Springer Series in

CHEMICAL PHYSICS 63

Springer
Berlin
Heidelberg
New York
Barcelona
Hong Kong
London
Milan
Paris
Singapore
Tokyo

Springer Series in
CHEMICAL PHYSICS

Series Editors: F. P. Schäfer J. P. Toennies W. Zinth

63 **Ultrafast Phenomena XI**
Editors: T. Elsaesser, J.G. Fujimoto, D.A. Wiersma, and W. Zinth

Volumes 1–62 are listed at the end of the book

T. Elsaesser J.G. Fujimoto D.A. Wiersma
W. Zinth

Ultrafast Phenomena XI

Proceedings
of the 11th International Conference,
Garmisch-Partenkirchen, Germany,
July 12–17, 1998

With 507 Figures

Springer

Professor Dr. Thomas Elsaesser
Max-Born Institut für Nichtlineare Optik
und Kurzzeitspektroskopie
Rudower Chaussee 6
D-12489 Berlin, Germany
e-mail: elsasser@mbi.berlin.de

Professor Dr. Douwe A. Wiersma
University of Groningen,
Faculty of Sciences & Mathematics
Ultrafast Laser and Spectroscopy Laboratory
Nijenborgh 4
9747 AG Groningen, The Netherlands
e-mail: wiersma@chem.rug.nl

Professor James G. Fujimoto
Massachusetts Institute of Technology,
Dept. of Electrical Engineering & Computer Science
77 Massachusetts Ave
Cambridge, MA 02139, USA
e-mail: jgfuji@MIT.edu

Professor Wolfgang Zinth
Universität München,
Institut für Medizinische Optik
Öttingerstr. 67
D-80538 München, Germany
e-mail: wolfgang.zinth@physik.uni-muenchen.de

Series Editors:

Professor F.P. Schäfer
Max-Planck-Institut für Biophysikalische Chemie
D-37077 Göttingen-Nikolausberg, Germany

Professor J.P. Toennies
Max-Planck-Institut für Strömungsforschung
Bunsenstrasse 10
D-37073 Göttingen, Germany

Professor W. Zinth
Universität München,
Institut für Medizinische Optik
Öttingerstr. 67
D-80538 München, Germany

ISSN 0172-6218
ISBN-13: 978-3-642-72291-2 e-ISBN-13: 978-3-642-72289-9
DOI: 10.1007 / 978-3-642-72289-9

Library of Congress Cataloging-in-Publication Data applied for

Die Deutsche Bibliothek – CIP-Einheitsaufnahme
Ultrafast phenomena ...: proceedings of the ... international conference ...
- Berlin ; Heidelberg ; New York ; Barcelona ; Hong Kong ; London ; Milan ; Paris ; Singapore ; Tokyo : Springer
Bis 5 (1986) Zusatz zum Hauptsachtitel: proceedings of the ... OSA top. meeting
Bis 3 (1982) u.d.T.: Picosecond phenomena
11. Garmisch-Partenkirchen, Germany, July 12–17, 1998. - 1998
(Springer series in chemical physics ; 63)
ISBN-13: 978-3-642-72291-2

Softcover reprint of the hardcover 1st edition 1998

Typesetting: Camera-ready copies by the editors
Cover concept: eStudio Calamar Steinen
Cover production: *design & production* GmbH, Heidelberg

SPIN: 10673635 57/3144 - 5 4 3 2 1 0 – Printed on acid-free paper

Preface

This volume contains papers presented at the Eleventh International Conference on Ultrafast Phenomena held at Garmisch-Partenkirchen, Germany, from July 12 to 17, 1998. The biannual Ultrafast Phenomena Conferences provide a forum for discussion of the latest advances in ultrafast optics and their applications in science and engineering. The Garmisch conference brought together a multidisciplinary group of 440 participants from 27 countries, including 127 students. The enthusiasm of this large number of participants, the high quality of the papers they presented and the magnificent conference site resulted in a successful and pleasant conference.

Progress was reported in the technology of generating ultrashort pulses, including new techniques for improving laser-pulse duration, tunability over broad wavelength ranges, output power and peak intensity. Ultrafast spectroscopy continues to provide new insight into fundamental processes in physics, chemistry, biology, and engineering. In addition to analyzing ultrafast phenomena, control of ultrafast dynamics now represents an important topic. Ultrafast concepts and techniques are being applied in imaging and microscopy, high speed optoelectronics, material diagnostics and processing, reflecting the maturing of the field.

Acknowledgements. Many people contributed to the success of the conference. Members of the international program committee provided general guidance, reviewed the papers and put the final program together. The local organizing committee deserves special thanks for making all the meeting arrangements and running the conference very smoothly. We would like to acknowledge the Deutsche Forschungsgemeinschaft, the Bayerische Staatsministerium für Unterricht, Kultus, Wissenschaft und Kunst, the Ludwig-Maximilians-Universität München, and the Optical Society of America for sponsorship as well as support from Coherent, and Spectra Physics.

Berlin, Germany
Cambridge, MA
Groningen, Netherlands
München, Germany
August 1998

Thomas Elsaesser
James G. Fujimoto
Douwe Wiersma
Wolfgang Zinth

Contents

Part VI: Coherence in Semiconductors

Part IX: Higher Harmonics, X-Ray Generation and Diffraction, High Intensity Physics

Part XI: Vibronic Excitations and Radiationless Processes

Part XII: Dynamics in Liquids and Solutions

Part XIII: Ultrafast Photochemistry

Part XIV: Biology and Medicine

Part I

Lasers for Ultrashort Pulses

Sub-5-fs Pulses: Generation, Characterisation, and Experiments

Maxim S. Pshenichnikov[1], Andrius Baltuška[1], Róbert Szipöcs[2], and Douwe A. Wiersma[1]

[1] Laser and Spectroscopy Laboratory, Department of Chemistry, University of Groningen, Nijenborgh 4, 9747 AG Groningen, The Netherlands
E-mail: M.S.Pshenichnikov@chem.rug.nl
[2] Research Institute for Solid State Physics, P.O. Box 49, H-1525 Budapest, Hungary
E-mail: szipoecs@sunserv.kfki.hu

Abstract. We present a sub-5-fs 1-MHz-repetition-rate all-solid-state laser system based on the fibre-compression of the output of a cavity-dumped Ti:sapphire oscillator. Frequency-resolved optical gating is applied to measure precisely the phase correction needed for the compression, and to characterise the 4.5-fs pulses. The applications of the developed laser in nonlinear experiments on liquid-phase dynamics are discussed.

1. Introduction

The way towards shorter pulses lies in the generation of white-light continuum (WLC) with sufficient spectral bandwidth. Furthermore, sub-5-fs pulses require sophisticated spectral phase correction across the bandwidth of several hundreds of nanometers which cannot be realised without a multistage pulse compressor [1,2]. To design such a compressor it is imperative to learn the spectral phase of the WLC which has to be compensated by the compressor. Therefore, a direct phase measurement of the WLC is invaluable for compressor design and optimisation in order to obtain the ultimate bandwidth-limited pulses. Moreover, a comprehensive knowledge of phase distortions in compressed pulses is crucial for most spectroscopical applications.

In this contribution we demonstrate how, starting from the 15-fs output of a Ti:sapphire laser, one can substantially reduce the pulse duration. In our experiments, characterisation of the chirped WLC by second-harmonic generation frequency-resolved optical gating (SHG FROG) has provided vital information to develop an appropriate pulse compressor. Subsequently, we applied SHG FROG to measure the compressed pulses that consist of only 2.5 optical cycles. We also address the spectroscopical applications of this sub-5-fs laser.

2. SHG FROG of ultrabroadband pulses

FROG yields the full intensity and phase of ultrashort pulses and guarantees the validity of the data through internal consistency checks [3]. A distinct advantage of FROG, compared to many other existing ultrashort pulse measurement techniques, is that it utilises the excite-probe geometry, common for most non-linear optical experiments. The second harmonic generation (SHG) version of FROG is the most appropriate technique for low-energy pulses as provided by an oscillator [4]. Furthermore, the second-order nonlinearity in transparent crystals is generally considered instantaneous and wavelength-independent, which is essential for the measurement of ultrashort pulses. This is in a favourable contrast to $\chi^{(3)}$-based techniques which have finite response time due to the Raman contribution to nonlinearity.

The FROG measurement of such broadband pulses, either spectrum-limited or temporally broadened, presents a considerable challenge. The main experimental difficulties result from the geometrical time-smearing due to non-collinear geometry and the limited phase-matching bandwidth of the non-linear crystal [4]. We re-examined the effect of geometrical broadening and found that in the case of 5-fs pulses it amounts to less than 5% of the duration provided the experimental geometry is carefully chosen. To estimate the influence of the phase-matching conditions, we performed numerical simulation of FROG traces using the full expression for the FROG signal that is applicable even to single-cycle pulses [5]. The results convincingly showed that despite the ultrabroad bandwidth of 5-fs pulses, the signal can still be decomposed to the product of an ideally phase-matched FROG and a spectral filter, originated from the finite conversion bandwidth due to phase matching. This allows us to run a commercially available FROG inversion software and, more importantly, to use the frequency marginal [4] to verify experimental data.

Our SHG FROG apparatus is a convertible collinear/noncollinear autocorrelator designed for extremely short pulses [1]. To provide adequate doubling efficiency around 600 nm, a 10-μm BBO Type I crystal with the central SHG wavelength of 700 nm was employed. Dispersive lengthening of a 5-fs pulse by such a crystal does not exceed 0.2 fs. Fundamental pulses are focussed with an r= 25-cm spherical mirror at near normal incidence to minimise astigmatism. Due to the low curvature of the mirrors delay variations within each beam are less than 0.1 fs. Two separate PC1000 (Ocean Optics) spectrometers are used to detect the fundamental and the SH spectra. The use of a piezo translation stage (Physik Instrumente) ensured precise actuation of the optical delay. He-Ne laser interference fringes from a Michelson interferometer served as time calibration marks.

3. Pulse compression and characterisation

The broad bandwidth needed for ultrashort pulse generation can be produced by a variety of methods such as self-phase modulation (SPM) in glass fibres [1,6] or gas-filled capillaries [2,7], harmonic generation in waveguides [8] and optical parametric amplification [9]. The distinct advantage of the first method is that it requires pulse energies of only few tens of nJ which can be generated at MHz-repetition rates. In our set-up the WLC is produced upon injection of ~13-fs, 35 nJ pulses from a "Millennia"-pumped cavity-dumped Ti:sapphire laser into a single-mode fused silica fiber [1]. Because of self-phase-modulation (SPM) in the fiber, the spectrum of the exiting pulse spans from 500 nm to 1.1 μm (Fig.1).

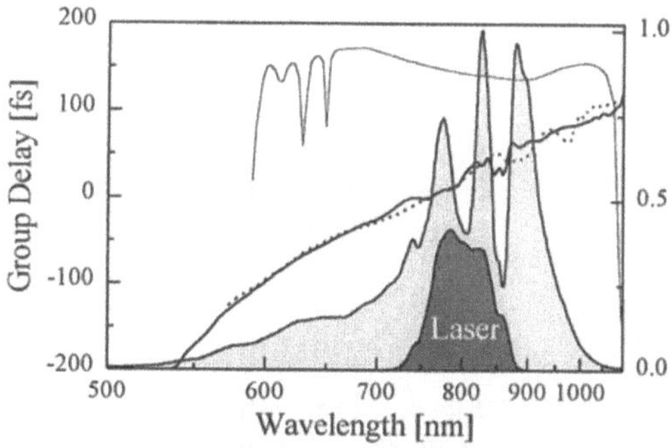

Fig. 1. The WLC characterisation and compressor design. The spectrum of the cavity-dumped Ti:sapphire laser entering the fibre is shown as a dark-shaded contour. WLC spectrum exiting the fibre is depicted as a light-shaded contour. The measured group delay (GD) of the WLC and the calculated GD of the pulse compressor are shown as solid and dashed curves, respectively. The thin line represents the overall compressor throughput.

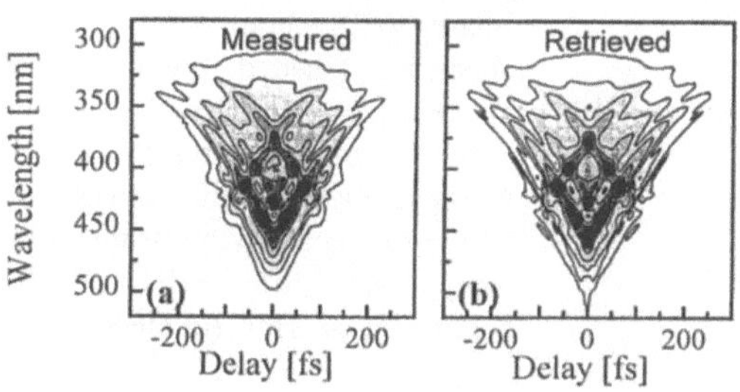

Fig. 2. Experimentally measured (a) and retrieved (b) SHG FROG traces of the chirped WLC. The traces are shown as a density plot with overlaid contours at the values 2%, 5%, 10%, 20%, 40%, 60%, and 80% of the maximum.

The measurement of strongly-chirped pulses like the one exiting the fibre turns out to be considerably more complicated compared with the case of compressed pulses. First, the upconverted signals are weaker due to the lower peak power. Second, a higher dynamic range of the measurements is required because of uneven temporal spread of spectral wings. Third, FROG inversion demands greater matrix sizes to provide adequate sampling in both time and frequency domains. Last, special care should be taken to collect all spatial components of the signal to prevent possible distortion of FROG data originating from delay-dependent change in the beam direction. The measured and retrieved SHG FROG traces of the chirped WLC are depicted in Fig.2. The combined action of SPM and dispersion leads to a nearly linear group delay (GD) over most of the spectrum (Fig.1, solid curve). We have also performed WLC characterisation for different input pulses. It turned out that the overall behaviour of the GD remains virtually unaffected, although the spectral width changes dramatically. This ensures efficient pulse compression under different experimental conditions.

Group delay measurements of the generated continuum served as a target function for the design of the three-stage, high throughput compressor, consisting of a quartz 45°-prism pair, broadband chirped mirrors and thin-film Gires-Tournois dielectric interferometers [1]. The phase characteristics of the compressor have been analysed using dispersive ray tracing and mapped onto the measured group delay of the continuum (Fig.1, dashed curve).

The experimental and retrieved FROG traces of compressed pulses are depicted in Fig.3. Figure 4 shows the retrieved intensity and phase in the time and frequency domains. The comparison of the FROG frequency marginal and the autoconvolution of the fundamental spectrum (Fig.4, right inset) indicates that no loss of spectral information has occurred. The obtained pulse duration is 4.5 fs

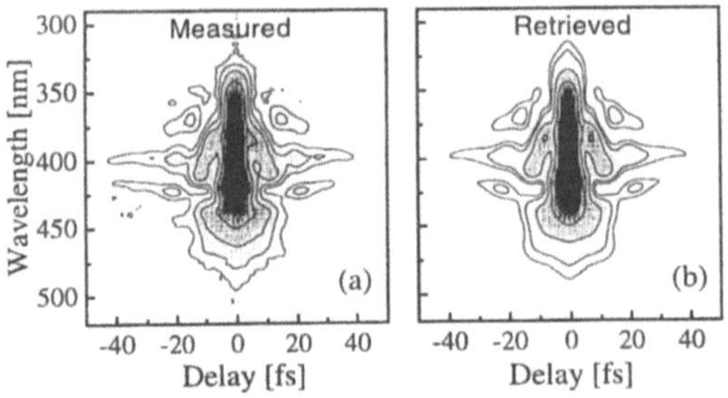

Fig. 3. Experimentally measured (a) and retrieved (b) SHG FROG traces of 4.5-fs pulses.

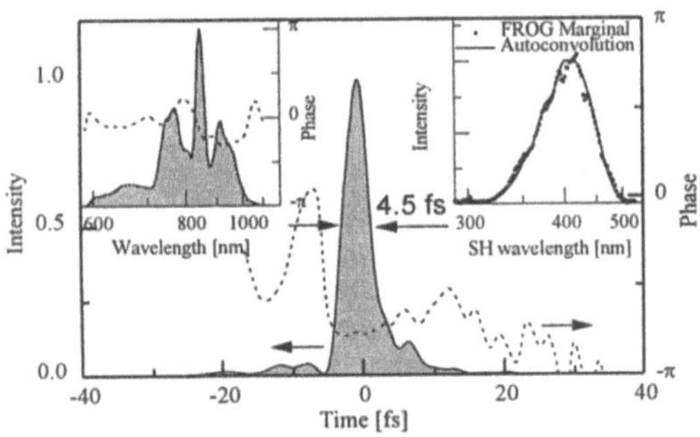

Fig. 4. The retrieved temporal pulse shape (shaded contour) and phase (dashed line). Inset left: The spectrum (shaded contour) and spectral phase (dashed line) of the 4.5-fs pulse. Inset right: Comparison of the autoconvolution of the fundamental spectrum (solid line) and the SHG FROG frequency marginal (circles).

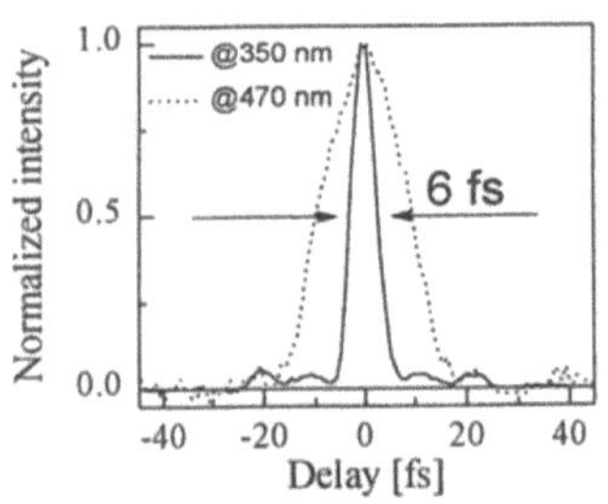

Fig. 5. Autocorrelation traces at 350 nm (solid curve) and 470 nm (dashed curve).

while variations of the spectral phase is less than $\pm\pi/4$ across the whole bandwidth. These results fully confirm our previous analysis based on the interferometric autocorrelation [1].

Figure 5 shows two autocorrelation traces derived from the complete FROG spectrogram at two separate wavelengths. The FWHM of the autocorrelation at 350 nm is merely 6 fs which is indicative of an ~4-fs pulse duration. However, the autocorrelation at 470 nm is three times broader. Such a difference clearly illustrates the effect of the spectral filtering in SH detection and/or nonlinear crystal on the autocorrelation width. This also underscores the importance of pulse characterisation by frequency-resolved (e.g., FROG) rather than non-frequency-resolved (e.g., intensity autocorrelation) methods if one deals with such broadband pulses.

4. Experiments

The high repetition rate, simplicity, and large usable spectral bandwidth (>400 nm) make the developed light source ideal for spectroscopic applications in condensed phase. With the unprecedented 5-fs resolution, the direct access to the

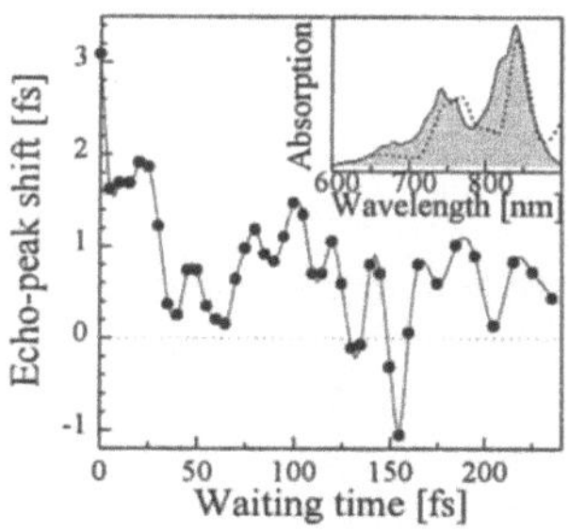

Fig. 6. Echo-peak shift experiment on TCNQ$^-$ in ethylene glycol. Inset shows the TCNQ$^-$ absorption spectrum (solid curve) and the spectrum of 4.5-fs pulses (dashed curve).

most controversial initial part of the hydrated electron dynamics has become possible [10].

Figure 6 summarises the results of stimulated photon echo experiments performed on TCNQ$^-$. This molecule is much smaller and more rigid than dye molecules conventionally used to study solvation dynamics. Also, 5-fs pulses excite most of the absorption band making it possible to investigate the effect of high-frequency vibronic transitions on solvation. Note that the echo-peak shift function [11] becomes negative at ~155 fs. This suggests that the energy-gap fluctuations become (partly) anticorrelated. Further experiments to explore this interesting system are underway.

5. Conclusions

We have constructed a 1-MHz repetition rate, sub-5-fs laser source that is highly suitable for condensed-phase spectroscopy. Thorough phase characterisation by SHG FROG at all stages of the system has enabled us to achieve pulse compression down to 4.5 fs. To the best of our knowledge, these are the shortest pulses whose phase and amplitude have been completely measured. The developed technology can be directly applied to produce even shorter pulses that comprise a single optical cycle. Undoubtedly, such pulses will find their way in many spectroscopic applications.

Acknowledgements. We gratefully acknowledge fruitful discussions with K.W. DeLong, R. Trebino, A. Rundquist, H.C. Kapteyn, M.M. Murnane, A. Gaeta, and V.V. Krasnikov.

References

1 A. Baltuška, Zh. Wei, M.S. Pshenichnikov, and D.A. Wiersma, Opt. Lett. **22**, 102, 1997; Appl. Phys. **B65**, 175, 1997.
2 M. Nisoli *et al.*, Opt. Lett. **22**, 522, 1997; Appl. Phys. **B65**, 189, 1997.
3 D.J. Kane and R. Trebino, IEEE J. Quantum Electron. **29**, 571, 1993
4 G. Taft *et al.*, IEEE J. of selected topics in **QE2**, 575, 1996
5 A. Baltuška *et al.*, Opt. Lett. **23**, 1998 (to be published)
6 R.L. Fork *et al.*, Opt. Lett. **12**, 483, 1987
7 M. Nisoli *et al.*, Appl. Phys. Lett. **68**, 2793, 1996
8 C. Durfee *et al.*, Opt. Lett. **22**, 1565, 1997
9 T. Wilhelm *et al.*, Opt. Lett., **22**, 1494, 1997; G. Cerullo *et al.*, Appl. Phys. Lett. **71**, 3616, 1997; A. Shirakawa *et al.*, Appl. Phys. Lett. **72**, 147, 1998; see also this book
10 M.F. Emde *et al.*, *Hydrated Electron Dynamics on a 5-fs Timescale*, in this book
11 W.P de Boeij *et al.*, Annu. Rev. Phys. Chem. **49**, 99, 1998

Generation of Intense Diffraction-Limited White Light and 4-fs Pulses

Z. Cheng, G. Tempea, T. Brabec, K. Ferencz[1], Ch. Spielmann, and F. Krausz

Abteilung Quantenelektronik und Lasertechnik, Technische Universität Wien, Gußhausstrasse 27-29, A-1040 Vienna, Austria

[1]Optical Coating Laboratory, Research Institute for Solid State Physics, H-1525 Budapest, POB 49, Hungary

Abstract. Sub-mJ 25-fs pulses produce, by weakly ionizing argon in a capillary, white-light continuum in a nearly diffraction-limited beam. The near-infrared portion of the spectrum can be compressed to bandwidth-limited 4-fs pulses using novel ultrabroadband chirped mirrors.

1. Introduction

Ever since its discovery [1] spectral continuum generation by ultrashort laser pulses has attracted much interest owing to its significance for ultrafast spectroscopy. Femtosecond continua produced by near-infrared pulses in various transparent media extend well into the violet with a characteristic nearly exponential decrease of the spectral intensity with decreasing wavelength.[2-4]. The emergence of a continuum was inherently linked to strong and for applications undesirable modifications of the diffraction-limited input beam, including break-up into filaments and conical emission.[2-4]. In this paper we report, for what is to our knowledge the first time, the generation of nearly diffraction-limited truly white-light femtosecond continuum exhibiting comparable spectral intensities all over the range of 400-1100 nm.

This performance could be obtained by propagating very short (~25 fs) pulses through a capillary (or hollow waveguide) filled with some noble gas as a nonlinear medium. The key advantage of waveguiding lies in the fact that energy transfer into higher-order transverse modes, and hence associated accumulation of spatial effects (e.g. self focusing/defocusing, filamentation) is confined to a relatively short propagation length over which higher-order modes dephase with respect to the fundamental one [5]. Spatially uniform self-phase modulation (SPM) of the fundamental mode, however, can accumulate over much longer distances limited merely by propagation losses and possible pulse broadening effects. As a result, hollow waveguides offer the potential for a much stronger nonlinear phase accumulation without degradation of beam quality as compared to bulk media.

2. White Light Generation

In our experiments, we used an argon-filled fused-silica capillary[6] with an inner diameter of 250 μm and a length of 95 cm. 25-fs pulses with energies ranging from 0.2 mJ to 1 mJ produced by a 1-kHz Ti:sapphire amplifier were passed into the waveguide. With the pressure optimized (in the range of 1 bar to 0.2 bar, respectively) for maximum spectral broadening in the absence of spatial effects, white-light continuum could be observed in the entire pulse energy range investigated. The full line in Figure 1(b) depicts the spectrum of the output pulse having an energy of E_{out} = 0.15mJ obtained with E_{in} = 0.22mJ and p = 1 bar. Spectral components having intensities comparable to those in the range of the seed spectrum (around 800 nm) extend to wavelengths as short as 400 nm. This supercontinuum is delivered in a nearly diffraction-limited beam ($M^2 < 1.5$) at all wavelengths. Preliminary calculations and computer simulations indicate that SPM induced by ionization provides a significant contribution to the observed spectral broadening and blue shift. In fact, both the plasma-induced and the Kerr-induced change in the refractive index have been estimated as $|\Delta n_{plasma}| \cong |\Delta n_{Kerr}| \cong 10^{-5}$ on the waveguide axis at the entrance of the waveguide. Numerical modeling of the phenomenon including SPM due to ionization and the Kerr effect as well as self-steepening and dispersion is under way.

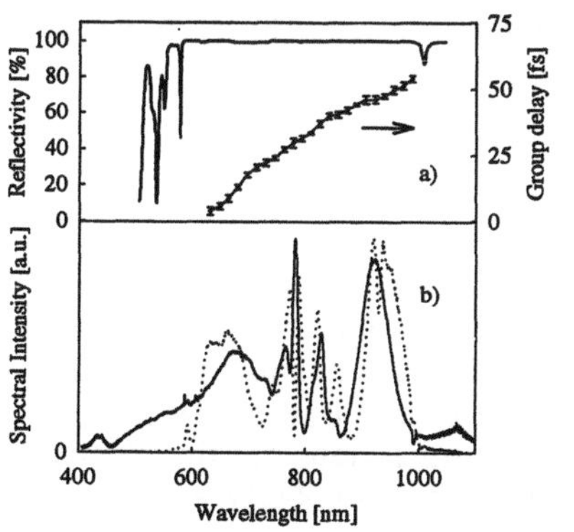

Fig. 1. (a) Reflectivity and group delay dispersion of ultrabroadband chirped mirrors consisting of 52 layers of SiO_2 and TiO_2. (b) White-light continuum emerging from the argon-filled capillary (full line) and output of the 2-stage compressor described in the text.

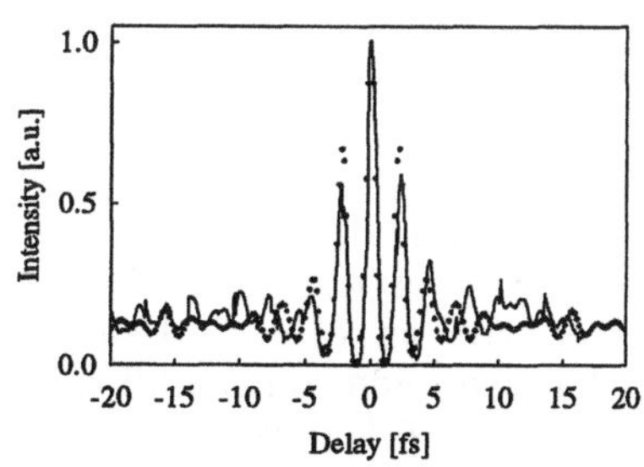

Fig. 2. Interferometric autocorrelation trace measured at the output of the 2-stage compressor described in the text (full line) and calculated (cross hairs) from the spectrum depicted by the dotted line in Fig. 1(b). The cross hairs imply a pulse duration (FWHM) of 3.7fs.

3. Pulse Compression

Recently developed ultrabroadband chirped mirrors[7] providing nearly constant negative group delay dispersion over the unprecedented bandwidth of 350 nm, as shown in Fig. 1(a), invite to test the linearity of the SPM-induced chirp carried by the output pulses. Attempts to compress the capillary output yielded pulse durations around 5 fs, some 20% longer than the Fourier limit. The compressed pulses with an energy of ≅120μJ have been gently focused in air to a beam waist of w_0≅140μm. The generated spectrum exhibited negligible variation across the beam profile. 5 reflections off the ultrabroad-band chirped mirrors and a pair of thin fused silica wedged plates were used for final dispersion control. A representative output spectrum, subject to some small variation in shape, and an interferometric autocorrelation signal produced by a 15μm-thick BBO crystal are shown by the dotted line in Fig. 1(b) and by the full line in Fig. 2, respectively. The reasonable agreement between the measured autocorrelation trace and that calculated from the measured spectrum by assuming constant spectral phase (depicted by cross hairs in Fig. 2 indicate the generation of nearly bandwidth-limited 4-fs pulses with a peak power of 30 GW. Developing chirped mirrors for the blue-green-yellow spectral range and a pair of carefully designed thin wedged plates for wavelength separation, the presented system will be capable of delivering synchronized sub-10fs pulses tunable throughout the visible and near infrared for a wide range of applications in physics, chemistry, and biology.

4. Conclusions

Sub-milijoule 25-fs pulses propagating in an Argon-filled hollow waveguide generate a white light continuum exhibiting comparable spectral intensities all over the range of 400-1100nm. The white light beam is exhibith excellent spatial coherence, with an $M^2<1.5$. Using exclusively novel ultrabrodband chirped mirrors for dispersion compensation, the near-infrared range of the spectrum can be compressed to 4-fs pulses, with peak powers of up to 30GW and focusable to peak intensities of 10^{16} W/cm^2.

Acknowledgements. This research was supported by the Austria Science Fund, grant Y44-PHY.

References

1 R. R. Alfano and S. L. Shapiro, Phys. Rev. Lett. 24, 584 (1970)
2 R. L. Fork et al., Opt. Lett. 8, 1 (1983)
3 P. B. Corkum et al., Phys. Rev. Lett. 57, 2268 (1986)
4 P. B. Corkum and C. Rolland, IEEE J. Quantum Electron. 25, 2634 (1989)
5 G. Tempea and T. Brabec, Opt. Lett. 23, 762 (1998)
6 M. Nisoli et al., Appl. Phys. Lett. 68, 2793 (1996)
7 G. Tempea et al., IEEE JSTQE 4, 193 (1998)

Broadband Operation of a Ti:Sapphire Laser Using Novel Double-Chirped Mirrors and Semiconductor Saturable Absorbers

D. H. Sutter[1], I. D. Jung[1], N. Matuschek[1], F. Morier-Genoud[1], F. X. Kärtner[1], V. Scheuer[2], M. Tilsch[2], T. Tschudi[2] and U. Keller[1]

[1]Ultrafast Laser Physics Laboratory, Institute of Quantum Electronics, Swiss Federal Institute of Technology, ETH Hönggerberg – HPT, CH-8093 Zürich, Switzerland
http://www.iqe.ethz.ch/ultrafast/
[2] Institute for Applied Physics, TH Darmstadt, Germany

Abstract. Pulses of 6.5 fs duration have been obtained using custom tailored double-chirped mirrors (DCMs) and a broadband semiconductor saturable absorber mirror (SESAM) that leads to reliable self-starting operation. With a single set of novel ultra-broadband DCMs we can tune the center wavelength of self-starting 30-fs pulses over 300 nm. The exchange of the cavity mirrors, which is usually neccessary when tuning over such a broad wavelength range, can thus be avoided.

1. Introduction

Within the last decade, Ti:sapphire became the most important gain medium for ultrashort laser pulses in the near infrared. The shortest pulses directly from an oscillator were generated using DCMs for higher order dispersion compensation [1, 2]. The important aspects in the mirror design are both the reflectivity as well as the influence on the phase φ as functions of the optical frequency. Only chirped mirrors show a high-reflectivity range that is broad enough and a controlled group delay dispersion ($GDD = d^2\varphi/d\omega^2$) that is smooth enough (with respect to wavelength) in order to support pulses below 8 fs. Reliable self-starting operation of dispersion-controlled lasers can be obtained using broadband SESAMs [3]. Moreover, the critical cavity alignment for pure Kerr-lens modelocking (KLM) becomes more relaxed when a SESAM is used.

2. Experimental Results

Broadband SESAMs. For the results presented here, we have used thin low-finesse antiresonant Fabry-Perot saturable absorbers. The design and fabrication issues of such SESAMs has been described before [3]. Our SESAMs can initiate modelocking with measured pulse build-up times down to 60 μs. For sub-10-fs pulses, the pulse formation is still dominated by KLM. In this case, SESAMs stabilize the pulse and relax the cavity design issues. However, SESAMs with a higher modulation depth but less bandwidth generated pulses as short as 13 fs with soliton modelocking over the full cavity stability range [4].

Double-Chirped Mirrors. DCMs can be custom designed to compensate for the higher order dispersion of the laser cavity [5]. For the shortest pulses directly from a laser oscillator, we have used DCMs that show a controlled *GDD* between 670 and 880 nm [2] (Fig. 1a). This limited the pulse formation to spectra below 900 nm. Using a single set of chirped mirrors in a commercial laser, 85 fs pulses have previously been demonstrated over a bandwidth of almost 300 nm [6]. We obtain significantly shorter pulses that are tunable over a slightly wider wavelength range, using novel ultrabroadband DCMs. The typical *GDD* fluctuations of broadband dielectric mirrors are minimized, however, some oscillations remain because it is extremely difficult to include a perfect broadband antireflection coating on top of the mirror structure. The ultrabroadband DCMs cover a much wider wavelength range than the mirrors used for the 6.5 fs pulses, at the expense of increased variations in the *GDD* (Fig. 1a).

Laser Performance. The Ti:sapphire laser, shown in Fig. 1b, is pumped by all lines of an argon ion laser (≈7 W). With a single set of DCMs, the Ti:sapphire cw output can now be tuned from 670 to 1010 nm due to the increased reflectivity range of the mirrors (R > 99.8% between 620 and 1020 nm). Self-starting 30-fs pulses were easily obtained over a range extending from 700 to 1000 nm, using two different SESAMs, one with a GaAs absorber for wavelengths $\lambda<870$ nm, the other with InGaAs for $\lambda>770$ nm (Fig. 2a). We have characterized the pulses using both interferometric autocorrelation and non-collinear frequency resolved optical gating (FROG). With strong KLM we have reached 7.3-fs pulses with the same set of ultrabroadband DCMs. For this result, we did not use any extra-cavity pulse shaping or dispersion compensation prior to the FROG set-up. The geometrical smearing error due to non-collinear interaction was also not substracted. However, the spectrum exhibited a considerable sub-structure due to the enhanced *GDD* oscillations (Fig. 1a). The trade-off between bandwidth and a smooth *GDD* limited the pulse generation to 7 fs in this case. In contrast, 6.5 fs pulses (Fig. 2b) have been obtained from a similar cavity with DCMs that exhibit a smoother *GDD* (Fig. 1b).

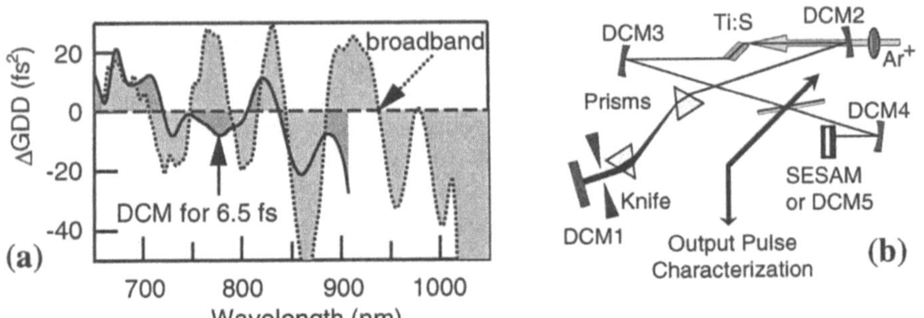

Fig. 1. (a) Measured deviation from desired *GDD* for a single reflection on the DCMs used to generate 6.5-fs pulses and for the ultrabroadband DCMs used for tunable 30-fs pulses. **(b)** Schematic of the cavity, which consists of a 2.3 mm Ti:sapphire crystal (doping 0.25 wt.%), a semiconductor saturable absorber mirror (SESAM), a fused silica prism pair with a 30 cm apex separation and double-chirped mirrors (DCMs). Knife edges are used for tuning, and a single-side AR coated glass plate for output coupling.

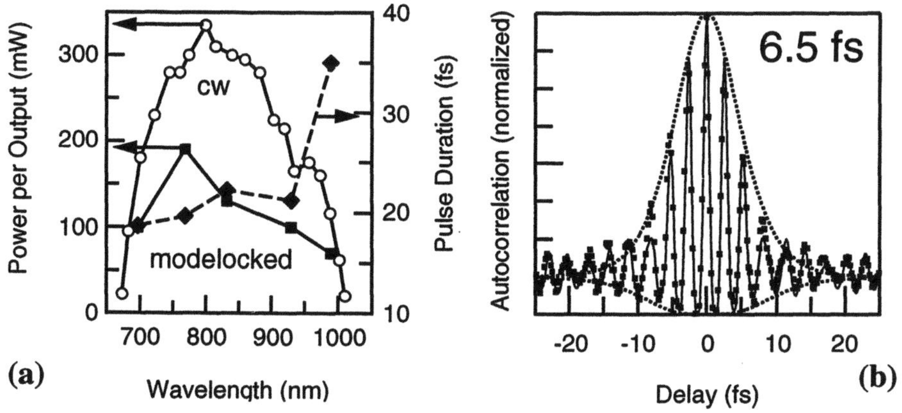

Fig. 2. (a) Tunability with broadband DCMs: power per output beam and durations (b) Interferometric autocorrelation of a 6.5-fs pulse

3. Conclusion and Outlook

A single set of DCMs supports ultrabroadband operation of Ti:sapphire lasers. Such mirrors can have a group delay that is smooth enough to support 6.5-fs pulses as well as broadly tunable pulses of 30 fs or less. SESAMs are the key components to obtain reliable, self-starting, "hands-off" operation. Future work will be devoted to the design of DCM combinations with less *GDD* oscillations. The challenge in SESAM technology will be to extend the wavelength range of broadband SESAMs to cover the whole gain spectrum of Ti:sapphire and to design SESAMs that support sub-10-fs pulses without KLM.

Acknowledgments. We would like to thank the Swiss National Science Foundation for support.

References

1 I. D. Jung, F. X. Kärtner, N. Matuschek, D. H. Sutter, F. Morier-Genoud, G. Zhang, U. Keller, V. Scheuer, M. Tilsch, T. Tschudi, *Opt. Lett.* **22**, 1009, 1997.
2 D. H. Sutter, I. D. Jung, F. X. Kärtner, N. Matuschek, F. Morier-Genoud, V. Scheuer, M. Tilsch, T. Tschudi, U. Keller, *IEEE JSTQE* **4**, 169, 1998.
3 I. D. Jung, F. X. Kärtner, N. Matuschek, D. H. Sutter, F. Morier-Genoud, Z. Shi, V. Scheuer, M. Tilsch, T. Tschudi, U. Keller, *Appl. Phys. B* **65**, 137, 1997.
4 F. X. Kärtner, I. D. Jung, and U. Keller, *IEEE JSTQE* **2**, 540, 1996.
5 F. X. Kärtner, N. Matuschek, T. Schibli, U. Keller, H. A. Haus, C. Heine, R. Morf, V. Scheuer, M. Tilsch, T. Tschudi, *Opt. Lett.* **22**, 831, 1997.
6 E. J. Mayer, J. Möbius, A. Euteneuer, W. W. Rühle, R. Szipöcs, *Opt. Lett.* **22**, 528, 1997.

Improved Broadband Double-Chirped Mirrors with Controlled Dispersion Characteristics

N. Matuschek[1], F. X. Kärtner[1], D. H. Sutter[1], V. Scheuer[2], M. Tilsch[2], T. Tschudi[2], and U. Keller[1]

[1]Ultrafast Laser Physics Laboratory, Institute of Quantum Electronics, Swiss Federal Institute of Technology, ETH Hönggerberg – HPT, CH-8093 Zürich, Switzerland
E-mail: nicolai@iqe.phys.ethz.ch
[2] Institute for Applied Physics, TH Darmstadt, D-64289 Darmstadt, Germany
E-mail: volker.scheuer@physik.th-darmstadt.de

Abstract. We present a theory for the analytical design of double-chirped mirrors (DCMs) with a controlled dispersion characteristic. Theoretical designs are derived from an analytical chirp law. These are then used as excellent starting designs for later computer refinement. The design procedure presented here results in improved broadband DCM designs for higher order dispersion compensation in sub-10-fs Ti:sapphire lasers.

1. Introduction

Chirped mirrors [1-3] have been established in various set-ups for the generation of ultrashort laser pulses. In ref. [1], chirping means that the Bragg wavelength is gradually increased along the mirror, producing a negative group delay dispersion (GDD). However, such a simple-chirped mirror shows large oscillations in the group delay and GDD. The physical origin of these oscillations were unclear at that time. Therefore, these oscillations have been minimized by computer optimization. Recently, we developed a theory of chirped mirrors which is based on an exact coupled-mode analysis [4]. This theory shows the origin of the oscillations in the group delay as a matching problem and gives advice on overcoming it. In order to avoid the oscillations, one has to use the double-chirp technique in combination with a broadband antireflection (AR) coating (Fig. 1) [3, 5]. In [6] we related the chirp of the Bragg wavelength to the resulting GDD. An inversion gives us the tool to find the chirp law for a desired GDD. Using this powerful chirp law we can design DCMs with smooth and custom tailored GDDs suitable for generating 6.5 fs pulses directly from a Ti:sapphire laser [7]. With similar DCMs,

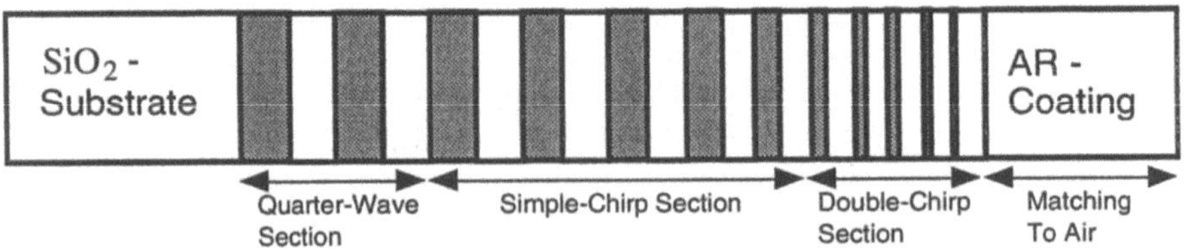

Fig. 1. Schematic picture of a double-chirped mirror (DCM) which can be considered to be composed of four different sections, each with a different task

we demonstrated ≈30 fs pulses tunable over a bandwidth of 300 nm, without changing the cavity mirrors [8]. The precise fabrication of the DCMs is guaranteed by the use of ion-beam sputtering with an active layer control during growth [9].

2. Analytical Design of Double-Chirped Mirrors

Universal DCM Structure. According to Fig. 1, a DCM is a multilayer interference coating that can be considered as a composition of four sections. Each of the sections serves a different task [6]. The layer materials considered here are SiO_2 and TiO_2. The first section is the AR coating, typically composed of 10-14 layers. It is necessary to match the first layer of the mirror to the ambient air. The other sections represent the actual DCM structure, as derived from theory. The double-chirp section is responsible for the elimination of the oscillations in the GDD. Double-chirping means that, in addition to the local Bragg wavelength, the local coupling of the incident wave to the reflected wave is independently chirped. This corresponds to an adiabatic matching of the impedance. In the simple-chirp section, only the Bragg wavelength is chirped, whereas in the quarter-wave section the Bragg wavelength is kept constant. This method produces a high reflectivity for the long wavelengths.

Analytical Chirp Law. For the design of a DCM with a custom tailored dispersion characteristic, a simple approach of linearly chirping the Bragg wavenumber along the mirror, is very often insufficient. A linear chirp produces only a positive average third order dispersion (TOD) for all wavelengths in the interesting high reflectivity region [5]. Thus, the GDD of such designs is far away from the desired GDD, especially when the desired TOD should be negative over the whole spectrum or parts of the spectrum [3, 5-8]. We are able to generate a DCM with the desired GDD by applying a proper chirp to the Bragg wavenumber, which automatically produces the right behavior of the GDD [6]. The theoretical DCM design, combined with a broadband AR coating, is used as an excellent starting design for a numerical optimization program. Since we start from a theoretical

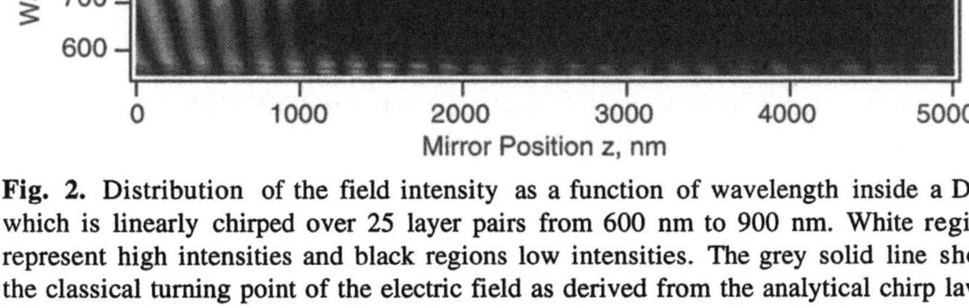

Fig. 2. Distribution of the field intensity as a function of wavelength inside a DCM which is linearly chirped over 25 layer pairs from 600 nm to 900 nm. White regions represent high intensities and black regions low intensities. The grey solid line shows the classical turning point of the electric field as derived from the analytical chirp law.

design that is close to the desired design goal, a local optimization using a standard gradient algorithm is sufficient.

Field Intensity. In a chirped mirror, the longer wavelengths penetrate deeper into the mirror structure than shorter wavelengths, because of the chirp of the Bragg wavelength to higher values. This results in a negative GDD, since the wavelength dependent group delay is essentially the time different spectral components spend inside the mirror. More precisely, we have shown that, for a DCM, the WKB solution of the group delay in zero order approximation is given by the optical path length for a wave travelling from the mirror surface to the classical turning point and back again [6]. Thus, we define the penetration depth of a DCM as the wavelength dependent physical distance from the mirror surface to the classical turning point. Fig. 2 shows the distribution of the field intensity for a typical DCM. No Ar coating is incorporated, because, for the theoretical design, we assume an ideal matching to the ambient medium (air). The horizontal axis gives the physical distance z from the mirror surface and the vertical axis gives the wavelength of the incident light. White regions of the standing-wave pattern correspond to high intensities and black regions to low intensities. Fig. 2 illustrates in an impressive way the larger penetration depth of the electric field for the longer wavelengths. The grey solid line shows the classical turning point of the electric field as derived from the analytical chirp law. Obviously, this curve is in excellent agreement with the positions where the exactly calculated field intensities start to decay exponentially.

3. Conclusion

Following the analytic design procedure described above combined with simple computer refinement techniques, we can design improved broadband DCMs with custom tailored dispersion characteristics useful for various applications [3, 5-8].

Acknowledgement. This work has been supported by the Swiss National Science Foundation.

References

1 R. Szipöcs, K. Ferencz, C. Spielmann, F. Krausz, *Opt. Lett.* **19**, 201, 1994.
2 G. Tempea, F. Krausz, Ch. Spielmann, K. Ferencz, *IEEE JSTQE* **4**, 193, 1997.
3 F. X. Kärtner, N. Matuschek, T. Schibli, U. Keller, H. Haus, C. Heine, R. Morf, V. Scheuer, M. Tilsch, T. Tschudi, *Opt. Lett.* **22**, 831, 1997.
4 N. Matuschek, F. X. Kärtner, U. Keller, *IEEE JQE* **33**, 295, 1997.
5 N. Matuschek, F. X. Kärtner, U. Keller, *IEEE JSTQE* **4**, 197, 1998.
6 N. Matuschek, F. X. Kärtner, U. Keller, submitted to *IEEE JQE*, 1998.
7 I. D. Jung, F. X. Kärtner, N. Matuschek, D. H. Sutter, F. Morier-Genoud, G. Zhang, U. Keller, V. Scheuer, M. Tilsch, T. Tschudi, *Opt. Lett.* **22**, 1009, 1997.
8 D. H. Sutter, I. D. Jung, N. Matuschek, F. Morier-Genoud, F. X. Kärtner, V. Scheuer, M. Tilsch, T. Tschudi, U. Keller, this issue, 1998.
9 M. Tilsch, V. Scheuer, T. Tschudi, *Conf. Proc. SPIE* **2253**, 414, 1994.

A New Type of Saturable Bragg Reflector with Record-Low Saturation Fluence and Broad Tuning Range

Jia-Min Shieh[1], T. C. Huang[2], K. F. Huang[2], Chi-Luen Wang[3], and Ci-Ling Pan[1*]

[1] Institute of Electro-Optical Engineering
[2] Dept. of Electrophysics,
National Chiao Tung University 1001 Ta-Hsueh Rd. Hsinchu, Taiwan 30010, R. O. C.
Tel: +886-3-5731921 Fax: +886-3-5716631
E-mail: clpan@cc.nctu.edu.tw
[3] Precision Instrument Development Center, Science-Based Industrial Park, Hsinchu, Taiwan 300

Abstract. The saturation fluence for a strained triple-quantum-well saturable Bragg reflector was as low as 12 $\mu J/cm^2$. Self-starting femtosecond pulses tunable from 768 to 804nm was realized.

1. Introduction

Semiconductor saturable-absorber mirrors (SESAM) have been successfully used for passive mode locking of solid state lasers.[1-3] Pulses as short as 6.5 fs were generated.[4] Low saturation fluence for the SESAM is important for low power diode-pumped solid state lasers. The other key issue for SESAM's is the tuning range. In this work, we report a new type of SBR using triple strained-layer quantum wells with separate and sequential absorption peaks as the absorbing layer. Record-low saturation fluence and broad tuning range are achieved.

2. Experimental Methods

Figure 1 illustrates the structure of the strained-layer saturable Bragg reflector (SSBR). A DBR with 15-pairs of high-low $\lambda/4$ layers of $AlAs/Al_{0.25}Ga_{0.75}As$ were grown by molecular beam epitaxy (MBE). An additional $\lambda/2$ layer of $Al_{0.25}Ga_{0.75}As$ was grown on top of the DBR mirror. Three strained quantum wells ($In_xAl_{1-x-y}Ga_yAs$, x=0.15 and y=0.6 for the center well) with separate absorption peak wavelength were inserted in this layer. The reflectivity of our SSBR was measured to be greater than 94% (770-800 nm) as shown in Fig.2. The photo-luminescence (PL) spectrum is also shown.

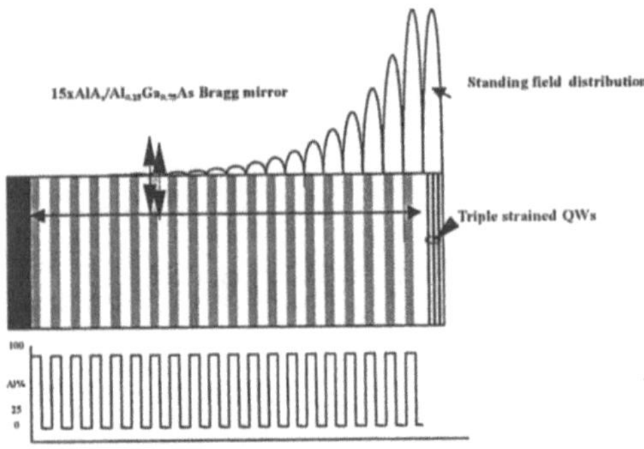

Fig.1. Structure of the strained-layer saturable Bragg reflector (SSBR) with a distributed Bragg reflector (DBR) and an additional λ/2 layer of $Al_{0.25}Ga_{0.75}As$; three strained quantum wells with different absorption peaks were inserted in this layer.

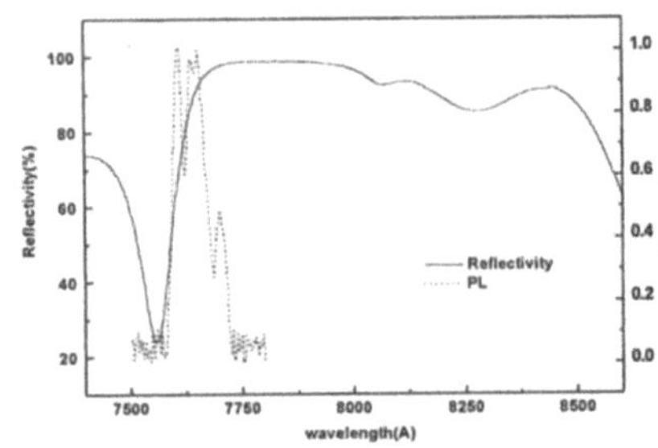

Fig.2. The reflectivity and the photo-luminescence spectra of the SSBR.

3. Results and Discussion

The SSBR was employed as the end mirror in the dispersion-compensation arm of a four-mirror X-folded Ti: sapphire cavity (without focusing on the SSBR). The beam spot size and optical fluence on the SSBR were about 1 mm and E ≈ 7μJ/cm^2 respectively. Self-starting mode-locking was achieved from 768 to 804 nm. Throughout this range, we are able to generate transform-limited sub-100fs Gaussian pulses. The laser can be mode-locked with output power as low as 40 mW. The corresponding threshold fluence was 1 μJ/cm^2. The saturation fluence of the SSBR is measured to be ≈ 12μJ/cm^2 (E_{sat} ≈ 150 μJ/cm^2 for the unstrained SBR). The normalized nonlinear reflectivity ΔR/R measured at λ = 755 nm (peak wavelength of the PL spectrum) as a function of average power P_{av} of the incident pulse on the SSBR sample is shown in Fig.3. We find that the ΔR/R increased from 1.0×10^{-3} to 2.0×10^{-2} as P_{av} changed from 5 mW to 60 mW. Obviously, ΔR/R was saturated for optical fluence larger than E ≈ 18 μJ/cm^2 (P_{av} ≈ 60 mW). This is consistent with independent measurement of the saturation optical fluence E_{sat} of the Ti: sapphire/SSBR laser, E_{sat} ≈ 12 μJ/cm^2 (P_{av} ≈ 40 mW). In Fig. 3, we also showed the same measurement at another wavelength, λ = 775 nm. The E_{sat} was found to be identical to that at λ = 755 nm. Due to the excitonic absorption bandwidth, however, ΔR/R at λ = 775 nm was smaller by about one order of magnitude. The transient second harmonic signal (both theoretical and experimental) for this laser is plotted as a function of delay time after the laser onset in Fig. 4. For the theoretical buildup curve, parameters such as the unsaturated loss, the fast decay time and the slow decay time of SSBR were measured to be 2 ~ 3×10^{-3}, 280 fs, and 40 ps, respectively. We find that the cw stage and the total buildup times to completely mode-locked pulse train were 60 us and 100 us, respectively. The fast buildup is consistent with the low self-starting threshold of the SSBR.

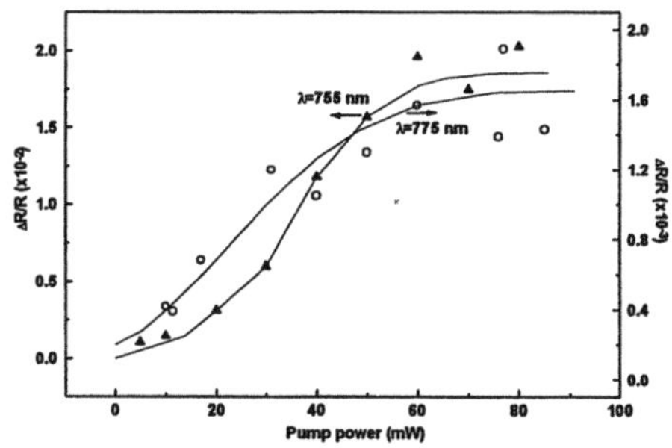

Fig.3. Normalized nonlinear reflectivity of the SSBR at λ = 755 nm and 775 nm are plotted as a function of incident average power.

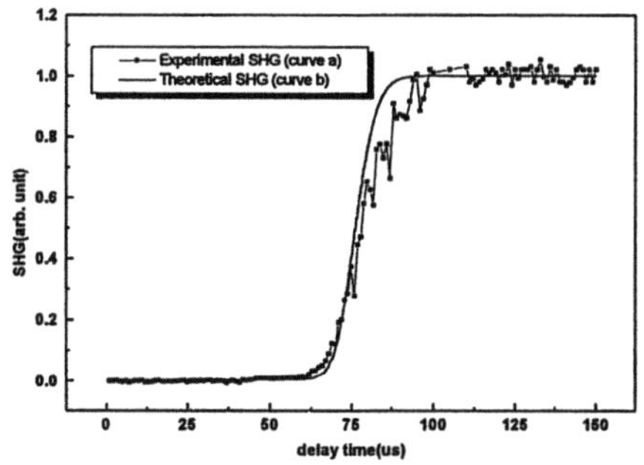

Fig.4. Theoretical and experimental buildup of the transient second harmonic signal as a function of delay time after the laser onset.

4. Conclusions

In conclusion, we have demonstrated a novel SESAM with record-low saturation fluence. Self-starting mode-locking range exceeding 40 nm was also demonstrated. This is presently limited by the bandwidth of the DBR ($\approx$ 45nm). On the other hand, the tuning range of SBR's with single and triple (with sequential absorption peaks) unstrained QW's were ~8 and ~30 nm, respectively. This work was supported in part by the National Science Council of the Republic of China

References

1. S. Tsudå, W. H. Knox, S. T. Cundiff, W. Y. Jan, and J. E. Cunninjham, IEEE Sel. Topics Quantum Electron. 2, 454. (1996).
2. M. J. Hayduk, S. T. John, M. F. Frol, C. P. Pollock, and R. P. Leavitt, Optics Comm., 137, 55 (1997).
3. U. Keller, K. J. Weingarten, F. X. Kartner, D. Kopf, B. Braun, I. D. Jung, R. Fluck, C. Honninger, N. Matuschek, and J. A. der Au, IEEE J. Sel. Topics Quantum Electron. 2, 435 (1996).
4. I. D. Jung, F. X. Kartner, Nl. Matuschek, D. H. Sutter, F. Morier-Genoud, G. Zhang, U. Keller, V. Scheuer, M. Tsilsch, and T. Tschudi, Opt. Lett. 22, 1009 (1997).

Time-Resolved Measurement of Electronic and Vibrational Nonlinear Refraction in Ti:sapphire and Cr:LiSGaF

Sylvia Smolorz and Frank Wise

Department of Applied Physics, Cornell University, Ithaca, NY 14853, USA
E-mail: sylvias@msc.cornell.edu

Abstract. We observe electronic and vibrational contributions in the femtosecond nonlinear response of Ti:sapphire and Cr:LiSGaF. The results suggest that nonlinear refraction is resonantly enhanced in Ti:sapphire relative to that of undoped sapphire. Cr:LiSGaF shows a substantial vibrational component and finite two-photon absorption. The implications of the nonlinear responses for short-pulse formation are discussed.

1. Introduction

In recent years, progress in modelocked Ti:sapphire lasers has led to the generation of pulses with bandwidths approaching the gain bandwidth [1, 2]. Jung *et al.* have observed pulses with a duration of 6.5 fs from a Ti:sapphire laser [1], and Sorokina *et al.* recently reported the generation of 14-fs pulses from a Cr:LiSGaF laser [3], which is a promising candidate for the generation of ultrashort pulses from an all-solid-state system [4]. Precise knowledge of the nonlinear refractive response of the gain medium is required for accurate modeling of lasers generating the shortest pulses. Despite its importance, we are not aware of a prior measurement of the nonlinear index n_2 of Ti:sapphire. The magnitude of n_2 is usually assumed to be identical to that of undoped sapphire, and the response is either assumed to be instantaneous or an exponential decaying in a few femtoseconds [5].

However, there may be a significant vibrational contribution to the nonlinear response. Both sapphire and colquiriite crystals, such as LiSGaF and LiSAF, have several Raman-active vibrational modes, which should be observable in the dynamic nonlinear response. Since LiSGaF has a smaller nonlinear index than sapphire [6], but a stronger Raman cross section [7], this vibrational component could be a significant fraction of its refractive nonlinearity.

Pulse shaping in KLM lasers is soliton-like, so the consequences of a vibrational contribution in the gain crystal should be similar to those observed in soliton propagation in fibers. Stolen *et al.* have shown that the delay due to the Raman response in silica fiber leads to a decrease of the effective nonlinearity, effectively leading to nonlinear chirp [8]. A significant vibrational contribution to n_2 in the gain medium would lead to an additional dispersion term in the propagation equation for a pulse in the laser cavity, while two-photon absorption (TPA) would decrease the stability.

2. Experimental Methods

Detection of small nonresonant nonlinearities requires a combination of high sensitivity and time resolution that has not been readily available. Kang *et al.* showed that very sensitive measurements of the magnitude and sign of third-order nonlinearities can be made using spectrally-resolved two-beam coupling (SRTBC) [9]. In addition, the dynamics of the nonlinear interaction are resolved, and the relative contributions of electronic and vibrational components are determined accurately in a single measurement.

We measured the nonlinear responses of Ti:sapphire (Ti^{3+} doping 0.15%) and Cr:LiSGaF (Cr^{3+} doping 2.5%) using SRTBC with 17-fs pulses from a modelocked Ti:sapphire laser. The bandwidth of the pulses (centered at 807 nm) exceeded the highest vibrational frequency of either material, so the measurements include all of the information about the Raman response. The samples were Brewster-cut crystals used in KLM lasers. The probe pulse propagated as it would in a laser ($E \parallel c$ and $k \perp c$) and the pump pulse with the same polarization propagated at a small angle to the probe. The pump intensity at the focus was ~20 GW/cm^2.

For materials with small TPA, n_2 can be determined by direct comparison to the signal from a reference material, SF59 glass ($n_2 = 3.1 \cdot 10^{-12}$ esu), without numerical fitting.

3. Results and Discussion

The experimental results are summarized in Table 1. The value of n_2 that we obtain for Ti:sapphire is 70% larger than the value customarily used ($1.25 \cdot 10^{-13}$ esu) [10]. We tentatively attribute the increase to some resonant enhancement arising from the linear absorption band at 800 nm, and we would also expect resonant enhancement of the vibrational contribution.

Table 1. Nonlinear index of refraction n_2, TPA coefficient β, and vibrational response time T_R of Ti:sapphire and Cr:LiSGaF. The uncertainty in the Raman frequencies is ±3 cm^{-1}.

	n_2(10^{-14} esu)	β(cm/GW)	T_R (ps)
Ti:sapphire, electronic	22±4	$<1 \cdot 10^{-4}$	
Ω=419 cm^{-1}	0.4±0.1	-	6±1
Ω=647 cm^{-1}	0.2±0.1	-	2.0±0.5
Cr:LiSGaF, electronic	5.5±1	$(9.2 \pm 2.3) \cdot 10^{-4}$	
Ω=233 cm^{-1}	0.07±0.02	$(2.4 \pm 0.8) \cdot 10^{-6}$	1.6±0.3
Ω=351 cm^{-1}	0.06±0.02	$(2.1 \pm 0.7) \cdot 10^{-6}$	0.6±0.1
Ω=554 cm^{-1}	1.16±0.35	$(3.7 \pm 1.1) \cdot 10^{-5}$	0.9±0.2

Figure 1 shows the vibrational signal from Ti:sapphire, together with the signal expected from the Raman spectrum. The agreement is very good and illustrates the excellent sensitivity and time-resolution possible with SRTBC. Both A_{1g} modes at 419 cm^{-1} and at 647 cm^{-1} contribute to the signal. The vibrational component of the response contributes ~2% to the total nonlinearity.

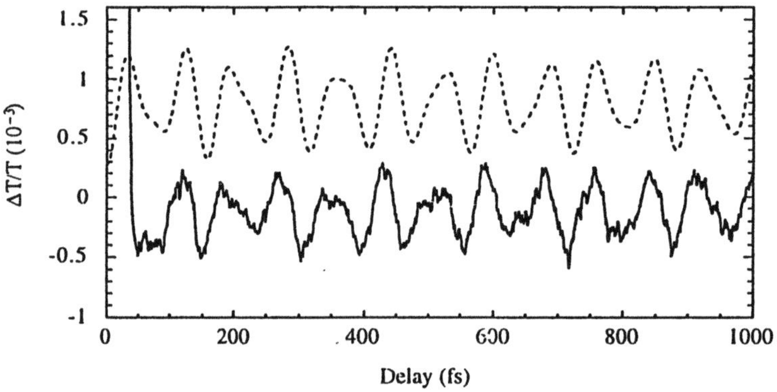

Fig. 1. Vibrational nonlinear refraction in Ti:sapphire

In Cr:LiSGaF (Figure 2), the electronic nonlinear index is a factor of 4 smaller than that of Ti:sapphire, and the vibrational component is much more pronounced, contributing ~20% of the total nonlinearity. The strong A_{1g} mode at 554 cm^{-1} dominates, and the vibrational signal shows a single damped oscillation. The inset shows nonlinear refraction. As a comparison shows, the absorptive nonlinearity of Cr:LiSGAF is small compared to the refractive part, but it is detectable.

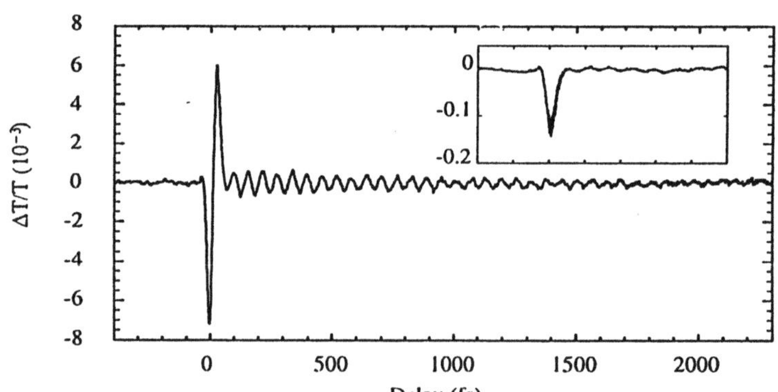

Fig. 2. Nonlinear refraction and TPA (inset) in Cr:LiSGaF

4. Conclusions

In conclusion, we observe the nonlinear refractive response of Ti:sapphire and Cr:LiSGaF directly in the time domain. These observations include what is to our knowledge the first measurement of n_2 of Ti:sapphire: it is 1.5 - 2.0 times the previously-accepted value. Aside from the dynamics of the electronic response, which this experiment cannot resolve, these data provide all the information necessary for accurate calculations of pulse formation in KLM lasers with these gain media. The detailed characterization of the material response will become most important as pulse durations approach fundamental limits. Since the vibrational nonlinear refraction in Ti:sapphire is quite small, it is unlikely that it will present a serious limitation to short-pulse generation. In Cr:LiSGaF, the relatively small electronic response and large vibrational response will result in a decrease of the nonlinear phase shift for short pulses, and this may become a limiting factor in the generation of ultrashort pulses. Also, the nonlinear absorption will be detrimental to the stability of the laser.

References

1 I. D. Jung, F. X. Kärtner, N. Matuschek, D. H. Suttner, F. Morier-Genoud, G. Zhang, U. Keller, V. Scheuer, M. Tilsch, and T. Tschudi, in *Opt. Lett.* Vol. 22, 1009, 1997.

2 A. Stingl, M. Lenzner, Ch. Spielmann, F. Krausz, and R. Szipöcs, in *Opt. Lett.* Vol. 20, 602, 1995.

3 I. T. Sorokina, E. Sorokin, E. Wintner, A. Cassanho, H. P. Jenssen, and R. Szipöcs, in *Opt. Lett.* Vol. 22, 1716, 1997.

4 Yanovsky, V.P., Wise, F.W., Cassanho, A., and Jenssen, H.P., in *Opt. Lett.* Vol. 20, 1304, 1995.

5 I. P. Christov, H. C. Kapteyn, M. M. Murnane, Ch.-P. Huang, and J. Zhou, in *Opt. Lett.* Vol. 20, 309, 1995.

6 A. Cassanho, V. K. Castillo, G. J. Quarles, H. P. Jenssen, S. Buchter, A. Bustamante, and D. Hagan, in *Conference proceedings LEOS '96*, Vol. 1, 96, 1996.

7 I. T. Sorokina, E. Sorokin, E. Wintner, A. Cassanho, and H. P. Jenssen, in *Avanced Solid State Lasers*, OSA Proceedings Series. Washington, DC: Optical Society of America, 400, 1998.

8 R. H. Stolen and W. J. Tomlinson, in *J. Opt. Soc. Am. B* Vol. 9, 565, 1992.

9 I. Kang, T. Krauss, and F. Wise, in *Opt. Lett.* Vol 22, 1077, 1997.

10 R. Adair, L. L. Chase, and S. A. Payne, in *Phys. Rev B* , Vol. 39, 3337, 1989.

Generation of amplified shaped pulses for highly adiabatic excitation

Matthew Fetterman, Debabrata Goswami, Dorine Keusters, June-Koo Rhee, and Warren S. Warren

Departments of Chemistry and Electrical Engineering, and the Center for Ultrafast Laser Applications, Princeton University, Princeton, NJ, 08544-1009

Abstract. Complex amplified pulses, including a 1 ps, 100 μJ tanh-swept sech pulse for adiabatic inversion, are generated experimentally. STRUT detection verifies the modulation and follows the dynamics induced by such pulses in Rb vapor. Applications to production of spin-polarized gases for medical imaging are discussed.

1. Introduction

Shaped femtosecond pulses have potential applications in areas ranging from optical communications, coherent control, and medical imaging to quantum computing. In this paper we demonstrate amplified pulse shaping with extremely high resolution at a wavelength of 800nm. Using an acousto-optic modulator (AOM) in a zero-dispersion line, we can impose both amplitude- and phase-modulation on the spectrum of a femtosecond laser pulse, allowing us to create an arbitrary pulse-shape. Specifically, we will discuss the uses of frequency swept laser pulses for generating arbitrary inversion profiles, and the application of such pulses to preparing large excited populations in optically dense media.

The pulse shaping system has been discussed in previous work[1]. A Mira mode-locked Ti-Sapph laser operating at 800nm and producing 150fs pulses provides the input for the pulse shaper, where both phase and amplitude modulation can be imposed on the pulse using the AOM. The output pulse shape is determined by the radio-frequency signal sent into the AOM. Following the pulse shaper, we have added a pulse amplification system, consisting of a Clark-MXR stretcher, regenerative amplifier, and compressor. The output pulse energy of the amplifier is 100μJ, at a repetition rate of 1kHz. The system has been optimized so that the pulse is amplified with high fidelity. An example of a highly amplitude-modulated pulse is shown in Fig 1. In this data set, a 200 MHz RF sine wave is amplitude modulated by a series of ones and zeros and applied to the AOM. Note that at this pulse energy the amplification process does not introduce crosstalk between the frequency components.

More complex amplified pulses can characterized using the Spectrally and Temporally Resolved Upconversion Technique, (STRUT) [2]. We use this advanced detection technique for the first time to characterize an arbitrarily generated pulse-shape, allowing us to demonstrate the full capabilities of the pulse shaper. STRUT is essentially a frequency-resolved autocorrelation-technique, where a narrow part of the spectrum in one of the arms is selected.

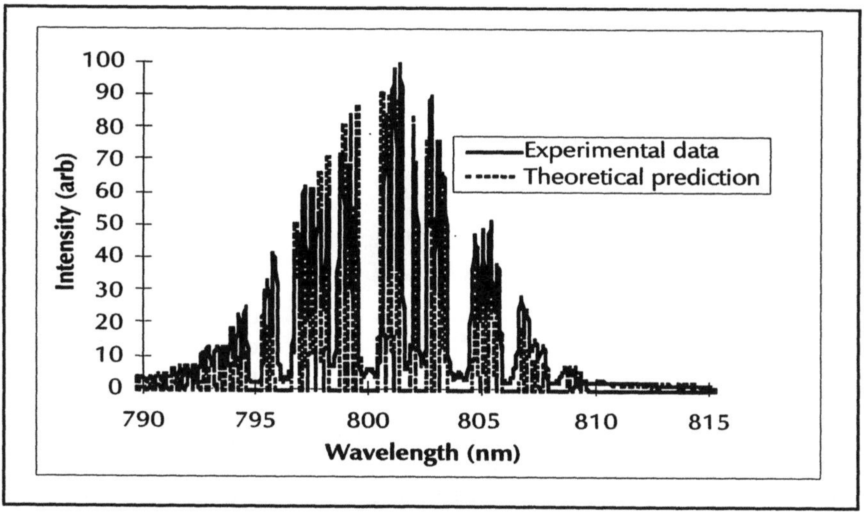

Figure 1. Measured spectrum of an amplified pulse produced by AOM femtosecond pulse shaping. In this case the pulse out of the oscillator was spectrally modulated with a series of ones and zeroes (the dotted function), then amplified to 100 μJ pulse energy. Note the absence of spurious frequency components, which correspond to intermodulation products produced by nonlinearities in the amplification.

This narrow pulse is then used to characterize the full pulse in the other arm of the autocorrelator. As an example of a pulse that is modulated both in amplitude and phase, Figs. 2 and 3 show the experimental and calculated shapes of a sech pulse with a tanh frequency-sweep. This modulation of this pulse can be described by the function $\mathcal{E}(t)=\mathrm{sech}(\alpha t)^{1+10i}$ (Figure 2, below). It has long been known theoretically that this type of pulse will give highly selective adiabatic inversion [3] and applications of the radiofrequency analogs of such pulses in magnetic resonance imaging are routine. The only previous experimental optical work [4], however, gave 10 ns resolution and 100 mW peak power, and was barely sufficient to achieve adiabatic inversion in I_2. The required peak power scales as the *inverse square* of the pulse length, hence amplified pulses are needed for ps or fs applications.

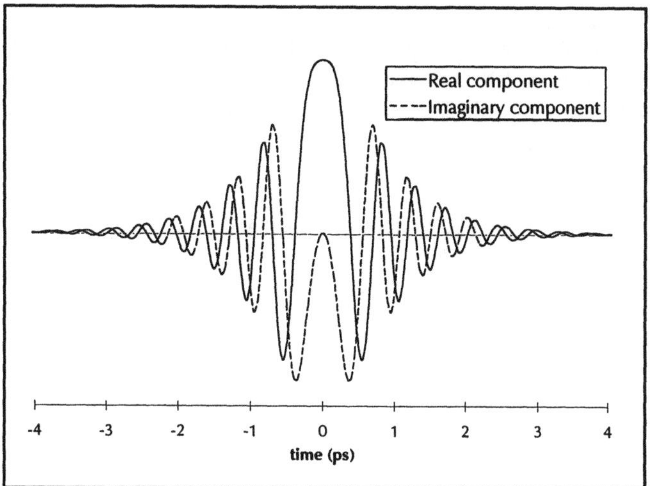

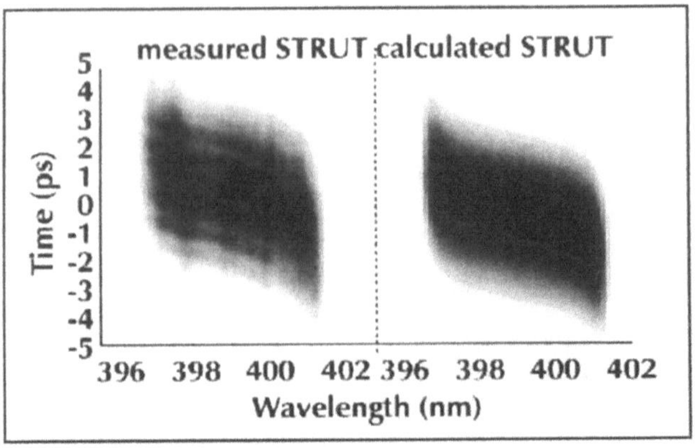

Figure 3. Measured and calculated STRUT for the sech(αt)$^{1+10i}$ pulse (100 μJ, 1 ps) illustrated above. Note the characteristic tanh frequency sweep (the pulse itself is centered at twice this wavelength). The clean sweep is an extremely sensitive test of linearity of the pulse amplification.

The peak power of approximately 25 MW is sufficient to easily achieve adiabatic inversion in common atomic and molecular systems. One interesting example of such a system, which we have been exploring, is to our knowledge the only chemical or medical "reagent" which is now prepared by laser excitation: spin polarized noble gases (helium-3 and xenon) for use in magnetic resonance imaging. The most successful preparation method to date involves optically pumping a cell containing 10^{11}-10^{14} atoms per cubic centimeter of Rb, the desired noble gas, and nitrogen to quench emission.[5] In a typical application, tens to hundreds of watts of resonant cw laser light continuously irradiate an optically thick pumping cell for many minutes, producing spin polarized gases with an overall efficiency substantially less than 1%. We have demonstrated that frequency swept pulses can produce dramatic alterations in the excitation dynamics even in optically dense cells. This permits far larger excited state populations than could be achieved by more conventional pumping methods, which may lead to a "shaped pulse" method for more efficient spin-polarized preparation.

References

1 J.X. Tull, M. A. Dugan, and W.S. Warren, Adv. Magn. Opt. Res., 20, 1997,1-63; M.A. Dugan, J.X. Tull and W.S. Warren, J. Opt. Sci. Am B, 14 (6), 1997, 1-11; C.W. Hillegas, J.X. Tull, D. Goswami, D. Strickland and W.S. Warren, Opt. Lett., 19 (10), 737-739,1994

2 J.-K. Rhee, T.S. Sosnowski, A.-C. Tien, T.B. Norris, J. Opt. Soc, Am. B., 13 (8), 1996, 1780-1785.

3 L. Allen, J.H. Eberly, Optical Resonance and Two Level Atoms, John Wiley & Sons Inc., New York, 1975

4 C. P. Lin, J. Bates, J. Mayer and W. S. Warren, J. Chem. Phys. 86, 3750 (1987).

5 T. G. Walker and W. Happer, Rev. Mod. Phys. 69, 629 (1997)

Transient Excited-State Absorption Measurement in Chromium Doped Forsterite

N. Zhavoronkov[1], V. Mikhailov[1], V. Petrov[2], and F. Noack[2]

[1] International Laser Center, Belarus State Polytechnical Academy, 65 F. Skoryna Ave., Minsk 220027, Belarus
e-mail: laser@org.by.

[2] Max-Born-Institute for Nonlinear Optics and Ultrafast Spectroscopy, Rudower Chaussee 6, Berlin D-12489, Germany
E-mail: petrov@mbi-berlin.de

Abstract. Excited state absorption (ESA) is one of the possible loss mechanisms reducing the efficiency in lasers based on chromium doped forsterite. In this work we study the kinetics of ESA with femtosecond temporal resolution using probe pulses in the ultraviolet (306 nm), visible (650 nm) and infrared (1500 nm), and estimate a nonradiative decay from the $^3T_1(^3F)$ to the $^3T_2$ metastable level of 3 ps.

1. Introduction

Tetravalent chromium (Cr^{4+}) ions are active laser centers in YAG, forsterite and other hosts [1, 2], in which tunable stimulated emission has been observed in the near infrared. The laser efficiency of Cr^{4+}-doped materials is, however, relatively low. Recently, ESA in the near infrared was shown to be an important source of losses [3]. ESA bands were observed from 1100 to 2100 nm in the infrared, from 600 to 800 nm in the visible, as well as from 200 to 400 nm in the ultraviolet. The main objective of the present study is to investigate the transient ESA behavior and to understand the role of the $^3T_1(^3P)$ and $^3T_1(^3F)$ energy levels in the ESA process.

2. Experimental Methods

The single crystals of Cr^{4+}:forsterite with ≈ 2 cm^{-1} absorption at 1.06 μm were grown by the Czochralski method. We used pump pulses of 100 fs duration either from a fs Ti:sapphire regenerative amplifier at 773 nm or from the frequency doubled signal output of a traveling wave optical parametric generator at 750 nm, pumped by the amplifier at 1 kHz repetition rate. Probe pulses of nearly equal duration were derived from the same optical parametrical generator. The pump beam was linearly polarized along the sample b-axis (E||b, P_{bmn} space group) and propagated along the a-axis for E||c probing and along the c-axis for E||a probing. Probe pulses at 306 nm (E||c), 650 nm (E||a) and 1500 nm (E||c) correspond to the maxima of the ESA observed previously in experiments with μs temporal resolution [3]. The kinetics of ESA is studied here by measuring the difference in the transmission ΔT with and without pumping the samples using lock-in modulation and averaging. The optical density variation ΔOD is calculated from the ΔT values and normalized to its maximum value. All measurements were performed at room temperature. To analyze the experimental results of ESA an energy level diagram shown in Fig.1 was assumed.

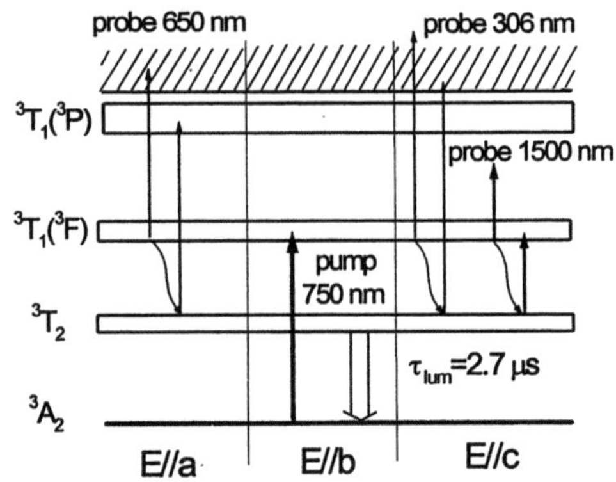

Fig. 1. An energy level diagram of Chromium doped Forsterite.

3. Results and Discussion

Recent estimations of the $^3T_1(^3F)$ level lifetime didn't have the necessary temporal resolution and provided only an upper limit of 15 ps [4]. Within this time the population reaches the $^3T_2$ metastable level by nonradiative decay. Probe pulses at 306 nm have enough energy for the charge transfer transition excitation both from the directly pumped $^3T_1(^3F)$ level and from the $^3T_2$ level. At this probe wavelength we observed fast rise time of the ESA corresponding to the absorption from the $^3T_1(^3F)$ level. This rise time was, however, identified as group mismatch between the pump and probe pulses. Pulses at 650 nm probe the ESA from the $^3T_1(^3F)$ band to the top of the $^3T_1(^3P)$ manifold and to the conductivity band. The ESA at 650 nm appears in the $^3T_2$-$^3T_1(^3P)$ channel during nonradiative decay to the metastable $^3T_2$ level and we have a combination of two ESA processes.

Using, however, probe pulses near 1500 nm it is impossible to excite the $^3T_1(^3P)$ level and ESA occurs only due to the $^3T_2$-$^3T_1(^3F)$ transition. In this case the measured group delay between the pump and probe pulses was negligible as compared to the pulse duration. The temporal evolution of the normalized optical density difference is shown in Fig.2 where the pump wavelength was 750 nm. The artifact at 0-time delay was identified as induced focusing which modified the spatial distribution of the probe beam on the detector. It served as a method to determine exactly the 0-delay point and to confirm the temporal resolution since the n_2-effect can be regarded as instantaneous. The $^3T_2$-$^3T_1(^3F)$ ESA rise time corresponds to the $^3T_1(^3F)$-$^3T_2$ relaxation time. The rate equations governing the population dynamics in the $^3T_2$, $^3T_1(^3F)$, and $^3A_2$ states were solved in order to study the evolution of the population density. The following approximations were used: the pump pulse is short compared to the $^3T_1(^3F)$ level lifetime, the radiative transition rate from the $^3T_1(^3F)$ band is negligible as compared to the nonradiative lifetime, ESA of the pump beam is ignored, and the pump beam is assumed to be Gaussian in time and „top-hat" in space. The difference of the optical density was calculated following Gayen et al. [5]. The best fit to the experimental data is shown in Fig.2 with a thick solid line and corresponds to a $^3T_1(^3F)$ level lifetime of 3 ps whereas the thin lines correspond to ±1 ps deviation from that estimation.

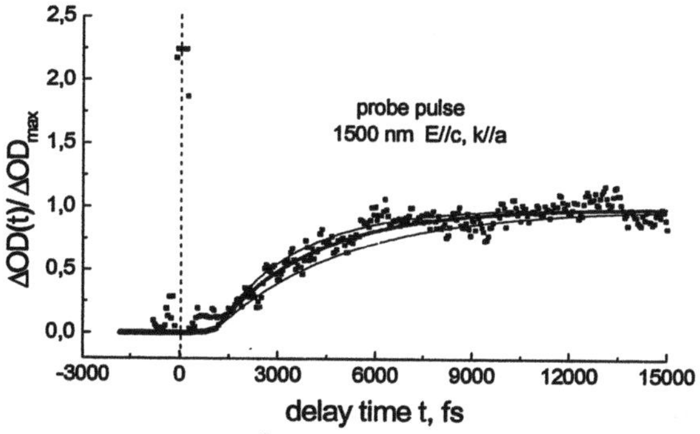

Fig.2. The temporal dependence of ESA for a probe pulse at 1500 nm.

4. Conclusions

Time resolved pump and probe measurements of ESA in Cr^{4+}:forsterite were performed with femtosecond resolution and the results were fitted using rate equations. The salient features of the normalized optical density curves are a rapid rise time followed by a slow decay. The lifetime for the nonradiative transition from $^3T_1(^3F)$ to the $^3T_2$ metastable level could be resolved for the first time yielding 3 ps, and the long decay of the curves observed reflects the radiative depopulation of the metastable $^3T_2$ level (2.7 μs).

References

1. V. Petricevic, S. K. Gayen, and R. R. Alfano, in *Appl. Phys. Lett.*, **53**, 2590, 1988.
2. N. Borodin, V. Zhitniuk, A. Okhrimchuk, and A. Shestakov, in *Izv. Akad. Nauk SSSR, Ser. Fiz.*, **54**, 1500, 1990.
3. N. V. Kuleshov, V. G. Shcherbitsky, V. Mikhailov, S. Hartug, T. Danger, S. Kueck, K. Petermann, G. Huber. in *J. of Luminescence*, **75**, 319, 1997.
4. K. V. Yumashev, N. V. Kuleshov, P. V. Prokoshin, A. M. Malyrevich and V. P. Mikhaillov , in *Appl. Phys. Lett.*, **70**, 2523, 1997.
5. S. K. Gayen, W. B. Wang, V. Petricevic, and R. R. Alfano, in *J. of Luminescence*, **47**, 181, 1991.

Ultrafast Fiber Lasers

Erich P. Ippen[1,2], David J. Jones[1], Lynn E. Nelson[1] and Hermann A. Haus[1]

[1] Department of Electrical Engineering and Computer Science, [2] Department of Physics, Research Laboratory of Electronics, Massachusetts Institute of Technology, Cambridge, MA 02139
E-mail: ippen@mit.edu

Abstract. Erbium-doped fibers are making it possible to develop compact, diode-pumped sources of ultrashort pulses for a variety of applications. Operating principles of these devices are reviewed and recent experimental progress in our laboratory on soliton and stretched-pulse lasers is described.

1. Introduction

Fiber lasers are attractive sources of ultrashort pulses for several reasons. They produce pulses at 1.55 μm fiber communication wavelengths; they can be frequency doubled to Ti:sapphire wavelengths; they are low noise; and they can be diode-pumped. Their operating characteristics, however, depend strongly on the choice of cavity design, the balance between fiber nonlinearity and dispersion, and internal spectral filtering. Results achieved with a variety of different systems have been reviewed in the literature [1,2,3,4]. In this paper we discuss recent work in our laboratory on soliton and stretched-pulse ring laser geometries. Soliton ring lasers can produce pulses with sub-500-fs durations and energies of tens of picojoules. When they are harmonically driven, they can produce stable pulse trains at greater than 10-GHz rep rates or act as memory rings to maintain pulse patterns (data packets) for long periods of time. Stretched-pulse lasers, on the other hand, are designed to avoid the pulse shortening and energy limitations inherent to soliton rings. They can produce sub-100-fs pulses and pulse energies as high as several nanojoules. Frequency-doubled, they are attractive as seed sources for femtosecond Ti:sapphire amplifier systems.

2. The Soliton Laser

Self-starting passive modelocking of fiber ring lasers is achieved with the artificial saturable absorber technique called Polarization-Additive-Pulse-Modelocking (P-APM) [5]. In lasers with net anomalous dispersion around the ring, conventional soliton formation determines much of the steady-state output character; but P-APM provides the necessary suppression of cw oscillation. A typical soliton ring cavity [6] is shown in Figure 1. The Faraday isolator ensures unidirectionality and fiber-loop polarization controllers provide the polarization bias needed for the P-APM. Traces in the lower part of Figure 1 show the 40-MHz, 450-fs pulses.

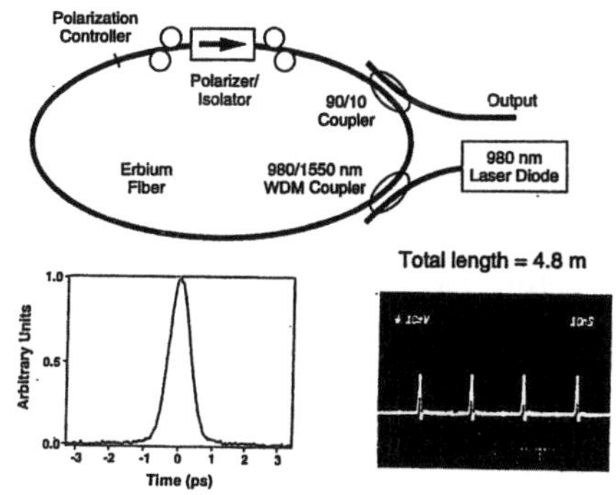

Fig. 1. All-fiber soliton laser; insets show pulse autocorrelation and an oscilloscope trace of the output

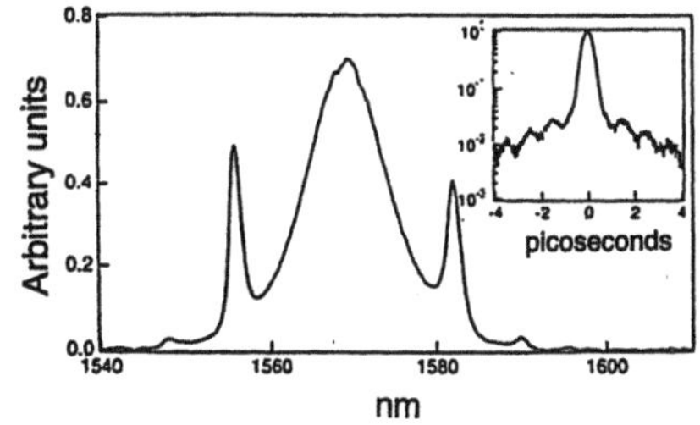

Fig. 2. Spectrum and SHG autocorrelation (inset) of pulses from a soliton laser

The output spectrum, shown in Figure 2, indicates that the spectral width, and therefore the minimum pulse duration, is limited by the spectral sidebands formed by roundtrip-periodic pulse perturbations [7]. This pulsewidth limit then also imposes a limit on the soliton energy via the soliton area theorem. With a spectral filter inside the cavity, the sidebands can be reduced dramatically; but, for shorter pulses and more single-pulse output power, one needs to make the cavity shorter. In all-fiber systems, however, the required length of gain fiber results in cavity lengths of 4 meters or more. One approach toward shorter cavity lengths is the use of more highly doped glass-waveguide amplifiers. A schematic of a laser incorporating such a waveguide amplifier into a fiber ring is shown in Figure 3 [8]. This laser generated 113-fs solitons with energies of 160 pJ at a rep rate of 130 MHz. Work toward shorter cavities, with perhaps semiconductor saturable absorbers for self-starting, is in progress. A potential advantage of such waveguide devices is the possibility of further integration and on-chip fabrication of other laser components.

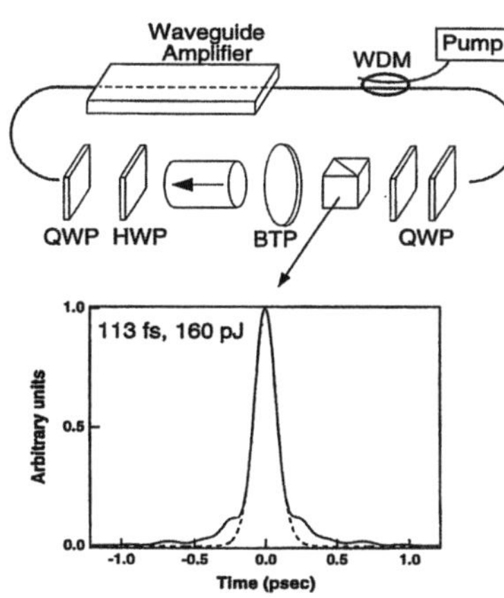

Fig. 3. Schematic and output pulse of an Er/Yb glass waveguide laser.

3. Harmonic Modelocking

With a fiber soliton laser, high (multi-GHz) repetition rates can be achieved by using an electro-optic modulator in the loop. The modulator is driven at a high harmonic of the cavity roundtrip rate to facilitate the formation of multiple solitons in the ring. Two conditions are required for stability: (1) that the roundtrip loss for the filtered solitons be less than that for low intensity pulses produced by active modelocking alone, and (2) that the modulator itself must not drive the solitons unstable. These stability conditions have been investigated using a polarization-maintaining, 45-m ring operating at 5 GHz. By varying modulation depth and filter bandwidth Jones et al [9] investigated a range of operating points. Figure 4a presents their parametric map, with lines indicating the boundaries proscribed by conditions (1) and (2). Operating points between the curves were characterized by stable pulses; those outside showed pulse pedestal build-up. Figure 4b illustrates this deterioriation with autocorrelation traces taken at the specific points indicated.

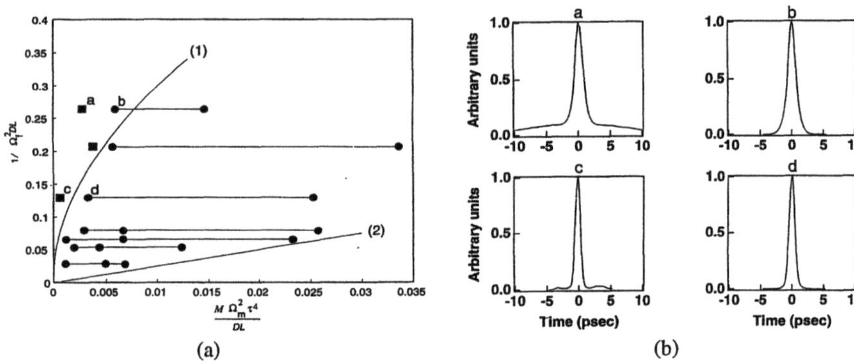

Fig. 4. (a) Stability diagram for solitons; and (b) SHG autocorrelations of selected points

4. An Optical Memory

Harmonic modulation and soliton stability also have important consequence for optical memories – storage rings in which patterns of pulses (optical data) circulate. Theory and experiments show that, with proper adjustment of nonlinearity and filtering as well as loop gain, patterns can be maintained essentially indefinitely. The storage of packets of up to 3.8 kbits of 50-Gbit/s data has been demonstrated [10]. Recently, we have explored a new mode of operation in which a phase modulator in the ring is driven at a frequency slightly different from that of the cavity harmonic [11]. This asynchronous modulation still provides significant noise suppression, but it places less stringent requirements on cavity length stability and the precise parameters of pulse packets being loaded into the memory. The experimental setup in shown in Figure 5. A 5-GHz data packet was produced by an externally-modulated, modelocked semiconductor laser. A control beam, produced by a gain-switched semiconductor laser and turned off when a packet was injected, was used to minimize relaxation oscillations. After each round trip in the buffer, the packet was monitored by a fast detector and sampling oscilloscope. Recordings of a packet before and after 450 circulations (44 μsec) are shown in Figure 6. The apparent widening of the pulses was caused by oscilloscope jitter which prevented pulse-resolved observation at

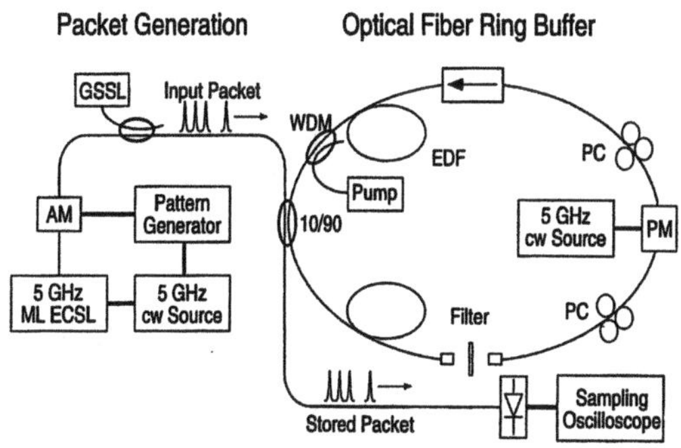

Fig. 5. Asynchronous optical buffer with connections for packet loading and monitoring

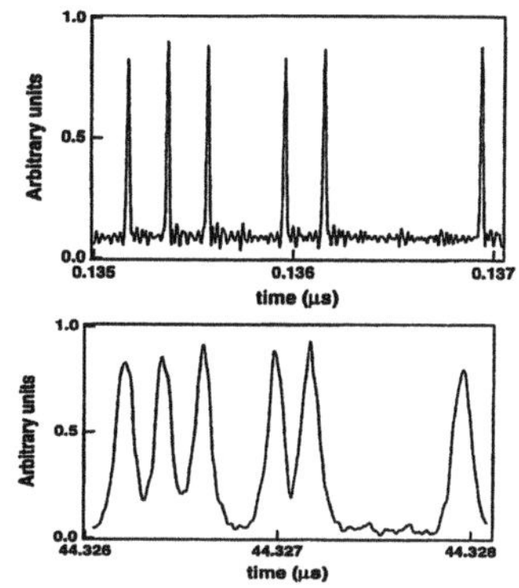

Fig. 6. Oscilloscope traces at time of loading (above) and after 44 µsec

longer times. Well-resolved packet envelopes were observed, however, for more than 1600 circulations (157 µsec), and even longer storage times seem possible.

5. The Stretched-Pulse Laser

To circumvent the pulse shortening limitation encountered in soliton lasers, Tamura et al [12] devised a ring comprised of two segments of fiber with strongly opposite dispersion. Modelocking is achieved by P-APM, as in the soliton laser; but, as the pulse gets shorter, it is chirped, alternately positively and negatively, by the two fiber segments. The shorter it gets, the more strongly it is stretched and

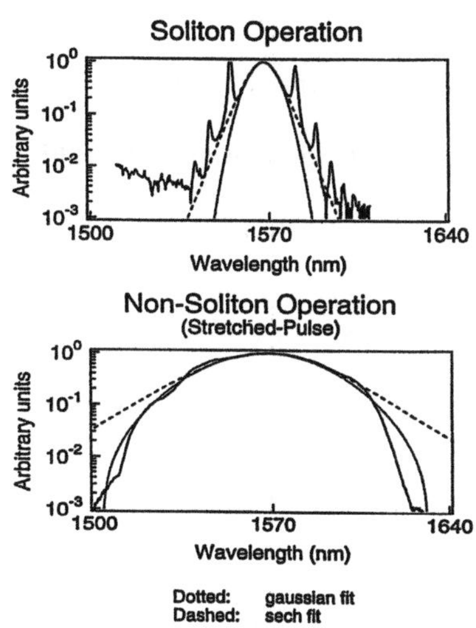

Fig. 7. Comparison of soliton and stretched-pulse laser spectra

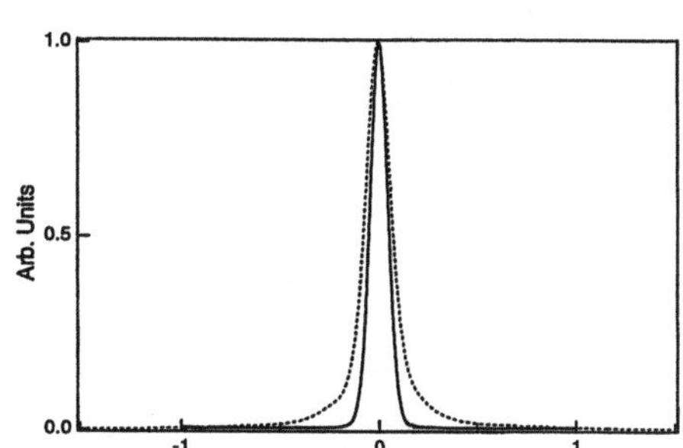

Fig. 8. Autocorrelations of frequency-doubled stretched-pulse laser pulses (solid curve) and of fundamental pulses (dashed curve)

the shorter is the distance over which its peak power is high. Thus, the effective nonlinearity is automatically adjusted in such a way as to prevent the peak power from being clamped. Sideband generation is also greatly diminished. As a result, dramatically higher energies and shorter pulses are produced. Soliton and stretched-pulse laser spectra are compared in Figure 7. The soliton case exhibits sidebands and a shape corresponding to a sech pulse (dashed theoretical curves); the stretched-pulse spectrum is broader, relatively free of sidebands and more gaussian like (dotted theoretical curves).

Stretched-pulse lasers, have produced pulses as short as 63 fs [13] and energies as high as 2.7 nJ [14]. Their broad bandwidths make it possible to create wavelength-tunable sub-picosecond pulses (by filtering) or to generate many wavelength channels simultaneously for dense wavelength-division-multiplexed (WDM) optical networks [15]. Their high peak powers permit efficient frequency doubling to 775 nm for femtosecond pulse amplification using Ti:sapphire amplifiers. As shown in Figure 8, the frequency-doubled pulses are less than 90 fs in duration and are pedestal free [14]. Their 270 pJ energies are more than sufficient to seed high-power amplifiers; so diode-pumped, frequency-doubled stretched-pulse lasers can be compact, less expensive replacements for argon-laser-pumped Ti:sapphire oscillators in some amplified systems.

Acknowledgements: This work was supported at MIT by AFOSR and DARPA. EPI also gratefully acknowledges support by the Alexander von Humboldt Foundation. LEN is presently with Lucent Technologies, Bell Laboratories.

References

1. I. N. Duling III, *Compact Sources of Ultrashort Pulses*, Cambridge University Press, Cambridge UK, 1995
2. M. E. Fermann, A. Galvanauskas, G. Sucha and D. Harter, Appl. Phys. B **65**, 259 (1997)
3. L. E. Nelson, D. J. Jones, K. Tamura, H. A. Haus and E. P. Ippen, Appl. Phys. B **65**, 277 (1997)
4. D. J. Jones, H. A. Haus, L. E. Nelson and E. P. Ippen, IECE Trans. Electron. **E81-C**, 180, (1998)
5. H. A. Haus, J. G. Fujimoto and E. P. Ippen, J. Opt. Soc. Am. B **8**, 2068 (1991)
6. K. Tamura, H. A. Haus and E. P. Ippen, Electron. Lett. **28**, 2226 (1992)
7. S. M. J. Kelly, Electron. Lett. **28**, 806 (1992)
8. D. J. Jones, S. Namiki, D. Barbier, E. P. Ippen and H. A. Haus, IEEE Photon. Tech. Lett. **10**, 666 (1998)
9. D. J. Jones, H. A. Haus and E. P. Ippen, Opt. Lett. **21**, 1818 (1996)
10. J. D. Moores, W. S. Wong and K. L. Hall, Opt. Lett. **20**, 2547 (1995)
11. D. J. Jones, K. L. Hall, H. A. Haus and E. P. Ippen, Opt. Lett. **23**, 177 (1998)
12. K. Tamura, E. P. Ippen, H. A. Haus and L. E. Nelson, Opt. Lett. 18, 1080 (1993)
13. K. Tamura, E. P. Ippen and H. A. Haus, Appl. Phys. Lett. **67**, 158 (1995)
14. L. E. Nelson, S. B. Fleischer, G. Lenz and E. P. Ippen, Opt. Lett. **21**, 1759 (1996)
15. L. Boivin, M. C. Nuss, W. H. Knox and J. B. Stark, Electron. Lett. **33**, 827 (1997)

First Experimental Observation of Polarization Locked Temporal Vector Solitons

B. C. Collings[1,5], S. T. Cundiff[2], N. N. Akhmediev[3], J. M. Soto-Crespo[4], K. Bergman[1], W. H. Knox[5]

Addresses: [1]Department of Electrical Engineering, Princeton University, Princeton, NJ 08544, USA. [2]JILA, University of Colorado and NIST, Boulder, CO 80309, USA. [3]Australian Photonics Cooperative Research Centre, Optical Sciences Centre, The Australian National University, Canberra, ACT 0200, Australia. [4]Insituto de Optica, Consejo Superior de Investigaciones Cientificas, Serrano 121, 28006 Madrid, Spain. [5]Bell Laboratories, Lucent Technologies, Holmdel, NJ 07733, USA.

Abstract. We experimentally observe polarization locked temporal vector solitons, i.e. solitons that preserve their polarization despite a birefringent environment, in a passively mode-locked fiber laser. The two intracavity orthogonal polarization modes of a 420 fs soliton pulse phase lock when the differential nonlinear phase shift arising from their unequal intensities compensates for the linear cavity birefringence. The stabilization mechanism involving the nonlinear coherent energy coupling between components is described.

Significant theoretical research has been devoted to the study of short pulse propagation in birefringent optical fiber. This has resulted in predictions of soliton solutions that propagate as a single particle-like unit, maintaining their shape and possibly state of polarization [1-4]. These classes of solitons represent examples of vector solitons due to their nonlinearly coupled, two mode composition [2]. Propagation of temporal vector solitons has generally been considered in the regimes of high and low birefringence. With high birefringence (unequal group velocities), the nonlinear coherent energy coupling (CEC), (also known as four-wave mixing) can be neglected and stable solitons preserving their pulse shape, but not their polarization, can fórm [2]. The two components, having different center wavelengths, are coupled through cross-phase modulation (CPM) and propagate with an equal group velocity. In low birefringence (standard telecommunication) fiber where the group velocity difference is negligible, solitons that preserve their pulse shapes are a natural consequence. However, since the CEC cannot be ignored for this case, a class of vector solitons, polarization locked vector solitons (PLVSs), with locked phase velocities and a stationary polarization state has been predicted [4,5]. Recently, experimental evidence of the nonlinear instability of the fast axis and stable elliptically polarized states has been observed in single pass, short pulse propagation experiments [6]. However, due to the randomness of the birefringence of standard single mode fibers hindering the clear isolation of orthogonal polarization modes, the experimental study of PLVSs in transmission systems has been difficult.

In this paper, we report the first experimental observations, to our knowledge, of PLVSs. In our previous work, we observed locking of the output polarization [7]. With the detailed characterization of the intracavity pulses presented here, we determine that PLVSs are forming within the cavity. The pulses are fundamental solitons of the averaged cavity nonlinearity and dispersion. Due to the weak perturbations (amplitude, nonlinear and dispersive) per round-trip (<5%), the laser approaches a conservative system. The laser cavity provides a periodic and controllable birefringent environment such that the orthogonal polarization modes along the principle axes of the cavity are easily defined and measured. Thus, the laser corresponds to a (infinitely) long periodic soliton propagation and the nonlinear polarization evolution of the pulses is tractable. Several theoretically predicted features of PLVSs are clearly observed, including stationary elliptical polarization states in the presence of small, non-zero linear birefringence.

The laser is shown schematically in Fig. 1 and described in Ref. [8]. Passively modelocked by a saturable Bragg reflector (SBR), it consists of standard single mode and Er/Yb fiber with a repetition rate of 48 MHz. The fundamental solitons are typically 420 fs in width (FWHM) with pulse energies around 90 pJ resulting in a soliton period of 4.5 m, slightly longer than the round-trip length of the cavity. For this reason, the soliton responds primarily to the averaged cavity parameters, rather than the local environment. For control of the cavity birefringence, two fiber polarization controllers (FPCs, 3 wraps around a 5 cm mandrel approximating $\pi/2$ radians of retardance) adjustable in angle are inserted near one end of the cavity. The total retardance of the cavity is the sum of the contribution from the two FPCs and any naturally occurring birefringence present in the fiber or fusion splices.

For this work, the total cavity retardance is low (<0.2 radians per round-trip) and the two orthogonal polarization components are defined along the principle cavity axes. The output of the laser is passed through a linear polarizer and detected by a fast photodiode connected to a rf spectrum analyzer [7]. When the orthogonal components of the soliton are not phase locked or the laser is operating CW, the output polarization evolves in time due to the beating between orthogonal components. The principle axes are determined by rotating the polarizer to null this beating. The beat frequency is used to determine the magnitude of the birefringence. All measurements of the cavity axes and retardance are performed under CW operation. The stationary output polarization state is measured by a HP 8509 polarization analyzer.

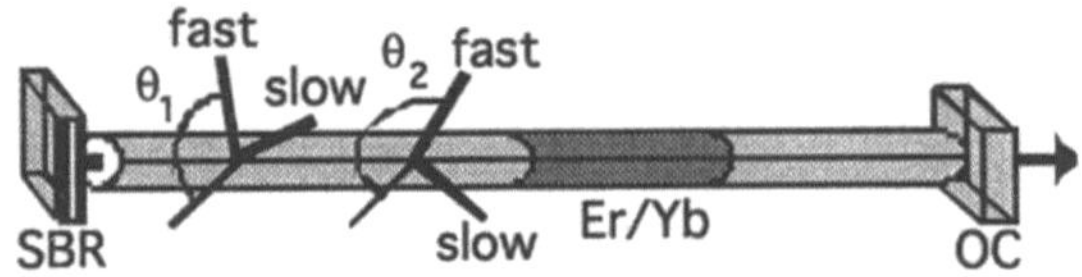

Fig. 1. Schematic of the laser cavity (OC, output coupler). The variable FPCs are indicated as fast and slow axes.

For small, yet non-zero, amounts of cavity birefringence, the two orthogonal polarization components become phase locked producing a stationary elliptical output polarization state. This corresponds to a PLVS in the laser cavity. This PLVS has a constant polarization state for the entire pulse since the intensity through a linear polarizer can be nulled. The range of cavity birefringence over which this locking occurs is pulse energy dependent as shown in Fig. 2a. The ratio of the intensities of the polarization modes and their relative optical phase is plotted as function of cavity retardance in Fig. 2b.

For the PLVS to maintain a stationary polarization state, each polarization component must receive an equal net phase delay, both linear and nonlinear, per round-trip. The unequal nonlinear phase shifts experienced by the two components of unequal intensities compensates the linear retardance. This is clearly depicted in Fig. 2b. Where the retardance is a minimum, the two components are nearly equal in intensity. However, as the retardance is increased, the ratio decreases.

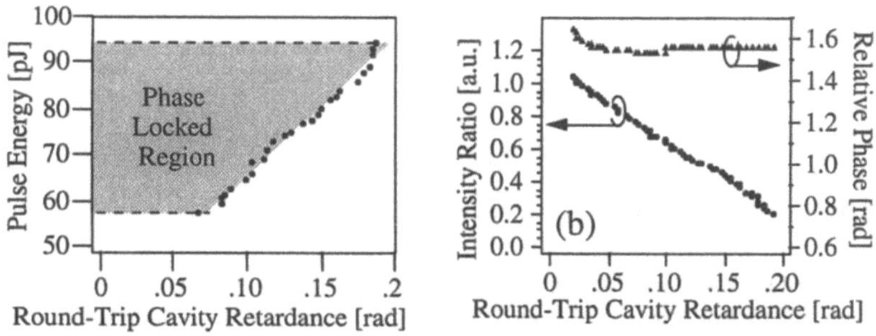

Fig. 2a. Plot of the phase locking region versus the round-trip cavity retardance illustrating the pulse energy dependence of the locking. **2b.** Plot of the ratio of the intensities in the polarization components and the relative optical phase between components.

This configuration alone is not stable. Stabilization occurs through the action of both the CEC and CPM. The direction and magnitude of the energy transfer between polarization components due to CEC dependents on their relative optical phase. Therefore, if a perturbation causes a change in the relative phase, energy is transferred to the proper component, increasing the nonlinear phase shift of that component and correcting the initial phase perturbation. Similarly, amplitude perturbations cause a nonlinearly induced phase perturbation which in turn causes energy transfer via CEC. For a stationary configuration, no net energy transfer can take place between components. For CEC, this occurs for a relative phase of $\pm\pi/2$ radians as experimentally measured (Fig. 2b). Numerical investigations also show the CEC to be critical to the stability of the PLVS [5].

In conclusion, we have experimentally observed (the first to our knowledge) temporal PLVSs in a passively modelocked fiber laser. The stable region of total cavity birefringence is energy dependent. The polarization state of the vector solitons is stationary and the relative optical phase between components is approximately constant over the entire birefringence range. The intensity ratio varies with cavity birefringence such that the net phase velocities (linear plus nonlinear) are equal. Stability is provided through the combined action of the CEC and CPM.

References

1. D. N. Christodoulides and R. I. Joseph, *Optics Lett.* **13**, 53 (1988).
2. C. R. Menyuk, *IEEE J. Quantum Electron.* **25**, 2674 (1989).
3. S. G. Evangelidies Jr., L. F. Mollenauer, J. P. Gordon and N. S. Bergano, *J. Lightwave Technol.* **10**, 28 (1992).
4. N. Akhmediev, A. Buryak and J. M. Soto-Crespo, *Optics Comm.* **112**, 278 (1994)
5. N. N. Akhmediev, J. M. Soto-Crespo, S. T. Cundiff, B. C. Collings and W. H. Knox, *Optics Lett.* **23**, 852 (1998).
6. Y. Barad and Y. Silberberg, *Phys. Rev. Lett.* **78**, 3290 (1997).
7. S. T. Cundiff, B. C. Collings and W. H. Knox, *Opt. Express* **1**, 12 (1997); www.osa.org.
8. B. C. Collings, K. Bergman, S. T. Cundiff, S. Tsuda, N. Kutz, J. E. Cunningham, W. Y. Jan and W. H. Knox, *IEEE Sel. Topics Quantum Electron.* **3**, 1065 (1997).

Time-Domain Jitter Measurement of Modelocked Fiber Lasers

Gregg Sucha, Martin. E. Fermann, and Donald. J. Harter

IMRA America,1044 Woodridge Ave.,Ann Arbor, MI 48105, USA
E-mail: gsucha@imra.com

Abstract: The timing jitter of modelocked Er:doped fiber lasers is measured using a new time-domain method involving cross-correlation. The measured jitter is compared with the quantum-limited timing jitter predicted by the theory of Haus & Mecozzi.

1. Introduction

Just as ever shorter pulsewidths have been a goal of laser development, so also will be the development of lasers with extremely low timing jitter. The recent development of low-noise modelocked lasers with nearly quantum-limited timing jitter will prove to be of great utility in high-precision measurement systems employing ultrafast lasers. Characterization of these low-jitter lasers requires improved jitter measurement techniques over conventional RF techniques.[1] Additionally, new techniques are required to measure relative timing jitter between two modelocked lasers, as rapid-scanning dual laser systems emerge as workhorse measurement systems. We are particularly interested in the time range of 10's of microseconds to several milliseconds as these time scales are the most relevant for rapid-scanning data acquisition. Here, we present time-domain jitter measurements[2] of passively modelocked fiber lasers. Using this new technique, we find a jitter growth rate of 0.25 ps/ms or 0.25 parts-per-billion (ppb). This is compared with theoretical predictions of timing jitter in passively modelocked lasers.

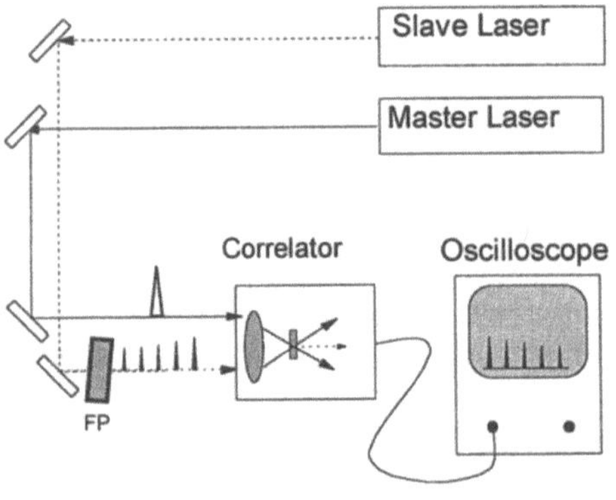

Figure 1. Jitter measurement system using cross-correlator and Fabry-Perot (rattle-plate) for timing calibration.

2. Experimental Method

Laser jitter is measured directly in the time domain by cross-correlating two identical modelocked fiber lasers via sum-frequency mixing (Fig. 1). This method gives much higher accuracy than frequency domain methods due to the very high timing accuracy which is inherently available with ultrashort lasers. The laser system consists of a pair of identical Er:fiber lasers wrapped on a common spool [3], each generating 600 fsec pulses, and having a repetition frequency of 4.6 Mhz. Co-wrapping the lasers greatly reduces the effects of environmental perturbations on timing jitter, and allows better access to the intrinsic laser properties. A cavity length mismatch is introduced which causes the lasers to scan through each other at an offset frequency of Δf ~5 Hz. A Fabry-Perot etalon is inserted into one beam to act as a rattle-plate, giving a series of timing calibration pulses. We used etalons with spacings of 50 ps and 500 ps. The peak positions in a sweep are measured to ±10 fs accuracy (within a 95% confidence interval) by fitting them with gaussian lineshapes. Nominally, the pulse spacings should be equal, but jitter causes slight deviations from a constant scan rate. Statisitcal analysis of these deviations gives the time-dependence of the jitter, shown in Fig.2. Jitter is measured over times ranging from 50 μs to 2.0 ms which are relevant time scales for data acquisition during a rapid scanning measurement. The jitter growth is found to be linear in time (Fig.2) with a jitter growth rate of 0.25 ps/ms or 0.25 parts-per-billion (ppb).

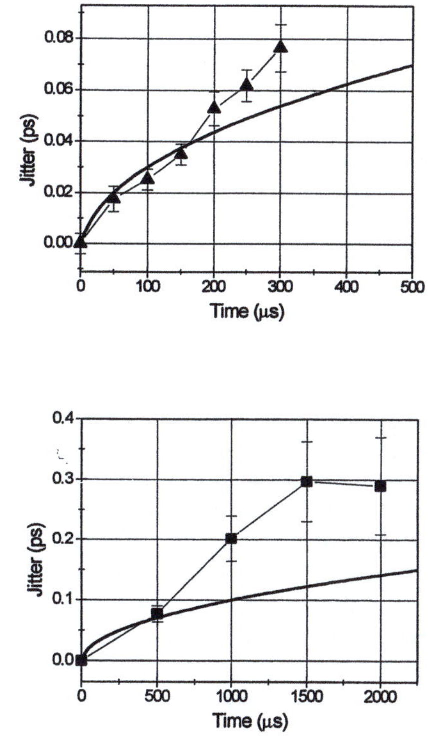

Figure 2. Timing jitter growth for modelocked fiber lasers. Theory (solid) and experimental (points). Time range up to 0.4 ms (upper) and up to 2.0 ms (lower).

3. Theory and Discussion

Neglecting other noise sources, the quantum-limited timing jitter of a passively modelocked laser is given by the Haus-Mecozzi theory[4]:

$$\Delta t_J(T) = 4D\sqrt{\frac{g\theta}{3}\frac{h\nu}{w_0}\Omega_g^3\tau^2}\left[\frac{T}{\tau_p} - 1 + \exp(T/\tau_p)\right]^{1/2}$$

The laser parameters are: gain (g = .7), enhancement factor ($\theta \sim 2$), bandwidth (Ω_g = 18 ps^{-1}), cavity dispersion (D = .04 ps^2), round-trip time (T_R = 217 ns), pulse energy (w_0 = 10 pJ), and pulsewidth (τ = .4 ps). These parameters give a jitter relaxation time of $\tau_p = (3/4g)(\Omega_g\tau)^2T_R = 10\ \mu s$. According to theory, jitter grows linearly for $T < \tau_p$ and grows as $T^{1/2}$ for $T > \tau_p$. The calculated jitter is plotted in Fig. 2 along with the measured jitter. The agreement is reasonably good for times T < .5 ms although the functional form is different. For longer times, the theory and the data diverge further. The linear growth of the jitter for times T < 1.5 ms differs from the random-walk calculations of quantum-limited timing jitter which predicts a $T^{1/2}$ dependence for times T > 10 μs suggesting that our laser jitter is not quantum limited for times T > .5 ms. This indicates the existence of an additional jitter noise source with a high frequency cutoff somwhere below 500 Hz, which could be due to a number of external factors including pump laser fluctuations.
Time domain jitter measurements provide the high accuracy needed to measure the small timing jitter of low-noise modelocked lasers. In addition, time-domain representation of the jitter provides very straightforward comparison with existing theories of timing jitter, and therefore enables us to identify and eliminate various sources of timing jitter.

References

[1]D. von der Linde, *App. Phys. B*, **39**, 201 (1986).
[2] G. Sucha, M. E. Fermann, and D. Harter, CLEO, San Francisco, CA, 1998.
[3] G. Sucha, M. Hofer, M. E. Fermann, F. Haberl, and D. Harter, *Opt. Lett.* **21,** 1570 (1996).
[4] H. A. Haus and A. Mecozzi, *IEEE J. of Quantum Electron.* **QE-29**, 983 (1993).

0.28 TW laser system at 1 kHz, Scaleable to 2 TW at 1 kHz

Sterling Backus, Charles G. Durfee III, Margaret M. Murnane, Henry C. Kapteyn

Center for Ultrafast Optical Science,
University of Michigan,
2200 Bonisteel BLVD, Ann Arbor, MI 48109-2099
Phone:(734) 936-0248, FAX:(734) 763-4876
e-mail:sbackus@eecs.umich.edu

Abstract. We demonstrate the full characterization of a high average power 1 kHz laser system, producing 4.5 mJ in 16 fs (0.28 TW). It uses cryogenic cooling of the power amplifier crystal to eliminate thermal distortions. The laser system is scalable to peak powers of > 1 TW, and we are currently constructing a 2 TW version, which uses diode pumped pump lasers, in a very compact design. This new laser system, will produce 15 fs pulses with an energy of 28 mJ at a 1 kHz repetition rate.

1. Introduction

High average power and high peak power ultrafast laser amplifiers have many applications in science and technology. For example, systems which can generate >14W of average power, with pulse durations of 5 - 20fs, have been used to generate ultrafast coherent ultraviolet and soft-x-ray pulses [1-4]. At the present time these high average power, high-repetition-rate, lasers are rapidly approaching the terawatt peak power level, which will make them useful for experiments usually reserved for low repetition rate systems (~10 Hz) [5]. As these laser systems approach pulse durations well below 20fs, difficulties arise because of high-order phase aberrations associated with the high damage threshold reflective dielectric optics which must be used. Thus far it has been possible to implement spectral phase correction to fifth-order in the Taylor expansion for phase. Another technical difficulty with the high average power systems are thermal problems as a result of the high heat load induced by the pump light with the > 30% quantum defect of the Ti:sapphire amplifier material.

In this paper we report the highest average/peak power laser system demonstarted to date. It represents a 3 times increase in peak power at 1kHz repetition rate than any other laser system demonstrated in this class. The combination of spectral phase control and cryogenic thermal management has alleviated many of the problems associated with high average power femtosecond laser systems. We present not only a state of the art laser system, but also the next generation of high average power/high peak power lasers, which will allow us to obtain ~ 2TW, with a variable repetition rates between 500Hz -1kHz.

2. Experimental Setup

Our laser system consists of a short pulse oscillator, an all-reflective stretcher, prism pair (for fourth-order dispersion compensation) a ring pre-amplifier with intra-cavity etalon, a cryogenic power amplifier, and a two-grating compressor with high 70% throughput [6-8]. The cryogenic power amplifier consists of a "cryotiger" cold head (APD) and compressor, and an evacuated cell with brewster windows. The cold tip contains a getter for trapping residual molecules, and ensures that the crystal face remains clear. The Ti:sapphire crystal is held at a temperature of 125K, and the cold head can remove up to 20W of thermal power at 125K. The cool-down time is relatively short at 20-min. The crystal is pumped from both sides with 18W of total absorbed power from two Quantronix Nd:YLF lasers. The pre-amplifier contains an intra-cavity etalon to suppress the peak of the gain profile, and reduce the effects of gain narrowing [5]. A bandwidth of 64nm is obtained for moderate modulation using a 2.5μm nitro-cellulose pellicle, angle tuned for maximum output bandwidth. The output of ~800 μJ from the pre-amplifier is then injected into the power amplifier, where it makes four passes through the gain medium. The output from the power amplifier is 6.5mJ (6.5W), giving an extraction efficiency of > 30%. The pulses are then recompressed to 4.5mJ, with 16fs pulse duration, corresponding to a peak power of 0.28TW.

3. Pulse Measurement and Model

For achieving short pulse durations it is important to keep the phase distortions of the entire system to a minimum. Multi-pass amplifiers are more suited for short pulse operation as they do not suffer from large material lengths and a high number of reflections from the dielectric high power optics which contain a certain amount of phase distortion. Characterizing ultrashort pulses has been made easier and highly accurate with the advent of FROG. This technique allows the full retrieval of the intensity and phase of an ultrashort pulse. This has allowed us to fully characterize and correct our amplifier system with high accuracy. Figure 1 shows the measured compressed pulse, using second harmonic FROG. The measured and predicted group delay of the system is shown in Fig 1(a), and the output pulse in Fig. 1(b). The model used to predict the output of the laser is quite comprehensive. It includes gain narrowing, gain dispersion, and spectral shaping due to the amplification of a chirped pulse. We also incorporate all the phase contributions from materials, reflective optics, gratings, prisms, and gain dispersion. This model has a high degree of accuracy in predicting the output from this laser system, as can be seen from Figure 1.

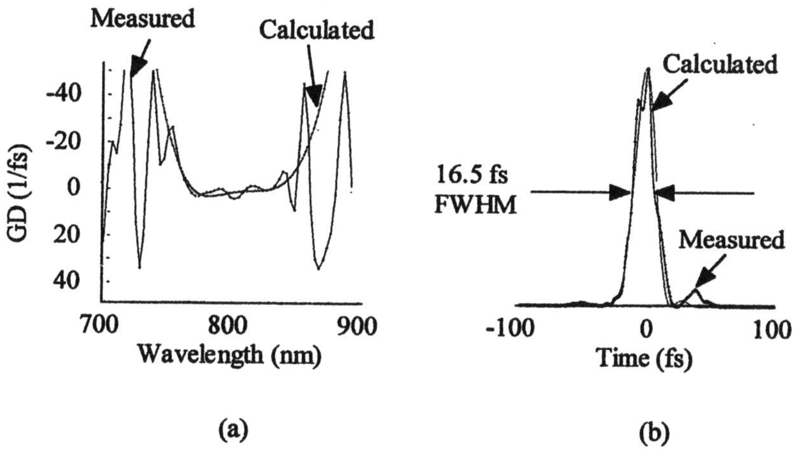

Fig. 1. Calculated and measured group delay of laser system (a), and measured and calculated pulses (b).

4. Conclusion

Under construction at present is a ~2TW laser system at a repetition rate of 1kHz. This system will use two 50mJ diode-pumped Nd:YAG slab lasers running at 1kHz for the pump source. This laser will also include a pulse shaper for arbitrary group delay correction and amplitude modulation. The laser output will be 28mJ, in 15fs, at 1kHz.

References

1 M. Nisoli, S. D. Silvestri, O. Svelto, *et al.*, Optics Letters **22**, 522 (1997).
2 S. Backus, C. Durfee, M. M. Murnane, *et al.*, Optics Letters **22**, 1256 (1997).
3 S. Backus, C. Durfee, M. Murnane, H. Kapteyn, Rev. Sci. Inst., 69, 1207 (1998).
4 S. Backus, J. Peatross, C. P. Huang, *et al.*, Optics Letters **20**, 2000 (1995).
5 C. Barty, G. Korn, F. Raksi, *et al.*, Optics Letters **21**, 219 (1996).
6 A. DeFranzo and B. Pazol, Applied Optics **32**, 2224 (1993).
7 W. Koechner, *Solid-State Laser Engineering* (Springer-Verlag, Heidelberg, 1996).
8 C. Hillimann, O. Seddiki, J. F. Morhange, *et al.*, Opt. Comm. **59**, 52 (1986).

Generation of 100 TW, Sub-20 fs Pulses at a 10 Hz Repetition Rate in Ti:Sapphire

Koichi Yamakawa[1], Makoto Aoyama[1], Shinichi Matsuoka[1], Teiji Kase[1], Yutaka Akahane[1], Hiroshi Takuma[1], David N. Fittinghoff[2] and Christopher P. J. Barty[2]

[1] Advanced Photon Research Center, KANSAI Research Establishment
Japan Atomic Energy Research Institute, Tokai, Naka, Ibaraki 319-11, Japan
E-mail: yamakawa@thsun.naka.jaeri.go.jp
[2] Institute for Nonlinear Science, University of California, San Diego
La Jolla, California 92093-0339, USA

Abstract. We have developed a compact three-stage Ti:sapphire amplifier laser system that produces a peak power in excess of 100 TW for a pulse duration of less than 19 fs and an average power of 19 W at 10 Hz repetition rate. The spatial beam quality is about twice the diffraction limited for the full amplified, compressed output pulse.

The technique of chirped-pulse amplification (CPA) has become a common approach for producing high peak power, ultrashort optical pulses [1, 2]. As a realistic laboratory tool for investigations requiring ultrahigh peak intensities (> 10^{20} W/cm^2), we have been developing an ultrahigh peak power, ultrashort pulse laser system with a capability of high repetition rate (~ 10 Hz) operation. In this paper we describe a Ti:sapphire amplifier system that has produced 100 TW laser pulses of less than 19 fs pulse duration at 10-Hz repetition rate with an average power of 19 W and a final amplifier efficiency reaching the theoretical quantum limit. To our knowledge, these results represent the highest peak power and average power pulses yet produced and the highest amplifier efficiency used in any Ti:sapphire CPA system.

The present 100 TW system is composed by adding a Ti:sapphire disk amplifier which is designed as an optimized power amplifier following our existing 10 TW, 16 fs Ti:sapphire laser system [3]. The 10 fs seed pulses from an all-solid-state mirror-dispersion-controlled Ti:sapphire oscillator were stretched to ~ 1.7 ns in a cylindrical-mirror-based pulse expander. The stretched pulse is then first amplified in a regenerative amplifier. The regenerative amplifier uses two 3 μm thick etalons as regenerative pulse shaping to counter gain narrowing. The 8 mJ output from the regenerative amplifier was then introduced into a 4-pass pre-amplifier. The output pulse energy from the amplifier was ~ 340 mJ with 700 mJ of 532 nm pump light. The output of the pre-amplifier is up collimated to an ~ 18 mm diameter with a Galilean telescope and then introduced into a final 4-pass power amplifier that is designed to produce > 3 J of radiation. This amplifier uses a water cooled 40 mm diameter 25 mm long Ti:sapphire crystal with anti-reflection coatings on both faces and is pumped with a custom built Nd:YAG laser which is capable of producing ~ 7 J of 532 nm radiation at 10 Hz. When the power amplifier was operated in the double-pass configuration, the output energy at high pump levels (~ 6 J/cm^2) was considerably less than expected. This was due to the presence of parasitic oscillation (PO) and amplified spontaneous emission (ASE) within the amplifier disk. According to these we have optimized the fluence and distribution of the pump laser and number of passes and small signal gain of the signal through the gain region in the power amplifier in order to suppress ASE and PO while simultaneously increasing the amplifier efficiencies.

The saturation characteristics for the power amplifier is shown in Fig. 1. The three curves are the calculated efficiencies of the power amplifier as a function of pump pulse fluence at different input pulse fluences based on Frantz-Nodvic simulation [4]. Squares, circles and a triangle represent the measured efficiencies in the power amplifier. With 6.4 J of pump light incident upon the crystal the amplifier has produced 3.3 J of 800-nm radiation. Under this condition, this amplifier has reached to the theoretical maximum conversion efficiency of 532-nm pump light to 800-nm radiation.

One important note in operating a Ti:sapphire amplifier with such high fluences is that the amplified pulse spectrum is reshaped and red-shifted due to saturation. Instead, a broad amplified bandwidth in a high energy output was obtained by shifting the spectrum of the input toward the short wavelength side of the desired output. This was accomplished by tuning the incidence angle of one of the etalons in the regenerative amplifier (Fig. 2). We have also checked the thermal lensing effect in the power amplifier under full power operation. The thermal lens in the horizontal plane is about f = + 25 m whereas it is f = + 9.5 m in the vertical plane. This difference between two planes may be caused by inhomogeneties of the Ti:sapphire crystal or thermal aberration in the crystal. To compensate for this lensing in the vertical plane, we placed a cylindrical mirror of f = - 9 m just after the pre-amplifier.

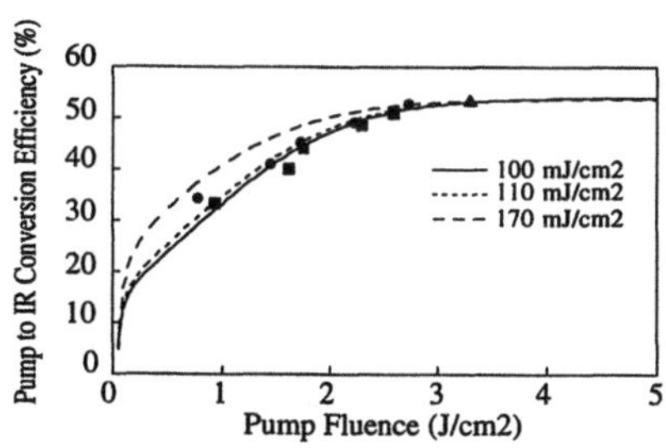

Fig.1 Saturation characteristics for the 4-pass Ti:sapphire power-amplifier.

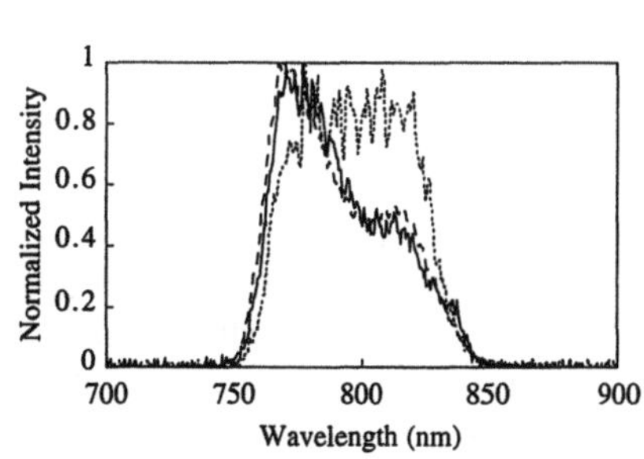

Fig.2 Measured amplified spectrum after the regenerative amplifier (solid line), the 4-pass pre-amplifier (dashed line) and the 4-pass power-amplifier (dotted line), respectively.

The output of the power amplifier then passes through the relay imaging optics with a magnification (M = 3.3) to provide spatially uniform beam onto the pulse compression gratings as well as collimating the beam diameter to an approximately 50 mm. A typical autocorrelation trace and an amplified spectrum after the compressor are shown in Fig. 3. The FWHM of the measured pulse duration is 18.7 fs. The duration of the transform limit, as calculated from the measured, amplified spectrum after the compressor is 17 fs. The high degree of agreement suggests that the compressed pulses are nearly transform limited. The transmission of the compressor was ~ 57%, yielding a compressed output pulse energy of ~ 1.9 J, which implies a peak power for the laser pulse of 102 TW. Because gain saturation in the amplifier stabilizes the pulse-to-pulse amplitude fluctuation, the energy stability is typically less than 2%.

The spatial beam quality is determined by focusing the attenuated output with a 3 m focal-length spherical mirror and measuring the spot size at the focus with a CCD camera. Under the full power conditions, the spatial quality of the beam was about 2 and 2.5 times diffraction limited in vertical- and horizontal-planes, respectively. We have determined that approximately 80% of the focused energy lies within the spot. With an f/3 off-axis parabolic mirror, focused intensities of ~ 3 x 10^{20} W/cm^2 should be possible with this beam quality.

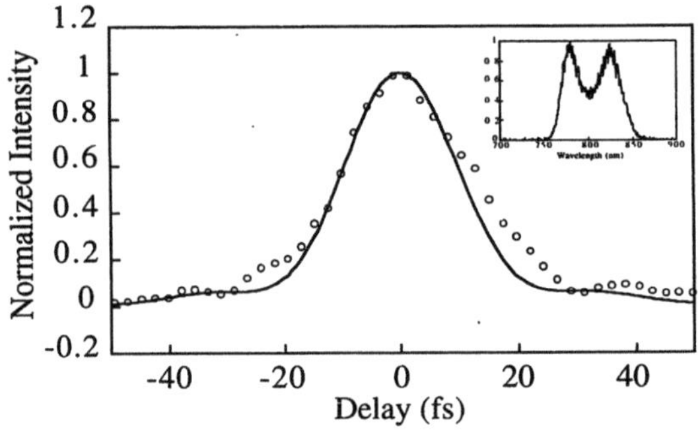

Fig.3 Measured and calculated autocorrelation. The solid line is the calculated, transform-limited autocorrelation based on the measured, amplified spectrum after the pulse compressor. The circles represent the measured autocorrelation of the 18.7 fs pulse. Inset : measured amplified spectrum after compression.

References

1 P. Maine, D. Strickland, P. Bado, M. Pessot, and G. Mourou, IEEE J. Quantum Electron. **QE-24,** 398, 1988.

2 J. D. Kmetec, J. J. Macklin and J. F. Young, Opt. Lett. **16**, 1001, 1991; J. Zhou, C.-P. Huang, M. M. Murnane, and H. C. Kapteyn, Opt. Lett. **20**, 64, 1995; C. P. J. Barty, C. L. Gordon III, and B. E. Lemoff, Opt. Lett. **19**, 1442, 1994.

3 K. Yamakawa, M Aoyama, S. Matsuoka, H. Takuma, C. P. J. Barty and D. Fittinghoff, Opt. Lett. **23,** 525, 1998.

4 L. M. Frantz and J. S. Nodvik, J. Appl. Phys. **34,** 2346, 1963.

Part II

Nonlinear Frequency Conversion and Pulse Shaping

Ultra-Broadband Optical Parametric Amplifiers in the Visible and in the Near-Infrared

G. Cerullo[1], M. Nisoli[1], S. Stagira[1], S. De Silvestri[1], O. Svelto[1], G. Valiulis[2], A. Varanavicius[2]

[1]INFM-Dipartimento di Fisica, CEQSE-CNR, Politecnico, Piazza L. da Vinci 32 – 20133 Milano, Italy E-mail: giulio.cerullo@polimi.it
[2] Laser Research Center, Vilnius University, Vilnius (Lithuania)

Abstract. We report on two different optical parametric amplifier (OPA) schemes for the generation of few-optical-cycles pulses. Pulses with 180-THz bandwidth and 2-μJ energy are generated by a noncollinear OPA in the visible; a portion of the amplified pulse spectrum is compressed down to 7.2 fs using a thin prism sequence. IR pulses tunable from 1.1 to 2.6 μm with duration as short as 14.5 fs are achieved in a type-II OPA pumped by 780-nm, 18-fs pulses generated by a hollow-fiber compressor.

1. Introduction

Femtosecond pulses with μJ level energy and broad tunability in the visible and in the near infrared are required for a variety of spectroscopic applications. Since the invention of Kerr Lens Mode-locking, the most widespread source of such pulses has been the optical parametric amplifier (OPA) pumped by an amplified Ti:sapphire laser system [1, 2]. Several schemes have been proposed, including seeding by parametric superfluorescence or by a white light continuum. Pumping with the fundamental wavelength (FW) or the second harmonic (SH) of the Ti:sapphire laser, it is possible to achieve tunability in the near-infrared and in the visible respectively. Until recently, however, the shortest pulses achieved from such systems have been in the 30-50 fs range, limited either by the narrow phase matching bandwidths or by the relatively long pump pulse durations.

In this work we report on two different OPA schemes which overcome these limitations and generate few optical cycles pulses in the visible as well as in the infrared. For the visible, we use long (≈150 fs) pump pulses and exploit noncollinear phase matching to generate ultrabroadband pulses, which are then compressed to sub-10-fs duration. For the infrared, we use sub-20-fs pump pulses obtained from an hollow fiber compressor to generate microjoule-energy-level pulses tunable from 1.1 to 2.6 μm with duration as short as 14.5 fs.

2. Visible OPA

Visible light pulses with a duration of less than 10 fs are required in several advanced ultrafast spectroscopy experiments, such as photon echoes and the study

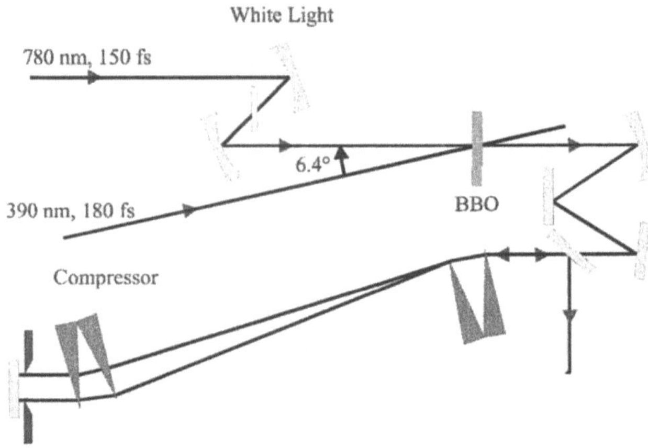

Fig. 1. Schematic of the noncollinear OPA and the pulse compressor.

of wavepacket dynamics. OPAs pumped by the SH of a Ti:Sapphire laser are a robust source of femtosecond pulses in the visible, but their durations are limited by the narrow phase matching bandwidths. Gale et al. [3] discovered a peculiar property of the β-barium borate (BBO) crystal for optical parametric amplification, when pumped at 400 nm: using noncollinear phase matching with a pump-signal angle of ≈ 3.7°, the phase matching angle becomes independent of wavelength over most of the visible range, so that very large gain bandwiths are possible. Recently, this property was exploited to build noncollinear OPAs tunable across the visible, generating sub-20 fs pulses [4-6].

In this work we demonstrate a noncollinear OPA, pumped by the SH of an amplified Ti:Sapphire laser, that generates ultrabroadband pulses in the visible: these pulses are shortened, using an optimized compressor, down to 7.2 fs.

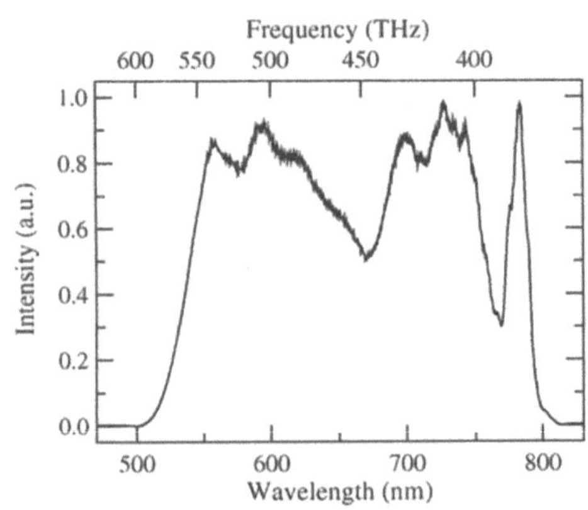

Fig. 2. Spectrum of the amplified pulses.

The scheme of the experimental setup is shown in Fig. 1. We start with an amplified Ti:Sapphire laser producing 150-fs pulses at 780 nm at 1 kHz repetition rate. A fraction of the light is frequency doubled in a 1-mm-thick LBO crystal, yielding 10-μJ pulses, used to pump the OPA. Another fraction of the light is focused into a 1-mm-thick sapphire plate to generate a stable white-light continuum, used to seed the OPA. The visible portion of the white light has a small linear chirp, with a group delay dispersion of ≈ 100 fs^2 [5];

to avoid any additional chirp, we employ reflective optics to guide the white light to the amplification stage. Parametric gain is achieved in a 1-mm-thick BBO crystal, cut at $\theta = 32°$, using type I phase matching; we use a single pass configuration to maximize the gain bandwidth. When illuminated by the pump pulse and aligned perpendicular to the pump beam, the BBO crystal emits a strong parametric superfluorescence in a cone with apex angle of 6.4°; if the white light is adjusted at exactly this angle, a very broad gain bandwidth is observed for optimum pump-seed delay. A typical amplified pulse spectrum, extending over most of the visible, is shown in Fig. 2; the amplified pulses, with a maximum energy of 2 μJ, maintain good spatial beam quality and amplitude stability, with peak to peak fluctuations of less than 7%. The amplified bandwidths in our system are considerably broader than those reported in [4], partly because of a lower white light chirp and partly because our longer pump pulses allow to temporally overlap a broader portion of the white light spectrum.

The compressor (see Fig. 1) consists of two pairs of fused silica prisms of small apex angle (20°): using small-angle prisms instead of the conventional Brewster-cut ones allows to substantially reduce the positive dispersion due to the path of the beams in the glass. As a result, the negative GDD required from the compressor is greatly reduced and, correspondingly, higher order phase errors are minimized. Reflective optics are used to collimate, steer and focus the amplified pulses, to minimize the dispersion of the compressor. The compressor cannot provide the required dispersion over the entire amplified pulse bandwidth: therefore we compressed only the short-wavelength portion of the spectrum, selected using a slit in the compressor (see Fig. 1). The optimum prism distance was found to be 1.9 m: the corresponding interferometric autocorrelation of the compressed pulse, measured using a 20-μm thick BBO crystal, is shown in Fig. 3: the pulse duration is 7.2 fs [7]. The Fourier transform of the entire pulse spectrum results in considerably shorter, sub-5-fs pulses: to reach this limit an accurate compensation of higher order dispersion over the entire pulse bandwidth, using chirped mirrors, will be required.

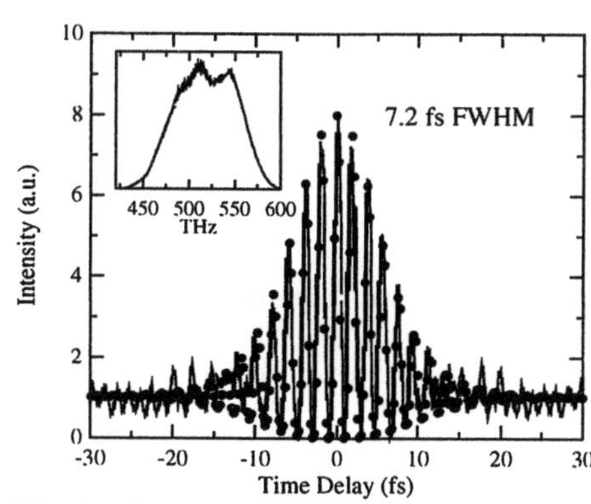

Fig. 3. Autocorrelation of the compressed pulse. Inset: spectrum.

3. Infrared OPA

Ultrashort high-energy pulses tunable in the infrared are strongly needed for applications in nonlinear optics and spectroscopy. Sub-100-fs infrared pulses were achieved by traveling-wave optical parametric amplification (OPA). The availability of high energy sub-20-fs pump pulses offers the possibility to investigate parametric processes in an ultrashort temporal regime. In the following we report on traveling-wave parametric conversion of 18-fs pulses at 780 nm in type-II β-BaB_2O_4 (BBO) crystals, yielding tunable infrared pulses as short as 14.5 fs with µJ-energy at 1-kHz repetition rate.

The choice of the phase-matching type is an important issue. In type-I BBO crystal signal and idler pulses sweep the pump pulse in the same direction in all the tuning range, so that, for long interaction lengths, a short spike with several satellites of different amplitudes is expected at the signal wavelength, while the idler pulse emerges as a broad and irregular background (see simulation in Fig.4(a)). In type-II BBO signal and idler pulses propagate in opposite directions with roughly the same relative velocities with respect to the pump pulse. In this case the temporal spreading of the parametric pulses is overcome by nonlinear compression for pump intensities larger than 0.4 TW/cm^2. At optimum pump intensities both signal and idler pulses are smooth and with negligible side-lobes in all the tuning range (see Fig.4(b)). On the basis of these considerations we have chosen type-II phase-matching.

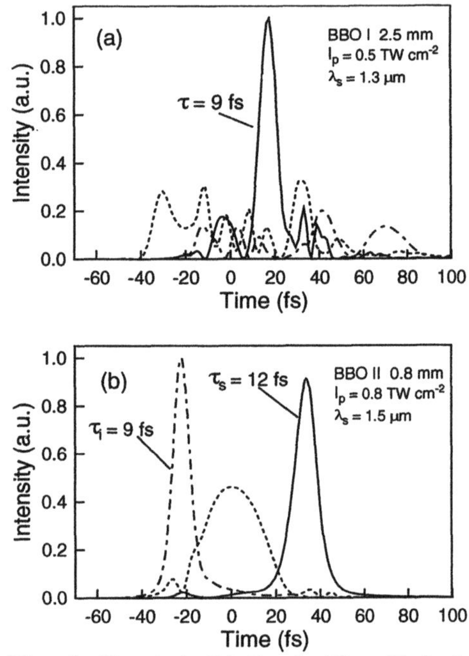

Fig. 4. Signal (solid lines), idler (dashed-dotted lines) and pump (dotted lines) pulses for type-I (BBO I) (a) and type-II (BBO II) (b) for 20-fs pump pulses at 780 nm. I_p is the pump peak intensity, τ_s and τ_i are the signal and idler pulse durations.

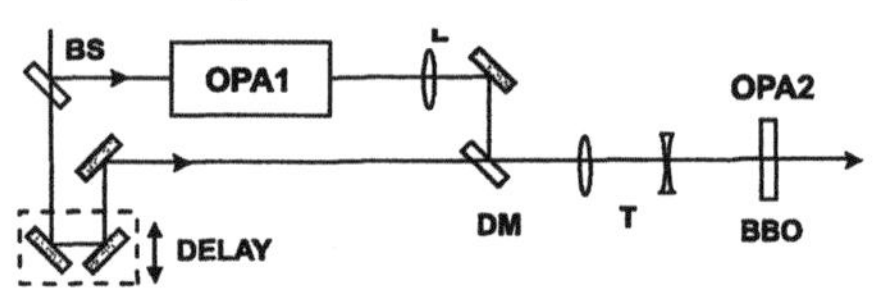

Fig. 5. Experimental setup: BS, beam splitter; DM, dichroic mirror; L, lens; T, telescope.

The experimental setup is shown in Fig.5. The pump pulses were obtained by compression of 140-fs, 500-µJ pulses from a 1-kHz CPA Ti:sapphire laser using the hollow fiber technique [8]. The beam was coupled into a 300-µm-diameter, 75-cm-long fused-silica hollow fiber filled with argon at a pressure of 3 bar. The frequency-broadened pulses were then compressed to 18 fs by double pass through a prism pair. Pump pulse energy was 200 µJ.

Parametric light conversion was accomplished in a three-pass OPA based on two BBO crystals. The seed pulse was formed in a generator/amplifier (OPA1) designed in two-pass configuration employing a single 5-mm-long type-II BBO crystal pumped by ~100-μJ pulses. The energy of the seed pulse was then boosted up to tens of μJ in the power amplifier (OPA2), based on a 1-mm-long type-II BBO crystal, pumped by ~100-μJ pulses. The signal pulses were tunable from degeneracy to 1.1 μm; the corresponding idler pulses tuning range was 1.56-2.68 μm. A maximum conversion efficiency of ~15% in OPA2 was measured at a signal wavelength around 1.4 μm. The peak to peak fluctuations were less than 15 %. Signal pulse duration was below 25 fs in all the tuning range.

Transform-limited pulses as short as 14.5 fs (FWHM) were generated at 1.5 μm, with an energy of 8 μJ [9]. Figure 3 shows a typical autocorrelation trace. These pulses are the shortest generated to date in this spectral region at microjoule-energy level. We point out, on the basis of the experimental results and of computer simulations, that the phase modulation in the nonlinear crystal due to third order nonlinearity is the main limiting factor for parametric amplification in the sub-20-fs temporal domain. With pump intensities close to 2 TW/cm^2 we observed the saturation of parametric energy conversion accompanied by severe degradation of output beam quality due to spatial self-phase modulation, though the crystal damage did not appear up to pump intensities of ~ 4 TW/cm^2.

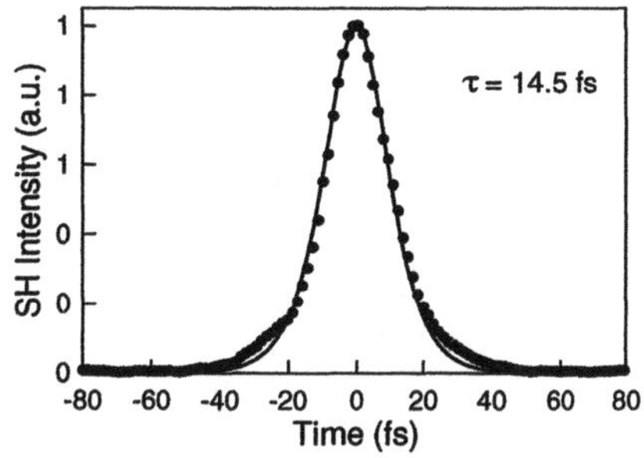

Fig. 6. Intensity autocorrelation function of the signal pulse at 1.5 μm.

In conclusion, we have demonstrated a traveling-wave optical parametric amplifier pumped by sub-20-fs pulses generated by a Ti:sapphire laser followed by a hollow-fiber compressor, that yields microjoule-energy-level pulses tunable from 1.1 to 2.6 μm with duration as short as 14.5 fs.

References

1 M. K. Reed, M.K. Steiner-Shepard, M.S. Armas, and D. Negus, J. Opt. Soc. Am. B **12**, 2229 (1995).
2 K. R. Wilson and V.V. Yakovlev, J. Opt. Soc. Am. B **14**, 444 (1997).
3 G. M. Gale, M. Cavallari, T.J. Driscoll, and F. Hache, Opt. Lett. **20**, 1562 (1995).
4 T. Wihlhelm, J. Piel, and E. Riedle, Opt. Lett. **22**, 1494 (1997).
5 G. Cerullo, M. Nisoli, and S. De Silvestri, Appl. Phys. Lett. **71**, 3616 (1997).
6 A. Shirakawa and T. Kobayashi, Appl. Phys. Lett. **72**, 147 (1998).
7 G. Cerullo, M. Nisoli, S. Stagira, and S. De Silvestri, Opt. Lett., in press.
8 M. Nisoli, S. De Silvestri, and O. Svelto, Appl. Phys. Lett. **68**, 2793 (1996).
9 M. Nisoli et al., Opt. Lett. **23**, 630 (1998).

Sub-5-fs Pulse Generation by Pulse-front-matched Optical Parametric Amplification

Akira Shirakawa, Isao Sakane, and Takayoshi Kobayashi

Department of Physics, University of Tokyo
7-3-1 Hongo, Bunkyo, Tokyo 113-0033, Japan
E-mail : takakoba@phys.s.u-tokyo.ac.jp

Abstract. A noncollinear optical parametric amplifier generates shortest 4.7-fs visible pulses or transform-limited sub-9-fs pulses tunable both in the visible (550-700nm) and near-infrared (900-1300nm) for the first time.

Noncollinearly phase-matched optical parametric conversion is a candidate to generate extremely-short pulses, as was pointed out by Dr. Gale and his co-workers in 1995 [1]. Recent efforts for noncollinear optical parametric amplification (NOPA) based on this signal-idler group-velocity-matching scheme are remarkable [2-4]. However, even sub-10-fs duration is expected from the broad signal spectra, only far from transform-limited (TL) pulses were reported so far. This is mainly due to the pulse-front tilting inevitable in the noncollinear interaction [2,5]. In this paper, the generation of TL sub-5-fs visible pulses from a NOPA with a tilted-pumped geometry is presented for the first time to our knowledge (Fig. 1). Tunable TL pulse generation both in the visible and near-infrared (NIR) by controlling the seed-chirp is also to be mentioned [6]. This NOPA is superior to a traditional continuum-compression method [7,8] in many aspects (tunability, pulse energy, flexible pulse-width and bandwidth, etc.), and is believed to be the most useful light source for ultrafast spectroscopy with the highest time-resolution.

The schematic of the tilted-pumped NOPA is shown in Fig.2(a). The second harmonics of a Ti:sapphire regenerative amplifier pumps a 1-mm-thick BBO crystal

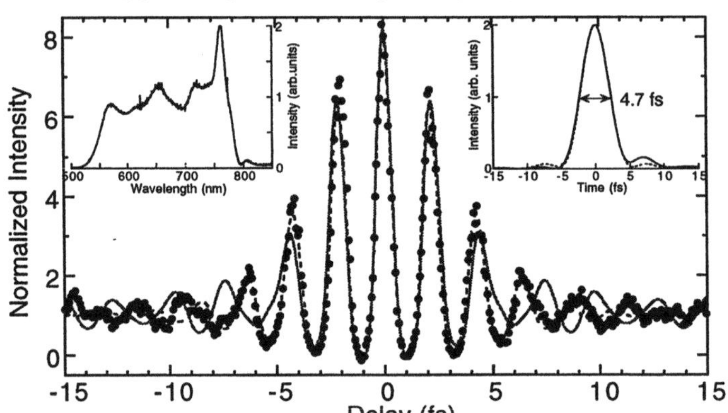

Fig. 1. Fringe-resolved autocorrelation (FRAC) trace of the full-bandwidth signal. The measured (full circles) dispersion-included Fourier-transforming fit (solid curve) and sech^2-fit (dashed curve) traces are shown. Each fit deduces the pulse width of 4.7fs and 4.1fs. The corresponding spectra and intensity profiles (solid curve : deduced profile, dashed curve : 4.5-fs TL profile) are also written in the inset on the upper-left and upper-right, respectively. The doubling crystal is a 10-μm-thick β-BaB_2O_4 (BBO).

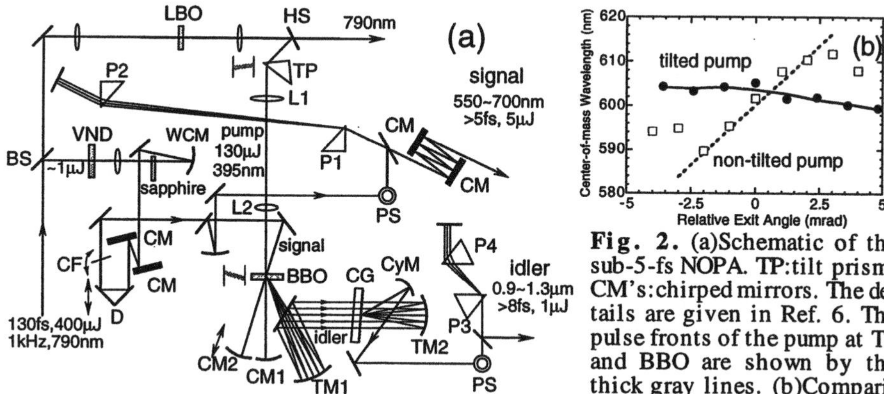

Fig. 2. (a)Schematic of the sub-5-fs NOPA. TP:tilt prism, CM's:chirped mirrors. The details are given in Ref. 6. The pulse fronts of the pump at TP and BBO are shown by the thick gray lines. (b)Comparison of the spatial chirp of the amplified signal.

and the single-filament continuum is noncollinearly amplified with the noncollinear internal angle α of 3.7° for the broadband amplification over 150THz. The pulse front of the pump beam is tilted 2.3° by a 45° fused-silica prism and then by the telescope the appropriate tilt angle which is equal to α in the BBO is attained so that the pulse volume is well overlapped with that of the signal. This tilt-matched interaction dramatically decreases the spatial chirp which was severe problem in a conventional non-tilted-pumped NOPA [2] (Fig. 2(b)), which indicates the non-tilted signal pulse generation [5] and realizes the effective pulse shortening. The final pulse energy is 5μJ after the 2nd-stage amplification with conserving the pulse-front matching. The seed-chirp is suppressed by the customized chirped mirror-pair (R&D Lezer-Optika) and a full-bandwidth amplification is attained. A prism-chirped mirror compressor is designed to compensate the measured group-delay (GD) across the whole spectral range.

The FRAC traces and spectrum are shown in Fig.1. The bandwidth is as broad as 250nm (150THz), and the sech^2-fit pulse width is 4.1fs. The fitting by Fourier-transformation of the spectrum with parameters of GD dispersion (GDD) and third-order dispersion [8] deduces 4.7-fs duration, which is fairly close to the TL width (4.5fs) and evidently indicates the importance of the pulse-front matching. As far as we know this is the first report of the visible sub-5-fs pulse. The error of the fitting is due to the insufficient bandwidth of the chirped mirror currently used, which is on being improved for the shorter pulse generation. One of the greatest merits of NOPA is the flexibility of the pulse width in the range longer than 5-fs by selecting some spectral portion by increasing the seed chirp [3]. For example, a tunable sub-10-fs generation is straightforward without the chirped mirrors or by inserting a glass plate before the crystal. The former case is shown in Fig. 3(a-c) where the tunable TL pulses with a 6 to 7-fs duration are generated from 550 to 700nm by scanning the delay-line of the seed [6].

The idler is generated with a large angular spectral dispersion to be phase-matched with a broad spectrum of the signal. This is a unique property of the group-velocity matching between the signal and idler. The all-reflective telescope accompanied with a grating suppresses the dispersion across the wide spectral range (800-1400nm) and a collimated idler beam is obtained with 1-μJ pulse energy. The measured negative chirp is well explained by the inversion of the seed-chirp through the parametric interaction (this case without seed-chirp suppression), which is compressed by a prism pair under a positive GDD region. A FRAC trace of the idler at 1100nm (Fig.

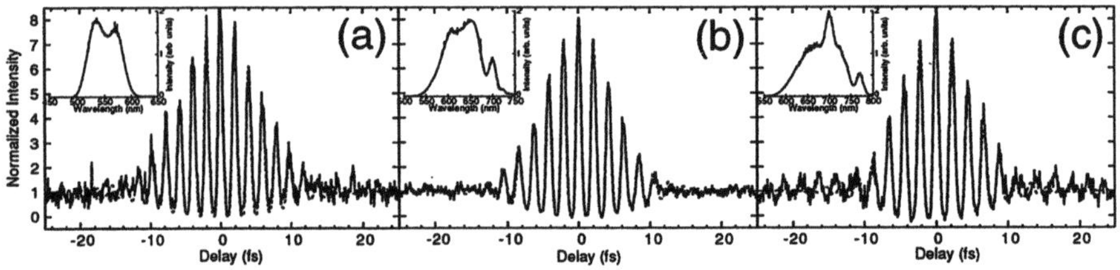

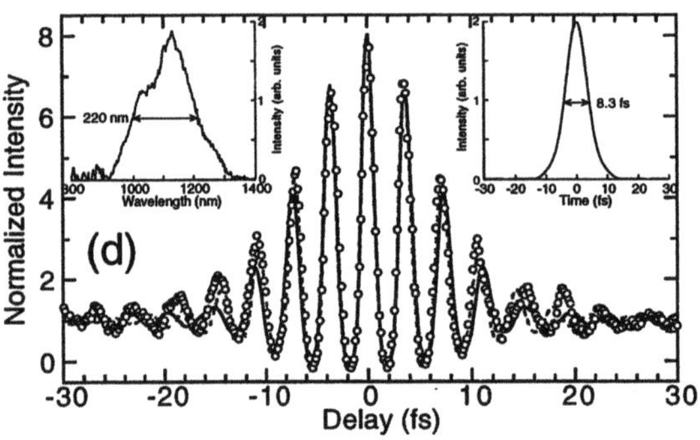

Fig. 3. (a)-(c)FRAC traces of the tunable signal. A prism-grating compressor is used in this case. The $sech^2$-fit (dashed curves) pulse widths are (a)7.1fs at 550nm, (b)6.2fs at 620nm, and (c)6.1 fs at 680 nm. Each spectrum is in the inset. (d)FRAC trace of idler at 1100nm. The calculated TL (solid curve, 8.3fs) and $sech^2$-fit (dashed curve, 8.4fs) traces are shown. The spectrum and TL intensity profiles are shown in the insets.

3(d)) also shows a nearly TL pulse shape, where the $sech^2$-fit pulse width of 8.4 fs is also reasonably matched with the TL width (8.3 fs). The tuning range is from 900 to 1300nm corresponding to the signal. A spatial light modulator is being introduced in the Fourier-plane of the angular-dispersion compensator for the more precise phase-collection, especially in the case of the full-bandwidth amplification (Fig. 1), and the sub-5-fs pulse generation is also being pursued in the NIR.

In conclusion, visible TL sub-5-fs pulses as well as tunable visible and NIR sub-9-fs pulses are generated from a novel tilted-pumped NOPA for the first time. These features are quite useful and will lead to a new regime in the ultrafast science. We are currently doing the study on the dynamics of the excited state of the hydrated electron or gap states of conjugated polymers with the highest time-resolution.

Acknowledgments. The authors are grateful to fruitful discussions with Drs. K. Misawa, R. Szipöcs, and V. V. Yakovlev. This work was supported by a Program of Research for the Future of Japan Society for the Promotion of Science.

References

1 G. M. Gale, M. Cavallari, T. J. Driscoll, and F. Hache, Opt. Lett. **20**, 1562 (1995); G. M. Gale, M. Cavallari, and F. Hache, J. Opt. Soc. Am. B **15**, 702 (1998), and references therein.
2 A. Shirakawa and T. Kobayashi, Appl. Phys. Lett. **72**, 147 (1998); IEICE Trans. Electron. **E81-C**, 246 (1998).
3 T. Wilhelm, J. Piel, and E. Riedle, Opt. Lett. **22**, 1494 (1997), this conference, WeB3.
4 G. Cerullo, M. Nisoli, and S. De Silvestri, Appl. Phys. Lett. **71**, 3616 (1997); this conference, WeB1.
5 R. Danielius, A. Piskarskas, P. Di Trapani, A. Andreoni, C. Solcia, and P. Foggi, IEEE J. Quantum. Electron. **QE-34**, 459 (1998), and references therein.
6 A. Shirakawa, I. Sakane, and T. Kobayashi, Opt. Lett., August 15 1998, in press.
7 A. Baltuska, Z. Wei, M. S. Pshenichnikov, D. A. Wiersma, and R. Szipöcs, Appl. Phys. B **65**, 175 (1997).
8 M. Nisoli, S. Stagira, S. De Silvestri, O. Svelto, S. Sartania, Z. Cheng, M. Lenzner, Ch. Spielmann, and F. Krausz, Appl. Phys. B **65**, 189 (1997).

Tunable Visible and NIR Parametric Amplifiers at 1 kHz and Pulse Lengths Down to 10 fs

S. Lochbrunner, T. Wilhelm, J. Piel, P. Huppmann, S. Spörlein, and E. Riedle

Institut für Medizinische Optik, Ludwig-Maximilians-Universität München, Oettingenstraße 67, D-80538 München, Germany
E-mail: Stefan.Lochbrunner@physik.uni-muenchen.de

Abstract. Noncollinear phase matching in a type I BBO optical parametric amplifier seeded by a high quality continuum is shown to produce pulses at 470 to 730 nm center wavelength with lengths down to 10 fs. Type II phase matching allows to access the 700 to 1100 nm range. The amplification preserves the phase of the seed light even at high efficiency. For characterization of the ultrashort visible pulses a SiC photodiode is found to be superior to thin nonlinear crystals.

1. Introduction

Recently we were able to demonstrate that optical parametric amplification (OPA) in a noncollinearly phase matched BBO crystal pumped by the second harmonic of a Ti:sapphire laser system leads to tunable visible pulses with sub-20-fs length [1]. Improved arrangements even allow the generation of sub-10-fs pulses [2,3].

In this paper we demonstrate, that the phase of the continuum seed light is not affected by the highly efficient amplification. The shortening from typically 100 fs pump pulses to the 10 fs duration of the visible pulses already occurs in the continuum generation.

For type I phase matching the ultrashort pulses cannot be tuned beyond the degeneracy point. Our calculations do, however, indicate that this is possible for type II phase matching and we experimentally prove this conjecture.

Both for the autocorrelation measurement of individual visible pulses on the 10 fs time scale and for the crosscorrelation of pulses from two OPAs the use of thin nonlinear crystals is found to be critical. Therefore we investigate the applicability of SiC photodiodes for this purpose.

2. Experimental Results

In our previously reported setup for the generation of tunable visible pulses of sub-20 fs duration the seed light generated in a 3 mm sapphire

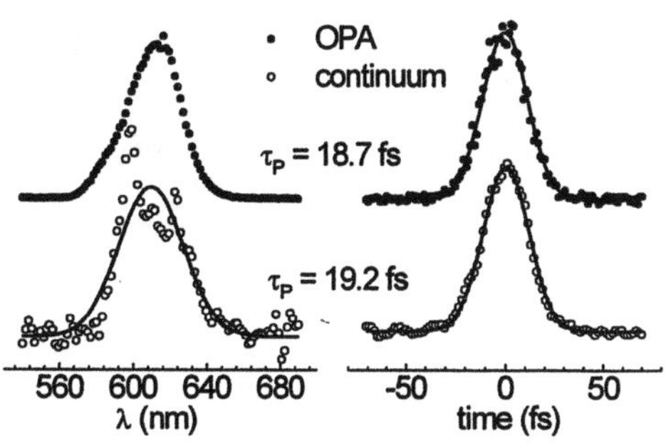

Fig. 1. Spectrum and autocorrelation of OPA-pulses and the corresponding part of the non amplified continuum.

disk was considerably chirped [1]. We have reduced this chirp by the use of fused silica lenses and thereby increased the spectral width of the pulses to greater than 48 THz. We are able to compress the pulses to less than 10 fs. The experimental bandwidth matches the prediction from calculations that only consider the phase matching. Due to the high pump intensities also other pulse shaping mechanisms can be imagined [4].

To characterize the amplification process amplified pulses with a spectrum around 610 nm (see Fig. 1) were compressed in a prism compressor to a duration of 19 fs. Their autocorrelation was measured via SHG in a 25 μm thick BBO crystal as depicted in Fig. 1. Then the pump beam was blocked and the continuum passed the OPA crystal without amplification. Since the directions of the amplified pulses and the continuum coincide nearly perfectly, only a very slight realignment was necessary to get the continuum beam on the same axis as the amplified one. A slit was inserted in front of the end mirror of the compressor to choose the corresponding part of the continuum spectrum (see Fig. 1). Without any realignment of the compressor the autocorrelation trace depicted in Fig. 1 was recorded. Because of the low energy of the continuum a 100 μm thick BBO crystal was used and 4000 scans were averaged. The autocorrelation of the amplified pulses and of the continuum agree remarkably well indicating that the spectral phase of the continuum matches that of the amplified pulses. The phase of the continuum is conserved in the amplification and no other pulse shaping mechanisms contribute significantly.

For spectroscopic applications we have set up two parallel OPAs. When their crosscorrelation is measured in thin BBO crystals, major complications arise. Both OPAs deliver output pulses that can be compressed to less than 20 fs duration (see Fig. 2a) and display a crosscorrelation of 29 fs when both pulses are optimally compressed (Fig. 2b). However, if both pulses are slightly chirped, the observed crosscorrelation width depends dramatically on the nonlinear medium used in the measurement. For identical settings of the compressors we find a correlation width of 27 fs in 100 μm BBO (Fig. 2 c)), 54 fs in 25 μm BBO (Fig. 2 d)) and 65 fs in a SiC photodiode (Fig. 2 e)). The usefulness of SiC photodiodes for autocorrelation measurements by two photon photoconductivity was recently demonstrated [5] and

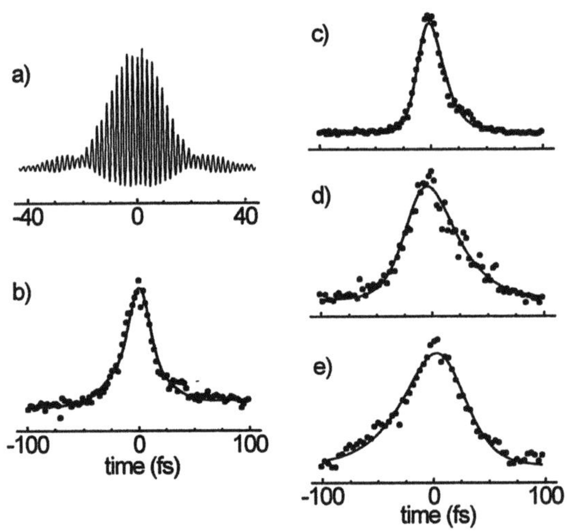

Fig. 2. Interferometric autocorrelation of 600 nm pulses (a) and crosscorrelation of unchirped 600 and 510 nm pulses (b), both measured with a SiC photodiode. The crosscorrelation of chirped pulses measured in a 100-μm-thick BBO crystal (c), a 25-μm-thick BBO crystal (d) and a SiC-photodiode (e).

we checked their applicability on the 10 fs time scale by interferometric autocorrelation measurements. The comparison of the three curves clearly shows that the acceptance bandwidth of even the 25 μm BBO crystal still is not sufficient to render a correct measurement. Since no phase matching is required in the SiC photodiode it is superior for the characterization of extremely short visible pulses. We therefore used it for the recording of the curves shown in Fig. 2 a) and b).

Broad bandwidth noncollinear amplification in a type I arrangement is limited to wavelengths well below the degeneracy point [1,2,3]. This is due to the fact that in BBO the group velocity increases monotonically for wavelength up to 1300 nm and therefore the projection of the off-axis visible group velocity cannot match the on-axis NIR one as required. This limitation does not exist in type II phase matching in BBO. Our calculations indicate that for an ordinary seed wavelength between 700 and 1100 nm, a corresponding extraordinary off-axis wave and an extraordinary blue pump the required noncollinear angle between pump and seed varies from 7 to 12 degrees. In a preliminary experiment we were indeed able to achieve efficient broadband amplification both at 960 and 750 nm. The spectrum generated in the OPA is shown in grey in the left part of Fig. 3 and its width of 37 THz corresponds to a pulse length close to 10 fs. A slightly narrower spectrum (black line) was compressed in a fused silica prism sequence and the pulse length measured to 29 fs assuming a $sech^2$ shape (see Fig. 3 right side). Further improvements should be readily possible.

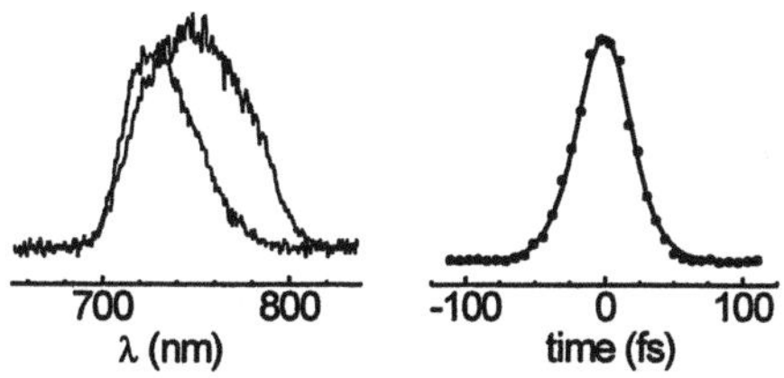

Fig. 3. Spectrum and autocorrelation of type II OPA. The pulse duration is 29 fs.

3. Conclusions

Noncollinear phase matching in type I and II BBO leads to blue pumped OPAs that can produce pulses tunable between 470 and 1100 nm with pulselengths down to 10 fs. Such pulses are extremley useful for spectroscopic experiments and indeed they were already used for the investigation of the azulene S_1 dynamics and the ultrafast primary electron transfer in photosynthesis.

Acknowledgements. We thank Wolfgang Zinth for continuous support and valuable discussions and the Deutsche Forschungsgemeinschaft for financial support.

References

1 T. Wilhelm, J. Piel, and E. Riedle, Opt. Lett. **22**, 1494, 1997
2 G. Cerullo, M. Nisoli, and S. De Silvestri, Appl. Phys. Lett. **71**, 3616, 1997
3 A. Shirakawa and T. Kobayashi, Appl. Phys. Lett. **72**, 147, 1998
4 S. Fournier, R. Lòpez-Martens, C. Le Blanc, E. Baubeau, and F. Salin, Opt. Lett. **23**, 627, 1998
5 T. Feurer, A. Glass, and R. Sauerbrey, Appl. Phys. B **65**, 295, 1997

Parametric chirped pulse microamplifier based on engineerable quasi-phase-matched $LiNbO_3$

A. Galvanauskas[1], A. Hariharan[1], D. Harter[1], M. A. Arbore[2] and M. M. Fejer[2]

[1]IMRA America, Inc., 1044 Woodridge Avenue, Ann Arbor, Michigan 48105-9774
E-mail: almisg@imra.com
[2]E. L. Ginzton Laboratory, Stanford University, Stanford, CA 94305-4085

Abstract. A new class of compact diode-pumped femtosecond amplifiers based on quasi-phase-matched parametric gain in $LiNbO_3$ can provide single-pass gain of up to 80 dB and arbitrarily broad, optically engineerable bandwidth.

For decades nonlinear interactions were exploited exclusively for wavelength conversion rather than for optical power and energy amplification. Traditionally, efficient optical parametric amplification required extremely high pump intensities. This rendered such power-amplification impractical due to the lack of simple pump sources with necessary peak powers.

Recently [1] we have demonstrated that the high nonlinearity of electrically poled $LiNbO_3$ enables to achieve very high parametric gain by pumping with practically any standard nanosecond-pulse Q-switched laser. Such an amplifier, used in conjunction with a chirped pulse amplification [2], provides a simple single-pass scheme for nanosecond-to-femtosecond energy conversion. This scheme takes advantage of the simplicity of conventional Q-switched lasers and overcomes the technological complexity and certain principal drawbacks of existing regenerative and multipass ultrashort-pulse amplifiers based on population-inversion based solid-state gain media.

Here we report a compact system for parametric chirped pulse amplification in electrically poled $LiNbO_3$. This diode-pumped, fiber laser seeded system produces femtosecond microjoule pulses using a microchip Q-switched nanosecond-pump laser and a compact chirped-pulse amplification arrangement. Furthermore, we demonstrate that the gain bandwidth of a quasi-phase-matched amplifier can be optically engineered to have any arbitrary width and to support even the shortest pulses.

The experimental set-up is shown in Fig. 1. A mode-locked fiber oscillator produces 180 fs seed pulses at 1560 nm. A diode-pumped actively Q-switched microlaser generates 1 ns 10 μJ pump pulses at 1.06 μm. Active Q-switching of a microlaser allows the synchronisation of 5 kHz repetition rate pump pulses to the mode-locked fiber oscillator pulses with the timing accuracy (jitter) of ~300 ps. Parametric amplification is accomplished in 5 cm long crystals of periodically poled lithium niobate (PPLN). The use of CPA is essential for this approach –

stretching of femtosecond pulses allows one to match pump and seed pulse durations and, consequently, to achieve efficient energy conversion [2]. A hybrid compact chirped pulse amplification (CPA) circuit is implemented using a chirped fiber grating and a chirped quasi-phase-matched (QPM) grating [3]. A 10 cm fiber grating stretches fiber-laser pulses to ~ 1 ns duration. Additional length of a highly-dispersive fiber spliced to the stretcher port of the fiber grating prevents complete recompression and, therefore, prevents nonlinear effects at the compressor port of the grating fiber. Final recompression occurs in a 5 cm long chirped-PPLN compressor producing 780 nm bandwidth-limited second-harmonic pulses. Such a hybrid arrangement enables one to produce recompressed-pulse energies significantly greater than with a fiber-grating alone.

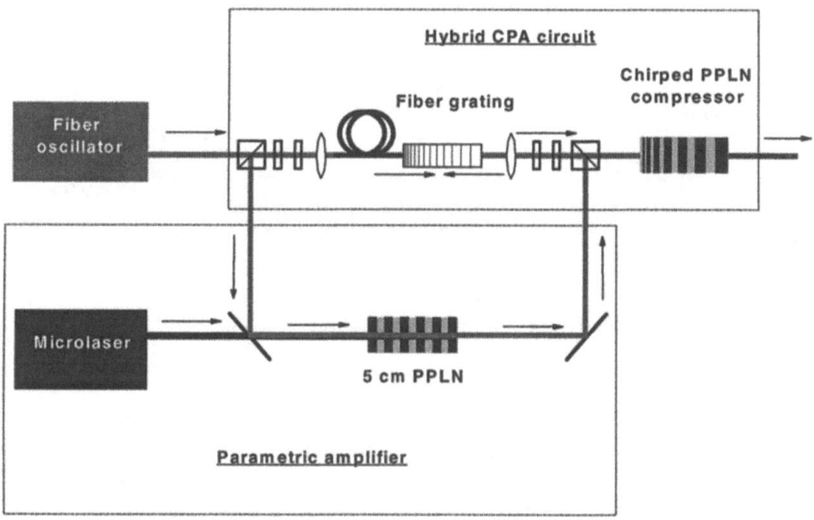

Fig. 1. Microlaser pumped PCPA system

The highest energy extraction and highest small-signal gain have been achieved for a uniform-period PPLN parametric amplifier. A single-pass gain of up to 80 dB and an energy conversion of up to 20 % has been obtained, producing 2 μJ amplified pulses. It is important to note that this high gain has been achieved with ~ 45 MW/cm^2 of pump intensity in the crystal, an intensity which is approximately two orders of magnitude below the surface damage threshold for nanosecond pump pulses.

Amplified and seed-pulse spectra are shown in Fig. 2. A uniform 5 cm long grating provides a narrow parametric gain spectrum of ~ 6 nm. To overcome this limitation we fabricated a chirped PPLN structure with a parametric gain bandwidth of ~ 20 nm in order to accommodate the ~ 23 nm seed spectrum. Such a selection of an engineered structure allowed us to shorten pulses after the amplification and SHG recompression from 500 fs to 200 fs. The overall transmission efficiency of the hybrid compressor is ~ 25 %, resulting from the coupling losses into the grating fiber, the limited grating reflectivity and the SHG conversion efficiency of a chirped-PPLN compressor.

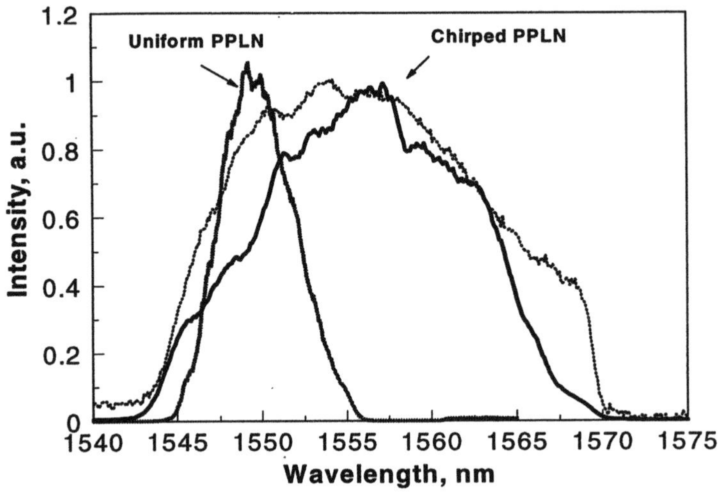

Fig. 2. Seed pulse (dotted) and amplified pulse (solid) spectra

To summarize, electrically poled $LiNbO_3$ is a practical parametric gain medium possessing unprecedented properties for ultrashort pulse amplification. Ultimate simplicity and compactness is available with such amplifiers and is significantly complemented by unique engineerability of the gain medium, with the potential for extremely broad gain bandwidths in amplifying few-femtosecond pulses.

References

1. A. Galvanauskas, A. Hariharan, M. A. Arbore, M. M. Fejer and D. Harter, CLEO 97, posteadline paper CPD7
2. A. Dubietis, G. Jonusauskas, and A. Piskarskas, Opt. Comm. **88**, 437 (1992)
3. M. A. Arbore, A. Galvanauskas, D. Harter, M. H. Chou and M. M. Fejer, Opt. Lett. **22**, 1341 (1997)

High Repetition Rate Mid-Infrared Femtosecond Optical Parametric Oscillator

S. Marzenell, R. Beigang and R. Wallenstein

Fachbereich Physik, Universität Kaiserslautern, Erwin Schrödinger Str., D 67663 Kaiserslautern, Germany
E-mail: beigang@rhrk.uni-kl.de

Abstract Operation of a cw mode-locked femtosecond optical parametric oscillator operating in the wavelength range from 2 μm to 10 μm using $AgGaSe_2$ is reported.

1. Introduction

Optical parametric oscillators (OPOs) are well suited for the generation of high power ps and fs light pulses in the near infrared region up to 4 μm using crystals such as KTP and its isomorphs. The spectral region of the mid infrared can, in principle, be covered using nonlinear crystals such as $AgGaS_2$, $AgGaSe_2$ and $ZnGeP_2$. The broad transmission range and birefringent properties of these materials allow the generation of radiation between 1.3 μm and 12 μm. This has been demonstrated with pulsed ns OPOs [1,2] and mode-locked ps-OPOs [3,4]. Here, we report what we believe to be the first demonstration of femtosecond continuous-wave (cw) mode-locked operation of an $AgGaSe_2$-OPO in the mid infrared spectral region.

2. Experimental Methods

The experimental setup consists of a cascaded OPO-system: a cw mode-locked Ti:sapphire laser synchronously pumps a Cesium-Titanyl-Arsenate (CTA)-OPO which in turn excites the $AgGaSe_2$-OPO. The CTA-OPO provides the required near infrared pump wavelength ($\lambda > 1.4$ μm) for the $AgGaSe_2$-OPO. CTA allows noncritical phase-matching (NCPM) between 1.3 μm and 2.2 μm by changing the pump wavelength of the fs Ti:sapphire laser. We used a CTA crystal cut at $\Theta = 90°$ and a mirror set for the e-polarized wave at 1.5 μm. The non-resonant o-polarized wave was coupled out via a polarizing beam splitter placed inside the OPO resonator. With this set up the o-polarized wave was tunable between 1.5 μm and 1.65 μm limitated by the spectral properties of the used polarizing beam splitter. An intensity autocorrelation of the o-wave indicated that the pulses were of 150 fs duration, while the pump pulses were 120 fs long. With an average pump power of 1.5 W up to 510 mW of usable o-wave power was generated, corresponding to a peak power of 41 kW.

The o-polarized radiation was focused by a f = 200 mm lens into the $AgGaSe_2$ crystal through one of the curved resonator mirrors. The $AgGaSe_2$ crystal (4x4x2

mm^3) was cut for type I phase-matching (e→o+o), at Θ = 70°. Due to the small walk-off angle of 0.4°, collinear pumping was used. The signal resonant cavity consisted of two spherical mirrors with a radius of curvature of 75 mm and two plane mirrors which defined a beam waist w_0=25 μm at the center of the crystal. Tuning can be accomplished by either rotating the crystal or changing the wavelength of the Ti:sapphire laser which results in a wavelength change of the pump radiation provided by the CTA-OPO. With a crystal cut at Θ = 70°, for example, the idler wave covers the range from 4 μm to 7 μm.

3. Results and Discussion

Fig. 1 shows the average output power of the signal wave at 2.35 μm (idler wavelength λ_i = 4.8 μm) as a function of the 1.58 μm pump power obtained with the crystal cut at Θ = 70° and three different output couplers. The threshold pump powers between 75 mW and 125 mW measured in front of the focusing lens (60 mW to 100 mW inside the crystal) were in good agreement with a calculated value of 70 mW, assuming a 2 mm long $AgGaSe_2$ crystal and round trip losses for the signal wave of 6 %. The nonresonant idler wave experienced a power loss of 24 % as a result of the crystal reflectivity and transmission of the resonator mirrors. At input powers of 400 mW the signal power exceeded 65 mW. The measured pump depletion of 40 % indicates that 160 mW of the pump power was converted into signal and idler power. Considering the Manley-Rowe relation, this corresponds to 108 mW signal and 53 mW idler power generated in the crystal. These values are in good agreement with the measured power, taking into account the losses for both waves.

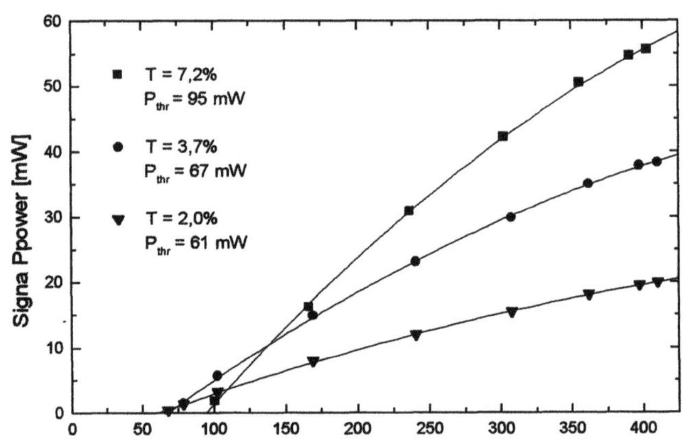

Fig. 2: Signal wave output power of the $AgGaSe_2$-OPO

The group velocity mismatch (GVM) between pump and resonant signal wave is about 140 fs/mm. As a consequence, with 150 fs long pump pulses only 1 mm of the 2 mm long crystal is used for parametric amplification before signal and pump wave are separated in time. The signal pulse length was significantly increased to 670 fs caused by group velocity mismatch and group velocity dispersion together with self phase modulation of the resonant wave. In addition a possible chirp of the pump pulses is enhanced in the $AgGaSe_2$ crystal which will lead to an additional pulse broadening. In contrast, for the idler wave self phase modulation is reduced (non resonant wave), the chirp enhancement is smaller and has the opposite sign. Together with a positive group velocity dispersion both a pulse length reduction and a reduction of the chirp compared to the signal pulses can be expected. The idler pulse length at 4.8 μm was measured to be 250 fs without a noticeable chirp in agreement with this assumptions (Fig.2).

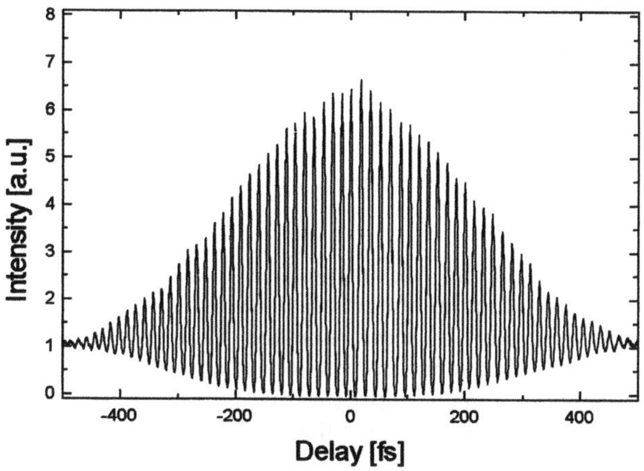

Fig. 2: Interferometric autocerrelation of the idler pulse at 4.8 μm

The $AgGaSe_2$ OPO was tuned from 1.95 μm to 2.5μm and from 4.2 μm to 7.9 μm for signal and idler wave, respectively. Water vapor absorption around 1.9 μm and 2.7 μm limited the possible tuning range.

4. Conclusions

Based on our calculations and first experimental results we expect this $AgGaSe_2$-OPO to produce pulses with pulse lengths well below 100 fs with average output powers in the order of several 100 mW over a wide tuning range (2 μm < λ < 10 μm) which can easily be reached by tuning the pump lasers.

References

1 R.C. Eckardt et al., Appl. Phys. Lett. **49**, 608 (1986).
2 J. Kirton, Opt. Commun. **115**, 93 (1995).
3 E.C. Cheung et al., Opt. Lett. **19**, 631 (1994).
4 Ch. Grässer et al., OSA TOPS Vol.**1**, 158 (1996)

Generation of High Repetition Rate Femtosecond Infrared Pulses Tunable from 9 to 18 μm

R. A. Kaindl[1], D. C. Smith[2], M. Joschko[1], M. P. Hasselbeck[1], M. Woerner[1], and T. Elsaesser[1]

[1] Max-Born-Institut für Nichtlineare Optik und Kurzzeitspektroskopie, Rudower Chaussee 6, D-12489 Berlin, Germany
E-mail: kaindl@mbi-berlin.de
[2] Department of Physics, Clarendon Laboratory, Oxford University, Oxford OX1 3PU, UK

Abstract. Nearly bandwidth-limited mid-infrared pulses as short as 110 fs are generated by phasematched difference frequency mixing within the broad spectrum of 20 fs Ti:sapphire pulses in GaSe.

Femtosecond pulses in the mid-infrared (3 to 20 μm) are important for many applications, e.g. for studying the ultrafast dynamics of vibrations in molecules or intersubband transitions in semiconductor nanostructures. Different schemes based on parametric frequency conversion of amplified femtosecond pulses have been used to generate tunable pulses in that range [1]. These are complex setups working at kilohertz repetition rates that make ultrafast experiments with a large dynamic range quite difficult. More stable sources such as optical parametric oscillators (OPOs) have been used to generate mid-infrared pulses at 76 MHz repetition rate and wavelengths up to 12 μm by difference frequency mixing in thin $LiIO_3$ and $AgGaS_2$ crystals [2]. In a simpler approach, a train of short mid-infrared pulses of large spectral width was produced by non-phasematched optical rectification of 15 fs Ti:sapphire laser pulses at the surface of GaAs [3]. This source, however, exhibits low average power and the pulse duration has not been characterized.

We report a versatile and compact source of femtosecond pulses tunable from $\lambda_{MIR} = 9$ to 18 μm at 88 MHz repetition rate [4]. Broadband 20 fs pulses from a mode-locked Ti:sapphire oscillator are used for difference frequency mixing in a GaSe crystal, in which different components of the input spectrum contribute to the phasematched parametric process. Both ordinary and extraordinary polarizations are obtained from a single near-infrared beam by orienting the crystal's rotation axis at an angle of 45° with respect to the incoming polarization. Type-I phasematching is employed with the crystal's y-axis perpendicular to the rotation axis. The wavelength of the mid-infrared pulses generated in this way is continuously tunable by varying the phasematching angle.

In Fig. 1, we present results of electric field correlation measurements [3]. Fig. 1 (a) shows an interferogram of two mid-infrared pulses at $\lambda_{MIR} = 15$ μm. The spectra shown in Fig. 1 (b) are derived from such data for different phasematching angles and exhibit bandwidths of 10 to 15 meV. They demonstrate the broad tunability of the pulses between $\lambda_{MIR} = 9$ and 18 μm. The tuning range is limited by the finite bandwidth of the 20 fs input pulses and the transparency cutoff of GaSe.

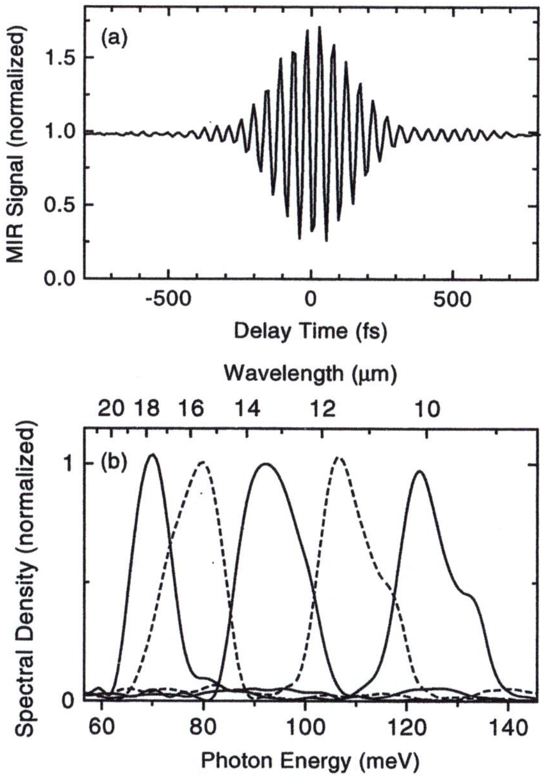

Fig. 1. (a) Interferometric electric field correlation of femtosecond mid-infrared pulses at λ_{MIR} = 15 μm. The signal is plotted as a function of delay time between the pulse and its replica. (b) Tuning range of the light source showing normalized spectra of the midinfrared pulses at different phasematching angles.

The bandwidth of the phasematched pulse is given by the spectral acceptance of the difference frequency mixing process, which is directly related to the effective interaction length L_{eff} and group velocity mismatch (GVM) between the mid-infrared and near-infrared pulses. For λ_{MIR} = 15 μm, we estimate L_{eff} = 150 μm from the GVM between the two interacting spectral components of the incident pulse, corresponding to a spectral width of $\Delta\lambda \approx 2$ μm, in good agreement with our experiment. Prechirping the incident pulse leads to a longer interaction length and thus higher infrared power. We obtain average powers of up to 1 μW, corresponding to a photon conversion efficiency of 2×10^{-4}. Such values are about two orders of magnitude higher than for infrared generation by optical rectification on a GaSe or GaAs surface.

The signal in Fig. 1 (a) displays the linear correlation function of the generated light field and thus provides a measure of the coherence length, but *not* the pulse duration. To determine the pulse duration, we perform pump-probe experiments

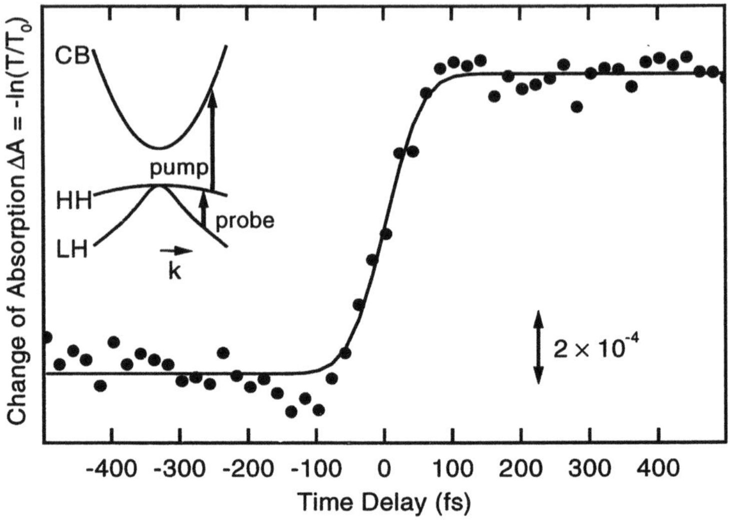

Fig. 2. Transient inter-valence band absorption of Ge at a probe wavelength of 13 μm after excitation by a 20 fs pulse at 830 nm. The change of absorption $\Delta A=-\ln(T/T_0)$ is plotted as a function of delay time between pump and probe (symbols, T_0,T: transmission of the sample before and after excitation). Solid line: calculated response. Inset: Schematic of the pump-probe scheme in the bandstructure of Ge.

with a bulk germanium sample at 300 K. Near-infrared excitation with 20 fs pulses creates an electron-hole plasma and the resulting absorption change is probed by the mid-infrared pulses at λ_{MIR} = 13 μm. In Fig. 2, the change of absorption is plotted versus the delay time between pump and probe. We observe a step-like increase of absorption that is dominated by intervalence-band absorption of photoexcited heavy holes. The observed absorption increase follows the time-integral of the excitation pulse convolved with the probe pulse. This is equivalent to the time-integrated cross-correlation function. The solid line in Fig. 2 represents the time integral of a Gaussian cross-correlation function with a duration of 110 fs (FWHM). Deconvolution yields a mid-infrared pulse duration close to 110 fs corresponding to a time-bandwidth product of 0.4.

References

1 F. Seifert, V. Petrov, and M. Woerner, Opt. Lett. **19**, 2009, 1994.
2 A. Lohner, P. Kruck, and W.W. Rühle, Appl. Phys. B **59**, 211, 1994.
3 M. Joffre, A. Bonvalet, A. Migus, and J.L. Martin, Opt. Lett. **21**, 964, 1996
4 R. A. Kaindl, D.C. Smith, M. Joschko, M.P. Hasselbeck, M. Woerner, and T. Elsaesser, Opt. Lett. **23**, 861, 1998.

A Proof of Principle Experiment of Sum Frequency Generation Using Chirped Pulses

Károly Osvay[1,2], Ian N. Ross[2]

1 Department of Optics and Quantum Electronics, JATE University, P.O.Box 406 Szeged H-6701, Hungary
E-mail: osvay@physx.u-szeged.hu
2 Central Laser Facility, Rutherford Appleton Laboratory, Chilton, Didcot, Oxon OX11 0QX UK
E-mail: I.N.Ross@RL.AC.UK

Abstract. Broadband and spectrally tuneable ultrashort pulses are efficiently generated by providing the input 50 fs pulses with appropriate chirps having opposite sign. The scheme may be of special interest in ultrashort pulse laser systems using the technique of chirped pulse amplification.

1. Introduction

The main limitation to efficient broad bandwidth frequency conversion originates from the dispersion of the non-linear crystal. To overcome such severe restrictions on efficiency, an alternative technique in theory [1,2] has been proposed in which the generating beams to be mixed in the nonlinear material are chirped at different chirp rates such that the phase matching condition can be met at all times.

For experiments scaling physical and chemical phenomena with pulse length, it would be extremely useful if the pulse duration is easily varied. Spectral shaping of ultrashort pulses provides this possibility at the expense of poor temporal contrast due to spectral diffraction on the mask.

In this paper we show experimentally that the bandwidth and the spectral position of a pulse generated by frequency mixing can be independently tuned by changing the chirp of and the delay between the generating pulses, respectively.

2. Theoretical considerations

In our description of the broadband scheme we consider a collinear type I mixing process, where the ordinary $\omega_1<\omega_2$ waves generate the extraordinary ω_3 wave. The frequencies are assumed varying with time. For perfect generation the total phase mismatch of the process $\Delta\psi=0$ should be satisfied in time and in space simultaneously. This requirement can be fulfilled by introducing a chirp (β_1, β_2) into the initial fields, so it can be interpreted as chirp-assisted group velocity matching (CGV) [1].

Narrow band operation can be described more easily on the basis of the above picture. If the two generating pulses are substantially stretched by approximately the same amount but with chirps of opposite sign, there are only very few frequency pairs satisfying the phase matching condition and this results in narrow generated bandwidth. Moreover, a change of delay between the input pulses gives rise to a spectral tuning of the generated field within the achievable largest generated bandwidth. It is worth noting that in this tunable narrow band case an increase of the crystal length leads to an increase of efficiency along with narrowing the generated bandwidth. This description of the CGV scheme applies equally in the case of type II processes.

3. Experiment

The theoretical predictions have been tested by an experiment carried out using sub-50 fs pulses from a Ti:S oscillator. The 744 nm beam is negatively stretched by a double pass prism stretcher, then split into two. One part of the beam is overcompressed by glass slabs while the other part remains unchanged in pulse duration but its polarisation is rotated. Recombining the two beams and focusing weakly into a 0.65 mm type II BBO crystal the generated 372 nm spectrum along with the generating 744 nm spectra have been recorded. Measurement of both generated and input powers established the efficiency.

The overall figure of merit of this new scheme can be defined as the product of the generated bandwidth and the efficiency. Figure 1 shows the essential summary of the series of measurements carried out at different chirps of the generating pulses. The ratio of the chirps β_1/β_2 have been calculated from the measured pulse durations. The length of the longer pulse (τ_2) unchanged in the course of each series is also shown. It can be seen that in agreement with the theory the peak of the bandwidth-efficiency product (i) increases with less stretched generating pulses; (ii) occurs at lower chirp ratio when the pulses are less stretched and (iii) there is a factor of 40 between the lowest and highest figure of merit obtained in the experiment [3]. The broadest bandwidth is generated in the vicinity of $\beta_1/\beta_2 \approx -3$ whereas the narrowest tuneable pulses are obtained at -1.

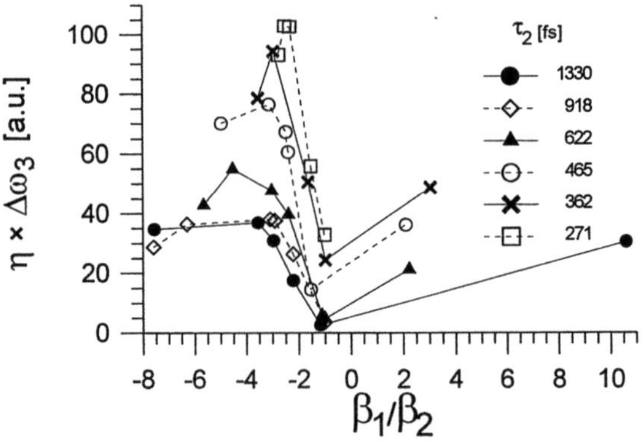

Fig. 1. The product of the generated bandwidth and the efficiency vs. the chirp ratio of the generating pulses.

An example of the spectral tuning achieved in our experiment is plotted in Figure 2. Since the process depends on the interacting bandwidths, the delay is given in units of the generating bandwidth $1/\Delta\omega_0$ which is ≈17 fs. The generated bandwidth is kept almost constant over the whole tuning range with the spectral tuning resolution of 2.4 cm^{-1}, which could be easily improved by several orders of magnitude by using a longer crystal and more highly stretched generating pulses.

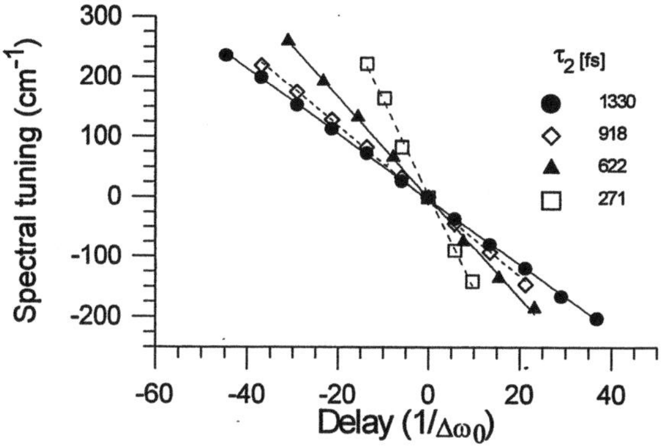

Fig. 2. Spectral tuning ranges from the experiment.

4. Conclusions

We have experimentally demonstrated a significant reduction of the effects of dispersion in frequency mixing using type II. SH generation of sub-50fs pulses. The product of generated bandwidth and efficiency varies by 40× according to the chirps of the generating pulses. The bandwidth and the spectral position of the generated pulse can be controlled by the chirp ratio of and the delay between the interacting pulses, respectively.

This CGV technique can be especially beneficial for high power CPA laser systems where the generating pulses are *ab initio* stretched. The smooth spectral profile and high temporal contrast of the generated pulse are adventegous for such laser systems while the extremely narrow band tuneable high power operation suffering no mode problems is essential in spectroscopy.

Acknowledgements. During this project Dr. Osvay has held a Magyary Zoltán Postdoctoral Scholarship. This work was partially supported by OTKA # F26065 and the Soros Foundation Hungary.

References

1 K.Osvay, I.N.Ross, J.Opt.Soc.Am.B **13**, 1431, 1996.
2 A.C.Boscheron, C.J.Sauteret, A.Migus, J.Opt.Soc.Am. B **13**, 818, 1996.
3 K.Osvay, I.N.Ross, submitted

A Novel Scheme for High-Intensity Ultrashort Pulse Generation

Gabriel Tempea and Thomas Brabec

Abteilung Quantenelektronik und Lasertechnik, Technische Universität Wien, Gußhausstrasse 27-29, A-1040 Vienna, Austria

Abstract. We analyze laser pulse propagation in a hollow waveguide filled with noble gas, at intensities where tunnel ionization occurs. A parameter range is identified where self-defocusing by the plasma lens is compensated by the guiding properties of the hollow waveguide. The plasma-induced self phase modulation introduces a strong spectral broadening and a nearly linear spectral chirp. Compensation of the linear component of the chirp yields sub-4-fs pulses with energies of a few milijoules.

1. Introduction

Recently, 5-fs pulses in the sub-mJ range were generated by using a hollow fiber filled with noble gas[1]. So far, it was assumed that a further increase in pulse energy is not possible because of the onset of self-focusing and ionization.

We show that the self phase modulation (SPM) due to the nonlinearity of a tunnel-ionizing gas can be used for pulse compression [2]. Our calculations predict the generation of sub-4-fs pulses with an energy of a few mJ. The predicted pulse energy is by more than one order of magnitude higher than the pulse energies which were obtained so far by using the Kerr nonlinearity.

2. Theory

We analyze pulse propagation in a gas-filled hollow fiber by means of coupled-mode theory. In the coupled-mode picture defocusing manifests as an energy transfer from the fundamental to higher-order modes. Our analysis reveals that the energy in the higher-order modes can grow only over the coherence length L_2. The coherence length L_2 is determined by the wave vector mismatch between the fundamental and the next higher-mode. If the plasma nonlinearity is small over L_2, the energy transfer to the higher-order modes will be weak and the pulse will propagate mainly in the fundamental transverse mode. The key-parameter to fulfill this condition is the gas density in the hollow fiber that must be chosen smaller than the critical density. In reference [2] we have determined the critical (highest permitted) density as a function of fiber radius and pulse-intensity.

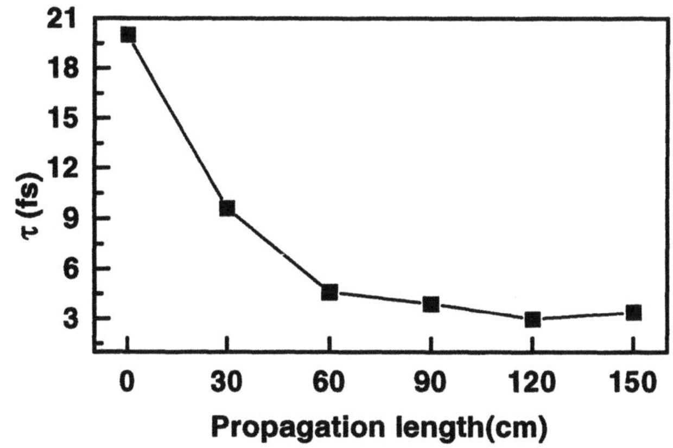

Fig. 2. Pulse duration as function of propagation length, for the parameters given in the text. The gas filling the fiber is Argon.

propagation distances do not lead to a further pulse shortening, however the energy loss is increased.

4. Conclusions

We have investigated a hollow-fiber pulse-compression scheme which utilizes the nonlinear response of an ionizing noble gas as a source for spectral-broadening. Our calculations have shown that the plasma nonlinearity has the potential to generate sub-4-fs pulses with energies of several milijoules.

Acknowledgements. We would like to thank Prof. A. J. Schmidt for his support. This work was supported by the Austrian FFWF grant P12631-PHY and by the Austrian Nationalbank grant 7178.

References

1 M. Nisoli, S. Stagria, S. De Silvestri, O. Svelto, S. Sartania, Z. Cheng, M. Lenzner, Ch. Spielmann, and F. Krausz, *Appl. Phys. B* 65, 189 (1997)

2 G. Tempea and T. Brabec, *Opt. Lett.* 23 (August 15, 1998)

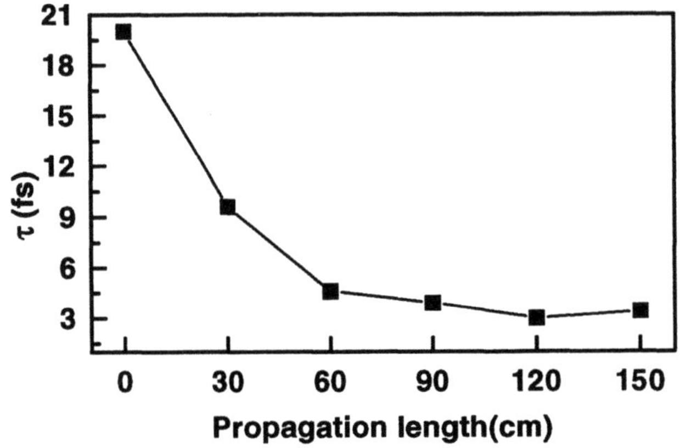

Fig. 2. Pulse duration as function of propagation length, for the parameters given in the text. The gas filling the fiber is Argon.

propagation distances do not lead to a further pulse shortening, however the energy loss is increased.

4. Conclusions

We have investigated a hollow-fiber pulse-compression scheme which utilizes the nonlinear response of an ionizing noble gas as a source for spectral-broadening. Our calculations have shown that the plasma nonlinearity has the potential to generate sub-4-fs pulses with energies of several milijoules.

Acknowledgements. We would like to thank Prof. A. J. Schmidt for his support. This work was supported by the Austrian FFWF grant P12631-PHY and by the Austrian Nationalbank grant 7178.

References

1 M. Nisoli, S. Stagria, S. De Silvestri, O. Svelto, S. Sartania, Z. Cheng, M. Lenzner, Ch. Spielmann, and F. Krausz, *Appl. Phys. B* 65, 189 (1997)

2 G. Tempea and T. Brabec, *Opt. Lett.* 23 (August 15, 1998)

Self-focusing in waveguides

Thomas Brabec and Gabriel Tempea

Abteilung Quantenelektronik und Lasertechnik, Technische Universtiät Wien
Gußhausstraße 27-29, A-1040 Wien, Austria
E-mail: brabec@rs6.iaee.tuwien.ac.at

Abstract. A theoretical analysis of self-focusing in waveguides is presented. Our analysis yields a critical power for self-focusing in waveguides which is by a factor of 5 larger than the respective critical power in bulk materials. The theory is used to obtain design criteria for multimode fiber-amplifiers and for hollow waveguide pulse compression systems.

1. Introduction

Recent developments in ultrafast sciences opened novel avenues towards the generation of high-energy ultrashort optical pulses. By using a hollow waveguide filled with noble gas high-energy optical pulses were compressed down to 5 fs [1]. The progress in fiber technology made it possible to reduce mode coupling in solid-core step-index multimode fibers so that single mode propagation over tens of meters can be realized [2]. This result has renewed the interest in multimode fiber-lasers and -amplifiers. In hollow-waveguide pulse-compressors as well as in multimode fiber-amplifiers the pulse peak power can become large enough so that self-focusing presents a severe limitation for the highest pulse energies which can be realized in these systems. We present, stimulated by these latest experimental developments, a coupled-mode theory of self-focusing in waveguides. Our analysis yields a critical power for self-focusing in waveguides which is by a factor of 5 larger than the critical power in bulk media. The phase mismatch is identified as the physical mechanism which counteracts self-focusing and which determines the critical power in a waveguide. Finally, a design criterium for hollow-fiber pulse-compressors and for multimode fiber-amplifiers is obtained.

2. Coupled-Mode Theory

The mode solutions of a hollow waveguide can be approximately described by a set of linearly polarized modes LP_{jp}, a wavevector β_{jp}, and a mode loss α_{jp},[3] where j and p are the mode indices. In the limit of weak guiding and far from cutoff, the eigenmodes of a multimode fiber are described by the same set of solutions [4] however, with $\alpha_{jp} = 0$. For the derivation of the coupled mode equations [3] it is assumed that the laser pulse is initially in the fundamental mode and that the coupling between the fundamental and the higher-order modes is weak. In this limit coupling occurs only between the fundamental mode and the modes with index j = 0. Therefore, the index j is omitted from hereon.

3. Results and Discussion

The Kerr nonlinearity of the fiber core material has two effects on the pulse evolution: (i) spectral broadening and (ii) self-focusing. In the mode picture self-focusing manifests as a transfer of radiation from the fundamental to the higher-order modes. Growth of radiation in the higher-order modes is only possible over the coherence length $L_p = \pi/\Delta\beta_p^{(0)}$, [3] where $\Delta\beta_p^{(0)}$ is the wavevector mismatch between the modes LP_1 and LP_p. As the coherence length $L_2 >> L_p$ $(p > 2)$, the dominant part of the radiation is transferred to the mode LP_2. The maximum ratio of the peak powers in the modes LP_1 and LP_2 is determined by the parameter κ^2, which is a good measure of the distortion introduced by self-focusing. The influence of self-focusing is weak, as long as

$$\kappa^2 = \left(\frac{P_0}{P_{cw}}\right)^2 << 1, \text{ where } P_{cw} \approx \frac{25\lambda^2}{8\pi n_2}. \tag{1}$$

Here, n_2 is the nonlinear index of refraction, λ is the center wavelength of the laser pulse, and P_0 is the initial peak power of the laser pulse. The parameter P_{cw} is the critical power around which self-focusing in a waveguide becomes significant and resembles the critical power for self-focusing in a bulk material, $P_{cb} = \alpha\lambda^2/8\pi n_2$. The parameter $\alpha = 4 - 6$ is a numerically determined correction factor [5]. Comparison of the critical powers gives the ratio $P_{cw} / P_{cb} = 4\text{-}6$.

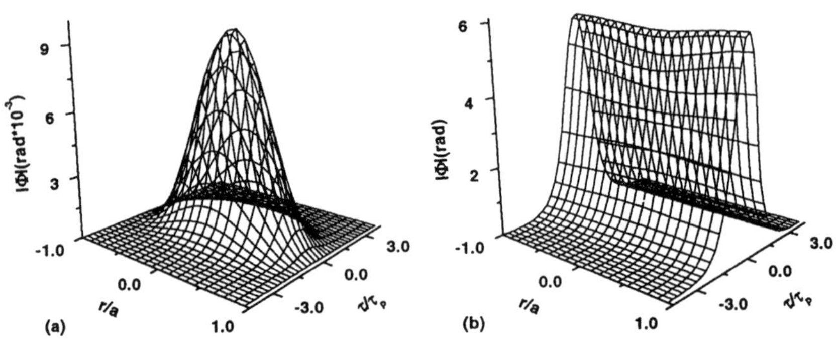

Fig. 1. depicts the pulse phase Φ due to the Kerr nonlinearity versus the radial (normalized to the fiber radius) and the temporal (normalized to the pulse duration) coordinates for $\kappa = 0.2$. The propagation distance is $z << L_2$ (a) and $z >> L_2$ (b).

From the coupled mode analysis the amplitude and the phase of the electric field in the waveguide is calculated. Figure 1 reveals the effect of the phase mismatch on self-focusing in waveguides. For distances $z<<L_2$ the phase mismatch plays no role and radiation is transferred into the higher-order modes. The Kerr nonlinearity introduces an equally strong temporal and spatial curvature of the pulse phase. For $z>>L_2$, the phase mismatch stops the energy transfer to the higher modes however, the temporal phase modulation keeps growing linearly with the propagation distance. This flattens increasingly the transverse variation of the nonlinear phase shift, see Fig. 1b.

The suppression of transversal effects due to self-focusing, as revealed by Fig. 1, is an important requirement. The transfer of energy to higher-order modes impairs the spatial quality of the laser pulse [2]. Therefore, self-focusing in waveguides must be controlled in order to avoid undesirable spatial effects. From Eq. (1) we find a criterium for the maximum peak power of a laser pulse, which can be propagated in waveguides before self-focusing becomes dominant. It is assumed that the transversal distortions of the pulse phase and of the pulse profile are tolerable as long as $\kappa < 0.3$. This parameter range which fulfills this requirement is found to be

$$P_0 \leq 0.3\frac{\lambda^2}{n_2} \tag{2}$$

The inequality (2) presents a design criterium for obtaining a good beam quality in hollow fiber pulse compression systems and in multimode fiber amplifiers and lasers. Equation (2) demonstrates that gas filled waveguides are more suitable for the compression of high energy pulses than multimode glass fibers, the Kerr index of which is by several orders of magnitude of higher.

4. Conclusions

We have identified a critical power for self-focusing in waveguides. From that a condition was derived which determines the optimum parameter regime for hollow-fiber pulse-compression systems and for multimode fiber-amplifiers and -lasers.

Acknowledgements. We would like to thank Prof. A. J. Schmidt for his support. This work was supported by the Austrian FFWF grant P12631-PHY and by the Austrian Nationalbank grant 7178.

References

1 M. Nisoli et al., Appl. Phys. B **65**, 189 (1997).
2 M. E. Fermann, Opt. Lett. **23**, 52 (1998).
3 G. Tempea and T. Brabec, Opt. Lett. **23**, 762 (1998).
4 D. Marcuse, *Theory of Dielectric Optical Waveguides,* Academic Press, New York, 1974.
5 M. Sheik-Bahae et al. Opt. Eng. **30**, 1228 (1984).

Observation of Raman Self-Conversion of fs-pulse Frequency due to Impulsive Excitation of Molecular Vibrations

Georg Korn, Oliver Dühr and Alexander Nazarkin

Max-Born-Institut für Nichtlineare Optik und Kurzzeitspektroskopie
Rudower Chaussee 6, 12489 Berlin, Germany
E-mail: korn@mbi-berlin.de

Abstract: We report on the observation of a Raman frequency conversion mechanism different from the usual Stokes (anti-Stokes) stimulated process. A continuous down-shift of a laser pulse frequency was measured for sufficiently short ($\leq$ 40 fs) pulses propagating in a gaseous SF_6. We attribute this behavior to the extremely nonstationary nuclear response of SF_6-molecules to the light field with a duration shorter than the period of the Raman-active vibration.

1. Introduction

Stimulated Raman scattering (SRS) of light has been studied in great detail both for the case of quasistationary interaction and for laser pulses comparable and shorter than the vibrational relaxation time of a Raman medium [1], but still much longer than the corresponding molecular vibrational period $T_v=2\pi\Omega_v^{-1}$. For fs-pulses of a duration $\tau_p < T_v$, the pulse spectrum initially contains Stokes and anti-Stokes frequency components satisfying the condition of Raman resonance, and single-pulse excitation of Raman-active vibrational-rotational transitions takes place. This mechanism (so called „impulsive SRS“) was recently used for real-time spectroscopic studies of individual molecular vibrations in solids and liquids and for investigation of rotational coherence in gases [2].

Here we report on first the results of experimental and theoretical studies of intense fs-pulse propagation in a Raman-active medium (SF_6) in the region of pulse durations $\tau_p < T_v$.

2. Experimental Results

The experiments were performed with frequency doubled pulses from a 30-fs - 1kHz Ti:sapphire laser. Frequency doubled 200µJ - 40fs at 400nm pulses were sent in a hollow waveguide [3,4] with a diameter of 120µm filled with SF_6. This technique allowed to enlarge the length of focused pulse interaction with the gas medium. Fig.1 shows the recorded spectra for different pressures of SF_6. Competing with pulse spectral broadening we have observed a well-pronounced shift of the pulse central frequency towards lower frequencies which increased

nearly proportionally to the gas pressure. In addition, a progressing asymmetry in the shape of the pulse spectrum was observed. With an increase of gas pressure, a noticeable suppression of the high-frequency wing of the spectrum occurred. The above features appeared for sufficiently short pulse durations (τ_p < 40fs).
The mechanism responsible for the observed behavior was examined by measuring the changes in the spectrum of a weak fs-laser pulse (τ_p < 20fs) delayed in respect to the strong pump pulse. The observed oscillation of the center of gravity of the probe pulse (see Fig.2) had a period of 43fs corresponding to the excitation of the symmetric vibrational mode A_{1g} of SF_6. Thus, an evidence for a fs-pulse frequency conversion in the regime of impulsive vibrational excitation was shown.

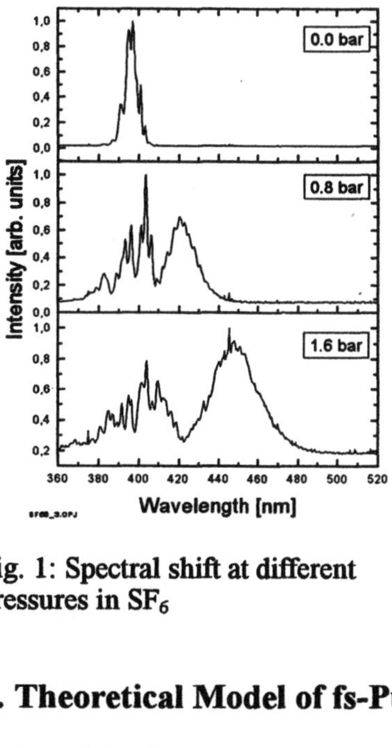

Fig. 1: Spectral shift at different pressures in SF_6

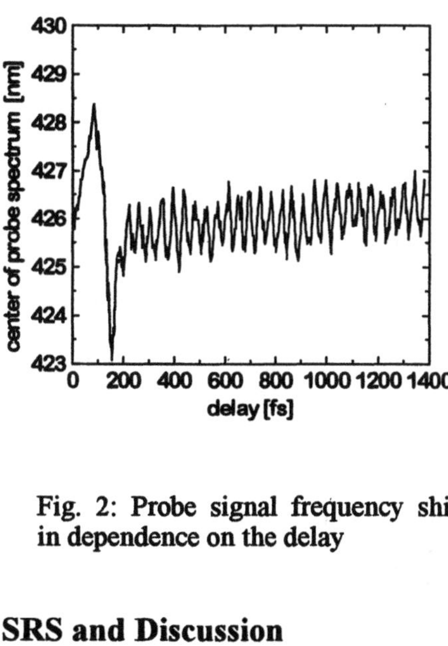

Fig. 2: Probe signal frequency shift in dependence on the delay

2. Theoretical Model of fs-Pulse SRS and Discussion

We explain the pulse dynamics on the basis of a self-consistent solution of the Maxwell equation for the pulse field and equations of molecule vibrational excitation. A highly symmetric SF_6 molecule exhibits no Raman rotational spectrum and its symmetric vibrational mode A_{1g} gives rise to a very strong (compared to the other modes) Raman line (775 cm^{-1}). To a good approximation, the molecule can be described by a model of a single mode Raman oscillator with a normal coordinate Q , an effective mass M, and a natural frequency Ω_v, which is driven by a force quadratic in a laser field $\mathcal{E}(t)$ [1]. The full medium polarization contains the linear part (material and waveguide dispersion) and the nonlinear part due to the electronic Kerr effect (P_{Electr}) and the Raman contribution $P_{Raman} = N\alpha' Q\, \mathcal{E}$, where α' is the derivative of the molecule electronic polarizability $\alpha(Q)$ in respect to the vibrational coordinate Q.

In the regime of ultrashort pulse-molecule interaction with $\tau_p < T_v = 2\pi\Omega_v^{-1}$, from the equation of molecular motion it follows that the molecule vibrational coordinate changes as

$$Q(t) \approx (1/M)\alpha' \int_0^t dt' \int_0^{t'} dt'' |E(t'')|^2 \quad (1)$$

where E(t) is the amplitude of the laser pulse. Using solution (1), for the initial stage of evolution of the pulse instant frequency $\dot{\varphi}(z,\tau)$ with the distance z we have found the following equation:

$$d\dot{\varphi}/dz = -\beta_{Electr}(\partial I/\partial\tau) - \beta_{Raman}\int_0^{\tau} I(\tau')d\tau' \quad (2)$$

where $\beta_{Electr} = 2\pi N\omega\chi^{(3)}/cn$, $\beta_{Raman} = 4\pi^2 N\omega(\alpha')^2/Mc^2n^2$, and $I(z,\tau)$ is the laser pulse intensity. As can be seen, the Raman contribution (the second term in Eq.(2)) appears as an additional red shift which increases monotonically from the pulse leading to its trailing edge in proportion to the laser energy passed through a given cross section. This results in the continuous down shift of the center of gravity of the pulse spectrum. Second, the combined action of the electronic Kerr effect (the first term in Eq.(2)) and the inertial Raman contribution results in an enhancement of the red and suppression of the blue spectral components which explains the asymmetry in the measured pulse spectrum.

The observed behavior seems to be a general characteristic of SRS-media on the fs-time scale because the frequency shift depends neither on the Raman vibrational frequency Ω_v nor on the gain line width, and because many types of matter exhibit Raman activity due to scattering by phonons, plasmons, spin and other elementary excitations. Our analysis of the pulse dynamics at higher gas pressures performed beyond the scope of the slowly-varying amplitude approximation shows that the observed down-conversion mechanism can lead to a generation of tunable IR pulses with only several cycles of electromagnetic wave [5]. This can be interesting for the generation of continuously tunable fs-radiation.

Acknowledgements:

We would like to thankProf. I.V. Hertel, Prof. T. Elsaesser and Prof. W. Zinth for fruitful discussions.

References

1 W. Kaiser and M. Maier, in *Laser Handbook*, ed F. Arecchi (Amsterdam,1970)
2 S. Ruhman, B. Kohler, A. G, Joly, and K. Nelson, IEEE J.Q.E.**24**, 470 (1988); M. Morgen, W. Price, L. Hunziker et al. Chem. Phys.Lett.**209**, 1(1993)
3 E. T. J. Nibbering, O. Dühr, and G. Korn, Opt. Lett. **22**, 1355 (1997)
4 M. Nisoli, S. De Silvestri, and O. Svelto, Appl. Phys. Lett. **68**, 2793 (1996)
5 A. Nazarkin and G. Korn, Phys. Rev. **A58**, R61(1998)

Spatio-temporal coherence properties of super-continuum single filament in bulk media

Emmanuel Baubeau, François Salin, and Catherine Le Blanc

Laboratoire d'Optique Appliquée, ENSTA - Ecole Polytechnique, Batterie de l'Yvette, 91761 Palaiseau Cedex, France
E-mail: salin@enstay.ensta.fr

Abstract. We study the spatio-temporal behaviour of the continuum produced by the propagation of an ultrashort pulse in a bulk media. We present a direct imaging of the continuum and a study of its temporal behaviour by spectral interferometry.

1. Introduction

The generation of a spectral continuum has been known and used since the first ultrashort laser sources were available [1]. This continuum is produced by an intense ultrashort pulse propagating in any medium including air. For sufficient power, it is accompanied by small scale self-focusing, the beam breaks into multiple filaments, destorting the transverse profile of the pulse. Recently, observations of long light filaments in the air have been reported [2] and several models have been proposed including ionisation, moving-focus and self-channeling [3,4]. In order to gather more precise information about this still not understood phenomenon, we performed several experiments on the spatial behaviour of the filaments produced in bulk media. These experiments were carried out with a high repetition rate Ti :sapphire oscillator amplifier chain delivering 50 fs, 800 μJ pulses around 800 nm at 1 kHz. Due to the spatial filtering introduced by the regenerative amplifier used in the system, the output beam is nearly gaussian.

2. Spatial study of continuum generation

In a first set of experiment, we focused a small part of the laser. with a 50 mm lens on a 6 mm-long Ti:sapphire crystal polished on every side. Inside the crystal, the beam self-focuses after about 1mm and propagates from that point as self-guided filament undergoing at the same time strong self-phase modulation. The blue part of this white spectrum is absorbed in the Ti:sapphire crystal and is converted into 800nm fluorescence which is emitted on 4π steradians. A microscope objective images the filament from the side onto a CCD camera. Both objective and camera could be translated parallel to the crystal and images of the filament were taken all along its propagation. A mosaic of images taken with a x20 objective is given in Fig. 1. The filament originates at the very left of the picture.

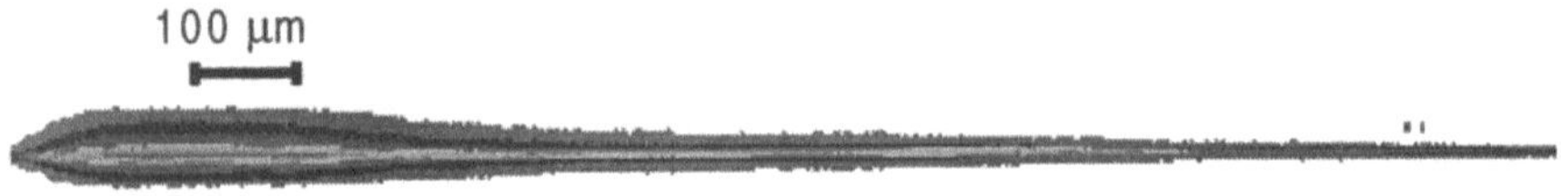

Fig.1: Side image of a single white light filament.

Several observations can be made from these images:
(i) the first 300 µm are the brightest part of the filament. In that region, the gaussian profile of the intensity sits on a large pedestral which seems to correspond to the conical emission (CE) already observed in solids and gases. The brightness of the first part could be caused by the absorption of the CE itself. It is interesting to note that 300 µm is the length over which the beam self-focuses in the moving focus model. We believe that the CE is produced by this moving focus via the Cherenkov effect and that the rest of the filament is related to self-guiding and not to a moving focus.
(ii)The filament exhibits an 8 µm gaussian-like shape along 1.5 mm, which represents more than 10 Rayleigh lengths. The strong absorption in the crystal did not allow us to take further images, but we saw no sign of divergence in the last images. A stable 3D self-guided propagation cannot be explained in the sole frame of the nonlinear Schrödinger equation but needs extra physical processes to exist. Preliminary calculations show that ionization could be a good candidate to stabilize the self-focusing effect.

In a second experiment, we investigated the transverse profile of the filament by an axial imaging. As the propagation inside the crystal is non-linear, one cannot make a direct optical imaging of the filament. We imaged the output of the crystal on a CCD camera and we controlled the length of the filament by changing the energy of the pulse. We checked by using side views that the profile of a filament was independent of the energy of the laser pulse. The side-imaging apparatus of the previous experiment is used to measure the length of the filament. Over the whole range the filament kept a gaussian profile, and the 8 µm diameter stays unchanged along 1.2 mm, which corroborates the results of the first experiment. After 1.2 mm the filament diverges with a half-angle of 10 mrad.

3. Study of the temporal coherence of the continuum

We used spectral interferometry [5] to investigate the behaviour of the spectral phase of the pulse during continuum generation. The pulse is sent in a Mach-Zender interferometer. A continuum is produced in one arm, the second arm is the reference. The output of the interferometer is sent to an spectrometer and we recorded the spectral interferences between the two beams. The spectrum gives then the spectral dephasing between the original 800 nm pulse and the part of the continuum corresponding to this spectral range. The spectral fringes were perfectly stable indicating that the continuum is temporally coherent with the incident pulse (fig.2). More surprising is the fact that the phase deduced from that spectral interferometry measurement did not exhibit the shot to shot fluctuations one can expect for this highly non-linear process. No trace of any strong non-linear dephasing linked to self phase modulation was observed.

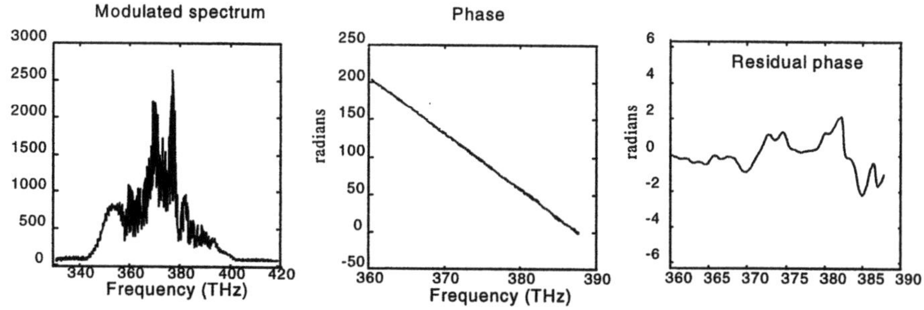

Fig. 2: Spectral interferences between the pump beam and the continuum

In a second time, two continua are separately produced from the same pulse in both arms of the interferometer. Once again the continuum, although produced in separated samples were perfectly coherent and phase locked over millions of shots. No evidence of difference in SPM among the two arms was observed although the intensities were not perfectly equal and the media different. This phase locking along with an insensibility of the phase to intensity fluctuation are reminiscent of soliton propagation.

4. Conclusion

Although the conditions for soliton propagation are not fullfil (positive GVD) we do see a stabilization of the spatial profile and spectral phase which are puzzling and cannot be explain with present theories of continuum generation.

References

1 R.R. Alfano and S. L. Shapiro, in Physical Review Letters, 24, 592 (1970)
2 A. Braun et all., in Optics Letters, vol. 20, n°1, 73 (1995)
3 A. Brodeur, O.G. Kosareva et al., in Optics Letters, vol. 22, n°5, 304 (1997)
4 E. T. J. Nibbering et al., in Optics Letters, vol. 21, n°1, 62 (1996)
5 C. Froehly et al., in Journal d'Optique, 4, 183 (1973)

Second Harmonic Pulse Compression

Jens Biegert, Vaclav Kubecek and Jean-Claude Diels

The University of New Mexico, Department of Physics & Astronomy, Albuquerque, New Mexico, U.S.A., phone/fax: +01 (505) 277-4026

Abstract. The coupled propagation equations are solved numerically in the frequency domain taking dispersion into account to all orders. Simulation and experimental data concur to show a 45-fold compression of the output of a Nd:YAG system down to 309 fs.

Second-harmonic pulse compression using the output of a Nd:YAG laser system to get a source of intense femtosecond radiation is a compact alternative to the much larger, more expensive and also more complicated Ti:Sapphire - regenerative amplifier systems. Q-Switched and mode-locked Nd:YAG laser systems are known to be the workhorse for many applications, routinely producing an output of pulses of 50 to 100 picosecond duration. However, pulses of 10 picosecond duration with energy fluctuations of less than 1% can be obtained from the same laser after the implementation of negative feedback [1]. These pulses are ideal to get significant pulse compression down to the femtosecond regime using group velocity dispersion in type II second-harmonic interaction inside a potassium dihydrogen phosphate (KDP) crystal as predicted by Wang and Dragila [2]. Heinz and Laubereau [3] obtained experimentally a 4-fold compression in KDP.

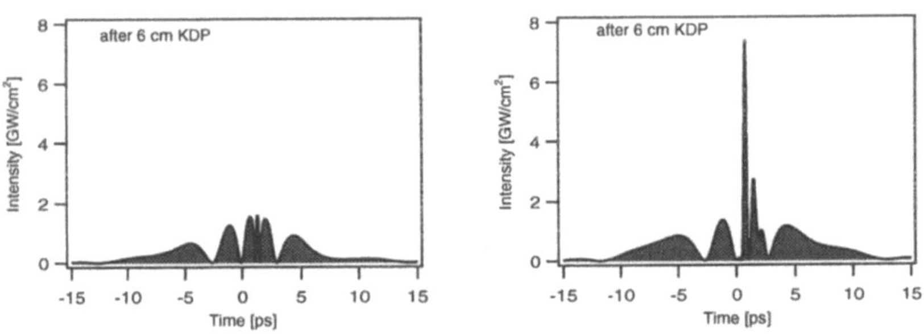

(a) neglecting group velocity dispersion

(b) full wavevectors

Figure 1: Second harmonic pulses generated in 6 cm of KDP. The initial e and o pulses are Gaussian, 14 ps FWHM with a peak intensity of 0.9 GW/cm^2 and 1.5 GW/cm^2. The fundamental pulses are depleted as they move into each other. The predelay was 14 ps.

We analyze the coupled wave interaction in the frequency domain, taking into account all dispersion effects to all orders. Solving the coupled differential equations in the frequency domain rather than the time domain allows us to use the exact functional dependence of the wave vectors. The common approach in the time domain instead uses a series expansion of differential operators hereby neglecting all higher order dispersion terms which account for pulse reshaping. Hence, the result of numerical simulations are even qualitatively different from those obtained by the standard approach in the time domain, where even group velocity dispersion is neglected. Experimental data, in agreement with the simulations, show the compression of 14 ps pulses down to 309 fs in the second harmonic. Figs. 1 show the difference between a calculation based solely on three different group velocities (a) compared to the exact solution (b). The dissimilarities between the two plots in Fig. 1 clearly illustrate that it is not justified to neglect higher order dispersion terms. The compressed pulse width of 230 fs in Fig. 1 (b) is very close to the limit of 220 fs given by the phase-matching bandwidth for 6 cm KDP.
We have also carried out the experimental investigation of pulse compression. The experimental arrangement shown in Fig. 2 is similar to the arrangement described by Umbrasas and Diels [4].

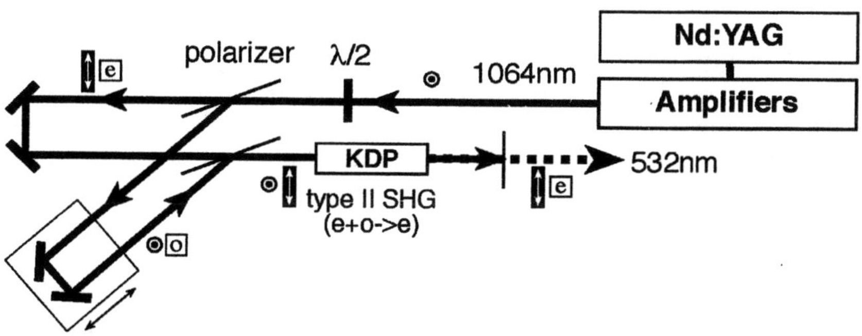

Figure 2: This simple setup makes it possible to control the ratio of the pulse energies in the different polarizations (using the waveplate) as well as the predelay (translation stage). After the second polarizer, the pulses are colinearly recombined and sent into the KDP crystal.

The linearly polarized output from the Nd:YAG laser passes through a $\lambda/2$ waveplate which rotates the plane of polarization by 45°. The beam is separated into two components by a polarizing beam splitter, and each component is being given a relative delay before being recombined at a second polarizing beam splitter and sent into the second harmonic generating crystal (6 cm KDP). The duration of the compressed pulse was measured with a single shot autocorrelator based on noncolinear second harmonic generation in a 2 mm KDP type I crystal. Fig. 3 shows a single-shot autocorrelation trace for the compressed 532 nm pulse.

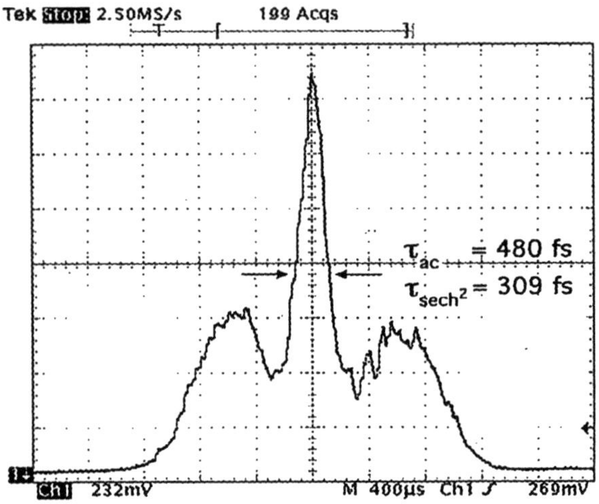

Figure 3: Single-Shot Autocorrelation of the compressed 532 nm pulse. The autocorrelation width was 480 fs. A sech^2 fit would give a temporal width of 309 fs. The modulation in the right pedestal results from scratches on a filter in one of the arms.

The total energy in the fundamentals (1064 nm) was 20 mJ, the pulse duration 14 ps, the resulting intensities $I_o = 1.5\ GW/cm^2$ and $I_e = 0.9\ GW/cm^2$ and the predelay 9.5 ps. This resulted in a 45-fold compression in the green and an energy conversion of 10 %. The temporal width of the compressed pulse was 309 fs. Broader pulses without satellites can also be obtained.

[1] A. Agnesi, A. Del Corno, P. Trapani, M. Fogliani, G. C. Reali, Jean-Claude Diels, Chao-Yung Yeh, Xin Miao Zhao, and V. Kubecek. Generation of extended pulse trains of minimum duration by passive negative feedback applied to solid-state q-switched lasers. *IEEE J. of Quant. Electr.*, 28:710–719, 1992.

[2] Y. Wang and R. Dragila. Efficient conversion of picosecond laser pulses into second-harmonic frequency using group-velocity dispersion. *Phys. Rev. A*, 41:5645–5649, 1990.

[3] P. Heinz, A. Laubereau, A. Dubietis, and A. Piskarskas. Fiberless two-step parametric compression of sub-picosecond laser pulses. *Lith. Phys. Review*, 33:314–317, 1993.

[4] A. Umbrasas, J. C. Diels, J. Jacob, and A. Piskarskas. Parametric oscillation and compression in KTP crystals. *Opt. Lett.*, 19:1753–1755, 1994.

Adaptive Control of Pulse Phase In A Chirped Pulse Amplifier

A. Efimov,[1] M. D. Moores,[1] N. M. Beach,[1] J. L. Krause,[2] and D. H. Reitze[1]

[1] Physics Department, University of Florida, Gainesville, FL 32611
E-mail: reitze@phys.ufl.edu
[2] Quantum Theory Project, University of Florida, Gainesville, FL 32611
E-mail: krause@qtp.ufl.edu

Abstract. By monitoring the magnitude of the spectral blueshift in a plasma breakdown experiment, we demonstrate that a femtosecond chirped pulse amplifier can use experimental feedback to adaptively find the phase that minimizes the laser pulse duration.

1. Introduction

Controlling ultrashort laser pulse parameters and their interactions with matter has long been a goal of the ultrafast laser community. Efforts in laser pulse control have been directed toward decreasing pulse durations, controlling the temporal and spatial profiles of the pulse, and increasing pulse energies. In coherent (or quantum) control of matter, the goal is to use optical fields that coherently interact with the dynamically evolving system and drive it to a specific desired final state. Recently, adaptive learning methods have been shown to be particularly suitable for experimental implementation of control [1]. These techniques directly couple computer-based learning algorithms with the experimental system under study together in a feedback loop such that the experiment 'learns' the optimal laser field for optimizing the laser-matter interaction. Demonstrations of adaptive control include the compression and shaping of femtosecond pulses from a Ti:sapphire laser [2], the optimization of fluorescence from organic molecules [3], and control of Rydberg wavepackets [4]. Here, we use a genetic algorithm-based adaptive learning scheme to compensate the phase of an amplified pulse. The goal of our experiments is to actively eliminate higher order phase that accumulates during the stretching, amplification, and compression process.

2. Experimental Methods

Our amplifier is seeded with pulses from a 2 nJ, 14 fs Ti:sapphire laser. Pulses are stretched to 10 ps in a dual-pass, dual 600 lines/mm grating configuration incorporating a programmable spatial light modulator (SLM), amplified to 1 mJ in a ten pass Nd:YAG pumped Ti:sapphire amplifier, and compressed in a double-pass grating compressor. In the absence of spectral phase applied to the SLM, the output pulse duration is determined by the amount of excess higher order phase on the pulse and the output bandwidth of the amplifier.

The feedback signal we 'lock' to is the magnitude of the spectral blueshift in a plasma breakdown experiment [5]. The magnitude of the spectral shift has been shown to be a sensitive probe of the pulse duration and shape. To generate the plasma, the pulses are focused using a 6X objective into air. The light emerging from the breakdown region is collected, spectrally filtered to pass wavelengths 400 nm < λ < 450 nm, and monitored by a photomultiplier tube. The feedback loop monitors the intensity of the blue light and iteratively updates the phase on the SLM. We use GENESIS, a genetic algorithm (GA), for performing the optimization [6]. The GA was programmed to simultaneously optimize cubic and nonlinear (NL) B-integral phase [7]:

$$\Phi(\omega)=\frac{1}{3!}\frac{d^3\Phi}{d\omega^3}(\omega-\omega_0)^3+\Phi_{NL}(\omega-\omega_0) \quad (1)$$

with 10-bit parameterization of the $d^{(i)}\Phi/d\omega^{(i)}$ and Φ_{NL}. An initial random population of 25 trials are tested and propagated via genetic operators to subsequent generations for subsequent evaluation.

3. Results and Discussion

Fig. 1 displays the evolution of the average and best blueshift magnitude for each generation. Within three generations, the GA has found a phase that produces the optimal blueshift signal. In the absence of phase correction, we observe no blueshifted component below 450 nm. Figure 2 displays logarithmic autocorrelations for both the optimized phase (solid line) and zero phase correction (dotted line). The optimized output pulse duration (30 fs FWHM) has decreased from the original pulse duration (37 fs FWHM), with a noticeable reduction in the wings of the pulse.

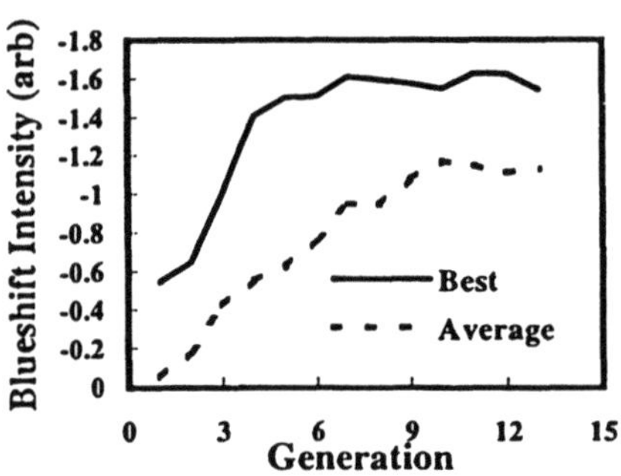

Fig. 1. The best evaluation (solid line) and average evaluation (dashed line) of the magnitude of the blueshifted light as a function of generation.

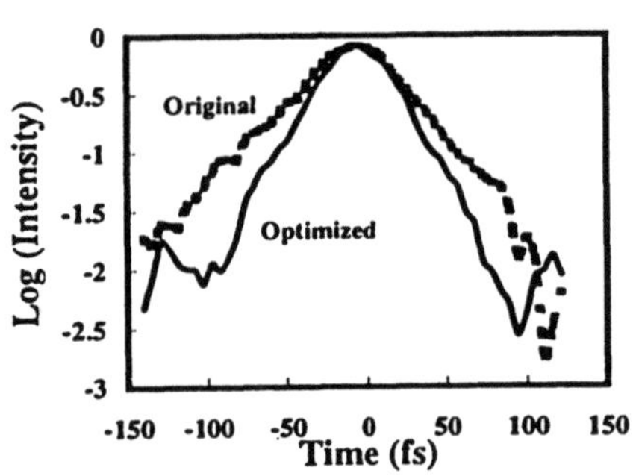

Fig.2. The autocorrelation (plotted logarithmically) of the optimized amplified pulse (solid line) and initial amplified pulse (dashed line).

Figure 3 plots the average value in each generation of the cubic and NL phases for each generation. Both the cubic and the NL phase values change significantly during the first few generations as the GA is searching the parameter space. By the fifth generation, however, both phases have converged to relatively constant values. At this point, the GA has tried roughly just 150 possible solutions.

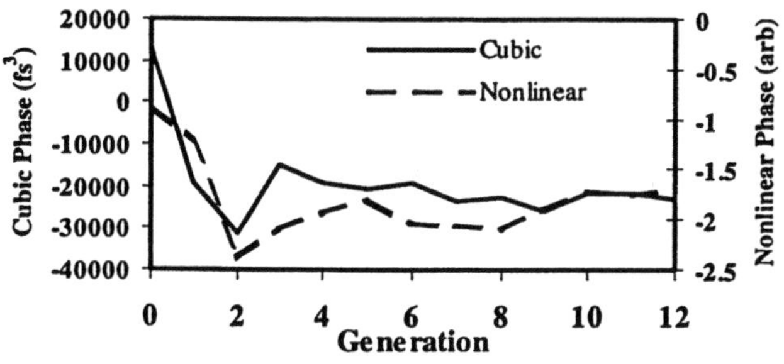

Fig. 3. The evolution of the average value of the cubic and nonlinear phase in each generation as a function of generation. Both the cubic and the nonlinear phase converge to well defined values after five generations.

4. Conclusions

We have demonstrated phase compensation of fs pulses in a chirped pulse amplifier using adaptive feedback. Using a genetic algorithm to reprogram the phase in pulse stretcher, we are able to achieve a significant reduction in pulse duration.

Acknowledgements. We acknowledge the support of NSF (National High Magnetic Field Laboratory) in support of this research. DHR gratefully acknowedges a Cottrell Scholarship from the Research Corporation.

References

1 R. S. Judson and H. Rabitz, Phys. Rev. Lett. **68**, 1500-1503 (1992).
2 D. Yelin, D. Meshulach, and Y. Silberberg, Op. Lett. **22,** 1793-1795 (1997).
3 C. J. Bardeen, V. V. Yakovlev, K. R. Wilson, S. D. Carpenter, P. M. Weber, W. S. Warren, Chem. Phys. Lett. **280**, 151-158 (1997).
4 T. C. Weinacht, J. Ahn, and P. H. Bucksbaum, submitted to Phys. Rev. Lett.
5 W. Wood, C. Siders, and M. C. Downer, IEEE Trans. Plasma Science **21**, 20 (1993).
6 J. D. Grefenstette, GENESIS v5.0, Copyright © 1990.
7 A. Braun, S. Kane, T. Norris, Op. Lett. **22**, 615 (1997).

Wavefront Correction of High Intensity Femtosecond Lasers

G. Chériaux[1], F. Druon[1], A. Maksimchuk[1], M. Nantel[1], G. Vdovin[2] G. Mourou[1].

1 Center for Ultrafast Optical Science, University of Michigan, 1006 IST building, 2200 Bonisteel Blvd, Ann Arbor, MI 48109-2099, USA.

2 Department of Electronic Instrumentation, Technical University of Delft, P.O. Box 5031, 2600 GA Delft, The Netherlands.

Abstract. We report the wavefront correction of high intensity laser beam, using deformable mirror. The B-integral and the thermal lensing wave-front distortions are characterized and compensated.

With the CPA technique, laser chains delivering up to multijoule energies in subpicosecond pulses can be focused to high intensities for laser-matter interaction experiments. Traditionally, increasing the energy and shortening the pulse duration are the methods used to achieve these intensities, but an equally important--and so far neglected--way is to produce an ideal wavefront for ultimate focusing quality. The wavefront of subpicosecond lasers can show distortions as high as several lambda [1], which we suppress with the proper use of deformable mirrors. Knowledge of the laser wavefront distortions is then necessary to tune the deformable mirror to the optimal shape, and to quantify the improvements thus achieved. We measure the wavefront with an Achromatic Three Wave Lateral Shearing Interferometer (**ATWLSI**) [1,2], which is particularly suitable for single-shot measurement of both phase and intensity of femtosecond pulses. Combining both the phase and intensity profiles of the beam, we have access to the focal spot size and its peak intensity through the normalized Strehl ratio (the ratio of the peak intensity at focus from a beam with a distorted wavefront to that of the same beam with a perfect wavefront). We show that, in fact, the Strehl ratio is a much better criterion for characterizing the focusing quality than the more widely used „diffraction limited“ focused spot. Indeed, the latter is far from reliable because of the sizeable energy that can be found *outside* the main diffraction spot in a beam with wavefront distortions.

The first experiment is performed with a 10-Hz, 100-fs Ti:Sapphire laser with 5-mJ pulses. The beam is focused into a 1.33-cm thick piece of glass to create wavefront distortions due to non-linear index of refraction and a spatial variation of the intensity in the beam. The measured distorted wavefront is shown on figure 1b. As expected this wave front is proportional to the intensity profile (figure 1a) with a peak value of 0.67λ which is equivalent of a B-integral of 4.1. The Strehl ratio for this distortion is 57%. The compensation given by a 10mm diameter deformable mirror controlled by 19 electrostatic actuators is shown in figure 1c, leading to the final corrected wave-front (figure 1d). The amplitude of distortions is now of 0.28λ and the Strehl ratio is increased to 88%. A net gain of a factor of 1.54 is obtained in the focused intensity.

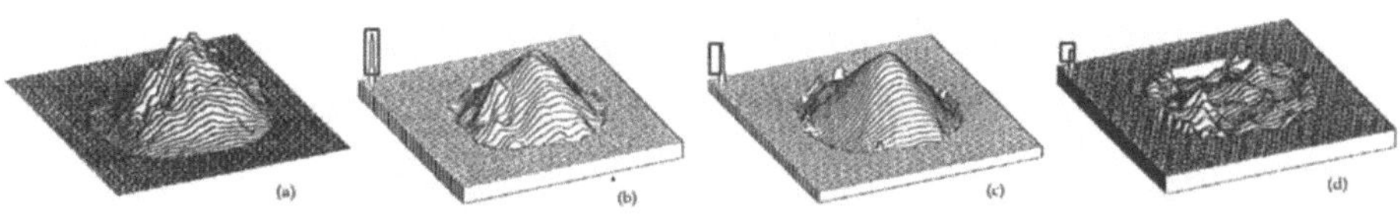

Figure 1: Experimental results for a distorted wavefront due to third-order non-linearity

The second experiment concerns the correction of the wave-front of a hybrid Titanium: Sapphire/Nd:Glass amplifiers laser that generates pulses of 2.5-J with duration of 400 fs. The correction will then have to take into account the wavefront distortions due to the large size of the optical components and the thermal lensing occurring at high energy. The wave-front correction of these multi-terawatt femtosecond pulses is realized after compression under vacuum, by the use of a 50mm diameter deformable mirror, whose surface can be modified by 37 actuators. The figures 2 show the experimental results for the non-corrected and the corrected 750-mJ, 400 fs beam. The peak value of the non-corrected wave-front distortions (figure 2b) is 2l leading to a Strehl ratio of 21% and to a spot size of 1.3 times the diffraction limit. This result shows the importance of the measurement of the wave-front in order to really quantify the peak intensity after focusing. The peak value of wave-front after the correction with the deformable mirror is 0.5l leading to a Strehl ratio of 87% and a focused spot size of 1.03 times the diffraction limit. A factor of 4.1 is gained in the focused intensity.

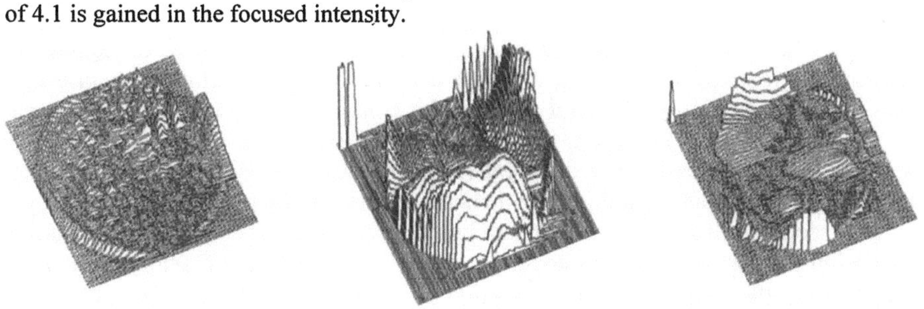

a b c

Figures 2: Intensity distribution (figure 2a) and associated wavefront (figure 2b) before that the correction is applied. Figure 2c is the corrected wave-front.

The single-shot wavefront correction of femtosecond laser pulses with a deformable mirror, after its characterization using the ATWLSI, is a powerful tool to avoid the distortions due to non-linear effects, thermal effects and large size optical components. This technique allows then to deposit almost all the energy of the pulse in the focused spot. The deformable mirror can also be very useful to correct the wavefront of femtosecond pulses with improved contrast, after passing through a second harmonic crystal.

References

1: J.C. Chanteloup & al, Opt. Let, Vol 23, p 621, 1998

2: J. Primot & al, J. Opt. Soc. Am. A, Vol. 12, p. 2679, 1995

Spatiotemporal Imaging, Spatiotemporal Pulse Shaping, and Spatiotemporal Coherent Control

Richard M. Koehl, Timothy F. Crimmins, Ciaran J. Brennan, and Keith A. Nelson

Department of Chemistry, Massachusetts Institute of Technology,
Cambridge MA 02139, USA
E-mail: kanelson@mit.edu

Abstract. Anharmonic phonon-polaritons are driven in polar crystals and their propagation is imaged spatially. Spatially and temporally shaped waveforms are used to control their behavior.

Three new advances have been combined to extend significantly the degree of optical control that can be exercised over coherent excitations in condensed matter. First, it is now possible to drive lattice vibrations in some materials well beyond the harmonic limit through impulsive stimulated Raman scattering (ISRS).[1] Anharmonic vibrational responses have been observed in several materials, providing insight into the corresponding anharmonic crystal potential energy surfaces whose characterization is generally elusive. Figure 1 shows an example of anharmonic responses in lithium tantalate crystal. The anharmonic response may also manifest itself in the crystal through the generation of wavevector overtones in a process similar to second-harmonic generation. The time dependence at wavevector overtones is very different from that at the fundamental and includes frequency components matching the dispersion curve at the overtone wavevector.

The second advance is real-space imaging of propagating phonon-polariton excitations in crystals. In contrast to earlier observations of phonon-polariton propagation through coherent scattering with probe pulses that were delayed temporally and separated spatially from the excitation pulse by specified amounts (the spatial separation increasing linearly with temporal delay),[2] in the present case a large probe spot which includes the excitation region and the entire surrounding region into which the excitation may propagate is used. The probe pulse is not incident at the phase-matching angle for coherent scattering, but simply passes through the coherent phonon-polaritons which act as a moving phase pattern. The transmitted probe light is imaged directly onto a CCD to reveal the phonon-polariton propagation pattern as shown in Figure 2 for the case of crossed-pulse excitation.[3] This technique may be used to observe apparent wavevector overtones in combination with the fundamental excitation.

The third advance is the extension of femtosecond pulse shaping to the spatial as well as temporal domain, in a manner compatible with automated control through a computer-interfaced 2-dimensional liquid crystal display (LCD).[4,5]

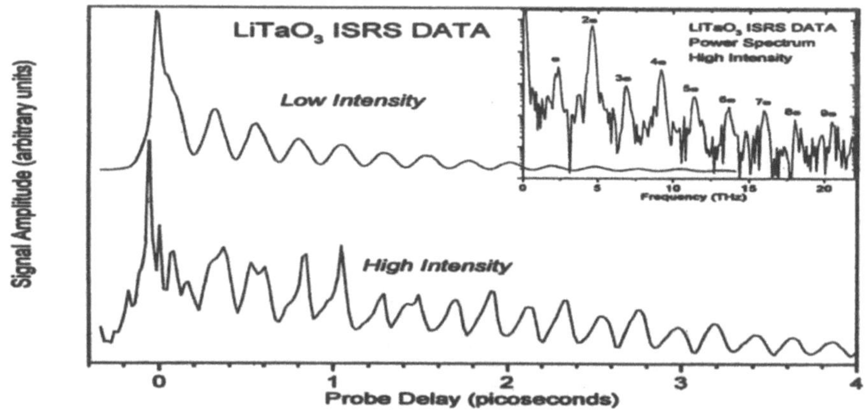

Figure 1. ISRS signal probed at the fundamental grating wavevector. With high intensity excitation pulses, lattice anharmonicity gives rise to frequency and wavevector overtones. Adapted from Ref. [1].

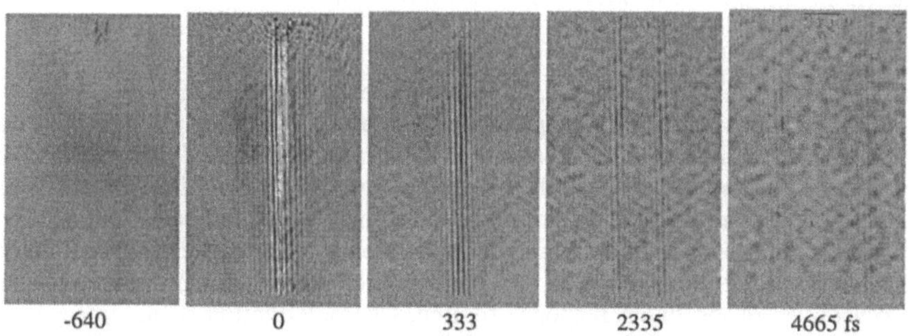

Figure 2. Polaritons imaged with femtosecond resolution, propagating at one-fifth the speed of light in air. Although only five images are shown, 220 images were taken in this experiment. The data illustrate the capabilities for terahertz imaging with visible light.

Through automated spatiotemporal pulse shaping, it is now possible to transform a single incident beam consisting of a single pulse into many spatially separated output beams, each one consisting of an independently specified waveform whose time-dependent profiles are controlled. The output is spatially as well as temporally phase-coherent, so options such as time-varying interference patterns and wavevector shaping are possible. The automated process is still rather crude due to limitations in the LCD hardware used, but already it is possible to specify spatially and temporally shaped waveforms to a computer and obtain them through its interface with the 2-D LCD. Figure 3 illustrates some automated spatiotemporal pulse shaping results.

The three advances discussed above can be combined to achieve coherent control over time and position-dependent sample responses, through specification of the time and position-dependent driving field and with feedback from real-

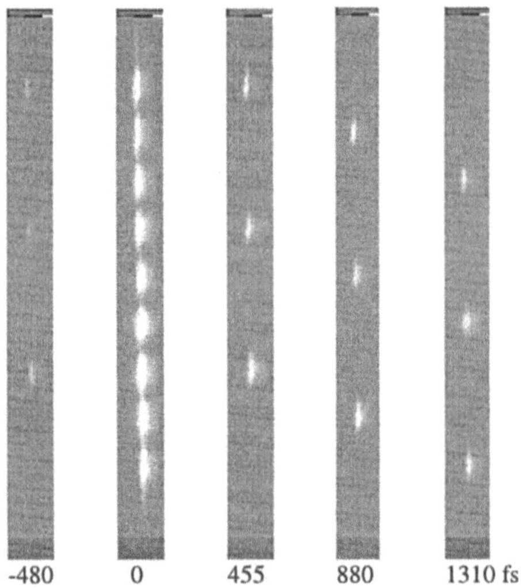

Figure 3. Automated spatiotemporal pulse shaping results. A single incoming pulse is transformed into nine beams with the specified waveforms.

space imaging of the responses. Preliminary experiments are under way with phonon-polaritons, and other propagating excitations including exciton-polaritons may also be manipulated by spatially and temporally shaped waveforms. The propagating excitation itself can thereby be "shaped" arbitrarily in a manner which has no analog in molecular or atomic spectroscopy, since direct optical access to different spatial locations into which the excitation moves can be achieved. Optical amplification, directional control, and frequency or wavevector filtering are among the objectives in the linear-response regime. Characterization of nonlinear lattice dynamics, soliton formation, and nonlinear polariton-polariton interactions are current goals of coherent control in the anharmonic regime.

References

1 Preliminary results are reported in "Direct time-resolved measurement of anharmonic lattice vibrations in ferroelectric crystals", C.J. Brennan and K.A. Nelson, J. Chem. Phys. **107**, 9691-9694 (1997).
2 "Electrooptic generation and detection of femtosecond electrical transients", D.H. Auston and M.C. Nuss, IEEE J. Quant. Electron. **24**, 184 (1988).
3 R. M. Koehl, S. Adachi, and K. A. Nelson, Journal of Chemical Physics (submitted).
4 "Time-domain images", M.C. Nuss and R.L. Morrison, Opt. Lett. **20**, 740-742 (1995).
5 "Multidimensional shaping of ultrafast optical waveforms", M.M. Wefers, K. A. Nelson, and A. M. Weiner, Opt. Lett. **21**, 1-3 (1996).

Part III

Pulse Diagnostics, Measurement Techniques, Nonlinear Optics

Infrared streak camera

L. D. Noordam, M. Drabbels, and C. W. Rella

FOM-Institute AMOLF, Kruislaan 407, 1098 SJ Amsterdam, The Netherlands
E-mail: noordam@amolf.nl, www.amolf.nl/external/wwwlab/atoms/femtophys

Abstract. An atomic streak camera has been constructed with a spectral range from 1 to 100 μm and a 2 ps time resolution. Gas-phase Rydberg atoms are used as a infrared sensitive photocathode. Prompt electron emission from the autoionizing Rydberg atoms is obtained when a red Stark states are selected.

1. Introduction: spectral limitation of streak cameras

The spectral range at which conventional streak camera systems operate is limited by the spectral response of the photocathode used. For most photocathode materials the sensitivity is limited to wavelengths shorter than 1.5 μm. Beyond this wavelength the quantum efficiency of the photocathode becomes negligible. Thus the temporal profile of mid- and far-infrared light pulses can not be measured directly using a conventional streak camera.

We have constructed an atomic streak camera [1-2] in which the photocathode of a conventional streak camera is replaced by a sample of gas phase Rydberg atoms for the conversion from photons to electrons. Due to the low binding energy of the electrons of Rydberg atoms, the IR photon energy is sufficient to photoionize the atoms and create electrons. The combination of a low ionization threshold with the high photoionization cross section makes a Rydberg atom photocathode ideally suited for use in an IR streak camera.

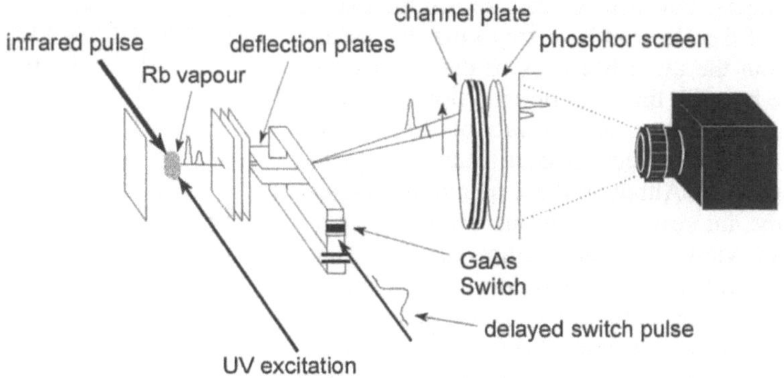

Fig. 1. Principal of operation of the infrared streak camera

2. Infrared streak camera: principle of operation.

The principle of operation of the atomic infrared streak camera is sketched in figure 1. Gas phase atoms are excited to a Rydberg state by a UV laser prior to exposure to the IR radiation. The number of photoelectrons, created by the IR photoionization of the Rydberg atoms, is proportional to the instantaneous IR intensity; therefore the temporal profile of the electron pulse mimics the profile of the IR radiation. Just as in a conventional streak camera, the electron pulse is accelerated towards a position sensitive detector. The electron pulse passes a set of deflection plates. The electric field of this condenser is ramped in time, causing a time-dependent deflection of the electrons. Hence the electron pulse is streaked over the position sensitive detector, and the profile on the detector is a direct measure of the profile of the IR pulse.

3. Limitations at long wavelengths

After the initial demonstration of the infrared streak camera at 2.6 μm [1], we investigated the response at longer wavelength. As a IR source the Free Electron Laser FELIX is used (located in Nieuwegein, The Netherlands). The FEL is tunable from 4 to 100 μm, and from the spectral bandwidth the pulse duration is estimated to be about 1 ps. In this particular case the FEL was tuned near 7 μm and the initial Rydberg state of potasium was n=10. Although the time resolution of the streak camera is excellent at 6.5 μm (<3 ps) the electron emission from the Rydberg atom is no longer prompt at longer wavelengths. For instance at 7.25 μm the excess energy of the electron is so small that the escape time from the atom becomes of the order of 10 ps. The beats in the electron emission are due to angular momentum oscillations of the electron [3, 4] while the electron is still attached to the ionic K+ core. Although these quantum beats are certainly interesting from an atomic physics point of view, they are rather disturbing for proper streak camera operation. Thus different initial Rydberg states have to be selected for different spectral regimes. So for λ=7.25 μm we start from n=11 rather than n=10. Moreover in the presence of the electric field each n state fans out in n Stark states. Some of these Stark states are better candidates for the IR streak camera than others....

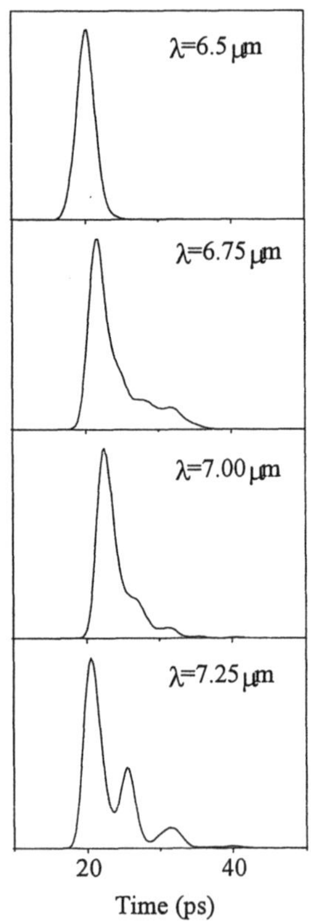

Fig.2

We investigated the ionization dynamics starting from different Rydberg Stark states as initial state: (1) blue shifted Stark states (the energy of the states increases with increasing field strength, hence the excitation laser has to be blue shifted to stay tuned) and (2) red shifted Stark states. From the results [5] we deduce that the red shifted Stark states with most of the wavefunction located at the down stream end of the potential give the sharpest electron emission peaks and are the states we will use for the streak camera operating at long wavelengths.

4. Results at 35 and 85 μm.

Selecting a red Stark state as an initial state we tested the streak camera in the far infrared [2]. Here results at 35 and 85 μm are shown. The single shot trace taken at 35 μm shows that even at these long wavelength the streak camera has a time resolution of about 1 ps. The streak trace at 85 μm shows the asymmetric pulse shape of the free electron laser at long wavelengths. Such an exponential growth of the laser intensity was predicted by theory but never actually observed. Note that the oscillation time of the electromagnetic field at 35 μm is already 0.12 ps and at 85 μm is 0.28 ps!

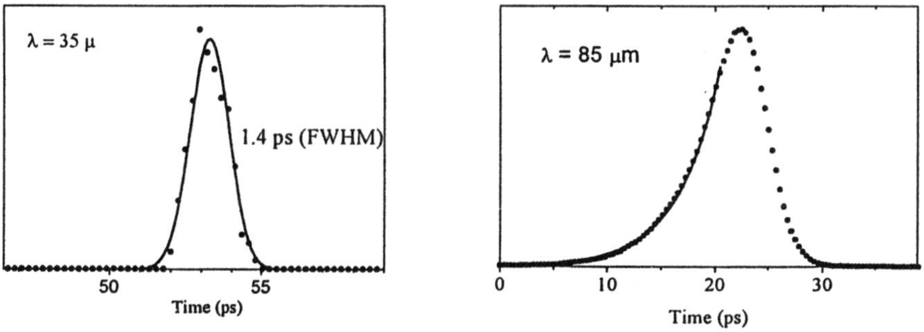

Fig. 3. Streak trace at 35 and 85 μm.

Acknowledgements. The work described in this article is supported by the Technology Foundation (STW), and the Stichting FOM, which is financially supported by NWO.

References

1. M. Drabbels, and L. D. Noordam, Opt. Lett. **22**, 1436 (1997)
2. M. Drabbels, G. M. Lankhuijzen and L. D. Noordam, IEEE-QE (1998)
3. G. M. Lankhuijzen and L. D. Noordam, Phys. Rev. Lett. **76**, 1784 (1996)
4. G. M. Lankhuijzen, F. Robicheaux and L. D. Noordam, PRL **79**, 2427 (1997)
5. G. M. Lankhuijzen, M. Drabbels, F. Robicheaux, and L. D. Noordam, Phys. Rev. A **57**, 440 (1998).

In-situ Waveform Retrieval and Display of Femtosecond Optical Pulses and Picosecond Electrical Pulses

Woopoung Kim[1], Heeseok Lee[1], and Joungho Kim[1]

[1] Dept of Electrical Engineering, KAIST
373-1 Kusong-dong, Yusong-Gu, Taejon, South Korea, 305-701
Email : yousu@eeinfo.kaist.ac.kr

Abstract. In-situ reconstruction of femtosecond optical pulses and picosecond electrical pulse waveforms was demonstrated using a neural network. A Fuzzy C-means clustering network is so flexible that adding the training data can improve the performance of the network, and also is insensitive to the noise. SHG FROGs are used to train the network and a experimental data from Ti:Sapphire laser is applied to retrieve its pulse.

1. Introduction

Measuring the correct waveforms of ultrafast optical and electrical pulses is an essential procedure for the study and the application of ultrafast science and technology. FROG (Frequency Resolved Optical Gating) [1] has been intensively used to retrieve femtosecond optical pulses. Also, photoconductive sampling has been used for measuring ultrafast electrical pulses.[2] Generally, iterative methods are used for retrieving the femtosecond pulse waveforms from the FROG. Also, an iterative method was introduced for waveform reconstruction of on-wafer sampling and terahertz pulse detection. [3] However, in-situ reconstruction and display of the measured waveforms are prohibited due to exhaustive computing time of the iterative algorithm being used.[4]

In this paper, Neural network is introduced as a useful tool for in-situ waveform recognition. The applications of the neural networks have been mostly confined to control and digital signal processing of vision and sound such as character recognition, face recognition, and so on. In 1996 the first application of neural network to FROG was published by Marco A. Krumbugel et al. [5] The computing time using the neural network was found to be less than 1% of that of the conventional iterative algorithm. In addition, the neural networks can be implemented into a CMOS chip design and fabricated in a single Si chip, so that a single chip can replace a computer in the FROG system and a single probe can integrate both a photoconductive probe switch and a signal processing unit. In this paper, Fuzzy C-means clustering neural network is applied to the retrieval of the FROG traces and autocorrelation data due to its flexibility and short training time. We trained the net with 108 FROG traces and tested.

2. Neural Network

A Fuzzy C-means clustering neural network[6] which is a kind of classification network is more stable and noise-insensitive than a backpropagation neural network. When the backpropagation network is used as a regression function, generalization and stability of the neural network must be proved. On the other hand, the Fuzzy C-means clustering neural network do not need good generalization and stability. It distinguishes among all the training data during the training process and calculates their centers. If a new FROG data is inserted into the network, the network computes all the distances from the new FROG data to all the centers, and the center which has the shortest distance is chosen by the neural network. We used the Fuzzy C-means program in the matlab fuzzy toolbox. The pulse parameter in [5] is used also.

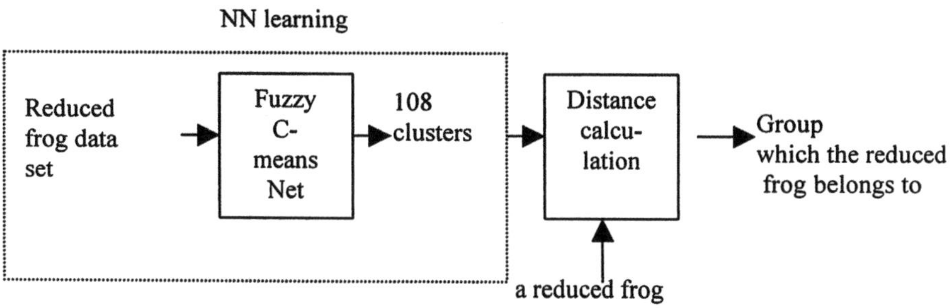

Fig. 1. Neural network for retrieval of ultrafast laser pulses from FROG traces

The structure of our net has 32 input nodes and 108 clusters. Reduced frog data are uniquely extracted from the FROG traces. Because the data dimension of a frog data is 32 x 32(= 1024) which is too large for neural network training, data reduction must be made. A contour line at the same height in a frog trace is unique according to the frog trace, and the horizontal distances between the contour line at each row are taken as the reduced frog trace(= 32).

3. Results and Discussion

130 epochs are enough to learn the Fuzzy C-means Network, and it takes about 2 minute in PC. From a experimental frog trace to the output through generating the reduced frog data and calculating distances, it takes about 8 second. As the number of clusters decrease, the time is reduced, but the result goes poor. The more is the number of the clusters, the higher is the performance of the neural net but the more time is needed.

After training, we tested the network performance with a validation set of new pulses not in the training set which is randomly chosen. Fig. 2 shows a measured FROG trace of Ti:Sapphire short pulse laser which is almost same with the retrieved one. The neural network produced very good estimates. Fuzzy C-means clustering net is so flexible that the precision of the net can be easily

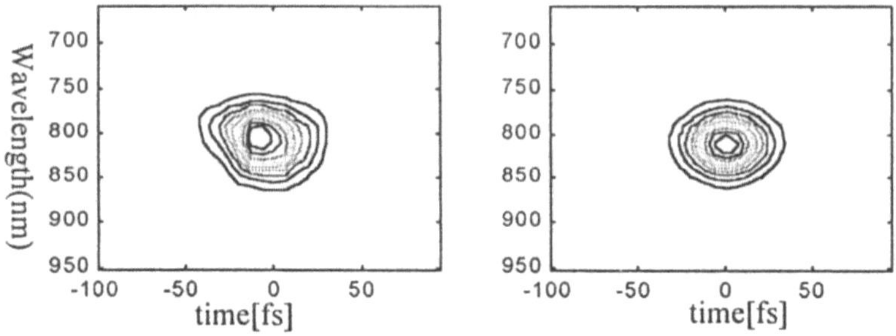

Fig. 2. A Measured SHG FROG trace(a), its neural network output(b)

improved by only increasing the number of clusters, and takes less time to train than the backpropagation net. The computing time required in the iterative algorithms increase steeply as data points increase. But required computing time of the neural networks is almost constant and is less than ten second.

4. Conclusions

Due to its flexibility and short training time, Fuzzy C-means clustering(FCM) neural networks are useful for in-situ reconstruction of femtosecond optical pulse and picosecond electrical pulse waveforms in replace of the slow iterative algorithm. Compared with backpropagation neural networks, FCM has slow retrieval time , but is insensitive to noise and trained fast.

References

1 Rick Trebino and Daniel J.Kane, *J. Opt. Soc. Am. A*, Vol. 10, No. 5, 1101(1993)
2 Joungho Kim, Steven Williamson, John Nees, Shin-ichi Wakana, and John Whitaker, *Appl. Phys. Lett.* 62,2268(1993)
3 Woopoung Kim, Joungju Lee, and Joungho Kim, International Workshop on *Contemporary Photonics Technology*(CPT'98), 233(1998)
4 K.W. Delong, C.L. Ladera, and R. Trebino, *Opt. Lett.* 20, 486(1995)
5 Marco A. Krumbugel, Celso L. Ladera, Kenneth W. DeLong, David N. Fittinghoff, John N. Sweetster, and Rich Trebino, *Opt. Lett.* 21, 143(1996)
6 J. C. Bezdek, *Pattern Recognition with Fuzzy Objective Function Algorithms*. New York: Plenum,1981

Spectral Phase Interferometry for Direct Electric Field Reconstruction of Ultrashort Optical Pulses

Chris Iaconis[1,2], Matthew E. Anderson[1], and Ian A. Walmsley[1]

[1] The Institute of Optics, University of Rochester, Rochester NY 14627 USA
E-mail: iaconis@optics.rochester.edu
[2] USAF Phillips Laboratory, PL/LIDA, Kirtland AFB, NM 87117

Abstract. We present a self referencing interferometric technique for measuring the time dependent intensity and phase of ultrashort optical pulses. The apparatus requires no moving components and the phase retrieval procedure is direct (noniterative). The technique can be used to characterize pulses from the ultraviolet to the infrared. We apply the technique here to characterize the second harmonic pulses generated from the output of a Ti:Sapphire regenerative amplifier.

1. Introduction

Interferometry offers a particularly simple means of measuring the time-dependent intensity and phase of ultrashort optical pulses. Although many interferometric methods require a well characterized reference [1-3], it is possible to use self referencing interferometry, such as direct optical spectral phase measurement (DOSPM) [4]. DOSPM infers a phase function from the rapid time-domain beating of discrete spectral components of the test pulse. In this paper we experimentally demonstrate a self referencing interferometric technique in which all measurements are made exclusively in the frequency domain. We refer to this method as spectral phase interferometry for direct electric field reconstruction (SPIDER). We apply SPIDER to characterize second harmonic pulses generated from the output of a Ti:Sapphire regenerative amplifier.

2. Theory of SPIDER

SPIDER is a version of spectral shearing interferometry [5-6]. Two replicas of the pulse to be characterized are coincident on an integrating detector. The pulses are identical except that they are shifted (sheared) in frequency with respect to one another. The resulting interferogram is related to the input pulse field spectrum, $\tilde{E}(\omega_C)$, by

$$S(\omega_C)=\left|\tilde{E}(\omega_C)\right|^2+\left|\tilde{E}(\omega_C+\Omega)\right|^2+2\left|\tilde{E}(\omega_C)\tilde{E}(\omega_C+\Omega)\right|\cos\left[\phi_\omega(\omega_C+\Omega)-\phi_\omega(\omega_C)+\omega_C\tau\right]$$

where Ω is the amount of spectral shear, τ is the temporal delay between the two replicas, and ω_C is the variable center passband frequency of the spectrometer. The final term in the above expression for the interferogram provides the spectral phase in the form of the phase difference between spectral components separated by the shear Ω. The spectral phase for a set of discrete frequencies separated by Ω is obtained from the spectral phase difference by simply adding up the appropriate phase differences. The spectral amplitude at these frequencies is obtained from the square root of an independently recorded pulse spectrum.

The SPIDER apparatus is shown in figure 1. A pair of replicas of the pulse to be characterized that are separated in time by the fixed delay τ are mixed with a stretched pulse in a nonlinear medium. The stretched pulse must be highly chirped so that each frequency occurs at a different time. Since the pulses in the test pair are delayed with respect to one another by τ, each is mixed with a different temporal, and hence spectral slice of the stretched pulse. The result is a pair of replicas of the pulse to be characterized that have been frequency shifted and are spectrally sheared by Ω. The frequency converted pulse pair is passed to a spectrometer and the spectral interferogram at the mixed frequency is recorded with a slow detector.

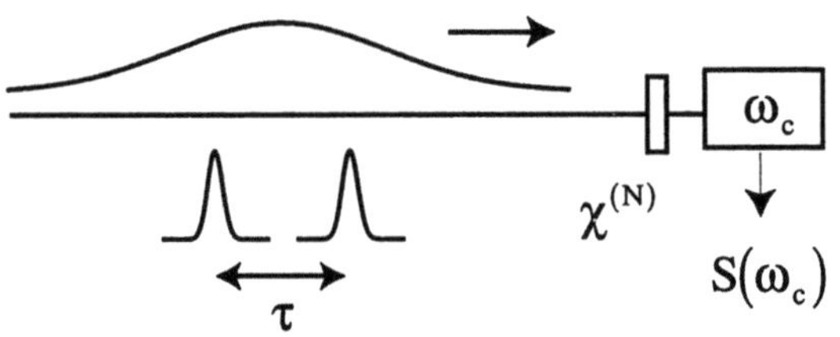

Fig. 1. The SPIDER apparatus.

3. Experimental Results

In our experiments we use a passively stabilized Michelson interferometer to generate the test replicas (reflections from an uncoated solid etalon also work well for this purpose), and a standard 1200 lines/mm grating compressor to generate the stretched pulse. These are frequency mixed in a Type II 250μm long BBO crystal.

In an earlier paper [7], we presented an experiment in which we used SPIDER to characterize the pulses output from a continuous wave modelocked Ti:Sapphire oscillator (7 nJ, 83 MHz). There is no ambiguity as to the direction of time even though a second order nonlinearity is used. Hence, SPIDER is particularly well suited for the measurement of oscillator strength pulses.

In the present experiment we characterize the second harmonic pulses generated from the output of a Ti:Sapphire regenerative amplifier (5 μJ, 2 KHz). The second harmonic test pulses leaving the Michelson arrangement are separated by appoximately 4 ps. A portion of the fundamental pulse is stretched to roughly 40 ps. The second harmonic pulse replicas are downconverted with the fundamental stretched pulse in the BBO crystal. The reconstructed pulse amplitude and phase in the time-domain are shown in figure 2. Notice that the phase retrieval procedure is capable of extracting the phase discontinuity between the main pulse and the satelite prepulse.

In this experiment we used downconversion so that the resultant interferogram

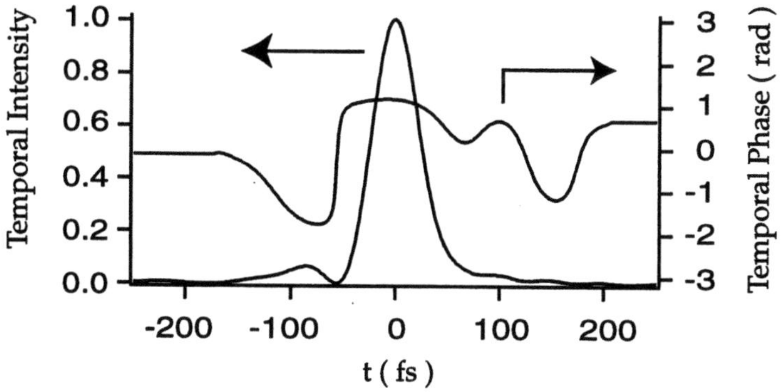

Fig. 2. Intensity and phase of second harmonic pulses measured with SPIDER.

is centered near the fundamental wavelength (nominally 830 nm) where it is easily detected with a silicon photodiode. Downconversion can be used in this arrangement to characterize uv pulses to about 200 nm. Of course, the ultimate limit is the transparency cutoff of the nonlinear material.

4. Conclusions

SPIDER is ideally suited for characterizing pulses in the ultraviolet by using downconversion and for characterizing pulses in the visibile and infrared by using upconversion. The collinear geometry with no moving components and the simple, direct, and rapid data inversion procedure make SPIDER an ideal candidate for single shot and real time applications. Also, with the use of a 2-D detector array, SPIDER is easily adapted for measuring both the temporal and spatial dependence of the field.

References

1 J. E. Rothenberg and D. Grischkowsky, Opt. Lett. **12**, 99 (1987).
2 L. Lepetit, G. Cheriaux and M. Joffre, J. Opt. Soc. Am B **12**, 2467 (1995).
3 D. N. Fittinghoff, et al., Opt. Lett. **21**, 884 (1996).
4 K. C. Chu, J. P. Heritage, R. S. Grant, K. X. Liu, A. Dienes, W. E. White and A. Sullivan, Opt. Lett. **20**, 904 (1995).
5 V. A. Zubov and T. I. Kuznetsova, Laser Physics **2**, 73 (1992).
6 V. Wong and I. A. Walmsley, Opt. Lett. **19**, 287 (1994).
7 C. Iaconis and I. A. Walmsley, Opt. Lett. **23**, 792 (1998).

Spatial-temporal field distribution of self-focused femtosecond pulses

Tai-Wei Yau[1], Chau-Hwang Lee[2], and Jyhpyng Wang[1,2]

[1] Graduate Institute of Electro-Optical Engineering, National Taiwan University, Taipei 106, Taiwan
[2] Institute of Atomic and Molecular Sciences, Academia Sinica, P.O. Box 23-166, Taipei 106, Taiwan
E-mail: jwang@ltl.iams.sinica.edu.tw

Abstract. We developed a 3-D phase-retrieval cross-correlation technique to measure the field distribution of femtosecond pulses, and studied its nonlinear propagation under strong self-focusing.

Self-focusing is an important nonlinear optical phenomenon. Its application covers a range as wide as passive mode-locking, *z*-scan spectroscopy, plasma generation, supercontinuum generation, guided lightning, micromachining, spatial soliton, and optical squeezing. Weak self-focusing can be treated as a perturbation to diffraction. More interesting and much less understood is self-focusing at well above the critical power. Under the scalar paraxial approximation [1], numerical simulations predicted that the beam would catastrophically collapse to a singular point, whereas some vector-field theories predicted differently [2]. By going beyond the paraxial approximation, it was predicted that the beam profile will oscillate [3]. When the effect of group velocity dispersion is included, temporal splitting in self-focused ultrashort pulses is also predicted [4,5]. To date, few of these theories have been verified by experiments. In this paper we study the process of self-focusing experimentally by measuring the spatial-temporal field distribution of self-focused femtosecond pulses.

To measure the field distribution of a sample pulse, we cross-correlated it in both the space and the time domain by second harmonic generation with a much weaker reference pulse of known field distribution. The experimental set-up is shown in Fig. 1. The sample beam passes through the BBO crystal without being focused, while the reference beam is focused to a 80-μm diameter. By scanning the position of interception we measure the spatial profile of the sample beam. The second-harmonic cross-correlation signal is frequency resolved, enabling the use of phase retrieval algorithms to reconstruct the filed distribution of the sample pulses [6].

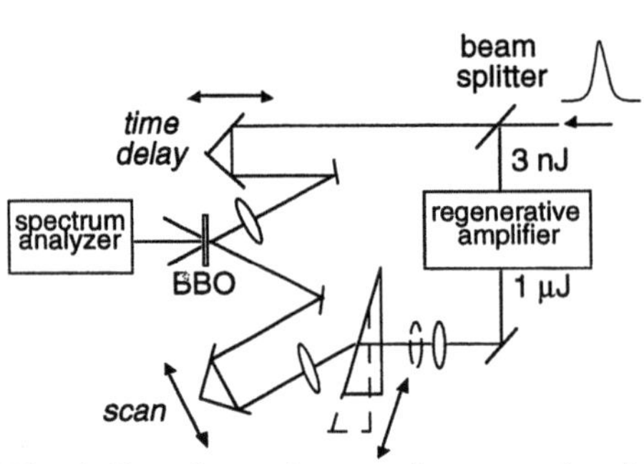

Fig. 1. Experimental set-up for measuring 3-D waveform of ultrashort pulse.

The phase-retrieval algorithm in cross-correlation measurement is much simpler and faster than those for autocorrelation frequency-resolved optical gating (FROG). In FROG, the waveform of the pulse $a(t)$ can be calculated from the measured spectrogram $|\tilde{S}(\omega,\tau)|$, by either applying the spectral constraint [7], or by minimizing the error function $Z=\sum_{t,\tau=1}^{N}|S(t,\tau)-a(t)a(t-\tau)|^2$ with respect to $a(t)$, where $S(t,\tau)$ is the Fourier transform of $|\tilde{S}(\omega,\tau)|$ [6]. The former is known to be less reliable [6], while the latter requires heavy computation. With cross-correlation, the error function becomes $Z=\sum_{t,\tau=1}^{N}|S(t,\tau)-a(t)a_r(t-\tau)|^2$, where $a_r(t)$ is the waveform of the reference pulse. Because $a_r(t)$ is known, minimization is reached in just one step by a simple algebraic relation: $a(t)=E^{-1}\sum_{\tau=1}^{N}S(t,\tau)a_r^*(t-\tau)$, where $E=\int|a_r(\tau)|^2d\tau$ is the energy of the reference pulse. The above equation greatly simplifies the iteration process, making this algorithm fast and reliable.

First we measured the output pulse of a 95-fs Kerr-lens mode-locked Ti:sapphire oscillator. From the 3-D field distribution we plot in Fig. 2 the diameter of the pulse for each time slice. The data clearly show the pulse center is focused more than the pulse wings by about 6%, in good agreement with the Kerr-lens mode-locking model. This gives a direct proof that self-focusing is the pulse-shortening mechanism.

Next we measured the field distribution of an amplified pulse after it passes through a Kerr medium made of a wedge of sapphire crystal. Fig. 3 shows the instantaneous beam diameters at different propagation distance for 180-fs, 0.8-μJ pulses focused to 28-μm diameter (2×1/*e*-radius) at the entrance face of the crystal. The confocal parameter is 0.76 mm. Self-focusing overcomes diffraction, reducing the beam diameter as the pulses propagate. The interesting point is that the pulse self-focuses almost uniformly, *i.e.*, it focuses as much in the wings as in the center. This is completely different from the behavior of pulses below the critical power. It shows clearly the break down of adiabatic (slowly varying amplitude) theories.

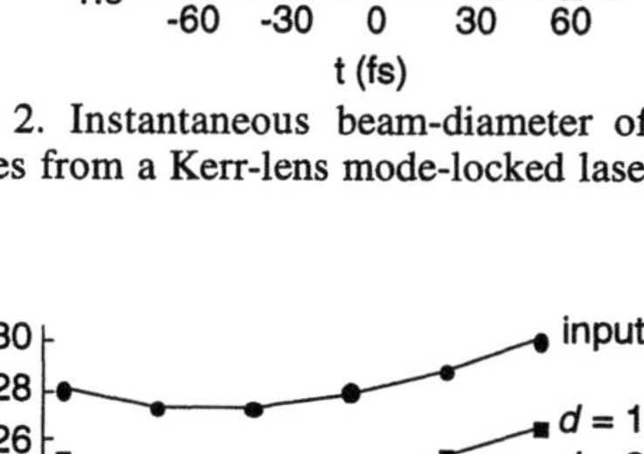

Fig. 2. Instantaneous beam-diameter of the pulses from a Kerr-lens mode-locked laser.

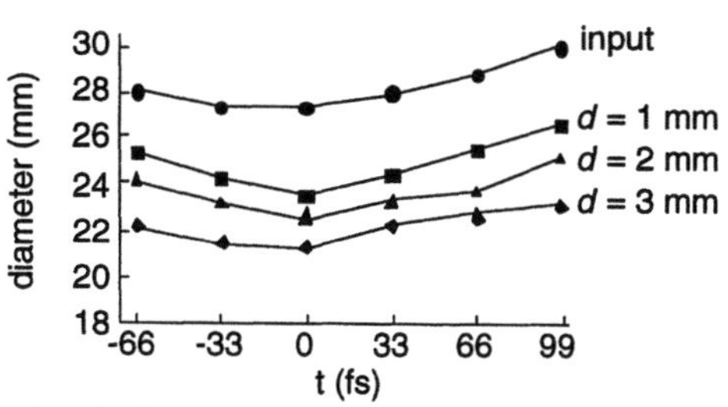

Fig. 3. Instantaneous beam diameter of self-focused high energy pulses. *d*: propagation distance.

At higher peak power and longer distance we observed beam trapping and pulse splitting. Fig. 4 shows the evolution of amplitude distribution for pulses with peak power equal to 2.8 times the critical power P_c. Self-focusing is obvious, with the pulse-center beam diameter (FWHM) shrinking from 44 μm to 20 μm. As the pulse propagates, it dissipates some energy into the outer region while maintaining a high intensity central core. After several millime-

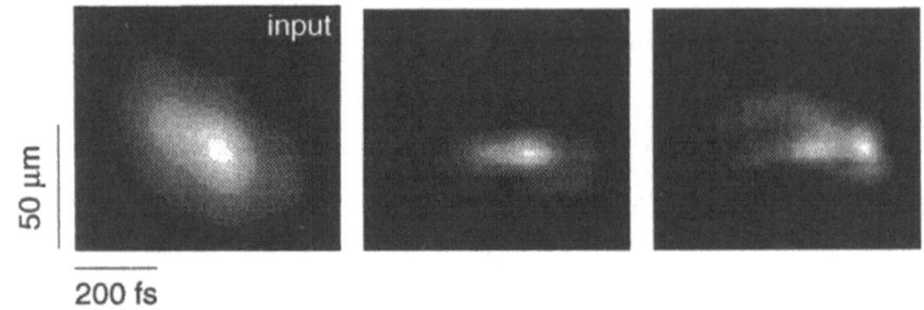

Fig. 4. Field evolution of self-focused femtosecond pulses.

ters, the chirped pulse splits, with each peak containing different frequency components, in agreement with [4,5].

In Fig. 5 we show the pulse-center beam diameter as a function of the propagation distance for different peak power. The 2.8-P_c beam is trapped to a 20-25-µm diameter all the way to 8 mm. The other two cases, which have a much larger peak power, also focus to about the same diameter. Due to spatial splitting, for these two cases the beam diameter after 4 mm is not well defined, therefore we show them in dashed curves.

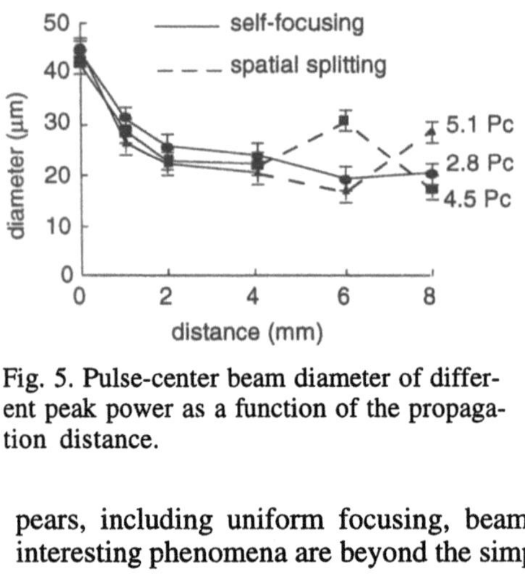

Fig. 5. Pulse-center beam diameter of different peak power as a function of the propagation distance.

It appears that within our experimental range of peak power, the beam always focuses to a 20–25 µm diameter, before it splits spatially. This result is consistent with our previous results published at CLEO'97, where the experiments were done with femtosecond dye lasers at 620 nm [8].

Our experiments show that 3-D phase retrieval cross correlation is a powerful technique for studying nonlinear ultrafast optics. Using this technique we provided the first direct proof of the Kerr-lens mode-locking mechanism, and found that at above the critical power, many new phenomena appears, including uniform focusing, beam trapping, and spatial splitting. These interesting phenomena are beyond the simple concept of self-focusing.

References

1. P. L. Kelley, Phys. Rev. Lett. **15,** 1005 (1965).
2. S. Chi and Q. Guo, Opt. Lett. **20,** 1598 (1995).
3. G. Fibich, Phys. Rev. Lett. **76,** 4356 (1996).
4. J. E. Rothenberg, Opt. Lett. **17,** 583 (1992).
5. J. K. Ranka, R. W. Schirmer, and A. L. Gaeta, Phys. Rev. Lett. **77,** 3783 (1996).
6. K. W. Delong, et al., J. Opt. Soc. Am. B **11,** 2206 (1994).
7. J. Paye, et al., Opt. Lett. **18,** 1946 (1993).
8. Y.-Y. Hwang, et al., in Conference on Lasers and Electro-Optics, 1997, Vol. 11 of 1997 OSA Technical Digest Series, paper CTuP22.

Pulse Characterization Techniques for sub-30 Femtosecond Terawatt Lasers

Csaba Tóth[1], David N. Fittinghoff[1], Barry C. Walker[2], Jeff A. Squier[3], Christopher P. J. Barty[1]

[1] Institute for Nonlinear Science, University of California at San Diego, 9500 Gilman Dr., La Jolla, CA 92093-0339, USA
E-mail: cxt@chem.ucsd.edu, dnf@chem.ucsd.edu, cpjb@chem.ucsd.edu
[2] Department of Chemistry and Biochemistry,
E-mail: bcw@chem.ucsd.edu
[3] Department of Electrical and Computer Engineering,
E-mail: jas@chem.ucsd.edu

Abstract. Single-shot characterization of spectral phase, pulse duration, and spectrum of femtosecond, terawatt pulses is described, using a specially designed, low-dispersion polarization-gate frequency-resolved optical gating arrangement.

1. Introduction

Chirped pulse amplification (CPA) of 10 fs Ti:sapphire seed pulses has led to the generation of multiterawatt optical pulses at unprecedented repetition rates – three to four orders of magnitude higher rates than similar peak power, large-scale Nd:glass or gas laser systems. These ultrafast systems now allow investigations of ultrahigh field laser/matter interactions and applications to be performed in university-scale laboratories, bringing the physics of the extreme to the table-top [1]. In order to build and characterize these systems it is important to obtain precise knowledge of the evolution of the optical pulse spectrum, duration and phase distortions during amplification [2,3]. For high peak power pulses of 50 fs and below, the most commonly used pulse characterization techniques have been single shot second harmonic generation (SHG) autocorrelation and time-integrated single shot spectroscopy [4,5]. Single shot autocorrelation, however, contains no information about the temporal asymmetry of the pulse or the phase content. Furthermore, in most ultrafast CPA systems, it is possible (and even likely) to have outputs which are accompanied by a significant component of ns-duration amplified spontaneous emission (ASE). In a time-integrated spectral measurement, this ASE content can alter the measured spectrum and lead to false conclusions about the potential duration of the amplified, short pulse. In this contribution, we discuss the considerations for an optimal measurement configuration and describe the results of a modified version of the commonly used frequency resolved optical gating technique which is ideally suited for measuring the intensity, phase and spectrum of 20 fs CPA systems.

2. Experimental Methods

Many techniques exist for measuring the intensity and phase of isolated femtosecond pulses. In particular frequency-resolved optical gating (FROG) [6] has become popular for its relatively straightforward experimental geometries and robust retrieval algorithms. The FROG geometry which has been used for pulses near or below 30 fs duration has generally been second-harmonic generation [7]. SHG FROG, however, has an ambiguity in the direction of time and restrictions on the thickness of the nonlinear crystal due to phase-matching for the large bandwidths needed to support such short pulses.

In large amplified systems, a polarization-gate (PG) FROG would be much more desirable. The large, thin and expensive crystals or semiconductors required for single-shot SHG generally have poor optical quality. PG FROG, however, uses thin, excellent optical quality, fused silica plates as the nonlinear material for Kerr gating. Also, large CPA systems can produce peak powers over 1 TW in unfocused beams, eliminating the need for a strong second-order nonlinearity. For such a system, a FROG based on a third-order nonlinear process can take advantage of the high power and eliminate the temporal ambiguity associated with SHG FROG. The main difficulty in using a PG FROG for very short (<50 fs) amplified pulses has been the dispersion of the calcite polarizers that have been used to produce the polarization gate. A good PG FROG device generally requires extinction ratios higher than 10^5. Calcite, while easily meeting these requirements on polarization purity, is very dispersive. Propagation in 1.0 cm of e-calcite will lengthen an initially transform-limited Gaussian pulse from 20 fs to over 50 fs. Furthermore, because of calcite's rather unusual higher order phase content [2], it

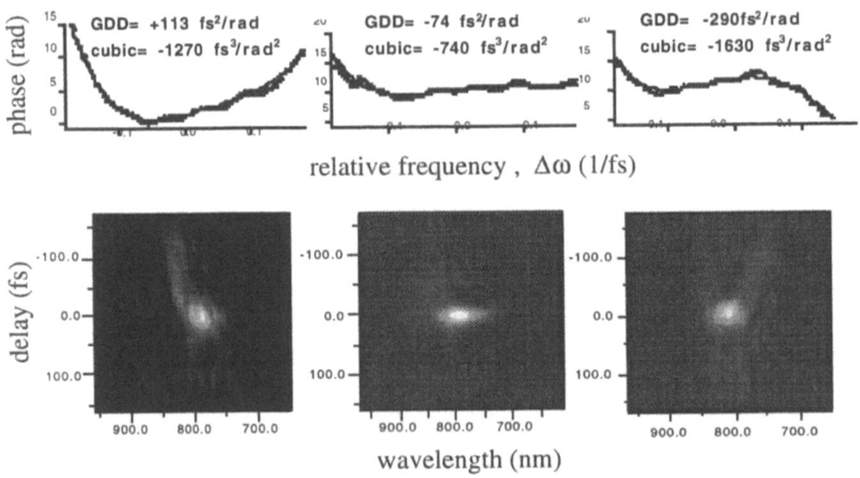

Fig. 1. PG-FROG traces (bottom panel) and polynomial fitting (top panel) of the retrieved spectral phase of amplified 22 fs pulses of a Ti:sapphire CPA laser for different relative positions (Δd) of the compressor gratings. (a): Δd = -190 μm, (b): Δd = 0 μm, (c): Δd = +220 μm.

is difficult to precisely precompensate for these polarizers in the laser system. To eliminate these problems, we replaced the first calcite polarizer in the normal PG FROG with a pair of broadband low-dispersion thin-film polarizers. This produces a low dispersion PG FROG for use on high peak-power ultrafast pulses. This geometry with <340 μm gate medium could be used to measure pulses as short as 10 fs with better than 0.1 % FROG error.

3. Results

As an example of the utility of the TFP based PG FROG, we have measured the sensitivity of the simplest dispersion parameter of a CPA system: the change of spectral phase vs. compressor gratings separation (Fig. 1). The quality of the FROG measurements and retrieval - characterized by FROG errors on the order of $3x10^{-3}$ - was excellent throughout our study. The higher order coefficients of the spectral phase were determined by polynomial fitting. Fig. 2 shows the derived second (group delay dispersion - GDD = $\partial^2\varphi/\partial\omega^2$) and third order (cubic dispersion = $\partial^3\varphi/\partial\omega^3$) coefficients as a function of the relative grating separation. The solid lines represent an analytical simulation. Consistent polynomial fits for orders higher than third are difficult [3]. To completely describe the higher order phase distortions, it may be necessary to define a spectrally weighted, average phase error as well.

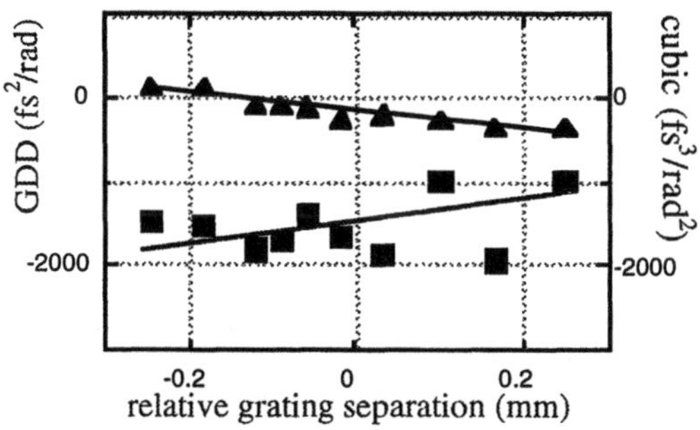

Fig. 2. Polynomial phase coefficients (triangles: GDD, boxes: cubic). The continuous lines show the expected variations based on a dispersion model [2] of the optical components of the CPA system.

Finally, it is important to note, that the same device can be used to measure the time gated spectrum of the pulse simply by broadening the gate pulse (e.g., by putting a piece of dispersive material in the gate arm) in this way discriminating the short amplified pulse from ASE, commonly produced in regenerative and multipass amplifiers. Spectra obtained in this manner do not require a retrieval algorithm.

References

1 G. A. Mourou, C. P. J. Barty, and M. D. Perry, Phys. Today, pp. 22, Jan. 1998.
2 D. N. Fittinghoff et al., IEEE J. Sel. Topics in Quant. Electron., **4**, 430, 1998.
3 A. Sullivan and W.E. White, Opt. Lett., **20**, 192, 1995.
4 C. P. J. Barty et al., Opt. Lett., **21**, 668, 1996.
5 P. F. Curley et al., Opt. Commun., **120**, 71, 1995.
6 D. J. Kane and R. Trebino, Opt. Lett., **18**, 823, 1993.
7 G. Taft et al, IEEE J. Sel. Topics in Quant. Electron., **2**, 575, 1996.

Single-Shot Third-Order Correlator for Femtosecond Terawatt Ti:sapphire Lasers

V.Sirutkaitis[1], R.Grigonis[1], A.Piskarskas[1], M.Grätz[2], A.Persson[2] and S.Svanberg[2]

[1] Laser Research Centre, Vilnius University, Sauletekio 10, 2054 Vilnius, Lithuania
E-mail: valdas.sirutkaitis@ff.vu.lt
[2] Division of Atomic Physics, Lund Institute of Technology, P.O. Box 118, S - 221 00 Lund, Sweden
E-mail: Sune.Svanberg@fysik.lth.se

Abstract. We discuss the design of a single-shot third-order correlator for pulse shape investigation of femtosecond Ti:sapphire lasers with a dynamic range of 6-7 orders. Measurements on the Ti:sapphire laser at the Lund Laser Center with this correlator is reported.

1. Introduction

An important quality factor of a femtosecond terawatt laser, apart from the peak power and duration, is the contrast between the main pulse and possible prepulses or a background level (pedestal). In laser-matter interactions, the pulse pedestal disturbs the material to be studied before arrival of the main short pulse. High-dynamic-range autocorrelation measurements must be used to characterize the temporal quality of the pulses generated by such systems. The conventional second-order autocorrelation is symmetrical and cannot determine pulse asymmetry. For this reason, in high-dynamic-range autocorrelation measurements third-order autocorrelators have been introduced [1, 2]. In most experiments it is important to know the pulse contrast in every shot. Thus, single-shot third-order correlation measurements are necessary. In this paper we discuss the design of a single-shot third-order correlator for pulse shape investigation of femtosecond Ti:sapphire lasers with a dynamic range of 6-7 orders. Measurements with this correlator on the terawatt Ti:sapphire laser at the Lund Laser Center are reported.

2. Construction of the Autocorrelator

The optical layout of the third-order autocorrelator is shown in Fig. 1. The Ti:sapphire laser beam falls, after passing the apertures A1 and A2, on the beam splitter BS, where 25 % of the fundamental beam is reflected to the arm, which comprises the aperture A3 and the third-harmonic crystal. The half wave plate HWP, the mirrors M1 and M2 and the retroreflector R are components in this arm. 75 % of the fundamental beam goes through the beamsplitter BS, then reflects from the mirrors M3 and M4 and enters the second-harmonic crystal SHC, where the second harmonic is generated.

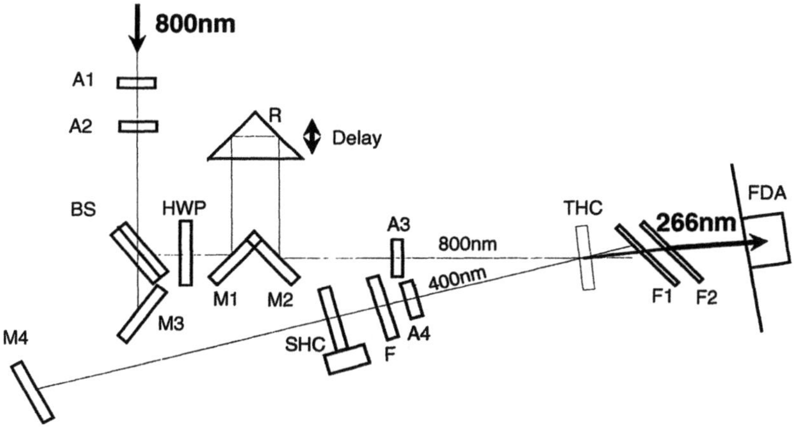

Fig. 1. Optical schematic of the third-order autocorrelator.

After the second-harmonic crystal (KDP), the fundamental beam is blocked by the filter F and the second-harmonic beam goes through the aperture A4 to the third-harmonic crystal THC, where it overlaps with the fundamental beam. The second-harmonic pulse is correlated with the fundamental pulse by sum-frequency generation in the third-harmonic crystal (KDP). The method used in such an autocorrelator is based on the non-colinear sum frequency (SF) generation, where the spatial distribution of the SF light contains information on the time behavior of the fundamental light pulse. This technique combines the zero background and the single-shot measurement capability and is analogous to the single-shot second-order autocorrelation [3]. The filters Fl-F2 are used for separation of the third harmonic with high contrast. A 1024-element thermoelectrically cooled CCD photodiode array with 12-bit analog to digital conversion is used for registration of the spatial distribution of the third-harmonic beam.

3. Results and Discussion

The single-shot third-order autocorrelator developed was tested with single femtosecond pulses from a chirped-pulse amplification table-top terawatt Ti:sapphire laser system. For single shots, the third-order correlation was recorded over a temporal range of about 3 ps. The corresponding correlation trace is presented in Fig. 2. As the dynamic range of the photodiode array was ~ 10^3, an increasing oversaturation of the array was used. The antiblooming properties of the photodiode array used allowed for larger than 400 times oversaturation. Even in this case the satellite pulses or pedestal near the main pulse could be registrated. Using calibrated attenuators for 266 nm radiation we were able to provide measurements over an intensity range of more than 6 orders of magnitude. For the measurements in a more wide temporal range the first harmonic pulse was delayed by ~3 ps steps in comparison with the second-harmonic pulse and in such steps measurements could be provided in the temporal range from - 60 ps to + 60 ps.

The composite autocorrelation curve is presented in Fig. 3. The measured autocorrelation curve shows more than one main peak. A few of them are related to the reflections in the beamsplitters, but there are three additional ones at the leading front of the pulse and one at the tail.

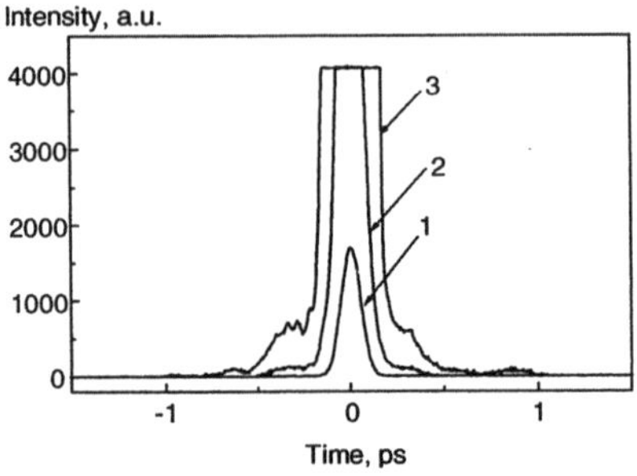

Fig. 2. Autocorrelation of the main pulse at different signal attenuation: 1 - 200 times; 2 - 10 times; 3 - without attenuation.

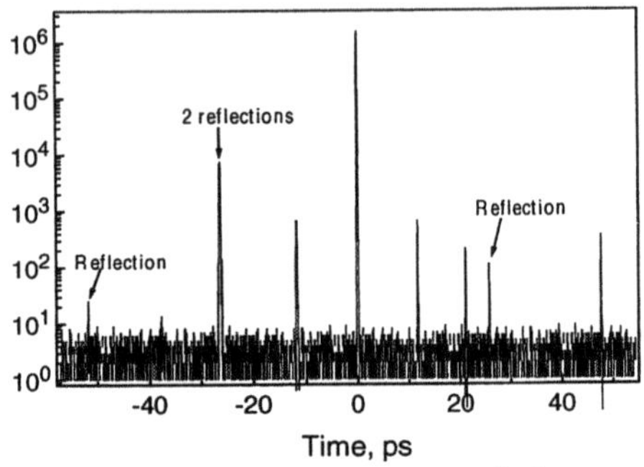

Fig. 3. The composite autocorrelation curve of the Ti:sapphire laser.

The statistical investigation of satellite pulse durations showed that they are produced as reflections from the different optical components of amplifiers. In our measurements the background was caused by noise of the photodiode array but not by the background of the laser pulse. This showed that the background of the amplified and compressed pulse of the system with the exception of the reflected satellites is at least lower than $5 \cdot 10^{-7}$.

4. Conclusions

In conclusion we have examined the possibilities of the single-shot third-order autocorrelator developed. It has been shown that at an input femtosecond pulse intensity ~ 25 GW/cm^2 the dynamic range of the autocorrelator is more than 6 orders.

References

1 G. Albrecht, A. Antonetti, G. Mourou, Opt. Commun., V. 40, p. 59 (1981).
2 Y. Yamakawa, H. Shiraga, Y. Kato and C.P.J. Barty, Opt. Lett., V. 15, p. 1593, (1991).
3 J. Jansky, G. Corradi and R.N. Gyuzalian, Opt. Commun., V. 23, p. 293 (1977).

Temporal evolution of selfguided femtosecond laser pulses in the presence of Kerr-nonlinearity and ionisation

H. R. Lange[1], J. -F. Ripoche[1], A. A. Chiron[2], B. Lamouroux[1], M. A. Franco[1], B. S. Prade[1], E. T. J. Nibbering[3] and A. Mysyrowicz[1]

[1] Laboratoire d'Optique Appliquée, École Polytechnique, Chemin de la Hunière, F - 91761 Palaiseau CEDEX, France
phone: +33-1-69 31 98 27, fax: +33-1-60 10 60 85, mysy@enstay.ensta.fr
[2] Dipartimento di Fisica, Universita di Pisa, Piazza Torricelli 2, I - 56100 Pisa, Italy
[3] Max-Born-Institut für Nichtlineare Optik und Kurzzeitspektroskopie, Rudower Chaussee 6, D - 12489 Berlin, Germany

Abstract. We study numerically and experimentally the dynamics of self-guided femtosecond laser pulses. Self-compression of the pulse duration is observed.

Self-channeling of intense femtosecond laser pulses in air has attracted considerable interest [1,2]. High-intensity filaments with diameters down to 80 μm have been shown to propagate over distances of more than one hundred meters. It has been assumed that the stable propagation is due to a dynamic balance between a positive Kerr-nonlinearity that leads to self-focusing, and a negative contribution due to ionisation, that causes diffraction. Up to now, the temporal aspects of the nonlinear propagation process have not been examined in experiments. Numerical studies however have predicted the effects of pulse narrowing and pulse splitting [3].

We report here on the temporal dynamics of these self-guided pulses. We show that a temporal self-compression occurs in the self-focusing stage, after which the propagation of the selfguided pulse is characterised by a splitting into shorter temporal fragments and an eventual stabilisation into a single shorter pulse.

The pulse duration of the central segment of the beam has been determined in the self-focusing region, before the onset of filamentation by second-order autocorrelation. Measurements have been performed as a function of initial pulse duration, position along propagation axis and initial pulse energy. In this ionisation-free region a spatio-temporal contraction of the pulse occurs, due to the spatial redistribution of different time slices of the pulse, as shown in Figure 1a).

Once the on-axis intensity is sufficient, multi-photon ionisation of the gas molecules steps in abruptly, preventing beam collapse and leading to the formation of a filament. The pulse duration in the filamentary region has been measured by a self-diffraction experiment. We divided the laser pulse into two beams of equal intensity. The two pulse replica generated two separate filaments which were made to overlap in a thin piece of fused silica. Figure 1b) shows the self-diffraction signal measured after onset of filamentation. One observes a compression of the pulse duration by a factor two.

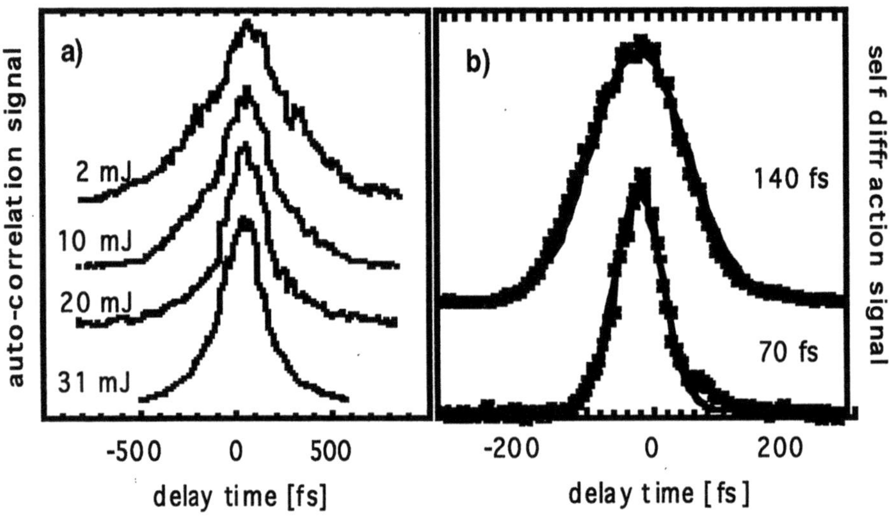

Figure 1. a) Pulse autocorrelation traces recorded before onset of filamentation as a function of pulse energy, b) self-diffraction signal of input pulse recorded before (upper trace) and 0.7 m after onset of filamentation (lower trace). The pulse durations are extracted assuming Gaussian laser pulses.

In order to understand the physics behind these results we have numerically solved the nonlinear Schrödinger-equation describing the field evolution in the presence of diffraction, instantaneous and non-instantaneous Kerr effect and multiphoton ionisation [2]. Key results of the simulation are: 1) the degree of ionisation remains always small and the pulse leaves a trail of unconnected plasmas segments (see Figure 2), 2) whereas the filament is stable on a macroscopic scale, closer inspection on a microscopic view reveals a complex dynamical evolution of temporal substructures, 3) the experimentally determined values for pulse durations correspond to spatially averaged mean values of the numerical solution.

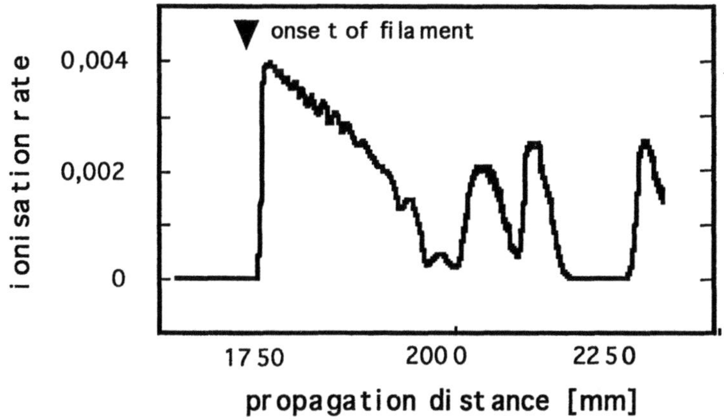

Figure 3. Ionisation rate at the onset of filamentation.

References

1 A. Braun, et al., *Opt. Lett.* 20, 73, 1995.
2 E. T. J.Nibbering, et al., *Opt. Lett.* 21, 62, 1996.
3 A. Brodeur et al., *Opt. Lett.* 22, 304, 1997.
4 L. Wöste, *Laser und Optoelektronik* 29, E2688, 1997.
5 O. G. Kosareva, et al., *Opt. Lett.* 22, 1232, 1997.
6 H. R. Lange, et al., *Opt. Lett.* 23, 120 (1998).
7 E. Esarey, et al., IEEE J. Quantum Electron. 33, 1879, 1997.

Femtosecond Lidar

L. Wöste[1], C. Wedekind[1], H. Wille[1], P. Rairoux[1], M. Rodriguez[1], B. Stein[1], R. Sauerbrey[2], H. Schillinger[2], F. Ronneberger[2], S. Niedermaier[2]

[1]Institut für Experimentalphysik, Freie Universität Berlin, Arnimallee14, D-14195 Berlin,
[2]Institut für Optik und Quantenelektronik, Friedrich Schiller Universität, D-07743 Jena

Abstract. We sent the slightly, negatively prechirped light of a terawatt femtosecond laser system vertically into the sky and observed the onset of an extended white light channel [1]. The generated plasma channel is highly sensitive to the pulse duration and chirp settings and is most efficient when the group velocity dispersion between laser and focus is fairly well precompensated. Scattered white light from altitudes beyond 12 kilometers was recorded; the spectra exhibit very pronounced features of atmospheric constituents.

1. Introduction

Lidar is a powerful and important tool for a range-resolved characterization of the atmosphere [2]. In standard systems the light of a pulsed laser is usually launched into the atmosphere, where it is elastically scattered by means of Rayleigh- or Mie- , and inelastically scattered by means of fluorescence- or Raman-scattering. A major drawback of the method is the fact that all these scatter processes exhibit almost isotropic angular distributions; therefore the intensity of the recorded backscattered light decreases quadratically with the observation distance. In addition, the measurement of fluorescence is restricted to altitude regimes, where collision quenching isn't dominant (>40 km). The small Raman-cross section, on the other hand, restricts Lidar-measurements to abundant atmospheric constituents like oxygen or nitrogen. As a result, the most commonly employed Lidar-systems are based on elastic scattering, and information about the investigated atmosphere is drawn from the spectral absorption of the light along its path. This, however, restricts the application of the method to such particles and atmosphere properties that can be characterized within the spectral range of the available laser source. In addition, the required light detection at the wavelength of the emitter demands great care in protecting the detector, when the laser is fired. An alternative approach of using nonlinear optical processes could avoid the aforementioned disadvantages and considerably improve the performance of Lidar systems. Their application has become feasible with the availability of high power femtosecond laser pulses.

2. Experiment

The experimental setup consists of a tabletop terawatt laser system. The laser operates at a wavelength of 790 nm with a maximum output energy of 200 mJ per pulse at a repetition rate of 10 Hz. The final pulse duration at optimum compressor settings amounts to 100 fs (FWHM), which was characterized by FROG techniques. A longer pulse duration with linear change of wavelength in time, a so-called positive or negative linear chirp can be introduced by the compressor alignment, which is incorporated inside the amplifier system of the laser. In this way the group velocity dispersion of the air between laser and focus can fairly well be compensated. The focus was placed at some distance away using a weak positive lens (f = 30 m). On the receiver side, the backscattered white light was collected with a 40 cm diameter f/3-telescope placed in a bistatic arrangement a few meters away from the emitted beam. The collected light was conducted with a fiber optic cable into the spectrometer. The recordings were performed with a gated optical multichannel analyzer, which allowed for detection of spectra from 350 to 850 nm. Range resolution was obtained by restricting the observed time window at a desired delay after the laser was fired.

3 Results and Conclusion

Depending on chirp settings a clearly visible, white light channel appeared in the air. The spectrum of this channel recorded at low altitude is shown in figure 1 (top) for two different chirp settings. Both curves exhibit large wings around the laser wavelength at 790 nm towards the UV and the IR, which is cut off, however, at 850 nm due to the sensitivity loss of the detector. Spectral integration indicates that the proportion of backscattered white light to laser light is over 40 %. The bandwidth of the laser light propagating through the atmosphere is plotted in figure 1 (buttom). It shows an almost linear increase beginning at 11 nm at the location of the laser and reaching 21 nm at a distance of 100 meters, after which this

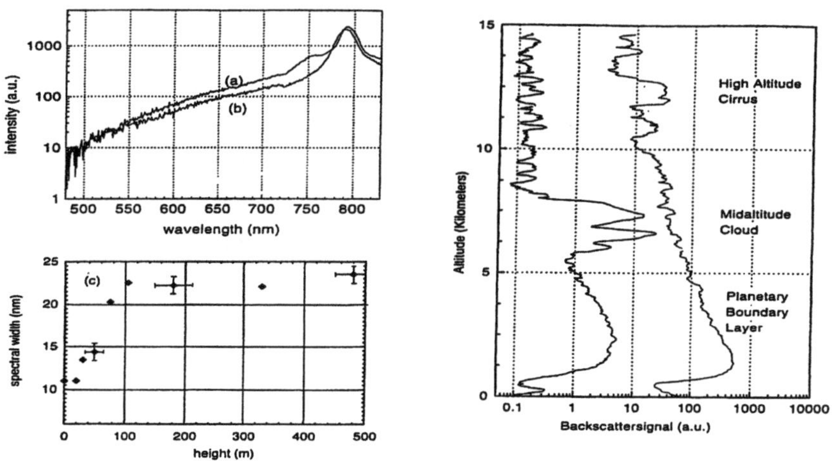

Figure 1: left: Spectra of the atmospheric white light channel recorded with an optical multichannel analyzer. (a) and (b) represent two chirp settings (top). Spectral broadening of the laser pulse along its propagation path (bottom).

right: Two different vertical backscatter profiles of the atmospheric white light

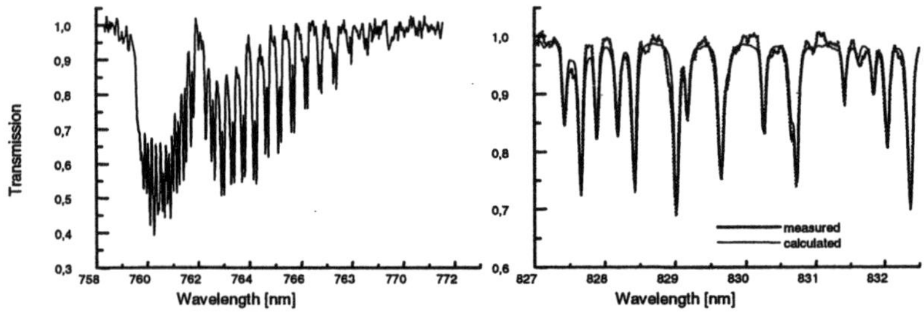

Figure 2: Recorded spectrum of the b $^3\Sigma^-_g \rightarrow$ b $^1\Sigma^+_g$ transition of oxygen (left), and of atmospheric water compared to calculations based on the HITRAN database (right).

value remains constant. This can be related to the geometrical, temporal and spectral channeling features of the propagating beam and indicates that the process of temporal self phase modulation ceases after this distance. Apparently the channel irradiates the atmosphere above and below in a narrow cone of white light. As a result white light signals could be recorded from the atmosphere at altitudes reaching above 12 kilometers as can be seen in figure 1 (right). It shows two curves of different atmospheric conditions exhibiting clear features of the planetary boundary layer, midaltitude cumulus and highaltitude cirrus clouds. Well resolved spectral sequences of the white light, after passing some distance across the atmosphere, are shown in figure 2. These results were obtained in an altitude between 600 and 850 meters. In the spectral range of 758 to 772 nm the progressions of the forbidden b $^3\Sigma^-_g \rightarrow$ b $^1\Sigma^+_g$ transition of O_2 appear. The cross-section of this strongly forbidden transition is only 2,5E-24 cm^2, so by scaling the sensitivity of this method to particles having typical cross sections of 10^{-16}cm^2, one can see that a detection limit in the ppb-range can be obtained. In the range of 826 to 836 nm well pronounced lines of the 000 $\rightarrow$ 211 band of H_2O appear. They compare extremely well with computations of the Hitran data base. The relative absorption between emitter and receiver allows one to deduce from this a relative humidity of 75 %, which is within the error of the ground measurements.

We have demonstrated the formation of very extended white light plasma channels in the atmosphere through the use of adequately well prechirped and shaped femtosecond laser pulses in the terawatt regime. The analysis of absorption spectra retrieved from this white light indicates that there is sufficient intensity coming from these channels so that range-resolved multicomponent analyses of the atmosphere can well be retrieved from this. In addition the plasma channel offers fascinating perspectives in the fields of aerosol-, cloud- and lightning research.

References

1 A. Braun, G. Korn, G. Mourou et.al. in *Optics Letters,* Vol 20, 73, 1995.
2 R.M. Measures *Laser Remote Sensing*, Whiley, New York, 1984
3 L. Wöste et al. in *Laser und Optoelektronik,* Vol. 29, 1997
4 A. Brodeur, S. L. Chin et. al. in *Optics Letters*, Vol. 22, 304, 1997

Superluminal Localized Waves in Femtosecond Optics

P.Saari

Institute of Physics, University of Tartu, Riia 142, Tartu EE2400, Estonia
E-mail: PSaari@fi.tartu.ee

Abstract. Theoretical and experimental study of optical implementations of various new nonspreading wave-packet solutions to linear wave equations is overviewed. Startling superluminal group velocity of pulsed Bessel beams and so-called X-type waves is discussed and mathematically described in terms of electromagnetic field of a tachyonic source.

1. Introduction

Several interesting classes of nonspreading wideband wave-packet solutions to the linear wave equations of mathematical physics ("focus wave modes", "Bessel-Gauss pulses", "electromagnetic bullets", "directed energy pulse trains," "X waves," etc.) have been discovered recently, which maintain their spatio-temporally localized shape in the course of propagation [1,2]. The feasibility of experimental launching has been tested in acoustics for the directed energy pulse train and for the X wave (see Refs. in [1,2]). The X-type waves are attracting growing interest, particularly, due to the outlook of application in medical ultrasonic imaging [2].

An optical version of the X wave, named the Bessel-X pulse, was treated theoretically and its propagation-invariant high localization (e.g. $\sim 10^{-3}$ mm, $\sim 10^{-15}$ sec) was shown by computer simulations [3-5] (see also figures below). Specially designed femtosecond-duration Bessel-X pulses not only maintain their lateral localization but also the longitudinal one in dispersive propagation media [3]. Such a possibility to suppress the temporal spread caused by the group velocity dispersion has been experimentally verified on subpicosecond laser pulses [4] by making use of a computer-generated holographic optical element designed for formation of the common Bessel beam.

First measurement [6] of the whole three-dimensional distribution of the field of optical X waves as well as study of terahertz pulses (see [7] and Refs. therein) have finally proved the promising prospects of various applications of electromagnetic localized waves in the field of ultrafast phenomena.

In this paper we present the main experimental result of Ref. [6] – recording of the characteristic shape of the Bessel-X wave in free space. We comment the startling superluminal group velocity of propagation-invariant beams and pulses. As a new result, we present a preliminary theoretical description of X-type pulses in terms of a coupled pair of retarded and advanced potentials of a tachyonic source.

2. Experimental Recording of Bessel-X wave

A simplest way to comprehend the nature of Bessel-X wave is to consider it as built up from interfering pairs of identical bursts of plane waves. Every plane wave pair makes up an X-shaped propagation-invariant interference pattern moving along the axis z with speed $v = c/\cos\theta$ which is both the phase and the group velocity of the wave field in the direction of the propagation axis z of the pattern. This speed is superluminal in a similar way as one gets a faster-than-light movement of a reflection/refraction stripe along surface of a dielectric medium when a plane wave light pulse is falling at the angle θ onto the surface plane. In the axisymmetrical superposition of all possible such pairs around z the only point of completely constructive interference is a point on the axis z, which becomes the bright center of the bullet-like pulse (Fig.1).

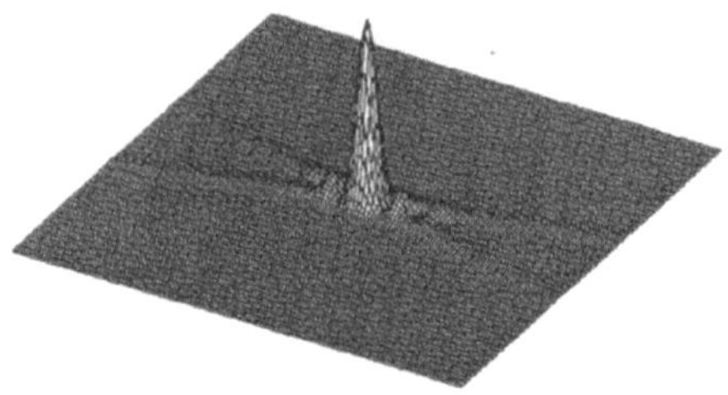

Fig.1. Intensity (upwards) of the Bessel-X pulse versus propagation distance $z - vt$ (to the right) and a lateral axis, computed for a 3-femtosec (FWHM) pulse with central wavelength 0.6 μm.

A straightforward method for a frame-by-frame recording of the field shape would use a CCD camera with a gate in front of it, which should possess a temporal resolution and a variable firing delay both in subfemtosecond range. As such a gate is not realizable, a workable idea of experiment has to resort to a field cross-correlation technique. Fortunately, one can take advantage of the superluminal speed of the Bessel-X pulse, which allows the latter to catch up with a reference plane wave pulse generated in the same optical scheme and propagating collinearly with the Bessel-X wave. If one generates (by a routine Bessel beam set-up) the wave with a very small conical angle θ (which is an half of the angle between the branches of X-like cross section), the velocity v exceeds only slightly c (e.g. for $\theta = 0.006$ *rad* $v/c = 1/\cos\theta = 1.000018$).

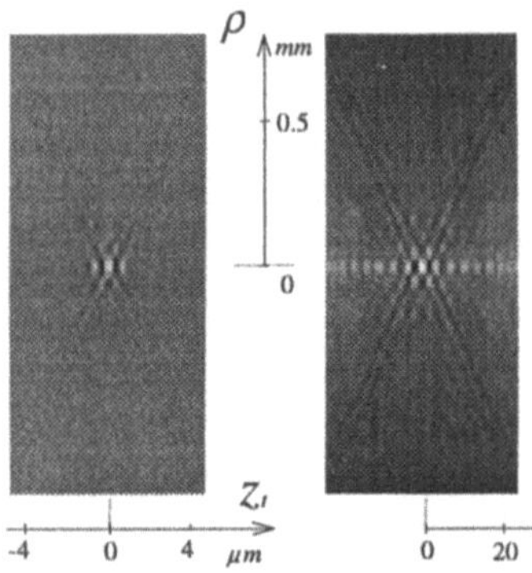

Fig. 2. Experimentally recorded cross-correlation (right panel) in comparison with theoretical field amplitude distribution of the Bessel-X pulse in a plane of the axis z (left panel).

Therefore, fortunately, the region where the plane wave and the faster Bessel-X wave intersect each other, is stretched along the flight direction over tens of centimeters and one can record mutual interference patterns corresponding to various relative sub-femtosecond delays, simply by placing a CCD camera at different distances along the flight directon.

A result of such experiment (see details in Ref. [6]) is shown in Fig.2.

3. Discussion

If one says that the superluminal Bessel-X pulse is not a real one but simply an interference pattern, one has to admit also, that there are no real pulses, say, in femtosecond lasers or wherever, but a superposition effect of (longitudinal) modes only. Shortly, the X-type EM field pulses belong to nontrivial (but allowed by the special theory of relativity) superluminal movements (for a review, see [8]).

As an intriguing remark we add that the Bessel-X pulse field is nothing but an "observable" shape of a tachyon (see [9] and Refs. therein) as well as a double version of the Mach cone of a shock wave. This could be demonstrated directly for a EM field of a superluminal source, if the calculation, instead of using the retarded Green function only – the common approach in field theory, is based upon the singular solution $D_0(\mathbf{r},t)$ to the homogeneous wave equation. The latter includes the advanced (incoming and collapsing at the origin) spherical wave in addition to the causal expanding one. As a result, the EM field can be expressed *via* radially symmetric scalar and vector potentials $\Phi(r,z,t)$ and $\mathbf{A}(r,z,t) = \Phi(r,z,t)\mathbf{v}/c$; $\mathbf{v} = (0,0,v)$ being the speed of uniform movement along the axis z and $\Phi(r,z,t)$ being given by the following expression ($Y(x)$ denotes the Heaviside step function)

$$\Phi(r,z,t)=\frac{2e}{\sqrt{(z-vt)^2+r^2(1-v^2/c^2)}}\left\{Y\left[-(z-vt)-r\sqrt{v^2/c^2-1}\right]-Y\left[(z-vt)-r\sqrt{v^2/c^2-1}\right]\right\} \quad (1)$$

The first of the two terms in the Eq.(1) gives nothing but a Vavilov-Cherenkov radiation of a superluminal electron in vacuum, first calculated by Sommerfeld [10] long before the Tamm and Frank's theory appeared, but forgotten after Einstein published his theory of relativity. The second term describes a leading and reversed Mach cone collapsing as if there is a superluminal sink flying along the axis. Hence, the pulse is created by a Huygens-type non-material source generated by the cone of incoming waves, what is not at variance with the special theory of relativity.

References

1 R. W. Ziolkowski, I .M. Besieris, A. M. Shaarawi, J. Opt. Soc. Am. A **10,** 75, 1993.
2 J. Lu, IEEE Trans. Ultrason. Ferroelec. Freq. Contr. **42**, 1050, 1995.
3 H. Sõnajalg and P. Saari, Optics Lett., **21**, 1162, 1996.
4 H. Sõnajalg, M. Rätsep, and P. Saari, Opt. Lett. **22**, 310, 1997.
5 P. Saari, K. Reivelt, M. Rätsep, and H. Sõnajalg, in *Ultrafast Processes in Spectroscopy (Proc. of the X Intern. Symposium held in Tartu, Estonia, 24-28 Aug. 1997)* Edited by R. Kaarli, A. Freiberg, and P. Saari, Inst. of Physics, University of Tartu, 184, 1998.
6 P. Saari and K. Reivelt, Phys. Rev. Lett., **79**, 4135, 1997.
7 S. Feng, H. G. Winful, and R. W. Hellwarth, Opt. Lett., **23**, 385, 1998.
8 V. L. Ginzburg, in *Progress in Optics*, Edited by E. Wolf, **32**, 267, 1993 .
9 E. Recami, Physica A, **252**, 586, 1998.
10 A. Sommerfeld, in *Göttinger Nachrichten*, 99, 1904.

Even-order nonlinear interaction of focused femtosecond beam with the bulk of chiral media.

A.V.Balakin[1], D.Boucher[2], A.S.Dunin, E.Fertein[2], N.I.Koroteev[1], P.Masselin[2], A.V.Pakulev[1] and A.P.Shkurinov[1]

[1] International Laser Center and Physics Department, M.V.Lomonosov Moscow State University, 119899, Moscow, Russia. E-mail : koroteev@nik.phys.msu.su
[2] Laboratoire de Physico-Chimie de l'Atmosphère, Université du Littoral-Côte d'Opale, 59140 Dunkerque, France. E-mail : masselin@pca3.univ-littoral.fr

Abstract We report on new applications of femtosecond laser pulses for studies of chiral media by nonlinear optical techniques.

1. Introduction

Second harmonic (SH) generation is a powerful tool for getting information about properties of a matter [1], but for isotropic materials possessing inversion symmetry the electric dipole SH generation in the bulk is forbidden. If isotropic media is mirror-asymmetric, second-order sum frequency (SF) and fourth-order SH generation processes because of electric dipole mechanism of two plane waves interaction [2] become possible in non collinear geometry scheme.

2. Theoretical Background and Experimental

An application of femtosecond laser pulses in nonlinear optics has huge potential not only due to possibility to study ultrafast processes occurring under photo-excitation. There are some other useful aspects of ultrashort pulses. First, due to their high peak power and low average one it is possible to detect processes caused by weak nonlinearities, for an example high-order ones, without demolition of the sample. Next because of well known relation $\Delta t \cdot \Delta \nu \sim 1$, the light emission of typical femtosecond Ti:Sapphire laser systems has wide enough spectral band $\Delta\lambda \sim 10$nm ($\Delta t \sim 100$fs, $\lambda \approx 800$nm). So, there is continuous frequency set inside of single laser pulse and femtosecond laser emission should be treated as non-monochromatic light if investigated process is forbidden for monochromatic one. Additional focusing of femtosecond laser light enriches a spatial spectrum of the emission and gives rise to effects which are forbidden for a single plane wave.

Taking into account the Gaussian spectral and spatial profile of the excitation pulse and expression for electric dipole second-order susceptibility $\chi^{(2)D}$ from [2] one gets next polarization dependence of quadratic response localized at SH frequency if the polarization vector of the fundamental beam is expressed as $\vec{e}_0 = \vec{e}_x \cdot \cos(\varphi) + \vec{e}_y \cdot i \cdot \sin(\varphi)$:

$$E_x^{2\omega} \propto i \cdot \sin(\varphi) \cdot e^{i\varphi}, \qquad E_y^{2\omega} \propto -\cos(\varphi) \cdot e^{i\varphi} \qquad (1)$$

For non-local second-order SH generation under the same conditions and by use of the expression for non-local second order susceptibility $\chi^{(2)N}$ from [3]:

$$E_x^{2\omega} \propto i \cdot \cos(\varphi) \cdot e^{i\varphi}, \qquad E_y^{2\omega} \propto -\sin(\varphi) \cdot e^{i\varphi} \qquad (2)$$

Expression for electric dipole fourth-order susceptibility $\chi^{(4)D}$ [2] and the same consideration for exciting lead to next dependence of SH signal on polarization of fundamental laser beam :

$$\begin{aligned} E_x^{2\omega} &\propto i \cdot \left[\sin\varphi + a \cdot \cos^2\varphi \sin\varphi - b \cdot \cos(2\varphi)\sin(2\varphi)\right] e^{i\varphi} \\ E_y^{2\omega} &\propto -\left[\cos\varphi + a \cdot \cos\varphi \sin^2\varphi + i \cdot b \cdot \cos(2\varphi)\sin(2\varphi)\right] e^{i\varphi} \end{aligned} \quad (3)$$

here a and b are complex values that characterize relative ratios and phase shifts between contributions with different polarization behavior into fourth-order signal at SH frequency.

We should note that it is possible to detect interference between second- and fourth-order processes in forming of total signal at SH frequency if contributions of eq(1), eq(2) and eq(3) have similar order of magnitude.

Here we present experimental results on "forbidden" SH generation into the bulk of isotropic optical active medium - fine prepared aqueous suspension of purple membranes *Halobacterium halobium*, containing bacteriorhodopsin protein. An average size of the membrane fragments in solution was determined to be equal to 50 nm. Laser system used in the experiments provided emission at wavelength 820 nm with precisely tunable energy ranging from 0 to 1.2 μJ/pulse into the sample, pulse duration of 300 fs and repetition rate of 200 kHz. Polarization state of laser beam was tuned according to $\vec{e}_0 = \vec{e}_x \cdot \cos(\varphi) + \vec{e}_y \cdot i \cdot \sin(\varphi)$. Besides we had a possibility to obtain additional laser beam (by means of optical parametric amplifier) with wavelength tunable from 530 to 670 nm, energy varied from 0 to 40 nJ/pulse, pulse duration of 150-200 fs and the same repetition rate. By precise delay line we delayed pulse of the additional beam relatively to the pulse at wavelength 820 nm. In experiments we measured vertical ($\vec{e}_x$) linear component of SH signal.

From measurements of SH signal energy dependencies on energy of exciting radiation [4] we have found the interference between second- and fourth-order nonlinear processes and determined effective fourth-order $(\chi^{(4)})$ to effective second-order $(\chi^{(2)})$ susceptibilities ratio as : $|\chi^{(4)}| / |\chi^{(2)}| \sim 4 \cdot 10^{-10}$ esu (out-of-resonance excitation $\lambda_{exc} = 820$ nm) and $|\chi^{(4)}| / |\chi^{(2)}| \sim 1.4 \cdot 10^{-8}$ esu (in-resonance excitation $\lambda_{exc} = 590$ nm). The same interfering interaction have been found from polarization dependence of the SH signal [4]. The experimental data may be explained only if local and non-local (eq. (1), eq. (2)) second- and fourth-order (eq. (3)) contributions are taken into account.

There is a spectral dependence of the excitation of the "forbidden" SH generation process in figure (1). One can see that the maximum efficiency of SH generation ($\lambda_{exc} = 610$ nm) is shifted relatively to the maximum of the sample absorption and the SH resonance curve (Fig. 1 dashed line) could not be interpreted in the term of second-order nonlinear interaction and one needs to take into consideration the role of fourth-order one, because $\chi^{(4)D}$ which describes it has specificity at 3ω [2]. This fact is in good agreement with previous experiments, where was shown that under resonance excitation the contribution from fourth-order nonlinearity is enhanced.

Because of bacteriorhodopsin (BR) molecule is regarded as photo-chromic molecule we carried out time resolved experiment to confirm the SH signal sensitivity to internal molecular structure. Figure (2) shows dependence of SH

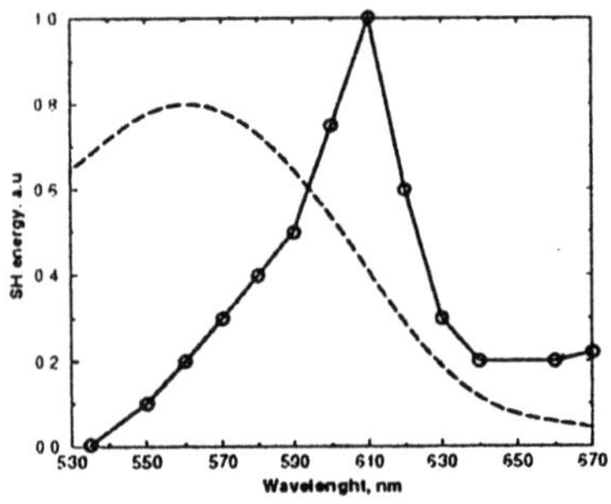

Figure 1:

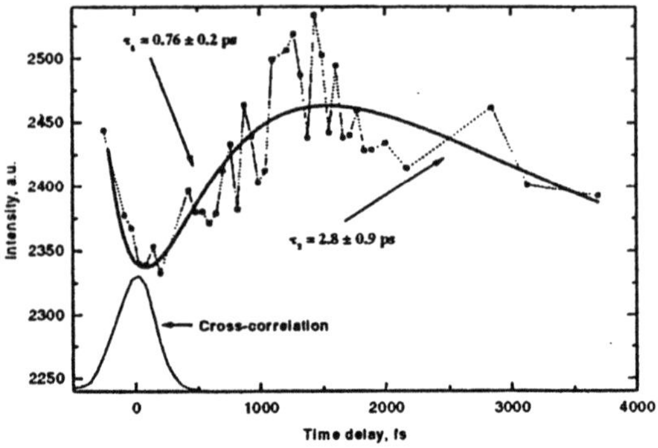

Figure 2:

signal intensity at 410 nm in presence of exciting beam ($\lambda_{exc} = 590$nm) versus the delay time between exciting and fundamental pulses. Taking into account times of BR photoproduct formation [5] which are in good agreement with times determined from time resolved experiment (figure 2), one gets additional confirmation of sensitivity of SH signal to internal molecular structure changes.

3. Conclusions

We predicted a possibility of rising new nonlinear optical effect under excitation of isotropic optical active media with single beam of femtosecond laser pulses and found out its polarization properties. Experimental results on investigation of this "forbidden" nonlinear signal confirmed its chiral-sensitivity and manifested the interference character of interaction between nonlinear optical processes of different orders (second and fourth ones).

Acknowledgments. This work was supported by Université du Littoral-Côte d'Opale and Russian Fundation of Fundamental Research (grant 93.0216596). A.V.B. also thanks Samsung Electronics Co. LPCA participates with the Centre d'Etudes et de Recherche Laser et Applications, supported by the Ministère de la Recherche, the Région Nord/Pas de Calais, and the Fond Européen de Développement Economique des Régions.

References

[1] Y.R. Shen, "*The Principles of Nonlinear Optics*" , Ed. Wiley, New York, 1984

[2] N.I. Koroteev, **JETP**, Vol 79, pp 681, (1994)

[3] Yu.I.Sirotin, M.P.Shaskol'skaya, "*Osnovy Kristallofiziki (Bases of Crystal Physics)*" , Nauka, Moscow, 1979 (in Russian)

[4] A.V.Balakin, D.Boucher, E.Fertein, N.I.Koroteev, P.Masselin, A.V.Pakulev, A.Yu.Resnianski, A.P.Shkurinov, **Opt. Comm.**, Vol 141, pp 343, (1997)

[5] A.V. Sharkov, A.V. Pakulev, S.V. Chekalin, Yu.A. Matveetz, **Biochem. et Biophys. Acta**, Vol 808, pp 94, (1985)

Complete Electric Field of Femtosecond Photon Echoes at Their Point of Origin

J. D. Hybl, A. W. Albrecht, S. M. Gallagher Faeder, and D. M. Jonas

Department of Chemistry and Biochemistry, University of Colorado,
Boulder, CO 80309-0215 USA E-mail: david.jonas@colorado.edu

Abstract. Three pulse scattering signals from solutions of the dye IR144 in methanol exhibit a positive chirp that varies with the delay between excitation pulses. Signal emission times, signal spectra, Wigner spectrograms of individual signals, and two dimensional electronic spectra correlating absorption frequency and signal frequency all suggest delayed echo-like signals do not exhibit a significant Stokes' shift.

1. Introduction

It has been suggested[1,2] that spectral interferometry could measure the electric field of nonlinear scattering signals, and three recent papers[3-5] used spectral interferometry to detect four-wave mixing signals. Spectral interferometry determines the phase difference $\Delta\phi(\omega)$ between temporally separated signal and reference pulses by measuring the frequency domain interference fringes which result from spectrally resolving and hence temporally broadening the pulses so that they overlap in time. The detected spectrum is

$$I(\omega) \propto E_{sig}^2(\omega) + 2E_{sig}(\omega)E_{ref}(\omega)\cos[\Delta\phi(\omega)] + E_{ref}^2(\omega) \qquad (1)$$

Since the phase difference $\Delta\phi(\omega)$ is determined at the point where the two pulses are combined, it includes phase distortions from both the sample and subsequent optics. Theory[6] suggests that photon echo chirp contains useful information especially on the time dependent Stokes' shift, but pulse propagation effects become increasingly severe for short pulses. We have developed a method for characterizing the echo field at the sample exit.[7] A tracer pulse[8] is sent along the signal path and characterized at the sample exit by FROG and measurement of the free induction decay. Spectral interferometry is then used to determine the tracer-reference phase difference $\Delta\phi_{tr}(\omega)$ and the signal-reference phase difference $\Delta\phi_{sr}(\omega)$. The undistorted spectral phase of the signal at the sample exit is given by $\phi_{signal}(\omega) = \Delta\phi_{sr}(\omega) - \Delta\phi_{tr}(\omega) + \phi_{tracer}(\omega)$.

2. Experimental Methods

Three-pulse scattering signals were generated by 250 pJ pulses of 24 fs duration from a cavity dumped Kerr lens mode-locked Ti:sapphire laser operating at 10 kHz. A double-pass through a prism pair pre-compensated for dispersion in the five beam interferometer which produces three excitation beams, a tracer, and a reference. Beams 1, 2, and 3 were positioned at three corners of a square and focused to a 40 diameter μm spot in a 100 μm path flow cell containing a 0.4 mM solution of IR144 in methanol to generate a four-wave mixing signal with wave-vector $k_s = k_3 + k_2 - k_1$. The tracer was focused into the sample along the path of signal emission ($k_t = k_s$). Second harmonic FROG traces recorded at the sample recovered the excitation and tracer pulse phases. The reference arrived at the beamsplitter ~750 fs before the tracer to produce a well-defined fringe pattern so the signal amplitude and phase can be recovered from a single interferogram.[2] Signal times are measured to ~0.3 fs accuracy and constant spectral phase shifts to within ~0.1 rad.

3. Results and Discussion

Three pulse scattering signals were measured as a function of $\tau \equiv (t_2 - t_1)$ for several values of $T \equiv \min[(t_3-t_1),(t_3-t_2)]$. For T=0, the maximum time integrated signal occurs at $\tau \sim 15$ fs (non-zero peak shift), the average time of signal emission increases slowly with τ for non-overlapping pulses, and the signal spectra blue shift as the pulse overlap decreases. The Wigner spectrograms[6] in Figure 1 show the instantaneous echo spectra as a function of the time t after the third pulse for six delays τ at T=0. All six spectrograms shift to higher frequencies as time increases, but this up-chirp becomes less pronounced for large τ where the spectra are blue shifted. As T is increased, the signals have less delayed echo character, less chirp, a red shift, and τ independent spectra.

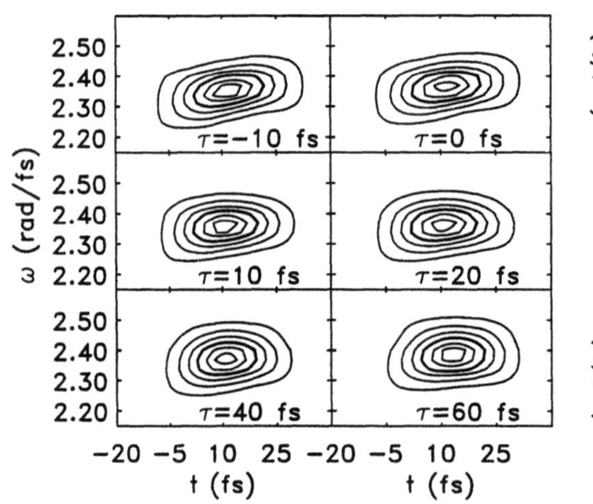

Fig. 1. Wigner spectrograms of three pulse scattering signal for six values of τ at T=0.

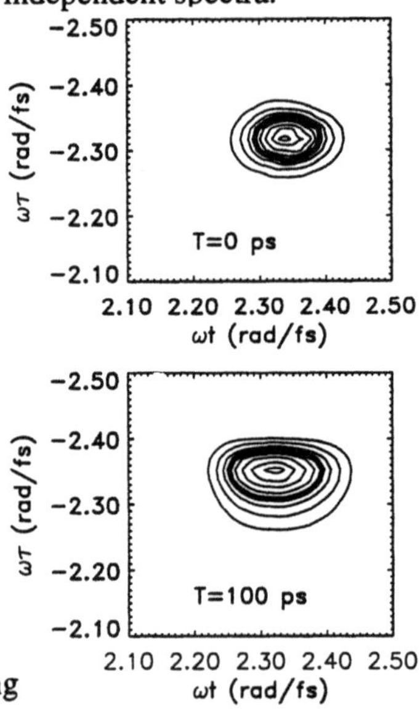

Fig. 2. Absolute value 2D electronic spectra for T=0 and T=100 ps.

The electronic frequency during the τ interval is encoded as a phase modulation ($-\omega_{eg}\tau$ referenced to t=0) of the four-wave mixing signal. A two dimensional (2D) Fourier transform[9] of the signal electric field with respect to t (ω_t) and τ (ω_τ) at constant T produces a 2D electronic spectrum which correlates signal absorption and emission spectra (Figure 2). For positive ω_t, there are no peaks outside the region shown in Figure 2. Unlike 2D Raman spectra of liquids[10] and a 2D map of the second order nonlinear response of KDP,[11] which both recovered difference frequencies within the laser bandwidth on 2D transformation, the 2D spectra shown here recover optical frequencies. To our knowledge, Figure 2 shows the first Fourier transform 2D electronic spectra. The peaks contain 2D absorption lineshapes in both frequency dimensions and a dispersion lineshape which widens the absolute value spectrum along ω_t. The frequency $-\omega_\tau$ reflects the initial electronic absorption frequency while ω_t reflects the emission frequency. Figure 2 shows that at T=0, the 2D spectrum peaks on the diagonal $\omega_\tau = -\omega_t$ within error while for T=100 ps, the 2D spectrum peaks off the diagonal with $|\omega_\tau| > |\omega_t|$, apparently reflecting an irreversible Stokes' shift in the signal emission. This contrasts sharply with 2D NMR where peaks from a single transition are centered on the diagonal.

4. Conclusions

Apparently, rephasing echo-like contributions to the three pulse scattering signals have a spectral peak near the initial absorption frequency while the signals are red-shifted in the absence of rephasing. This suggests that echo-like signals result from molecules in which the Stokes' shift is either absent or partially reversible.

Acknowledgments. This work was supported by the National Science Foundation, the Packard Foundation, the Dreyfus Foundation, and the Petroleum Research Fund administered by the American Chemical Society.

References

1 L. Lepetit, G. Chériaux, and M. Joffre, J. Opt. Soc. Am. B **12,** 2467, 1995.
2 D. N. Fittinghoff, et al., Opt. Lett. **21,** 884, 1996.
3 J.-P. Likforman, M. Joffre, and V. Thierry-Mieg, Opt. Lett. **22,** 1104, 1997.
4 M. F. Emde, et al., Opt. Lett. **22,** 1338, 1997.
5 X. Chen, et al., Phys. Rev. B **56,** 9738, 1997.
6 S. Mukamel, C. Ciordas-Ciurdariu, and V. Khidekel, IEEE J. Quantum Electronics **32,** 1278, 1996.
7 S. M. Gallagher, et al., J. Opt. Soc. Am. B , in press, 1998.
8 P. Vöhringer, et al., Chem. Phys. Lett. **237,** 387, 1995.
9 R. R. Ernst, G. Bodenhausen, and A. Wokaun, *Principles of nuclear magnetic resonance in one and two dimensions* Oxford University Press, Oxford, 1987.
10 A. Tokmakoff, et al., Phys. Rev. Lett. **79,** 2702, 1997.
11 L. Lepetit and M. Joffre, Opt. Lett. **21,** 564, 1996.

Characterization of the Complex Third-order Susceptibility of Myoglobin using Diffractive Optics

Gregory D. Goodno[1], Vladimir Astinov[2], and R. J. Dwayne Miller[2]

[1] Department of Physics and Astronomy, University of Rochester, Rochester, New York 14627, USA
[2] Department of Chemistry and Department of Physics, 60 St. George St., University of Toronto, Toronto, Ontario M5S 1A7, Canada
E-mail: dmiller@lphys.chem.utoronto.ca

Abstract. We measured ultrafast structural relaxations in horse carboxymyoglobin (MbCO) using heterodyne-detected transient grating spectroscopy. The anisotropic component of the protein response to photodissociation was isolated by acquiring all nonzero elements of the complex third-order nonlinear susceptibility ($\chi^{(3)}$) tensor.

1. Introduction

In order to determine the relationship between a protein's structure and its biological function, it is crucial to first understand the mechanical details of the conformation changes that occur upon transfer between ligated and unligated states. Despite the efforts of many workers, a clear picture of the response of the heme protein MbCO to photodissociation of the heme-ligand bond is still far from complete, especially on ultrafast timescales. This is primarily because changes in the protein conformation do not directly couple to optical or Raman transitions, so it is not easy to find experimentally accessible observables that contain unambiguous information.

Previously, a sub-20-ps volume change in MbCO was detected through its effect on the photogenerated acoustics using transient grating (TG) spectroscopy [1], which is one of the few experimental methods that is sensitive to structure changes in the protein. While photoacoustic spectroscopy can be extended to time resolutions of ~10 ps [2], this is not fast enough to directly measure the contribution of global collective motions of the protein convolved to the bond dissociation coordinate. Furthermore, the direct-detected TG signal is proportional to the modulus-square of the nonlinear susceptibility $\chi^{(3)}$ of the material, so contributions from structure changes cannot be easily isolated from effects arising from electronic or vibrational relaxation.

We have used a recently-developed method of optically heterodyne-detected (OHD) TG spectroscopy [3] to reduce the ambiguity associated with interpreting the TG signals on ps timescales. This technique linearizes the observed signals in $\chi^{(3)}$, eliminating cross-terms between different processes that complicate the direct-detected signal. In addition, absorptive changes can easily be distinguished from changes in the real refractive index through this method.

2. Experimental Methods

Our experimental apparatus is shown in Fig. 1. A 400-nm pump pulse is overlapped with a variably-delayed 800-nm probe pulse at the surface of a diffractive optical element, which splits each into two (± 1 orders). The resulting four pulses are achromatically imaged onto the sample by a spherical mirror at a 1:1 conjugate ratio. This geometry provides automatic Bragg phase matching for the TG experiment, as well as a passively phase-stabilized (~5° rms phase noise) reference beam for OHD [3].

Demonstrative OHD signals from impulsive stimulated Raman scattering in liquid CS_2 with 800-nm pump and probe pulses are shown in Fig. 2. The different traces were obtained by tilting a 200-μm thick cover slip in the probe beam path to change the phase of the probe field with respect to the other input fields. The real (Re) and imaginary (Im) signal components were clearly identified based on their characteristic features in the time domain. For the MbCO studies, the Re and Im components were identified based on the ns-timescale response, where it is expected that photoacoustic oscillations arising from solvent density fluctuations should only be present in the Re component [2]. Both tensor elements χ_{1111} and χ_{1122} of MbCO were measured by varying the polarizations of the input fields. The anisotropic element χ_{1212} could not be experimentally detected with OHD as there are no photoacoustics to provide a basis for separation of its Re and Im components, so it was constructed based on the symmetry relation for isotropic media [4], $\chi_{1212} = (\chi_{1111} - \chi_{1122})/2$. Each OHD component was taken to be the difference of 2 sets of data acquired with a π-phase shift between them. This eliminates both residual direct-detected signal contamination as well as a pump-probe modulation of the reference beam.

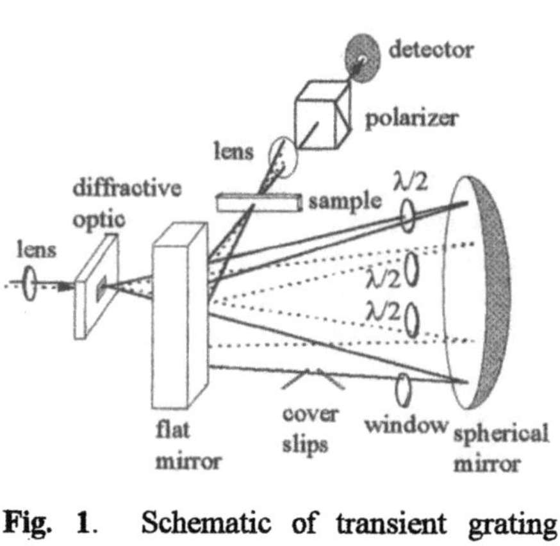

Fig. 1. Schematic of transient grating experiment using a diffractive optic. The dashed line represents the pump beams and the solid line the probe/reference beams.

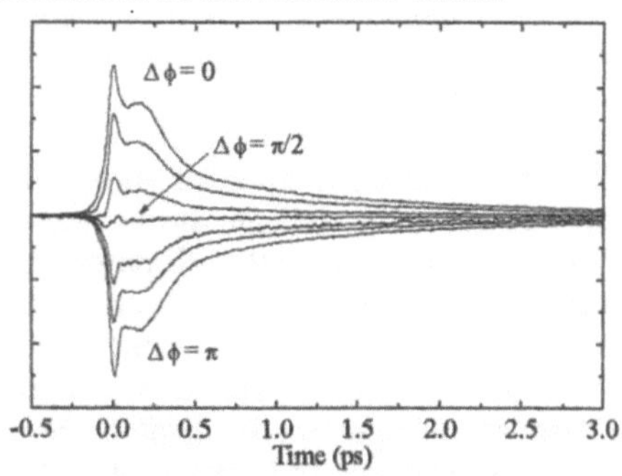

Fig. 2. Transient grating signals in liquid CS_2 for a series of values of the phase difference $\Delta\phi$ between the signal and reference fields (χ_{1111} tensor element). These data demonstrate the passive phase stability of the optical setup.

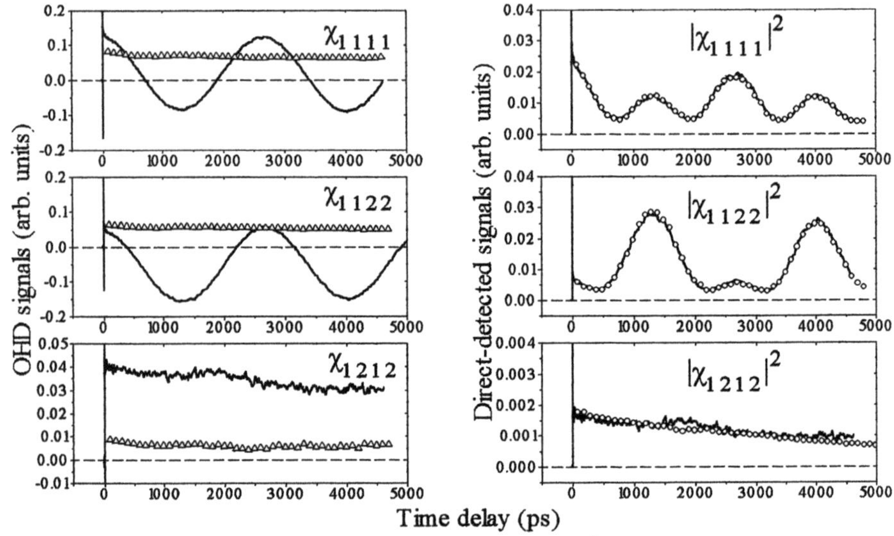

Fig. 3. Left side: OHD signal for the Im($\chi^{(3)}$) (triangles) and Re($\chi^{(3)}$) (solid line) response of MbCO, all normalized to the same scale. Right side: comparison of the scaled direct-detected experimental signal (circles) to $[\text{Im}(\chi^{(3)})]^2 + [\text{Re}(\chi^{(3)})]^2$ (solid line).

3. Results and Discussion

The OHD response of MbCO is displayed in Fig. 3 on acoustic timescales, along with a comparison to the direct-detected signals. This demonstrates that the Im and Re components are linear in $\chi^{(3)}$ and in quadrature with one another.

The χ_{1212} element is proportional to the anisotropic response of the protein. The Im component contains information about the electric dipole properties of the excited state, and an initial anisotropy of 0.09 ± .01 can be inferred from these data, in agreement with pump-probe measurements [5]. The Re component contains both dipole and structural contributions, and the anisotropy is much larger than for the Im. This observation illustrates that the modes involved in the conformation change are highly anisotropic with respect to the heme plane. These anisotropic mode displacements drive specific motions integral to the protein's function and represent a direct observation of mode-specific coupling [6]. Future work will be directed towards quantifying this effect.

References

1 L. Genburg et al., Science **251**, 1051 (1991).
2 G. Dadusc, G.D. Goodno, H.L. Chiu, J. Ogilvie, and R.J.D. Miller, sub. for publ.
3 G.D. Goodno, G. Dadusc, and R.J.D. Miller, J. Opt. Soc. Am. B **15**, 1791 (1998).
4 R. W. Hellwarth, Prog. Quant. Electr. **5**, 1 (1977).
5 A. Ansari et al., Biophys. J. **64**, 852 (1993).
6 J. Deak, H.L. Chiu, C. Lewis, and R.J.D. Miller, J. Phys. Chem. (in press) (1998).

Time domain separation of the nuclear and electronic contributions to the third order nonlinearity in glasses

S. Montant[1], A. Le Calvez[1], E. Freysz[1], A. Ducasse[1], M. Couzi[2]

[1] Centre de Physique Moleculaire Optique et Hertzienne
[2] Laboratoire de Physico-Chimie Moleculaire, UMR 5803
Universite Bordeaux I, 354 cours de la Liberation 33405 TALENCE CEDEX, France
E-mail: montant@cribx1.u-bordeaux.fr

Abstract. The electronic and nuclear contributions to the third order nonlinearity of glasses are separated using 100 fs pulses with a time-resolved heterodyne optical Kerr effect technique. The time evolution of the nuclear contribution is in very good agreement with results deduced from the Raman spectra. A direct estimate of the relative strengths of electronic and nuclear contributions is made. The ratio between the two response function is ~5/1 in Tellurium oxide glass sample. The validity of our method to separate these two contributions has been confirmed in different glasses.

Knowledge of the temporal response of materials with large third order nonlinearities is important for optical switching applications, and is particularly important in understanding the propagation of short pulses in optical fibers. In this context, the unavoidable nuclear phenomenon is consequential since it induces a thermal effect and may modify the temporal structure of the pulses, thereby limiting the speed of response of the device. By analyzing Raman scattering spectra along with intensity-induced polarization changes, Hellwarth et al. were able to estimate the relative electronic and nuclear contributions to the third order nonlinearities of a number of glasses [1]. The nuclear contribution has to be associated with a slow temporal response. However in usual glasses, previous experiments performed in the time domain with short pulses (~ 100 fs) failed to demonstrate such a contribution [2, 3]. Very recently, using very short pulses (~ 35 fs) and exciting large and well defined stretching modes of original glasses, Kang et al. were able to evidence such a contribution [4]. However, their experimental approach is useful only for some particular materials.

The purpose of this article is to demonstrate that using 100 fs pulses with a Time Resolved Heterodyne Optical Kerr Effect (TRHOKE) technique, it is generally possible to evidence a nuclear contribution of glasses in the time domain. The comparison of our TRHOKE signal with the differential Raman scattering cross-section in glass allows us to estimate the electronic and nuclear contribution to the third-order nonlinearity. Below the glass transition temperature T_g, the amplitudes of these contributions are not affected by temperature change but present a clear evolution with respect to the glass composition.

The TRHOKE set-up has been described elsewhere [5]. It is based on a self mode-locked Ti:Sapphire laser (Coherent Mira 900) yielding 10 nJ 100 fs Fourier transform limited pulses at 800 nm with a 76 MHz repetition rate. The TRHOKE signal is expressed as

$$S^{\mathrm{TRHOKE}}(t_D) = \sigma G^{(2)}(t_D) + \int_{-\infty}^{\infty} dt\, b(t) G^{(2)}(t - t_D) = S_E^{\mathrm{TRHOKE}}(t_D) + S_N^{\mathrm{TRHOKE}}(t_D) \tag{1}$$

where σ and b(t) are respectively the electronic and nuclear contributions, $G^{(2)}(t_D)$ is the intensity autocorrelation function, t_D is time delay between pump and probe beam. According to Hellwarth et al. the differential Raman scattering cross-section is closely related to the nuclear responses b(t) [1]. The b(t) components are associated to the elements $\chi^{(3)}{}_{1221}(\Delta) = \chi^{(3)}{}_{1212}(\Delta)$ of the third-order nonlinear tensor by a Fourier transform. According to the fluctuation dissipation theorem the imaginary part of the third order susceptibility is linked to the depolarized differential Raman scattering cross-section. Practically, the real part of the third order susceptibility is deduced through the usual Kramers-Kronig integrals.

The absolute value of the time resolved nonlinear refraction change of the nonlinear 20 % Nb_2O_5 – 80 % TeO_2 glass sample is shown in figure 1 in a linear-log scale. The phase of the signal is shown in to the inset. The π phase change around t = 0.5 ps indicates a change of the sign of the recorded signal.

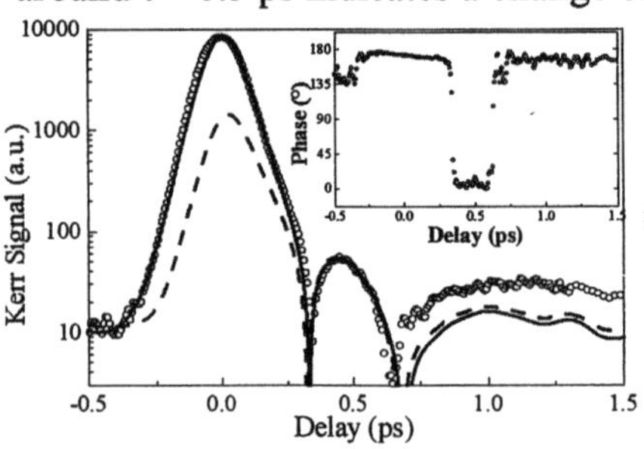

Fig. 1. (o) TRHOKE signal of a 20 % Nb_2O_5 – 80 % TeO_2 glass.

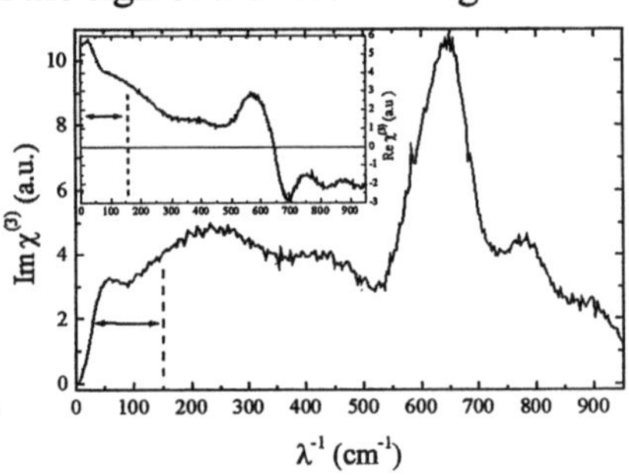

Fig. 2. Imaginary part of the corrected Raman spectra.

Therefore, in addition to a sharp spike near zero delay, there is a damped a oscillation. The signal around the zero time delay is mainly due to the electronic nonlinear response and has a width about equal to the pulse autocorrelation. Three decades in the signal to noise ratio are needed to record the damped oscillation. It was not recorded in the previous homodyne experiments since six decades would have been necessary [2,3]. To deduce the TRHOKE nuclear signal from the Raman spectrum, we used the following procedure. We weighted the corrected Raman spectrum (figure 2) by the Fourier transform of the experimental autocorrelation function $G_2(t_D)$. By using a Fourier transform of this weighted nuclear third-order susceptibility, we deduced the TRHOKE nuclear response function of the glass $S_N{}^{\mathrm{TRHOKE}}(t_D)$ (dashed line in figure 1). By adding to the nuclear response $S_N{}^{\mathrm{TRHOKE}}(t_D)$ and the instantaneous electronic response $S_E{}^{\mathrm{TRHOKE}}(t_D)$, calculated according to the measured autocorrelation of the used pulses, one obtains a very good agreement between computed and experimental results (straight line in figure 1). Since the electronic and nuclear contributions to the nonlinear response can be separated in the time domain, we can directly estimate their relative strengths. By integrating in the time the nuclear contribution of this sample, we found 20 % of the nonlinear refractive index is due to the nuclear motion.

With 100 fs pulses, our TRHOKE experiments mainly probe the low frequency modes of glasses ($\lambda^{-1} < 100$ cm^{-1}). In this frequency range, the glass spectrum exhibits a characteristic feature, the so-called Boson peak. These modes correspond primarily to quasi localized modes of 20 or more atomic masses, secondly, in usual temperature range (T > 300 K), to relaxation processes of structural « defects » of the glass. To evidence the contribution of these later modes, we measured the TRHOKE signal between 300 to 460 K. It was almost unaffected by the temperature variation. This results is not surprising since we noticed temperature mainly affects very low wave number modes which have a negligible contribution in the TRHOKE signal. However, as expected the structure of damped oscillation depends on the Nb_2O_5 content in the glass (figure 3a) and is perfectly reproduced according to the Raman spectra (figure 3b). More generally, we also recorded the TRHOKE signal in differently doped TeO_2 glasses and in commercial SF Schott glass samples (SF1, SF2, SF4, SF6, SF57, SF58, SF59).

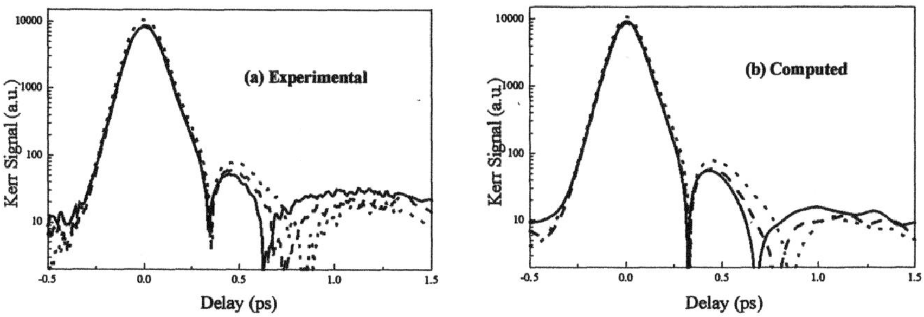

Fig.3: Experimental (a) and computed (b) Kerr signal for 20 % Nb_2O_5 – 80 % TeO_2 (____),15 % Nb_2O_5 – 85 % TeO_2 (------),10 % Nb_2O_5 – 90 % TeO_2 (......) glasses.

The signals had the same characteristic temporal behavior in both types of glass and were fairly well reproduced according to their Raman spectra

In conclusion, we have shown that the nuclear contribution to the optical nonlinearities of many different glasses can be evidenced using 100 fs pulses and TRHOKE set-up. An influence of the glass composition on the nuclear response function is observed and is closely related to the evolution of the Raman spectra

References

1 R. Hellwarth, J. Cherlow, T.T. Tang, Phys. Rev. B 11, 964 (1975).
2 I. Thomaseau, J. Etchepare, G. Grillon, A. Migus, Opt. Letters. 10, 223 (1985).
3 I. Kang, T. D. Krauss, F. W. Wise, B. G. Aitken, N. F. Borrelli, JOSA B 12, 2053 (1995).
4 I. Kang, T. D. Krauss, F. W. Wise, B. G. Aitken, N. F. Borrelli, Phys. Rev. B 54, 641 (1996).
5 A. Le Calvez et al., Chem. Phys. Letters, 258, 620 (1996).

Multiple-Pulse Control and Bispectral 2D Raman Analysis of Nonlinear Lattice Dynamics

Richard M. Koehl,[1] Satoru Adachi,[2] and Keith A. Nelson[1]

[1] Department of Chemistry
Massachusetts Institute of Technology
Cambridge, MA 02139 USA
E-mail: kanelson@mit.edu

[2] Department of Materials Science
Himeji Institute of Technology
1475-2 Harima Science Garden City
Hyogo 678-1297, Japan

Abstract. Anharmonic lattice dynamics are investigated with a heterodyne bispectral 2D Raman technique.

Multiple-pulse ISRS excitation of lattice vibrations has been conducted for some time.[1] Very recently, ISRS excitation of highly anharmonic, nonlinear lattice vibrational responses has been demonstrated with the use of intense excitation pulses.[2] Excitation pulses crossed at difference wavevector $\mathbf{q}$ generate lattice vibrational responses at the fundamental frequency $\omega(\mathbf{q})$ and at high frequency harmonics $2\omega(\mathbf{q})$, $3\omega(\mathbf{q})$, etc., and also vibrational responses at high wavevector harmonics $2\mathbf{q}$, $3\mathbf{q}$, etc., up to $6\mathbf{q}$, all clear signatures of strongly anharmonic vibrational responses. The crystals studied are ferroelectrics such as LiTaO3, and the lattice vibrations are phonon-polaritons which propagate through the crystals at about 1/5 the speed of light in air. Real-space imaging of propagating phonon-polaritons has been demonstrated[3] and is a form of Terahertz imaging carried out with all-optical wavelengths.

The frequency and wavevector overtones we observe can be thought of as resulting from sum-frequency (and sum-wavevector) mixing among the anharmonically interacting phonon-polaritons at the fundamental (ω, $\mathbf{q}$). Anharmonic mode-mode interactions in molecular liquids have recently been studied using 2D Raman spectroscopy,[4-7] which is a form of bispectral or multispectral analysis used in many applications involving nonlinear systems, e.g. nonlinear signal processing,[8] nonlinear and chaotic dynamics,[9, 10] and system identification for control over macroscopic nonlinear systems.[11]

We have performed experiments involving multiple-pulse excitation and bispectral analysis of anharmonic lattice vibrations. Two pairs of crossed excitation pulses reached the sample with variable delay t_1 with the spatial phases of the grating patterns maintained through the use of a corrugated phase mask to produce the beam pairs.[12] A probe and reference beam for heterodyne detection[13] at the wavevector fundamental or overtones were also produced by the phase mask and incident at the sample after variable delay time t_2. In this manner it is possible to probe the three-time correlation functions of the fundamental and overtone wavevectors. In this case we are not interested in dephasing processes, but in nonlinear lattice couplings between modes. The phase-sensitive experimental response (Figure 1) appear consistent with qualitative calculations using a simple classical model of a driven fundamental mode nonlinearly coupled to a second mode by two-phonon and three-phonon lattice processes.

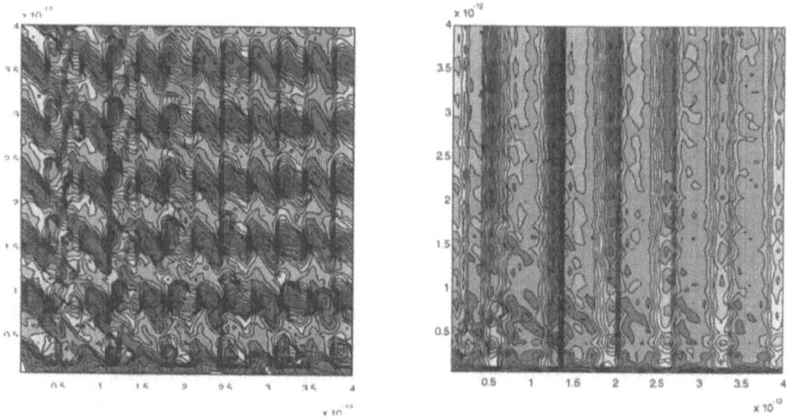

Figure 1. (a) 2D data probed at the fundamental wavevector. Two pairs of pulses, created by a corrugated phase mask to preserve spatial phase relationships, are variably delayed by a time t_1. A probe for wavevector q and heterodyne reference pulse are also produced by the phase mask, and are delayed by a time t_2. (b) 2D data probed at the lowest wavevector overtone. The same pump sequence as in Fig. 1a is used, but the probe is now at 3q.

The response of a system probed at **q** (Fig. 1a) shows oscillations with frequency ω(**q**) along the probe time delay axis. These oscillations are separated along the pump delay axis by a period equivalent to the oscillations of frequency ω(**q**). The response of a system probed at 3**q** (Fig. 1b) shows oscillations along t_2 with frequency ω(3**q**) (close to 3ω(**q**) at this point on the dispersion curve) which appear to be separated on the $t_1$axis by a period corresponding to a frequency ω(**q**), *the frequency that is seen in the fundamental.* This behavior is qualitatively reproduced by the calculations. The results show that wavevector overtones are not arising from multiple or nonlinear scattering of the crossed excitation pulses, as opposed to genuine lattice anharmonicity. The results are also consistent with a recently completed theoretical analysis that shows how our data may be used to describe the anharmonic potential energy surface.[14]

In summary, bispectral, multi-wavevector, heterodyne 2D Raman analysis of nonlinear lattice responses excited through multiple-pulse, spatially phase-locked ISRS has been conducted to reveal the anharmonic mode-mode interactions in the crystal. The novel experimental results that have been brought to bear on the problem are

(1) multiple-pulse excitation beyond the harmonic limit;
(2) multiple-pulse grating excitation with phase masks to ensure reliable spatial phase;

(3) broadly applicable, experimentally simple heterodyne detection of transient grating signals;
(4) real-space imaging of propagating phonon-polaritons; and
(5) 2D Raman spectroscopy and bispectral analysis, similar to that applied recently to molecular liquids, now applied to fundamental and overtone wavevectors of lattice responses.

Acknowledgements. We thank T. F. Crimmins and A. A. Maznev for stimulating discussions. This work was supported in part by NSF Grant No. CHE-9713388.

References

1 A.M. Weiner, D.E. Leaird, G.P. Wiederrecht, and K.A. Nelson, Science **247,** 1317, 1990.
2 C. J. Brennan and K.A. Nelson, Journal of Chemical Physics **107**, 9691, 1997.
3 R.M. Koehl,S. Adachi, and K.A. Nelson, Journal of Chemical Physics (submitted).
4 Y. Tanimura and S. Mukamel, J. Chem. Phys. **99**, 9496 (1993).
5 T. Steffen and K. Duppen, Physical Review Letters, **76**, 1224, 1996.
6 A. Tokmakoff, M.J. Lang, D.S. Larsen, G.R. Fleming, V. Chernyak, and S. Mukamel, Physical Review Letters **79**, 2702, 1997.
7 A. Tokmakoff and G.R. Fleming, Journal of Chemical Physics **106**, 2569, 1997.
8 C. L. Nikias and A.P. Petropulu, *Higher-Order Spectral Analysis: A Nonlinear Signal Processing Framework*, PTR Prentice Hall, Englewood Cliffs, NJ, 1993.
9 K. Hasselmann, *Bispectra of Ocean Waves*, in *Time Series Analysis*, Edited by M. Rosenblatt, John Wiley and Sons, New York, 1963.
10 M. D. Miller, Physical Review B, **34**, 6326, 1986.
11 M. Schetzen, *The Volterra and Wiener Theories of Nonlinear Systems*. Wiley, New York, 1980.
12 A. A. Maznev, T. F. Crimmins, and K.A. Nelson, Optics Letters (accepted).
13 A. A. Maznev, K.A. Nelson, and J. A. Rogers, Optics Letters (in press).
14 V. Romero-Rochin, R. M. Koehl, and K.A. Nelson, in preparation.

Engineering Quantum Indistinguishability in Ultrafast Parametric Downconversion

David Branning[1], Warren Grice[2], Reinhardt Erdmann[3], and Ian Walmsley[3]

[1] Department of Physics and Astronomy, University of Rochester, Rochester, NY 14627, USA
[2] Department of Physics, Southern Illinois University, Edwardsville, IL, 62026-1654, USA
[3] The Institute of Optics, University of Rochester, Rochester, NY 14627, USA
E-mail: walmsley@optics.rochester.edu

Abstract. Ultrashort optical pulses provide timing information for entangled photons generated by downconversion. This information destroys two-particle nonlocal quantum interference, but enables multiparticle interference. We demonstrate a general technique for restoring the nonlocal correlation of the photon pairs without compromising the source brightness.

1. Introduction

Correlated photon pairs produced by spontaneous parametric downconversion have proved their utility in a variety of experiments exploring the foundations of quantum theory. The entangled nature of signal and idler photons has been exploited in violations of local realism, measurements of the tunneling time for light, "quantum eraser" demonstrations of wave/particle complementarity, and in numerous other optical realizations of what might once have been called gedankenexperiments.[1] In the past year ultrafast pulsed pump sources have gained increasing attention as a means of synchronizing the emission and detection of photons from two or more independent downconversion processes. When a parametric downconverter is pumped by a train of ultrafast optical pulses, the pump also acts as a master clock against which the arrival of the other photons in the experiment may be gauged. The high timing resolution gained in this way appears to be a necessary requirement for implementing so-called "GHZ" demonstrations of nonlocality with multiple entangled particles, and was also recently put to use in the first experimental demonstration of "quantum teleportation," a process which may ultimately play a role quantum computation.[2]

Useful though it may be, this timing information comes at a price. For example, if one wishes to maintain the interferometric stability of a copropagating arrangement, then the difference in dispersion between the two crystal axes of a Type-II parametric downconverter that is necessary for phase matching also leads to reduced entanglement of the two photons.

2. Indistinguishability in two-photon interferometry

When two photons are coincident simultaneously on different input ports of a beamsplitter, they always exit at the same output port. Thus two detectors located at these ports never register a photon simultaneously - there is a complete lack of coincidence counts. This counter intuitive quantum feature of photon pairs is intimately related to the distinguishability of two-photon pathways leading to a coincidence count, and is exploited in the Hong-Ou-Mandel (HOM) interferometer. When the inputs to the two ports are the signal and idler photons generated by parametric downconversion, the lack of coincidence counts is a measure of the entanglement of the photon pair. We exploit this effect in our experiments to demonstrate a rather general technique for preserving maximal entanglement of the photon pair without postselection using an ancillary measurement.

The interference of photon pairs in a HOM interferometer shows reduced visibility as the pump pulse duration decreases, as shown in Ref.[3] for type-II downconversion in BBO using 100 fs-duration pump pulses at 400 nm. The reason for this is that a broad pump spectrum produces orthogonally polarized signal and idler photon pairs which satisfy the phase-matching conditions over a larger range of emission frequencies than in the case of cw pumping, but this range is not increased in equal measure for each photon as the pump bandwidth increases, owing to the dispersion characteristics of the (birefringent) downconversion medium. Therefore, the signal and idler emerge from such a process with markedly different bandwidths, even in the case where their center frequencies are identical. This bandwidth difference is a kind of "spectral information" which can serve to distinguish one photon from the other in principle, and thus to destroy quantum interference. The visibility can be restored either by restricting the pump bandwidth, or by limiting the range of frequencies seen by the detectors, so that the pairs of photons with the most asymmetric frequency distributions are not admitted -- in effect, selecting only the most indistinguishable photons for detection. Both methods reduce the timing resolution and counting rates at the detectors.

Another scheme for restoring the quantum interference is to eliminate the spectral distinguishability at the source, by overlapping two processes for parametric downconversion with the axes of polarization switched between the two. In this symmetrized configuration, both the signal and the idler photons carry identical superpositions of both the broad and the narrow spectra leading to an optimally entangled state. When these photon pairs are passed through the HOM interferometer, the interference visibility is restored, as shown in Fig.1, without recourse to spectral filters, and without reduction of the counting rate or loss of timing information. The relative phases of the superpositions are critical and the coincidence rate may be set to zero or to twice the background rate depending on whether the phase is 0 or π. In either case, this implies the paths are nearly completely indistinguishable.

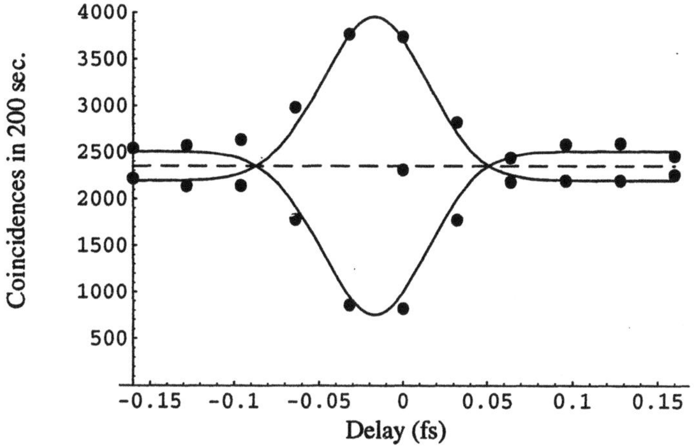

Fig. 1. Restored interference visibility in the HOM interferometer using photon pairs from a source that has been "engineered" to produce indistinguishable ultrashort entangled photon wave packets. Points: experimental data, lines; theory, for phase settings of 0 and π.

4. Conclusions

Information and interference are complementary phenomena. The ability to inhibit the generation of distinguishing information in the production of photon pairs is therefore critical in maximizing the entanglement of the pair. The method of symmetrizing the probability amplitudes of the pair with respect to the frequency of each (or the time of emission) allows the generation of well-entangled photon pairs at a high rate. Since it allows one to combine the best features of both Type-I and Type-II sources, and may be directly generalized to multi-particle sources, we anticipate that the technique will be of broad applicability in quantum information processing.

Acknowledgements. We thank Professor L. Mandel for many useful conversations.

References

1 See, for example, L. Mandel, Ann. NY Acad. Sci., 575, 1, 1995.
2 D. Bouwmeester *et al.*, Nature, 390, 575, 1997.
3 W. Grice, D. Branning, R. Erdmann and I. A. Walmsley, Phys. Rev. A, 57, R4219, 1998.
4 A. M. Steinberg, P. G. Kwiat, and R. Y. Chiao, Phys. Rev. Lett., 68, 2421, 1992.

Part IV

Imaging and Microscopy

Ultrafast Laser Technology Applied to Whole-Field Fluorescence Lifetime Changing

K. Dowling[1], R. Jones[1], M. J. Dayel, P. M. W. French[1], H. Ikeda-Gerling[2], M. J. Lever[2], J. D. Hares[3], A. K. L. Dymoke-Bradshaw[3], M. A. A. Neil[4], R. Juskaitis[4] and T. Wilson[4]

[1] Femtosecond Optics Group, Physics Department, Imperial College, Prince Consort Road, London, SW7 2BZl, U.K. E-mail: paul.french@ic.ac.uk, http://www.lsr.ph.ic.ac.uk/femto/
[2] Department of Biological and Medical Systems, Imperial College,
[3] Kentech Instruments Ltd. Unit 9, Hall Farm Workshops, South Moreton, Didcot, Oxfordshire, OX11 9AG, U.K., http://www.kentech.co.uk/
[4] Department of Engineering Science, University of Oxford, Parks Road, Oxford, OX1 3PJ.

Abstract. We describe an ultrafast fluorescence lifetime imaging system that can image lifetime differences < 10ps. Preliminary results concerning autofluorescence lifetime images of whole and composite tissue samples are presented. Using structured illumination light, we have also demonstrated optically sectioned fluorescence lifetime imaging.

1. Introduction

Laser induced fluorescence is a powerful modality for chemically specific imaging, with potential biomedical applications ranging from optical biopsy to functional imaging. As well as revealing fluorophore distribution, fluorescence imaging can also yield information concerning the local chemical and/or structural environment, in as much as this impacts the quantum efficiency via the non-radiative decay rate. In principle, quantum efficiency may be determined by absolute intensity measurements, but the heterogeneity of tissue and background autofluorescence usually make this extremely difficult *in vivo* or *in vitro* with thick tissue samples. Alternatively, the fluorescence decay time, τ, which also depends on the non-radiative decay rate, may be determined for any fluorophore using only relative intensity measurements [1]. Fluorescence lifetime imaging (FLIM) provides a robust means of acquiring spatially resolved information regarding the local environment of a distributed fluorophore in biological tissue and other heterogeneous or turbid media. Fluorescence lifetime probes have already been demonstrated for the measurement of e.g. [Ca^{2+}], [O_2] and pH.

2. FLIM Apparatus

The experimental FLIM set-up is shown in Figure 1. Fluorescent samples were typically illuminated by ~10 ps pulses at 5 kHz repetition rate with ~1 μJ energy at 415 nm. These pulses were derived from a commercial Ti:sapphire laser and amplified in a tunable Cr:LiSAF regenerative amplifier [2]. We note that this could be replaced by an all-solid-state diode-pumped system [3]. After each excitation pulse, the emitted fluorescence distribution was imaged onto a time-gated image intensifier (Kentech Instruments Ltd., Gated Optical Imager, GOI), which provided an effective (measured) exposure time gate of < 90 ps.

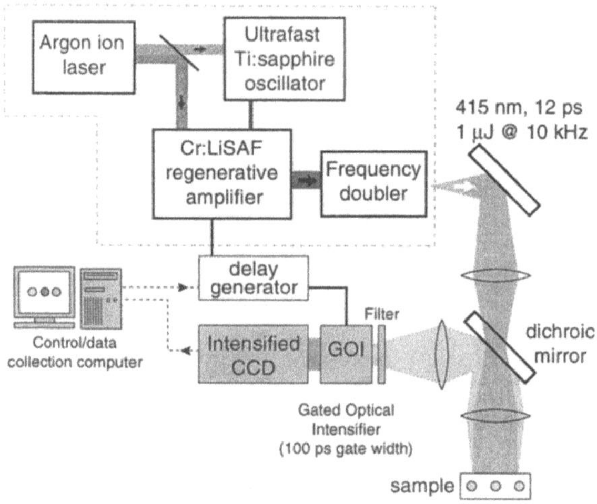

Fig. 1. Experimental FLIM setup

The GOI was triggered from the regenerative amplifier's Pockels cell. This trigger signal was synchronized to within ~ 20 ps of the 80 MHz photodiode signal from the oscillator, using a trigger resynchronization unit (Kentech Instruments Ltd., JE1). By using an electronic delay to vary the detector trigger position with respect to the excitation pulse, a series of time-gated images could be recorded. The temporal decay of the fluorescence was thereby acquired simultaneously for each pixel in the field of view. A non-linear least-squares fitting algorithm was then used to determine the fluorescence lifetime for each individual pixel. The data collection and subsequent processing typically took a few minutes to produce fluorescence lifetime maps. Low resolution FLIM maps, of fluorophore distributions fitted to single exponential decay profiles, could be acquired and displayed with an update time of just 3 seconds.

Most previously demonstrated FLIM systems have been limited to temporal responses of a few nanoseconds by the availability of excitation sources and detectors. Using an ultrafast laser system and time-gated image intensifier, however, permits a FLIM system with a temporal system response of less than 100 ps [4] to be implemented. This system is suitable for imaging subtle changes in fluorophore environment [5], as is illustrated in Figure 2, which shows the FLIM map of four pipettes containing the dye DASPI in solutions of varying viscosity. The reproducibility of this measurement over 100 minutes is also shown, indicating that the recorded lifetime differences vary by only a few picoseconds. By imaging DASPI solutions of very similar viscosity, we have demonstrated the ability to repeatedly image differences in lifetime of < 10 ps.

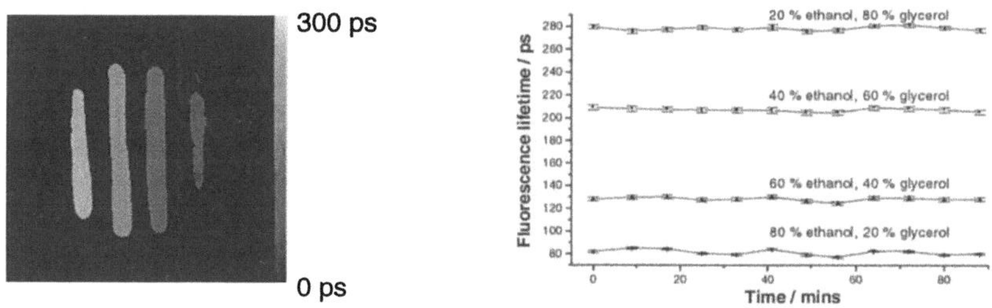

(a) (b)

Fig. 2. (a) FLIM map of four pipettes of DASPI in ethanol/glycerol solvent mixtures of varying viscosity and (b) the variation of measured fluorescence lifetimes of each pipette for a series of acquisitions repeated over 100 minutes

3. Biological FLIM

Streak camera measurements have already demonstrated that fluorescence lifetime can be used to distinguish between different tissue types [6] and work at Imperial College is underway applying FLIM to biological tissue in vitro. First results suggest that, for excitation with 10 ps pulses at 415 nm, the fluorescence profiles of collagen, elastin and other tissue components are reasonably well fitted by double exponential decays.

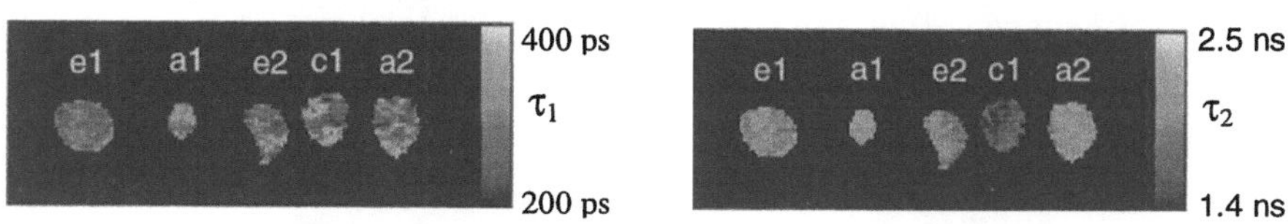

(a) (b)

Fig. 3. FLIM maps of (a) short-lived component (τ_1), and (b) long-lived component (τ_2) of fluorescence decay of samples of elastin (e1 and e2); aorta (a1 and a2) and collagen (c1)

Figure 3 shows the FLIM images corresponding to the first and second temporal components of the fluorescence decay of samples of elastin (e1 and e2), collagen (c1) and aorta (a1 and a2), extracted from rat. It is apparent that, while the short-lived component of the fluorescence (τ_1) does not provide a strong contrast, the FLIM map of the long-lived component (τ_2) clearly provides a means to distinguish, not only the collagen and elastin, but also the elastin and aorta. For elastin and collagen this contrast between different tissues is to be expected. The difference between the FLIM images of elastin and aorta is more surprising as the elastin was prepared by boiling aorta in 0.1 M NaOH for 30 minutes. The FLIM contrast therefore suggests a potential technique for imaging relatively subtle changes in biological tissue e.g. *in situ*, during therapy.

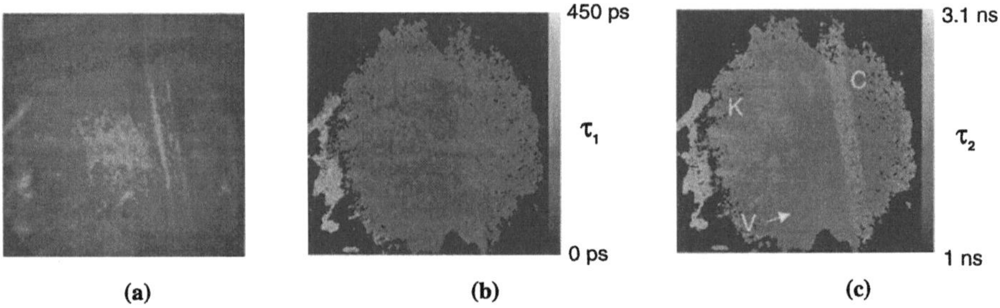

Fig. 4. (a) Time-gated fluorescence image immediately following excitation of rat ear, together with FLIM maps of (b) short-lived (τ_1), and (c) long-lived decay component (τ_2).

Whole tissue samples have also been investigated *in-vitro* to determine whether fluorescence lifetime may provide contrast between different tissue types. Figure 4 shows the fluorescence intensity image of a 10 μm thick sample of rat ear captured immediately after excitation, together with the FLIM maps corresponding to the two fitted exponents to the fluorescence decay profiles. In this case the second decay constant (τ_2) can be seen to provide contrast between hair follicles and epidermal tissue (mainly keratin, K), a blood vessel (V) and cartilage (mainly collagen, C).

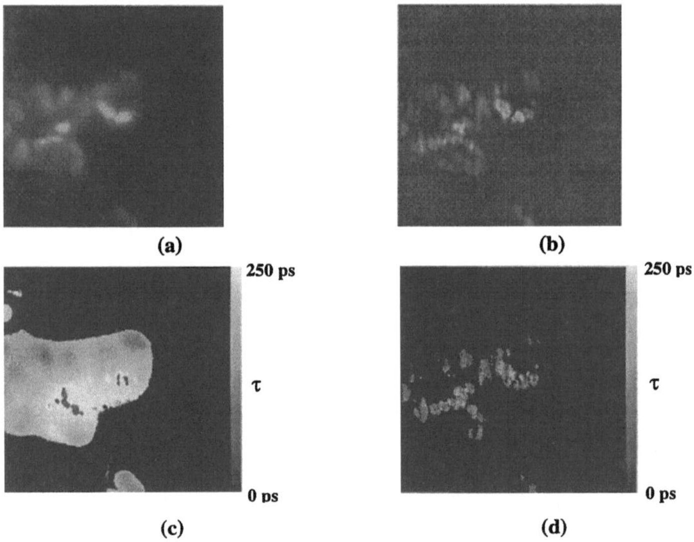

Fig. 5. (a) conventional and (b) optically sectioned fluorescence microscope images of 10 μm moss spoors recorded directly after excitation, together with FLIM maps obtained from (c) conventional and (d) optically sectioned time-gated fluorescence images.

4. Optically Sectioned FLIM

Our time-gated FLIM system can be used in conjunction with the whole-field optical sectioning technique pioneered by Neil et al. [7] to record optical sectioned FLIM maps with a depth resolution comparable to confocal microscopy. The optical sectioning is achieved by using a structured excitation intensity distribution, such as a grid pattern, at the sample. In a proof of principle experiment, moss spores were imaged by a microscope with 250x magnification. A grating of 50 line pairs/mm was projected onto the sample using the 415 nm pulses and a conventional and sectioned image were obtained using the method outlined in reference [7]. Figures 5 (a) and (b) show the conventional and sectioned images of the fluorescence from ~ 10 μm diameter moss spoors directly after excitation. By recording images at different delays after excitation, conventional and optically sectioned FLIM images were made, as shown in Figure 5 (c) and (d) respectively. The sectioned images have been determined to have a depth resolution of ~ 1.5 μm.

5. Conclusions

We report a FLIM system that can image lifetime differences of ≤ ~ 10 ps. This has been applied to tissue autofluorescence *in-vitro* with 415 nm excitation, which has been shown to exhibit double exponential decay profiles. The decay constants have been shown to provide contrast between different tissue constituents. Using structured excitation, optically sectioned FLIM images have been obtained of 10 μm diameter moss spores with a depth resolution of ~ 1.5 μm. This work has been supported by the UK Engineering and Physical Sciences Research Council.

References

[1] J. R. Lakowicz, "Topics in Fluorescence Spectroscopy, vol. 1- Techniques", Plenum Press, New York (1991)

[2] S. C. W. Hyde, N. P. Barry, R. Mellish, P. M. W. French, J. R. Taylor, C. J. van der Poel and A. Valster, *Opt Lett* **20**, pp 160-162 (1995)

[3] R. Mellish, N. P. Barry, S. C. W. Hyde, R. Jones, P. M. W. French, J. R. Taylor, C. J. van der Poel and A. Valster, *Opt Lett*, **20** pp 2312-2314 (1995)

[4] K. Dowling, M.J. Dayel, M.J. Lever, P.M.W. French, J.D. Hares, A.K.L. Dymoke-Bradshaw, *Opt. Lett.* **23,** pp 810-812 (1998)

[5] K. Dowling, M. J. Dayel, S. C. W. Hyde, J. C. Dainty, P. M. W. French, P. Vourdas, M. J. Lever, A. K. L. Dymoke-Bradshaw, J. D. Hares and P. A. Kellett, *IEEE JSTQE* **4**(2), pp 370-375 (1998)

[6] B.B. Das, F. Liu and R.R. Alfano, *Rep. Prog. Phys.*, **60**, pp. 227-292 (1997)

[7] M. A. A. Neil, R. Juskaitis and T. Wilson, Opt. Lett. **22** (24), pp 1905-1907 (1997)

Optical Coherence Tomography using Femtosecond Lasers

Stephen A. Boppart[1], Costas Pitris[1], Brett E. Bouma[1], Mark E. Brezinski[1,2], and James G. Fujimoto[1]

[1] Department of Electrical Engineering and Computer Science, Research Laboratory of Electronics, Massachusetts Institute of Technology, 77 Massachusetts Avenue, Room 36-345, Cambridge, MA 02139, USA
[2] Massachusetts General Hospital, and Harvard Medical School, Boston, MA 02114

Abstract. Optical coherence tomography (OCT) using femtosecond lasers enables subcellular resolutions of < 5 μm and real time imaging at several frames per second. Catheter-endoscope imaging and medical applications are discussed.

Optical coherence tomography (OCT) is a recently developed imaging technique for performing high resolution cross-sectional imaging of internal microstructure in biological and other systems [1]. OCT imaging is analogous to ultrasound except that it uses light rather than sound and constructs cross-sectional images by measuring the echo delay and intensity of backscattered light. An axial backscattering profile is measured at several transverse positions to yield a two dimensional data set which can be displayed as a gray scale or false color image. Because the velocity of light is extremely high, echo delay times cannot be measured directly and interferometry using short coherence length cw or ultrashort light pulses is used. The axial resolution is determined by the coherence length of the light while the transverse resolution is governed by beam focusing.

OCT can perform optical biopsy, the in situ cross sectional imaging of tissue morphology [1-3]. Coupled with catheter, endoscopic, or laparoscopic delivery, OCT holds the promise of enabling the screening and diagnosis of a wide range of diseases including cancerous and precancerous tissue changes without the need for excisional biopsy and histological processing. In conjunction with conventional microscopy, OCT enables the imaging of internal structures in living specimens without the need for sacrifice and histology.

Femtosecond solid state lasers are enabling for performing OCT because they have short pulse duration and short coherence lengths [4-6]. Femtosecond lasers are also attractive because they have high average powers and can enable high speed imaging with high sensitivity. KLM modelocked $Ti:Al_2O_3$ and Cr^{4+}:Forsterite lasers are especially interesting for performing OCT. The $Ti:Al_2O_3$ laser features a broad tuning range and can generate pulses of less than 10 fs duration. Cr^{4+}:Forsterite operates in the ~ 1.3 micron wavelength range and can produce pulse durations of less than 50 fs. Higher resolutions can be achieved by broadening the spectral bandwidth using self phase modulation in fibers. The Cr^{4+}:Forsterite is well suited for imaging in biological tissues because scattering effects which limit imaging depth are reduced at these wavelengths.

Figure 1 shows a schematic of the OCT system. The system consists of a self-phase modulated KLM Cr^{4+}:Forsterite laser coupled into a fiber optic Michelson interferometer [6]. The laser produces an output power of 30 mW with a Gaussian

FWHM spectral bandwidth of 75 nm, centered at 1280 nm. These parameters yield a measured free space axial resolution of 10 μm and a SNR of 110 dB. A rapidly scanning optical delay line is constructed using a grating phase control device, similar to that used to generate programmable pulses [7]. The grating phase control device is attractive because both the phase and group velocity of the scan can be independently controlled. Since the delay is scanned by varying the mirror angle, extremely high scan speed is possible. The scanner can scan a 1 kHz triangle wave, corresponding to 2000 axial scans per second or an imaging speed of 8 frames per second for a 250 pixel transverse image.

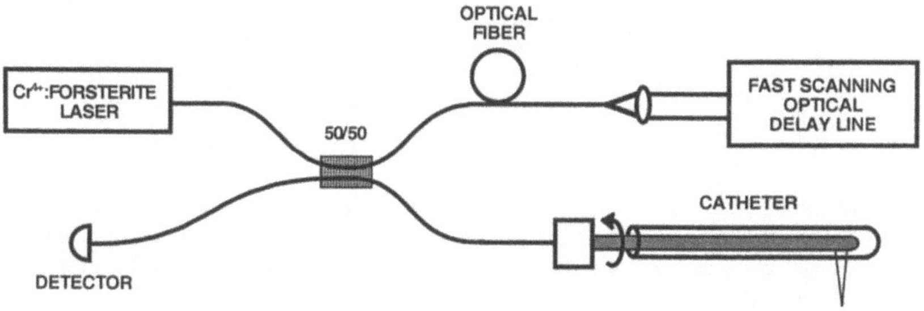

Fig. 1. Schematic of OCT instrument for high speed catheter-endoscope based imaging. The laser source is a KLM Cr:Forsterite laser producing short pulses at 1.3 μm. Rapid scanning of the optical delay is performed using a grating phase control delay line.

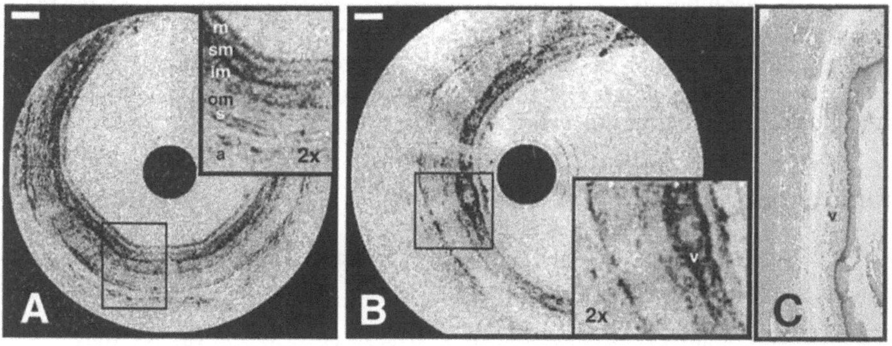

Fig. 2. A. OCT image in vivo of the of a New Zealand White Rabbit. The image clearly shows the esophageal layers including the mucosa (m), the submucosa (sm), the inner muscular layer (im), the outer muscular layer (om), the serosa (s) and the adipose and vascular supportive tissues (a). B. A blood vessel (v) can be seen within the sub-mucosa of the esophagus. C. Corresponding histology for Figure 3B. The bars represent 500 μm.

In medical diagnostics, OCT can be used in conjunction with a fiber optic catheter-endoscope to perform imaging of internal organ systems [8]. Using fiber optics, a catheter-endoscope with an outer diameter of 2.9 French or 1.0 mm has been constructed. A single mode optical fiber runs the length of the catheter and the distal end consists of a GRIN lens and a microprism to direct the OCT beam radially. The fiber and distal optics are rotated so that the OCT beam scans an angular, radar like, pattern and image cross sectionally through internal organs.

The high speed catheter-endoscope OCT system enables the acquisition of in vivo images of internal organ systems. Figure 2 shows an example of a cather-endoscope OCT image of the gastrointestinal tract of a rabbit in vivo. In vivo imaging of the pulmonary tract, gastrointestinal tract, and urinary tract as well as arterial imaging have also been performed. These studies demonstrate the feasibility of performing OCT imaging of internal organ systems.

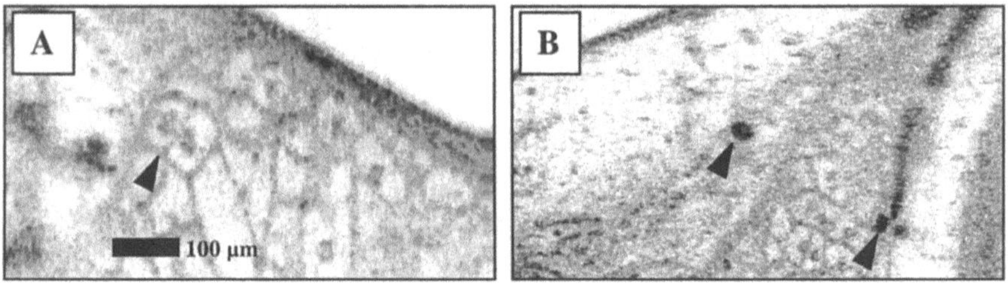

Fig. 3. High resolution OCT images of a Xenopus laevis (African Frog) tadpole in vivo. Figure 2A mesenchymal cells immediately following cell division with two daughter cells (arrow). Cell membranes and individual cell nuclei are apparent. Figure B melanin-laden neural crest cells which migrate during development.

Figure 3 show an example of high resolution OCT images of a *Xenopus laevis* (African frog) tadpole specimen, used in developmental biology [9]. The OCT beam was focused to a 9 µm diameter spot (100 µm confocal parameter). The free space axial resolution was 5.1 µm. Assuming an average index of 1.35 for these specimens, the *in vivo* axial resolution was ~ 3.8 µm.

The ability to image tissue, *in vivo*, at high resolutions and high speed can enable identification of normal and diseased tissue without having to biopsy a specimen for histological processing.

References

1 D. Huang, E. A. Swanson, C. P. Lin, J. S. Schuman, W. G. Stinson, W. Chang, M. R. Hee, T. Flotte, K. Gregory, C. A. Puliafito and J. G. Fujimoto, Science **254**, 1178, 1991.
2 J. M. Schmitt, M. Yadlowsky and R. F. Bonner, Dermatology **191**, 93, 1995.
3 M. E. Brezinski, G. J. Tearney, B. E. Bouma, J. A. Izatt, M. R. Hee, E. A. Swanson, J. F. Southern and J. G. Fujimoto, Circulation **93**, 1206, 1996.
4 X. Clivaz, F. Marquis-Weible and R. P. Salathe, Electron. Lett. **28**, 1553, 1996.
5 B. E. Bouma, G. J. Tearney, S. A. Boppart, M. R. Hee, M. E. Brezinski and J. G. Fujimoto, Opt. Lett. **20**, 1486, 1995.
6 B. E. Bouma, G. J. Tearney, I. P. Bilinsky, B. Golubovic and J. G. Fujimoto, Opt. Lett. **21**, 1839, 1996.
7 A. M. Weiner, J. P. Heritage and E. M. Kirschner, J. Opt. Soc. Am. B **5**, 1563, 1988.
8 G. J. Tearney, M. E. Brezinski, B. E. Bouma, S. A. Boppart, C. Pitris, J. F. Southern and J. G. Fujimoto, Science **276**, 2037, 1997.
9 S. A. Boppart, M. E. Brezinski, B. E. Bouma, G. J. Tearney and J. G. Fujimoto, Dev. Biol. **177**, 54, 1996.

3D-microscopy using third-harmonic generation at interfaces in biological and non-biological specimens

Jeff A. Squier[1], Kent R. Wilson[1], Michiel Müller[2], and G.J. Brakenhoff[2]

[1]University of California, San Diego, LaJolla, CA 92093-0339 USA
[2]BioCentrum Amsterdam, Institute for Molecular Cell Biology, Kruislaan 316, 1098 SM Amsterdam, The Netherlands.

Abstract. 3D imaging of the structure of transparent biological and non-biological samples is demonstrated using third harmonic generation at interfaces within the specimen.

Recently Barad et al. [1] have shown that the third harmonic generation (THG) at the interface of a transparent sample can be used to map the surface of optical glass fibers in index matching fluid. In this paper we extend the demonstrated principle of the technique and show the potential of three-dimensional third harmonic volume imaging of both biological and non-biological specimens. We show THG imaging is significant for a variety of reasons including 1) it is a background free imaging technique requiring no additional staining, 2) it is non-fading in nature, 3) it is sensitive to the orientation of interfaces within the specimen. Further, since two-photon microscopy is often done with excitation wavelengths around 800 nm (or below), these results imply significant amounts of 265 nm radiation (or shorter) are routinely generated within the sample. Thus, these results may have important implications in choosing the excitation wavelength in two-photon absorption imaging in order to minimize radiation damage.

Tunable femtosecond pulses in the 1.0 - 2.0 μm range are generated at 1 kHz in an optical parametric amplifier (OPA). The output of the OPA is focused by a cylindrical lens to create a line cursor which is then focused by a high numerical aperture objective into the specimen. By scanning the cursor and detecting the third harmonic in the forward direction, we are able to optically section the specimen. The sectioning effect is a result of the production of third harmonic exclusively at the focus, and only when there is an interface which has a component perpendicular to the direction of light propagation [1, 2]. A series of optical sections of an optical fiber in immersion oil taken in this manner is shown in Figure 1.

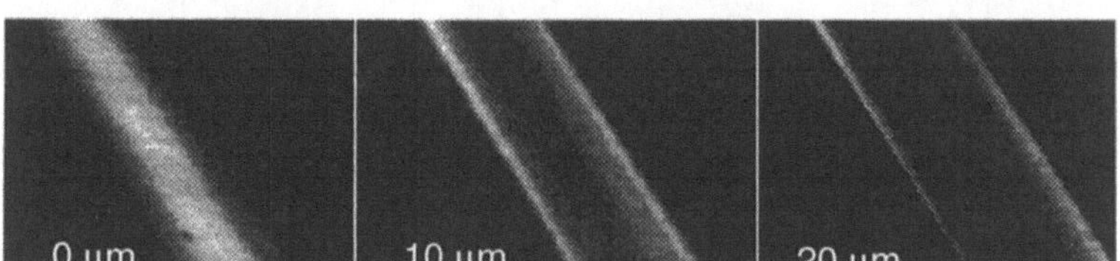

Figure 1. The cross sectional field was 340 x 340 μm. Each successive panel is 10 μm deeper within the fiber, starting at the top of the fiber.

In Figure 2, a series of cross sectional images of the optical fiber is stacked together to render a full three dimensional volume. The k-vector dependence of the third harmonic radiation is especially dramatic in this figure, as the image virtually disappears in those optical sections where the fiber interfaces are parallel to the propagation direction of the excitation beam.

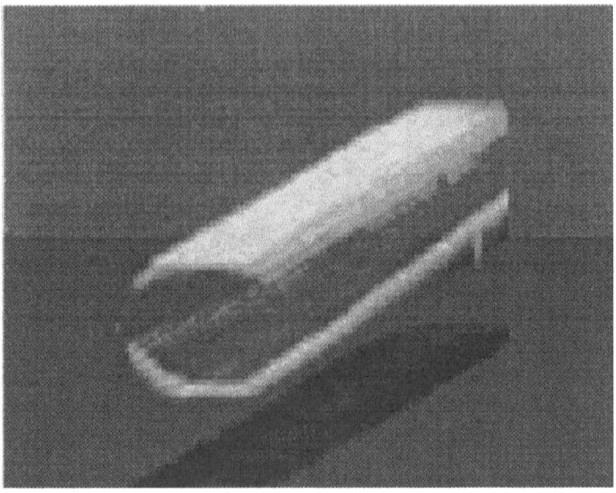

Figure 2. 3-D reconstruction of the fiber.

In Figure 3, a THG optical section of a plant leaf is shown. The resistence to fading and sample damage of this imaging technique was tested by continuously exposing the leaf sample to the excitation beam, and recording a new image every 5 minutes for 1 hour. The specimen was never translated either laterally or axially, so the same specimen volume was continuously exposed. After one hour the image showed no reduction in intensity or structural change, indicating a lack of fading and visible damage, in contrast to fluorescence microscopy. This can be understood in terms of the nonresonant nature of the frequency conversion process, and the minimal absorption at both the fundamental and third harmonic wavelengths.

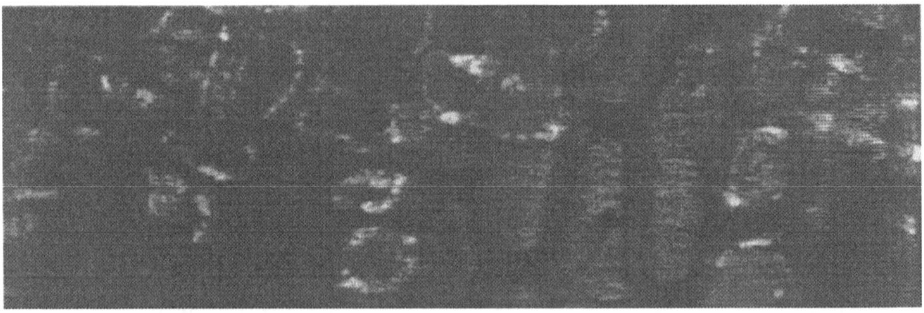

Figure 3. Cross sectional THG image of the plant leaf Herba Mollicoma Extramurana

An interesting opportunity presents itself in combining THG imaging with three-photon absorption (3PA) imaging. In this case the former is detected in the forward direction, whereas the latter can be detected in the backscattering direction. Such a set-up would provide the opportunity for relative localisation of fluorescence labeling dyes with respect to interfaces. Both THG and 3PA provide inherent three-dimensional imaging potential and can be done simultaneously. Since the sectioning in both techniques is determined by the excitation process, co-localising the THG and 3PA images is straight-forward.

References

1. Y. Barad, H. Eisenberg, M. Horowitz, Y. Silberberg, Appl. Phys. Lett. **70** 922-924 (1997).
2.T. Tsang, Phys. Rev. **A 52**, 4116-4125 (1995).

Ultrafast STM-tip Localized Responses from Nanostructured Surfaces

Yish-Hann Liau,[a] Mark Feldstein[b], Norbert F. Scherer[a,c]

a) Department of Chemistry and James Franck Institute, University of Chicago, 5735 S. Ellis Avenue, Chicago, IL 60637 Tel. (773) 702-7069, nscherer@rainbow.uchicago.edu
b) Naval Research Laboratory (Code 6910), 4555 Overlook Ave, SW, Washington, DC 20375 (202) 404-6129
c) Interdivisional Institute for Molecular and Cellular Biophysics, University of Chicago

Abstract: Spatially correlated reactivity and femtosecond time-scale responses from mesoscopic rough films are reported. The STM-tip spatially-localized transient SHG dynamics of nanometer scale metallic structures are used to establish localized surface plasmon dynamics.

The relationship between structure and functionality (or reactivity) is an underlying principle for the molecular biological and chemical sciences. The structure–function paradigm also applies to observables related to excitation of collective modes of materials as exemplified in the sensitivity of surface plasmon polaritons (SPP) to the structural and chemical nature of the interface.[1] The coupling of photons into the SPP mode of a material and the resultant large local electromagnetic fields are central to realizing surface enhanced spectroscopy (SES). Experimental support for the types of surface topographies that give rise to the largest predicted field enhancements came through measurements of SERS signals from structured surfaces.[2-4] Rough surfaces result in greater signal enhancement over smooth ones;[4] metal islands resulting from the formation of regular features (*i.e.*, hemisphereoids or spheroids) in quasi-periodic patterns yield greater enhancements still.[2] By contrast, theoretical studies of SES have established the local field enhancements for particular geometries on model surfaces.[3,5] These high local electromagnetic fields affect the local propensity for higher order processes like multi-photon ionization (MPI) and harmonic generation. MPI from metal films is, therefore, a representative measure of the electromagnetic contribution to SES processes since, under appropriate conditions, it is a direct result of SPP excitation.[4,6,7]

This paper reports on two types of measurements that directly correlate metal structural features with surface plasmon-induced responses. The first, correlated optical reactivity and scanning tunneling microscopy (CORSTM), measures SPP-induced reactivity initiated with femtosecond pulses and the near-field local plasmon dynamics on femtosecond time scales and nanometer length scales. A spatially localized probe (*i.e.*, STM) is used for simultaneous characterization of local surface topography and detection of the local field enhancement via SPP-mediated MPI from the metal surface. The second approach is the STM-tip localized excitation of nanometer Au features probing the SHG response as a function of tip position or interferometric time delay.

In both cases, the excitation geometry, with the pump and probe beams from the femtosecond pulse laser and STM interrogating the surface from opposite sides, facilitates excitation of SPP-modes. The optical source is a home-built cavity-dumped Ti:Sapphire laser that produces 25-30 fs pulses, centered at 780 nm, with energies up to 30 nJ. Focusing conditions yield peak intensities at the Au and Ag films of 30 MW/cm^2 and 1 GW/cm^2 for t he SHG and CORSTM experiments, respectively. The short duration pulses and high peak powers are critical for successful simultaneous measurement of the surface topography and reactivity.

The surface topography of the rough Ag films was determined by measuring the STM DC tunneling current. Simultaneous determination of the photo-induced signal component is achieved by lockin amplifier demodulation of the output of a second pre-amp located in the STM head. Topography photo-reactivity correlations[8] are found for three qualitatively different surface structures: from tens of nanometer length scale roughness, the edges of oblate ellipsoids, and from the gaps between prolate ellipsoids in close parallel proximity. These experimental results confirm the predictions of SPP-reduced electromagnetic field enhancements. Spatial resolution in these unique photo-induced surface reactivity measurements is achieved through the favorable interaction of several factors: the close proximity (~1 nm) of a sharp metallic tip to the surface, tip-surface field focusing, and the highly nonlinear nature of the process being studied.[6]

The temporal response associated with SPP-mediated MPI can be established using a pump-probe scheme; *i.e.*, detection of the local ionization response as a function of the temporal separation between two excitation pulses. The spatially localized pump–probe response shown in Figure 1 is dominated by the ultrafast pulse width-limited feature at the zero of time with a smaller amplitude component extending well beyond the pulse duration that decays to a baseline level of ~325 fA (that was subtracted in the figure). The ultrafast peak results from an "instantaneous" three photon ionization response. The longer decay component is exponential with time constants of 500-700fs. By comparison with an all-optical measurement that established that the thermalization of hot electrons created by SPP excitation in Ag occurs with a time constant of 670(70) fs,[9] the decays in Figure 1 are attributed to the relaxation of hot electrons by inelastic electron-phonon collisions. The asymmetry of the decay times is attributed to small asymmetries in the intensities and spatial overlap of the beams relative to the STM tip.

A second approach for examining spatial and time-resolved electronic dynamics of mesoscopically-structured films employed tip-localized SHG responses of individual mesoscopic scale metal particles. The surface feature selection scheme utilizes the local field enhancement that results from the plasmon resonance from the STM tip–material junction, where local field enhancements of 100-fold are possible.[10] We have already observed tip- enhanced SHG signals from structured Au and Ag films with ~100nm size particles. Tip-localized 2-pul interferometric measurements of localized plasmon dynamics have been obtained, as shown in Figure 2. Since the electronically interacting particles (and interactions with the Au or Ag tip) give rise to another plasmon resonance that is on-resonance with the fundamental wavelength, temporally broadened, fringe-resolved autocorrelation are obtained. Although position-dependent variations in the response are observed, spatial drift of the tip results in large signal changes that have prevented

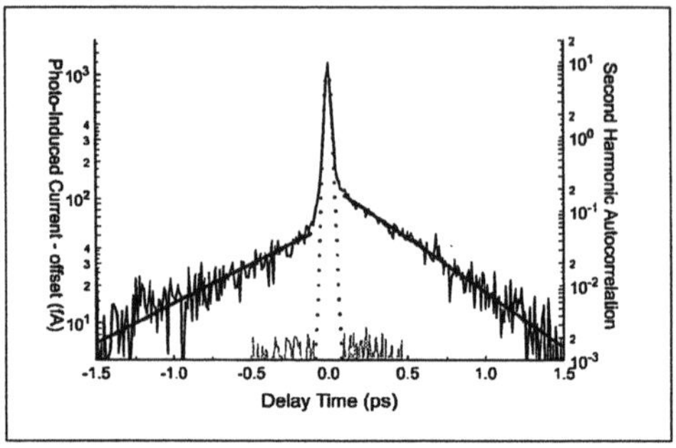

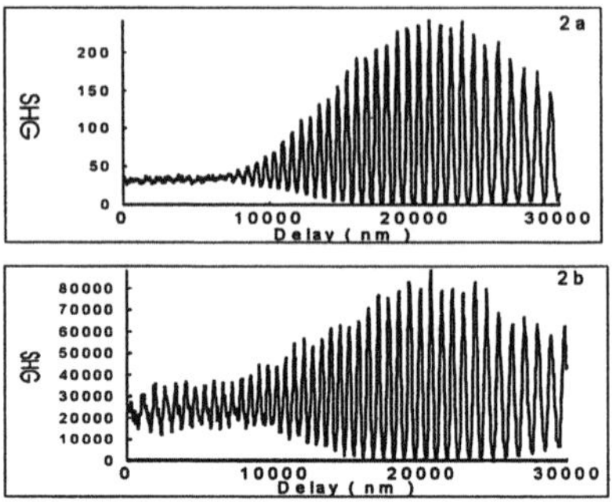

Figure 1. Semi-log plot of temporal SPP-MPI data from Ag films (solid lines) and SHG autocorrelation (dashed lines). Exponential component temporal decay constants of 680 and 490 fs. Signal was obtained with 50 KHz pulse repetition rate.

Figure 2a Interferometric SHG response from 100 μm KDP crystal.

Figure 2b Interferometric SHG from Ag particle film. Spatial variations of this signal are observed.

systematic analysis. Nevertheless, temporal broadening of the autocorrelation response suggests particle plasmon dephasing times of about 30 fs for Ag but much shorter for Au.

CORSTM measurements have been used to directly image the surface plasmon local electric field on rough Ag surfaces and establish localized plasmon and hot electron relaxation dynamics. The tip-localized SHG responses from Ag and Au nanoparticle films indicate particle plasmon resonance dephasing times of more than 30 fs for Ag nanoparticles.

Acknowledgement We acknowledge the NSF, Packard Foundation and the University of Chicago MRSEC for financial support.

References

(1) H. Raether, *Surface Plasmons*, Springer-Verlag, Berlin, (1988).
(2) R. P. Van Duyne, J. C. Hulteen, D. A. Treichel, *J. Chem. Phys.* **99**, 2101, (1993).
(3) M. Moskovits, *Rev. Mod. Phys.* **57**, 783, (1985) and references therein.
(4) C. Douketis, V.M. Shlaev, T.L. Haslett, Z. Wang, M. Moskovits, *J. Elec. Spec. Rel. Phen.* **64/65**, 167, (1993).
(5) J. Gersten, A. Nitzan, *J. Chem. Phys.* **73**, 3023, (1980); U. Laor, G.C. Schatz, *J. Phys. Chem.* **76**, 2888, (1982); H. Metiu, P. Das, *Ann. Rev. Phys. Chem.* **35**, 507, (1984) and references therein.
(6) M. J. Feldstein, P. Vöhringer, W. Wang, N. F. Scherer, *J. Phys. Chem.* **100**, 4739, (1996).
(7) T. Tang, T. Srinivasan-Rao, J. Fischer, *Phys. Rev. B.* **43**, 8870, (1991).
(8) M. J. Feldstein and N.F. Scherer, Phys. Rev. Lett., submitted, (1998).
(9) R.H.M. Groenveld, R. Sprik, A. Lagendijk, *Phys. Rev. Lett.* **67**, 784, (1990).
(10) L. Novotny, R. X. Bian and X. S. Xie, *Phys. Rev. Lett.* **78**, 1339 (1997).

Ultrafast Scanning Tunneling Microscopy Using a Photoexcited Low-Temperature-Grown GaAs Tip

Giovanni P Donati[1], George Rodriguez[2] and Antoinette J Taylor[3]

[1] Los Alamos National Laboratory, Los Alamos NM 87545, USA
E-mail: giove@lanl.gov, giove@frontiernet.net
[2] Los Alamos National Laboratory, Los Alamos NM 87545, USA
E-mail: rodrigeo@lanl.gov
[3] Los Alamos National Laboratory, Los Alamos NM 87545, USA
E-mail: ttaylor@lanl.gov

Abstract We report ultrafast scanning tunneling microscopy using a low-temperature-grown GaAs tip photoexcited by 100-fs, 800-nm pulses. We use this tip to detect picosecond transients on a coplanar stripline and demonstrate a temporal resolution of 1.7 ps. A dependence of the transient signal upon spatial position of the tip is revealed, indicating that the signal arises from areas on the sample smaller than ~20nm

1. Introduction

In the quest for atomic spatial and picosecond temporal resolutions, several groups [1-5] have integrated an STM tip with an ultrafast optoelectronic switch that gates the tunneling current from the tip. We report a novel ultrafast STM tip consisting of a cleaved GaAs substrate with a 1-μm thick epilayer of low-temperature grown GaAs (LT-GaAs) deposited on the face. Since LT-GaAs has a carrier lifetime of 1 ps, photo-excitation of the tip with an ultrafast above-bandgap pulse provides carriers for the tunneling current and photoconductively gates the current from the tip with picoseconds time resolution. We report a characterization of the tip under ultrafast photoexcitation and use this tip to detect picosecond voltage transients on a coplanar stripline.

2. Experimental Setup

A mode-locked Ti:sapphire laser provides 100-fs, 800-nm optical pulses at a repetition rate of 82 MHz. The output of the laser is split into a pump beam and a time delayed probe beam. The probe beam is focused on the LT-GaAs tip tunneling above a stripline. The pump beam is mechanically chopped and generates the voltage transients by optically switching the LT-GaAs epilayer between the stripline. The transient tunneling current is then revealed via lock-in detection for each value of delay τ.

The stripline consists of 50 μm wide, 7 mm long platinum lines sputtered 10 μm apart on a 1 μm LT-GaAs epilayer and is held at a voltage difference of 15 V. The tip consists of a LT-GaAs square of 0.04 mm^2 100 μm thick bonded to a tungsten wire using a back gold contact and conducting epoxy. The dark resistance of the tip is typically of 3 GΩ.

3. Results and Discussion

Figure 1a shows the IV characteristic for an LT-GaAs tip under different laser fluence at constant tip-sample separation. Since the tip energy bands are bent upwards the surface field separates the photo-generated carriers and a non zero current flows at zero bias due to electrons tunneling into surface photo-holes. When the ultrafast pulse passes under the tip the sample Fermi level shifts changing the number of electrons involved in the photo-enhanced tunneling.

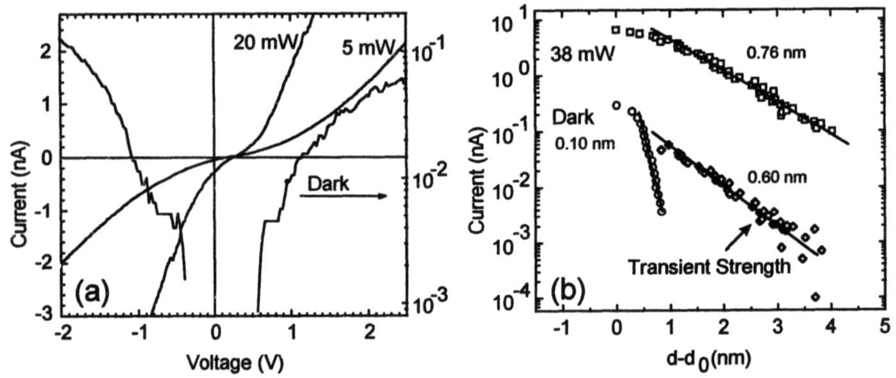

Fig. 1 DC-Current versus voltage (a) and versus tip/sample separation (b) for the LT-GaAs with 38 mW power on the tip at zero STM bias and in dark at 3V. The dependence of the transient strength on the separation is also shown.

Figure 1b shows the tunneling current as a function of tip-sample separation ($d-d_0$) with and without photoexcitation. For a clean tip and sample the expected work function would be ~5 eV. Although the difference in the apparent barrier with the tip in dark (0.84 eV) can be explained by native oxide contamination and the amorphous metal sample, the small value for the photoexcited tip (0.02 eV) is unclear but may be caused by photo-activation of surface states. The peak transient current also exhibits an exponential decay on separation with a decay constant close to that measured for the photoexcited DC tunneling current.

Fig 2a shows the transient current as a function of the delay. The first waveform is the correlation of the signal on the stripline. Deconvolving the 4.0 ps pulse yields a duration of 2.8 ps. The second pulse at 27 ps is a reflection off the end of the stripline. Unlike previous results [1-5], the waveforms from the tip in contact and in tunneling mode have the same shape and are free from other spurious signals. The width of the first pulse in tunneling regime is 3.3 ps indicating a temporal resolution of 1.7 ps after deconvolution.

Fig. 2b is a spatio-temporal scan of the transient current spanning 200 nm over the stripline and 45 ps in the delay. Features of 20 nm size due to differences in the sample conductivity are observed and indicate that the transient current arises from areas on the sample smaller than these dimensions.

The tunneling current can be described by a function I=I(x,t,s,V) where x is the position, t the delay, s the tip sample separation and V the voltage which includes the dependence of the voltage transient. Since the transient signal in Fig. 2b is the amplitude of the modulated tunnel current induced by the modulating voltage at the chopping frequency of the pump beam, it equals $\delta I = \partial I / \partial V \cdot \delta V(t)$ where $\delta V(t)$ is the transient voltage amplitude. Since the total derivative of the current along the delay is zero being the STM operated in constant current mode, also the conductivity $\partial I / \partial V$ is a constant of the delay. The scan in Fig. 2b can then be modeled knowing the voltage pulse waveform and the conductivity along the distance with the assumption that the voltage pulse does not depend upon the position. The fit obtained from this model is in agreement with the measured signal of Fig 2b.

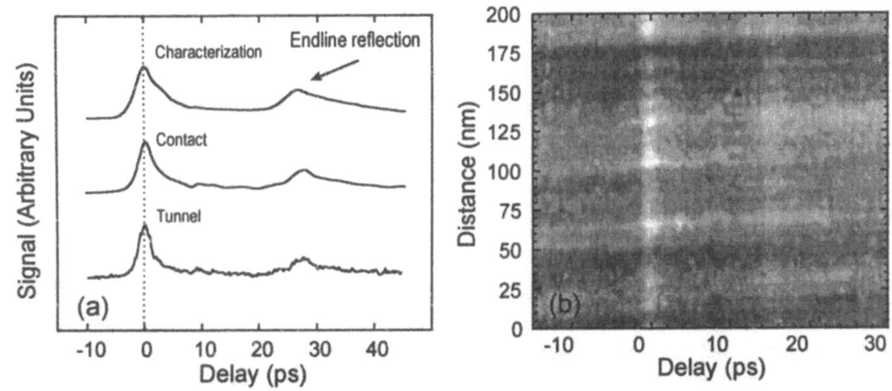

Fig. 2 (a) Transient signal as a function of the delay as detected from device characterization, with the tip in contact and with the tip in tunneling regime. (b) Spatio-temporal scan of the transient signal on the stripline acquired in constant current mode.

4. Conclusions

We have shown the performances of a novel LT-GaAs tip under photoexcitation and used it to detect picosecond electrical transients on a coplanar stripline. The time dependent tunneling current has shown a temporal resolution of 1.7 ps and is free from other spurious signal. A spatio-temporal scan of the time dependent signal was presented and has revealed that the transient tunneling current arises from areas on the sample smaller than 20 nm.

References

1 S.Weiss, D.F.Ogletree, D.Botkin, M.Salmeron, and D.S.Chemla, *Appl. Phys. Lett.* **63**, 2567 (1993).

2 R. H. M. Groenveld and H. van Kempen, *Appl. Phys. Lett.* **69**(15), 2294 (1996)

3 D. Botkin, J. Glass, D. S. Chemla, D. F. Ogletree, M. Salmerson and S. Weiss, *Appl. Phys. Lett.* **69**(9), 1321 (1996)

4 U. D. Keil, J. J. Jensen, and J. M. Hvam, *J. Appl. Phys.* **81**, 2929 (1997).

5 M. W. J. Prins, R. Jansen, R. H. M. Groeneveld, Ap. P. van Gelder, and H. van Kempen, *Phys. Rev. B* **53** (12), 8090 (1996).

Photoconductive sampling with a scanning tunneling microscope

O. Gielkens[1], L.M.F. Kaufmann[2], E. Smalbrugge[2], M.R. Melloch[3], R.H.M. Groeneveld[1], Th. Rasing[1], and H. van Kempen[1]

[1] Research Institute of Materials, University of Nijmegen, Toernooiveld 1, NL-6525 ED Nijmegen, The Netherlands
E-mail: olafg@sci.kun.nl
[2] COBRA Interuniversity Research Institute, Electronic Devices Group, Eindhoven University of Technology, PO Box 513, NL-5600 MB Eindhoven, the Netherlands
[3] School of Electrical Engineering, Purdue University, West Lafayette, Indiana 47907-1285, USA

Abstract. An experimental equivalent of the lumped circuit model of a photoconductively gated scanning tunneling microscope (PG-STM) is studied. This model can be used as long as the distance between the tunnel-junction and the pc-switch is small. For suitable parameters it implies a response-time below the carrier lifetime in the pc-switch.

1. Introduction

The scanning tunneling microscope (STM) is a widely used surface probe with atomic-scale resolution. Fundamental research on for example ultrafast dynamical processes in nanometer-scale objects and applied research on for example ultrafast submicrometer electronic devices, both demand much larger bandwidths than possible with a standard STM.

In recent years efforts have been made to integrate photoconductive sampling with an STM, resulting in a photoconductively gated STM (PG-STM). Picosecond and subpicosecond resolution have been reported [1,2,3], but the spatial resolution remains a point of controversy [4]. Both model and experiment suggest that the main contribution to the measured signal is coupled through the geometrical capacitance of the tip-sample junction [3,4].

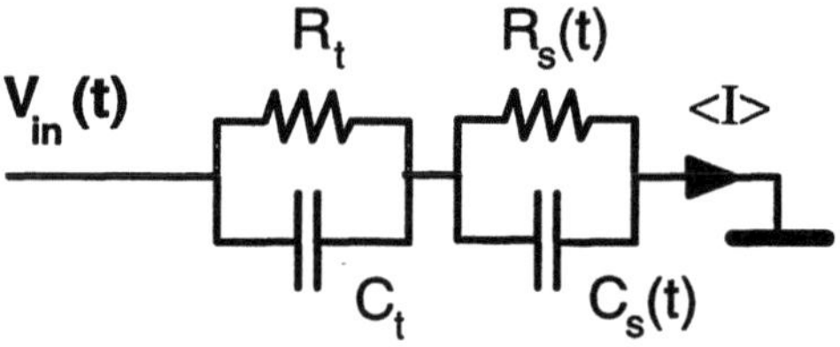

Figure 1. Lumped circuit model for a PG-STM. The tunnel junction (Rt, Ct) and photoconductive switch (Rs(t), Cs(t)) are represented by their resistance and capacitance.

This model (see fig.1) also predicts that a suitable choice of resistances and capacitances will result in a time resolution better than the carrier lifetime in the PC-switch. Comparing the theoretical model with experiments requires however unrealistic small values for the tunnel resistance.

2. Experiments

To overcome this problem an experimental equivalent of the PG-STM is studied, which allows for the use of small tunnel resistances. The tunnel resistance can be represented by a pc-switch illuminated by a cw-laser. Tuning the power of the laser tunes the resistance of this pc-gap.

Fast electrical transients on a coplanar stripline are measured with this geometry and with standard pc-sampling. Figure 2 shows how these two sampling methods are integrated into one sample together with the cps.

The pulsed laser illuminates the sample with 100 fs pulses at a repetition rate of 76 MHz and a wavelength of 765 nm. The generating laser beam (1) is chopped at 1.2 kHz and the average current is measured with a lock-in amplifier as a function of the delay-time between the pump and probe pulse.

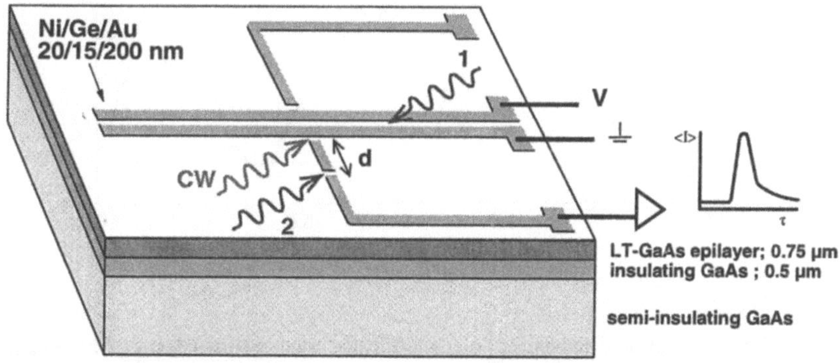

Figure 2. Schematic representation of the sample. Laser beam 1 generates the signal on the cps, while laser 2 switches the pc-switch. The cw laser illuminates the pc-gap representing the tunnel-junction.

3. Results

The influence of the distance between the two gaps, and the tunnel-resistance are studied. A typical result for low tunnel-resistance is shown is figure 3. For gap separations comparable to the wavelength of the signals on the cps, reflections between the two gaps can clearly be resolved. Here the model fails and an extra impedance should be introduced between the two gaps.

For small gap separation (50 μm) the detected signal has a shorter rise-time and smaller FWHM than the conventional sampling method. Due to the short distance, the individual reflections are not observable any more and cause a tail. The tail is however smaller than the tail of the conventional sampling signal.

For larger tunnel resistances, the signal-amplitude decreases rapidly and the reflections are hardly detectable any more. An increase in resolution, as predicted, has however not yet been observed.

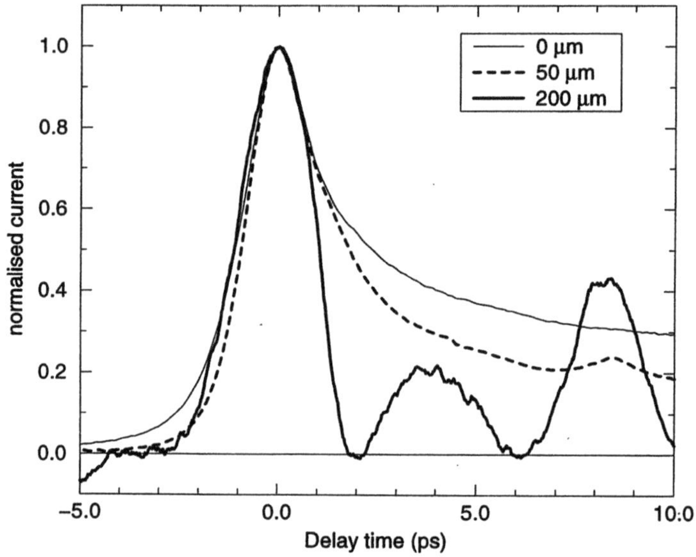

Figure 3. Transient signal measured on a CPS for two geometries of the PG-STM equivalent and normal PC sampling (thin solid line). The geometries used here have distances of 50 μm and 200 μm respectively (thick dashed and solid line)between the tunnel junction and the PC switch. The 'tunnel resistance' Rt = 33 kΩ.

References

1 D. Botkin, J. Glass, D.S. Chemla, D.F. Ogletree, M. Salmeron, and S. Weiss, Appl. Phys. Lett. **69**, 1321 (1996)

2 R.H.M. Groeneveld, Th. Rasing, L.M.F. Kaufmann, E. Smalbrugge, J.H. Wolter, M.R. Melloch, and H. van Kempen, Physica B **218**, 294 (1996) and Journ. Vac. Sci. & Technol. B. **14**, 1321 (1996).

3 U.D. Keil, J.R. Jensen, and J.M. Hvam, Appl. Phys. Lett. **70**, 2625 (1997)

4 R.H.M. Groeneveld and H. van Kempen, Appl. Phys. Lett. **69**, 2294 (1996)

Femtosecond Nonlinear Microscopy of Photodetectors

P. Dorn, J. Jasapara, J. Zeller, W. Rudolph, and M. Sheik-Bahae

University of New Mexico, Department of Physics and Astronomy, 800 Yale Blvd NE, Albuquerque, NM 87131, USA
email: wrudolph@unm.edu

Abstract. Diffraction and bandwidth limited pulses from 10 to 100 femtoseconds in the focal plane of microscope objectives are used to perform multiphoton current and luminescence microscopy of nonlinear photodetectors. Information on material parameters and the electric field distribution is obtained.

1. Introduction

High resolution nonlinear microscopy requires sources with high intensities at low mean power in addition to diffraction limited focusing. This can be achieved using ultrashort pulses and optimum focusing geometries. Nonlinear microscopy based on femtosecond pulse illumination has found attractive applications [1] resulting from its increased three-dimensional imaging capabilities [2]. We report on nonlinear microscopic imaging of ZnSe detectors under multi-photon excitation using a Ti:sapphire laser with pulse durations as short as 10 fs in the focal plane of a high-resolution microscope objective. These detectors have recently received attention for their application in femtosecond laser pulse diagnostics [3].

Laser pulses shorter than 10 fs duration are now routinely produced from Ti:sapphire laser oscillators with prism or mirror dispersion control [4,5]. After careful pre-chirping and beam-shaping using a combination of a prism sequence and a chirped mirror (third-order dispersion [5]), we obtained 11.2 fs pulses at nearly diffraction limited spot sizes of less than 0.53 μm in the focal plane of a common microscope objective. Figure 1. (left) shows the measured FWHM of the line spread function versus the numerical aperture (NA) while Figure 1. (right) depicts an interferometric autocorrelation at NA=0.85 based on two-photon fluorescence in a Rh6G dye. The measured spot sizes are close to the diffraction limit. We also notice that the pulse duration increases with NA. This behavior is expected due to the residual chromatic aberration of the objective and the microscope cover slide (0.15 mm), both introducing a radius dependent group delay. It is worth mentioning that with the above focusing, a laser oscillator with an average power of 1 W can produce intensities that exceed 10^{14} W/cm^2. This opens attractive possibilities for nonlinear optics, nonlinear imaging and material structuring using a single oscillator without additional pulse amplification.

A laser scanning microscope illuminated by a fs Ti:sapphire laser was used to image a biased ZnSe photodetector [3]. Figure 2. (left) shows the confocal reflection image, the two-photon luminescence image, and the current image. An area where the metal fingers are detached from the ZnSe substrate can clearly be identified in the two-photon current images, an information that is not contained in the confocal reflection image. Figure 2. (right) shows the confocal and two-photon current signal when the longitudinal position of the focus is scanned while keeping its lateral position fixed.

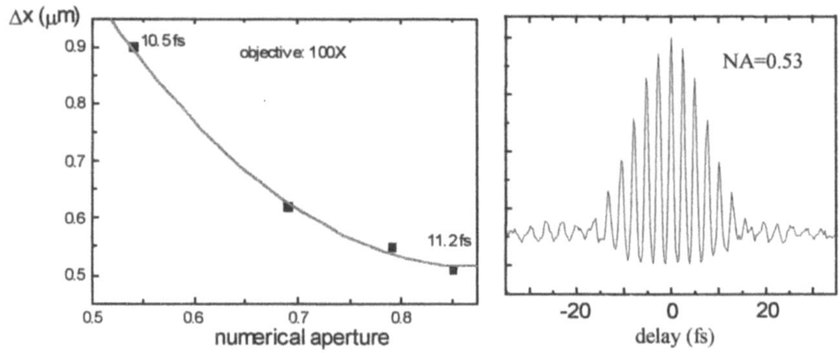

Fig.1. left: Width of the line spread function obtained after focusing 10 fs pulses using a 100x microscope objective as a function of the numerical aperture. The actual pulse duration in the focal plane is indicated for two apertures. Right: Interferometric autocorrelation (two-photon fluorescence in Rh6G) in the focal plane for NA=0.85.

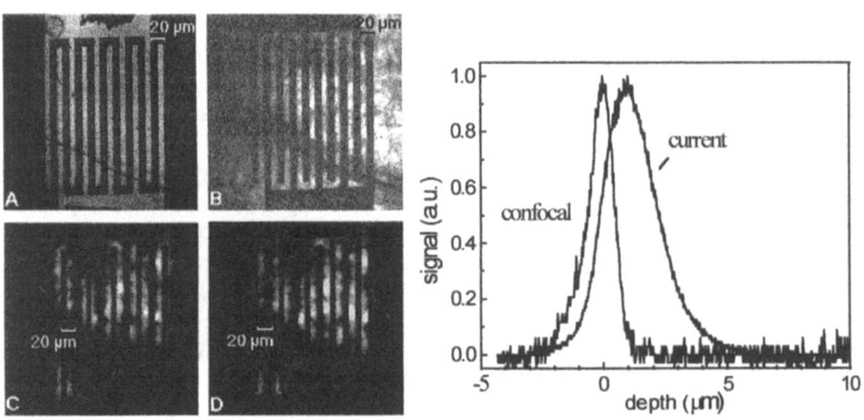

Fig.2. left: Confocal reflection signal (A), multi-photon luminescence image (B), two-photon current signal with the positive electrode on the bottom (C) and the positive electrode on top (D) of the ZnSe structure. Bright regions indicate a high signal. right: depth scan of the two-photon current and the confocal reflection signal..

The two-photon current signal between any pair of electrodes is nonuniform and has a maximum near the anode. Since the current is dominated by electrons, due to their higher mobility, the recombination time of the electron must be faster than the drift time. From the measurements, the latter can be estimated to be about 1 ns. Figure 2. (right) reveals that the active (fringe-field) layer is only 3 μm thick indicating that dispersion does not affect autocorrelation measurements of pulses as short as 5 fs. From the confocal reflection signal, the depth resolution of this measurement can be estimated to be ≈1.5 μm. To first order, the current distribution describes the electric field distribution in the ZnSe making the nonlinear microscope a sensitive electric potential detector with micron spatial and femtosecond temporal resolution. The ZnSe sample was polycrsystalline, a fact that is not obvious from the linear reflection image but is clearly visible in the two-photon current image. For comparison the corresponding images of a single-crystalline sample of a linear array of 32 detector pixels are shown in Figure 3 (left). The uniform two-photon current response suggests the application of such an array for femtosecond pulse autocorrelation measurements in a single shot or in compact correlators without moving components. An example of such a correlation is shown in Figure 3 (right).

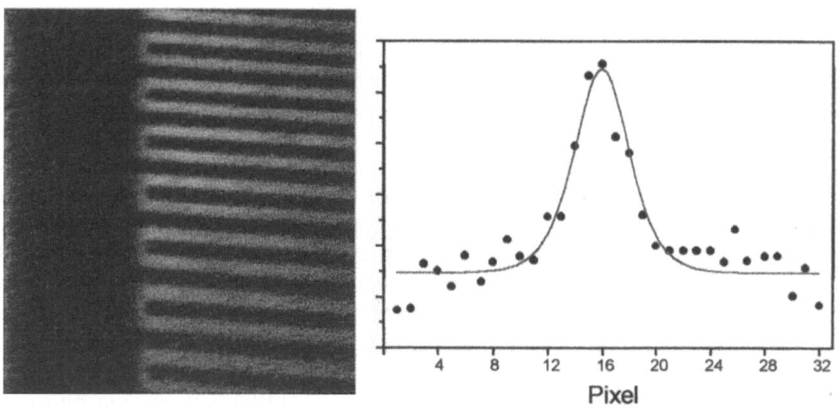

Fig.3. left: Two-photon current image of a detector array based on monocrystalline ZnSe. right: Autocorrelation of a 70 fs pulse taken with the array detector in a correlator without moving parts.

Acknowledgements. This work was supported by NSF (#9625532, #9601890) and the W. M. Keck Foundation.

References

1. W. Denk, J. H. Strickler, W. Webb, Science **248** (1990)73
2. J. Deitche, M. Kempe, W. Rudolph, J. of Microscopy **174** (1994)69
3. W. Rudolph, M. Sheik-Bahae, A. Bernstein, L. Lester, Opt. Lett. **22** (1997)313
4. J. Zhou, G. Taft, C.-P. Huang, M. M. Murnane, H. C. Kapteyn, I. P. Christov, Opt. Lett. **19** (1994)1149
5. A. Stingl, M. Lenzner, Ch. Spielmann, F. Krausz, R. Szipocz, Opt. Lett. **21**(1996)1259

Time-Resolved Near-Field Spectroscopy of the Carrier Dynamics in single GaAs Nanostructures

Ch. Lienau[1], A. Richter[1], T. Elsaesser[1], M. Ramsteiner[2], R. Nötzel[2], and K. H. Ploog[2]

[1] Max-Born-Institut für Nichtlineare Optik und Kurzzeitspektroskopie, Rudower Chaussee 6, D-12489 Berlin, Germany,
E-mail: lienau@mbi-berlin.de
[2] Paul-Drude-Institut für Festkörperelektronik, Hausvoigteiplatz 5-7, D-10117 Berlin, Germany

Abstract. We report the first time-resolved near-field luminescence study of excitonic real-space transfer into *single* GaAs quantum wires. Excitons generated by local optical excitation in a 250 nm spot undergo diffusive transport over lengths of microns and are subsequently trapped into the quantum wire by optical phonon emission.

Direct studies of the dynamics of photogenerated electron-hole pairs in single semiconductor nanostructures require both ultrahigh spatial and temporal resolution. Near-field scanning optical microscopy (NSOM), offering subwavelength spatial resolution in the 100 nm range, in combination with time-resolved detection schemes is a particularly promising technique for such experiments. Here, we report on the first time-resolved near-field spectroscopic investigation of the carrier dynamics in single GaAs/(AlGa)As sidewall quantum wires.

Quantum wires (QWR) with a lateral width of 50 nm and a thickness of up to 13 nm embedded in a 6 nm GaAs quantum well (QW) were grown by molecular beam epitaxy of GaAs/(AlGa)As multilayer structures on patterned GaAs (311)A substrates along the sidewall of 15-20 nm high mesa stripes oriented along the [01-1] direction [1]. The quasi-one-dimensional (1D) confinement potential of the QWR is directly extracted from near-field photoluminescence excitation spectra at cryogenic temperatures [2] (Fig. 1). These experiments reveal shallow asymmetric energetic barriers with a height of about 18 meV on the mesa top and 14 meV on the mesa bottom in the vicinity of the QWR, separating QWR and embedding QW.

The lateral bandgap variations on a subwavelength scale have a pronounced influence on the real-space transfer within the 2D continuum and thus on the carrier trapping dynamics into the 1D QWR states as is directly revealed using time-resolved near-field spectroscopy. In these experiments, electron-hole pairs are created by transmitting a fs-Ti:Sapphire laser pulse at 1.614 eV through an NSOM fiber probe with an aperture diameter of about 200 nm. The laser spectrum overlaps with the excitonic absorption band of the GaAs QW. QWR luminescence at 1.54 eV is detected in the far-field as a function of tip position using time-correlated single photon counting detection.

In Fig. 2, the intensity of the time-resolved QWR emission at a sample temperature of 100 K is shown as a function of tip position and delay time as the tip is

Fig. 1. Lateral confinement potential and subband structure of the QWR as obtained from near-field spectroscopy. The QWR is surrounded by asymmetric energetic barriers. A schematic of the QW embedded QWR is shown in the inset.

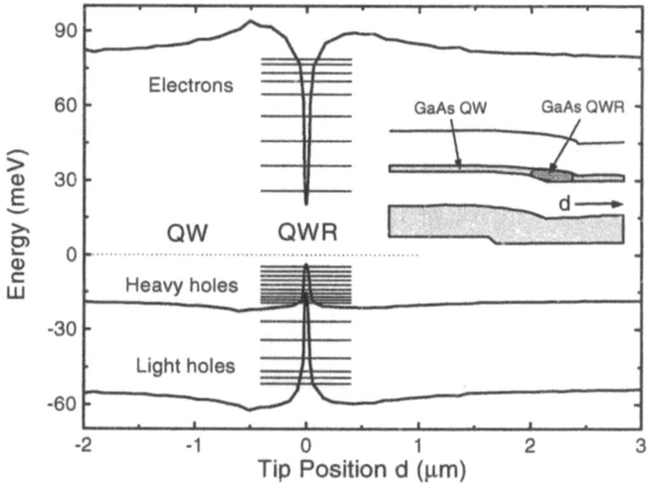

scanned perpendicular to the QWR, located at d = 0. For localized excitation in a narrow region around the QWR, i.e. between the two energetic barriers separated by about 700 nm, carrier capture into the QWR and relaxation to the bottom of the QWR potential by optical phonon emission is fast, leading to an onset of QWR luminescence within the time resolution of the present experiment. The temporal dependence of the QWR emission is dominated by the QWR recombination lifetime of 1.5 ns and in this region independent on excitation position.

For excitation outside the barriers, however, the onset of QWR luminescence is delayed by several hundred ps. This delay increases with increasing distance bet-

Fig. 2. Time-resolved QWR luminescence for spatially resolved excitation of the absorption band of the GaAs QW as a function of tip position at a sample temperature of 100 K.

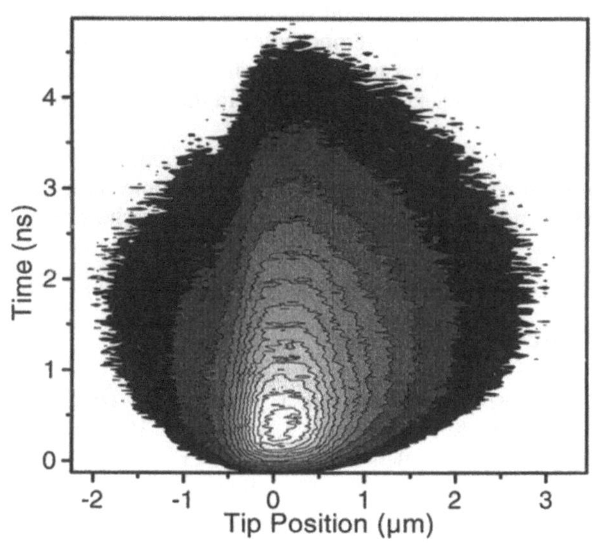

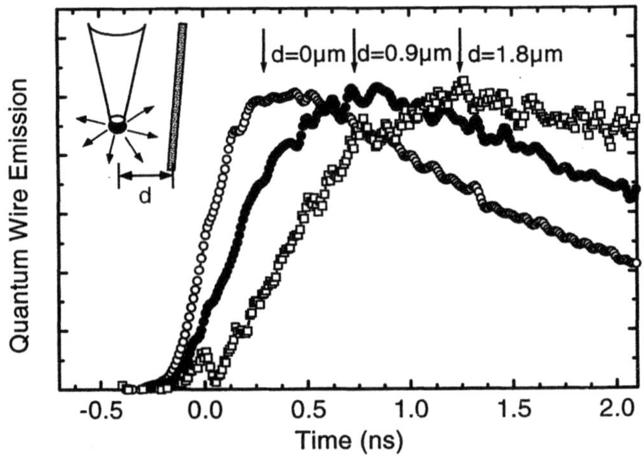

Fig. 3. Cross sections through Fig. 2 at fixed excitation positions d with respect to the QWR at d = 0.

ween excitation tip position and QWR location (Fig. 3). The delayed occurrence of QWR emission reflects the real-space transfer of electron-hole pairs within the embedding QW and carrier trapping into the QWR, i.e. the finite diffusion time of excitons. The carrier dynamics is quantitatively described within a model that accounts for (a) the diffusive carrier transport induced by the spatial gradient of the exciton concentration and (b) a drift term induced by the action of the local band gap gradient on the excitonic center of mass motion. The fraction of the carriers that is trapped into the QWR depends strongly on the local bandgap gradient and thus is markedly different on both sides of the QWR, as is evidenced by the asymmetric luminescence intensity and difference in dynamics on both sides of the QWR. An analysis of the data within this model allows us to extract the local exciton mobility of 2.000 cm^2/Vs at 100 K.

The influence of the lateral band-gap gradient on the local carrier transport increases with decreasing temperature as is revealed from time-resolved experiments at various temperatures between 10 and 100 K. At the lowest temperatures, the trapping of carriers generated within the QW is suppressed due to the presence of the barriers.

A complete description of the carrier trapping dynamics for excitation inside the barriers requires subwavelength spatial and subpicosecond temporal resolution. Here, near-field pump-probe experiments shall give important new information.

References

1 R. Nötzel, M. Ramsteiner, J. Menniger, A. Trampert, H.-P. Schönherr, L. Däweritz, and K. H. Ploog, Jpn. J. Appl. Phys. 35, L297 (1996).

2 A. Richter, G. Behme, M. Süptitz, Ch. Lienau, T. Elsaesser, M. Ramsteiner, R. Nötzel, and K. H. Ploog, Phys. Rev. Lett. **79**, 2145 (1997).

Time resolved near field imaging and diffraction with sub-wavelength far-infrared dipole sources

I. Brener[1], S. Hunsche[2], Y. Cai[1], M.C. Nuss[2], J. Wynn[1], J. Lopata[1], and L. Pfeiffer[1]

[1] Bell Laboratories, Lucent Technologies, 700 Mountain Ave, Murray Hill, NJ 07974, USA
E-mail: igal@bell-labs.com
[2] Bell Laboratories, Lucent Technologies, 101 Crawfords Corner Rd, Holmdel, NJ 07733, USA

Abstract We image objects in the far-infrared with a spatial resolution of ~λ/10 by scanning samples within the near-field of optically excited terahertz dipole sources. We also show time resolved data that illustrate the near field dipole-sample interaction.

Near field imaging with sub-wavelength resolution is typically impaired by poor coupling efficiency and a complicated sample-source interaction. Advances in these and related aspects are improving our understanding of the capabilities and limitations of near field optical imaging. In this work, we present a detailed study of both spatially and temporally resolved near field imaging using far infrared dipole sources.

We solve the coupling efficiency problem by scanning the objects to be imaged directly in the near field region of the electromagnetic emission, as it was recently suggested in the microwave regime [1]. Our far-infrared dipole sources are shown in Fig. 1 and consist of subpicosecond photoconducting switches that radiate terahertz (THz) radiation after being excited by femtosecond laser pulses. Structures a) and b) were fabricated on an optically opaque substrate of low temperature GaAs. In order to bring the emitting dipole close to a sample, the substrate was thinned to 100 μm. The emitted radiation on the back surface of the structure is localized to a diameter of 60 μm by total internal reflection in structure (a). This area was further reduced by a 25x25 μm aperture in structure (b). The emitter (c) is fabricated on ion-implanted silicon on sapphire and the gating laser beam is focused through the substrate. The emitter design is optimized according to Ref. [2]. The THz radiation is collected and detected in the standard way. [3]

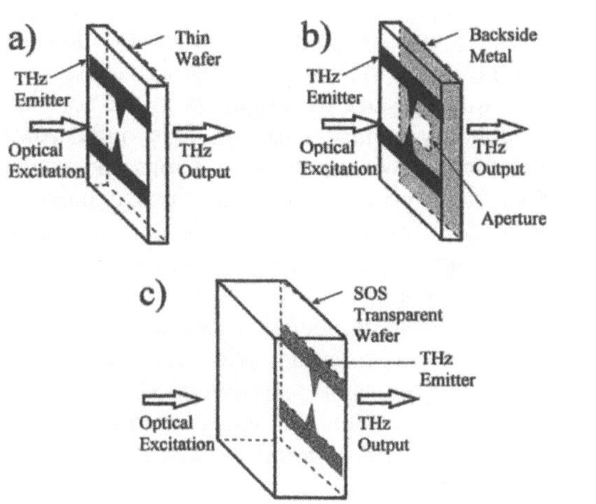

Fig. 1: Different terahertz near-field emitters used in this work.

When we scan a sample almost in contact to the THz emitters (and thus in the near field region of their emission), we invariably attain a resolution of 50-60 μm when the average wavelength is ~600 μm. This corresponds to a relative resolution of ~λ/10 (but is significantly larger than the actual size of the emitter area: ~5 μm).

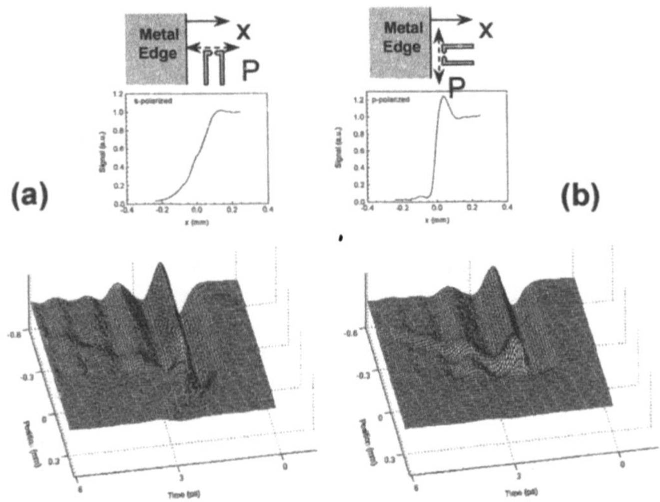

Fig. 2: Time-resolved THz near-field diffraction from a scanned metal edge. a) dipole perpendicular to the edge; b) dipole parallel to the edge. Insets: frequency integrated scan.

Time-resolved THz waveforms as a function of the sample position are shown in Fig. 2, using the scheme (c) in Fig. 1, and for dipole orientation perpendicular (a) and parallel (b) to the metal edge (the geometry and the frequency integrated scan are shown in the insets). Superimposed on the "direct" THz signal that peaks at a fixed time, a "diffracted" contribution appears on diagonals in the t-x plane, i.e., the sample structure acts as a secondary radiation source. This signal contribution appears with a different phase for the two dipole orientations and extends with significant amplitude to positive blade positions in 2(a). Consequently, the resolution is about a factor 4 worse than in 2(b). It is interesting to note that we operate in a particular regime where the time resolution is comparable to the light period and the spatial resolution is subwavelength.

The phase or sign reversal of the induced signal can be understood by considering a dipole close to a metal surface. For perpendicular orientation to the metal surface, it will induce an image dipole with the same orientation, and for parallel orientation one with opposite orientation. Therefore, secondary radiation that appears to arise from the image dipole will have either equal or opposite phase with respect to the field radiated by the initial dipole.

However, we note that the polarization dependence of the spatial resolution is also observed with a nonmetallic sample, as shown in Fig. 3. The sample consists of a GaAs wafer with three rectangular grooves etched to a depth of 90 μm. The ridges between the grooves are 140 μm wide. The solid line shows the integrated THz signal for a scan with the dipole parallel to the sample structures, the dashed line is the signal with both sample and scan direction rotated by 90 degrees.

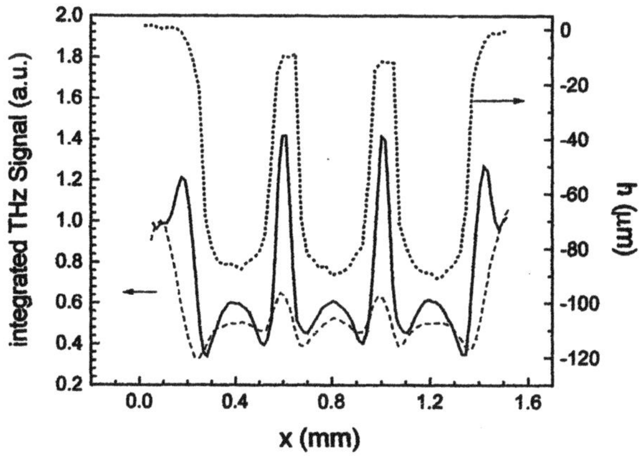

Fig. 3: Near-field scan over etched grooves in a GaAs surface. Solid line: dipole parallel to grooves, dashed: dipole perpendicular to grooves; dotted: topographic data from THz pulse delay.

Clearly, the resolution of the outer edges and the ridges is significantly better in the first case. However, in neither case does the shape of the signal directly reflect the structure of the sample, but shows additional features. A straightforward image interpretation is possible for this type of sample by measuring the temporal delay of the main THz signal, which directly maps the surface topology of the sample. The dashed curve shows the depth of the etched grooves as measured by the pulse delay and using the index difference between GaAs and air. This simple analysis gives accurate values at the center of the grooves without any calibration.

In summary, we have demonstrated a resolution of better than $\lambda/10$ in a novel and highly efficient far-infrared imaging system. We have also shown first "snapshots" of dipole-sample interaction in the near field directly observed in the time-domain. We conclude that polarization-dependent diffraction and edge enhancements significantly influence near-field imaging and that the achievable resolution, even with an almost ideal sub-wavelength dipole source, depends mainly on the sample-source interaction rather than the actual size of the source.

References

1 B. Knoll, F. Keilmann, A. Kramer, and R. Guckenberger, Appl. Phys. Lett. 70, 2667 (1997); R. D. Grober, R. J. Schoelkopf, and D. E. Prober, Appl. Phys. Lett. **70**, 1354 (1997).

2 I. Brener, D. Dykaar, A. Frommer, L.N. Pfeiffer, J. Lopata, J. Wynn, and K. West, Opt. Lett. **21**, 1924 (1996).

3 D. M. Mittleman, R. H. Jacobsen, and M. C. Nuss, IEEE J. Selected Topics in Quant. Electr. **2**, 679 (1996)

Part V

High Speed Optoelectronics and Thz Radiation

Ultrafast All-Optical Signal Processing at Over 100 Gbit/s with Mode-Locked Semiconductor Diode Lasers

H. Yokoyama[1, 2]

[1] Optoelectronics and High-Frequency Device Research Laboratories, NEC Corporation 34 Miyukigaoka, Tsukuba 305-8501, Japan
[2] Institute of Physical Science, University of Tsukuba Ten-nodai, Tsukuba 305-0006, Japan
E-mail: yoko@obl.cl.nec.co.jp

Abstract. The novel all-optical signal processing functions of mode-locked laser diodes in addition to ultrashort optical pulse generation are demonstrated for applications in ultrahigh-speed optical communication technologies.

1. Introduction

Development of the optical time division multiplexing (OTDM) system at over 100 Gbit/s is one of the most challenging subjects in optical communication technologies. Here, the key is the realization of novel kinds of optical devices that make possible some simple but ultrafast signal processing functions, including the generation of stable ultrashort light pulses, the all-optical demultiplexing (DEMUX) and the all-optical 3R (reshaping, retiming, regenerating), beyond the speed limit of the present electronics. It should be noted that these all-optical signal processing functions necessary for ultrafast OTDM systems are also beneficial in present long-haul optical communication systems incorporating wavelength division multiplexing (WDM) technologies. If simple and reliable all-optical 3R technologies are available, these may improve the limit of transmission schemes with Er-doped fiber amplifiers (EDFAs) that have only a regenerating function. In this article, we describe the novel potential of mode-locked laser diodes (MLLDs) for ultrafast all-optical signal processing applications.

2. Coherent Optical Pulse Generation

A remarkable feature of the MLLD is recognized, in general, that the direct generation of optical pulse trains at ultrahigh repetition rates is possible. Actually, tera-hertz rate sub-picosecond optical pulse generation experiments were reported [1, 2].

In the optical communication use, however, the requested most fundamental function of MLLDs is to generate transform-limited picosecond optical pulses with low timing jitter at a precisely controlled wavelength and repetition frequency. Although many efforts have been made till present towards the development of monolithically integrated MLLDs, it is still difficult to satisfy all the required operation characteristics. Instead, excellent quality optical pulses were generated with rigid external-cavity module configurations [3-5]. The MLLD device-chip used in our MLLD module had a simple bi-section electrode structure involving a saturable absorber (SA) section which is controlled by electric reverse-bias, and the chip was antireflection (AR) coated at the gain-section facet. Figure 1 shows the appearance of the module (PhotoOscillator). An SHG correlation trace and an RF spectrum in Fig. 2 indicates the transform-limited < 2-ps optical pulse generation with a very low timing jitter (0.5 ps) under a hybrid mode-locking condition. The mode-locking frequency tunability between 9.5 - 10.5 GHz, and the wavelength tunability within the bandwidth of a conventional EDFA (1.53 - 1.57 μm) were obtained with this MLLD module [5]. Presently, many sets of this MLLD modules are being used for ultrafast optical communication research purposes, in place of mode-locked Er-doped fiber lasers.

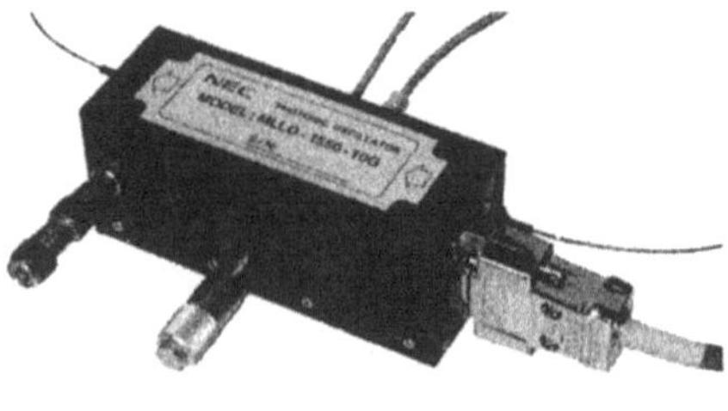

Fig. 1. An external-cavity MLLD module (PhotoOscillator) that can produce wavelength-tunable stable picosecond coherent optical pulses. The optical pulse repetition frequency is variable between 9.5 - 10.5 GHz, and the wavelength is tunable between 1.53 - 1.57 μm.

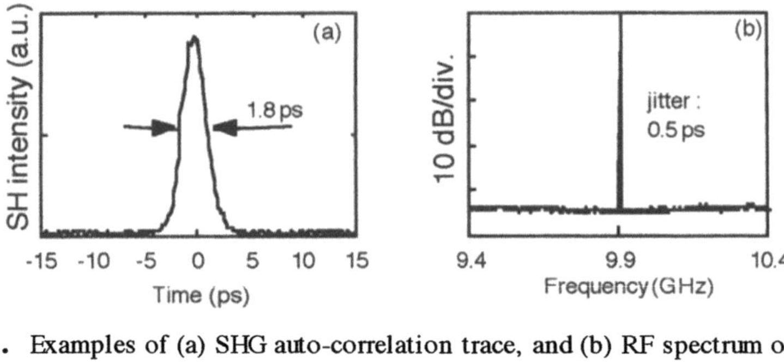

Fig. 2. Examples of (a) SHG auto-correlation trace, and (b) RF spectrum of optical pulses generated from a hybrid mode-locked external-cavity MLLD. The time and bandwidth product is 0.4, and the timing jitter is estimated to be < 0.5 ps.

3. All-Optical Clock Recovery

We also drew out new potentials of MLLDs which enable ultrafast all-optical signal processing. Here, "signal processing" means some simple functions providing interfaces between ultrafast optical signals and electronic signals of practical speed. Seeing all-optical 3R and DEMUX, both functions can be realized with combining the optical clock recovery and the optical gate switch, as shown in Fig. 3 (100- Gbit/s data rate case is depicted in Fig. 3, as an example). However, optical clock pulses should be extracted for 3R with a frequency same to the input data rate, while a subharmonic-frequency optical clock pulses are necessary for the DEMUX.

We demonstrated that these two different types of clock-extraction functions were possible with optical data injection into an MLLD at bit rates of over 100-Gbit/s with very low excess timing jitter (< 0.1 ps). The present all-optical clock extraction is based on the timing synchronization of MLLD output to the injected optical data pulse [6]. In this scheme, the extracted clock frequency was determined by the mode-locking frequency of the MLLD.

Although initial clock extraction experiments are demonstrated with a monolithic bi-section MLLD [5], we recently fabricated a specially designed external cavity MLLD also intending the precise control of the mode-locking frequency, the operation wavelength and the spectral width[7].

4. All-Optical Gate Switching

Furthermore, we obtained an optical gating function with an AR-coated MLLD structure. This was due to the absorption saturation and the following picosecond carrier sweepout from the SA section in a quantum-well MLLD [5, 8].

Through a pump and probe measurement, it was found that a 100-μm long SA section shows an absorption saturation energy of 1-2 pJ and an absorption recovery time of 7-8 ps. It should be noted that a very fast response of the optical gating device is important for all-optical 3R processing in which instantaneous material response to the optical input is required. This feature can cover the application for DEMUX operation which is also possible with a combination of a sub-nanosecond response material (generally LD amplifier) and an proper interferometer configuration [4, 9].

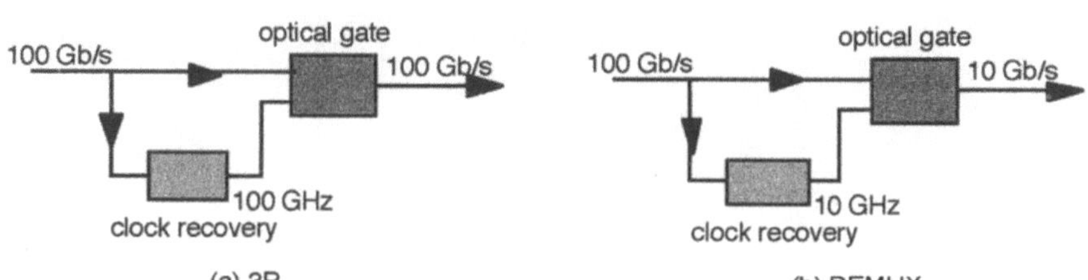

Fig. 3. Diagrammatic representations of all-optical 3R repeater (a) and demultiplexer. Both the functions are realized with a similar combination of clock recovery and optical gate switch, but with different frequencies for extracted optical clock pulses.

5. Error-Free All-Optical DEMUX Operation

By combining all these MLLD functions, 40-Gbit/s to 10-Gbit/s all-optical demultiplexing was realized at a receiver site [2] (the experimental configuration is similar to the one shown in Fig. 4). Recently, we have confirmed an error-free OTDM soliton transmission combined with this DEMUX configuration [10].

For over 100-Gbit/s operation, however, we utilized a nondegenerated four wave mixing (FWM) process in an LD device with amplifier-mode for the optical gate function. This is because the absorption recovery time of 8-ps in the present SA is still not fast enough to suppress optical data pulse neighboring the gated ones. Employing FWM with the configuration shown in Fig. 4, an error-free 200-Gbit/s to 10-Gbit/s DEMUX operation was also obtained. The bit error rate measurement data in Fig. 5 indicate this good DEMUX performance. Here, it should be noted that the SA effect was also incorporated to reduce spontaneous-emission-noise components and to improve the S/N [11, 12]. This indicates an another potential application of the waveguide SA for high-speed optical transmission systems.

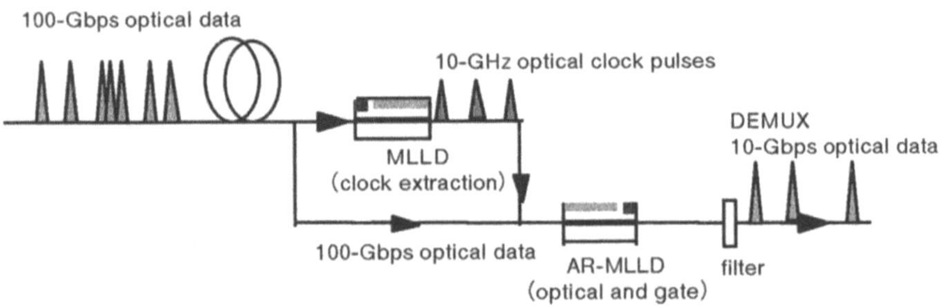

Fig. 4. Experimental setup for all-optical DEMUX using MLLDs as an optical pulse source, an optical clock extraction device, and an optical gate.

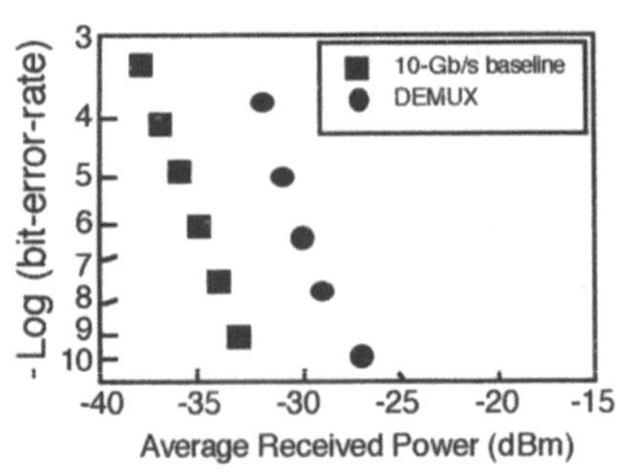

Fig. 5. Bit error measurement data showing error-free 200- to 10-Gbit/s DEMUX.

6. Conclusions

We described in this article the novel potential of MLLDs for all-optical signal processing functions that are required in ultrahigh-speed OTDM communication. It should be noted that the remarkable features demonstrated with simple and compact configurations are mainly due to strong nonlinear-optic properties of the saturable absorber which dominates the mode-locking. However, many other phenomena should be studied to find fast and large nonlinear effects useful for a practical optical gate switching device having a real response time of a picosecond or less. On the other hand, from the device technology point of view, we must realize fully integrated MLLDs that can generate communication grade coherent optical pulses.

We expect that further efforts to develop simple and reliable all-optical signal processing technologies will also make it possible to obtain the 3R-function, and will be beneficial in many respects for ultrahigh-capacity optical communication systems.

Acknowledgments. The author is grateful to H. Kurita, T. Shimizu, I. Ogura, and Y. Hashimoto for their experimental collaboration. We also wish to thank M. Yamaguchi, H. Yamada, T. Torigai, T. Sasaki, Y. Yano, T. Ono, K. Yoneyama, K. Kasahara, and K. Kitayama for their support in fabricating laser diodes and helpful discussions on ultrahigh-capacity optical communication systems.

References

1 S. Arahira, Y. Matsui, T. Kunii, S. Oshiba, and Y. Ogawa, IEEE Photon. Tech. Lett. **5**, 1362 (1993).
2 T. Shimizu, I. Ogura, and H. Yokoyama, Electron. Lett. **33**, 1868 (1997).
3 H. G. Weber, R. Ludwig, S. Diez, W. Pieper, H. J. Ehrke, C. M. Weinert, International Topical Workshop on Contemporary Photonic Technologies (CPT'98), Tokyo, Jan. 1998, Mb-01.
4 R. Ludwig, S. Diez, A. Ehrhardt, L. Küller, W. Pieper, and H. G. Weber, IEICE Trans. Electron. **E81-C**, 140 (1998).
5 H. Kurita, I. Ogura,, H. Yokoyama, IEICE Trans. Electron. **E81-C**, 129 (1998).
6 H. Yokoyama, T. Shimizu, T. Ono, Y. Yano, Opt. Rev. **2**, 85 (1995).
7 T. Shimizu, H. Kurita, and H. Yokoyama, to be presented at the 1998 Annual Meeting of IEEE Lasers and Elecro-Optics Society (LOES'98), Orland, Dec. 1998.
8 I. Ogura, Y. Hashimoto, H. Kurita, T. Shimizu, and H. Yokoyama, IEEE Photon. Tech. Lett. **10**, 603, 1998.
9 J. P. Sokoloff, P. R. Pruncnal, I. Glesk, and M. Kane, IEEE Photon. Technol. Lett. **5**, 787 (1993).
10 K. Yoneyama et al., to be presented at the 24th European Conference on Optical Communication (ECOC'98), Madrid, Sep. 1998, ThC13.
11 H. Kurita, Y. Hashimoto, H. Yokoyama, CPT'98, Tokyo, Jan. 1998, Pc-14.
12 H. Yokoyama, Y. Hashimoto and H. Kurita, Conference on Lasers and Electro-Optics (CLEO'98), San Francisco, May. 1998, FB-5.

Characterization of Interferometric Switching Devices for Ultrafast All-Optical Signal Processing

Enno Hilliger, Stefan Diez, Reinhold Ludwig, and Hans-Georg Weber

Optical Signal Processing, Photonics Department, Heinrich-Hertz-Institut für Nachrichtentechnik Berlin GmbH, Einsteinufer 37, D-10587 Berlin, Germany
E-mail: hilliger@hhi.de

Abstract. The temporal characteristics of interferometric switches based on semiconductor optical amplifiers are determined from picosecond pump-probe measurements on the solitary amplifiers.

1. Introduction

All-optical switching devices are of interest for many signal processing applications, such as optical time-division demultiplexing, sampling and data regeneration [1]. Very promising switching devices are semiconductor-optical amplifiers (SOAs) in interferometers. Their switching function is achieved by cross-phase modulation in the SOA due to gain saturation. This results in a differential phase shift in the two interferometer branches.

2. Method

To test the suitability of these devices for high-capacity optical communication systems, experiments with short optical pulses (≈ 1 ps) and high repetition frequencies (> 1 GHz) have to be carried out. Hereby, the investigation of the temporal switching behaviour (i.e. the switching window) is of particular importance. However, this requires to set up the complete interferometric switch.

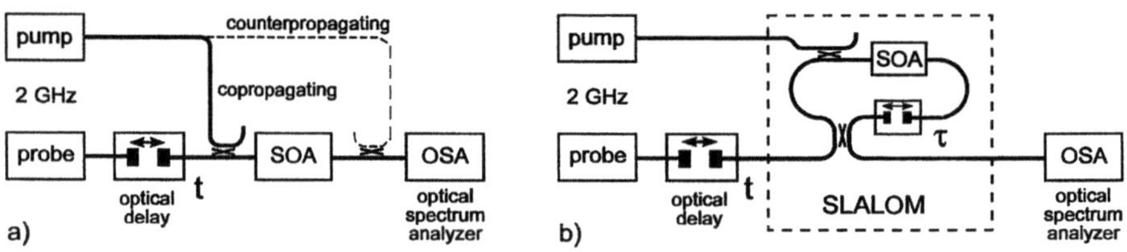

Fig. 1. Experimental setup: a) pump-probe measurements on a solitary SOA, b) switching window measurements on a SLALOM all-optical switch

This is very time consuming. It is therefore desirable to determine the overall switching function of the interferometer from measurements on the solitary SOAs already. Investigating an all-optical switch based on a hybrid Sagnac-Interferometer configuration (also named SLALOM or TOAD) [2], we calculate the switching windows from pump-probe measurements on the solitary SOA device and compare the results with direct measurements on the interferometric switch. The experimental setup is shown in Fig. 1. Two external cavity modelocked lasers [3] with a pulse repetition rate of 2 GHz were used as sources for pump (1550 nm) and probe (1538 nm). The temporal width of the optical pulses after amplification by an EDFA and subsequent filtering (not shown in the setup) was 1.5 ps FWHM as determined by autocorrelation measurements. The SOA under investigation was an InGaAsP-bulk SOA with a length of 980 μm.

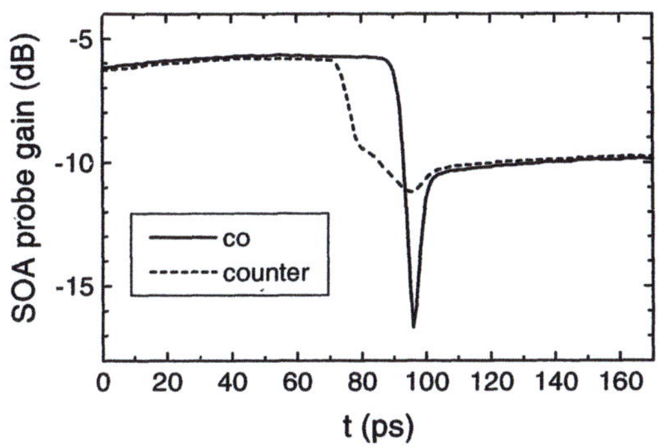

Fig. 2. Pump-probe curves for co- and counterpropagating pump and probe pulses in the solitary SOA

For the SOA a rather low injection current of 40 mA was chosen to avoid pulse distortions in the SOA. The average optical powers at the SOA input were 2 mW for the pump and 7 mW for the probe wave. Results of pump-probe measurements are shown in Fig. 2. These data can be used to obtain the clockwise G_{cw} and counterclockwise gain G_{ccw} of the SLALOM. For different assymmetries τ (difference of arrival time at the SOA for clock- and counterclockwise propagating pulses, see Fig. 1b) the transmittivity $T_{SLALOM}(t,\tau)$ can then be calculated by the following equation (using a linewidth enhancement factor α = 6).

$$T_{SLALOM}(t,\tau) = \frac{G_{CW}(t+\tau)+G_{CCW}(t)}{4} - \frac{\sqrt{G_{CW}(t+\tau)\cdot G_{CCW}(t)}}{2} \cdot \cos\left(\frac{\alpha}{2}\ln\frac{G_{CCW}(t)}{G_{CW}(t+\tau)}\right)$$

3. Results and Conclusion

The different curves in Fig. 3 result from different settings of τ using the same SOA with the same wavelengths and input powers at the SOA in both experiments. The good agreement of the calculated and the measured curves demonstrates that the SLALOM switching window can be successfully calculated from pump-probe measurements on the solitary SOA. It is thus possible to select suitable SOAs and to optimize the operation conditions for interferometric switches very efficiently before building the interferometer finally.

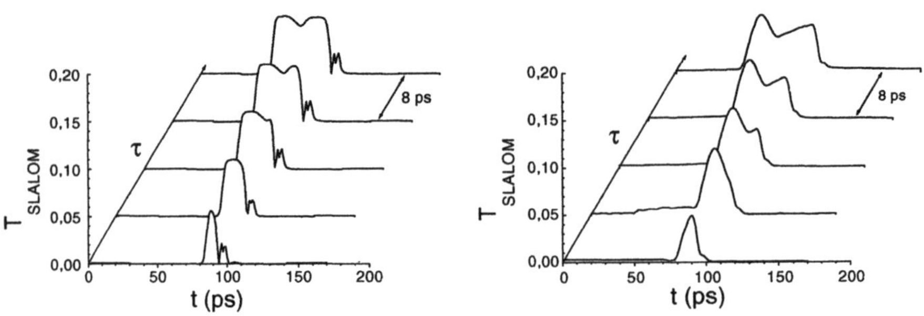

Fig. 3. Switching windows of a SLALOM: a) calculated from pump-probe measurements on the solitary SOA, b) measured directly in the interferometric switch.

References

1 Kawanishi, in *Optpelektronics*, Vol. 10, No. 4, pp. 447-460, 1995.
2 Eiselt et al., in *IEEE JLT*, Vol. 13, pp. 2099-2112, 1995.
3 Ludwig et al., in *Electronics Letters*, Vol. 31, pp. 1165-1167, 1995.

Measurement of 1.4 THz Beat Frequencies from a Dual Wavelength Gain–Switched Laser Diode using Frequency–Resolved Optical Gating.

L.P. Barry, J.M. Dudley, B.C. Thomsen and J.D. Harvey

Department of Physics, University of Auckland, Private Bag 92019, Auckland
NEW ZEALAND
Email: j.harvey@auckland.ac.nz

Abstract. The 1.4 THz optical beat frequency from a dual wavelength self–seeded gain–switched Fabry–Perot laser diode has been characterised using frequency–resolved optical gating.

1. Introduction

The generation of optical radiation modulated at THz frequencies has recently attracted much attention because of potential applications in mm–wave optical communication systems and in far–infrared THz spectroscopy [1, 2]. A convenient technique for the generation of THz–modulated optical radiation is the construction of dual–wavelength semiconductor lasers that operate on two longitudinal modes, and recent experiments have demonstrated both CW [1] and pulsed [3] output. The application of these laser sources will be dependent on their output intensity and phase characteristics, and in this paper we perform frequency–resolved optical gating (FROG) characterisation [4] of the pulsed output from a dual–wavelength gain–switched Fabry–Perot (FP) laser diode. The FROG measurements show the direct modulation of the intensity at a beat–frequency of 1.4 THz and the presence of a large frequency chirp.

2. Experimental Methods

Figure 1 shows the experimental set-up consisting of a 1.5 μm FP laser diode, gain-switched around 500 MHz, together with an all–fibre based external cavity comprising either one or two fibre Bragg gratings for single or dual wavelength operation respectively. The gratings used have central reflection wavelengths of 1541 nm and 1552 nm with bandwidths of 0.3 nm. Gain–switching without reinjection resulted in the generation of 12 ps pulses (FWHM) with a multi–mode spectrum. Self–seeded gain–switched operation was achieved by adjusting the laser temperature and the gain–switching frequency so that the external cavity feedback coincided with the build–up of the optical pulse due to the gain–switching cycle. The reinjection level was varied by adjusting the polarisation of the reflected signal using a polarisation controller (PC). The output pulses were characterised for both single and dual wavelength operation.

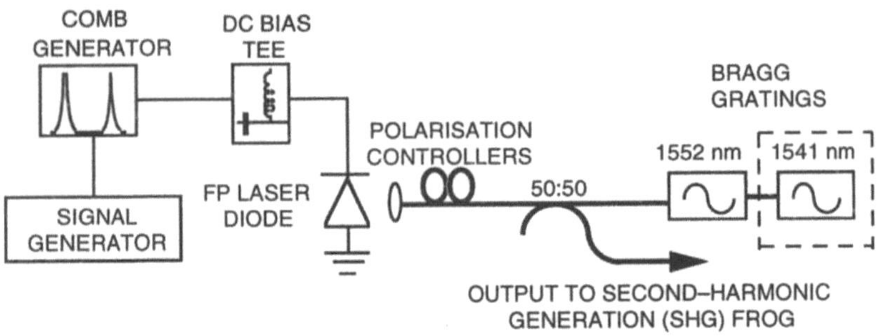

Figure 1: Experimental set-up for the gain-switched, self-seeded FP laser

3. Results and Discussion

Figure 2 shows typical results for single-wavelength operation. Fig. 2(a) shows the measured FROG trace, and Fig. 2(b) and 2(c) show respectively the spectral and the temporal intensity and phase of the retrieved pulse. Note that the independently-measured spectrum and autocorrelation were in excellent agreement with those derived from the retrieved field. The pulses are not transform-limited, with a large chirp arising from variations in carrier density in the gain region during the gain-switching cycle.

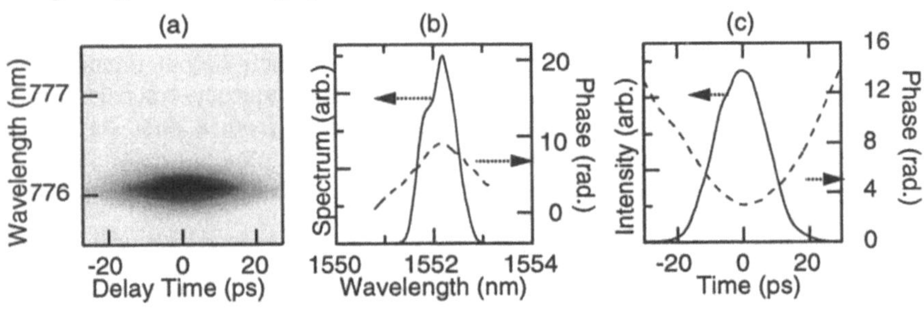

Figure 2. Results for single-wavelength operation showing (a) the SHG FROG trace, (b) the spectral intensity (solid line, left axis) and phase (dashed, line right axis) and (c) the temporal intensity (solid line, left axis) and phase (dashed line, right axis)

Figure 3 shows experimental results for dual-wavelength operation. Fig. 3(a) shows the measured FROG trace, Fig. 3(b) shows the spectral intensity and phase of the retrieved pulse, and Fig. 3(c) shows the retrieved temporal amplitude and phase. The FROG trace shows appreciable signal at the two SHG wavelengths corresponding to each oscillating mode, but also clearly shows a temporally modulated signal at a wavelength corresponding to inter-modal sum frequency mixing. Fig. 3(c) shows the corresponding oscillatory structure in the retrieved field amplitude, and note also that π phase discontinuities have been removed from the retrieved phase. The output intensity is modulated at frequency of 1.4 THz, corresponding to the beating between the two longitudinal modes separated by 11.2 nm.

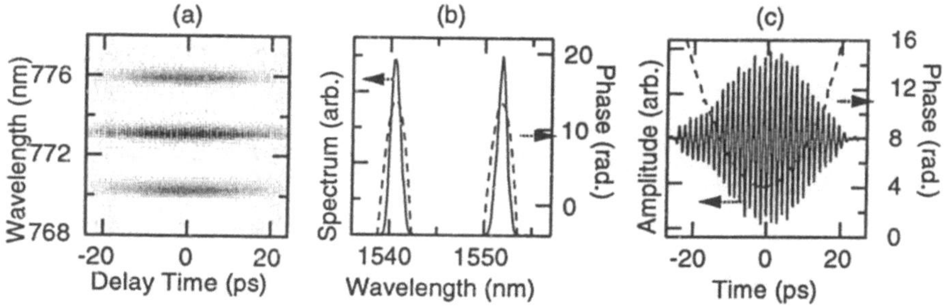

Figure 3. Results for dual–wavelength operation showing (a) the SHG FROG trace, (b) the spectral intensity (solid lines, left axis) and phase (dashed lines, right axis) and (c) the temporal amplitude (solid line, left axis) and phase (dashed line, right axis).

4. Conclusions

The FROG measurements clearly allow the retrieval of both this THz oscillatory structure, and the large residual frequency chirp on the superposition field generated by the dual wavelength gain switched laser. We expect further experiments will provide additional insight into the operation of these sources for applications in THz frequency generation.

Acknowledgements. We acknowledge the support of the New Zealand Foundation for Research Science and Technology

References

1. T. Hikada et al., Electron. Lett. **33**, 2039 (1997).
2. S. Matuura et al, Appl. Phys. Lett. **70**, 559 (1997).

Spectral Holography For Dynamic Dispersion Compensation

Y. Ding[1], D. D. Nolte[1], M. R. Melloch[2] and A. M. Weiner[2]

[1] Department of Physics, Purdue University, 1396 Physics Building, West Lafayette, IN 47907-1396, USA
E-mail: yiding@physics.purdue.edu
[2] School of Electrical and Computer Engineering, Purdue University, West Lafayette, IN 47907, USA

Abstract. Spectral holograms written in a photorefractive quantum well device convert a phase-distorted femtosecond pulse into its autocorrelation, which suppresses the pulse broadening due to dispersion. Photorefractive semiconductor quantum wells have a dynamic adaptibility bandwidth that extends from quasi-steady-state up to several kHz, enabling dynamic compensation of most environmentally induced distortions and random time drifts of femtosecond pulses that occur in practical applications.

1. Introduction

A serious issue in applications exploiting the large bandwidth of ultrafast optical pulses, such as femtosecond pulse processing, high-rate fiber optical communications and high-power chirped pulse amplification, is the dispersion in the ultrafast pulses. In pursuing data-rate of 10-100 Gbit/s in fiber communications, several means for dispersion compensation have been proposed and demonstrated, such as inserting fibers with opposite dispersion coefficient, using chirped fiber Bragg gratings, or exploiting optical phase conjugation. For data rates reaching Terabit/s, optical pulses on the sub-picosecond or femtosecond time scale are extremely sensitive to the dispersion, which can severely reduce the signal-to-noise ratio and make long distance data transmission impossible. Dispersion control is also a key technique in high-power femtosecond chirped pulse amplification. Femtosecond spectral holography [1] has been proposed and demonstrated to be an effective method to process femtosecond pulses [2]. As dynamic holographic media, semiconductor photorefractive quantum wells (PRQWs) have been demonstrated to be an appropriate material for dynamic femtosecond pulse processing [3]. In this paper, we take the advantage of this material and demonstrate a spectral holographic method to compensate the dispersion of femtosecond pulses **dynamically** with a compensation bandwidth up to kHz, which accommodates most environment-induced phase variations.

2. Experimental Methods

The dynamic dispersion compensation process is based on holography in the frequency domain [1,2] with a PRQW device as the dynamic holographic medium [3]. As shown in Fig. 1, the spatially spread frequency components of the signal and reference pulse $E_s(\omega)$ and $E_r(\omega)$ write a hologram with a strength proportional to $E^*_s(\omega)E_r(\omega)\exp(-i\boldsymbol{K}\cdot\boldsymbol{r})+c.c.$, where $\boldsymbol{K} = \boldsymbol{k}_s - \boldsymbol{k}_r$ is the grating vector with the propagation vector $\mathbf{k}_s$ and $\mathbf{k}_r$ of the signal and reference pulse, respectively. When the hologram is read by a test pulse $E_{test}(\omega)$, one of the first-order diffraction amplitudes is proportional to $E_{test}(\omega)E^*_s(\omega)E_r(\omega)\exp(-i\boldsymbol{k}_{out}\cdot\boldsymbol{r})$ in the direction of $\boldsymbol{k}_{out} = \boldsymbol{k}_{test} - \boldsymbol{k}_s + \boldsymbol{k}_r$. If the test beam is identical to the signal beam except for the incident direction, the output is then $E_s(\omega)E_s^*(\omega)E_r(\omega)\exp(-i\boldsymbol{k}_{out}\cdot\boldsymbol{r})$, which cancels the phase variations in $E_s(\omega)$ because of the term $E_s(\omega)E_s^*(\omega)$. In the time domain, it gives the autocorrelation of the signal pulse $e_t(t)$ propagating in the $\boldsymbol{k}_{out}$ direction, assuming the reference pulse is much shorter then the signal pulse. The autocorrelation compensates the dispersion in the signal pulses and compresses chirped pulses to the bandwidth limit.

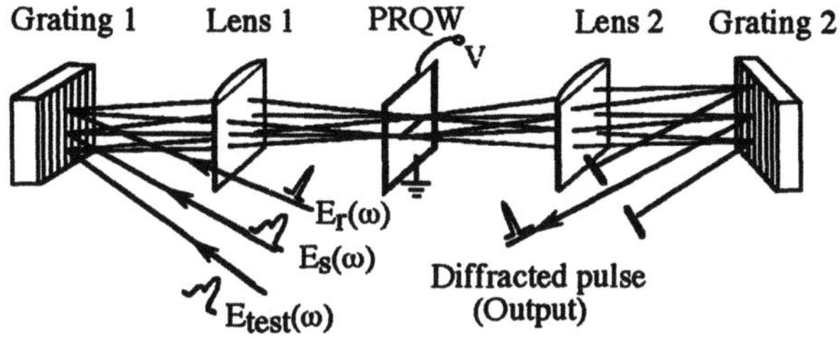

Fig. 1. Schematic experimental setup for dispersion compensation using dynamic spectral holograms

The femtosecond light source in this work is a Ti-Sapphire laser with a nominal pulse width of 130 fs and a central wavelength of 836 nm. The signal beam $E_s(\omega)$ is from a femtosecond pulse shaping apparatus and its spectral phase can be modified dynamically. The test beam is part of the signal beam. The power of each beam is about 400 μW. The temporal shape of the pulse is measured by electric-field cross-correlation. A dc-electric field of 7 kV/cm is applied on the GaAs/AlGaAs PRQW device in the transverse geometry. The photorefractive grating spacing is 22.5 μm and the input diffraction efficiency is 10^{-3}.

3. Results and Discussion

Fig. 2 shows the electric field cross correlation data and calculated envelopes of the femtosecond pulses involved in the experiments. When the input signal pulse with a Gaussian shape is not phase distorted (Fig. 2a), the diffracted output from the hologram is nearly Gaussian with a FWHM of 410 fs (Fig. 2b). The pulse broadening at this stage is caused by the bandwidth limit of excitonic nonlinearity in the PRQW device [4]. When a phase distorter, here a sapphire wafer with a rough spherical surface, is introduced in the frequency components of the input signal pulse, the pulse is severely broadened and distorted with a FWHM of more then 530 fs (Fig. 2c), whereas in the output pulses no broadening is observed (Fig.

2d), showing the ability of the spectral holography for dispersion compensation. The spectral phase of pulses in each case was measured experimentally by spectral interferometry. The data show that the strong phase distortion induced by the phase distorter in the input signal pulse is completely eliminated by the spectral holography. Because the hologram is dynamic, the compensation is in real-time limited only by the hologram response time. By interfering the output pulse with a reference pulse, we measured the phase changes and time shifts of the output pulse as a function of the phase modulation frequency of the input signal pulse. The experimental data show that at an intensity level of 5 mW/cm^2, phase variations and time shifts up to 1 kHz are completely compensated, which is fast enough for most environment-induced phase changes.

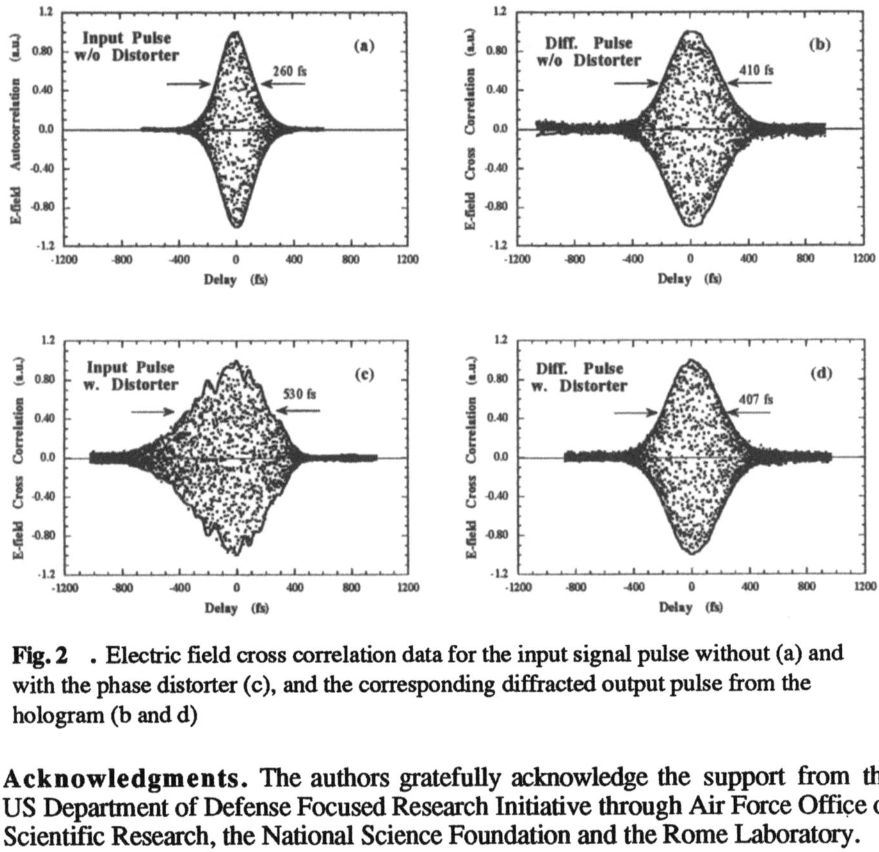

Fig. 2 . Electric field cross correlation data for the input signal pulse without (a) and with the phase distorter (c), and the corresponding diffracted output pulse from the hologram (b and d)

Acknowledgments. The authors gratefully acknowledge the support from the US Department of Defense Focused Research Initiative through Air Force Office of Scientific Research, the National Science Foundation and the Rome Laboratory.

References

1 Y. T. Mazurenko, Appl. Phys. **B50**, 102 (1990).
2 A. M. Weiner *et al.*, IEEE J. Quantum Electron. **28**, 2251 (1992).
3 Y. Ding *et al.*, IEEE J. Selected Topics Quantum Electron. **4**, 332 (1998)
4 R. M. Brubaker *et al.*, IEEE J. Quantum Electron. **33**, 2150 (1997)

Femtosecond Optical Pulse Generation from CW Laser Using Electrooptic Phase Modulator Featuring Lens Modulation

T. Khayim, M. Yamauchi, D.-S. Kim and T. Kobayashi

Division of Advanced Electronics and Optical Science, Faculty of Engineering Science, Osaka University, Toyonaka, Osaka 560-8531 Japan
E-mail: tattee@laser.ee.es.osaka-u.ac.jp

Abstract. We report a novel phase modulator featuring lens modulation and its application to the electrooptic chirping compression method for pedestal free ultrashort optical pulse generation.

The electrooptic chirping compression method can generate and control electrically ultrashort optical pulses from a narrow linewidth laser, while pulse compression of the passively mode-locked laser output is limited to a few kinds of lasers. We have developed a quasi-velocity-matched electrooptic phase modulator of high frequency with periodic domain inversion [1], succeeded in generating terahertz sidebands and formed 560 fs optical pulses using a grating pair as an anomalous dispersive element in the previous experiment [2].

However, in this method, a residual CW optical power between adjacent pulses after compression is not sufficiently small. This problem is effectively alleviated by using the novel phase modulator featuring lens modulation being reported here.

The basic structure of lens modulation consists of an electrooptic lens that has electrical controllability of the focal distance, a Fourier transform lens and a slit as shown in Figure 1. As well as performing phase modulation, at the same time, the electrooptic lens makes the focal point of a Fourier transform lens move sinusoidally. The modulated light has some parts cut periodically as if being amplitude modulated when a slit with suitable aperture is placed around the movement range of the optical focus. Therefore, it is possible to properly select the light which matches the frequency chirping from every modulation period by adjusting the position of a slit.

A bulk type $LiTaO_3$ electrooptic phase modulator featuring lens modulation was fabricated with parabolic-shaped periodic domain inversion, as shown in Figure 2, to perform lens modulation and make quasi-velocity-matching possible at a high frequency. For the thickness of a $LiTaO_3$ crystal $t = 0.5$ mm, the width of microstrip line W=1.0 mm, the wavelength of light $\lambda = 514.5$ nm, the phase velocity of microwave $v_m = 4.95 \times 10^7$ m/s and the group velocity of light $v_0 = 1.24 \times 10^8$ m/s, a half period of domain inversion is calculated to be 2.62 mm. A microwave resonator at 16.25 GHz is formed by an open-circuit terminated silver strip line.

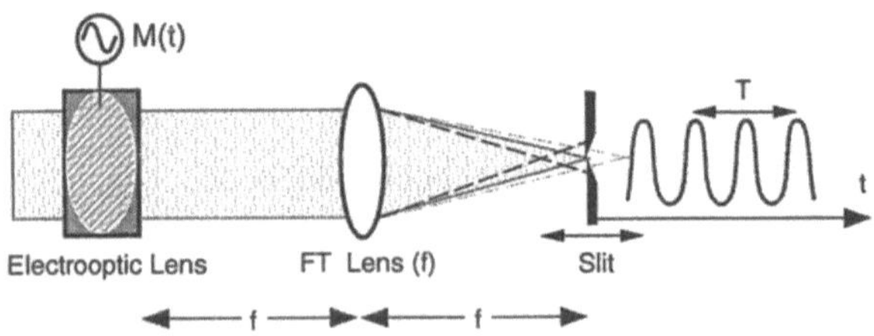

Figure 1: The basic structure of electrooptic lens modulation consisting of an electrooptic lens, a Fourier transform lens and a slit

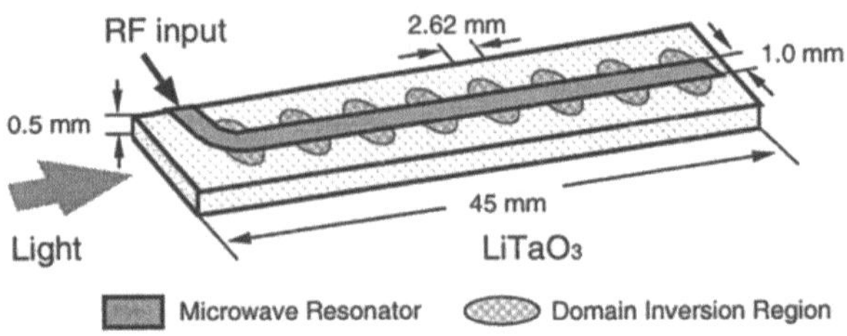

Figure 2: A bulk type electrooptic phase modulator featuring lens modulation

Figure 3 shows the experimental setup for ultrashort pulse generation. A 514.5 nm CW Ar laser was used as a light source, and a 16.25 GHz pulsed magnetron was used as a modulating microwave source. The light is phase modulated and lens modulated simultaneously, then a slit is regulated to allow only up-chirp frequency light to pass, and gate out others. Finally, an output light after the slit passes through the grating pair which provides the optimum group delay dispersion of 18.1 fs/GHz. As a result, pedestal free optical pulses of 550 fs with 16.25 GHz repetition rate were obtained as shown in Figure 4(a), and hence the extinction ratio is improved relatively to the previous result. In addition, down-chirp frequency light can also be selected instead for pulse compression by normal dispersion. Figure 4(b) obviously shows that the shape of the entire sideband spectral intensity becomes much flatter compared to a Bessel function form in the case of normal electrooptic phase modulation.

In conclusion, we proposed a novel electrooptic phase modulator featuring lens modulation and succeeded in generating transform-limited pulses of 550 fs from a CW laser without background light remaining by utilizing it with the electrooptic chirping compression method which uses electrooptic phase modulation instead of self-phase modulation. Moreover, generation of several tens of femtosecond optical pulses will be possible with the electrooptic method in the very near future.

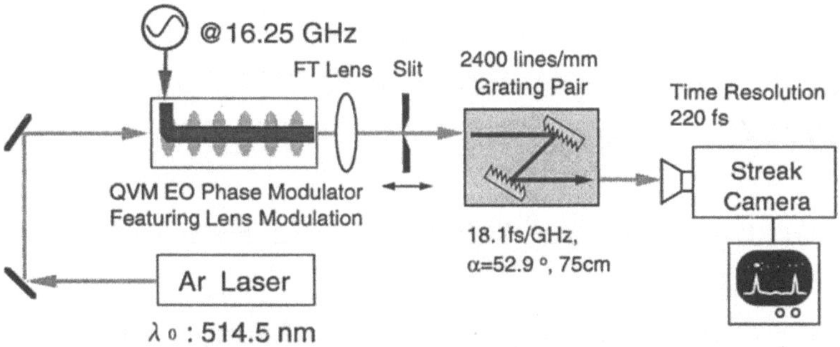

Figure 3: The experimental arrangement for ultrashort optical pulse generation

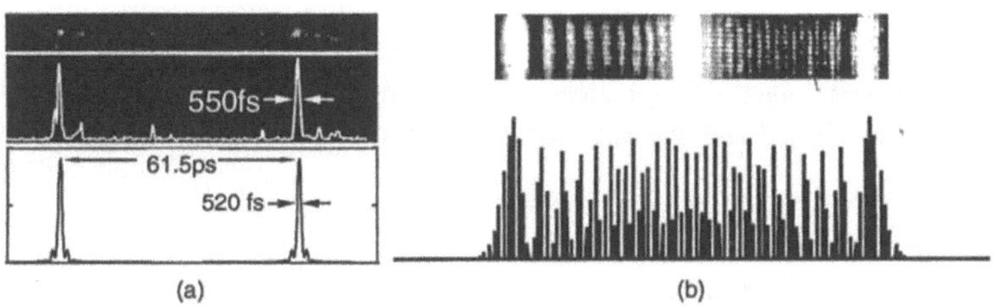

Figure 4: (a) The compressed pulses and (b) the sideband spectra generated by using a novel electrooptic phase modulator featuring lens modulation

Acknowledgements. The author would like to thank Prof. A. Morimoto of the College of Science and Engineering, Ritsumeikan University, for his kind and helpful advice.

References

[1] A. Morimoto, E. Saruwatari and T. Kobayashi: CLEO'94 (1994) CME5, 21.

[2] D.-S. Kim, M. Arisawa, A. Morimoto and T. Kobayashi: IEEE J. Select. Top. Quantum Electron. **2** (1996) 493.

Characterization of > 300 GHz Transistors Using a Novel Optoelectronic Network Analyzer

Nabil Sahri[1], Tadao Nagatsuma[1], Taiichi Otsuji[2], Naofumi Shimizu[2] and Makoto Yaita[1]

[1] NTT System Electronics Laboratories, 3-1 Morinosato Wakamyia, Atsugi, Kanagawa, 243-0198 Japan
E-mail: sahri @aecl.ntt.co.jp
[2] NTT Optical Network Systems Laboratories, 1-1 Hikarinooka, Yokosuka, Kanagawa, 243-0487 Japan

Abstract. A novel practical optoelectronic network analyzer with over 300-GHz bandwidth was developed and was successfully used to measure over 100-GHz HEMT S-parameters.

1. Introduction

Ultrafast electronic and photonic devices operating beyond 100 GHz are today commonly available [1] and are already being routinely incorporated in ICs [2]. For the characterization of such high-frequency devices, commercially available all-electronic network analyzer (NA) systems are limited to small signal diagnosis at frequencies up to 120 GHz. Optoelectronic techniques are capable of bandwidths greater than 300 GHz [3]. However, difficulties in instrumentation and poor detection sensitivity were major limitations for the practicability in the previously reported systems. This work addresses a novel technique to build over 300-GHz bandwidth practical optoelectronic network analyzer (ONA) based on electrooptic sampling (EOS). Using this system, a HEMT transit frequency (f_T) over 100 GHz is, to our knowledge, measured for the first time.

2. Optoelectronic Network Analyzer Configuration

Our approach to build an ultra-broadband ONA is shown in Fig.1. It consists of an integrated probe-type EOS-based NA. All the functions of stimulus, sampling and bias for the device under test (DUT) are integrated into a probe-head structure dedicated to two-port network measurement of devices with coplanar waveguide (CPW) type access electrodes. This configuration frees from the numerous parasitics due to the bonding between the DUT and the approaching transmission lines, and guarantees excellent stability and reproducibility of the measurements. One of the key devices in our ONA is a Uni-Traveling-Carrier photodiode (UTC-PD) [4] with a 50-Ω load resistor and a bias circuit. It can generate 2.2-ps electrical pulses with up to 1-V peak amplitude. The laser used is an extremely stable 1.55-μm sub-picosecond mode-locked fiber laser. This system allows us to perform small-and-large-signal network analysis in a wide frequency range of beyond 300 GHz. The UTC-PD is followed by 4.35-mm long, 50-Ω CPW on a quartz substrate, which assures perfect time-windowing with a bandwidth wider than 300 GHz. The full system geometry and the effect of connecting materials (Ag-Epoxy) were specified and designed on the basis of FDTD simulation, resulting in an excellent return loss of less than -20 dB throughout the electrical transmission paths.

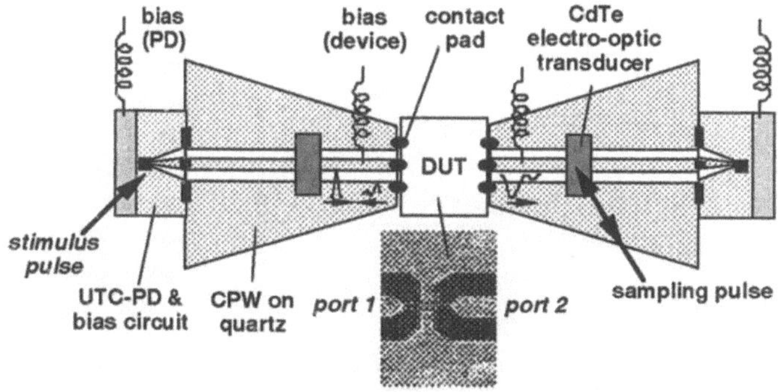

Fig.1. Configuration of all 1.55 μm-based optoelectronic network analyzer

3. HEMT Characterization

The device we measured is a 0.1-μm gate length HEMT [1]. First, its S-parameters were measured by conventional NA (CNA) up to 50 GHz as a reference for the DC gain of each S-parameter component. Optoelectronic network measurements were then realized by attaching two ONA probe heads to the gate node (port 1) and to the drain node (port 2) (inset of Fig. 1). Typical measured time-domain waveforms of the stimulus and the HEMT response are shown in Fig. 2.

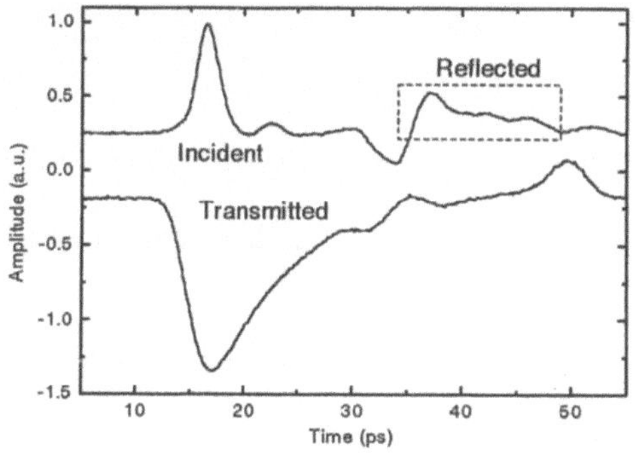

Fig. 2. Typical time-domain excitation and HEMT responses

Incident, reflected and transmitted signals are perfectly separated in time. The magnitudes of the device's S-parameters, calculated using Fourier transform are shown in Fig. 3. A good agreement exists with the CNA measurements. The f_T value of 152 GHz (inset of Fig. 3.) is also in agreement with the one obtained from the extrapolation of the CNA data. We should point out that these data include the effect of the parasitics due to the DUT's access transmission lines and the interconnection between the DUT and the ONA probes. Future on-wafer contact-probe-type ONA should allow to use conventional de-embedding techniques using passive devices.

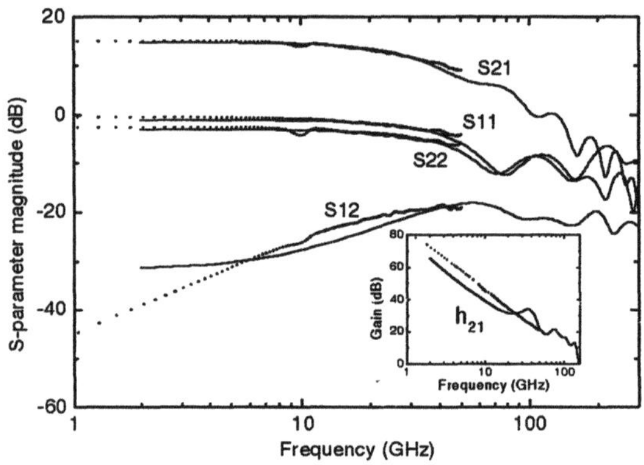

Fig. 3. Comparison between the magnitudes of the S-parameters calculated from the ONA measurements (solid lines) and the 50 GHz-bandwidth-CNA ones (dotted lines). The inset shows the respective calculated h_{21}

4. Conclusion

We developed a new practical approach to perform network analysis measurements over 300 GHz by optoelectronic means. S-parameters of a HEMT chip were measured within this bandwidth. A non-de-embedded transit frequency of 152 GHz was measured. This approach should be implemented in an external probe-type wide bandwidth NA for on-wafer measurements.

Acknowledgements. The authors would like to thank Drs. T. Ishibashi and Y. Umeda for their fruitful discussion and comments.

References

1 T. Enoki, H. Ito, K. Ikuta and Y. Ishii, in *Proc. 1995 Int'l. Conf. on IPRM*, 81, 1995.
2 T. Otsuji, K. Murata, T. Enoki and Y. Umeda, in *Tech. Dig. IEEE GaAs IC Symp.*, 183, 1997.
3 M. Y. Frankel, in *Opt. Quantum Electron.* **28**,783, 1996.
4 N. Shimizu, N.Watanabe, T. Furuta and T. Ishibashi, in *Tech. Dig. 55th Annual Research Conf.*, 164, 1997.

Free-Space Electro-Optic Detection of THz Radiation with Chirped Optical Beam

Zhiping Jiang, and X.-C. Zhang
Physics Department, Rensselaer Polytechnic Institute, Troy, NY 12180-3590 USA
Email:zhangx2@rpi.edu

Abstract. We report electro-optic detection of a freely propagating THz field with chirped optical probe pulses. We also present the single-shot measurement and one-dimensional THz field imaging with an unprecedented data acquisition rate.

1. Introduction

Conventional electro-optic sampling measures a repetitive electrical waveform, such as an electrical signal in an integrated circuit [1,2] or a THz beam in a free-space environment [3-5]. However, due to the nature of the repetitive measurement of the sampling system, for each time-delay between the electrical signal and a synchronized optical probe pulse, only a small portion of the temporal signal (about laser probe pulse duration) is measured. The entire waveform is reassembled by sequentially plotting the signal versus the time-delay. The data acquisition rate is very limited. Until now, there has been no report of a successful measurement of a picosecond electrical pulse on a single-shot basis.

Using a linearly chirped optical probe pulse in the electro-optic measurements, a temporal waveform of an electric field is linearly encoded onto the frequency spectrum of the probe pulse, and then decoded by dispersing the probe pulse from a grating to a detector array [6,7]. Real-time and single-shot acquisition of picosecond THz field pulses without using a mechanical time-delay device is achieved [8].

2. Temporal Measurement with a Chirped Beam

Figure 1(a) schematically illustrates electro-optic detection with a chirped optical probe beam. A mode-locked Ti:sapphire laser provides 250 fs pulses with a center wavelength at 820 nm, spectral bandwidth 10 nm, and pulse energy 3 μJ. The laser beam is split into pump and probe beams. The pump beam is used to generate THz beam from an emitter, then the THz beam is focused onto the electro-optic sensor by a polylens. The probe beam is frequency chirped and temporally stretched by a grating pair. When the chirped probe beam and a THz pulse co-propagate in the electro-optic crystal, different portions of the THz pulse modulates the different wavelength components of the chirped pulse through the Pockels effect. Therefore, after the optical analyzer, the THz waveform is encoded onto the wavelength spectrum of the probe beam. A spectrometer and a detector array are used to measure the spectral distribution.

Figure 1(b) shows the spectral distributions of the chirped probe pulse measured by the linear diode array (LDA) with and without a co-propagating THz field pulse.

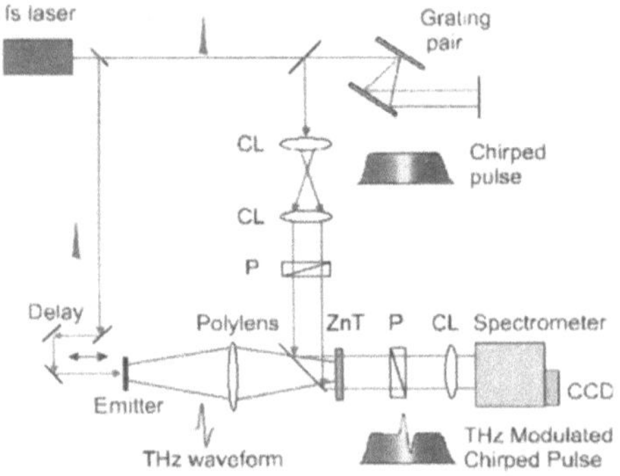

Fig. 1(a). Schematic of experimental setup of electro-optic measurement with a chirped optical probe beam.

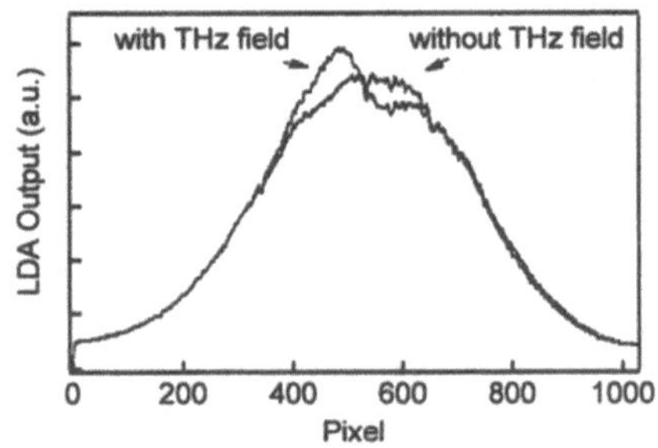

Fig. 1(b). Spectral distribution of the chirped probe pulse with & without a co-propagating THz field pulse.

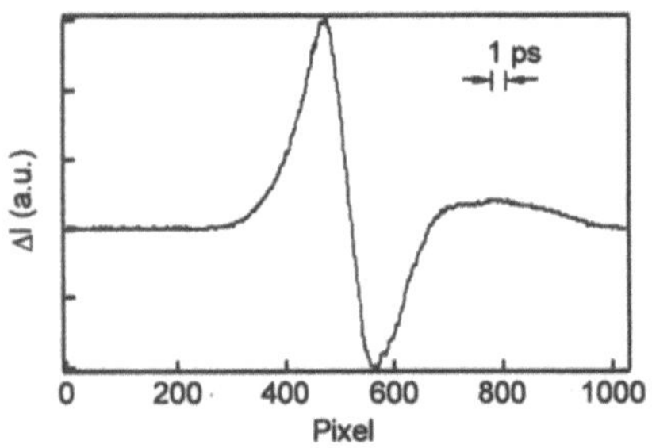

Fig. 1(c). The differential spectrum from Fig. 1(b).

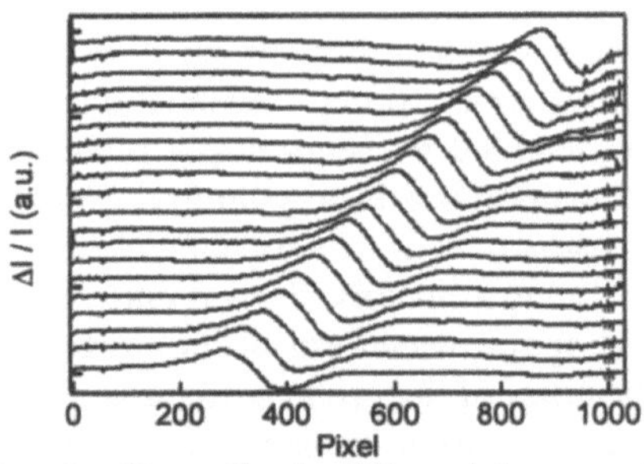

Fig. 2. Normalized differential spectral distribution (ΔI/I) by adjusting the fixed delay line at a step of 1.3 ps.

Figure 1(c) is a plot of the differential spectrum distribution (ΔI) from Figure 1(b). This differential distribution reconstructs both the amplitude and phase of the temporal waveform of the THz pulse. The differential spectrum (ΔI) shifts horizontally by adjusting the fixed time-delay line. Moving the fixed delay line is equivalent to placing the terahertz field in a different portion of the probe beam spectrum and it can be used as a marker to calibrate the time scale. Figure 2 shows the normalized differential spectrum distribution (ΔI/I) by adjusting the fixed time delay at a step of 1.3 ps.

High data acquisition rate in the chirped pulse measurement makes real-time 1D spatio-temporal THz imaging feasible. In imaging experiment, a CCD camera replaces a linear diode array. A spectrometer (SPEX 500) and a CCD camera

(Princeton Instruments, Inc., CCD-1242E) are used to measure the spectral distribution of the probe beam (wavelength-division-demultiplexing). The 0.5-meter spectrometer has a spectral resolution of 0.05 nm. The camera has 576x1242 pixels, dynamic range of 18 bits, and a full well capacity greater than 500,000 electrons per pixel. The THz emitter is an 8-mm wide GaAs photoconductor with a biased voltage of 3 kV. The electro-optic crystal is a 4-mm thick <110> ZnTe.

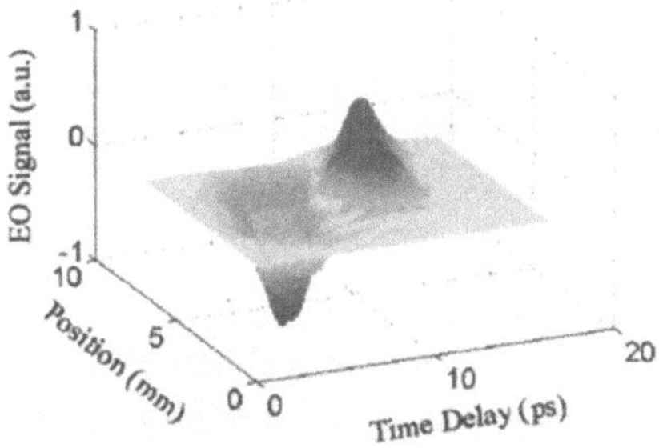

Fig. 3(a). One-dimensional spatio-temporal THz imaging of a dipole emitter.

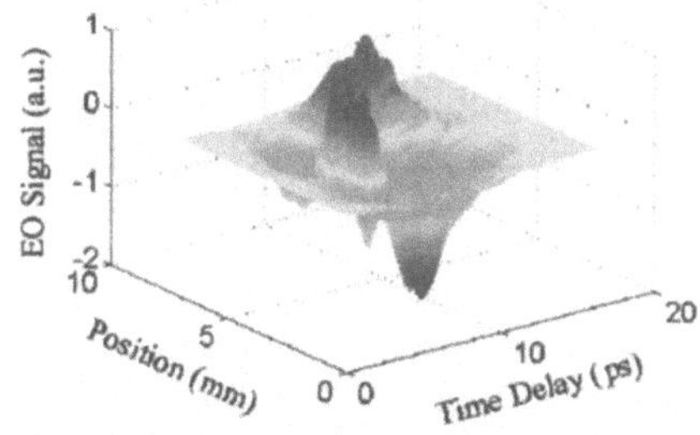

Fig. 3(b). One-dimensional spatio-temporal THz imaging of a quadrupole emitter.

Figures 3(a) and (b) show THz field images (x position versus time) emitted from a dipole and a quadrupole emitters, respectively. We achieved 26 frames/s by using a CCD camera.

3. Single-Shot Measurement

The electro-optic measurement with a chirped optical pulse is capable of a single-shot measurement of a full THz waveform. This technique can be used in the measurement of a single event, such as an explosion or breakdown of the electrical field in dielectrics.

To get a single laser pulse, low repetition rate of 10 kHz is used. An optical modulator with a 1/220 on/off ratio reduces the repetition rate to 45 Hz. The exposure time of the CCD camera was set to 10 ms. In this case, the camera could receive only one optical probe pulse during this single-shot exposure time at the pulse repetition of 50 Hz, where the adjacent pulses are 20 ms apart. Therefore, if the exposure time of the detector is less than 22 ms, only one pulse is captured for single exposure. The minimum 10 ms shutter-time of the camera satisfies this condition.

Figure 4(a) plots a single-shot differential spectral distribution waveform of a THz pulse. The measured signal has a 2-ps rise time (10% to 90%) with a signal-to-noise ratio better than 60:1. In this system, the optimal temporal resolution equals to the square root of the product of the original optical pulse duration and the chirped pulse duration. Figure 4(b) shows 3 individual single-shot THz pulses measured by adjusting the fixed time delay with a step of 3.3 ps.

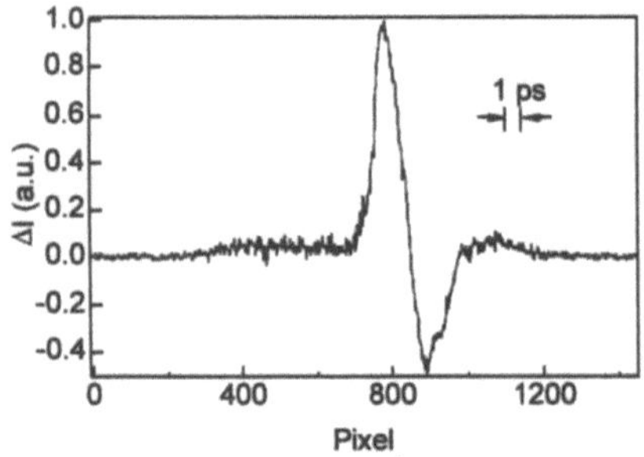

Fig. 4(a). A single-shot spectral waveform of a THz pulse measured by a chirped optical probe pulse. It reconstructs the temporal waveform of the THz pulse.

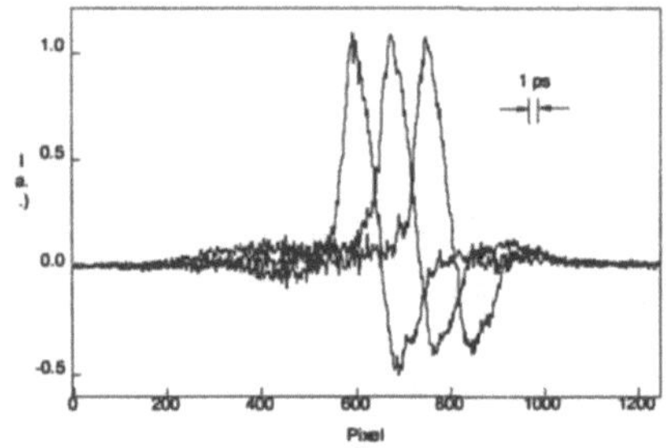

Fig. 4(b). Three THz pulses (single-shot) measured by adjusting the fixed time delay with a step of 3.3 ps.

A 1-D image of the field spatial distribution across the THz emitter can be obtained by focusing the probe beam into a line with a cylindrical lens [9]. For a spatio-temporal imaging, the size of the probe beam on the ZnTe sensor is about 0.4 mm by 10 mm. Figure 5(a) shows a simulated spatio-temporal THz pulse in the chirped pulse system, while Figure 5(b) is a plot of a single-shot image measured from a GaAs photoconductive dipole antenna. The dipole length is 7 mm and the bias voltage is 5 kV. The 1-D spatial distribution across the dipole and its temporal THz waveform are obtained simultaneously in a single laser pulse. The size of the spatio-temporal image is 10-mm by 25-ps. In this single-shot measurement, the background light per pixel on the CCD camera is about 200 counts, while that of the modulated probe pulse is about 50 counts. Typical oscillating features and the symmetric spatial distribution of the pulsed THz field are obtained. The spatial resolution in the imaging system is better than 1 mm.

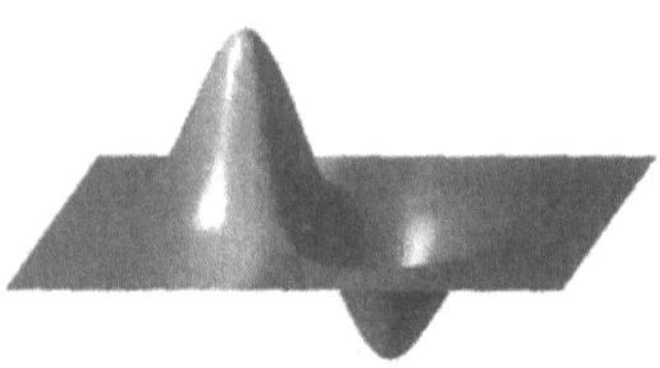

Fig. 5(a). Computer plot of a spatio-temporal THz image from a dipole (far field).

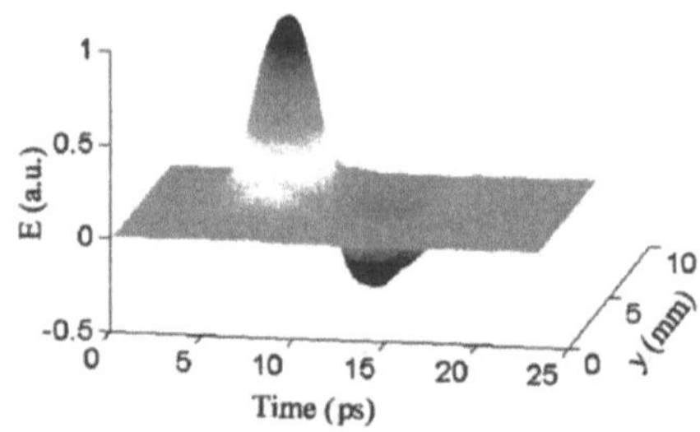

Fig. 5(b). Measured single-shot spatio-temporal image of a THz pulse from a dipole emitter.

In the measurement of Figure 5(b), since the x-position of the CCD camera is used for the wavelength measurement, this imaging system only provides a y-direction spatial measurement. It is a major challenge to use a CCD with 3 parameters (the x- and y-positions and the amplitude) for the measurement of THz field with 4 variables (the wavelength, the two dimensions (x and y), and the amplitude). It is possible to use an optical fiber array (2-D to 1-D converter) to perform a 2-D spatio-temporal THz field imaging.

4. Conclusion

A chirped optical pulse, without a mechanical moving element for the time delay, measures the temporal waveform of a THz pulse. The single-shot measurement capability allows us to capture a THz pulse with an unprecedented data acquisition rate. The conventional time-domain electro-optic sampling technique is modified for wavelength-division-processing by using a chirped probe pulse. Unparalleled by other THz sampling techniques, this single-shot method provides the highest possible data acquisition rate. Our results demonstrate the feasibility of spatio-temporal THz imaging of a single event with electro-optic detection.

Acknowledgements. This work was supported by the Army Research Office and the National Science Foundation.

References

1. B. H. Kolner and D. M. Bloom, *IEEE J. Quantum Electron.*, QE-**22**, 69, 1986.
2. J. A. Valdmanis and G. A. Mourou, *IEEE J. Quantum Electron.*, QE-**22**, 79, 1986.
3. Q. Wu and X.-C. Zhang, *Appl. Phys. Lett.*, **67**, 2523, 1995.
4. A. Nahata, D. H. Auston, T. F. Heinz, and C. J. Wu, *Appl. Phys. Lett.*, **68**, 150, 1996.
5. P. U. Jepsen, C. Winnewisser, M. Schall, V. Schyja, S. R. Keiding and H. Helm, *Phys. Rev. E*, **53**, pp. R3052 - R3054, 1996.
6. J. A. Valdmanis, *Proceeding of Ultrafast Phenomena V, Springer-Verlag*, 82, 1986; *Solid State Technology / Test & Measurement World*, Nov. S40, 1986.
7. Galvanauskas, J. A. Tellefsen, Jr., A. Krotkus, M. Oberg, and B. Broberg, *Appl. Phys. Lett.*, **60**, 145, 1992.
8. Zhiping Jiang and X.-C. Zhang, *Appl. Phys. Lett.*, **72**, 1945, 1998.
9. Zhiping Jiang and X.-C. Zhang, *Optics Lett.*, July 15 issue, 1998.

Spectrum control of intense THz-radiation from InAs under magnetic field irradiated with stretched femtosecond laser pulses

H. Ohtake, Z. Liu, S. Izumida, S. Ono, and N. Sarukura
Institute for Molecular Science (IMS), Myodaiji, Okazaki, 444-8585 Japan

Abstract: We report significant enhancement of THz-radiation from InAs under magnetic field irradiated with femtosecond pulses. Furthermore, the radiation spectrum is found to be controlled by the excitation pulsewidth, chirp direction of the excitation pulse, and the magnetic field.

Various THz-radiation sources have been intensively studied including photo conductive switches irradiated with ultrashort optical pulses. [1,2] For applications to sensing or imaging [3] and time-resolved spectroscopy in the far infrared region, an intense, compact, and simple light source is required. Zhang et al. reported quadratic dependence of the laser induced THz-radiation on the magnetic field. [4] Previously, we have reported the significant enhancement of THz-radiation power from InAs under magnetic field irradiated with femtosecond optical pulses. The power reaches sub-mW level in a 1.7-T magnetic field with 1.5-W excitation power. [5] In this paper, we report the spectrum control of this intense THz-radiation by the width of the excitation pulses and magnetic field.

A Ti:sapphire laser delivered 100-fsec pulses at 800 nm with an 82-MHz repetition rate using 1.0-W average power for the excitation. The sample was non-doped bulk InAs with a (100) surface. The InAs sample itself is highly reflective. The reflectivity of the THz radiation was measured to be approximately 70% for a 45 degree incidence angle. Therefore, the THz radiation was totally generated in the reflection direction. A liquid-helium-cooled InSb bolometer was provided for detection. The InAs sample was placed 45 degrees to the excitation laser beam path. The magnetic field was applied parallel to the sample. In this geometry, the THz radiation was detected even without the magnetic field, [5] and the polarization of the THz radiation was horizontal. The THz-radiation power exhibits almost quadratic dependence on the excitation power and higher magnetic field similar to the previous geometry. [5]

The spectrum of the THz radiation was obtained by a Polarizing Michelson interferometer. Typical spectrum obtained in the different magnetic field is shown in Fig. 1. The spectral shape differ not only to the magnetic field but also to the field direction. This difference on the magnetic field direction is due to the initial carrier acceleration direction due to the Lorentz force [4] as observed in time resolved measurement of the electric field of the THz-radiation as shown in Fig. 2.

For the control of the spectrum of this THz radiation, we introduced a pulse stretcher and compressor consisted of a SF-6 prism pair separated 120 cm for large negative dispersion and a closely placed SF-6 prism pair to provide adjustable positive dispersion material. With this pulse stretcher and compressor, the pulse duration could be changed between 100 fsec and 300 fsec with positive or negative chirp ($sech^2$ pulse shaped assumed). The center frequency of THz-radiation can be controlled with different excitation pulse width as shown in Fig. 3(a). The longer

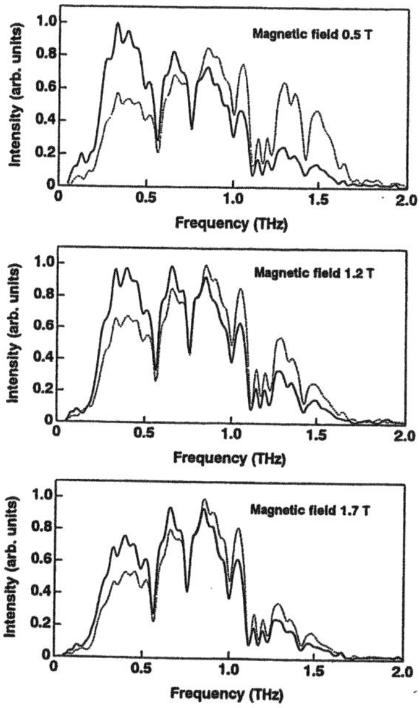

Fig. 1 Frequency spectrum of THz radiation from InAs under different magnetic field. Solid line: negative field; Dotted line: positive field.

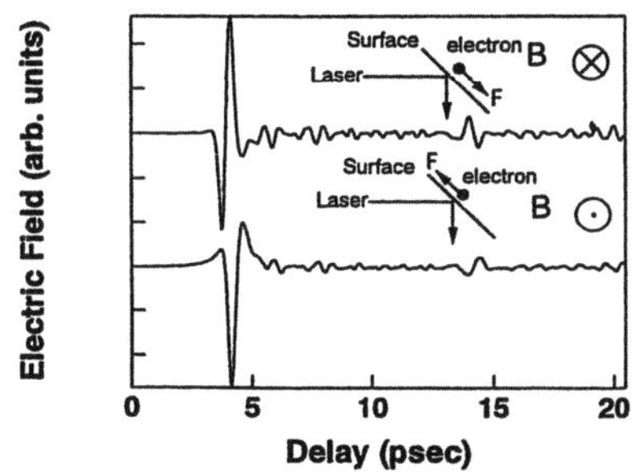

Fig. 2 The time-resolved measurement of electric field of the THz-radiation for the different direction of the magnetic field.

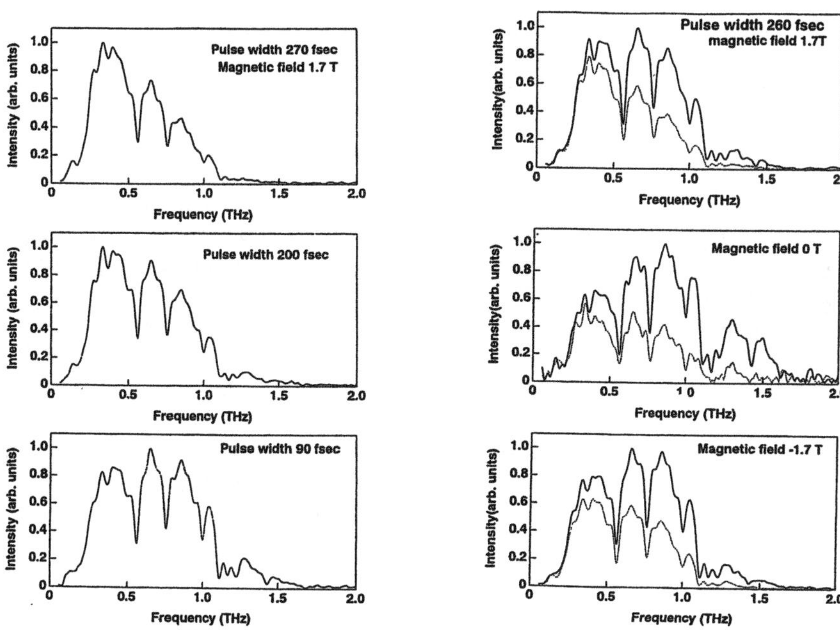

(a) (b)

Fig. 3 (a) Frequency spectrum of THz radiation from InAs under 1.7 -T magnetic field with the different excitation pulse duration with negative chirp. (b) Frequency spectrum of THz radiation from InAs with the different excitation chirp and magnetic field. The excitation pulse duration is 260 fsec in all case. Solid line: positively chirped-pulse excitation; Dotted line: negatively chirped-pulse excitation.

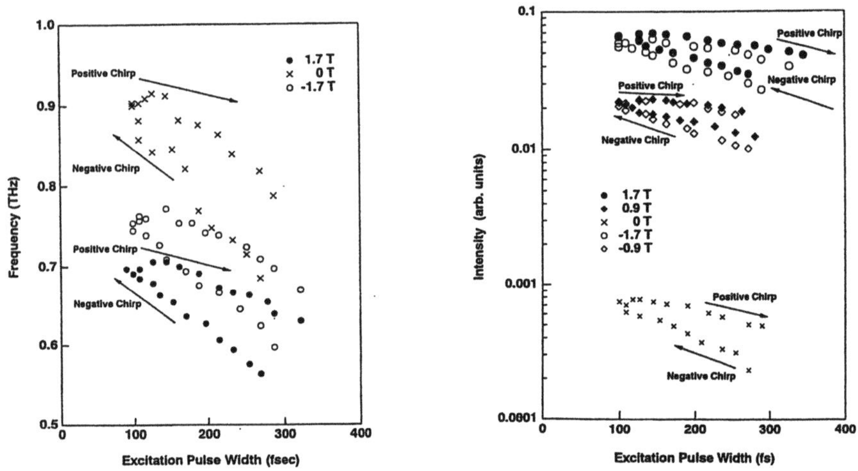

(a) (b)

Fig. 4 (a) Center frequency spectrum dependence of THz radiation with the different excitation chirp, pulse duration and magnetic field. (b) The dependence of THz radiation intensity with the different excitation chirp, pulse duration and magnetic field.

excitation pulse width resulted in lower center frequency (Fig. 4(a)), this is due to the corruption of the impulse response of the semiconductor in the longer pulse width region. Furthermore, the spectral shape of the radiation strongly depends on the chirp direction as shown in Fig. 3(b) This difference for the excitation chirp direction might be attributed to the difference of the photo-carrier relaxation process in the conduction band with differently chirped-pulse excitation. The excitation of the carrier from lower energy side (positively chirped-pulse excitation) will be more efficient to create carriers and less sensitive to carrier-scattering effect considering the carrier-relaxation in the conduction band, that would be favorable for the intense and higher frequency THz-radiation generation. This hypothesis is partly supported by the fact that the positively chirped-pulse excitation yielded higher power of the radiation as shown in Fig. 4(b).

In conclusion, we have demonstrated a new, simple, and intense THz-radiation source just using bulk InAs owing to the quadratic relationship to the magnetic field and excitation laser power. [5] Furthermore, the radiation spectrum is well-controlled by the excitation pulsewidth, chirp direction of the excitation pulse, and the magnetic field. Development of such simple and intense radiation sources with spectral controllability will open up new applications of THz radiation.

1. For example, D. H. Auston, Appl. Phys. Lett. 43, 713 (1983).
2. For example, I. Brener, D. Dykaar, A. Frommer, L. N. Pfeiffer, J. Lopata, J. Wynn, and K. West, and M. C. Nuss, Opt. Lett. 21, 1924 (1996).
3. D. M. Mittleman, S. Hunsche, L. Boivin, and M. C. Nuss, Opt. Lett. 22, 904 (1997).
4. X. -C. Zhang, Y. Lin, T. D. Hewitt, T. Sangsiri, L. E. Kingsley, and M. Weiner, Appl. Phys. Lett. 62, 2003 (1993).
5. N. Sarukura, H. Ohtake, S. Izumida, and Z. Liu, J. Appl. Phys., vol. 84, no. 1, 1998.

Dithered-edge sampling of terahertz pulses: fast detection using slow photoconductive receivers

Jake Bromage, Ian A. Walmsley, and C. R. Stroud, Jr.

Institute of Optics, University of Rochester, Rochester NY 14627

Abstract. We introduce dithered-edge sampling (DES) as a new technique for ultra-wideband measurement of terahertz pulses using photoconductive antennas. The terahertz pulse is sampled by first passing it through a photoconductive gate whose fast attenuation edge (limited only by the duration of the optical gating pulse) is dithered in time. A slow photoconductive receiver then measures the component of the terahertz electric field that is modulated at the dither frequency. When used alone, the receiver blurs the measured terahertz pulse width to 1.3 ps. However, the increased time resolution of the DES system (gate + receiver) allows one to measure source-limited terahertz pulse widths (400-fs in this case) [D.You *et. al.*, Opt. Lett. **18**, 290 (1993)]. In addition, we have used the DES technique to make what we believe to be the first direct measurement of a photoconductive receiver's temporal response.

1. Introduction

The major technologies for terahertz pulse measurement are photoconductive sampling [1] and electro-optic sampling [2]. For both, there is a trade-off between sensitivity and bandwidth. In photoconductive sampling, one controls this trade-off by choice of material carrier lifetime and the antenna geometry. In electro-optic sampling, the crystal length is the dominant parameter.

We have developed a technique for terahertz pulse measurement which decouples the sensitivity and bandwidth requirements. This decoupling is made possible by the use of two separate components. The first is a photoconductive

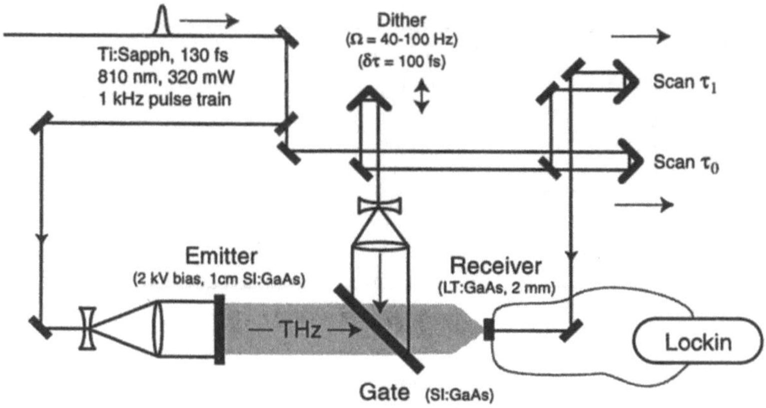

Fig. 1. The apparatus used for both the DES and PCS measurement schemes

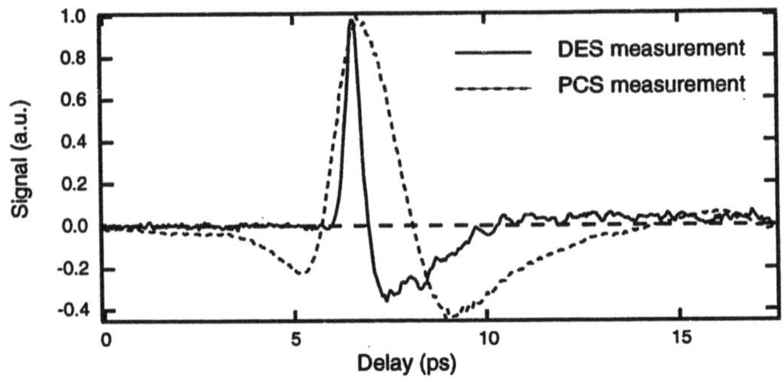

Fig. 2. Comparison of the THz pulse as measured by the DES and PCS schemes

gate which provides the bandwidth. The second is a standard photoconductive receiver which is responsible for the sensitivity. Because the receiver is not responsible for the temporal resolution of the detection system, one can instead use the system to measure the receiver's temporal response. This is the first direct measurement of the temporal response of a photoconductive receiver.

2. Techniques and Results

The experimental set-up is shown in Fig. 1. Both the emitter and receiver are described in detail in Ref. [3]. The gate is made from a SI:GaAs wafer. After it is illuminated with a 50 μJ optical pulse, transmission of THz power drops by 85%. This fall in transmission is quick—on the order of the 130-fs optical pulse width. This sharp edge in the transmission of the gate provides the temporal resolution.

Other attempts to measure the terahertz pulse shape have used the sharp edge of a photoconductive gate to provide the time resolution [4-6]. These schemes, however, measured the *integral* of the THz pulse. To obtain the electric field, a numerical derivative has to be performed which places tight constrants on the signal-to-noise ratio. In our scheme, the instrumentation takes the derivative directly. This is done by modulating the delay of the optical pulse which activates

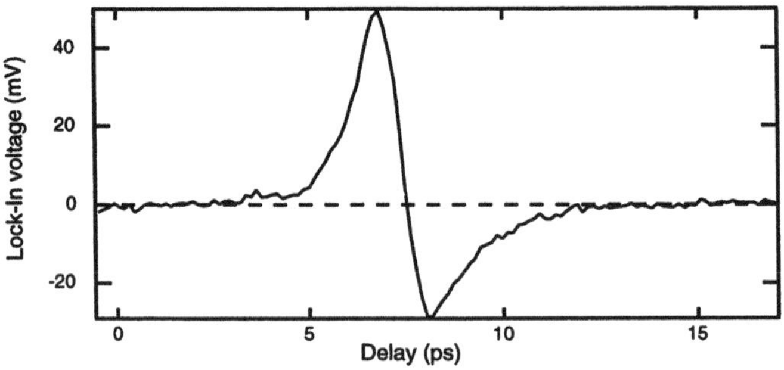

Fig. 3. The temporal response of the receiver as measured by the DES technique

the gate ($\delta\tau = 100\,\text{fs}$, $\Omega = 40 - 100\,\text{Hz}$), and locking to the signal from the receiver at the dither frequency. This signal, $S_\Omega(\tau_0, \tau_1)$, is a function of the relative delays between activating the emitter, gate and receiver, and depends on the terahertz pulse electric field (E) and the receiver's response (R) as given by

$$S_\Omega(\tau_0, \tau_1) \propto \delta\tau\, E(\tau_0)\, R(\tau_0 - \tau_1) \tag{1}$$

2.1 Measuring the terahertz pulse

We can see how to make the DES measurement of the THz pulse by examining Eq. (1). If $\tau_0 - \tau_1$ is fixed to some value that maximizes R, scanning τ_0 produces a signal which is proportional to the THz pulse electric field. Using the apparatus shown in Fig. 1, the THz pulse could also be measured using the receiver in the standard photoconductive sampling (PCS) mode [1] by blocking the gate's optical pulse. The signal that is measured in this case is a cross-correlation of the THz electric field and the receiver's response, as shown in the Eq. (2).

$$S(\tau) = \int_{-\infty}^{\infty} E(t)\, R(t - \tau)\, dt \tag{2}$$

In Fig. 2 we can see the comparison of the THz pulse as measured by the DES and PCS techniques. There are three points to note: firstly, the pulse measured by the DES scheme has a FWHM of 400 fs, compared to 1.3 ps for the PCS scheme. Secondly, the DES pulse shape is in agreement with the Current Surge Model for large-aperture photoconductive emitters [7], in that it does not have an exponential leading edge. Finally, oscillations due to water absorption are faithfully reproduced in the DES measurement.

2.2 Measuring the receiver's temporal response

As shown in Eq. (1), if τ_0 is fixed to maximize E, scanning τ_1 directly measures R. The measured receiver response is shown in Fig. 3. This response function is affected by the material carrier lifetime and the antenna geometry. The THz pulse calculated from the DES measurements of E and R using Eq. (2) is in good agreement with the pulse measured using the PCS scheme.

Acknowledgements. This work was supported by the US Army Research Office.

References

1. N. Katzenellenbogen and D. Grischkowsky, Appl. Phys. Lett. **58**, 222 (1991).
2. X. C. Zhang, SBMO/IEEE MTT-S conference proceedings, p.5 (1997).
3. J. Bromage, S. Radic, G. P. Agrawal *et al.*, Opt. Lett. **22**, 627 (1997).
4. G. Mourou, C. V. Stancampiano *et al.*, Appl. Phys. Lett. **38**, 470 (1981).
5. F. G. Sun, G. A. Wagoner, and X. C. Zhang, Appl. Phys. Lett. **67**, 1656 (1995).
6. D. You, R. R. Jones, P. H. Bucksbaum *et al.*, in Opt. Lett. **18**, 290 (1993).
7. P. K. Benicewicz, J. P. Roberts and A. J. Taylor, JOSA B **11**, 2533 (1994).

Driving Intersubband Transitions With THz Pulses

R. Kersting[1], J.N. Heyman[2], G. Strasser[1], and K. Unterrainer[1]

[1] Institut für Festkörperelektronik und Mikrostrukturzentrum der TU Wien, Floragasse 7, A-1040 Wien, Austria
E-mail: kersting@fkeserver.fke.tuwien.ac.at
[2] Department of Physics and Astronomy, Macalester College, St. Paul, MN 55105, USA
E-mail: heyman@macalester.edu

Abstract. Few-cycle THz pulses are used to drive electrons in semiconductor heterostructures. Time-resolved THz experiments show the collective intersubband oscillation of electrons and the decay of their coherent motion.

1. Introduction

The response of a two-level system following a driving field is one of the fundamentals of quantum mechanics. Semiconductor heterostructures provide an attractive material system to investigate this response time-resolved since the intersubband transitions can be tailored by modern semiconductor growth techniques. In the recent years the decay of the interband [1,2] and intraband polarization [3] of such extended electronic states in solids has been intensively investigated. In this work we present a study on electrons performing an intersubband transition. Our time-resolved THz experiments show the full coherent response of the electrons even during the oscillation of the driving THz pulse.

2. Experiment

In our experiments we transmit few-cycle THz pulses through semiconductor heterostructures which modulate the incident THz pulse. The generation of the THz pulses is performed on n-doped GaAs by excitation with 80fs laser pulses [4]. Additionally, we generate for the time-resolution nearly half-cycle THz pulses at a LT-GaAs epitaxial layer. Time-resolution of the transmitted signal is obtained by linear mixing of the transmitted and the half-cycle THz beams. The superposition is detected with a 4.2K bolometer as a function of the delay between the pulses.

The investigated heterostructures are a conventional multiple quantum well structure (MQW) and a parabolic quantum well (PQW). The MQW structure consists of 10 periods of 540Å GaAs wells separated by 1600Å $Al_{0.3}Ga_{0.7}As$ barriers. The wells are modulation doped to give a sheet density of $n_e=2.8*10^{10}cm^{-2}$. The PQW structure consists of a 2000Å wide PQW embedded in $Al_{0.3}Ga_{0.7}As$ and is modulation doped with a sheet density of $2.7*10^{11}cm^{-2}$.

Since the intersubband transitions couple only to electric fields parallel to the growth direction we use two techniques to drive the transitions: i) coupling the THz pulses into the cleaved edge of the structures. ii) incidence on the front surface and coupling by a 16μm gold grating on top of the structures. Schottky gates on top of the structures enable us to deplete the wells from electrons applying a negative bias.

3. Results

Figure 1a) shows the exciting THz pulse which is used to drive the electron intersubband transitions of the quantum structures. The transmitted pulse is recorded in cleaved edge geometry. It contains the signature of the intersubband transition since the polarization of the electrons counteracts the driving THz field. However, this pulse is strongly chirped due to group velocity dispersion in the GaAs bulk. This masks the signal of the coherent response of the electrons in the quantum well.

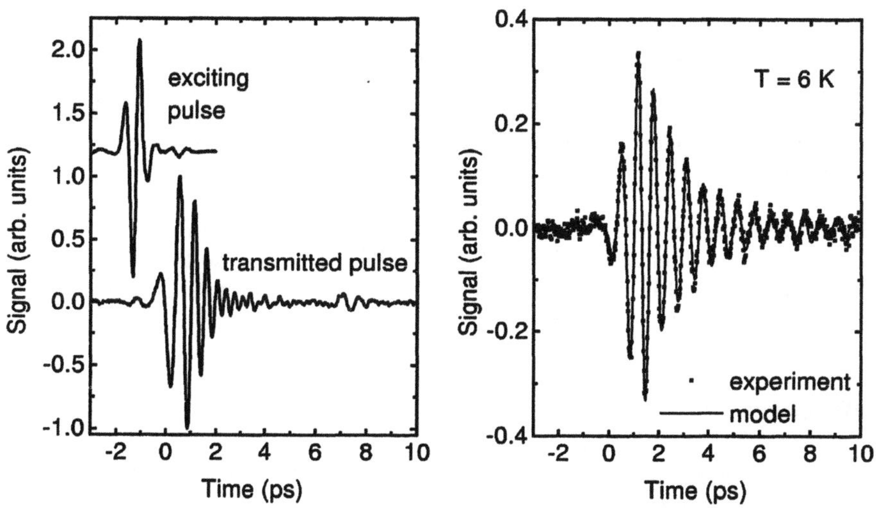

Fig. 1. a) Time-resolved THz pulse before and after transmission through the MQW structure in cleaved edge geometry. b) The dots show the signal recorded by modulation of the carrier density. The line shows the model calculation of the coherent polarization of the electrons in the quantum well.

Modulating the gate bias of the semiconductor structure and thus the electron density in the QW suppresses the chirp introduced by the GaAs [5] as shown in Fig. 1b). The recorded signal is directly proportional to the polarization of the driven electrons in the quantum well. It arises because the incident THz pulse excites carriers into a coherent superposition of states of the first and second subband. During the first picoseconds the amplitude of the polarization grows with each oscillation of the driving pulse. The resulting coherent oscillation of the electrons radiates along the direction of the incident beam. When the exciting THz pulse is over the carriers continue to radiate and the coherent signal is damped out by their free induction decay.

The calculation shown in Fig. 1b) is performed using the transient of the incident THz pulse, the intersubband frequency and a phenomenological damping time [5]. The excellent fit to the experimental data confirms the single-oscillator model. This enables us to determine the dielectric function of the isolated electron gas in the QW.

Temperature dependent measurements performed on the PQW structure give insight into the damping processes of the coherent oscillation of the electrons. Below 40K we observe a constant damping rate which can be attributed to alloy scattering. Above 40K the damping rate shows a strong increase which is due to the activation of optical phonon scattering.

4. Conclusion

We have performed time-domain measurements of electron intersubband oscillations in GaAs quantum structures. The data directly show the coherent response of the electrons. The results can be understood in terms of a single oscillator model. Temperature dependent measurements give insight into the different damping mechanisms.

Acknowledgements. We would like to acknowledge financial support by the Austrian Science Foundation (START Y47-PHY) and the Gesellschaft für Mikroelektronik (GMe). J.H. acknowledges support from a Cottrell College Science Award of Research Corporation.

References

1 E.O. Göbel, K. Leo, T.C. Damen, S. Schmitt-Rink, W. Schäfer, J.F. Müller and K. Köhler, *Phys. Rev. Lett.*, Vol. 64, 1801, 1990.

2 K. Leo, J. Shah, E.O. Göbel, T.C. Damen, S. Schmitt-Rink, W. Schäfer and K. Köhler, *Phys. Rev. Lett.*, Vol. 66, 201, 1991.

3 T. Dekorsy, P. Leisching, K. Köhler, and H. Kurz, *Phys. Rev. B*, Vol. 50, 8106, 1994.

4 R. Kersting, K. Unterrainer, G. Strasser, H.K. Kauffmann, and E. Gornik, *Phys. Rev. Lett.*, Vol. 79, 3038, 1997.

5 J.N. Heyman, R. Kersting, and K. Unterrainer, *Appl. Phys. Lett.*, Vol. 72, 644, 1998.

Part VI

Coherence in Semiconductors

Quantum Interference Control of Free Carriers in Semiconductors

H. M. van Driel, D. Côté, J. M. Fraser, N. Laman and J. E. Sipe

Department of Physics
University of Toronto
Toronto, Canada
M5S-1A7
Email:vandriel@physics.utoronto.ca

Abstract: Optically-induced quantum interference is used to generate and control electrical currents and free carrier populations in bulk semiconductors at room temperature. In a macroscopic picture this coherent response can be described in terms of nonlinear optical susceptibilities. We demonstrate current control in GaAs by simultaneous single and two photon absorption processes involving picosecond or femtosecond pulses via a divergent piece of a $\chi^{(3)}$ process. Using a polarization-controlled single beam and a $\chi^{(2)}$ process we demonstrate coherent control of current in CdSe, a wurtzite structure material. Finally we partially control carrier populations via a $\chi^{(2)}$ process.

1. Introduction

Manykin and Afanas'ev [1] were apparently the first to recognized that optical phase could be used along with interference between quantum mechanical pathways to control physical processes. Coherence control (CC) of matter has now been demonstrated for atomic and molecular processes [2,3]. Coherence effects have also been used to produce spatially asymmetric ejection of photoelectrons from atoms [4], and from AlGaAs quantum wells at 80 K [5] and to control exciton populations in AlGaAs quantum wells at 5K [6] using phase-related femtosecond pulses of the same wavelength. More recently, Wehner et al. [7] have used CC to reversibly control electron-phonon interactions in GaAs.

Until recently CC experiments were mainly confined to effective two-level system. However, a number of years ago we proposed [8] and later observed [9] CC of photocurrents via two beam *interband* transitions in a bulk semiconductor such as GaAs or low-temperature-grown GaAs (LT-GaAs). A description of CC can also be formulated within nonlinear optics since nonlinear susceptibilities determine the macroscopic coherent response of matter. Photocurrent CC processes are then obtained as previously unrecognized, physically divergent pieces of $\chi^{(2)}$ and $\chi^{(3)}$. We illustrate some of these photocurrent as well as population control effects here.

2. Coherently Controlled Currents in Semiconductors

The generation of electrical current can occur via simultaneous single and two photon absorption of coherently related beams of frequency ω and 2ω, for the case where $\hbar\omega < E_g < 2\hbar\omega$ with E_g being the semiconductor band-gap energy. The interference of quantum mechanical pathways[9] leads to an electron and hole (e,h) current *injection* process with the current evolution governed by

$$\frac{dJ_{e,h}}{dt} = \ddot{\eta}_{e,h} : E^{\omega} E^{\omega} E^{-2\omega} + c.c. - J_{e,h}/\tau_{e,h} \quad . \qquad (1)$$

Here $\tau_{e,h}$ is the current relaxation time and $\ddot{\eta}_{e,h}$ is a purely imaginary, fourth-rank tensor. Two beam CC currents are therefore possible in centrosymmetric and non-centrosymmetric materials. For both beams polarized along the crystalline x axis, the current can be related to $(\eta_{e,h})_{xxxx}$ and to the phase parameter $\Delta\phi = 2\phi_{\omega} - \phi_{2\omega}$.. It can be shown that, neglecting lifetime effects, the $\ddot{\eta}_{e,h}$ is related to a doubly divergent piece of a $\chi^{(3)}$ nonlinearity. Hence, in spite of the high order of this process, large effects can occur. For GaAs $(\eta_{e,h})_{xxxx} \sim 20\ s^{-2}mCV^{-3}$.

The generated current differs from a conventional current which occurs in a semiconductor when *pre-existing* or optically injected carriers move in response to a DC electric field. In the case of a CC current, carriers are *optically injected* into band states in a polar fashion with a speed determined by 2ω, E_g and the carrier effective masses respectively. The quantum interference in the two pathways leads to a classical-like state in which a majority of electrons initially move effectively in the same direction with a speed which is 100's of kms^{-1} for a 100 meV electron-hole excess energy. Surprisingly large peak currents up to ~ 1 $kAcm^{-2}$ are possible for a carrier density of only $\sim 10^{14}\ cm^{-3}$.

We have observed CC currents using harmonically related femtosecond, picosecond and nanosecond pulses with the fundamental beam having a wavelength of 1.3-1.6 μm. Fig. 1 shows the dependence of the current on the phase parameter $\Delta\phi$ in the case of illumination of an LT-GaAs (001) metal-semiconductor -metal (MSM) structure with a 50 μm gap. The source consists of 1 ps 1.55 μm and 775 nm pulses with nJ and pJ energy respectively. The data show the expected phase dependence when both beams have their optical polarization across the gap (crystal (100) axis), while no phase dependent signal is observed in the case of both beams polarized along the gap.

It is also possible to achieve CC effects when the two photon absorption process is nondegenerate. We have demonstrated this using 1400 and 1800 nm 150 fs pulses along with the sum frequency beam at 800 nm. In this case, as Eq. 1 gives and Fig. 2 shows, the induced CC current and steady state voltage on the MSM structure is linearly proportional to the sum frequency electric field strength.

In the above process the current magnitude and directionality is controlled by two optical beams which act externally to break the symmetry of the crystal. It is also possible to induce a coherently controlled photocurrent using a *single*

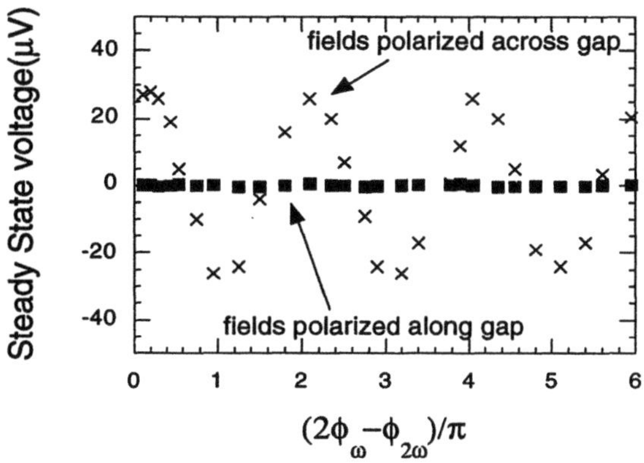

Fig. 1 Phase dependent steady-state voltage observed from an LT-GaAs wafer for two polarization configurations. A small background signal has been removed.

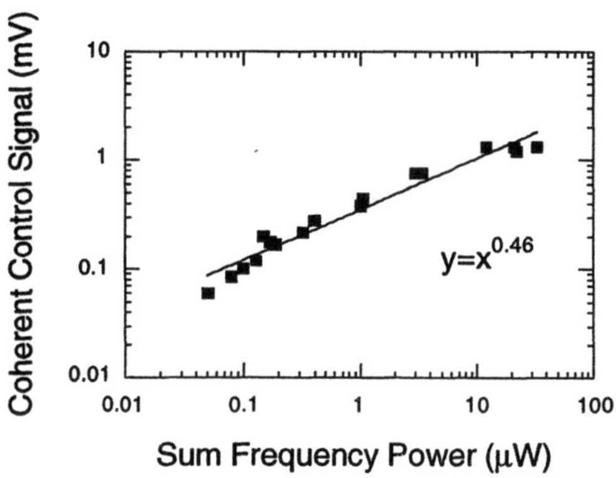

Fig. 2. Scaling of CC current with optical power .

optical beam but where the current directionality is controlled by the *internal* crystal symmetry and the relative phase of orthogonal beam components. When the photon energy is above the band-gap it is possible for certain crystal classes to allow interference between the absorption pathways of the two components leading to current generation. One seeks current injection terms of the form:

$$\frac{dJ_i}{dt} = \eta_{ijk}^{(2)} E_j^{-\omega} E_k^{\omega} = 2 \mid \eta_{ijk}^{(2)} \mid E_j^{-\omega} \mid\mid E_k^{\omega} \mid cos(\phi_j - \phi_k) \qquad (2)$$

for which the injected is linear with the beam intensity. Obviously one requires a crystal which is noncentrosymmetric. Symmetry requirements also dictate that the purely imaginary $\vec{\eta}$ tensor must be antisymmetric in the last two cartesian indices. The wurtzite family satisfies this condition and we have employed CdSe in preliminary experiments to observed this CC current. We calculate η_{yyz} to be ~10^8 $AV^{-2}s$, indicating that even a low power source such as a HeNe laser can induce a current. Figure 3 shows the steady state voltage induced across electrodes deposited onto a CdSe crystal oriented with its optic axis (z) along the gap. The signal shows the expected dependence on the relative phase between the y and z beam components, with the maximum voltage induced by right- or (with a sign change) left-handed circularly polarized light. This type of CC may provide an efficient method of producing THz radiation using ultrashort light pulses.

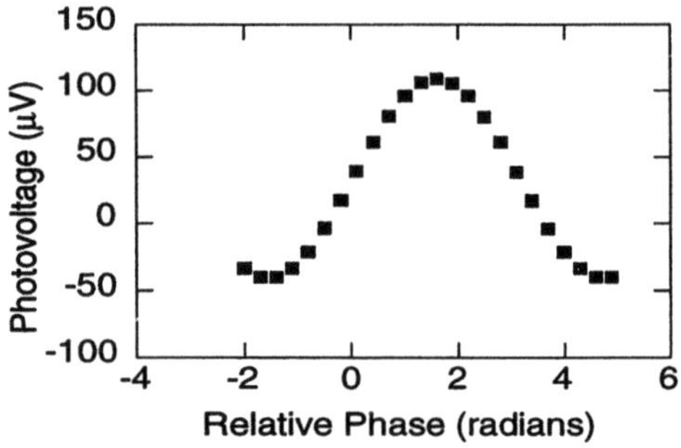

Fig. 3 Photovoltage generated by a phase controlled He-Ne laser beam; the relative phase is that between orthogonal polarization components. An induced DC photovoltage associated with contact effects shift the modulated curve.

3. Coherently Controlled Carrier Populations in GaAs

Coherent control of population has been demonstrated via discrete states in quantum well semiconductors [6]. To what extent can similar effects be observed in bulk semiconductors based on band-band transitions? We have attempted to control a population of carriers using harmonically related pulses. By considering the power deposition associated with coherently related beams one can show that interference effects lead to population control terms involving single- and two-photon absorption processes. The interference power deposition is:

$$P_{inter}^{\omega,2\omega} = \omega\varepsilon_0 (Im\,\chi_{ijk}^{(2)}(-2\omega;\omega,\omega)E_i^{-2\omega}E_j^{\omega}E_k^{\omega}\cos(\phi_i^{2\omega}-\phi_j^{\omega}-\phi_k^{\omega}) \quad (3)$$

For GaAs, the only nonzero element of $\chi^{(2)}$ is $\chi^{(2)}_{xyz}$. Therefore for beams normally incident on GaAs (001), there is no density modulation. For a GaAs (111) sample, however, the effective $\chi^{(2)}$ is $0.4\chi^{(2)}_{xyz}$. Experiments were carried out with 150 fs pulses with peak irradiance of 800 $MWcm^{-2}$ (1550 nm) and 20 $MWcm^{-2}$ (775 nm) respectively. In a 0.7 μm thick sample the peak carrier density is ~10^{18} cm^{-3}. The transmission of the 775 nm beam, whose total absorption is dictated by its relative phase with the 1550 nm beam is a measure of the total density induced.
We have inferred a modulation in the carrier density of a few % under these conditions. This low value is partly related to the small effective $\chi^{(2)}$

susceptibility for the geometry used as well as sample symmetry and dephasing effects.

4. Summary

Quantum interference processes have been used to induce and control electrical currents in bulk semiconductors at room temperature. Overall, coherent control points to new applications in optoelectronics and phase-controlled THz generation. As well it opens up new fundamental studies related to the generation and probing of previously unrealizable electron-hole distributions in semiconductors.

Acknowledgements. We are grateful for support from NSERC (Canada) and Photonics Research Ontario. HMvD acknowledges support from the Killam Fellowship Program of the Canada Council.

References

1. E.A. Manykin, and A.M. Afanas'ev, Sov. Phys. JETP **25**, 828 (1967); (J.Exptl. Theor. Phys. (USSR) **52**, 1246 (1967)).
2. P. Brumer and M.Shapiro, Accts. of Chem. Res. **22**, 407 (1989).
3. W.S. Warren, H. Rabitz, and M. Dahleh, Science **259**, 1581 (1993).
4. Y.Yin, C. Chen, D.S. Elliott, and A.V. Smith, Phys. Rev. Lett. **69**, 2353 (1992).
5. E. Dupont, P.B.Corkum, H.C. Liu, M. Buchanan, and Z. R. Wasilewski, Phys. Rev. Lett. **74**, 3596 (1995).
6. A.P. Heberle, J.J. Baumberg, and K. Köhler, Phys. Rev. Lett. **75**, 2598 (1995).
7. M. U. Wehner, M. H. Ulm, D.S. Chemla and M. Wegner, Phys. Rev. Lett. **80**, 1992 (1998).
8. R. Atanasov, A. Haché, J. L. P. Hughes,H. M. van Driel, and J. E. Sipe, Phys. Rev. Lett. **76**, 1703 (1996).
9. A. Haché, Y. Kostoulas, R. Atanasov, J. L. P. Hughes, J. E. Sipe, and H. M. van Driel, Phys. Rev. Lett. **78**, 306 (1997); A. Haché, J. E. Sipe, and H. M. van Driel, H. M., I.E.E.E. J. Quant. Elec. **34**, 1144 (1998).

Optical Control of Precession of Mn^{2+} Moment in $CdTe/Cd_{1-x}Mn_xTe$ Quantum Wells

R. Akimoto, K. Ando, F. Sasaki, S. Kobayashi, and T. Tani

Electrotechnical Laboratory , 1-1-4 Umezono, Tsukuba, Ibaraki 305, JAPAN
E-mail: akimoto@etl.go.jp

Abstract. We demonstrated the ultrafast optical manipulation of the Larmor precession of localized Mn^{2+} moments in $CdTe/Cd_{1-x}Mn_xTe$ quantum wells by two successive optical short pulses with duration of 200 fs.

In recent years, control of a quantum mechanical system by means of a sequence of optical short pulses, has attracted considerable interest in the field of condensed matter physics as well as optoelectronics aimed at ultrafast all-optical switching devices and chemistry pursuing unusual nonequilibrium states which may not appear in the usual way[1-3]. In this paper, we demonstrate the optical control of the Larmor precession of Mn^{2+} moments in $CdTe/Cd_{1-x}Mn_xTe$ quantum wells(QWs, x=0.35) by inputting two successive optical short pulses with its duration of 200 fs. The heterostructures based on diluted magnetic semiconductors[4], such as $Cd_{1-x}Mn_xTe$, have received a lot of attention due to the inherent strong s,p-d exchange interactions between mobile carriers and embedded magnetic ions as well as the controllability of the interacting volume between them by tailoring the confinement potential energy of carriers. In $CdTe/Cd_{1-x}Mn_xTe$ QWs structure, carriers penetrate slightly into the magnetic barrier layers and interact with the Mn^{2+} ions through s,p-d exchange interactions. Thus, we can obtain the dynamical information of Mn^{2+} spin at the vicinity of the interface between the well and barrier layer selectively.

Figure1(a) shows the time-resolved Kerr rotation signals at 3 Tesla for the sample with the well width (L_z) of 40 Å, when the sample is excited by only single σ^+ pump pulse. We observe pronounced oscillations of the Kerr rotation signal, which have two different periods of oscillations. As shown in earlier studies[5,6], the oscillations, which decay faster and have shorter periods, come from the Larmor precession of conduction electron moments around the external field and the frequency is amplified by s-d exchange interaction. The comparison with the rapidly decaying oscillations for the sample with wider well width in Fig. 1(b) also substantiates the above interpretation. Meanwhile, the heavy hole spin is quantized along a growth direction due to a large heavy- and light hole splitting of 60 meV in the present sample, so that the heavy-hole can not precess but simply decays. On the other hand, the long-persisting signal is attributed to the Larmor precession of Mn^{2+} moments around the external field and its free induction decay. The hole exchange field generates transient torques for Mn^{2+} moments and rotates them away from the external field[5,6].

Figure 1(b) shows the result of the comparable measurement for the sample with L_z=66 Å at 5 Tesla. Although a clear oscillatory signal originating from the Larmor precession of conduction electron moments is observed, a signal from the Mn^{2+} Larmor precession is not detected. As mentioned before, carriers penetrate slightly into the magnetic barrier layers and interact with the Mn^{2+} ions through s,p-d exchange interactions. Here, note that both samples have the identical thickness of magnetic barrier layer. Thus, the disappearance of Mn^{2+} beat for the sample with L_Z=66 Å is due to less penetration effect of hole wavefunction into barrier, that is, the number of Mn ions, which interacts with a hole, is not enough to observe the Mn^{2+} beat signal. Further, it also suggests that the Mn^{2+} beat signal observed in the sample with L_Z=40 Å reflects the dynamical information of Mn^{2+} spin not at the entire region of the barrier layer but at the vicinity of the interface between the well and barrier layer.

Figure 1(c) shows the field dependence of the maximum tipping angle measured for the sample with L_Z=40 Å and calculated from the Bloch model[6] including transient hole exchange field as a driving mechanism for the Mn^{2+} Larmor precession. The calculation can reproduce well the external field dependence on the tipping angle which decreases with increasing the external field. The best value of the initial hole exchange field to fit the experimental data is determined to be 0.6 mT. The obtained exchange field is in reasonable agreement with the effective exchange field estimated from p-d exchange Hamiltonian including the excitation density effect[6]. This agreement also supports the main mechanism described above is reasonable, that is, the Mn^{2+} moments are coherently rotated by the transverse exchange field of the heavy-hole spin.

Figure 2(a) and (b) shows the time-resolved Kerr rotation signals, when the sample with L_Z=40 Å is excited by a sequence of two σ^+ pump pulses. As shown in Fig. 3(a), the amplitude of the Larmor precession of Mn^{2+} moments is

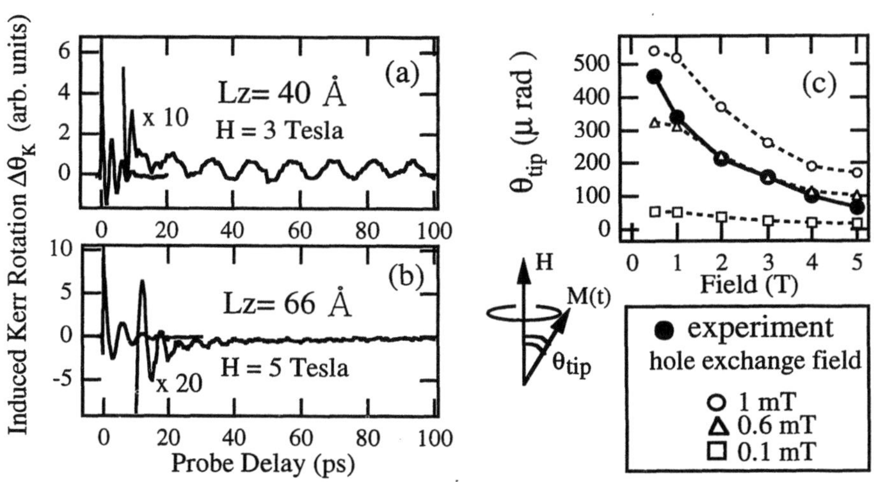

Fig. 1. Time-resolved Kerr rotation signals excited by single σ^+ pump pulse for samples with (a) L_Z=40 Å at 3 Tesla and (b) L_Z=66 Å at 5 Tesla. The temperature of both samples is 5 K. (c)The field dependence of the maximum tipping angle of Mn^{2+} moment obtained experimentally (solid circles, L_Z=40 Å), and the tipping angle calculated from the Bloch equation when the initial exchange field of the heavy-hole spin takes 0.1, 0.6, and 1 mT.

doubled, when the time separation between two pump pulses is $\tau=T_L=12$ ps, where T_L corresponds to the oscillation period of the Larmor precession of Mn^{2+} moment at 3 Tesla. Meanwhile, the amplitude of the Larmor precession is completely suppressed, for $\tau=1(1/2)T_L= 18$ ps. With a reverse of the polarization of the second pump pulse to σ^-, the Larmor precession is completely suppressed for $\tau=T_L$, while the amplitude of the Larmor precession is doubled for $\tau=1(1/2)T_L$ as shown in Fig. 2(c) and (d). A reverse of the circular polarization of the second pump pulse causes the reverse of the direction of the transient hole exchange field, because the direction of heavy hole spin depends on the circular polarization of the pump pulse. Thus, the specific time separation between two pump pulses to double or suppress the amplitude of the Larmor precession of Mn^{2+} moments exchanges each other.This result confirms that we can amplify the amplitude of the Larmor precession of Mn^{2+} moments by the repetitive excitation which is synchronizing with the period of the Mn^{2+} Larmor precession.

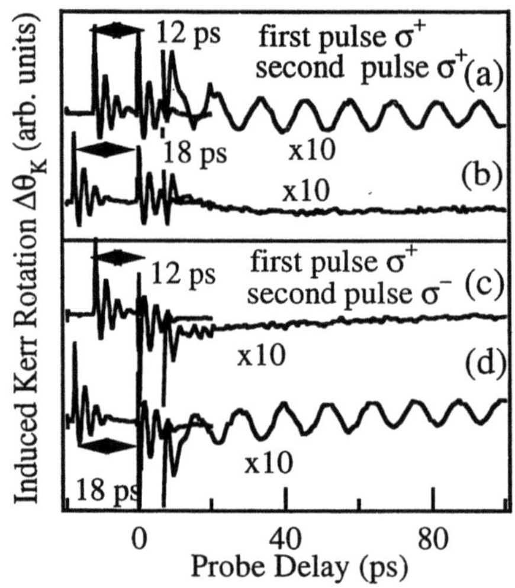

Fig. 2. Time-resolved Kerr rotation signals for the sample with L_Z=40 Å excited by double pump pulses(σ^+_{first}, σ^+_{second}) at the pulse separation of (a)12 ps and (b)18 ps, double pump pulses(σ^+, σ^-) at the pulse separation of (c)12 ps and (d)18 ps at 5 K and 3 Tesla.

References

1 P. C. M. Planken, I. Brener, M. C. Nuss, M. S. C. Luo, and S. L. Chuang, Phys. Rev. B**48**, 4903 (1993).

2 A. P. Heberle, J. J. Baumberg, and K. Kohler, Phys. Rev. Lett. **75**, 2598 (1995).

3 A. M. Weiner, D. E. Leaird, G. P. Wiederrecht, and K. A. Nelson, J. Opt. Soc. Am. B**8**, 1264 (1991).

4 Diluted Magnetic Semiconductors, edited by J. K. Furdyna and J. Kossut (Academic, Tokyo, 1988) Vol. 25. p.295.

5 S.A. Crooker, J.J. Baumberg, F. Flack, N. Samarth, and D.D. Awschalom , Phys. Rev. Lett. **77**, 2814 (1996).

6 R. Akimoto, K. Ando, F. Sasaki, S. Kobayashi, and T. Tani, Phys. Rev. B**57**, 7208 (1998).

Ultrafast Time-Resolved Quantum Beats in the Polarization State of Coherent Emission from Quantum Wells

Arthur L. Smirl, X. Chen, and O. Buccafusca

Laboratory for Photonics and Quantum Electronics, 138 IATL, University of Iowa, Iowa City, IA 52242
Phone: (319) 335-3460 Fax: (319) 335-3462 E-mail: art-smirl@uiowa.edu

Abstract. Dual-channel spectral interferometry is used to measure the ellipticity, the orientation of the polarization ellipse, and the sense of rotation of the four-wave-mixing signal from a GaAs-AlGaAs multiple quantum well. Each is observed to oscillate dramatically at the heavy-hole-light-hole quantum beat frequency. During each beat period (in the strong quantum beat regime), the ellipticity oscillates twice between linear and almost circular polarization, the orientation of the polarization ellipse rotates through a complete 180°, and the sense of rotation changes from left to right circular polarization.

Conventional four wave mixing (FWM) techniques have provided considerable information about the spectral and temporal amplitudes of ultrafast coherent radiation emitted by semiconductors and multiple quantum wells (MQWs), but they typically provide no information about the temporal or spectral phases. In addition, conventional techniques have been used to address the dependence of the magnitude of the FWM signal (or its spectrum) on the relative polarizations of the incident pulses, but there have been few attempts to measure the polarization state of the FWM signal itself. Recently, however, we have used direct time-resolved ellipsometric[1,2] and spectrally-resolved interferometric techniques[3] to characterize the polarization state of the FWM signal. These studies [1,2] have established the sensitivity of polarization-sensitive techniques for studying many body and biexcitonic processes. However, to date, our experiments have been limited to investigations of the emission from the heavy hole (hh). That is, the laser was purposefully tuned slightly below the hh exciton to ensure that we excited as few light holes (lh) as possible

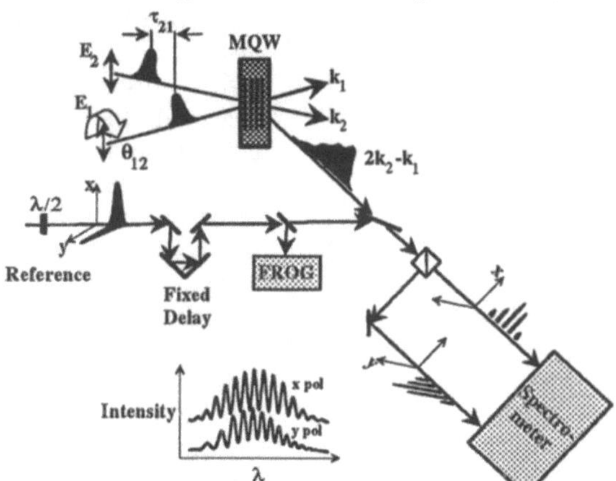

Fig. 1: Dual-channel measurement of the amplitude, phase and polarization state of FWM signal. Typical spectral interferograms are shown in inset.

and to avoid the complication of quantum beating between the lh and hh.

Here, we describe measurements of the dynamics of the amplitude, phase and polarization state of the emission from a GaAs-AlGaAs MQW when the excitation spectrum is tuned so that both hh and lh are excited and strong quantum beats are observed. We find that all of the parameters that describe the polarization state (including the ellipticity, the orientation of the polarization ellipse, and the sense of rotation) oscillate at the hh-lh beat frequency. To our knowledge, this is the first observation of time-resolved quantum beats in the polarization state. Quantum beats have, of course, been studied previously ([4] and references therein). In fact, it has been demonstrated that the quantum beats produced by incident pulses with parallel polarizations are exactly out of phase with the beats produced by orthogonally polarized pulses[5]. However, the polarization state of the emitted light was not measured.

We use a dual-channel spectral interferometric technique (Fig. 1) to completely characterize the emission[3]. In this technique, the FWM mixing signal is allowed to interfere with a well characterized time-delayed reference pulse (150 fs) that has equal x and y components. The combined signal is separated into x and y components, and each component is separately dispersed by a spectrometer. The spectral amplitude and the phase of each component of the signal can then be recovered from the corresponding spectral interferogram (inset Fig. 1), and the temporal amplitude and phase can be obtained by inverse Fourier transformation. Using this technique, we have systematically measured the amplitude, phase and polarization state of the FWM emission as the polarization of the E_1-pulse is rotated through an angle θ_{12} with respect to a constant linear E_2 polarization. These measurements were then repeated for selected time delays between the pump and probe pulses and as a function of the

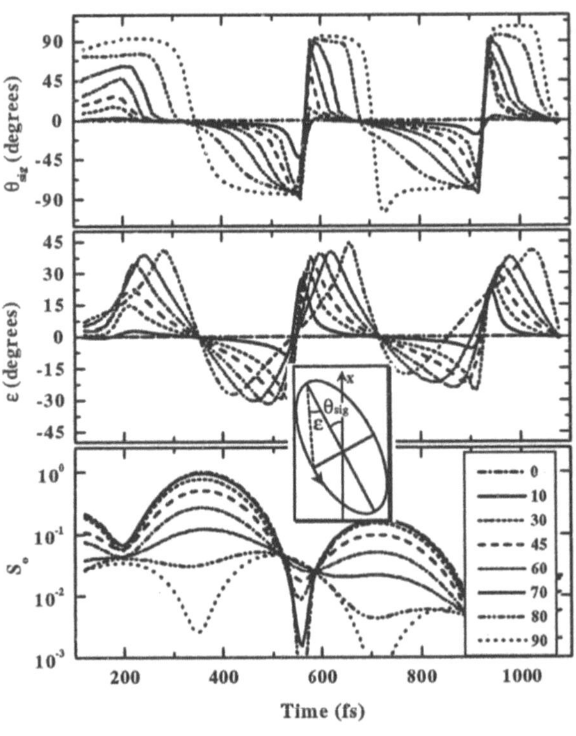

Fig, 2. Representative time-resolved measurements of the azimuthal angle θ_{sig}, ellipticity ε, and the total intensity S_0 of the FWM intensity for selected angles θ_{12} between the two linearly polarized incident pulses.

detuning of the excitation spectrum.

Typical measurements of the parameters that directly determine the time-resolved polarization state of the FWM emission from a MQW are shown in Fig. 2 for selected input polarizations. The sample consists of 10 periods of ~14-nm-wide GaAs wells alternating with 17-nm-thick $Al_{0.3}Ga_{0.7}As$ barriers, and the excitation spectrum was tuned such that the ratio of lh-to-hh emission was ~0.7. The corresponding temporal behavior of the polarization ellipse for approximately one beat period and for an angle between the two input polarizations of θ_{12}=60° is sketched in Fig. 3. Notice that the orientation of the ellipse rotates through a full 180° in one beat period. The ellipticity goes from linear to highly elliptical and then back to linear twice per beat period. And, finally, the sense of rotation changes from clockwise (left circular) to counter clockwise (right circular) each period.

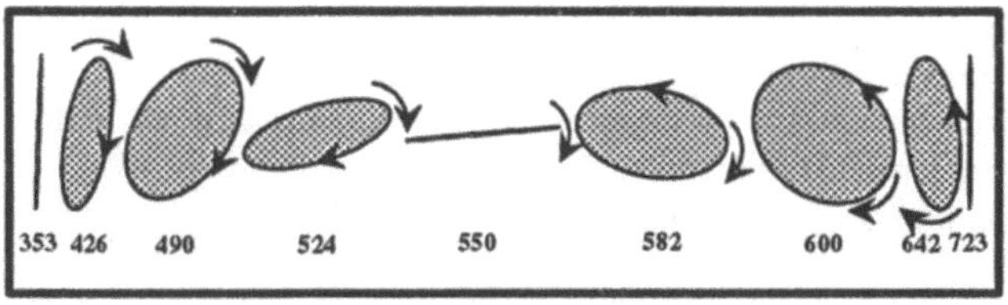

Fig. 3. Schematic of the temporal evolution of the polarization ellipse corresponding to the data in Fig. 2 for θ_{12}=60°.

The behavior illustrated in Fig. 3 is typical of that observed at this time delay (+300 fs) for hh to lh ratios near unity and for large angles between input polarizations. For other time delays, these tendencies vary significantly. Moreover, as expected, the beats become negligible when the laser is tuned well below the hh, so that few lh are excited. In the latter regime, our previous data[1,2] are recovered. The features in the polarization beating shown here (and others not shown) can be qualitatively described by a single-particle density matrix description[5] without the inclusion of many body effects; however a detailed description requires the inclusion of such effects.

Finally, we emphasize that a knowledge of the temporal phases of both the x and y components of the radiation are necessary for determining the vectorial dynamics (i.e., the polarization state) portrayed in Fig. 2 and 3 and that this information is not provided by conventional techniques.

Acknowledgments. This work was supported by the U. S. Army Research Office and the National Science Foundation.

References

1 J. A. Bolger, A. E. Paul, and A. L. Smirl, Phys. Rev. B 54, 11666 (1996).
2 A. E. Paul, J. A. Bolger and A. L. Smirl, J. Opt. Soc. Am. B 13, 1016 (1996).
3 W. J. Walecki, D. N. Fittinghoff, A. L. Smirl, R. Trebino, Opt. Lett. 22, 81 (1997)
4 A. E. Paul, W. Sha, S. Patkar, and A. L. Smirl, Phys. Rev. B 51, 4242 (1995).
5 S. Schmitt-Rink, D. Bennhardt, V. Heuckeroth, P. Thomas, P. Haring, G. Maidorn, H. Bakker, K. Leo, D. S. Kim, J. Shah, K. Köhler, Phys. Rev. B 46, 10460 (1992).

Femtosecond Coherent Response of Spin-orbit Split-off Excitons in InP

G. R. Allan, K. C. Hall and H. M. van Driel
Physics Department, University of Toronto
60 St. George St. Toronto, Canada, M5S 1A7
and
Walter Pötz
Physics Department, University of Illinois at Chicago
Chicago, IL 60607, USA

Abstract. From spectrally resolved femtosecond four-wave mixing (FWM) in InP at 5 K we observe excitons associated with the spin-orbit split-off gap at 1.52 eV. The FWM spectrum reveals Fano interactions of these excitons with continuum excitations when co-polarized pulses are used. For perpendicularly polarized pulses only weak excitonic features are observed and the spectrum is dominated by the continuum photon echo response.

1. Introduction

Coherent nonlinear optical techniques such as femtosecond four-wave mixing (FWM) have provided much insight into the properties of excitons associated with the fundamental (E_0) gap in bulk materials and related subbands in quantum wells [1]. However, to our knowledge no FWM experiments have been reported for excitons associated with higher energy gaps in bulk materials and indeed such excitons are usually not observable, even in linear optical experiments. Here we

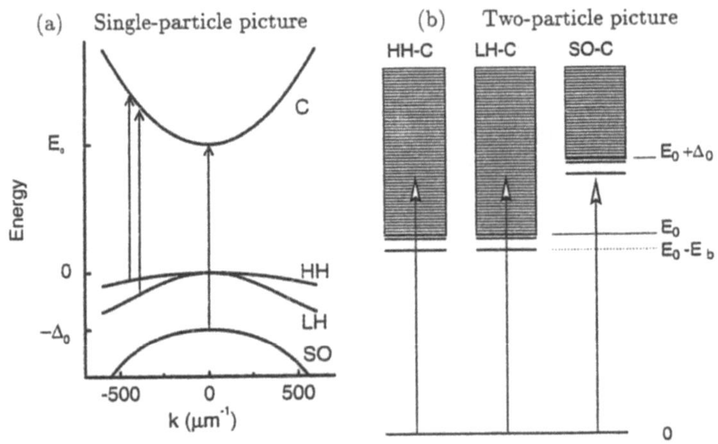

Fig. 1 a) Schematic band band-structure of InP. The energy axis is not to scale to show the valence bands more clearly; b) Energy level diagram of the excitation spectrum in the exciton picture.

offer results of spectrally resolved FWM experiments in InP (E_0 =1.41 eV) near the conduction band/ spin-orbit-split off (SO) band gap ($E_0+\Delta_0$= 1.52 eV) and clearly demonstrate the existence of such excitons. Fig. 1 shows the single and two-particle states in the vicinity of the fundamental gap. For optical excitation near the SO edge, excitons associated with the SO-conduction band (CB) are energetically resonant with transitions connecting light hole (LH) and heavy hole (HH) states to the CB. Because Δ_0 is only 110 meV a Fano interaction may also be anticipated since the SO and LH/HH states will be adjacent in k-space and Coulomb exchange effects are not negligible.

2. Experimental Results and Discussion

The FWM experiments used 35 fs, 1.52 eV (820 nm) pulses from a Ti:sapphire laser in the two-beam, self-diffraction geometry with co- (‖) or perpendicularly (⊥)

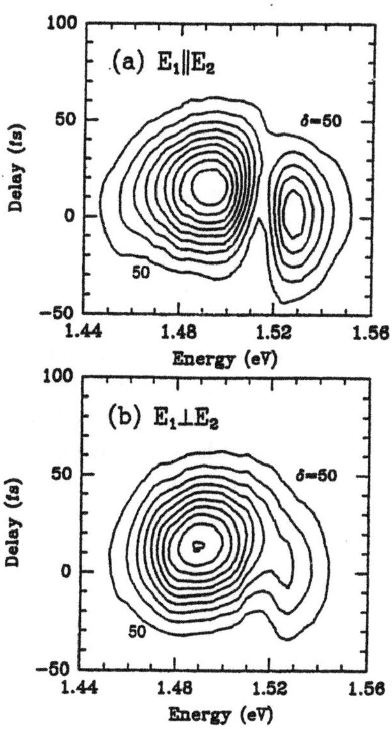

Fig.2: Contour plots for ‖ FWM (a) and ⊥FWM (b) signals at N= 1.5×10^{17} cm^{-3} with a spacing of 50 units.

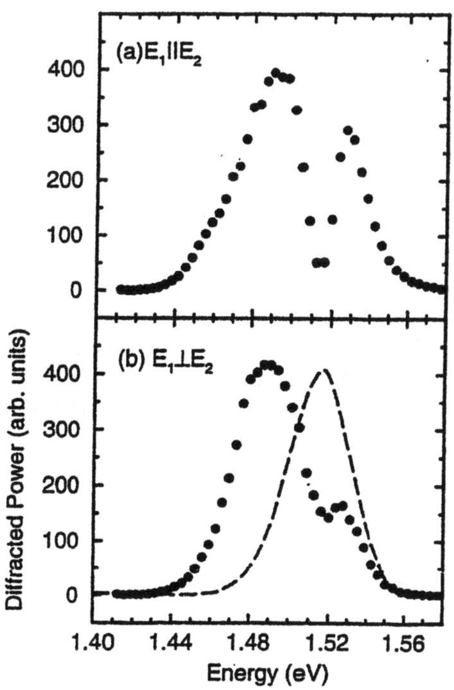

Fig. 3: Spectra of the ‖ (a) and ⊥ (b) FWM signals for zero delay corresponding to Fig. 1; the dashed curve indicates the laser spectrum.

polarized pulses. The sample is a 0.3 μm thick piece of bulk, intrinsic InP held at 5 K. Fig. 2 shows the ||FWM and ⊥FWM spectra for an excitation (electron-hole) density of 1.5×10^{17} cm^{-3} as a function of time delay between the two pulses. Fig. 3 shows the spectra for zero delay. For the || case, the broad emission which peaks at 1.49 eV is characteristic of a continuum photon echo response. The high energy side of the spectrum also has a peak at 1.525 eV near zero delay and a deep valley at 1.515 eV. These features, unlike the photon echo signal, do not shift as one tunes the laser. For the (⊥) case the structure near 1.52 eV is much weaker (more recent results of ours obtained with higher bandwidth waveplates indicates that essentially only the photon echo response is observed). The asymmetrical spectral characteristics near 1.52 eV are suggestive of Fano interference [2]. Prior observations of Fano effects in semiconductors using linear optical and/or FWM techniques were made under conditions of reduced dimensionality, e.g., using quantum wells [3] or magnetic fields [4]. For our bulk material there is no evidence for Fano effects in a linear transmission spectrum (only a weak shoulder appears as one tunes across the $E_0+\Delta_0$ gap).

To interpret our results a set of semiconductor Bloch equations for the three valence and one conduction band are developed. We include Coulomb interactions within the Hartree-Fock approximation [5] such as exchange coupling terms leading to exciton formation, but also the *intervalence band* exchange terms which couple SO and HH/LH states. The latter terms can yield a Fano interaction in the picture of Fig. 2 as discussed by Reining and Egry earlier [6]. From the model, in the linear regime only a very weak excitonic feature on a broad background is obtained consistent with Ref. [6]. In 3rd order there are two contributions to the FWM signal. The first contribution produces equal amplitude photon echoes for the || and ⊥ configurations, consistent with the data. The second contribution, from the coupled SO exciton-continuum, yields a Fano-like emission peaked at zero-delay. The optical polarization sensitivity of this contribution is contained within the matrix elements which establish the third-order electronic polarization density. This can occur both through the SO-LH/HH exchange interaction and also excitation-induced dephasing-like (EID) terms [1] wherein the density grating (and hence diffraction efficiency) vanishes for the ⊥ case. EID effects have likewise been attributed to explain polarization sensitivity in FWM studies of fundamental gap excitons.

References:

1. J. Shah, *Ultrafast Spectroscopy of Semiconductors and Semiconductor Nanostructures* (Springer Verlag, Berlin, 1996).
2. U. Fano, Phys. Rev. **124**, 1866 (1961).
3. D. A. Broido and L.J. Sham, Phys. Rev. B **34**, 3917 (1986).
4. S. Glutsch, U. Siegner, M.-A. Mycek and D.S. Chemla, Phys. Rev. B **50**, 17009 (1994).
5. W. Pötz, Phys. Rev. B **54**, 5647 (1996).
6. L. Reining and I. Egry, Sol. State Commun.,**68**, 981 (1988)

Femtosecond Dynamics and Coherence Properties of Resonant Secondary Emission from Excitons in GaAs Quantum Wells

Michael Woerner[1,2], Jagdeep Shah[1]

[1] Bell Laboratories, Lucent Technologies, Holmdel, NJ 07733, USA
[2] Permanent Address: Max Born Institut, Rudower Chaussee 6, D-12489 Berlin, Germany
Email: woerner@mbi-berlin.de

Abstract. Resonant excitation with two phase-locked pulses and subsequent femtosecond upconversion provides new insights into dynamics of various components of secondary emission with different coherence properties.

Resonant secondary emission (RSE) from quantum wells is a subject of considerable current interest. The initial dynamics of RSE in GaAs quantum wells was first investigated by Wang et al. [1] and later extended to lower intensities by Haacke et al. [2]. The emission was interpreted in terms of resonant Rayleigh scattering (RRS) from static disorder for low excitation intensities and in terms of incoherent photoluminescence (PL) for high intensities. Gurioli et al. [3] investigated the problem using a pair of excitation pulses which provide information about the ratio of the time-integrated RRS and PL similar to cw experiments [4]. A direct time-resolved separation of the two components is still missing.

We have taken a different approach in which the sample is resonantly excited by a pair of phase-locked pulses, and the emission is detected by femtosecond upconversion. This novel technique allows us to study the femtosecond dynamics as well as the coherence properties of the excitonic RSE simultaneously. We identify for the first time three different contributions to RSE with different coherence properties in space and time.

The experimental setup is shown in Fig.1. Two phase-locked femtosecond Ti:Sapphire laser pulses from an actively stabilized Michelson interferometer (separation τ_M and relative phase ϕ) excite excitons in a high quality GaAs/AlGaAs quantum well sample (17 nm wells, temperature 10 K). The subsequently emitted RSE from the excitons is upconverted in a nonlinear $LiIO_3$ crystal using 200-fs (λ=1530 nm) gating pulses from a synchronously pumped OPO. The gate delay τ_G is measured from the second excitation pulse.

Typical transients of the RSE intensity as a function of τ_G are shown in Fig. 1 for two different Michelson delays τ_M. Solid and dashed lines represent curves for opposite phases ϕ between the excitation pulses. The data exhibit two peaks at the positions of the excitation pulses that are due to nonresonant surface-scattering. The transient intensity shows a fast bell-shaped rise and decay within 10 ps followed by a slower decay at later times. As we vary the phase Φ between the excitation pulses the intensity of RSE follows a sinusoidal fringe pattern which is also present at τ_G=0. At this gate delay the surface-scattered light of the second excitation pulse constructively or destructively interferes with RSE triggered by the first pulse as visualized in the inset of Fig. 2. This means that there is a optically coherent component of RSE; i. e. it has a macroscopic electric field with a defined phase relation to the field of the first excitation pulse. The transient intensity of this fully coherent contribution can be calculated with $I_{coh}=[(I_{max})^{1/2}-(I_{min})^{1/2}]^2/4$ and is plotted as a function of time in Fig. 2(a).

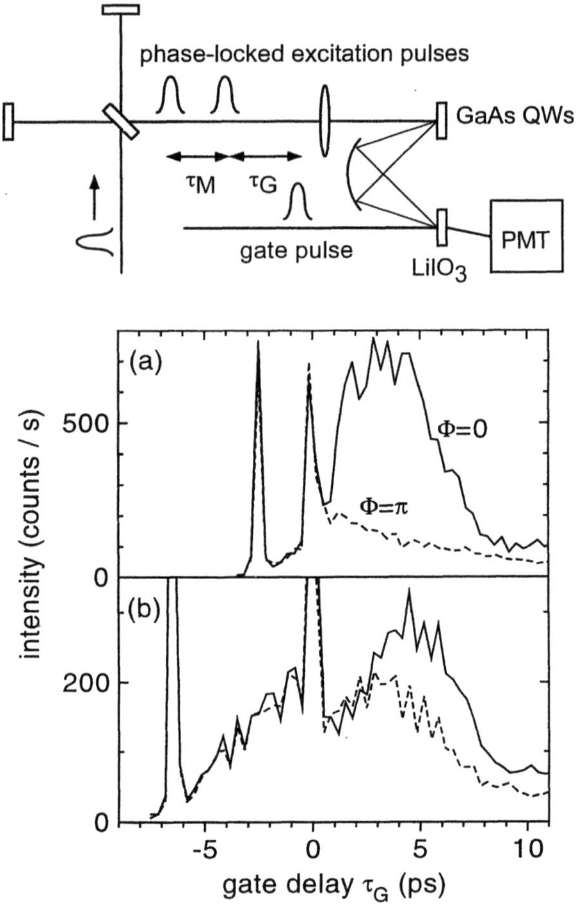

Fig. 1. Experimental setup for resonant excitation with a pair of phase-locked pulses (separation τ_M) and subsequent gated detection of secondary emission at τ_G after the second pulse. The intensity is plotted as a function of τ_G for opposite relative phases between the excitation pulses for pulse separations of (a) τ_M = 2.5 ps and (b) 6.5 ps.

Our experiments allow us for the first time to identify and separate three different contributions to RSE with different coherence properties: (i) As discussed in the previous section our experiments demonstrate that the RSE has a optically coherent part with an electric field in a defined phase relation to the excitation pulses. This is a new component of RSE which was never considered in previous studies [1-4].

Now, we discuss the optically incoherent components of RSE whose transient intensity still has some phase memory (coherence) of the excitation process. (ii) For very long gate delays τ_G>15 ps the fringe contrast defined by $R=(I_{max}-I_{min})/(I_{max}+I_{min})$ does not change as function of τ_G and has to be attributed solely excitation process. In this case RSE has no longer a phase relation to the excitation pulse and is consequently incoherent luminescence. As we plot the fringe contrast as a function of τ_M [solid symbols Fig. 2(b)] we obtain direct information about the excitonic absorption process [3]. The decay of R as a function of τ_M corresponds to the inverse linewidth of the excitonic absorption band. In this case RSE consists of two independent (incoherent) steps, first absorption and then emission.

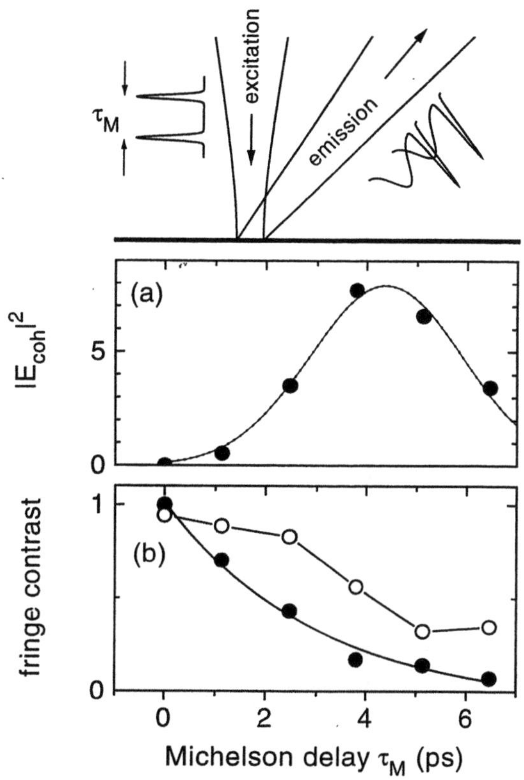

Fig. 2. (a) Interference of the surface-scattered light of the second pulse (τ_G=0 ps) with the spatially coherent part of RSE triggered by the first pulse. The transient intensity I_{coh} of the optically coherent RSE has been calculated from the fringe pattern using $I_{coh}=[(I_{max})^{1/2}-(I_{min})^{1/2}]^2/4$.
(b) Filled symbols: fringe contrast as a function of τ_M for large gate delays (τ_G > 15 ps). Solid line: exponential fit curve. Open symbols: highest fringe contrast observed in the time interval $0 < \tau_G < 12$ ps.

(iii) The most complex situation occurs in the interval of gate delays $0<\tau_G<12$ps. Here, even when the fringe contrast has decayed at long gate delays the intensity of RSE is still controlled by the phase of the excitation pulses. The open symbols in Fig. 2(b) show the highest fringe contrast in that interval of gate delays. The RSE in this time interval is a mixture of incoherent and coherent components governed by both the dynamics of excitonic populations (luminescence) and coherent polarizations (Resonance Rayleigh scattering).

Acknowledgements. We acknowledge L. N. Pfeiffer for the growth of the high-quality quantum well sample used in these studies and many useful discussions with D. Birkedal, E. Ippen, R.A. Kaindl, Ch. Lienau and T. Elsaesser.

References

1 H. Wang, J. Shah, T. C. Damen, L. N. Pfeiffer, Phys. Rev. Lett. **74**, 3065, 1995.
2 S. Haacke, R. A. Taylor, R. Zimmermann, I. Bar-Joseph, B. Deveaud, Phys. Rev. Lett. **78**, 2228, 1997.
3 M. Gurioli, F. Bogani, S. Ceccherini, M. Colocci, Phys. Rev. Lett. **78**, 3205, 1997.
4 J. Hegarty, M. D. Sturge, C. Weisbuch, A. C. Gossard, W. Wiegmann, Phys. Rev. Lett. **49**, 930, 1982.

Femtosecond Interferometric Measurements of Secondary Emission from Resonantly Excited Quantum Wells

Dan Birkedal and Jagdeep Shah

Bell Laboratories, Lucent Technologies
101 Crawfords Corner Road, Holmdel, NJ 07733, USA.

Abstract. We report on the first interferometric measurements of resonant Rayleigh scattering from quantum well excitons. Using ultrafast spectral interferometry we observe high-contrast spectral fringes resonantly enhanced over the exciton emission line. These studies allow us to determine the time evolution of Rayleigh scattering unambiguously. We discuss the implications of our observations for the scattering process and the nature of the quantum well disorder in high quality samples.

The nature of secondary emission from quantum well excitons following ultrafast resonant excitation has been intensively debated recently [1-3]. It is now commonly accepted that resonant secondary emission (RSE) is composed of luminescence and Rayleigh scattering. While luminescence is due to an inherent incoherent process, Rayleigh scattering is in principle fully coherent and an interferometric measurement technique could provide a clear separation of the two contributions to RSE. Using ultrafast spectral interferometry [4], or TADPOLE [5], where the RSE is cross-correlated with a reference laser pulse, we present the first interferometric measurements of RSE. The full amplitude and phase information provided by the interferometric technique allows us to recover the time evolution of the coherent part of the RSE, and provides new insights into the nature of Rayleigh scattering and disorder in quantum wells.

The sample under investigation is a molecular beam epitaxially grown 20 period $GaAs/Al_{0.3}Ga_{0.7}As$ quantum well structure with 130 Å wells and 150 Å barriers. The sample is mounted in a cold finger liquid He cryostat and excited by ≈120 fs transform-limited pulses from a self mode-locked Ti:Sapphire laser focused to a spot size of ≈ 40 μm. At the lowest excitation levels the sample shows a heavy-hole (hh) exciton emission line of full width at half maximum of 0.6 meV. The RSE is collimated and overlapped with a reference pulse on a variable beam splitter that allows adjustment of the relative intensities of the reference beam and the RSE. To assure maximum fringe contrast a 1 mm^2 aperture is placed in the output beam after the beam splitter to reject the part of the RSE which is not overlapped with the reference beam. The signal is dispersed by a 32 cm focal length spectrograph and recorded by a LN_2 cooled multichannel detector. A suitable number of spectral fringes over the resonances under investigation is achieved by adjusting the path difference between the emission and the reference beams.

We show in Fig. 1 the spectral interferogram and the time integrated RSE spectrum for excitation resonant with the hh exciton at an exciton density of $4\cdot10^8$ cm^{-2} and a sample temperature of 10 K. The interferogram has well resolved fringes resonantly enhanced at the hh and light-hole (lh) exciton transitions at 807.0 nm

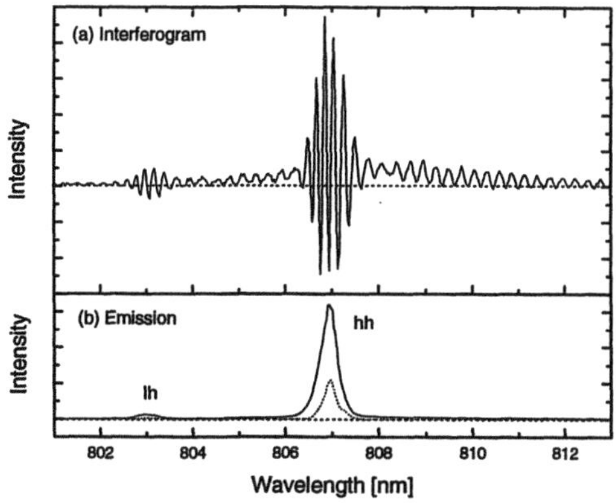

Fig. 1. Interferogram (a) and total time integrated secondary emission (b). Also shown in (b) is the spectrum of the coherent part of the RSE retrieved using Fourier-transform interferometry.

and 803.0 nm, respectively. The interferogram was obtained by first recording the RSE spectrum by blocking the reference beam, then the reference spectrum by blocking the RSE path and finally the spectrum resulting from interfering the RSE and reference beams. Subsequently, the interferogram was produced by subtracting the first two spectra from the last spectrum. The observation of high-contrast fringes shows that the phases of different spectral components are correlated; i.e. the emission is fully coherent.

The interferogram also shows fringes at wavelengths different from the resonant wavelengths. These fringes are due to non-resonantly surface-scattered light. The coherent part of the RSE has a smaller fringe spacing than the surface-scattered light. This indicates that the coherent part of the RSE is delayed in time with respect to the surface-scattered light. We also observe a phase shift and a node of the fringes near the resonance, indicating a phase shift between the surface-scattered light and the coherent part of the RSE.

A Fourier-transform of the spectral interferogram shown in Fig. 1(a) yields the temporal convolution of the fields of the coherent part of the RSE and the reference pulse [4,5]. We shown in Fig. 2 (a) the result of such a transform. The results display an initial narrow peak originating from the non-resonantly surface scattered light. This allow us to determine the time origin and the experimental resolution. The coherent part of the RSE is seen to display a delayed rise peaking at ≈ 4 ps in addition to a modulation due to the hh-lh exciton splitting. The time dependence of the

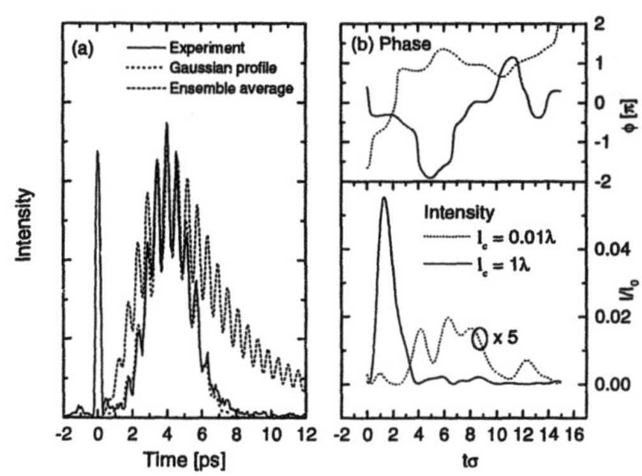

Fig. 2. Temporal profile of the RSE with line shape fits (a) and results of a model calculation of emitted field from correlated disorder (b).

coherent part of the RSE is found to be relatively independent of excitation density ($1 - 100 \cdot 10^8$ cm^{-2}) and temperature (10-70 K). Using the methods of spectral interferometry [4,5] we recover the spectrum of the coherent part of the RSE as shown in Fig. 1(b). This allows us to compare the fraction of the emission due to coherent and incoherent processes. We find that more than 25% of the total time integrated emission is fully coherent for excitation densities up to $50 \cdot 10^8$ cm^{-2}.

The experimental data in Fig. 2 (a) follows a Gaussian to a good approximations as shown by the dotted curve. We have also shown in Fig. 2(a) a fit to the data using a recent theory for resonant Rayleigh scattering [2,6]. As seen from the figure, this theory does not fit the data well. In particular the rise and decay of the data is poorly described. The theory fails to describe our data because the phase information associated with the microscopic scattering processes of the inhomogeneously broadened ensemble of excitons is ignored, and only the incoherent sum of scattered intensities is considered. In fact, any theory such as this, based on ensemble averaging, ignores phase correlations, and would predict the absence of a finite coherent field in contrast to our measurements.

To further investigate the nature of the scattering process itself we have performed model calculations on a linear chain of 500 dipoles with a Gaussian distribution of resonance frequencies. The dipole frequencies have been spatially correlated with a variable correlation length l_c to investigate to influence of spatial correlation on the emitted field. The field is calculated for a fixed angle with respect to the primary emission direction as a function of time and the intensity and phase are shown for two different correlation lengths ($l_c = 0.01\lambda$ and 1λ) in Fig. 2(b).

The calculation for $l_c = 1\lambda$ shows a single temporal peak with nearly constant phase during the emission, whereas the calculation for $l_c = 0.01\lambda$ shows fluctuating intensity and phase of the emitted field. The fluctuating phase for $l_c = 0.01\lambda$ would destroy the fringes in the spectral interferogram. In this case, ensemble averages might provide meaningful experimental predictions. However, for $l_c = 1\lambda$ the near constant phase of the emitted field shows that well-defined fringes would be present in spectral interferograms and that ensemble averaging is inapplicable to calculations of resonant Rayleigh scattering from such highly correlated disorder.

Acknowledgements. The authors would like to acknowledge L.N. Pfeiffer for sample preparation and M. Woerner for many useful discussions. D. B. acknowledges financial support from the Danish Natural Sciences Research Council.

References

1 H. Wang, J. Shah, T. Damen, and L.N. Pfeiffer, Phys. Rev. Lett, **74**, 3065 (1995).
2 S. Haacke *et al.*, Phys. Rev. Lett, **78**, 2228 (1997).
3 M. Gurioli, F. Bogani, S. Ceccherini, and M. Colocci, Phys. Rev. Lett. 78, 3205 (1997).
4 L. Lepetit, G. Cheriaux, and M. Joffre, J. Opt. Soc. Am. **12**, 2467 (1995).
5 D. Fittinghoff *et al.*, Opt. Lett. **21**, 884 (1996).
6 R. Zimmermann, Il Nuovo Cimento, **17 D**, 1801 (1995).

Quantum coherence dynamics of 2D excitons

P. Le Jeune, X. Marie, T. Amand, F. Romstad, J. Barrau, M. Brousseau

Laboratoire de Physique de la Matière Condensée, CNRS-UMR 5830, INSA, Complexe scientifique de Rangueil, 31077 Toulouse Cedex, France
E-mail : marie@insa-tlse.fr

Abstract. Optical and quantum interferences are distinguished in a coherent control experiment which relies on the linear optical response of the crystal.

We perform temporal coherent control of the optical orientation of excitons in a high quality GaAs/AlGaAs multiple quantum well (L_W=10 nm) at T=10 K. A sequence of two phase-locked 1.4 ps optical pulses of equal intensities, from a mode-locked Ti:Sapphire laser, resonantly excites the heavy-hole excitons at energy E_{XH}. The two pulses are linearly polarized, σ^X and σ^Y respectively. The Secondary Emission (SE) kinetics are recorded by up-converting the optical signal in a $LiIO_3$ non-linear crystal with the output from an optical parametric oscillator synchronously pumped by the same Ti:Sa laser. In fig.1, the time delay Δt between the two excitation pulses is zero and precisely adjusted so that the interference of the two pulses results in a single circularly polarized excitation pulse. The total SE dynamics (bold line) is characterized by two very different decay times which correspond successively to the coherent and incoherent parts of SE, i.e. Resonant Rayleigh Scattering (RRS) and luminescence respectively [1].

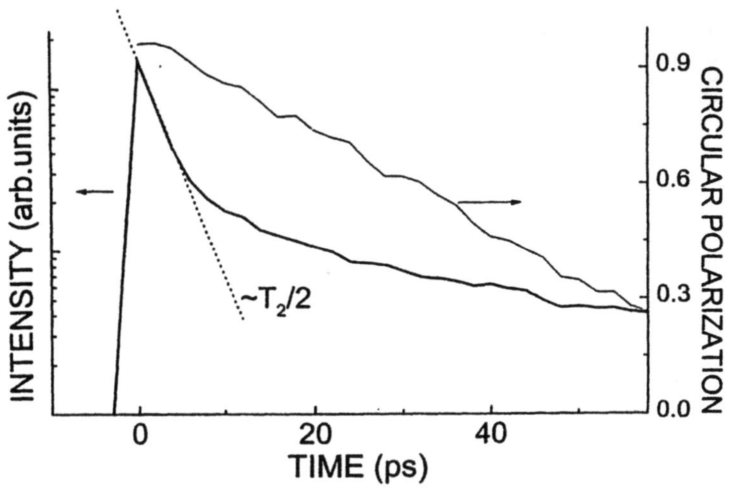

Fig. 1. Time-resolved intensity and circular polarization of the secondary emission after a circularly polarized excitation pulse. The photogenerated density is ~ 10^9 cm^{-2}.

The circular depolarization dynamics (solid line) is characterized by a single decay time, the so-called « longitudinal spin relaxation time » T_{s1} of XH excitons [2] : we measure T_{s1}=40 ps. This longitudinal spin relaxation time is generally much longer than the optical dephasing time T_2 of excitons [3].

Now, the time delay between the two excitation pulses is Δt = 4 ps (there is no temporal overlap) and their phase difference is precisely adjusted to an odd multiple of $\pi/2$. Fig.2 displays the right I^+ (□) and left I^- (O) circularly polarized SE components and the corresponding circular polarization (bold line). We observe that the excitation with the second laser pulse results in a sharp rise of the circular polarization of the excitonic SE. The inset displays the circular polarization oscillations (pulsation $\omega=E_{XH}/\hbar$) versus the fine temporal separation between the two pulses when the recording is taken just after the second pulse (~ 1 ps). As shown in [4], the amplitude of these oscillations decays as a function of the time delay Δt between the two excitation pulses. The associated damping time is the optical dephasing time T_2 of excitons : we measure T_2 = 6 ±1 ps. It is interesting to observe that the RRS emission decay in fig.1 is exponential and precisely fitted with a decay time equal to half the measured dephasing time (dotted line).

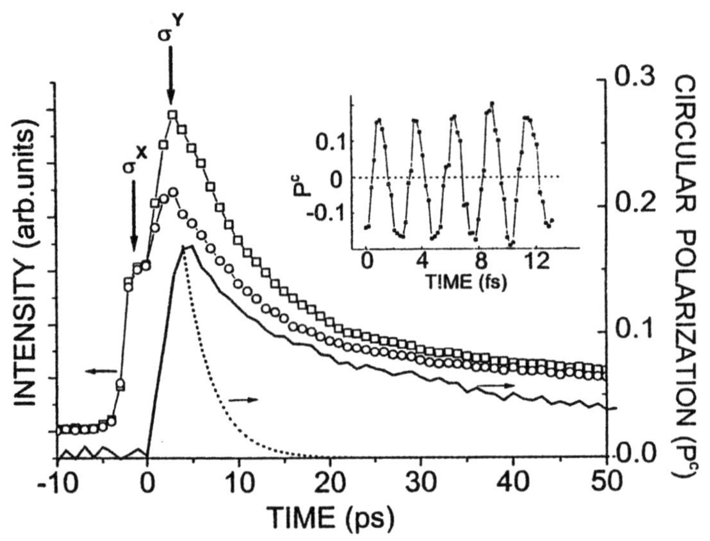

Fig. 2. The time delay between the two σ^X, σ^Y excitation pulses is Δt=4 ps . The dotted line represents the expected circular polarization decay in the OI hypothesis. Inset : circular polarization oscillations versus the fine temporal separation between the two pulses (the recording is taken just after (~1 ps) the second pulse)

The question addressed here is fundamental : do the observed phenomena in fig.2 result from Optical Interference (OI) or Quantum Interference (QI) ? In the OI case, the circular polarization observed just after the second pulse originates from the optical interference of the coherent secondary radiations from the independent $|X\rangle$ and $|Y\rangle$ excitons, created successively by the two pulses. In the QI case, the circular polarization is the polarization of the secondary radiation of elliptic excitons (a coherent superposition of $|X\rangle$ and $|Y\rangle$ states) resulting from the quantum interference between the excitation wavepackets created by the two pulses.

- In the *OI* hypothesis, the circular polarization decay should be the consequence of the optical interference extinction resulting from the optical dephasing of both $|X\rangle$ and $|Y\rangle$ excitons : the decay time of the circular polarization would then be equal to $T_2/2$ (such a decay is plotted as a dotted line in fig.2) [5].
- In the *QI* hypothesis, the circular polarization decay should be the consequence of the spin depolarization of the elliptic excitons characterized by the longitudinal spin relaxation time T_{s1}, much longer than $T_2/2$ [5].

The answer to the addressed question is given in fig.2. This figure shows very clearly that the circular polarization decays with T_{s1} and not $T_2/2$; this demonstrates the *QI* hypothesis. The result is in agreement with the observation reported in [7] : the Faraday rotation of the reflected probe after an excitation sequence (σ^X, σ^Y) proves that a quantum superposition of exciton states is indeed achieved. Time resolved four wave mixing experiments, which are based on the third order non-linear polarisability dynamics, allow also in principle to distinguish between *QI* and *OI* [8].

In conclusion, we have demonstrated here that the distinction betwen *QI* and *OI* can be made from an experiment which relies on the first order polarisation. It is interesting to note that a clear distinction between *QI* and *OI* was also made recently in atomic system with a temporal coherent control experiment based on two photons transition [9].

References

1 S. Haacke *et al*, Phys. Rev. Lett. **78**, 2228 (1997)
2 M.Z. Maialle *et al.*, Phys. Rev. B, **47**, 15776 (1993)
3 A. Vinattieri *et al*, Phys. Rev. B **50**, 10868 (1994)
4 X. Marie *et al*, Phys. Rev. Lett., **79**, 3222 (1997)
5 A theoretical analysis of the circular polarization decay in both cases, based on the consideration of the time evolution of the density matrix operator of the system is developped in [6]
6 X. Marie *et al*, Proceedings of the International workshop on coherent control of charge carrier dynamics in semiconductors, Kluwer Academic Publishers (1998)
7 A.P. Heberle *et al*, IEEE J. Quantum Electron. **2**, 769 (1996)
8 M. Koch *et al*, Phys. Rev. Lett. **69**, 3631 (1992)
9 V. Blanchet *et al*, Phys. Rev. Lett. **78**, 2716 (1997)

Femtosecond Two Color Four-Wave Mixing Experiments on GaAs Quantum Wells Using Two Independently Tunable Lasers

D. S. Kim[1], J. S. Yahng[1], Y. H. Ahn[1], J. Y. Sohn[1], T. Meier[2], and S. W. Koch[2]

[1] Department of physics, Seoul National University, Seoul 151-742, Korea
E-mail: denny@phya.snu.ac.kr
[2] Fachbereich Physik und Wissenschaftliches Zentrum Für Materialwissenschaften, Phillips-Universtät Marburg, 35032 Marburg, Germany

Abstract. Femtosecond four-wave mixing experiments on GaAs/AlGaAs quantum wells are performed with two independently tunable lasers. Many new results are found, including the heavy-hole-light-hole beating with these two mutually incoherent lasers.

Most of femtosecond four-wave mixing studies in semiconductors thus far have been one-color in nature. To perform femtosecond two color wave mixing experiments with a wide range of tunability, one needs to synchronize two femtosecond lasers with timing jitters as short as the pulse width. In this study, we have developed a simple external jitter compensation mechanism with "sample-and-hold" electronic switch, which improves the time resolution of a commercial, partially synchronized system from ~8 ps to 100 fs (Fig. 1).

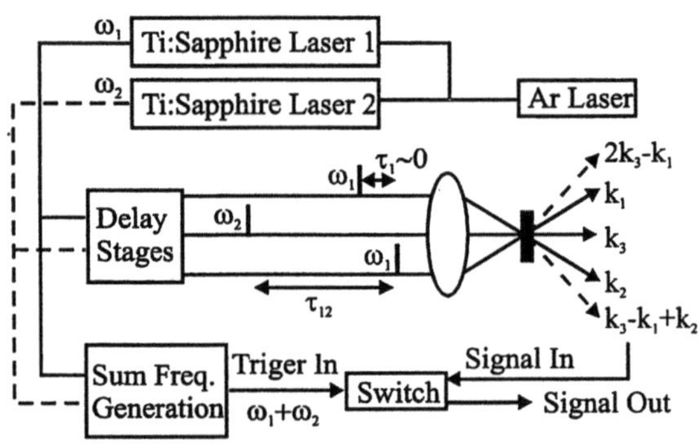

Fig. 1. Schematics of the experimental set-up

With this novel scheme, we have performed femtosecond two beam and three beam four-wave mixing experiments on GaAs quantum wells. Some of our exciting results are as follows:

(1) Heavy-hole and light-hole quantum beating is observed with these two essentially incoherent lasers after external jitter compensation in two beam experiment (Fig. 2).

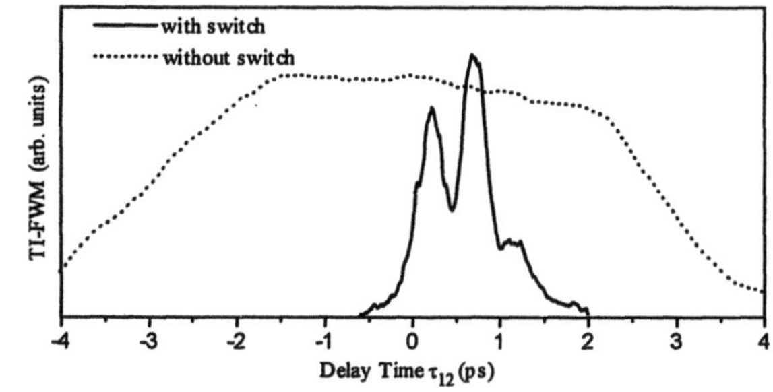

Fig. 2. Two beam FWM signal with and without the switch. Both lasers were tuned near the HH and LH excitons. The sample was a 30 period GaAs/AlGaAs multiple quantum well (x=0.25; 100 Å / 100 Å)

(2) Unlike the two beam experiments where the spectral overlap is essential in getting any signal [1], in three beam FWM, a wide range of spectrum can be diffracted off the population grating both above and below the band gap whether or not there exists spectral overlap between ω_1 and ω_2 (Fig. 3).

(3) When ω_1 is at the exciton resonance, surprisingly, the diffracted intensity of ω_2 below the band gap is stronger relative to that at the free carrier continuum for the equivalent detuning. We believe that the weak signal in the continuum states is mainly from the destructive interference among them [2]. In addition, short dephasing time for continuum states might further decrease the diffraction efficiency [3].

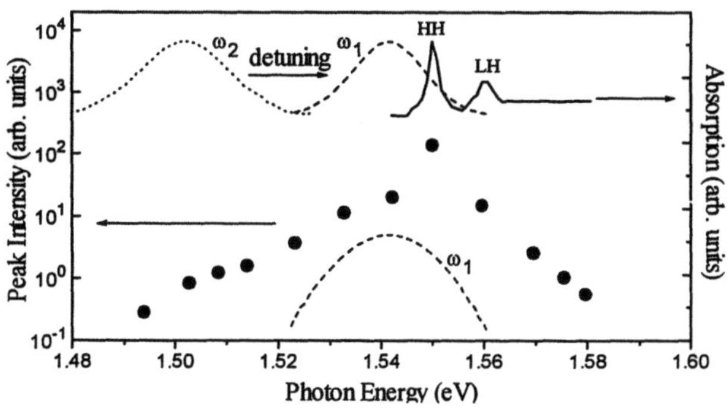

Fig. 3. Peak three beam FWM signal at $\tau_1 = 0$ ps and $\tau_{12} = 10$ ps as a function of the center photon energy of ω_2. Absorption curve is represented by solid lines.

(4) Spectrally resolved three beam FWM signal is dominated by the diffraction of the laser spectrum (Fig. 4a) while the excitonic resonances play an important role when ω_2 is close to them. This is very general, as we have verified this in

several other samples. The laser spectrum dominates even when ω_2 is CW as shown in the inset. The model calculations based on the semiconductor bloch equations (SBE) show the presence of strong off-resonant signal (Fig. 4b). The inset displays the peak FWM signal as a function of detuning, which qualitatively reproduces the experimentally observed decrease of about 2 orders of magnitudes between 10 and 60 meV of detuning below the HH exciton.

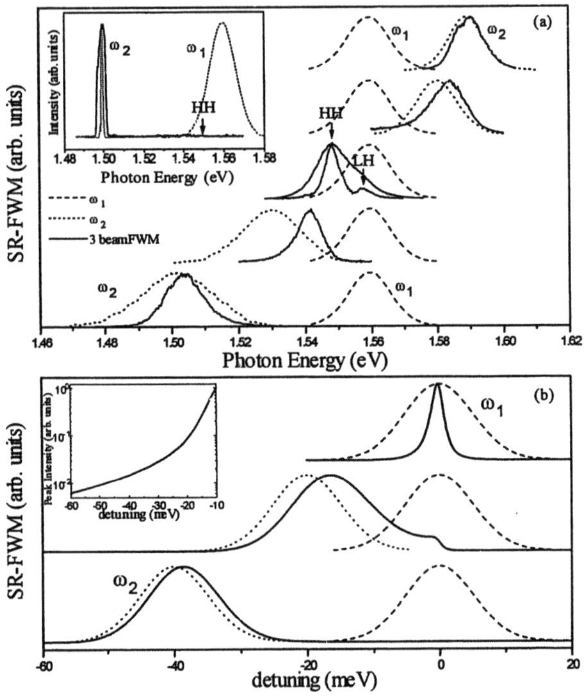

Fig. 4. (a) Three beam SR-FWM signals (solid lines) for a fixed ω_1 for various ω_2's at $\tau_1 = 0$ ps and $\tau_{12} = 10$ ps. In the middle curve, the SR-FWM at lower density is represented by dotted lines. (Inset) SR-FWM for CW ω_2 (b) Computed FWM spectra using the semiconductor bloch equations. (Inset) Peak FWM signal as a function of detuning below the HH exciton.

References

1 S. T. Cundiff et al., Phys. Rev. Lett. 77, 1107, 1996.
2 F. Jahnke, M. Kira, and S. W. Koch, Z. Phys. B 104, 559, 1997.
3 D. S. Kim et al., Phys. Rev. Lett. 68, 1006, 1992

Temporal and Spectral Characteristics of Radiatively Coupled Excitons in Multiple Quantum Wells

Dirk Ammerlahn[1], Jürgen Kuhl[1], Martin Hübner[1], Bernd Grote[2], Tineke Stroucken[2], Stephan W. Koch[2], Galina Khitrova[3], and Hyatt Gibbs[3]

[1] Max-Planck-Institut für Festkörperforschung, D–70569 Stuttgart, Germany
E-mail: ammer@servix.mpi-stuttgart.mpg.de
[2] Department of Physics and Materials Science Center, Philipps University, D–35032 Marburg, Germany
[3] Optical Sciences Center, University of Arizona, Tucson, AZ 85721, USA

Abstract. Interwell exciton-photon coupling in detuned multiple quantum well Bragg structures results in an ultrafast radiative decay of the coherent exciton polarization. Remarkably different temporal and spectral distributions of backward and forward degenerate four-wave mixing signals are found for detuned Bragg samples.

1. Introduction

Simultaneous carrier and photon confinement in multiple quantum well (MQW) structures leads to coupling between excitons and photons and results in an enhanced radiative decay of the coherent exciton polarization [1]. This coupling depends on the number of quantum wells N, their respective spacing d and the angle of the incident laser beam with respect to the growth direction. It becomes most efficient for MQW Bragg structures with $d = \lambda/2$, where λ is the wavelength of the heavy hole (hh) exciton transition. In a perfect Bragg structure, only one mode is active, the superradiant mode. All other $N-1$ modes remain optically dark. In the presence of disorder, however, part of the oscillator strength is transferred to the forbidden (subradiant) modes.

In this paper, we present a detailed study of the temporal and spectral behavior of the third-order nonlinear polarization of the hh exciton transition in wedge-shaped (In,Ga)As/GaAs MQWs. We investigate the effect of barrier thickness variations and discuss the role of sample disorder.

2. Experiment and Theory

To study the dynamics of the coupled exciton-photon modes, we performed time-integrated (TI) and spectrally-resolved (SR) degenerate four-wave mixing (DFWM) with 700 and 200 fs pulses both in reflection (RG) and transmission geometry (TG). Our sample consists of 30 (In,Ga)As/GaAs QWs with a width of 8 nm and an indium content of 3 % and was grown by molecular beam epitaxy on semi-insulating GaAs substrate. The barrier thickness decreases from the growth center to the edge of the wafer. Thus different sample positions correspond to different barrier thicknesses and the strength of the exciton-photon coupling can be studied in terms of different detunings from the Bragg condition by varying the sample position.

The optical response of the macroscopic interband polarization is computed from the semiconductor Bloch equations (SBE) in Hartree-Fock approximation. Maxwell's equations are solved to determine the self-consistent optical field which contains the externally-applied laser field and the retarded time derivative of the induced interband polarizations within different QWs. In order to compute DFWM signals, we make the slowly-varying envelope approximation, solve the semiconductor Maxwell Bloch equations (SMBE) iteratively up to third order in the electric field and pick out the contributions to the experimentally-studied DFWM signals [2].

3. Results and Discussion

Figure 1 shows in the upper line theoretical DFWM traces for different detunings. In the case of Bragg condition, the theory predicts an ultrafast mono-exponential decay with a decay time of 120 fs. Deviations of d from $\lambda/2$ reveal more complex forms of decay: while 2 % detuning seems to result in a biexponential decay, the DFWM trace with 4 % detuning shows quantum beating. In the lower line of Fig. 1, DFWM signals for different sample positions are depicted. The positions correspond to the above-mentioned detunings and the experimental data are in good agreement with the theoretical traces. For perfect Bragg condition, only the superradiant mode is active. At slight detuning, we find after the ultrafast decay of the superradiant mode a slow decay which is due to subradiant modes. In the experimental curves, subradiant modes occur also at Bragg condition because of disorder in the sample. The quantum beats in the 4 % detuning curve can be explained by different energies of the subradiant modes.

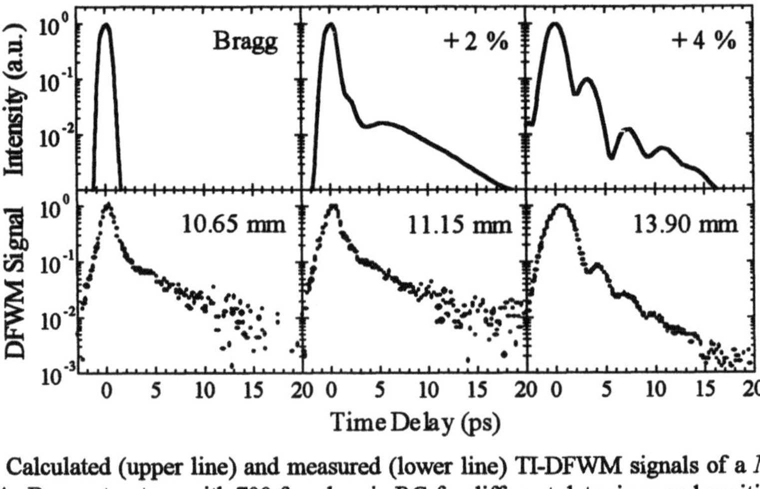

Fig. 1. Calculated (upper line) and measured (lower line) TI-DFWM signals of a $N = 30$ (InGa)As Bragg structure with 700 fs pulses in RG for different detunings and positions.

At resonance, we find for the superradiant mode with 200 fs pulses a decay time $\tau_{dec}(sup)$ = 220 fs and for the inhomogeneously distributed [3] subradiant modes $\tau_{dec}(sub)$ = 4 ps which yields T_2^{sup} = 440 fs and T_2^{sub} = 16 ps, respectively.

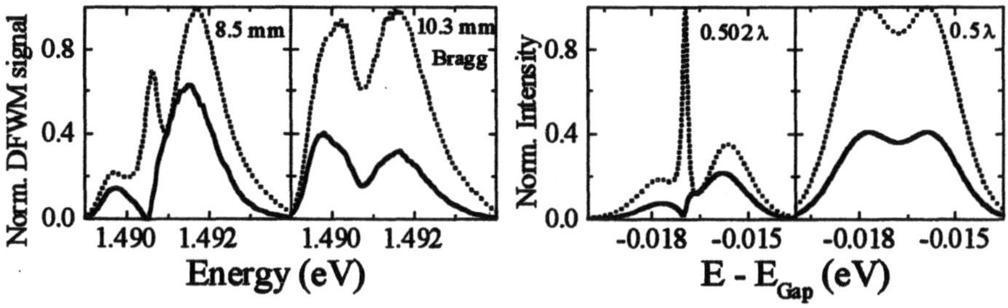

Fig. 2. Spectrally-resolved DFWM on a $N = 30$ (InGa)As Bragg sample for two sample positions in reflection (*solid lines*) and transmission (*dotted lines*) geometry (left) and calculated DFWM spectra for the corresponding detunings (right).

The different exciton-photon modes in a MQW Bragg structure are also observed in SR-DFWM signals presented in Fig. 2 both in RG and TG [3]. The DFWM spectra for two positions (left) correspond to calculated spectra for $d = \lambda/2$ and $d = 0.502\ \lambda$ (right). At resonance, the signal shape is identical for the two geometries. The different amplitude is due to reflection of the backward signal into the forward direction at the sample-air interface for the cladding layer thickness of $\lambda/2$. The dip is due to partial overlap of the two pulses and destructive interference. The FWHM reaches a value of 3.2 meV in accordance with the ultrafast decay of 0.22 ps. For the detuned case, differences between RG and TG occur both in the theoretical and measured spectra. In the DFWM signal in TG, a sharp peak appears, whereas in RG the DFWM signal drops to zero. This is due to subradiant modes which occur at slight detunings. They are hardly coupled and therefore the signal is preferentially emitted in forward direction.

4. Conclusions

Optical dephasing in MQW Bragg structures is dominated by radiative recombination. At slight detunings, DFWM signals show remarkable differences in RG and TG due to subradiant modes which are preferentially emitted in forward direction. The excitation of subradiant modes is forced by both detuning and sample disorder. Our data are well reproduced by solutions of the SMBE.

References

1 M. Hübner et al., Phys. Rev. Lett. **76**, 4199, 1996.
2 T. Stroucken et al., Phys. Rev. B **53**, 2026, 1996.
3 J. Kuhl et al. in *Festkörperprobleme/Advances in Solid State Physics* **38**, Ed. B. Kramer, Vieweg, Braunschweig/Wiesbaden, 1998.

Pulse Breakup due to Coherent Self-Induced Transmission on a Free Exciton Resonance

Harald Gießen[1,2], Andreas Knorr[1], Stephan W. Koch[1], Stefan Linden[2], Jürgen Kuhl[2], Michael Hetterich[3], and Michael Grün[3]

[1] Fachbereich Physik, Philipps-Universität, D-35032 Marburg, Germany
E-mail: giessen@physik.uni-marburg.de
[2] Max-Planck-Institut für Festkörperforschung, Heisenbergstraße 1, D-70569 Stuttgart, Germany
[3] Institut für Angewandte Physik, Universität Karlsruhe, Kaiserstraße 12, D-76128 Karlsruhe, Germany

Abstract. Theoretical and experimental studies of femtosecond pulses propagating on a free exciton resonance show coherent pulse breakup due to self-induced transmission at long propagation distances at pulse areas above π.

1. Introduction

In atomic two-level systems, it has long been known that pulses with an area of 2π can propagate lossless on resonance through an optically thick absorber [1]. The underlying coherent transient of this *self-induced transparency* (SIT) is the electric field of the pulse driving the Bloch vector of the system from 0 to π to 2π, leaving the system therefore in the ground state after transit. The pulse breakup observed at higher intensities corresponds to multiple rotations of the Bloch vector. Pulses with areas that are larger than 2π will break up into multiples of 2π. At least two conditions have to be fulfilled to observe this coherent pulse-breakup: Dephasing processes have to be long compared to the duration of the Bloch vector rotation because they reduce the coherence of the system, destroying the phase relationship between the Bloch vector and the electric field. Furthermore, the propagation length has to be sufficient in order to observe the desired breakup. It turned out that several Beer absorption lengths usually fulfill this requirement [2]. Recently, SIT has been found on atomic-like bound excitons in semiconductors [3]. However, ideal self-induced transparency (with 100% transmission) should not occur on the free exciton resonance in semiconductors because scattering of carriers into different k-states would lead to a loss for the reamplification of the propagating pulses, therefore inhibiting completely lossless propagation [4]. The key difference between atomic two-level systems and semiconductors are therefore the many-body effects. However, Rabi oscillations in the carrier density were predicted in theoretical studies [5] and should manifest in long-distance propagation in semiconductors.

2. Experimental Methods

We propagated 180 fs and 900 fs pulses resonantly with the A-exciton through bulk CdSe platelets of $\alpha L=1.7$ and 6.8, respectively. The intensities were in the range of several 10 MW/cm^2, giving pulse areas around π and multiples of π. The pulses were crosscorrelated with a 50 fs Fourier-limited pulse.

3. Results and Discussion

The 180 fs pulses show polariton propagation beats in the short sample at low intensities which are vanishing at increasing intensities. Around a pulse area of π, the transmitted pulse resembles the input pulse. Further increase of the input intensity leads to a leading and trailing shoulder in the pulse shape. The maximum transmitted intensity at a pulse area of 4π is more than 90%. Theoretical calculations employing the semiconductor Maxwell-Bloch equations show agreement for both the temporal pulse shapes and the spectra.

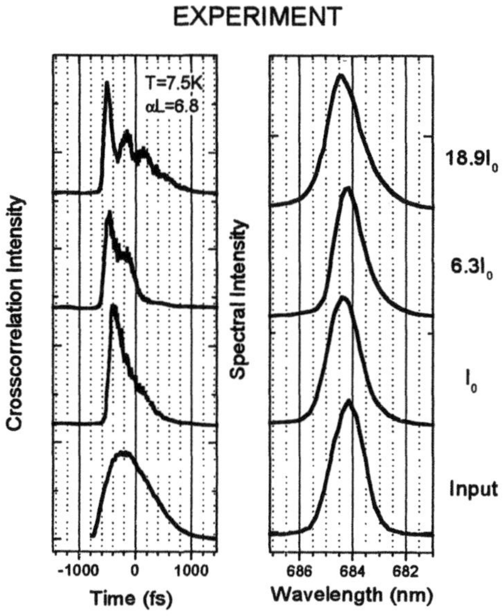

Fig. 1. Experimental result: Propagation of 900 fs pulses (crosscorrelation and transmitted spectra) through a CdSe bulk sample with $\alpha L=6.8$ at increasing intensities. $I_0=5$ MW/cm^2.

The theory predicts that the modulation in the pulse shape should increase and clear pulse breakup due to coherent interaction should be visible for propagation over long distances. Therefore, we investigated a thicker sample of strained epitaxial CdSe. Due to depth-dependent strain, the exciton was strongly inhomogeneously broadened. However, we were able to circumvent this problem by using a narrow-band spectral filter and relatively long pulses (900 fs), so that only a relatively small ensemble of excitons within the inhomogeneous distribution participated in the interaction process. Figure 1 shows the experimental results of propagation under increasing intensity through the sample: A breakup into two, three and four pulses becomes clearly visible, and the separation between the peaks decreases with increasing power, as one would expect from a faster polarization oscillation. The spectra show a slightly asymmetric broadening towards the blue side. Figure 2 shows the result of a semiconductor Maxwell-Bloch calculation, taking into account the dephasing processes microscopically. The agreement between theory and experiment for both temporal traces and spectra is very good. The Rabi oscillations in the polarization density predicted by the theory are causing the temporal pulse breakup. Carriers are coherently excited and destroyed, similar to coherent control experiments in the linear regime [6]. This effect manifests in the oscillations in the propagating pulse, but in contrast to experiments with tailored phases, in our case the electric field of the pulse itself in combination with the propagation is the reason for the breakup.

Incoherent scattering processes certainly play a role, as complete transparency is not observed. Despite the relatively long pulses, dephasing does not seem to destroy the coherence before the pulses do break up. Purely incoherent effects such as bleaching would rather effect the leading edge of the pulses, which was not observed in our experiment. The fact that relatively long pulses are used facilitates the observation of the breakup with crosscorrelation. Self-phase modulation can be ruled out as a cause for the pulse breakup as it takes places at much higher intensities [7].

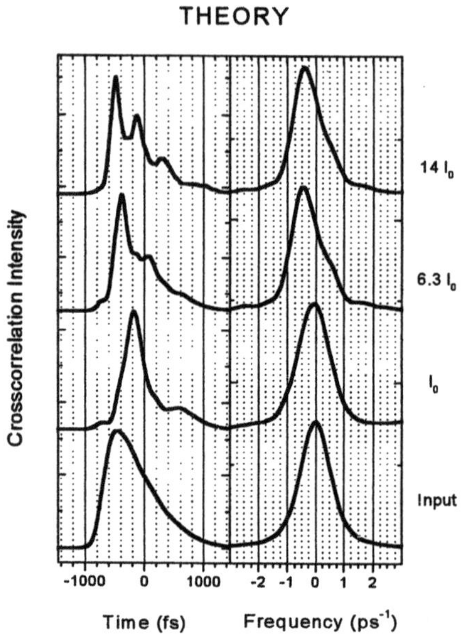

Fig. 2. Theoretical calculations with the parameters of the experiment of Fig. 1.

4. Conclusions

In conclusion, we have shown coherent self-induced transmission on a free exciton resonance in CdSe that manifests itself as propagation-induced, intensity dependent pulse breakup.

References

1 R.E. Slusher and H.M. Gibbs, Phys. Rev. A **5**, 1634 , 1972.
2 G.L. Lamb, Jr., Rev. Mod. Phys. **43**, 99, 1971.
3 M. Jütte, H. Stolz, and W. von der Osten, J. Opt. Soc. Am. B **13**, 1205, 1996.
4 A. Knorr et al., Phys. Rev. A **46**, 7179, 1992.
5 R. Binder et al., Phys.Rev. Lett. **65**, 899, 1990.
6 A. Heberle et al., IEEE J. Selected Topics in Quantum Electronics **2**, 769, 1996.
7 H. Giessen et al., phys. stat. sol. (b) **206**, 27, 1998.

Excitonic Quantum Beating in Space- and Time-Resolved Pump and Probe Experiments

S. Grosse[1], R. Otremba[1], G. von Plessen[1], M. Koch[1], J. Feldmann[1],

V.M. Axt[2], T. Kuhn[2], R. Rettig[3], and W. Stolz[3]

1 Lehrstuhl für Photonik und Optoelektronik, Sektion Physik, Ludwig-Maximilians-Universität, Amalienstr. 54, D-80799 München, Germany
E-mail: jochen.feldmann@physik.uni-muenchen.de
2 Westfälische Wilhelms-Universität, Wilhelm-Klemm-Straße 10
D-48149 Münster, Germany
3 Wissenschaftliches Zentrum für Materialwissenschaften, Philipps-Universität, Renthof 5, D-35032 Marburg, Germany

Abstract. In spatially and temporally resolved pump and probe experiments we observe a spatial beating which reflects the coherent dynamics of an excitonic wave packet.

1. Introduction

Recently efforts have been made to combine the ultrahigh spatial resolution provided by near-field scanning optical microscopes with the ultrahigh temporal resolution readily obtained in pump and probe experiments with femtosecond optical pulses. In semiconductors these spatio-temporally resolved studies aim to monitor the coherent and incoherent electron-hole pair and exciton dynamics including relaxation and exciton formation as well as ballistic and diffusive transport. One difficulty when performing time-resolved nonlinear optical experiments with a scanning near-field microscope is that the excited area and therefore the signal intensity is very small. An alternative approach which provides slightly less spatial resolution but is more flexible and furthermore does not suffer from low signal intensity uses a microscope objective to focus laser radiation as tightly as possible [1;2]. By spatially displacing the tiny focus of a pump beam versus that of a probe beam it is possible to observe signatures of heating and cooling of optically created electron-hole pairs with non-vanishing center-of-mass motion [1,2].

This paper focuses on the spatio-temporal dynamics of an electron-hole polarization and population directly after its optical creation by an ultrafast laser pulse. In particular we aim to observe coherent signatures in the spatial dynamics which arise from a macroscopic nonlinear polarization. We find that the spatial profile of the signal from the excited area oscillates in time. This *spatio-temporal beating* reflects the dynamics of an excitonic wave packet [3].

2. Experimental Setup and Sample

The experimental setup for spatially resolved pump and probe experiments is shown schematically in Fig. 1. Two laser pulses with variable delay which are perpendicularly polarized with respect to each other are focused onto the sample by a microscope objective. After the sample the pump beam is suppressed by means of a polarizer. By measuring time traces for different displacements ξ between the two laser spots the induced spatial expansion of the signal profile can be monitored [2]. This setup provides a spatial resolution of 1.5 μm limited by the spot size and a temporal resolution of 150 fs limited by the duration of the laser pulses. Chopping both pulse trains with different frequencies in the MHz range with acousto-optic modulators and detecting signals at the difference frequency suppresses interference effects in the region of temporal pulse overlap and, hence, allows access to the coherent time regime.

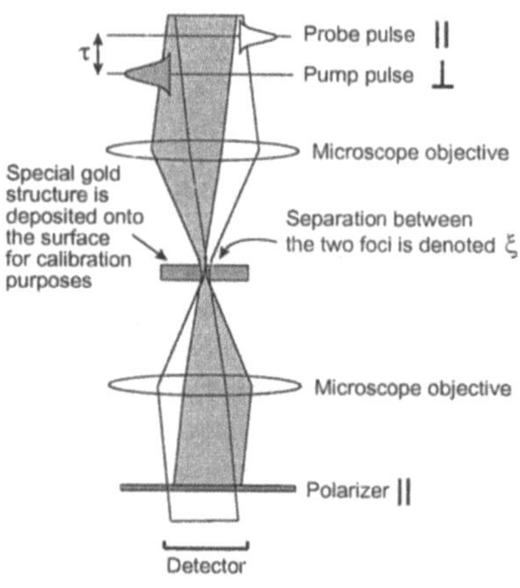

Fig. 1. Experimental setup for a space- and time-resolved pump and probe experiment.

As sample we choose a structure containing 50 $In_{0.11}Ga_{0.89}As/Ga_{0.75}AsP_{0.25}$ quantum wells in which the light hole transition is considerably blueshifted due to compressive strain. In this way we ensure that only heavy hole excitons are excited and complications due to light hole excitons are avoided.

3. Results and Discussion

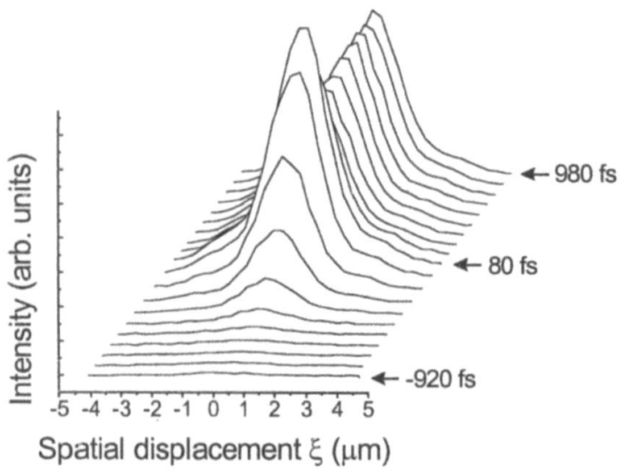

Fig. 2. Differential transmission profile ΔT(ξ) as a function of the time delay between the two pulses.

The differential transmission profiles ΔT(ξ) obtained for different time delays τ are displayed in Fig. 2. They can be well approximated by Gaussians, yet a close inspection reveals slight deviations from the Gaussian shape due to small side maxima which arise from diffraction of the beams by the aperture of the microscope objective. The most prominent feature in Fig. 2 is a modulation of the signal amplitude as a function of time delay.

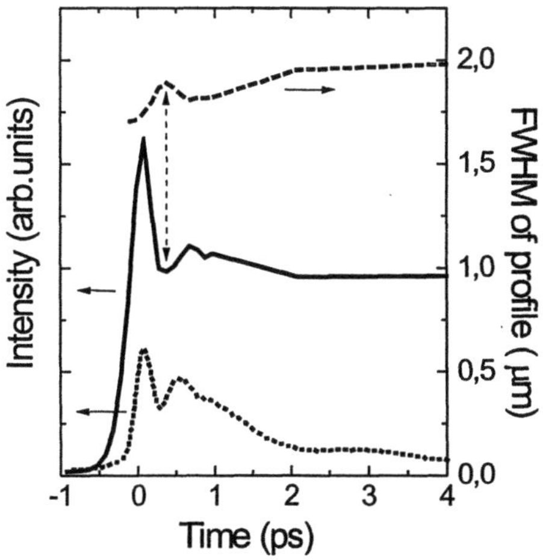

Fig. 3: Pump and probe transient for ξ=0 (solid) in comparison with the spatial width (FWHM) of the profile (dashed) and a four-wave mixing transient (dotted).

This oscillation can be seen more clearly in Fig. 3, where we plot the time trace obtained for ξ=0 (solid line). The full width at half maximum (FWHM) of the differential transmission profile (dashed line) also oscillates in time; a minimum in the pump and probe signal corresponds to a maximum in the spatial width of the profile, and vice versa.

We identify this „spatial beating" as a signature which arises from the temporal dynamics of an excitonic wave packet [3]. Due to the collinearity of the pump and the probe beams the detected signal is not only the usual pump and probe signal but contains also four wave mixing like signal contributions. For comparison a four-wave-mixing-transient obtained in the usual non-collinear geometry with large laser spots is also plotted in Fig. 3 (dotted line). The spatial profiles of the different signal components are given by a convolution which involves four or six spatial envelopes of the laser field. Both, the signal components which scale to leading order with E^4 in the laser amplitude and give a broader profile and those which scale which E^6 and have a more narrow profile oscillate in time; with the oscillation period roughly given by the inverse excitonic binding energy. Hence the relative weight of these components varies so that $\Delta T(\xi)$ finally exhibits a spatio-temporal beating. This interpretation is confirmed by numerical calculations based on the semiconductor Bloch equations.

At longer time scales we observe the transition to the incoherent regime in which center of mass momentum scattering becomes important. In particular it is possible to follow heating and cooling of electron-hole pairs with non-vanishing center of mass motion and the formation of excitons out of correlated electron-hole pairs [2].

Acknowledgements. This work has been financially supported by the Deutsche Forschungsgemeinschaft through the program *Quantenkohärenz in Halbleitern (Fe 323/5-2)* and by the European Union through the network *Ultrafast Quantum Optoelectronics (FMRX-CT97-0134).*

References

1 H. W. Yoon, D. R.Wake, J. P.Wolfe, H Morkoc, Phys. Rev. **B46**, 13461 (1992)
2 S. Grosse, et al., phys. stat. sol. (b) **204**, 147, 1997.
3 J. Feldmann, et al., Phys. Rev. Lett. 70, 3027 (1993).

Ultrafast phase relaxation of confined excitons in CuCl nanocrystals : Non-Markovian behaviors

Arao Nakamura and Hideki Ohmura

Department of Crystalline Materials Science, Graduate School of Engineering and Center for Integrated Research in Science and Engineering, Nagoya University, Chikusa-ku, Nagoya 464-8603, Japan
E-mail: nakamura@cirse.nagoya-u.ac.jp

Abstract. We have investigated femtosecond coherent transients of confined excitons in CuCl nanocrystals with radii in the range 2.7−5.3 nm by means degenerate four-wave mixing with a two-beam configuration. When the crystallite radius is smaller than 4 nm, the non-Markovian behavior is observed in the short time region comparable to the reservoir correlation time.

1. Introduction

In the short time region comparable to a reservoir correlation time τ_c, the relaxation can not be described by a constant relaxation rate and a reservoir memory effect, i.e. a non-Markovian effect plays an important role in relaxation process. Phase relaxation provides us with information on the interaction between the relevant electronic system and vibrational degrees of freedom (thermal reservoir) of condensed phases. Non-Markovian behaviors of phase relaxation have been studied both theoretically and experimentally[1-6]. In semiconductor nanocrystals, both phonons and excitons are confined and their density of state is transformed from a continuum to a series of discrete levels by size quantization. The confinement effects lead to changes of the electron-phonon interaction and the correlation time. In this paper, we report on non-Markovian phase relaxation of confined excitons in CuCl nanocrystals. We have investigated size and temperature dependent behaviors of phase relaxation by means of a femtosecond degenerate four-wave mixing (DFWM) with a 2-beam configuration.

2. Experimental

The samples used in this study are CuCl nanocrystals embedded in glass with radii ranging from 2.7 nm to 5.3 nm. Details of preparation method were described in Ref[7]. DFWM experiments with a two-beam configuration were carried out using a second harmonic light of a femtosecond mode-locked Ti:sapphire laser operating at 1 kHz. The pulse duration, the spectral width and the laser power density were 200 fs , 20 meV and 50 kW/cm^2, respectively.

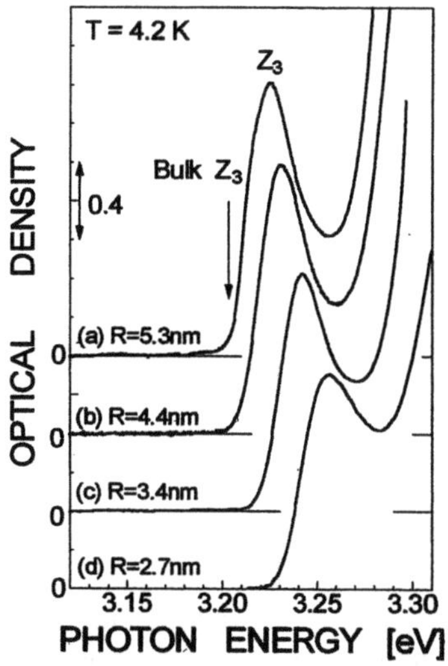

Fig.1. Absorption spectra at 4.2 K.

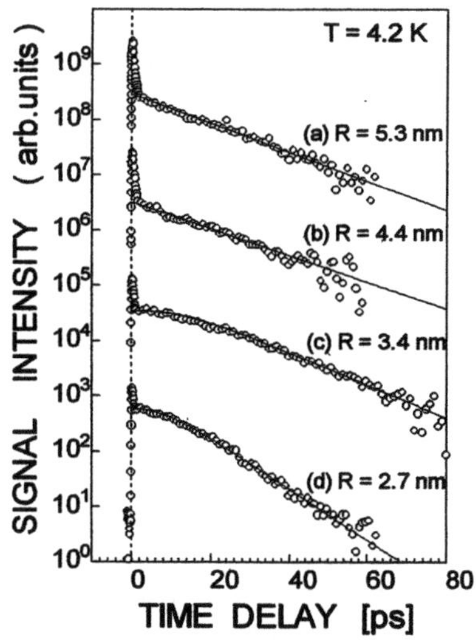

Fig.2. Correlation traces of DFWM signal.

3. Results and Discussion

Figure 1 shows absorption spectra of CuCl nanocrystals with mean radii of R = 5.3 nm(a), 4.4 nm(b), 3.4 nm(c), and 2.7 nm(d) at 4.2 K. The absorption peak due to Z_3 excitons is shifted to the higher energy side with decreasing crystallite radius, which indicates quantum confinement effects. In this size range, the exciton Bohr radius (0.68 nm) is much smaller than the crystallite radius, and thus the translational motion of excitons is confined [7]. The spectral width of exciton band is about 20 ~ 30meV, and is governed by inhomogeneous broadening due to crystallite size distribution.

Figure 2 shows correlation traces of DFWM signals at 4.2 K for CuCl nanocrystals with various crystallite radii. For R = 5.3 nm and 4.4 nm, the signal exhibits a two-component exponential decay. The time constants (homogeneous width Γ_h) of slow and fast decay components are 17 ps (19 μ eV) and 0.6 ps (0.5 meV) for R = 5.3 nm, and 18 ps (18 μ eV) and 0.5 ps (0.7 meV) for R = 4.4 nm, respectively. Similar two-component decay behaviors have been observed in dye-doped polymers[8]. Comparing the Fourier cosine transform of the echo signals with the persistent hole-burned spectra, the fast and slow components have been assigned to the phonon-side band and the zero-phonon band, respectively[8]. The zero-phonon line and its acoustic phonon wing have been observed in the hole-burned spectra of CuCl nanocrystals[9]. Therefore, We attribute the slow component seen in Fig.2 to the zero-phonon band and the fast one to the acoustic-phonon-side band of confined excitons.

For R = 3.4 nm and 2.7 nm, however, the slow component of the correlation trace does not exhibit an exponential behavior, which suggests that the phase relaxation cannot be simply described by the transverse relaxation time T_2. In a stochastic model taking into account a finiteness of τ_c, the DFWM signal I is written by[3]

$$I(t_{12}) \propto \exp\left[-4p^2\left\{\frac{t_{12}}{\tau_c}+\exp(\frac{-t_{12}}{\tau_c})-1-\frac{1}{2}\left(\exp(\frac{-t_{12}}{\tau_c})-1\right)^2\right\}\right] \quad (1)$$

where t_{12} is the time delay between two excitation pulses and $p = D\tau_c$. D is the root-mean-square amplitude of the fluctuations. The solid curves show the least-squares fits using adjustable parameters of τ_c and p. Taking $\tau_c = 8.5$ ps and p = 0.37 for 3.4 nm and $\tau_c = 8.0$ ps and p = 0.46 for R = 2.7 nm, the non-exponential decay behavior can be well fitted to Eq (1). These results indicate the reservoir memory effect, i.e. the non-Markovian phase relaxation on the time scale of 10 ps.

The temperature dependence of phase relaxation was also investigated. The homogeneous width Γ_h defined as $\Gamma_h = 2\hbar D^2 \tau_c$ is governed by the exciton scattering with confined acoustic phonons whose frequency is proportional to R^{-1}. The phonon confinement gives rise to a cut off of the phonon dispersion and narrowing of the width of density of state. As τ_c is inversely proportional to the width of the density of state, τ_c is lengthened for CuCl nanocrystals with smaller radii, and the non-Markovian behavior can be observed.

4. Conclusions

We have observed non-Markovian behaviors in phase relaxation of confined excitons in CuCl nanocrystals with radii smaller than 4 nm. The non-exponential decay of the transient DFWM trace was well fitted to the theoretical decay curve taking into account the finiteness of τ_c. The obtained values of τ_c at 4.2 K are 8.0 ps and 8.5 ps for 2.7 and 3.4 nm, respectively.

References

1 M. Aihara, Phys. Rev. B25, 53(1982).
2 R. F. Loring and S. Mukamel, Chem. Phys. Letters 114, 426(1985).
3 J.-Y. Bigot et al ., Phys. Rev. Lett. 66, 1138(1991).
4 E. T. J. Nibbering, D. A. Wiersma, and K. Duppen, Phys. Rev. Lett. 66, 2464(1991).
5 S. Saikan, J. W-I. Lin, and H. Nemoto, Phys. Rev. B46, 11125(1992).
6 L. Banyai et al., Phys. Rev. Lett. 75, 2188(1995).
7 T. Kataoka, T. Tokizaki, and A. Nakamura, Phys. Rev. B48, 2815(1993).
8 S. Saikan et al., Phys. Rev. B41, 3185(1990).
9 Y. Masumoto, T. Kawazoe, and N. Matsuura, J. Lumin. 76&77, 189(1998).

Study of hexagonal GaN by femtosecond-excitation correlation measurement and degenerate four-wave-mixing

S. Pau[1], J. Kuhl[1], F. Scholz[2], C. Haerle[2], M. A. Khan[3] and C. J. Sun[3]

[1] Max-Planck-Institut, Heisenbergstrasse 1, D-70569 Stuttgart, Germany
Email: spau@servix.mpi-stuttgart.mpg
[2] Universitat Stuttgart, 4. Physikalisches Institut, Kristallabor, D-70550 Stuttgart, Germany
[3] APA Optics Inc., 2950 N. E. 84th Lane, Blaine, Minnesota 55449, USA

Abstract. Carrier and exciton dynamics in bulk hexagonal GaN is studied spectrally and temporally by correlation and degenerate four-wave-mixing techniques for different excitation density, wavelength and temperature.

Recent advances in GaN based light emitters have bought tremendous interest in the study of the intrinsic properties of GaN. Accurate knowledge of the material parameters and emission mechanisms are essential in the design and improvement of the next generation of GaN devices. We report here a low temperature spectral and temporal study of carriers and excitons dynamics of GaN by femtosecond excitation correlation (FEC) measurement and degenerate four-wave-mixing (DFWM). FEC spectroscopy is a powerful technique to study incoherent emission dynamics with temporal resolution limited only by the excitation pulse width [1,2] and DFWM is a complementary technique which allows accurate determination of the exciton dephasing time, i.e. the coherent properties of the material transitions. For our studies, we use two nominally

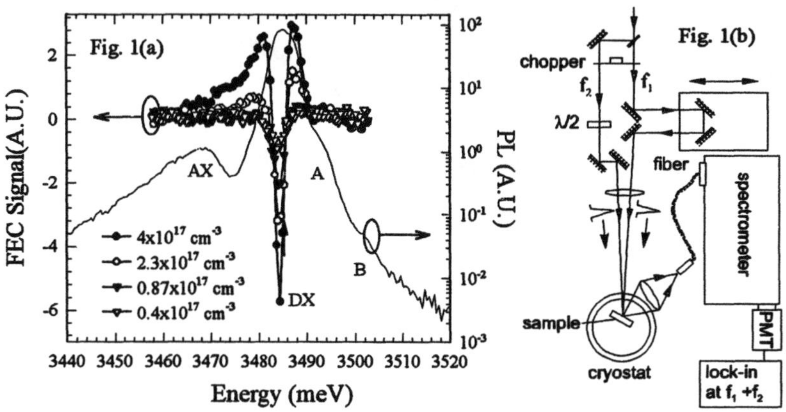

Figure 1. (a) PL and FEC spectra of GaN excitons at T=10K. (b) FEC experimental setup.

undoped samples of hexagonal GaN on sapphire grown by MOCVD by two independent groups.

Fig. 1 shows the typical photoluminescence (PL) spectrum of our GaN samples at T=10K. The emission is dominated by donor bound (DX), acceptor bound (AX) and free excitons (A and B). In the FEC experiment, the two cross polarized excitation beams (pulse width 150fs), chopped at two different frequencies, are set to be E_b=25 meV (A exciton binding energy) above the free A exciton energy to generate cold e-h pairs at the A bandedge, and the spectrally resolved photoluminescence at the A and DX exciton energies are detected by lock-in technique at the sum frequency as a function of time delay, τ, between the two pulses. The FEC spectrum for τ=0 and different pair densities are also shown in Fig. 1. A negative FEC signal at the DX energy signifies a sublinear variation of DX photoluminescence with increasing pump intensity [3].

Fig. 2 shows FEC traces taken at the DX exciton energy for different pair densities. At all excitation densities, we observe a dip in the signal at $\tau = 0$ which recovers after ~ 35 ps. We attribute the dip predominately to the nonlinear processes of bimolecular formation of exciton and the saturation of the donor concentration. The signal subsequently decays on a time scale of ~ 120 ps which roughly equals the DX exciton population lifetime. A fit of the FEC results to a rate equation model gives an estimate of the value of the A exciton formation coefficient to be $C = 1.2\times10^{-18}$ cm^3/ps [3].

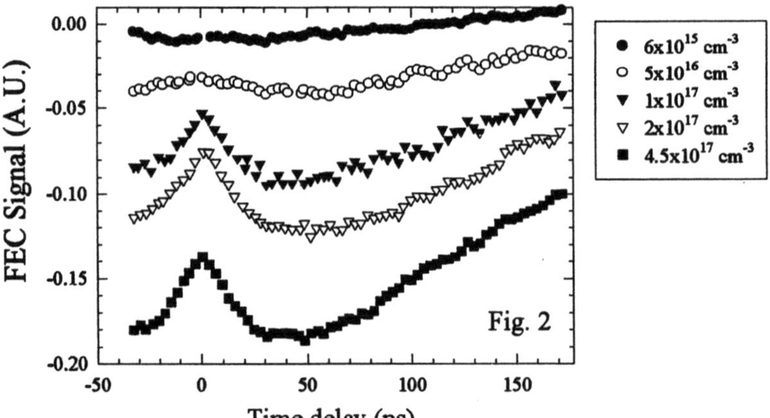

Figure 2. Time resolved FEC trace measured at DX energy for different pair densities at T=10K.

In addition to the FEC measurement, we performed DFWM on GaN as a function of energy, excitation density and temperature. The homogeneous linewidth of the free exciton to first order can be described by the relation

$$\Gamma(N,T) = \Gamma(0,0) + \beta_{xx} a_B^3 E_B N + \beta_{ph} T \qquad (1)$$

where N is the density, T is the temperature, a_B = 3.6 nm is the A exciton Bohr radius. The coefficient β_{xx} is a dimensionless parameter which measures the strength of the exciton-exciton scattering, whereas the coefficient β_{ph} is a parameter which measures the exciton-acoustic-phonon interaction. The summary of our results are listed in Table 1. Note that the linewidth is related to the dephasing time by $\Gamma_h = h/\pi T_2$. A comparison of the linewidth between the two samples shows that the measured β parameters are sample independent and we believe our results give an accurate value of the intrinsic exciton dephasing time in GaN. The residual linewidth of the exciton, Γ(0,0) ~ 4 ps, is caused predominately by exciton-impurity scattering. The material parameters of GaAs are also listed in Table 1 and are similar to that of GaN, provided that the exciton-exciton scattering coefficient is normalized with respect to the exciton binding energy and Bohr radius[4]. Finally, by measuring the scattering efficiency of both the 3rd and 5th order diffracted beams, we estimate the value of the effective nonlinear susceptibilities at the A exciton energy at 10 K to be $|\chi_{eff}^{(3)}| = 2 \times 10^{-11}$ esu and $|\chi_{eff}^{(5)}| = 0.5 \times 10^{-17}$ esu [5].

	Γ(0,0) (meV), T_2	*β_{ph} (μeV/K)*	*β_{xx}*
GaN, APA-C	0.29, 4.5 ps	9	6.8
GaN, HB1975	0.31, 4.2 ps	11	5.1
theory	----	7	---
bulk GaAs	0.18, 7.3 ps	17	10

Table 1: Summary of homogeneous linewidth and broadening parameters of GaN and GaAs

Acknowledgements. We thank Klaus Rother and H. Klann for technical support and T. Ruf, K. Syassen and H. J. Queisser for discussion.

References

1 D. von der Linde, J. Kuhl, and E. Rosengart, J. Lumin. **24/25**, 675, 1981; R. Stroebel, R. Eccleston, J. Kuhl and K. Koehler, Phys. Rev. **B43**, 12564, 1991.
2 J. L. A. Chilla, O. Buccafusca, J. J. Rocca, Phys. Rev. **B48**, 14347, 1993.
3 S. Pau, J. Kuhl, M. A. Khan and C. J. Sun, Phys. Rev. B, in press.
4 S. Pau, J. Kuhl, F. Scholz, V. Haerle, M. A. Khan and C. J. Sun, Appl. Phys. Lett. **72**, 557, 1998.
5 W. Schmid, T. Vogtmann, M. Schwoerer, Optics. Comm. **121**, 55, 1995.

Evidence of Trion and Carrier Localization in Modulation Doped Quantum Wells by fs and ps FWM Studies

P. Gilliot[1], D. Brinkmann[1], J. Kudrna[1], B. Hönerlage[1], A. Arnoult[2], J. Cibert[2], and S. Tatarenko[2]

[1] Institut de physique et de chimie des matériaux de Strasbourg, Groupe d'optique nonlinéaire et d'optoélectronique, (UMR 7504 CNRS – ULP – EPCM)
23, rue du Lœss, B.P. 20 CR, F-67037 Strasbourg CEDEX, France
E-mail: gilliot@ipcms.u-strasbg.fr
[2] Laboratoire de Spectrométrie Physique, (UMR 5588 UJF Grenoble I – CNRS)
B.P. 87, F-38402 Saint-Martin-d'Hères CEDEX, France
E-mail: jcibert@cea.fr

Abstract. The localization of holes and charged excitons is evidenced in modulation p-doped CdTe/CdZnMgTe quantum wells by fs and ps four-wave mixing experiments.

1. Introduction

Charged excitons, also called trions, are bound states of an exciton with a supplementary electron or hole. Their observation is only possible in quasi-two dimensional quantum structures [1] with excess carrier population, where their binding energy is drastically increased [2] compared to the 3D case. Up to now, most investigations on trions have dealt with the dependence of their properties on parameters like carrier density or magnetic field. We report here, to our knowledge, the first direct study of their dephasing dynamics. Four-wave mixing (FWM) experiments are performed in modulation doped CdTe/CdMgZnTe multiple quantum wells (QW). A layer of acceptors in the barriers produces a hole (h) gas in the wells with a density of almost $10^{11}cm^{-2}$. A resonance associated with a positive trion (X^+) can be observed [3] (Fig. 1(A)).

As the free carrier gas would screen a Coulomb potentiai and prevent the formation of trions and excitons, it was proposed that the fluctuations of the electrostatic potential due to the doping impurities in the barrier could localize the carriers [1, 4]. Our experiments support this hypothesis: they show that collisions with and between charged particles (X^+ and h) are quite inefficient in the dephasing of excitons and trions and suggest that the trions are localized like the holes.

2. Experiments and discussion

Spectrally resolved FWM with femtosecond pulses. Our degenerate FWM experiments are performed in the usual two-beam self-diffraction configuration. The sample temperature is kept at 5 K. It is excited by two subsequent laser pulses of 80 fs-duration from a tunable self-mode-locked Ti:Sapphire laser. The laser pulses have a spectral width of about 20 meV so that the exciton and the trion resonances can be coherently excited. The FWM signal is dispersed in a spectrometer in order to analyze separately the trion and exciton dynamics and detected as a function of the delay time between exciting pulses using a photomultiplier and a lock-in amplifier.

FWM traces of Fig. 1B show a periodic modulation superimposed on an exponential decay. The 1.4 ps oscillation period corresponds to an energy of 2.8 meV in good agreement with the trion-exciton separation (Fig.1A). We attribute the main contribution to the oscillations to quantum beats between the two resonances. This shows that X and X^+ are two states of the same electron-hole pair and are not created in separate regions on the QWs.

The dephasing times we find for the exciton are close to those measured in undoped samples [5]. The trion dephasing is even slower. This indicates that the holes do not interact strongly neither with neutral excitons, nor with charged trions and thus should be localized in the fluctuation of the electrostatic potential

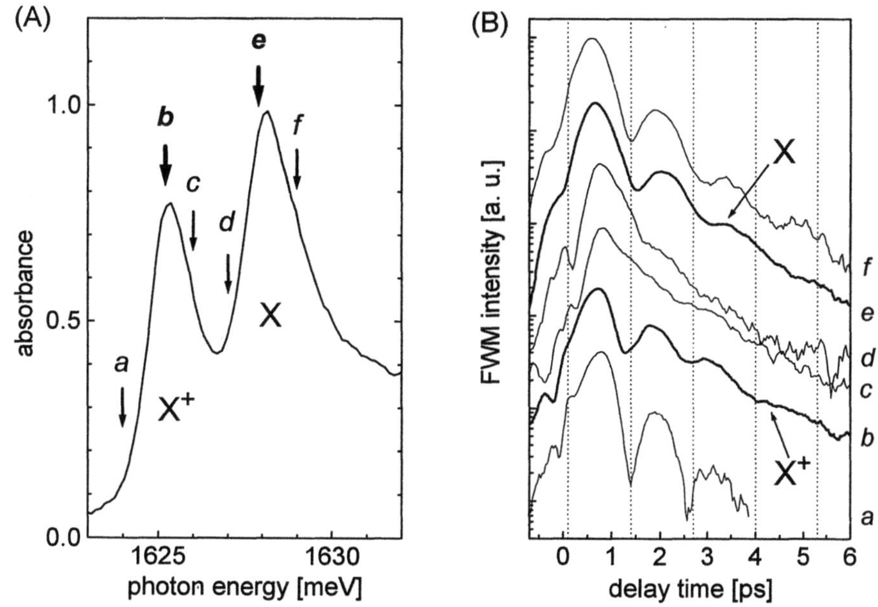

Fig. 1. (A) Absorption spectrum at 5 K with the X and X^+ peaks. **(B)** Femtosecond FWM traces for different detection photon energies shown by the arrows on (A). The bold curves correspond to the trion (X^+) and to the exciton (X) photon energies

FWM with spectrally narrow picosecond pulses. Using picosecond pulses in the same experimental conditions, we can determine separately the dephasing rates of trions and excitons without exciting the other resonance or the continuum states. We can then analyze the different collision processes: As seen in Fig. 2, trions are less sensitive to collisions than excitons. This is contradictory with the fact that the trions are charged and would strongly interact mutually . Trions can indeed be also localized and considered as excitons bound to localized holes.

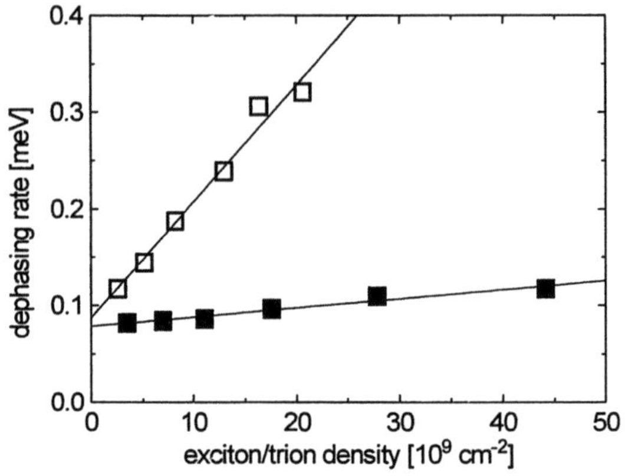

Fig. 2. Dephasing rates $2\hbar/T_2$ for trions (full squares) and excitons (open squares) measured with spectrally selective ps-excitation, as a function of the X and X^+ densities.

3. Conclusion

We demonstrate that FWM experiments can help determining the degree of localization of trions in QWs as a function of carrier density and disorder. In our modulation p-doped CdTe/CdZnMgTe QW sample, we evidence a localization of trions and holes in the fluctuations of the impurity induced electrostatic potential by measuring relatively long trion dephasing times.

References

1 K. Kheng, R.T. Cox, Y. Merle d'Aubigné, F. Bassani, K. Saminadayar, S.Tatarenko, Phys. Rev. Lett. **71**, 1752, (1993)
2 B. Stébé and A. Ainane, Superlattices Microstruct. **5**, 545 (1989)
3 A. Haury et al., 9th Int. Conf. Superlattices, Microstructures and Microdevices, Liège 1996, Proceedings to appear in Superlattices Microstruct.
4 G. Finkelstein, H. Hadas, and I. Bar-Joseph, Phys. Rev. Lett. **74**, 976 (1995)
5 E. J. Mayer, N.T. Pelekanos, J. Kuhl, N. Magnea, and H. Mariette, Phys. Rev. B **51**, 17263 (1995)

Ultrafast pump-probe signals at the band edge of bulk GaAs: Excitonic versus free carrier contributions

M. Joschko[1], M.P. Hasselbeck[1], M. Woerner[1], T. Elsaesser[1], R. Hey[2], H. Kostial[2], and K. Ploog[2]

[1] Max-Born-Institut für Nichtlineare Optik und Kurzzeitspektroskopie, Rudower Chaussee 6, D-12489 Berlin, Germany
E-mail: joschko@mbi-berlin.de
[2] Paul-Drude-Institut für Festkörperelktronik, D-10117 Berlin, Germany

Abstract. Polarization selection rules allow an experimental separation of free carrier from excitonic heavy-light hole quantum beats in fs-pump-probe signals at the absorption edge of GaAs.

Coherent polarizations govern the nonequilibrium carrier dynamics of semiconductors in the initial phase of ultrafast photoexcitation. Impulsive excitation of a coherent superposition of heavy (HH) and light hole (LH) states has allowed observation of quantum beats associated with transitions to excitonic states in quantum wells [1] and recently in the free carrier continuum of bulk GaAs [2]. At the band edge of GaAs, femtosecond pump-probe signals are influenced by both excitonic and free carrier contributions. A direct separation of these two components is important for a complete understanding of the pump-probe interaction.

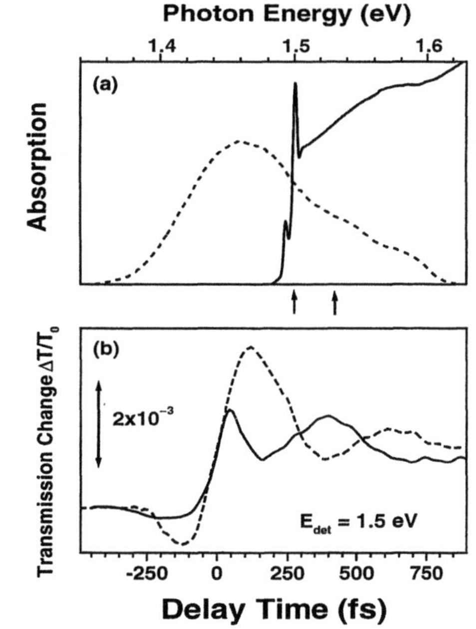

Fig. 1 (a) Linear absorption spectrum of the strained GaAs sample at T = 8 K (solid line) and spectrum of the 20 fs pump and probe pulses covering both excitonic and free carrier transitions (dashed line). Arrows are detection energy positions (E_{det}). (b) Spectrally dispersed transmission changes at $E_{det} = 1.5$ eV for an excitation density of 1 x 10^{15} cm^{-3}. $(T-T_0)/T_0$ is plotted as a function of delay time between pump and probe (T, T_0: transmission with and without excitation). Solid line: parallel-linear polarizations of pump and probe. Dashed line: perpendicular-linear polarizations.

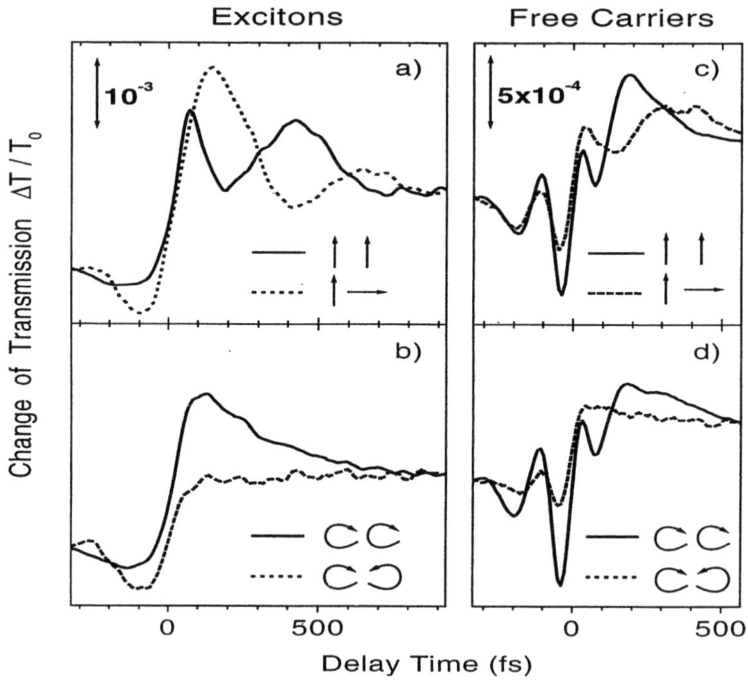

Fig. 2. Excitonic (a,c: E_{det} = 1.5 eV) and free carrier (c,d: E_{det} = 1.530 eV) heavy-light hole quantum beats measured in spectrally and temporally-resolved pump-probe experiments. The transmission change is plotted vs delay between pump and probe. Insets: Relative polarization orientation of pump and probe pulses.

We present an experimental study of HH-LH quantum beats at the band edge of a slightly strained bulk GaAs sample (thickness: 500 nm). The low temperature, linear absorption spectrum in Fig. 1a (solid line) reveals a splitting of excitonic transitions (ΔE_X = 8 meV) due to strain that develops during sample preparation. Spectrally and temporally-resolved measurements are performed with 20 fs pulses from a Ti:Sapphire laser arranged in four different polarization orientations. The spectrum of 20 fs pump pulses (dashed line) is centered below the bandgap and causes transitions into both excitonic and free carrier final states. Probe pulses at the same wavelength are spectrally resolved after passing through the sample. The transients in Fig. 1b (detection energy E_{det}=1.5 eV at exciton peak) show two components: (i) a transmission increase due to state-filling by photoexcited electron-hole pairs and (ii) oscillations attributable to HH-LH quantum coherence. The phase of the beats changes by π when the polarization of the pump-probe geometry is switched from parallel-linear (solid line) to perpendicular-linear (dashed line).

In the following, we concentrate on differential transmission signals for positive delay times where the influence of detuning oscillations and perturbed free induction decay are negligible (Fig. 2). We observe distinct differences in amplitude, frequency, and phase of the beats depending on E_{det} and the polarization of pump and probe pulses as indicated in the insets of Fig. 2. For E_{det} at the exciton peak (Figs. 2a, 2c), we find oscillations for linear-parallel and linear-perpendicular polarizations, but no oscillations for co-circular and counter-circular polarizations. The small strain polarizes the p-orbitals of HH and LH states resulting in polarization selection rules (PSR) similar to quantum wells [3,4]. In the free carrier continuum, strain is much less effective. The k-vector acts as quantization axis resulting in PSR of the bulk crystal [2,4]. Data obtained in the free carrier continuum (Figs. 2b, 2d) show beats for positive delay times with amplitude and phase exactly predicted by the PSR and distinctly different than observed at the exciton resonance. We detect strong oscillations for parallel-linear and co-circular polarizations, but very weak beats for perpendicular-linear and counter-circular arrangements. The variation of oscillation frequency with E_{det} in the continuum agrees with the previously characterized behavior of quantum beats [2], whereas the excitonic beats are fixed at frequency corresponding to $\Delta E_X = 8$ meV. Our experiments clearly demonstrate that free carrier HH-LH quantum beats are important in spectrally resolved pump-probe signals even in the vicinity of the absorption edge, where the role of excitonic states are usually assumed to be dominant.

References

1 K. Leo, T. C. Damen, J. Shah, E. O. Göbel, and K. Köhler, Appl. Phys. Lett. **57**, 19,1990.

2 M. Joschko, M. Woerner, T. Elsaesser, E. Binder, T. Kuhn, R. Hey, H. Kostial, and K. Ploog, Phys. Rev. Lett. **78,** 737, 1997.

3 S. Schmitt-Rink et al., Phys. Rev. **B 46,** 10460, 1992.

4 P. Enders, A. Bärwolff, M. Woerner, and D. Suisky, Phys. Rev. **B 51,** 16695, 1995.

Femtosecond Coherent Dynamics of Free Carriers in Semiconductors

Cornelius Fürst, Alfred Leitenstorfer, and Alfred Laubereau

Technische Universität München, Physik-Department E11, D-85748 Garching, Germany
E-mail: cfuerst@physik.tu-muenchen.de

Abstract. Using two-color femtosecond pump-probe spectroscopy, we study quantum coherent effects in the dynamics of nonthermal carrier distributions in semiconductors. At very low excitation densities, energy non-conserving transitions and memory effects are observed during the spontanous emission of LO phonons in GaAs. For strong excitation, we present the first experiment demonstating Rabi-oscillations of the carrier density in the absorption continuum of InP

1. Introduction

The recent development of femtosecond laser sources for ultrafast spectroscopy has given access to novel coherent phenomena in the dynamics of optically excited semiconductors. For weak excitation the evolution of the photogenerated carrier distributions is governed by carrier-phonon scattering [1]. In the past, most of the experiments investigating such scattering processes have been explained within the semiclassical Boltzmann equation. In this model, energy and momentum conservation holds for every single collision. In our experiments at very low carrier densities we show that this picture breakes down for ultrashort timescales [2].
At high excitation densities we demonstrate that despite ultrashort dephasing times the coherence of the excited polarizations persists long enough to allow for Rabi-oscillations of the occupation density in the interband continuum [3]. This coherent phenomenon originates from the resonant coupling of the coherent light wave with the quantum mechanical transitions.
In our experiments, differential transmission changes are measured in a shot-noise limited detection scheme, using synchronized femtosecond pulses of different spectral shape out of our two-color Ti:sapphire laser [4].

2. Weak Excitation: LO Phonon Quantum Kinetics

In Fig.1, we present time and energy resolved transmission changes $\Delta T/T$ in a bulk GaAs sample (T_L=15K). Laser pulses of a duration of 120fs excite 8×10^{14} electrons per cm^3 out of the valence band into the conduction band. The dynamics is probed by broadband 15fs pulses. Both pump and probe pulses are circularly

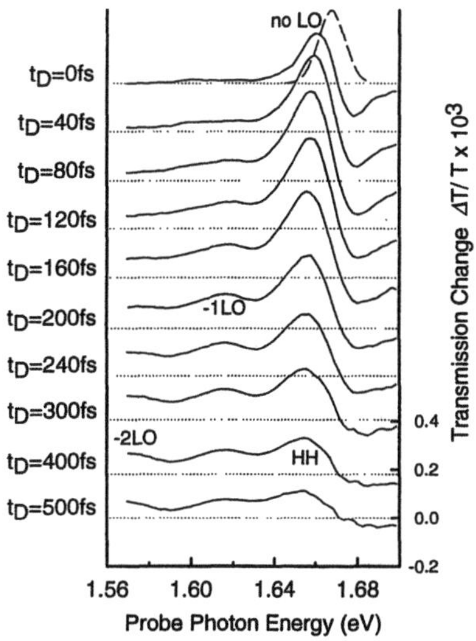

Fig. 1: Memory effects in the electron-phonon interaction at low excitation densities of $8x10^{14}cm^{-3}$ in GaAs. The pump and probe pulses have a duration of 120fs and 15fs, respectively.

polarized. In this geometry the transmission changes are dominated by the carriers generated involving the heavy hole band [2]. At a delay time of t_D=0fs, the photo-excited carrier distribution is reflected as a well pronounced peak in the spectrum. For delay times of 40fs and 80fs a broad shoulder appears between 1.59eV and 1.64eV representing electrons that have emitted one LO phonon. This feature is energetically much broader than the initial bleaching peak, demonstrating that energy needs not to be conserved for each scattering event. A minimum between the first maximum and the satellite peak cuts in after 100fs. This time scale coincides with the duration of one LO phonon oscillation period (115fs). After 200fs the system has remembered its history: Due to *intraband* quantum interference the width of the phonon satellite narrows towards the shape of the unrelaxed peak. These phenomena are repeated for the second satellite (-2LO) at later time delays.

Results from an analytical model treating the memory effects exactly are in agreement with our experimental observations [2]. More elaborate simulations of the experiment can be found in Refs. 5 and 6. We expect these memory effects to be important also in other dynamical systems whenever transition rates are of the same order of magnitude as the oscillation frequencies of the energy quanta involved.

3. Strong Excitation: Rabi-Oscillations

The coherent features of the optical response completely change for high intensity excitation. Carrier-carrier scattering now leads to a fast dephasing on a time scale substantially below 100fs. Nevertheless we show that the coherence between the light wave and the excited *interband polarizations* even at carrier densities around $10^{19}cm^{-3}$ persists long enough to allow for a driven recombination of the carriers due to the strong laser pulse. Such Rabi oscillations are demonstrated for the first time in the absorption continuum of a semiconductor. Electrons are excited from the heavy and light hole bands into the conduction band of InP (T_L=15K) by bandwidth limited pulses of a duration of 120fs. The dynamics of the electron population is probed by 15 fs pulses tuned to the transitions starting from the split-off band. In this way, we avoid coherent artifacts due to the coupling of the pump and probe pulses. Fig. 2 presents measured transmission changes versus time delay for an electron excess energy of 20 meV. The pump pulse intensity is varied from 0.2 GW/cm^2 to 17 GW/cm^2 resulting in excitation densities exceeding 1.5×10^{18}cm^{-3}. At peak intensities above 4 GW/cm^2 we find strongly damped Rabi flopping: Up to 20% of the carriers are recombined coherently after the maximum of the pump pulse. Two Rabi cycles are observed at 17 GW/cm^2. Stimulated recombination only occurs after the second cycle, whereas the first appears as a shoulder in the bleaching signal. This feature clearly shows that the dephasing rate decreases with time. Two effects account for the finding: (i) Pauli blocking of the states at the bottom of the band hinders carrier-carrier scattering (ii) The quantum kinetic build-up of the screening reduces the dephasing rates within the time interval of one plasmon period which is in the order of the duration of the pump pulses.

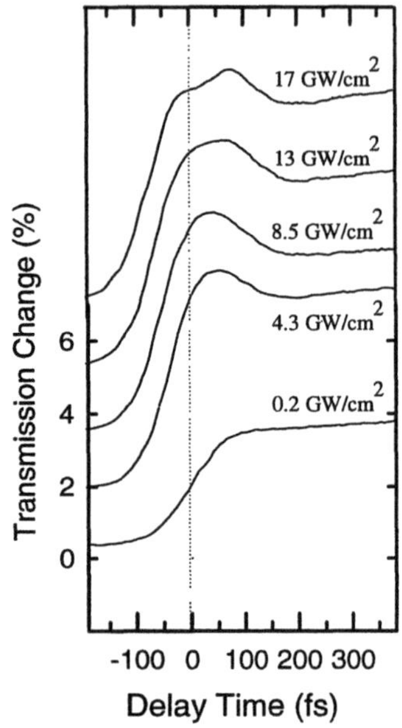

Fig. 2: Interband Rabi oscillations in InP for different pump pulse intensities. Electrons are excited 20 meV above the band gap by 120 fs pulses. The dynamics is probed using 15 fs pulses.

References

1 A.Leitenstorfer, C.Fürst, A.Laubereau, W.Kaiser, G.Tränkle, and G.Weimann, Phys.Rev. Lett. 76, 1545 (1996)

2 C. Fürst, A. Leitenstorfer, A. Laubereau, and R. Zimmermann, Phys. Rev. Lett. 78, 3733 (1997)

3 C. Fürst, A. Leitenstorfer, A. Nutsch, G, Tränkle, A. Zrenner, phys. stat. sol. (b) 204, 20 (1997)

4 C. Fürst, A. Leitenstorfer, A. Laubereau, IEEE J. of Select. Top. in Quan. Electr. 2, 473 (1996)

5 R. Zimmermann, J. Wauer, A. Leitenstorfer, C. Fürst, J. Lumin. 76&77, 34 (1998)

6 A. Schmenkel, L. Bányai, and H. Haug, J. Lumin. 76 & 77, 134 (1998)

Intraband Coherence after Energy Relaxation

F. Wolter[1], G. Bartels[1], R. Martini[1], P. Haring Bolivar[1], H. G. Roskos[2], R. Hey[3], H. T. Grahn[3], and H. Kurz[1]

[1] Institut für Halbleitertechnik II, RWTH Aachen, D-52056 Aachen, Germany
E-mail: wolter2@iht-ii.rwth-aachen.de
[2] Physikalisches Institut, Joh. Wolfg. Goethe-Universität, D-60054 Frankfurt, Germany
[3] P. Drude-Institut f. Festkörperelektron., Hausvogteiplatz 5-7, D-10177 Berlin, Germany

Abstract. An experimental and theoretical analysis of optically excited Bloch oscillations proves intraband coherence after LO-phonon emission. It is demonstrated that this relaxation process even can be an essential prerequisite for the development of coherent charge oscillations.

1. Introduction

Apart from being attractive sources of coherent Terahertz (THz) radiation, optically excited Bloch oscillations in semiconductor superlattices remain a field of active research, as they represent a model system for the general study of ultrafast coherent electronic dynamics within energy bands in semiconductors. The scope of this presentation is the experimental and theoretical demonstration that energy relaxation of electrons by LO-phonon emission can preserve the intraband coherence of an electronic ensemble performing Bloch oscillations. In addition, for certain excitation conditions, this relaxation process is even necessary to create a macroscopic coherent charge oscillation.

2. Experimental Results and Discussion

The superlattice sample consists of 35 periods of 9.7 nm-wide GaAs wells and 1.7 nm-wide AlGaAs barriers embedded in a pin mesa structure. AlGaAs buffer layers are avoided to prevent static screening effects [1]. The combined miniband width is $\Delta = 19$ meV. The sample is excited at 10 K with 150 fs-pulses from a Ti:sapphire laser. The photoexcited carrier density is kept constant at approximately 5×10^{9} cm^{-2}. The emitted coherent THz radiation is detected in a standard setup for THz emission spectroscopy [2].

Figure 1 displays transients of the coherent THz emission of carriers performing Bloch oscillations. The applied field corresponds to the observed frequency of 1.9 THz. The figure shows characteristic transients for various excitation energy positons. The energy denotes the excess energy with respect to the lower miniband edge at 1.54 eV. Three distinct observations can be derived from the data: i) The decay of the coherent intraband signal is independent of the excess energy. The fitted decay time constant of about 980 fs agrees with

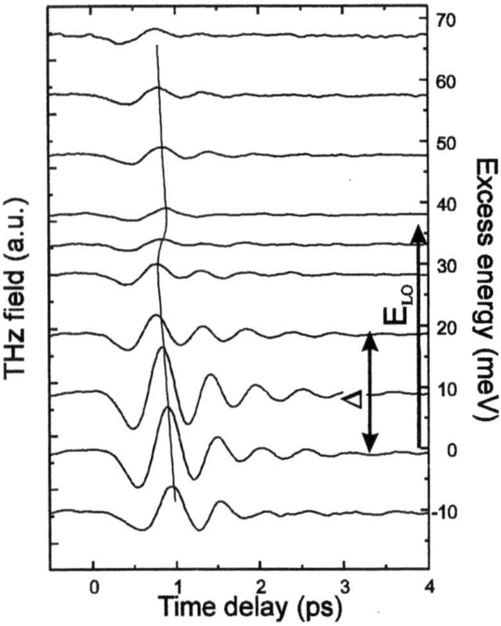

Fig.1. THz emission for different initial carrier excess energies. Δ marks the miniband width, E_{LO} the onset for LO-phonon emission

experimental carrier-carrier scattering times and in particular it does not change when the threshold for LO-phonon emission at E_{LO}= 36 meV is crossed. This is surprising, since pump-probe experiments have demonstrated that the LO-phonon emission time in super-lattices is much shorter and agrees with bulk values below 200 fs [3]. ii) The amplitude of the THz emission as a function of excess energy shows a maximum in the middle of the miniband and decreases for higher energies. Surprisingly, at the onset for phonon emission, the amplitude rises again. iii) A change of the phase of the oscillations at E_{LO} is observed. This phase jump of about 180° when E_{LO} is crossed is shown in more detail in Figure 2, which represents THz transients at different frequencies. Intuitively these observations can be understood as follows. As the excitation energy increases, the initial carrier distribution in 3D k-space spans an increasing k_z region of the mini-Brillouin zone. If the energy exceeds the miniband maximum at Δ, the carrier distribution spreads homogenously over k_z, which corresponds to phases from -π to π. As a consequence, a net carrier oscillation is suppressed as the contributions of all carriers cancel out [3]. The situation is similar to a completely filled electronic band, where current transport is blocked. A net carrier oscillation can thus only arise if the initial carrier distribution prepared by the pulse is redistributed by the emission of LO-phonons and if this process preserves intraband coherence. The constant dephasing even above E_{LO} (i) and the observation of Bloch oscillations for excess energies exceeding Δ (ii) confirm this intuitive picture The coherent redistribution is particularly evident from the observed phase jump (iii).

This interpretation is analyzed and conclusively confirmed by numerical simulations in a density matrix approach as employed e. g. in [4], using a 3D superlattice model, which additionally allows to gain insight into the relevant microscopic processes. Bulk LO-phonon interaction is taken into account in the Markov approximation. The calculations allow to separate different contributions to the total intraband carrier dipole moment: Interestingly both the carriers which have relaxed by LO-phonon emission, as well as the asymmetry of the remaining initially homogenous carrier distribution (the *intraband holes*) are found to contribute to the phonon relaxation-induced Bloch oscillation. For high excess energies the relaxed carriers are found to dominate the total

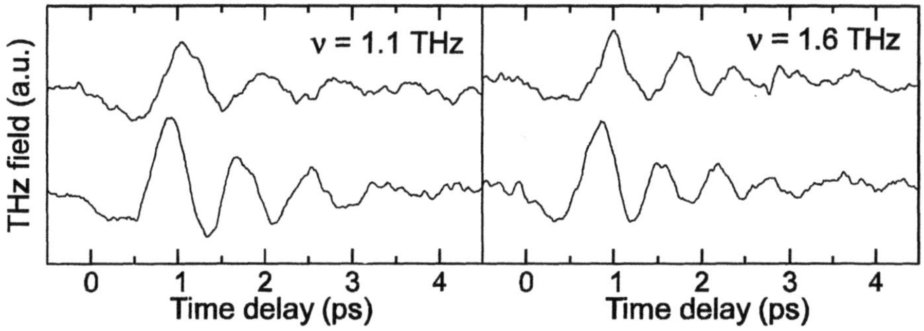

Fig. 2: THz transients for different frequencies in the vicinity of E_{LO}.

intraband dipole moment. The calculated dynamics agrees with the experimental observations and confirms both the intraband coherence conservation and the generation of a coherent charge oscillation by the LO-phonon emission process.

4. Conclusions

The emission of LO-phonons preserves intraband coherence and under special conditions is even an essential prerequisite for the generation of a coherent charge oscillation. This fact is derived from the investigation of Bloch oscillations excited with a high excess energy. The initially symmetrical phase distribution, which results in destructive interference, is broken due to redistribution in k-space.

Acknowledgements. We thank A. Stahl for fruitful discussions and gratefully acknowledge financial support by the Deutsche Forschungsgemeinschaft in the Schwerpunkt *Quantenkohärenz in Halbleitern.*

References

1 R. Martini, G. Klose, H. G. Roskos, H. Kurz, H. T. Grahn, and R. Hey, Phys. Rev. B **54**, R14325 (1996).
2 C. Waschke, H. G. Roskos, R. Schwedler, H. Kurz, and K. Köhler, Phys. Rev. Lett. **70**, 3319 (1993).
3 F. Wolter, H. G. Roskos, P. Haring Bolivar, G. Bartels, H. Kurz, K. Köhler, H. T. Grahn, and R. Hey, phys. stat. sol (b) **204**, 83 (1997).
4 T. Kuhn in *Theory of Transport Properties of Semiconductor Nanostructures*, Edited by E. Schoell (Chapman and Hall, London, 1998).

Subpicosecond Carrier-Phonon Interactions in $GaAs/Al_xGa_{1-x}As$ MQW': Plasmon-Phonon Coupling Mediated by Well-to-Well Coupling in MQW's.

K. J. Yee[1], D. S. Kim[1], D. S. Yee[1], T. Dekorsy[2], G. C. Cho[2], H. Kurz[2], and D. H. Woo[3]

[1] Department of Physics, Seoul National University, Seoul 151-742, Korea
E-mail: kedroi@phya.snu.ac.kr
[2] Institut fur Halbleitertechnik II, RWTH Aachen, Templergraben 55, D-52056 Aachen, Germany
E-mail: DEKORSY@iht-ii.rwth-aachen.de
[3] KIST, Cheongryang, Seoul 136-791, Korea

Abstract. We study coherent phonon oscillations in $GaAs/Al_xGa_{1-x}As$ MQW's with varying x and fixed well and barrier widths of 100Å. We have observed large plasmon-phonon coupling in x=0.2 MQW, although it is generally believed that the 2D nature of plasmon makes the plasmon-phonon coupling very difficult. One possible explanation for this phenomenon is the bulk-like, growth direction plasmon oscillations mediated by the anomalously large well-to-well coupling for small x (x<0.35).

1. Introduction

Coherent optical phonons are generated by the electric field induced acceleration of free carriers generated by an ultrafast laser pulse. One of the direct ways of studying the generation and dephasing of such phonon population is the use of time-resolved reflectivity change. One interesting topic in this area is the screening of phonon by photo-excited carriers, i.e. plasmon-phonon coupling. In bulk GaAs, shifts of LO phonon frequency to TO phonon frequency at high excitation density occurs as a result of this plasmon-phonon coupling. And the TO frequency component is roughly proportional to the photo-excited carrier density [1].

In narrow well $GaAs/Al_xGa_{1-x}As$ MQW's which is quasi 2D case, growth-direction plasmon-phonon coupling is very difficult because electrons can not move freely in the growth direction [2]. As the well width become large, bulk-like coupling becomes possible as in 3D. This transition from 2D to 3D behavior occurs when the inter-subband energy separations are near LO phonon frequency [2].

It was revealed that phonons can propagate through $GaAs/Al_xGa_{1-x}As$ MQW's when GaAs LO phonon and GaAs-like LO phonon in $Al_xGa_{1-x}As$ barrier overlap in dispersion curve for small x [3]. And anomalously large well-to-well coupling of carriers was also observed for barriers with small x. Our paper concerns the plasmon-phonon coupling in GaAs/$Al_xGa_{1-x}As$ MQW's with varying x.

2. Experimental Methods

We performed the standard reflective electro-optic sampling experiments varying pump beam intensity. The laser was a mode-locked Ti:sapphire laser with pulse duration of sub-60fs and photon energy of 1.61eV. All experiments were done at room temperature. We used a series of GaAs/$Al_xGa_{1-x}As$ MQW's with varying x and the same well and barrier widths of 100 Å in p-i-n structure. Bulk GaAs was also used for reference in the same set-up.

3. Results and Discussion

Figure 1. shows coherent phonon signal obtained in GaAs/$Al_xGa_{1-x}As$ MQW with x=0.2. Fig. 1(a) is the time domain signal and Fig. 1(b) is the corresponding Fourier transform data. At high intensity we can see nice beating between LO and TO phonon frequencies, and TO components decrease with pump intensity. This TO phonon frequency peak is a result of the growth direction plasmon-phonon coupling as is generally observed for bulk GaAs, but not expected in our sample because of large subband separation compared with phonon energy (E_{12}~100meV, E_{LO}~37meV) [2].

Figure 2. is coherent phonon data for MQW with x=0.5. There exists nearly no TO frequency component that is in agreement with the simple theory of ref.[2].

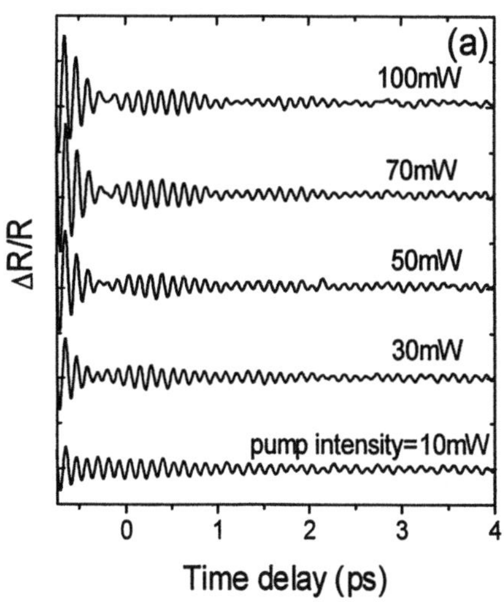

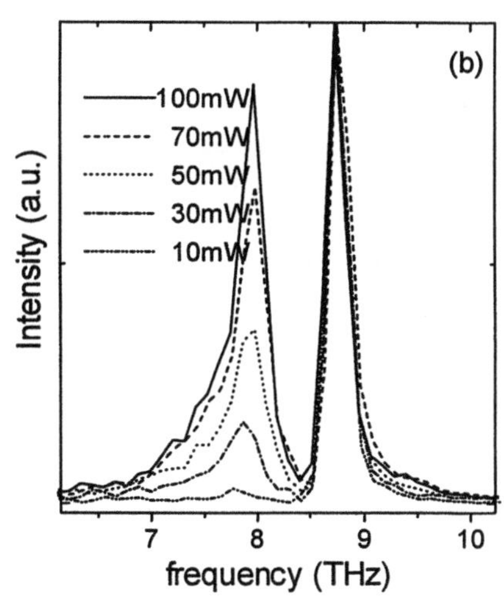

Fig. 1. (a) Coherent phonon signal in time domain and (b) Fourier transform of the MQW with x=0.2.

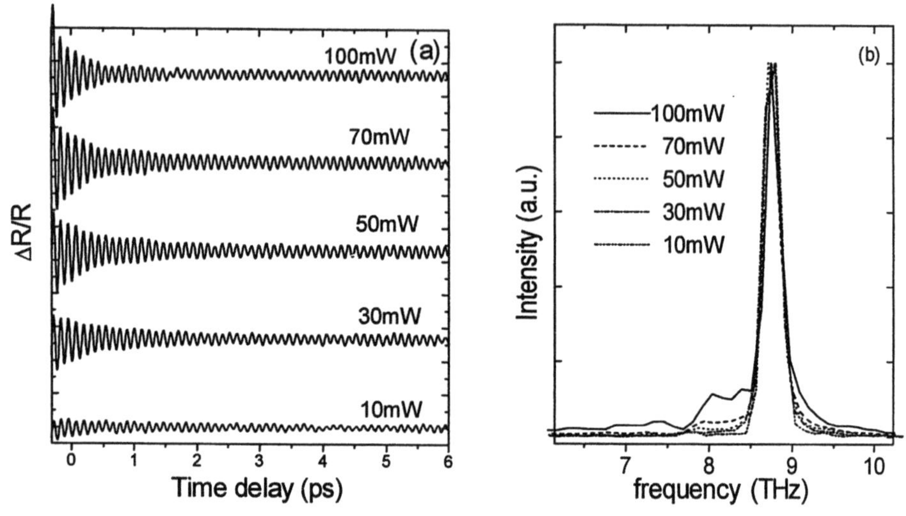

Fig. 2. (a) Coherent phonon signal in time domain and (b) Fourier transform of the MQW with x=0.5.

What is surprising is the appearance of growth-directional plasmon-phonon coupling in x=0.2 MQW with narrow well which is expected to behave as 2D case in the mean field approximation. This interesting phenomenon might originate from the anomalously large well-to-well coupling between barriers with small x (<0.35) [4]. In other words, anomalously large well-to-well coupling mediate growth-direction plasmon-phonon coupling in GaAs/$Al_xGa_{1-x}As$ MQW.

4. Conclusions

We studied coherent phonon oscillations in GaAs/$Al_xGa_{1-x}As$ MQW's. We could observe plasmon-phonon coupling in MQW with x=0.2 in spite of narrow well width. We contend that the growth-direction, bulk-like plasmon oscillations mediated by the anomalously large well-to-well coupling might screen the LO phonon oscillations.

References

1 W. A. Kutt, A. M. T. Kim, G. C. Cho, H. Kurz, A. V. Kuznetsov, and A. Forster, *IEEE. J. Quantum Electron.* Vol. 28, 2434, 1992.
2 T. Dekorsy, A. M. T. Kim, G. C. Cho,H. Kurz, A. V. Kuznetsov, and A. Forster, *Phys. Rev. B*, vo. 153, 1531, 1996.
3 D. S. Kim, Abdellatif Bouchalkha, J. M. Jacob, J. F. Zhou, and J. J. Song, *Phys. Rev. Lett.* Vol. 68, 1002, 1992.
4 D. S. Kim, et. al. *Phys. Rev. B*, vol. 54, 14580, 1996.

Study of coherent folded acoustic phonons in GaAs/AlAs superlattices by two-color pump-probe technique

K. MIZOGUCHI[1], M. HASE[1], S. NAKASHIMA[1], and M. NAKAYAMA[2]

[1] Department of Applied Physics, Osaka University, Yamadaoka 2-1, Suita 565-0871, Japan
E-mail: mizoguch@ap.eng.osaka-u.ac.jp
[2] Department of Applied Physics, Osaka City University, Sugimoto, Sumiyoshi-ku, Osaka 558-8585, Japan
E-mail: nakayama@a-phys.eng.osaka-cu.ac.jp

Abstract. We have observed coherent folded longitudinal acoustic phonons in periodic superlattices by one- and two-color pump-probe measurements. The time-resolved spectrum with violet pump and near-IR probe beams indicates that the coherent folded acoustic phonons generated near the surface propagate to a substrate and are observed only when the acoustic phonons propagate in the superlattice layer.

1. Introduction

The dynamics of coherent oscillations by using ultrashort laser pulses has attracted much attention since the time evolution of carriers and phonons was directly observed in time-domain [1,2]. Recently, coherent folded longitudinal acoustic (FLA) phonons in periodic superlattices (SLs) have been observed by a reflection-type pump-probe method using ultrashort pulse lasers [3]. However, there is no report on the propagation of the coherent FLA phonon. For use of two-color (2C) light beams, the pumping and probing regions are different because of the difference in the penetration depth: the phonon generation by the violet pump light near the surface and the detection by the near-IR probe in the SL. Then the 2C pump-probe experiment enable us to obtain the information on the propagation of the coherent FLA phonons in the SL. In the present paper, we report on the coherent FLA phonons propagating in a GaAs/AlAs periodic SL observed by using the 2C method.

2. Experimental Methods

The GaAs/AlAs SL under investigation has a periodic heterostructure, a $(GaAs)_{24}(AlAs)_{24}$ SL on a (001) GaAs substrate which consists of 50 periods, where the subscript denotes the number of the GaAs or AlAs monolayers (thickness of one monolayer is equal to 0.283 nm). The total thickness of the SL layer is about 680 nm. The Reflection-type pump-probe experiments were performed by using a mode-locked Ti:sapphire laser delivering 100 fs pulses. In the 2C measurement, the SHG light of the Ti:sapphire pulse laser was used as pump light (407 nm), and the fundamental laser light tuned to the first interband transition of the SL was used as probe light (814 nm). The pump (probe) power was adjusted to 70 mW (5 mW). The probe beam was delayed by a variable optical delay line over 200 ps.

3. Results and Discussion

Figure 1 shows the time derivative traces of reflectivity changes for $(GaAs)_{24}(AlAs)_{24}$ SL obtained by the one-color (1C) and 2C pump-probe measurements. Oscillatory structures with beats are observed in both measurements. The coherent oscillations obtained by the 1C pump-probe method are observed in the time range longer than 200 ps, while the oscillations observed by the 2C pump-probe technique almost disappear after about 140 ps.

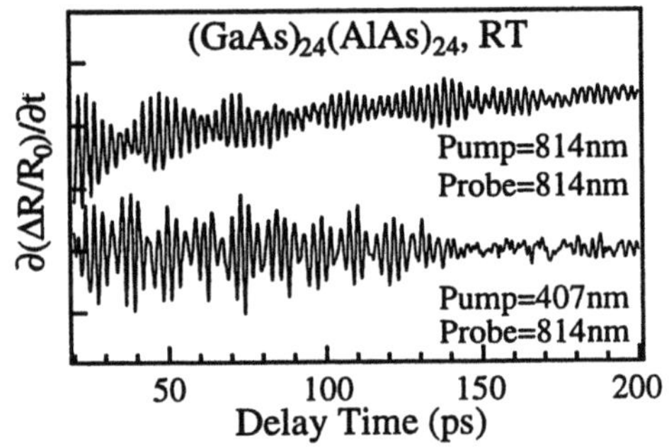

FIG. 1 Time derivative traces of reflectivity changes for $(GaAs)_{24}(AlAs)_{24}$ SL observed by using 1C and 2C pump-probe techniques.

In the 2C experiment with the 407-nm pump and 814-nm probe, it is to be noted that the penetration depth of the 407-nm pump pulses is several 10 nm. The pump pulse produces high-density electrons and holes near the surface region, and then it is considered that the surface band-bending due to the photoexcited carriers instantaneously occurs and generates a phonon pulse through the deformation potential. When the phonon pulse propagates through the SL layer, the dielectric constant of the SL layer is modulated. Then the coherent oscillations can be detected only during the propagation of the phonon pulse in the SL layer because the 814-nm light pulse penetrates entirely the SL layer. The observable time for the coherent oscillations in the 2C pump-probe experiment corresponds to the propagation time of the FLA phonons from the surface to the end of the SL layer. The propagation time is estimated to be about 133 ps from the total thickness of the SL layer and the average sound velocity, which is calculated from the thickness ratio of the GaAs and AlAs layers (v_s=5.1x10^5 cm/sec in $Ga_{0.5}Al_{0.5}As$ [4]). The estimated propagation time agrees with the time range of the large amplitude oscillations in the 2C pump-probe experiment. In order to confirm the propagation of the FLA phonons, we have measured the coherent acoustic phonon in other SLs capped at the surface with AlGaAs cladding layers. Similar results have been observed in the SLs with the cladding layer.

For more detailed analysis of the frequencies of the observed coherent phonons we performed Fourier-transform of the time-domain data (Fig.2). The Fourier transform (FT) spectrum obtained by the 1C pump-probe technique shows that a strong sharp peak accompanies weak satellite peaks at the lower- and higher- frequency sides. The frequencies of the strong sharp peaks are 0.39 and 0.78 THz, and the frequencies of the weak satellite peaks at lower frequency side are 0.35 and 0.43 THz. The FT spectrum obtained by the 2C pump-probe technique shows only two peaks at 0.35 and 0.43 THz. The strong sharp peak at 0.39 THz is not observed.

From the comparison between the observed frequencies and the phonon dispersion curve [5], the peaks at 0.39 and 0.78 THz observed in the 1C pump-probe experiment are attributed to the first and second FLA modes at the zone center (q=0), respectively. Moreover, the peaks at 0.35 and 0.43 THz are assigned to the first FLA modes near the zone center (q≠0). For the 2C pump-probe spectrum, only the FLA phonons with q≠0 which propagate in the SL layer are observed. If coherent phonons would be generated via impulsive stimulated Raman scattering predicted by Merlin [2], the wavevectors of the FLA phonons with q≠0 should change with the wavelength of the pump pulse. Our observation is not affirmative to this prediction. A possible explanation is that the coherent vibrations with various wavevectors are generated by the pump pulse and the coherent phonons with a certain wavevector are detected by the probe pulse.

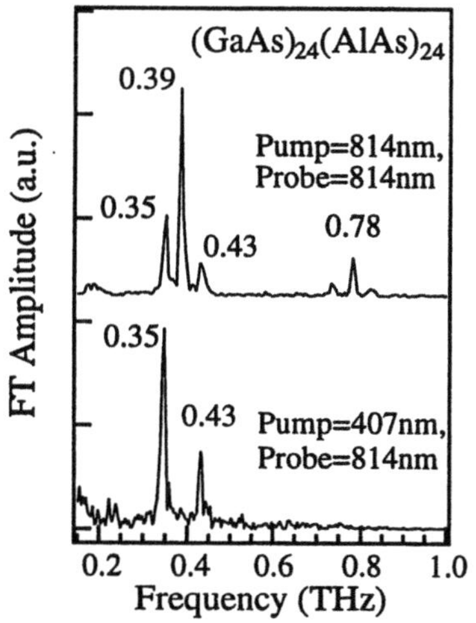

FIG. 2 Fourier-transform spectra of the time-domain signals in $(GaAs)_{24}(AlAs)_{24}$ SL obtained by using 1C and 2C pump-probe techniques.

3. Summary

Using the 2C pump-probe technique, we have measured the coherent FLA phonons in the SL which are generated near the surface and propagate through the SL layer. The FLA phonons with q≠0 have been observed only during propagating in the SL layer.

Acknowledgements. The samples were grown by the MBE facilities of Kwansei-Gakuin University, Japan. The works was partially supported by a Grant-in-Aid for the Scientific Research from the Ministry of Education, Science, and Culture of Japan.

References

1 W. A. Kütt, et al, IEEE, J. Quantum Electron., Vol.28, 2434, 1992; H. J. Zeiger, et al, Phys. Rev. B, Vol.45, 768, 1992.
2 R. Merlin, Solid State Commun., Vol.102, 207, 1997.
3 A. Yamamoto, et al, Phys. Rev. Lett. Vol.73, 740, 1994; K. Mizoguchi, et al, Physica B, 1998, in press.
4 S. Adachi, J. Appl. Phys., Vol.58, R1, 1985.
5 S. M. Rytov, Akust. Zh., Vol.2, 71, 1956.[Sov. Phys. Acoust. Vol.2, 68 (1956)]

Coherent Acoustic Phonon Oscillations in Superlattices

T. Dekorsy[1], A. Bartels[1], H. Kurz[1], and K. Köhler[2]

1 Institut für Halbleitertechnik, RWTH Aachen, Sommerfeldstr. 24, D-52056 Aachen, Germany
E-mail: dekorsy@iht-ii.rwth-aachen.de
2 Fraunhoferinstitut für Angewandte Festkörperphysik, D-79108 Freiburg, Germany

Abstract. Resonant impulsive stimulated Raman scattering is shown to be the driving force for coherent acoustic phonons in semiconductor superlattices. Three modes are observed for the first two backfolded phonon branches, which are assigned to q=0 and $q=2k_{laser}$ excitation via forward and backward scattering, respectively. Modes of different backfolded branches are controlled coherently.

The investigation of low-energy elementary excitations in semiconductor heterostructures is driven by their relevance as the final state in the process of energy relaxation. In particular, acoustic phonons play a dominant role for dephasing processes at low lattice temperature and for heat transport in general. However, these excitations with energies in the energy range of some meV are hardly accessible by means of coherent time-resolved spectroscopy, since Brillouin scattering at the wavevector of light can only couple to phonons of μeV energy. Semiconductor superlattices exhibit zone-folding of the acoustic branches within the mini-Brillouin zone due to the artificial periodicity. Thus a series of backfolded phonon branches is generated, enabling the coupling of light to phonons in the frequency range covered by the bulk acoustic phonon dispersion. The detection of coherent zone-folded acoustic mode was first demonstrated in GaAs/AlAs superlattices in a time-derivative detection scheme, where a single frequency of the first order backfolded mode was observed [1]. Here we present the first observation of several distinct back-folded modes. The appearance of several peaks allows the clear identification of the excitation process as resonant impulsive stimulated Raman scattering in both forward and backward direction, leading to modes with wavevectors **q**=0 and $\mathbf{q}=2\times\mathbf{k}_{laser}$, respectively.

The experiments are performed on GaAs/AlAs superlattices consisting of 19 monolayers of each constituent. The laser source is a Kerr-lense modelocked Ti:sapphire laser delivering pulses of 50 fs. The excitation and detection wavelengths are in resonance with the first interband resonance. The experimental challenge is to maintain a high signal-to-noise ratio for the detection of relative reflectivity changes $<10^{-6}$ in for time delays > 100 ps, which is approximately the dephasing time of acoustic modes at room temperature. This requirement is achieved in a step-and-repeat mode with a fast-scanning time-delay (≈80 Hz) .

Fig. 1 depicts the time resolved reflectivity changes due to coherent acoustic modes from the superlattice at room temperature. The exponentially decaying electronic background of the order 10^{-3} $\Delta R/R_0$ has been subtracted. The oscillations exhibit a complicated structure which stems from the superposition of 6 different frequencies. These frequencies belong to two triplets around 450 GHz and 900 GHz, i.e. the first and second order backfolded acoustic modes. The frequencies within one triplet agree well with the calculated **q**=0 and **q**=2×$\mathbf{k}_{laser}$ frequencies of the backfolded LA. The **q**=0 mode is clearly identified as the branch having A_1 symmetry within the superlattice point group, i.e. the lower branch in first order the upper in second order. The **q**=0 modes in the other branches of the dispersion have B_2 symmetry and are Raman forbidden. The **q**=2×$\mathbf{k}_{laser}$ modes have mixed symmetry and are thus Raman allowed. The high resolution determination of all frequencies and their identification with the calculated dispersion relation allows us to determine accurately the width of the acoustic gap at **q**=0, which is sensitive to interface imperfections.

The dephasing times of the modes are in excess of 100 ps. From long time-delay scans (>200 ps) we obtain linewidths as narrow as 2 GHz, which is well beyond linewidths achieved in high-resolution CW Raman spectroscopy. Since the detection is based on the acoustic deformation potential-exciton coupling, the temporal development of the amplitude is influenced by the transient changes of the excitonic resonances, which is broadened and bleached by the excited carriers within the first 50 ps. This observation is confirmed at low lattice temperatures (10 K), where the difference between the unperturbed exciton resonance and the excited resonance is stronger than at room temperature.

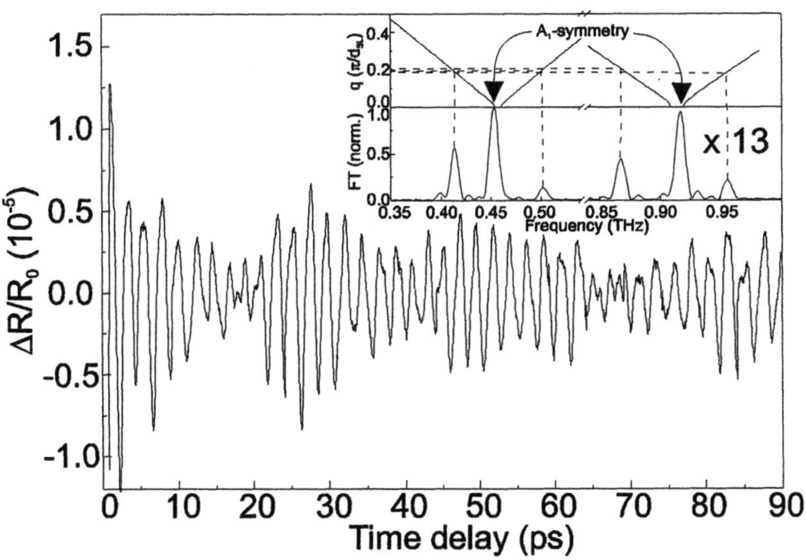

Fig. 1. Acoustic phonon oscillations from a $(GaAs)_{19}(AlAs)_{19}$ superlattice at room temperature. The insert shows the numerical Fourier transform with the LA phonon dispersion relation (dashed line) in the upper part.

Due to a constant average sound velocity up to phonon frequencies of 2 THz the different backfolded modes appear approx. at multiple integers of the frequency of the first backfolded mode. This fact can be exploited for the coherent control of the relative amplitudes of different backfolded modes. Fig. 2 shows that by application of successive pump pulses the first order mode can be completely silenced while the amplitude of the second order mode is enhanced. The experiment is performed by setting the time delay between two successive pump pulses to half the period of the **q**=0 first order backfolded mode, i.e. 1098 fs. The intensities of the two pulses are adjusted in order that both pulses provide the same driving force of opposite sign to first order the acoustic modes. Such coherent control experiments allows an additional identification of the excited modes and the selective study of their dephasing times [2].

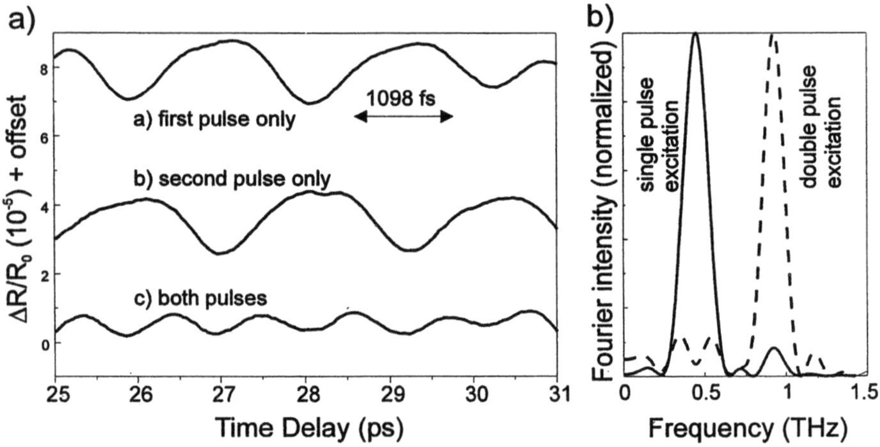

Fig. 2. a) Time resolved acoustic phonon traces of a $(GaAs)_{19}(AlAs)_{19}$ superlattice under single pulse (a and b) and double pulse (c) excitation. b) The Fourier spectrum of the single pulse excitation (solid line) and of the double pulse excitation (dashed line).

Acknowledgements. This work is supported by the Deutsche Forschungsgemeinschaft.

References

1 A. Yamamoto, T. Mishina, Y. Masumoto, and N. Nakayama, Phys. Rev. Lett. **73**, 740 (1994).

2 A. Bartels, T. Dekorsy, H. Kurz, and K. Köhler, Appl. Phys. Lett. **72**, 2844 (1998).

High-Field Transport in Semiconductors Studied via Ultrabroadband THz Sampling

A. Leitenstorfer, S. Hunsche, J. Shah, M. C. Nuss, and W. H. Knox

Bell Labs, Lucent Technologies, Holmdel, NJ 07733, USA
E-mail: aleitens@physik.tu-muenchen.de

Abstract. Ultrafast carrier transport in GaAs and InP is investigated with an unprecedented time resolution of 20 fs and for electric fields up to 130 kV/cm. We provide quantitative values for the maximum velocity overshoot, depending on material and electric field. The data are in agreement with the semiclassical theory even for very high fields.

The non-equilibrium transport properties of semiconductor materials represent a key issue in a wide field ranging from fundamental research in many-body physics to applications in the engineering of modern high-speed devices. Due to the lack of quantitative data on the relevant femtosecond time scales, our present understanding of this topic relies mainly on theoretical calculations based on the semiclassical Boltzmann equation.

We have performed the first direct investigation of high-field transport in GaAs and InP with a time resolution of 20 fs: Electron-hole pairs are excited in large-area p-i-n diodes at 300 K with 12 fs laser pulses of a central photon energy of 1.49 eV. Carrier densities are as low as 5×10^{14} cm^{-3}. Bias fields up to 130 kV/cm have been verified via the Franz-Keldysh effect. In the far zone, the time-dependent dipole moment that arises due to the rapid separation of the carrier pairs generates an electromagnetic field with an amplitude proportional to the momentary charge acceleration. We have developed a special geometry in which the THz transients of each elementary dipole arrive synchronously at a point 12 mm from the diode. Here, the signal is detected via ultrabroadband electro-optic sampling [1]. We have used <110> oriented ZnTe and GaP detector crystals as thin as 8 μm and 15 μm. The bandwidth of our setup exceeds 50 THz.

Integrating the measured transients twice in time yields a quantity that is proportional to the carrier displacement: After a few ps, the double integrals saturate at a known value, namely the width of the intrinsic region of the p-i-n devices. In this way, the time dependent displacements and its derivatives, the velocities and accelerations are calibrated quantitatively.

On the left hand side of Figure 1, the electro-optic THz signals from the GaAs sample measured with the GaP sensor are depicted for various bias fields E. The transients have to be corrected for the frequency response of the detector given by linear [1] and nonlinear [2] optical effects. The resulting time dependent accelerations are shown on the right hand side of Figure 1.

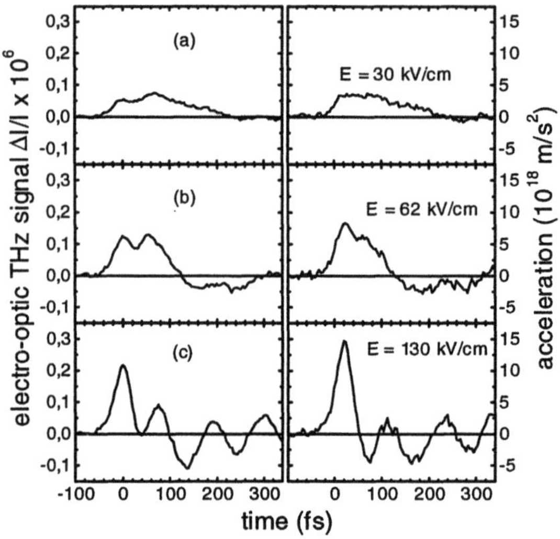

Fig. 1. Electro-optic signals (left) from the GaAs diode and carrier accelerations corrected for the detector response (right) at a temperature of 300 K and different electric fields E.

The data are dominated by the electron dynamics due to the low effective mass at the global minimum of the conduction band. For a bias field of 30 kV/cm [Fig. 1 (a)], the acceleration shows an instantaneous onset within the time resolution of our system and remains constant for approximately 100 fs. A decrease in acceleration follows since carriers have gained enough energy to scatter from the Γ-valley into the low mobility L- and X side minima of the conduction band. At 62 kV/cm [Fig. 1 (b)], the acceleration period shortens. A negative THz amplitude between 120 fs and 300 fs indicates strong deceleration: The electron velocity exhibits a large overshoot in the Γ-valley before equilibrium drift dominated by the high effective mass and strong scattering in the side valleys is established. These findings are in good agreement with Monte Carlo simulations of femtosecond transport within the framework of the semiclassical Boltzmann equation [3]. No theoretical calculations are found in the literature for fields above 100 kV/cm [Fig. 1 (c)]. In this regime, quantum phenomena might start to influence high-field transport. However, our results are still in qualitative agreement with the semiclassical model assuming an intervalley transfer rate of $(20\ \text{fs})^{-1}$.

The polarization originating from the displacement of the electron-hole pairs leads to a small but ultrafast change of the internal electric field. The ion cores respond by accelerating towards their new equilibrium positions and the polar crystal lattice becomes excited coherently. As a result, a component oscillating with the LO phonon frequency is superimposed on the signal connected to free carrier transport. When the bias is increased to 130 kV/cm, the acceleration–deceleration cycle becomes comparable to the oscillation period of the LO phonon (115 fs in GaAs) and the generation of coherent phonons is resonantly enhanced.

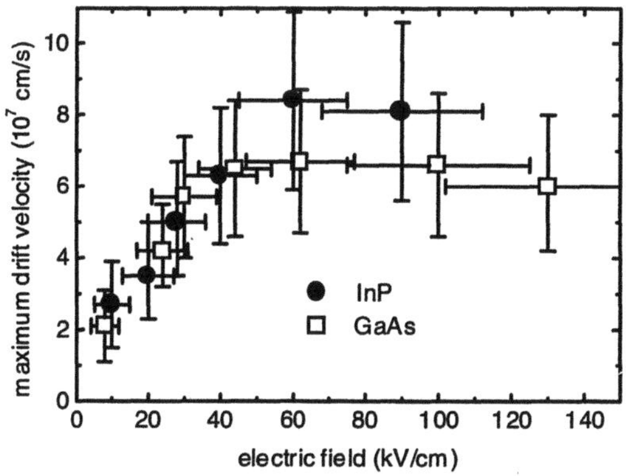

Fig. 2. Maximum drift velocities (velocity overshoot) in GaAs and InP versus electric field.

Figure 2 shows the maximum transient drift velocities measured during the non-equilibrium regime as a function of electric field. The velocity overshoot is similar in both materials for bias fields below 40 kV/cm. At stronger fields, InP allows higher transient velocities because of the larger energy splitting between the Γ- and L-minima in the conduction band (600 meV as compared to 330 meV in GaAs). The saturation behavior of the velocity overshoot indicates a regime of electric fields where momentum loss due to polar optical interaction with LO phonons within the central valley is negligible and the carriers accelerate ballistically to the intervalley threshold. The transition to a saturated maximum velocity occurs at lower fields in GaAs due to the less polar character and consequently weaker Fröhlich interaction in this compound.

In conclusion, we have developed a powerful technique giving direct and quantitative access to the ultrafast transport properties of semiconductors. The high-field dynamics of GaAs and InP has been studied with a resolution of 20 fs and in a previously unexplored regime of electric fields up to 130 kV/cm. Influences of various momentum scattering processes and band structure effects are identified. Coherent excitations of the crystal lattice due to the coupling to the electron-hole polarization via the linear dielectric function are observed. Our results are of substantial importance for the understanding of non-equilibrium electronic conduction and for the engineering of future high-speed electronics. This work was supported by the NEDO Femtosecond Technology Project.

References

1 Q. Wu and X.-C. Zhang, Appl. Phys. Lett. **70**, 1285 (1997) and Appl. Phys. Lett. **70**, 1784 (1997)
2 W. L. Faust, C. H. Henry, and R. H. Eick, Phys. Rev. **173**, 781 (1968)
3 G. M. Wysin, D. L. Smith, and A. Redondo, Phys. Rev. **B 38**, 12514 (1988)

Kinetics of Four-Wave Mixing for a Magnetoplasma in a Quantum Well

M. W. Wu and H. Haug

Institut für Theoretische Physik, J.W. Goethe Universität Frankfurt, D-60054 Frankfurt a. M., Germany
Email: wu@mandala.th.physik.uni-frankfurt.de

Abstract. Femtosecond kinetics of an optically excited 2D magnetoplasma is theoretically studied and the magnetic field dependence of the dephasing is calculated.

Femtosecond pulse excitation in semiconductors induces transient charge fluctuations, which can be studied through nonlinear-optical effects to elucidate many-body phenomena, such as time-dependent Coulomb correlations. Numerous experimental and theoretical studies have been devoted to this problem in the magnetic-field-free case in the last ten years[1-3]. Studies of femtosecond laser spectroscopy in the presence of a strong magnetic field are relatively rare and are mainly focused on magnetoexcitons[4-6], for low excitation densities. With strong resonant laser pulses which excite a dense carrier system in a strong magnetic field, one can study the kinetics of a magnetoplasma. While the plasma-density dependent spectra for a quasi-equilibrium 2D magnetoplasma have been calculated[7], the relaxation and dephasing kinetics of a 2D-magnetoplasma has not been treated either experimentally or theoretically.

We present a kinetic study of this regime in a quantum well (QW) which is further restricted by an additional weak lateral confinement which lifts the degeneracy of the Landau levels partially and broadens them into Landau subbands. The additional confinement is described by a harmonic oscillator potential. For the pulses shorter than the build-up time of the Coulomb screening, the dephasing and relaxation of the excited magnetoplasma is solely due to the bare Coulomb scattering. We calculate the time-resolved and time-integrated (TI) four-wave mixing (FWM) signals by taking into account three Landau subbands in both the valance band and the conduction band. Differing from an earlier theory for semiconductors in a magnetic field with low density excitations[4] which describes the collision terms by means of phenomenological relaxations, we treat them by Boltzmann-kinetics. Following the same scheme as for the magnetic-field-free case in Ref. 3, we developed the mean-field Bloch equations for the QW in a strong perpendicular magnetic field B in the rotating-wave approximation together with delayed quantum kinetic scattering integrals for the particle distributions and the interband polarization functions. We further assumed electron-hole symmetry ($i.e.$ $m_e = m_h$) which simplifies the numerics greatly.

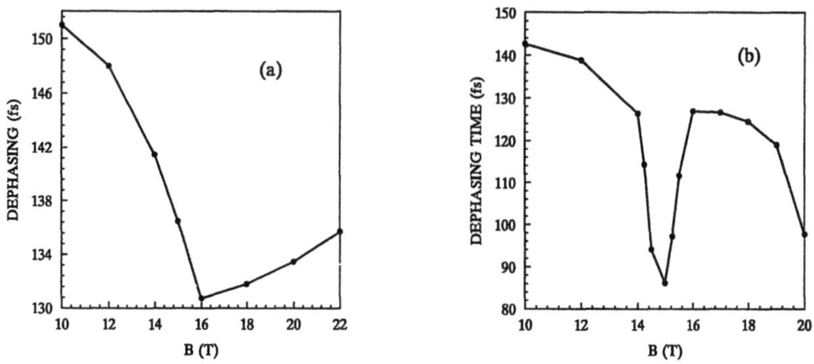

Figure 1: Dephasing time as a function of B. (a): Intermediate Density case; (b) High density case.

We performed numerical studies of our Bloch equations for a FWM setup with two pulses of 50 fs duration with an excess energy of 26.4 meV which in the energy spectrum may include up to 2 Landau subbands in both the conduction and the valence bands when B is larger than 10 T. We calculated both TR- and TI-FWM signals under different magnetic fields for intermediate- and high-density excitations. We observe quantum beats with frequency $2\omega_c$ for both signals under magnetic fields when the polarizations of the lowest two Landau subbands contribute to the FWM signals. The effective dephasing times are calculated from TI-FWM signals and are plotted as functions of magnetic field in Fig. 1 (a) and (b) with the former for the intermediate density case and the later for high density case.

For a fixed pulse, several effects compete with each other when the magnetic field increases. On one hand, the number of Landau subbands which contribute to the Coulomb scattering kinetics decreases. In particular the contributions to the dephasing from the intra- and inter-subband scattering of the higher Landau subbands as well as the inter-subband scattering between the higher and lower subbands decrease. For large populations in one subband the Pauli blocking may further reduce also the intra-band scattering rates. All these effects increase the dephasing time. On the other hand, with increasing B field the degeneracy of Landau subbands increases and the matrix elements of the Coulomb scattering become larger. Both of them reduce the dephasing time. Our results show when the increase of the Coulomb matrix elements and the number of states in a given Landau subband dominant, the dephasing time decreases with B field. In the situation when the loss of scattering channels exceeds these increasing effects, one gets a slight increase at the dephasing time. However, details of the strongly modulated scattering kinetics depend sensitively on the detuning, the plasma density, and the spectral pulse width relative to the cyclotron frequency.

1. Proceedings of the Third International Workshop on Nonlinear Optics and Excitation Kinetics in Semiconductors, Bad Honnef, Germany [Phys. Status Solidi B **173**, 11 (1992)].

2. J. Shah, *Ultrafast Spectroscopy of Semiconductors and Semiconductor Microstructures* (Springer, Berlin) 1996.

3. H. Haug and A.P. Jauho, *Quantum Kinetics in Transport and Optics of Semiconductors* (Springer, Berlin) 1996.

4. C. Stafford, S. Schmitt-Rink, and W. Schaefer, Phys. Rev. B **41**, 10000(1990).

5. S. Glutsch and D.S. Chemla, Phys. Rev. B **52**, 8317 (1995).

6. U. Siegner, S. Bar-Ad, and D.S. Chemla, Chem. Phys. **210**, 155 (1996).

7. G.E.W. Bauer, Phys. Rev. Lett. **64**, 60 (1981) and Phys. Rev. B **45**, 9153 (1993).

Part VII

Relaxation in Semiconductors, Processing of Semiconductors and Dielectrics

Ultrafast Heating and Cooling of Electrons in GaAs at Room Temperature

Jean-François Lampin[1], François-Xavier Camescasse[1], Véronique Thierry-Mieg[2], and Antigoni Alexandrou[1]

[1] Laboratoire d'Optique Appliquée, Ecole Polytechnique-Ecole Nationale Supérieure des Techniques Avancées CNRS UMR 7639, Chemin de la Hunière, F-91761 Palaiseau Cedex, France
E-mail: lampin@enstay.ensta.fr
[2] Laboratoire de Microstructures et de Microélectronique, CNRS UPR 20, 196 av. Henri Ravera, F-92225 Bagneux, France

Abstract. We isolate the heating of electrons photoinjected with small excess energy using a nondegenerate pump-probe configuration and compare it to the cooling observed for large excess energies.

1. Introduction

Relaxation and thermalization of nonequilibrium photoexcited carriers in semiconductors have been extensively investigated because of the insight they provide into the fundamental properties of semiconductors as well as because of their importance for the performances of high-speed optoelectronic devices. In most experiments, electron-hole pairs are injected with a large excess energy with respect to the band gap and their cooling towards the bottom of the conduction band is observed. Very few experiments have focused on the opposite case where carriers are injected with a small excess energy and have to gain energy in order to reach equilibrium with a warm lattice. The latter case has been studied using time-resolved photoluminescence in the 10-100 ps range at relatively low temperatures where the carrier heating is slow [1]. Hole heating was also monitored by measuring the transient absorption saturation in a pump-probe experiment at a wavelength where the signal is dominated by the hole dynamics [2]. However, to our knowledge, the heating of electrons in a room-temperature lattice has not been investigated before and its dynamics may be important for the functioning of new types of devices such as quantum-cascade lasers.

2. Experimental Methods

In this work, we have chosen a nondegenerate pump-probe scheme which yields a signal proportional to the full electron distribution function [3]: the pump pulse excites electrons from the heavy-hole (HH) and light-hole (LH) valence bands while the probe pulse tests the absorption saturation of the interband transition from the split-off (SO) valence band to the conduction band (C) [see insets of Fig.

1]. Due to the large spin-orbit splitting in GaAs (340 meV), no holes are present in the SO band and the differential absorption signal is proportional to the electron distribution only (if excitonic and band-gap renormalization effects are neglected). We have thus been able to follow the ultrafast heating of a nonequilibrium electron distribution injected with small excess energy in a GaAs layer at room temperature and compare it to the relaxation and cooling of nonequilibrium electrons injected with a large excess energy.

We used a setup consisting of a Ti:sapphire mode-locked oscillator and a 200-kHz regenerative amplifier system (Coherent RegA). The probe pulses were obtained by spectral continuum generation and were compressed down to 40 fs using a combination of prisms and gratings. The pump had a duration of 150 fs (15 meV FWHM) and was obtained either directly as the output of the regenerative amplifier with an energy of 1.544 eV (120 meV excess energy with respect to the band gap) or after parametric amplification (energy: 1.434 eV; 10 meV excess energy). The sample was an intrinsic GaAs layer of thickness d=0.65 μm antireflection-coated on both sides and was held at room temperature.

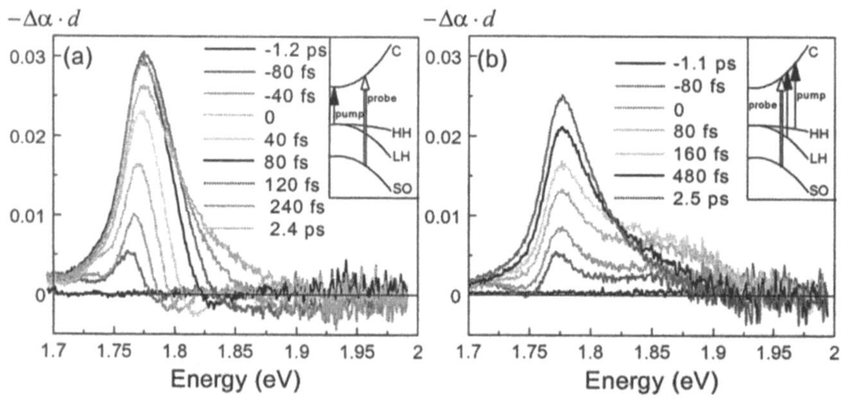

Fig. 1. Differential absorption spectra in undoped GaAs for a pump energy with a small [case(a)] and a large [case(b)] excess energy with respect to the band gap.

3. Results and Discussion

The differential absorption spectra obtained for different pump-probe delay times are shown in Fig. 1 for photoexcitation close to the band gap (a) and for photoexcitation with large excess energy (b) for the same total density of photoexcited carriers of about $8 \cdot 10^{16}$ cm^{-3}. The spectra of Fig. 1(b) show a nonequilibrium electron population at early times with a spectral hole burning signal due to the unresolved initially injected populations from the HH and LH valence bands (expected to appear at 1.911 and 1.860 eV, respectively). A characteristic sharpening of the Fermi edge at longer delay times indicates a reduction of the electron temperature down to the lattice temperature of 300K.

The opposite is observed in Fig. 1(a) where the electron temperature increases up to the lattice temperature mainly due to LO-phonon absorption. While LO-phonon emission is also responsible for the cooling in Fig. 1(b), an important electron energy loss mechanism in this case is electron-hole collisions where the holes injected with small excess energies exchange energy with the energetic electrons. The reverse process (heating of electrons through collisions with hot holes) is not operative in the case of excitation close to the band gap, at least at early delay times, since then both types of carriers are injected with small excess energies. In both cases of excitation with small and large excess energy the lattice temperature of 300 K is attained about 2 ps after the pump pulse and the differential spectra for all the subsequent delay times are identical. This is not surprising, even though the amount of energy to be given away in case (b) is much larger than the amount to be gained in case (a), because LO-phonon emission is 3-4 times more efficient than LO-phonon absorption.

We have also used the heating configuration in a three-pulse experiment in order to investigate the role of pre-injected carriers and of their temperature on the relaxation of nonequilibrium electron populations. The pre-pump pulse was of low excess energy and the pump pulse was of large excess energy. By varying the time delay between these two pulses, we observed that a cold pre-injected population accelerates the relaxation of non-equilibrium carriers (short delay between pump and pre-pump). There is no such effect with the same density of room-temperature pre-injected carriers (large delay between pump and pre-pump, the carriers induced by the pre-pump have thermalized). This effect was predicted by Scott *et al.* [4], and was ascribed to a strongly increased carrier-carrier scattering rate due to acoustic plasmon undamping.

4. Conclusions

We have studied the heating and cooling of nonequilibrium *electrons* in room-temperature GaAs. These measurements can be compared to theory without having to resort to an averaging of the electron and hole heating or cooling as is the case for results obtained from time-resolved photoluminescence or conventional pump-probe measurements. The information obtained on the heating dynamics allowed us to demonstrate the importance of the temperature of pre-injected carriers in a three-pulse experiment.

References

1 W. W. Rühle and H.-J. Polland, in *Physical Review B*, Vol. 36, 1683, 1987.

2 N. Del Fatti, P. Langot, R. Tommasi, and F. Vallée, in *Applied Physics Letters*, Vol. 71, 75, 1997.

3 F. X. Camescasse, A. Alexandrou, D. Hulin, L. Bányai, D. B. Tran Thoai, and H. Haug, in *Physical Review Letters*, Vol. 77, 5429, 1996.

4 D. C. Scott, R. Binder, and S. W. Koch, in *Physical Review Letters*, Vol. 69, 347, 1992.

Effects of carrier screening and state filling on the carrier scattering of Si doped GaAs

Weizhu Lin, Jinhui Wen, Yingyu Chen, Chun Huang, Zhiren Qiu, and Weiliang Li
State Key Laboratory of Ultrafast Laser Spectroscopy, Department of Physics
Zhongshan University, Guangzhou 510275, P.R. China
email: stslwz@zsu.edu.cn

Abstract Femtosecond spectroscopy studies on highly Si-doped GaAs show that the excited carrier initial scattering is reduced by the enhanced carrier screening and state filling.

Recent developments of high speed optoelectronic semiconductor devices have encouraged the studies of ultrafast carrier dynamics not only in intrinsic but also in n-type or p-type doped semiconductors. It has been reported that interaction between the photoexcited hot carriers and the doped or injected cold electron-hole plasma in GaAs accelerates the initial scattering or the thermalization of the as-excite carriers with the excess energy higher than the background Fermi level [1-3]. However, most of the previous researches carried out with higher exciting photon energies (2 eV). Intervalley scattering (between Γ and L , X valleys) and interband transition (from the split-off valance band to conduction band) complicate the interpretation of the experimental results. In this paper we report the investigations of the ultrafast carrier dynamics at the excited states near and above the Fermi level of the background electron distribution in highly Si doped GaAs using various photon energy femtosecond pulse pump-probe techniques. The experimental results present a comprehensive understanding of the influences of carrier screening, plasmon emission and state filling on the excited carrier initial relaxation.

The femtosecond absorption saturation measurements were performed at room temperature on Si doped GaAs film which was grown by MBE with Si dopant concentration of 2×10^{18} cm^{-3} , a typical concentration for semiconductor devices. The pump and probe pulses are generated from a Kerr lens mode-locked tunable Ti:sappire laser with a pulse repetition rate of 90 MHz and pulse duration of 60 fs. The center wavelength of the pulses are chosen , respectively, with photon energies above and near the doped electron Fermi level (~ 90 meV). The photoexcited carrier densities are about 1×10^{18} cm^{-3} . The transient absorption saturation traces of Si:GaAs for pump-probe photon energies of 1.65, 1.57 and 1.52 eV are shown in Fig.1 (a), (b) and (c), respectively. For comparison, the traces of intrinsic GaAs are also displayed in Fig.2. In all the traces, the artifact components arising from the coherent coupling of the parallel polarized pump - probe pulses have been subtracted by subtracting the pump-probe pulse intensity autocorrelation equal to the half of the amplitude of the pump-probe signal at the zero delay from the raw traces.

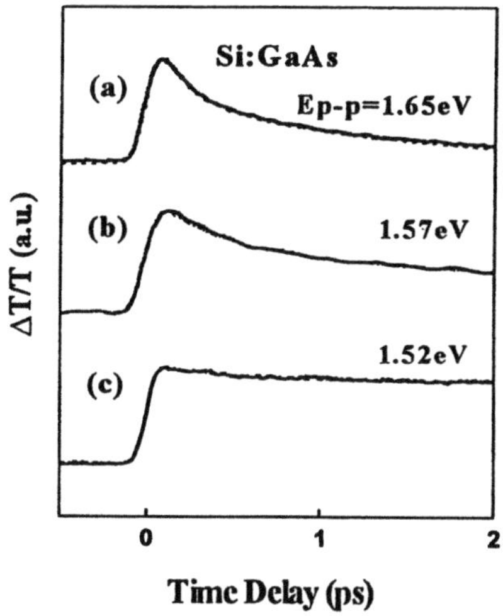

Fig.1 Femtosecond absorption saturation traces of Si:GaAs for various pump-probe photon energies .

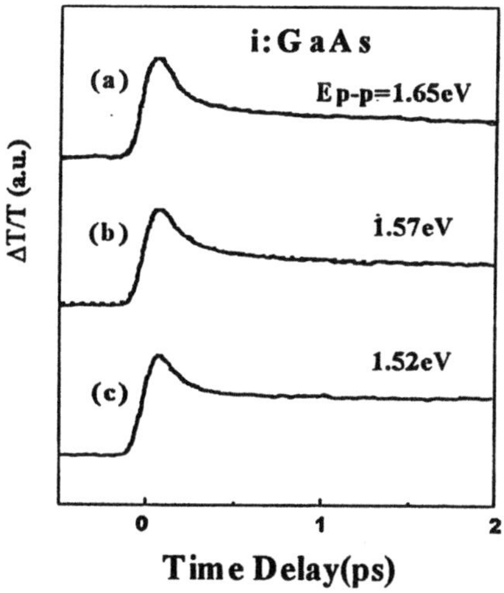

Fig.2 Fs absorption saturation traces of i:GaAs for various photon energies.

The trace (a) as well as (b) of Fig.1 shows a fast bleaching component which characterizes the ultrafast excited carrier initial scattering process. However, this fast component decreases significantly in trace (c). Two fast initial carrier scattering time of 150 , 180 fs and a much slower one of 500 fs are deduced, respectively, from the traces (a), (b) and (c) by theoretical fit (dashed curves in Fig.1). In contrast, much faster carrier initial scattering times of 85, 100 and 115 fs in i:GaAs are measured, respectively, from the traces (a), (b) and (c) of Fig.2. It is predicted by a theoretical approach of Boltzmann-like kinetic equations following Collet[4] that the carrier-carrier, carrier-phonon and carrier-plasmon interactions in i:GaAs give rise to an ultrafast initial carrier scattering in time scale of about 100 fs for excited carrier density of $10^{18}cm^{-3}$ and excess carrier energy of 100- 200 meV. The much slower initial carrier relaxation of Si:GaAs shown in Fig.1 (c) indicates that the carrier initial scattering is significantly reduced. For photon energy of 1.52 eV the excited electrons from the hh and lh valence bands to the conduction band are in the vicinity of the doped electron Fermi level, LO phonon emission ($\hbar\omega_{LO} \sim$ 36 meV)and plasmon emission ($\hbar\omega_p \sim$ 70 meV for a doped electron density of 2×10^{18} cm^{-3}) are inhibited. Electron-electron scattering in conduction band becomes less efficient. Conduction band electrons remain quasithermalized at the bath temperature. The initial carrier scattering processes are dominated by electron-hole and hole-hole scattering in the electron-hole plasma. Since the photogenerated holes are often cold (the excess energy of heavy hole is about 10 meV for photon energy of 1.52 eV), a carrier scattering time of 400 fs is calculated in static screening approximation and plasma density of $10^{18}cm^{-3}$, which is well consistent with the measured value.

Although for photon energies of 1.65 and 1.57 eV the excess energies of the excited electrons in conduction band of Si:GaAs are larger enough for plasma and LO phonon emission, the initial carrier scattering times are still longer than those of i:GaAs at the same photon energies. It seems that the enhanced carrier initial scattering arising from the interaction between the as-excited hot electrons and the cold dense electron plasma is overcompensated by the enhanced state filling. In addition, increased screening at higher carrier densities can also reduce the electron-electron scattering rate. We have also examined the influences of the photoexcited carrier densities on the carrier initial scattering in Si:GaAs. A slower rather than faster carrier initial scattering is observed in Si:GaAs for lower carrier density of about $10^{17}cm^{-3}$ and photo energy of 1.65 eV as predicted by the theoretical approaches. A similar reduced carrier scattering due to the enhanced state filling by the pre-injected carriers has been measured in a three pulse pump-probe experiment of GaAs for photon energy of 2 eV and carrier density of $6 \times 10^{16}cm^{-3}$ at 300 K [5]. These phenomena indicate that the enhanced state filling plays an important role in the carrier initial scattering processes of n-doped GaAs.

The slow decay components of the traces (a) and (b) of Si:GaAs correspond to the cooling of the hot electron-hole plasma to the Fermi equilibrium distribution through LO phonon emission in a time scale of 1.5 ps which is slightly longer than that of i:GaAs (~ 1.2 ps). No carrier cooling process presents in the trace (c) of Si:GaAs. This is because the conduction band electrons are in quasithermalization as mentioned above. The experimental results show that the carrier cooling times of Si:GaAs for excess energies of 100-200 meV are insensitive to the total energy initially injected into the plasma and essentially reflect the LO-phonon dephasing time of 2.1 ps[6]. The slight longer carrier cooling time in highly n-doped GaAs may also reflect the effects of screening of LO-phonon emission.

In conclusion, the femtosecond spectroscopy studies on highly Si-doped GaAs show that for the excited states closed to the doped electron Fermi level, the inhibited LO phonon emission and plasmon emission reduces significantly the carrier initial scattering; while for the higher excited states the enhanced state filling results in a slower carrier initial scattering process than that of i:GaAs.

This work was supported by the National Science Foundation of China.

References:

[1] U.Hohenester et al., Phys. Rev. B, 47, 13 233 (1993).
[2] R.Rodrigues-Herzog et al., Appl. Phys. Lett. 62, 264 (1995).
[3] T. Gong et al. Semicond.Sci.Technol. 9, 459 (1994).
[4] J. H. Collet, Physical Review B, 47, 10 279 (1993).
[5] Y. Kostoulas et al., Semicond.Sci.Technol. 9, 462 (1994).
[6] P. Langot et al., Physical Review B, 54, 14487, (1996).

Ultrafast Nonlinear Absorption in an Exciton Bragg Reflector

Kensuke Ogawa* and Yasuhiro Matsui

The Femtosecond Technology Research Association, 5-5 Tokodai, Tsukuba, Ibaraki 300-2635, Japan
*E-mail: kenogawa@festa.or.jp

Abstract. Absorption saturation associated with quantum well excitons periodically distributed in a Bragg reflector was investigated by pump-probe reflectance spectroscopy. Femtosecond spectrogram of reflected probe pulses yielded frequency chirping due to cross-phase modulation. The frequency chirping, which was obtained as 3.2 THz for a time interval of 370 fs, gives rise to an optical phase shift as large as 1.6π. The phase shift evidences the enhancement of excitonic nonlinearity by photon confinement in the Bragg reflector.

1. Introduction

Ultrafast response of nonlinear refractive index generates the phase modulation of optical pulses propagating in nonlinear optical media. The phase modulation plays a significant role in high-speed photonics technology such as optical pulse shaping and all-optical switching. Ultrafast nonlinear refractive index is caused by nonlinear absorption induced by photoexcited carriers in semiconductors [1]. High efficiency of the phase modulation is thus achieved by large magnitude of ultrafast nonlinear absorption. Excitons in quantum wells (QWs) are essential for this application since the QW excitons yield large magnitude and ultrafast response of absorption saturation [2, 3].

In this paper, we have investigated temporal and spectral characteristics of absorption saturation associated with excitons in QWs distributed in a Bragg reflector by transient reflectance spectroscopy. In a Bragg reflector showing photon confinement, optical absorption was enhanced [5]. Excitonic optical nonlinearity is enhanced and large phase modulation of optical pulses is achieved in such a Bragg reflector.

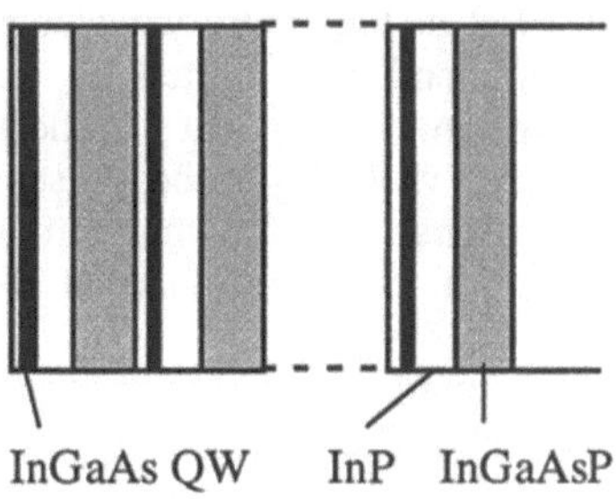

Fig. 1. Layer design of the exciton Bragg reflector (EBR).

2. Exciton Bragg Reflector

An exciton Bragg reflector (EBR) was grown by metallo-organic vapour phase epitaxy to investigate excitonic nonlinearity in Bragg reflector. A unit cell of the EBR consisted of a QW, a low refractive index (n_L) layer and a high refractive index (n_H) layer as the illustrated cross-section in Fig. 1. The QWs were 7-nm $In_{0.53}Ga_{0.47}As$, designed for the exciton resonance at a wavelength of 1500 nm. The QWs were located at the anti-nodes of the optical field at a wavelength of 1500 nm. The n_L layers were InP while the n_H layers were $In_{0.67}Ga_{0.33}As_{0.72}P_{0.28}$. Each pair of n_L/n_H layers had an optical pitch equivalent to a quarter of 1500 nm.

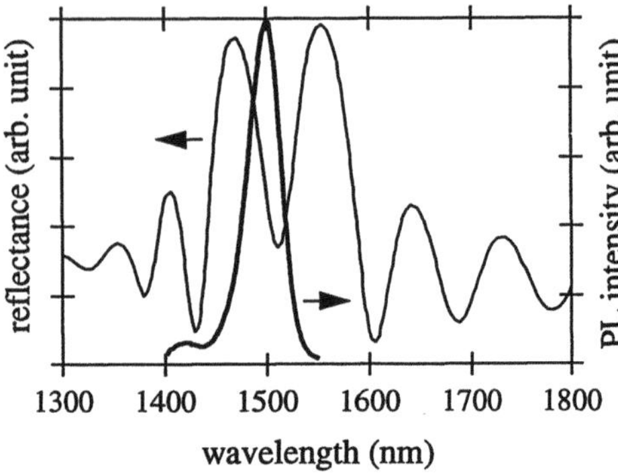

Fig. 2. Reflectance and photoluminescence (PL) spectra of the EBR.

Reflectance and PL spectra in Fig. 2 indicate the enhancement of optical transition at the centre wavelength of the EBR stopband. The reflectance spectrum has a dip at 1500 nm due to the enhancement of linear absorption. The PL intensity was nearly 5 times enhanced and the PL linewidth was narrowed by 0.7 compared with the PL spectrum of the QWs *not distributed in Bragg reflector*.

3. Femtosecond Reflectance Spectroscopy

Transient reflectance was measured in femtosecond pump-probe experiments to characterise absorption saturation in the EBR. The pump and probe pulses were transform-limited pulses generated from a synchronously pumped optical parametric oscillator. The pulse duration and the spectral width were 120 fs and 20 nm, respectively. The pump and probe beams were chopped by the dual acousto-optic modulators at different chopping frequencies. The reflected probe intensity was spectrally resolved through a spectrometer and detected with a lock-in amplifier at the difference of the dual chopping frequencies. Since the reflectance of the EBR at the centre wavelength of the stopband was limited by optical absorption, transient reflectance measured with the probe pulses was proportional to the magnitude of absorption saturation induced by the pump pulses.

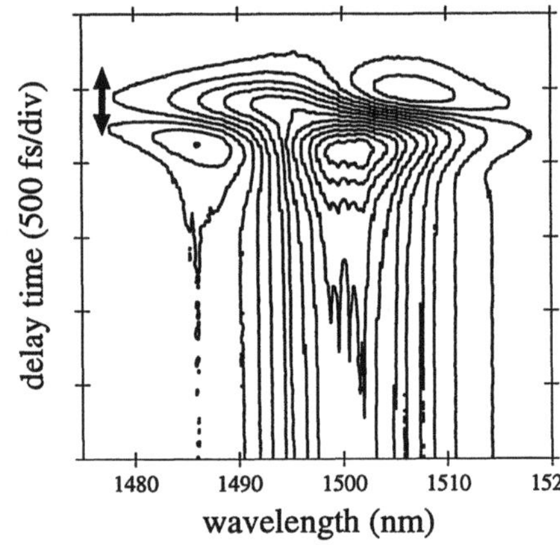

Fig. 3. Contour plot of pump-probe reflectance spectrogram of the EBR.

A femtosecond spectrogram of transient reflectance of the EBR is presented as a contour diagram in Fig. 3. The vertical axis is probe delay time which increases from top to bottom along the axis. The pump and the probe pulses overlapped within the delay time range indicated with the vertical arrow. In the delay time range of the pulse overlap, transient reflectance shows asymmetric change in terms of wavelength and time. In shorter delay time, transient reflectance increases at wavelengths shorter than 1500 nm, while it decreases at the other wavelength side. The opposite behavior is observed in the longer delay time. These characteristics indicate reflected probe pulses were frequency chirped due to cross-phase modulation.

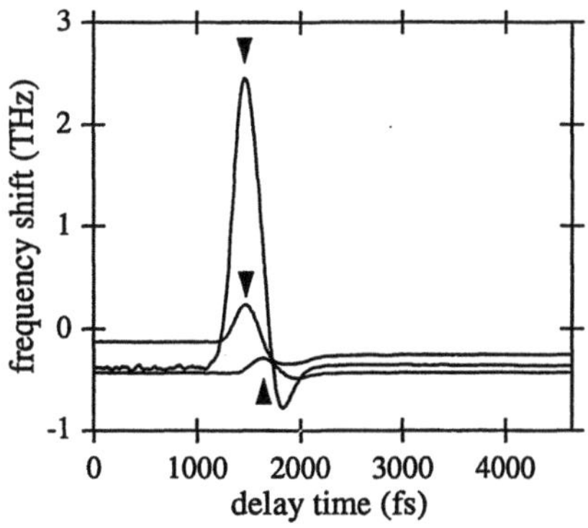

Fig. 4. Frequency chirping obtained from transient reflectance characteristics measured with different pulse energies.

To evaluate frequency chirping and resultant optical phase shift, time evolution of optical frequency shift in the transient reflectance was analyzed with the frequency shift curves plotted in Fig. 4. The frequency shift was given as a shift from the average frequency of the incident probe spectrum by the conversion of average wavelength in the transient reflectance to optical frequency. The three curves in Fig. 4 were obtained from the transient reflectance characteristics with pump/probe pulse energies, 1.3 nJ/1.0 nJ, 1.3 nJ/0.48 pJ and 21 pJ/0.48 pJ, respectively. The largest chirping, obtained from the characteristics shown in Fig. 3, was 3.2 THz for a time interval of 370 fs. This frequency chirping leads to an optical phase shift as large as 1.6π. The phase shift is an experimental evidence of the enhancement of excitonic nonlinearity in the EBR due to photon confinement. The peak of the frequency shift curves in Fig. 4 appear at longer delay time with decreasing pulse energy as indicated with the vertical arrows. This behavior is characteristic of the optical pulse shaping based on absorption saturation: the leading edge of incident pump pulses were attenuated by absorption according to the ratio of incident pulse energy against saturation energy [1].

Acknowledgements. This work was performed under the management of The Femtosecond Technology Research Association supported by New Energy and Industrial Technology Development Organization.

References

1 J-C. Diels and W. Rudolph, *Ultrashort Laser Pulse Phenomena*, Academic Press, San Diego, 1996.
2 D. A. B. Miller, D. S. Chemla, D. J. Eilenberger, P. W. Smith, A. C. Gossard and W. T. Tsang, Appl. Phys. Lett. **41**, 925, 1982.
3 W. H. Knox, R. L. Fork, M. C. Downer, D. A. B. Miller, D. S. Chemla, C. V. Shank, A. C. Gossard and W. Wiegmann, Phys. Rev. Lett. **54**, 1306, 1985.
4 K. Ogawa, Y. Matsui, T. Itatani and K. Ouchi, Appl. Phys. Lett. **72**, 155, 1998.

Super-Radiant Decay in Two-Dimensional Layered Semiconductors

S. Xu[1], S.J. Diol[1], A. Makinen[1,2], M.G. Mason[3], A.A. Muenter[3], L. Rothberg[1], Y. Gao[1,2], and R.J. D. Miller[4]

1. Center for Photoinduced Charge Transfer, Department of Chemistry, University of Rochester, Rochester, NY 14627
2. Department of Physics and Astronomy, University of Rochester, Rochester, NY 14627
3. Research Laboratories, Eastman Kodak, Rochester, NY 14652
4. Departments of Chemistry and Physics, 80 St. George Street, University of Toronto, Toronto, Ontario M5S 1A1, CANADA

Geometry confinement of excitons in quasi-two dimensional semiconductors, leads to a shrinkage of the exciton Bohr radius, resulting in an increase of the oscillator strength and binding energy.[1,2] This effect is evidenced by the nonlinear absorption and the strong luminescence observed in multiquantum wells (MQW) and semiconductor superlattices,[3,4] and harbors a large number of potential applications.[5] The theoretical study of Hanamura[6] shows that the formation of excitonic polaritons in two-dimensional (2D) systems leads to extremely fast superradiant decay due to the 2D confinement, and dynamics on the order of a picosecond for GaAs MQW and subpicosecond for CdS MQW were predicted.[6] The 2D confinement and fast decay makes the excitons deviate from ideal bosons. The resultant increase of the third-order susceptibility thus extends the potential application of quasi-2D semiconductors to fast response nonlinear optical devices.[6]

In this paper, we present time-resolved electron photoemission studies on the layered semiconductor SnS_2. This crystal has a CdI_2-type hexagonal closely packed layered structure. Each layer consists of a plane of tin sandwiched between two planes of sulfur, and has a thickness of 0.59 nm. The bonding between

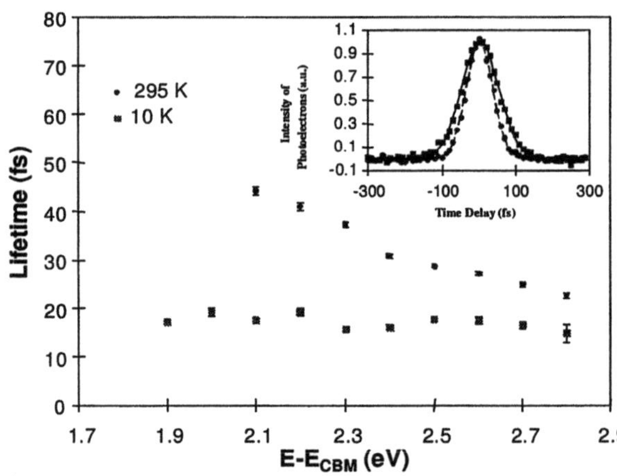

Figure 1 The lifetimes of the optical excitations in SnS_2 at 295 K (round dots) and at 10K (filled squares) versus the excitation energy referenced to the VBM. Insert: Time-resolved pump-probe spectrum at the excitation energy of 2.4 eV above the VBM at T= 10 K as an example of the time resolution [filled squares are experimental data, round dots are instrument response function (re: ref. 9)].

layers, by van der Waals force, is very weak and the electrons and excitons are confined in the layer. Hopping between layers is negligible such that this system forms an ideal model for the study of 2-D confinement.

The excited state dynamics of SnS_2 can be inferred from the time-resolved two-photon photoemission measurements shown in Figure 1. The lifetimes observed for different kinetic energy electrons are plotted with respect to the valence band maximum (VBM). It is obvious that the lifetimes at 10 K are quite different from the observations at 295 K. Although the lifetimes at 295 K increase with a decrease in excess energy relative to the CBM (2.2 eV at 295 K), as observed for all semiconductors to date, the lifetimes at 10 K show no energy dependence and significantly shorter decays (18 ± 5 fs) relative to the 295 K results. At the same time, when the temperature goes below 100 K, we observed strong luminescence from the sample. In contrast, no luminescence is observed from the sample at room temperature. When the temperature is cooled down just below 100 K, we observed a red dot near the laser spot. Further cooling of the sample below 35 K leads to a bright uniform glow across the whole 8 mm x 8 mm area. The luminescence spectrum is shown in Fig. 2 with images of the spatial distribution of emission from the sample at 83 K and 34 K.

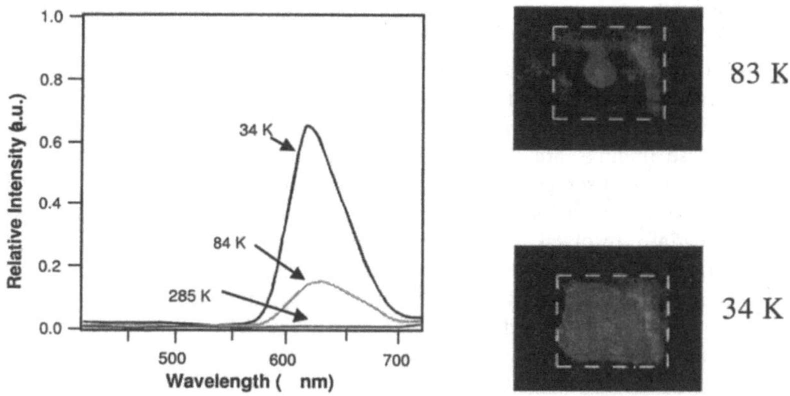

Figure 2 (a) The Photoluminescence spectra of SnS_2 at different temperatures. (b) Image of the sample at two different temperatures with the laser focused (50 μm spot size). At 83 K, the luminescence only came from the small area (red dot) near the laser spot. At 34 K, the luminescence came uniformly from the whole sample

From these findings, it is difficult to believe that the quasi-particle lifetime of 18 fs measured using electron photoemission is the same state which gives rise to the luminescence. In order for the quasi-particle with an 18 fs lifetime to transport from the center of the sample to the edge, its speed would have to be more than 10^9 m/s, which is unreasonable. Hence, these two states must be different. Both theoretical and experimental studies show that excitons in semiconductors can further interact with photons to form excitonic polaritons, which have two modes, a photon-like mode and an exciton-like mode. In 3D semiconductors, both of them are stable and do not show any decay.[7] However, the polaritons in quasi-2D semiconductors exhibit quite different behaviour.[7] The polariton with $k_{//} < \omega_0 n/c$ (c is the light speed, and n is the refractive index), called the photon-like mode,[7] is very unstable.[6] When it reaches the boundary of the two-dimensional layers, it will transform to a photon.[6,7] This decay process is usually called superradiance.[6,7] On the other hand, the mode with $k_{//} > \omega_0 n/c$, the exciton-like mode, is very stable.[7] It undergoes total internal reflection (TIR) at crystal boundaries and can transport to very far distances until it is scattered by phonons or impurities, where it becomes a radiative exciton and gives luminescence. Hence, we believe the observed radiation is from the exciton-like mode. Furthermore, unlike the photon-like mode, the exciton-like mode cannot couple with the external light. Consequently, in the time-resolved two-photon electron photoemission experiment, the probe beam photon can only access the photon-like mode, and the lifetime we measure is the superradiant decay time.

These results were compared to the theoretical model of Hanamura[6] using: an estimate for the oscillator strength of SnS_2 based on the electron effective mass;[5] the dielectric constant of SnS_2; and the reduced mass from band structure calculations[8] to estimate the exciton Bohr radius. This analysis gives an energy independent superradiant decay time of 76 fs. This comparison is within a factor of 3-4 the experimental observations and qualitatively captures the observed features at low temperatures. As the temperature is raised to ambient conditions, the excitons become unbound and optical excitation simply creates an electron-hole plasma. Ordinarily, these carriers would also have much longer decay times than in 3-D materials but due to the formation of a layered 2-D plasmon band the decays are accelerated through additional decay processes mediated through plasmon coupling.[9]

In summary, we have directly measured the decay time of excitonic polaritons in SnS_2 with time-resolved two-photon photoemission spectroscopy at 10 K. The experimental observations agree within a factor of 3-4 with the superradiant treatment of Hanamura. These findings are, to our knowledge, the shortest exciton lifetimes ever measured. We also observed extraordinary apparent exciton mobility due to the waveguide mode of the polaritons. The seemingly contradiction between the temporally resolved and spatially resolved measurements gives evidence for the existence of two excitonic polariton modes, a photon-like and an exciton-like (or wave-guide) mode in layered 2-D semiconductors, analogous to previous findings in other systems.[7]

This work was supported by the NSF (CHE 912001) and NSERC (R.J.D.M.). The samples were provided by B.A. Parkinson and L.I. Sharp.

1. G. Bastard, E.E. Mendez, L.L. Chang and L. Esaki, Phys. Rev. **B 26,** 1974 (1982).
2. B. Deveaud, F. Clerot, N. Roy, K. Satzke, B. Sermage and D.S. Katzer, Phys. Rev. Lett. **67,** 2355, (1991).
3. D.S. Chemla, D.A.B. Miller, P.W. Smith, A.C. Gossard, and W. Weigmann, IEE J. Quantum Electronics. **20,** 265 (1984); D.S. Chemla and D.A.B. Miller, J. Opt. Soc. Am B **2,** 1155 (1985).
4. E.O. Gobel, H. Jung, J. Kuhl and K. Ploog, Phys. Rev. Lett. **51,** 1588 (1983).
5. Elias Burstein and Claude Weisbuch, *Confined Electrons and Photons, New Physics and Applications,* Plenum Press, New York (1995).
6. E. Hanamura, Phys. Rev. B **38,** 1228 (1988).
7. J. Aaviksoo, J. Lumin. **48,** 57 (1991).
8. C.Y. Fong and M.L. Cohen, Phys. Rev. B **5,** 3095 (1972).
9. S. Xu, C.C. Miller, S.J. Diol, Y. Gao, D.A. Mantell, M.G. Mason, A.A. Muenter, L.I. Sharp, B.A. Parkinson and R.J.D. Miller, Chem. Phys. Lett. **272,** 209 (1997).

Spectral Hole Burning in Intersubband Absorption Bands of Quantum-Well-Structures

A. Seilmeier, J. Kaiser

Institute of Physics, University of Bayreuth, D-95440 Bayreuth, Germany
E-mail: alois.seilmeier@uni-bayreuth.de

Abstract. Transient hole burning experiments on a picosecond time scale give a convincing proof of inhomogeneous broadening of intersubband absorption bands for the first time. At excitation intensities of 400 kW/cm^2 population holes of ~60 cm^{-1} widths are found in narrow n-modulation doped GaAs/$Al_{0.35}Ga_{0.65}As$ quantum well structures. At intensities of a few MW/cm^2 „holes" in the absorption spectra due to a $\chi^{(3)}$ effect are observed.

1. Introduction

Generally, quantum well structures exhibit broad intersubband absorption bands in the infrared which are attributed to a transition from the lowest to the first excited subband. The largest spectral widths are observed in modulation doped structures with narrow quantum wells. While the frequency positions of the transition lines can be explained easily, the spectral widths and shapes are not understood very well. Interface roughness, and the dependence of the effective mass on the energy have been discussed as relevant broadening mechanisms in a single particle picture. Recent investigations have given evidence that the spectral properties may strongly depend on the collective nature of the intersubband transition [1]. Even the question, whether the absorption band is homogeneously or inhomogeneously broadened, is not answered finally. This point can be clarified by a hole burning experiment. However, if holes can be observed at all, the short lifetime of the upper subbands only allow transient holes on an ultrafast time scale. In this paper we report on the first observation of transient population holes in an intersubband absorption profile.

2. Experimental

The data are taken in the following way: A first infrared pulse of 2 ps duration and ~10 cm^{-1} spectral width excites electrons to the first excited subband. The subsequent change of the intersubband absorption is measured time and frequency resolved by a second infrared pulse [2].

The data presented here are taken on a multiple GaAs/$Al_{0.35}Ga_{0.65}As$ quantum well structure with 5.9 nm wide GaAs quantum wells embedded in between 36 nm thick $Al_{0.35}Ga_{0.65}As$ layers in which the central 8 nm are doped by silicon. The doping concentration is 1.4×10^{12} cm^{-2}. The sample shows an exceptionally broad absorption line of $\Delta\tilde{\nu}$=90 cm^{-1} at $\tilde{\nu}$=1200 cm^{-1} as indicated in Fig. 1 (dashed line).

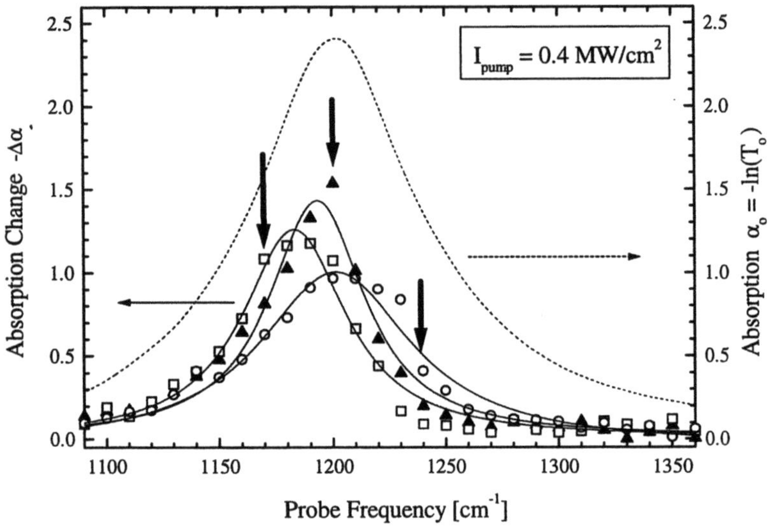

Fig.1:Intersubband absorption spectrum (dashed line), and spectra of transient holes (solid lines) after excitation at $\tilde{\nu}$=1170 cm^{-1} (□), 1200 cm^{-1} (▲) and 1240 cm^{-1}(O) (see arrows)

3. Results and Discussion

In a first type of experiments, population holes in the intersubband absorption band are observed. An infrared picosecond pulse of relatively low intensity of ~400 kW/cm^2 excites electrons to the upper state. The spectral width of 10 cm^{-1} of the pulses is considerably narrower than the width of the absorption line. A second tunable pulse measures the change in absorption $\Delta\alpha=\alpha-\alpha_0$ at a delay time of 0.5ps (α_0 is the absorption without excitation). $\Delta\alpha$ is shown in Fig. 1 for three excitation frequencies, $\tilde{\nu}$=1170 cm^{-1}, 1200 cm^{-1}, and 1240 cm^{-1} which are indicated by arrows. Bleaching of the absorption is observed close to the pump frequency. The linewidth of the bleaching spectra of ~60 cm^{-1} is clearly smaller than the linewidth of the intersubband absorption band. The observed shift between the excitation frequency and the frequency position of the maximum of the bleaching spectrum is expected for hole burning in steeply rising wings [3].

It should be mentioned that the intersubband transition is weakly saturated at excitation intensities of 400 kW/cm^2. Numerical fits taking into account the saturation give a homogeneous linewidth of ~30 cm^{-1}. Assuming an inhomogeneous broadening of ~65 cm^{-1} we obtain the spectral width of the absorption spectrum shown as dashed line in Fig. 1. The strong excitation is believed to reduce the collective character of the intersubband transition [4] allowing a discussion of the results in a single particle picture. An interface roughness of ± one atomic layer can explain the estimated inhomogeneity in the 5.9 nm wide quantum wells.

Experiments at considerably higher pump intensities of 2 MW/cm^2 show a different type of „holes" in the absorption spectra. In Fig. 2, not the change in ab-

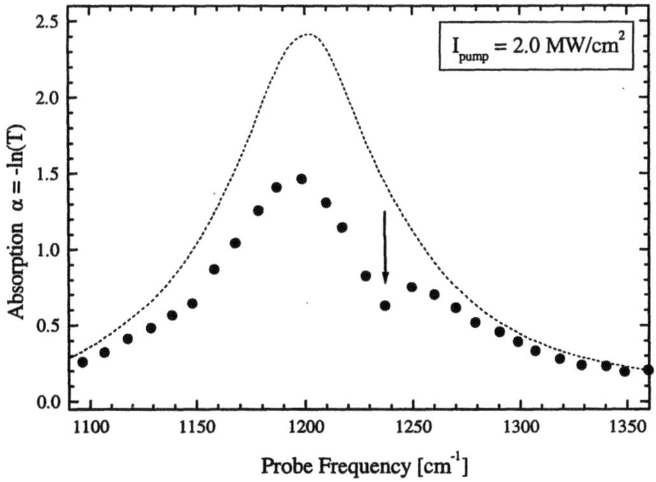

Fig. 2: Intersubband absorption (dashed line) and transient absorption (●) after strong excitation at $\tilde{\nu}$=1240 cm^{-1}

sorption but a transient absorption spectrum $\alpha(\tilde{\nu})$ taken at a delay time of 0.5 ps is shown. A hole is observed exactly at the frequency position of the excitation pulse ($\tilde{\nu}$=1240 cm^{-1}) indicated by an arrow. In this case, the spectral width is comparable to the width of the IR pulses and the spectral feature disappears with the correlation function of the laser pulses. This effect is interpreted as a coherent $\chi^{(3)}$ - process. The pump pulse is scattered by a polarization grating generated by the pump and the probe pulse resulting in a signal that is simultaneously measured with the transmitted probe pulse signal.

In conclusion we want to emphasize that these findings are not specific for the sample discussed here. Very recent experiments on a sample with considerably lower doping concentration and only 60 cm^{-1} wide absorption bands show population holes of comparable width but a less pronounced inhomogeneous broadening of the intersubband transition.

Acknowledgements. Financial support by the Deutsche Forschungsgemeinschaft is gratefully acknowledged.

References

1 R. J. Warburton, K. Weilhammer, J. P. Kotthaus, M. Thomas, H. Kroemer, Phys. Rev. Lett. **80,** 2185, 1998

2 J. Kaiser, A. Seilmeier, Phys. Stat. Sol (b) **204,** 205, 1997

3 H. Graener, G. Seifert, Chem. Phys. Lett. **185**, 68, 1991

4 K. Craig, B. Galdrikian, J. N. Heyman, A. G. Markelz, J. B. Williams, M. S. Sherwin, K. Campman, P. F. Hopkins, A. C. Gossard, Phys. Rev. Lett. **76**, 2382, 1996

Ultrafast Carrier Capture into InGaAs/GaAs Quantum Wells

L.V. Dao, M. Gal, H.H. Tan* and C. Jagadish*

School of Physics, University of New South Wales, Sydney, Australia 2052
*·Dept. Electr. Mat.Eng.,Research School of Physical Sciences and Engineering, IAS
Australian National University,Canberra, Australia 0200

Abstract. We have inverstigated the dynamics of photoexcited carriers in InGaAs/GaAs quantum wells using the photoluminescence up-conversion technique. In this study we describe the observation of an extremely fast (600fs) capture process in shallow InGaAs/GaAs quantum wells, resulting from resonant tunnelling from impurities in the GaAs barrier.

1.Introduction

In most modern optoelectronic devices, the upper limit to the modulation frequency is limited by the dynamics of the excited carriers. In quantum well (QW) based devices this is determined by carrier thermalisation and capture into the quantum wells, and the overall capture time is typically of the order of 10-50 ps. These capture times correspond accurately to the known diffusion constants and given barrier widths of the samples inverstigated. Photoluminescence up-conversion allows the study of the dynamics of photoexcited carriers with sub-picisecond resolution. In this paper, we describe the observation of an extremely fast (<600fs) capture process into shallow InGaAs/GaAs QW's, studied using photoluminescence (PL) up-conversion technique. We shall show that this exceptionally fast capture of carriers is attained through resonant tunnelling from impurities in the GaAs barrier.

2. Experiment

The PL up-conversion experiments [1](shown in Fig. 1) were performed using a femtosecond self-mode locked Ti:sapphire laser. The laser was tunable between 750 and 900 nm, the pulse width was 80fs, the repetition rate 85 MHz and the output power was 200 mW at λ =780 nm. The output was divided by a

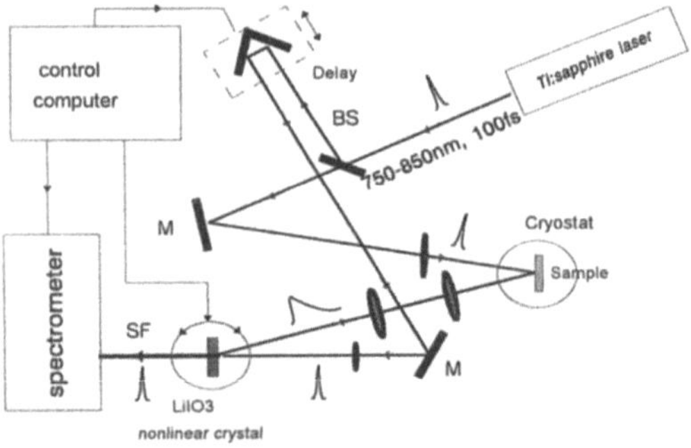

Fig.1. Schematic of experimental arrangement (M- mirror, BS- Beamspliter)

beamspliter (BS) into two beams. One beam excited the sample, the other was delayed by a variable delay stage. The luminescence from the sample and the delayed laser beam were focused on a $LiIO_3$ crystal. Noncollinear phase matching was used to generate the sum frequency (SF) at a wavelength determined by the angle of the $LiIO_3$ crystal to the incoming beams. The up-converted photons were dispersed by a spectrometer and detected by a photomultiplier. The measurement was controlled by a computer. The time resolution of our system was approximately 200fs. The optically excited carrier concentration was $4x10^{10}$carriers/cm^2. The $In_{0.2}Ga_{0.8}As/GaAs$ QW samples were mounted in a variable temperature He cryostat. The samples consisted of four single quantum wells having well widths equal to L_z=2, 4, 6 and 10nm, separated by 50nm thick GaAs, and grown on GaAs substrate.

3. Results and discussion

In terms of PL intensity and line-width, the emission from the four different QW's are similar (see insert in Fig. 2),however, the time evolution of the PL signals, shown in Fig. 1, indicate a marked difference between the shallowest QW (Lz=2nm) and the other QW's. The PL rise-time, τ , defined as the time required for the luminescence to reach its maximum, is less than 1ps for the shallowest well, but more than 100ps for the three deeper wells. We found that for the shallowest QW, between 10 and 150K, the initial very short rise-time was independent of temperature and was, within the experimental uncertainty, τ =600fs.PL rise-times of 100 ps's have been observed by several researchers[2] in similar systems, and have been attributed to the combination of carrier thermalisation, capture into the wells and exciton formation in the QW. Thus, the PL rise-times observed for the three deeper QW's are analogous to that observed by others. What we found surprising is the very short rise-time observed in the QW with the narrowest (2nm) well width. Although very short capture times (1ps) have also been reported previously[3], such 'quantum capture' processes only occur when the energy difference between the excited and ground state are separated by an integer number of LO phonon energies. We searched for LO phonon resonances by varying the exciting photon energy above the band gap of GaAs, but found the

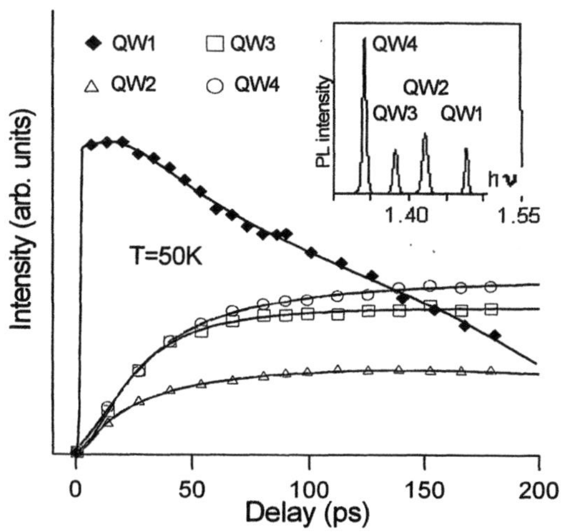

Fig. 2. Time evolution of the PL signal from four QW's. The insert shows the CW PL signals from the same sample

rise-time to be independent of the excitation photon energy. Therefore, the observed very short PL rise-time is not the result of quantum capture but some other rapid and efficient capture process. The temperature independent nature of the fast PL rise-time and the specific energy of the QW indicated that resonant tunnelling could be taking place between the GaAs barriers and the (Lz=2nm) QW. As the emission energy from an impurity in the GaAs barrier coincides with the emission from the QW, we postulated that the QW ground state energy is resonant with an impurity level, so carriers/exciton captured by the impurities in the barrier can efficiently tunnel into the QW.To test this possibility, we studied the PL properties of the GaAs barriers, using both time dependent and time integrated PL spectroscopies, and compared it to that observed for the QW sample. As the barrier luminescence in the QW samples was weak, we studied the GaAs emission in a reference sample, containing no QW's, but in all other respects similar to the other samples. The time evolution of the impurity related emission in the GaAs sample is shown in Fig. 3 together with that of the Lz=2nm QW. As can be seen, the initial very fast rise-time is identical in the two systems, but the decay times are radically different. The fact that the fast rise-times in the QW and in the barrier are identical supports our assumption that the impurities rapidly capture the photoexcited carriers/excitons which then resonantly tunnel into the QW. The observed different decay times indicate that while the capture is common, the decay process is distinct.

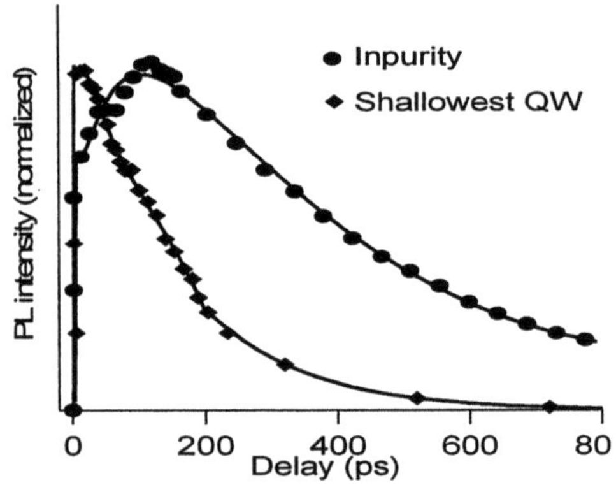

Fig. 3. Time evolution of the PL signal of the impurity related emission (h =1.48eV) in the GaAs barriers, and that of the shallowest QW (Lz=2nm), showing the similar PL rise-times but different decay times.

References:

1.L.V.Dao, M. Gal, G..Li, and C. Jagadish, Appl. Phys. Lett .**71**, 1849 (1997)

2.R. Spiegel, G.Bacher, K.Herz, A. Forchel, T. Litz, A.Waag, G.Landwehr, Phys. Rev. B**53**, 4544 (1996). M.R.X. Barros, P.C Becker, D. Morris, B. Deveaud, A. Regreny, and F. Beisser Phys. Rev. B **47**, 10951 (1993) and S.Lutgen, R.A. Kaindl, M.Woerner, T. Elsaesser, A. Hase, H. Kunzel, M. Gulia, D. Meglio and P. Lugli, Phys. Rev. Lett. **77**, 3657 (1996).

3. B. Deveaud, D. Morris, A. R. X. Barros, P. Becker, J. M. Gerard, Opt. and Quantum Electr. **26**, S679 (1994) and references therein

Femtosecond Dynamics in Semiconductor Lasers: Dark Pulse Formation

M. Kauer[1], A. P. Heberle[2], J. R. A. Cleaver[1], J. Allam[2], and J. J. Baumberg[2]

[1] Microelectronics Research Centre, University of Cambridge, Cavendish Laboratory, Madingley Road, Cambridge CB3 0HE, United Kingdom
E-mail: mk225@cam.ac.uk
[2] Hitachi Cambridge Laboratory, Cavendish Laboratory, Madingley Road, Cambridge CB3 0HE, United Kingdom

Abstract. We report on an ultrafast upconversion experiment for semiconductor lasers to track their dynamic response after injection of a femtosecond optical pulse. Due to high sensitivity and improved signal-to-noise ratio, both continuous-wave emission and evolution of the recirculating femtosecond pulse can be tracked on timescales in excess of 1 ns. We show that for GaAs quantum-well semiconductor lasers the injected pulse evolves into an ultrafast 'dark pulse'.

1. Introduction

Time-domain studies of optical pulse propagation in semiconductor diode lasers (LDs) and semiconductor optical amplifiers (SOAs) enable dynamical characterisation of these devices. This is important for applications, and furthermore probes the dynamics of the electron and photon systems and their interplay very sensitively. For example, the roundtrip propagation delay of femtosecond optical pulses has been used to determine group-velocity dispersion in LDs [1]. In SOAs, optical pulse propagation has been used to investigate pulse shaping [2] and coherent and incoherent many-body effects [3].

Here, we report on a high-dynamic range femtosecond pulse propagation experiment on continuous-wave (cw) LDs which tracks *both* cw and pulse evolution on timescales of over 1 ns.

2. Experimental Setup

A schematic of the experiment is shown in Fig. 1. A single femtosecond optical pulse centered on the lasing wavelength of the LD is coupled into the LD and the output is then time-resolved using an upconversion technique. Measurements on both 770 nm AlGaAs and 1.5 μm InGaAsP edge-emitting quantum-well LDs have been performed.

The source for the optical input pulses is a self-mode-locked Ti:sapphire laser which produces 120 fs (autocorrelation full width at half maximum (FWHM)) pulses near 800 nm with 82 MHz pulse repetition rate. For measurements on GaAs LDs, the Ti:sapphire laser output is split into a weak excitation (<30 μW) and a strong gating pulse. For measurements on InGaAsP devices, a synchronously pumped parametric oscillator (OPO) provides output pulses at 1.5 μm. The FWHM of the OPO pulses is 220 fs, while as much as 300 mW of the Ti:sapphire beam initially fed into the OPO is available for the gating pulse.

The coupling of pulses into the LD can be performed in either of two ways. In the transmission configuration, the excitation pulse is coupled into the LD front facet using a microscope objective. The output is collected at the rear facet by a second microscope objective. In the reflection configuration [Fig. 1], the excitation pulse is coupled into the front facet by a microscope objective and the output then collected from the same facet. A 20% beamsplitter between LD and nonlinear crystal brings in the excitation pulse. The advantage of this configuration is that access to only one LD facet is needed, therefore it is suitable for measuring packaged LDs and devices with a high reflecting facet. Furthermore, the coupling efficiency into the device can be monitored, since the reflected light from the front facet is also upconverted. However, the beamsplitter in the output beampath reduces the total signal.

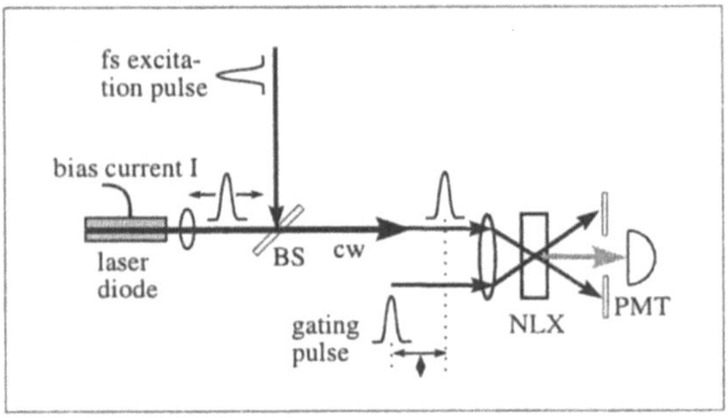

Fig. 1. Experimental scheme for the reflection configuration (BS=Beamsplitter, NLX=nonlinear crystal, PMT=photomultiplier)

In both configurations, the output signal is cross-correlated with the time-delayed 800 nm gating pulse in a nonlinear crystal. We use antireflection-coated 1 mm thick crystals of BBO (β-barium borate) for use with GaAs LDs and LBO (lithium triborate, 1 mm thickness, antireflection coated) crystals for use with InGaAsP LDs. The upconverted signal is detected by a photomultiplier.

The minimum detectable perturbation with our setup was 5 nW on a cw background of 10 mW. Therefore, care must be taken to avoid any spurious reflections in the exciting pulse; less than 0.1 % was achieved. At the same time, peak pulse powers in excess of 1 W can be detected, demonstrating the large dynamic range of our experiment.

3. Dark Pulse Formation in Semiconductor Lasers

A study of pulse propagation in quantum-well AlGaAs LDs [4] serves to illustrate the sensitivity and temporal resolution of our experiment. A single femtosecond pulse centred on the lasing wavelength of 773 nm is injected in the LD under cw bias above threshold. In the data presented both the cw and input pulse fields are TE polarized. The input pulse is amplified as it propagates in the cavity and is partially reflected at the uncoated front and rear facets of the LD, giving rise to a pulse train with a separation corresponding to the cavity round-trip time of 10.55 ps. Due to the high dynamic range, more than 100 round-trips of the circulating pulse can be observed. In this fashion, both optical pulses and the cw emission dynamics of the LD are tracked simultaneously at different points along otherwise impractical path lengths of several centimetres.

In particular, the relaxation oscillations in the cw intensity initiated by the input pulse can be seen clearly. The pulses retain their original shape until they vanish at $I_{bias} \approx I_{th}$ [Fig. 2, 60 mA]. However, pulse and cw emission start to influence each other strongly as the current is increased [Fig. 2, 85 mA]. Initially, the cw power drops after the pulse, but with increasing propagation time the drop steepens and becomes progressively shorter lived. This 'dark pulse' in the cw emission continues to propagate for hundreds of picoseconds after the original 'bright pulse' has died away. Preliminary results from a laser-diode model replicate this behaviour.

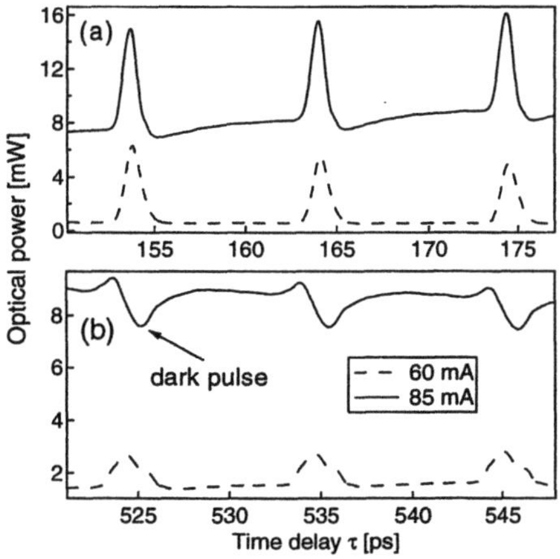

Fig. 2. Output pulse train produced by a single 85 fs pulse (P=2 μW) incident on the LD at $I_{bias}=1.3I_{th}=60$ mA and $I_{bias}=1.9I_{th}=85$ mA after 15 (a) and after 50 (b) round-trips in the cavity

4. Conclusions

In conclusion, we report a femtosecond upconversion experiment on semiconductor lasers which provides direct access to the interaction of pulse propagation and cw emission dynamics in these devices. The measurements presented reveal new dynamics following the injection of a femtosecond pulse, which evolves into an ultrafast 'dark pulse' in the cw emission.

Acknowledgements. The authors would like to thank Dr. Okai and his colleagues at HCRL for providing the laser devices. M. Kauer acknowledges financial support from EPSRC and German Academic Exchange Service (DAAD HSP III).

References

1 K. L. Hall, G. Lenz, and E. P. Ippen, in *Journal of Lightwave Technology*, Vol. 10, 616, 1992.

2 P. A. Harten, J. P. Sokoloff, F. Brown de Colstoun, S. G. Lee, R. Jin, E. M. Wright, G. Khitrova, H. M. Gibbs, S. W. Koch, and N. Peyghambarian, in *Physical Review Letters*, Vol. 69, 852, 1992.

3 R. A. Indik, R. Binder, M. Mlejnek, J. V. Moloney, S. Hughes, A. Knorr, and S. W. Koch, in *Physical Review A*, Vol. 53, 3614, 1996.

4 M. Kauer, J. R. A. Cleaver, J. J. Baumberg, and A. P. Heberle, in *Applied Physics Letters*, Vol. 72, 1626, 1998.

Ultrafast Dynamics of Field-Induced Charge Generation in Conducting Polymers

G. Cerullo[(1)], M. Nisoli[(1)], S. Stagira[(1)], S. De Silvestri[(1)], G. Lanzani[(2)], W. Graupner[(3)], E.J.W. List[(3)], G. Leising[(3)]

[1]INFM-Dipartimento di Fisica, CEQSE-CNR, Politecnico, Piazza L. da Vinci 32 – 20133 Milano, Italy E-mail: giulio.cerullo@polimi.it
[2] Istituto di Matematica e Fisica – INFM, Università di Sassari, Italy
[3] Institut für Festkörperphysik, Technische Universität Graz, Austria

Abstract. We performed field-assisted time-resolved femtosecond pump-probe experiments in light emitting diodes of laddertype poly-(para-phenylene). Field-induced charge carriers generation is directly observed monitoring the characteristic polaron absorption band at 1.91 eV. Charge carriers are directly formed by singlet exciton dissociation in the picosecond time domain.

1. Introduction

Optical absorption in organic semiconductors creates tightly bound electron-hole pairs, known as excitons. Charge carriers photogeneration is an indirect process that occurs when excitons break into positive and negative polarons. This process is crucial both for understanding the fundamental physics of these materials, and for applications in photovoltaic devices and photodetectors. Since exciton dissociation can be affected by an external electric field, field-induced (FI) spectroscopy is a powerful tool to study carrier generation. Moreover, FI dissociation, being a non radiative decay channel, is relevant to the electroluminescence efficiency of organic light emitting diodes (LEDs). The temporal evolution of FI charge generation in optically excited organic semiconductors was previously investigated either by transient quenching of photoluminescence [1] and photobleaching [2], or by photocurrent measurements . Such techniques are indirect probes and cannot monitor the state produced by the exciton dissociation, the nature of which is currently under debate [1, 3].

The target of this work is the direct observation of FI charge generation in conducting polymers with femtosecond time resolution. As a model system we have chosen the methyl substituted ladder type poly(para-phenylene) (m-LPPP) because of its low defect content, the availability of LEDs and the well-characterised photophysics [4]. The narrow linewidths of m-LPPP, both in the ground and excited state absorption spectra, allow to clearly resolve and identify its photoexcitations.

2. Experimental Methods

We used LED structures consisting of thin films of m-LPPP (~100 nm thick) sandwiched between aluminum and transparent indium tin oxide (ITO) electrodes. Pump pulses (~150 fs) at 3.18 eV were obtained by frequency doubling a CPA Ti:Sapphire laser, probing in the visible region was achieved by continuum generation in a thin sapphire plate. Pump and probe pulses reached the m-LPPP through the ITO electrode and were reflected back by the aluminum layer. With no electric field (F=0) applied to the diode, we measured conventional transient pump/probe spectra, named hereafter ΔT/T. In order to detect small field-induced changes of ΔT, we applied a modulated electric field (MF) to the device and referenced our detection at the field modulation frequency. In this way we are able to sensitively detect only field-induced changes in the ΔT/T spectra, which we call $(\Delta T/T)_{MF}$.

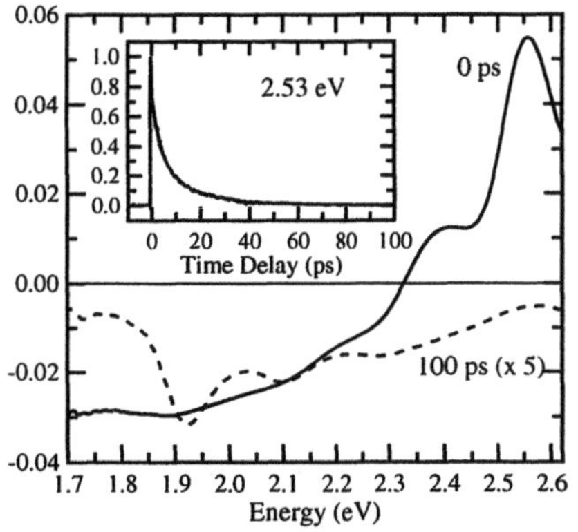

Fig. 1. (ΔT/T) spectra at 0 ps (solid line) and 100 ps (dashed line) probe delay. Inset: stimulated emission decay at 2.53 eV.

3. Results and Discussion

Conventional ΔT/T for m-LPPP films are shown in Fig. 1 for two different pump-probe delays (τ_D). At τ_D = 0 ps we observe both stimulated emission (SE) and photoinduced absorption (PA) of the singlet exciton, peaking at 2.53 eV and 1.4 eV (not shown in the figure) respectively [4]. The singlet exciton population decays on a time scale of 40 ps (see inset of Fig.1). In addition, for τ_D = 100 ps, we observe two additional long-lived PA bands peaking at 1.9 and 2.1 eV. Based on different cw spectroscopic techniques, such as doping-induced absorption and charge injection spectroscopy, we can

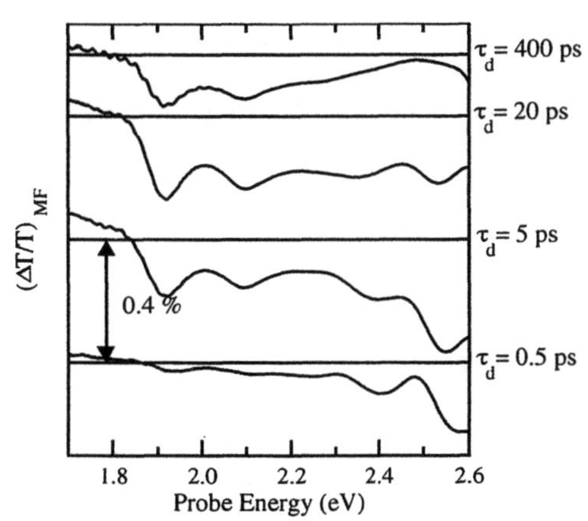

Fig. 2. ΔT_{MF} spectra for a positive bias of 13 V at different probe delays.

assign the PA bands to charged species (polarons). The appearance of these well resolved features is a fingerprint of charge generation in m-LPPP which allows to optically address the FI charge generation in LEDs.

In Fig. 2 we show $(\Delta T/T)_{MF}$ spectra for a LED positively biased at 13 V and different pump-probe delays. Two important phenomena take place in the presence of the field : quenching of singlet exciton features (decrease of the SE at 2.53 eV and of the PA at energies lower than 1.8 eV) and enhancement of PA at 1.9 and 2.1 eV. The former corresponds to a field induced reduction in the number of emitting states, i.e. exciton breaking; the latter is a new result which directly indicates the kinetics of polaron formation. Figure 3 shows the $(\Delta T/T)_{MF}$ signal at 1.9 eV as a function of pump-probe delay: at $\tau_D = 0$ ps the signal is zero, meaning that there is no significant exciton dissociation on the timescale of the pulse duration.The signal presents an initially fast rise time, growing to 50% of its final value in 2 ps, followed by a slower increase on the 40 ps time scale and a plateau on the 400 ps timescale. To highlight the time-dependence of the charge generation mechanism, we employed a simple phenomenological rate equation model in which the singlet exciton breaks directly into a polaron pair with a rate constant $\gamma(t)$. A plot of $\gamma(t)$ is reported in the inset of Fig.3: the rate shows an initial fast time dependence, settling down after ~3 ps to a constant value. We associate the initial fast dynamics of $\gamma(t)$ to a nonequilibrium process due to inter-chain dissociation during exciton migration within the disorder induced density of states. Once relaxation is completed, an additional time-independent dissociation mechanism becomes evident: we assign this process to on-chain dissociation of thermalized excitons.

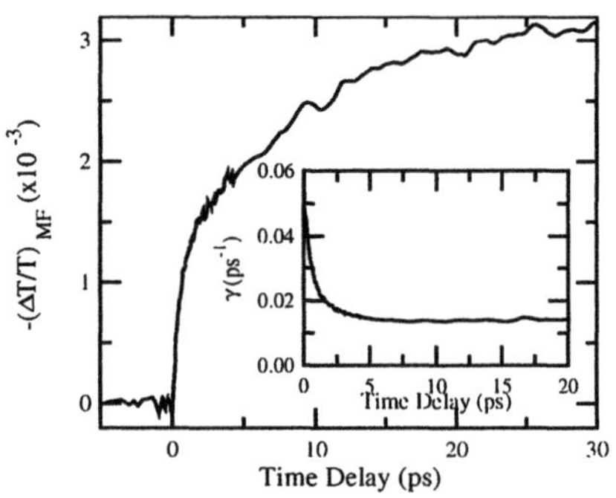

Fig. 3. ΔT_{MF} at 1.9 eV vs. delay; the inset shows a plot of $\gamma(t)$.

In conclusion, for the first time we have directly observed the dynamics of carrier generation from field-induced exciton breaking in an electroluminescent polymer. The data are interpreted in terms of direct exciton dissociation. We find two different regimes for this process, which are attribute to inter-chain and intra-chain dissociation respectively.

References

1 R. Kersting, U. Lemmer, M. Deussen, H.J. Bakker, R.F. Mahrt, H. Kurz, V. I. Arkhipov, H. Bässler, E.O. Göbel, Phys. Rev. Lett. **73**, 1440, (1994).

2 G. E. O'Keefe, J.J.M. Halls, C.A. Walsh, G.J. Denton, R.H. Friend, and H.L. Anderson, Chem. Phys. Lett. **276**, 78 (1997).

3 M. C. J. Vissenberg and M. J. M. de Jong, Phys. Rev. Lett. **77**, 4820, (1996).

4 G. Cerullo, S. Stagira, M. Nisoli, S. De Silvestri, G. Lanzani, G. Kranzelbinder, W. Graupner, G. Leising, Phys. Rev. B **57**, 12806 (1998).

Ultrafast Phase Transition in GeSb Films Triggered by Femtosecond Laser Pulses

J. Solis[1], K. Sokolowski-Tinten[2], J. Bialkowski[2], J. Siegel[1], C.N. Afonso[1] and D. von der Linde[2]

[1] Instituto de Optica, Consejo Superior de Investigaciones Científicas (CSIC)
Serrano 121, 28006-Madrid, SPAIN
Tel: +34-91-5616800, E-mail: iodjs37@pinar1.csic.es
[2] Institut für Laser- und Plasmaphysik, Universität-GHS-Essen
D-45117 Essen, GERMANY
Tel: +49-201-1832570, E-mail: kst@ILP.Physik.Uni-Essen.DE

Abstract. Time-resolved optical microscopy during femtosecond laser irradiation of amorphous GeSb films provides evidence for a non-thermal phase transition occurring on a subpicosecond time scale.

1. Introduction

Most of the work reported up to date on ultrafast, non-thermal, laser induced phase changes has concentrated on the solid-to-liquid transition in group IV- and III-V-semiconductors [1]. However non-thermal processes due to strong electronic excitation might be also important in triggering transitions between different solid phases of a material. It has been recently shown that the threshold fluence required for crystallizing amorphous GeSb thin films under pulsed laser irradiation decreases substantially for pulses shorter than one picosecond [2], indicating the relevance of electronic effects. In this work we have analyzed the dynamics of phase transformations induced in amorphous $Ge_{0.04}Sb_{0.96}$ films upon irradiation with 100 fs laser pulses at 620 nm by means of reflectivity measurements with fs-time and µm-space resolution (time-resolved optical microscopy) [3]. The spatial information contained in the reflectivity snapshot images of the surface provides directly the fluence dependence of the induced changes through the spatial intensity distribution of the pump laser beam.

2. Results and Discussion

The time dependence of the reflectivity upon irradiation can be seen in Fig.1A where it has been plotted for two different locations inside the irradiated spot, corresponding thus to two different local pump fluences. In both cases, the evolution of the reflectivity in the ps to ns time scale shows a plateau with a value of 0.69 which corresponds to the molten phase of the material [4] (melting threshold: 5 mJ/cm^2). For fluences above 19 mJ/cm^2 the reflectivity reaches a final ($t = \infty$) value of 0.71, consistent with the formation of crystalline material [4],

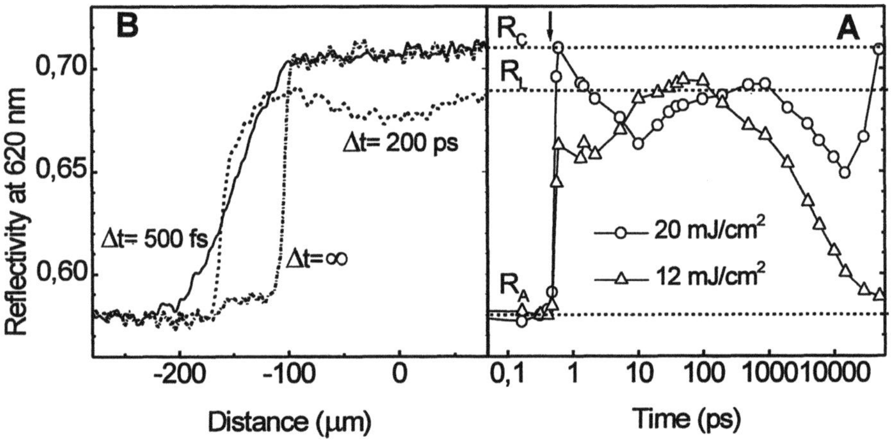

Figure 1. Reflectivity of the irradiated films at 620 nm: (A) As a function of time for two different fluences. R_C, R_L and R_A denote the reflectivities of the crystalline, liquid, and amorphous phases. The arrow marks the time of arrival of the pump pulse. The rightmost experimental points corresponds to t = ∞ (several seconds). (B) As a function of the distance to the center of the irradiated spot for three different delays after the pump pulse

whereas at lower fluences the material reamorphizes upon solidification. Figure 1B shows cross-sections of reflectivity images obtained for three different delay-times between the pump and the probe pulses (Δt= 500 fs, 200 ps, ∞). In the trace for Δt = ∞, the zone in the center of the spot with a constant reflectivity of 0.71 corresponds to the finally crystallized region. The trace for Δt = 200 ps, with top reflectivity values of 0.68-0.69, shows the extension of the region undergoing melting which is clearly bigger than that one finally crystallized. The trace for Δt = 500 fs shows the formation of a constant reflectivity region with a value of 0.71 at the center of the spot.

The sub-picosecond reflectivity increase which is observed in Fig.1A for fluences below 19 mJ/cm^2 can be related to the creation of a dense electron-hole plasma. It is followed by a slower increase to the reflectivity level of the liquid. At higher fluences, the reflectivity shows an abrupt increase to a maximum value of 0.71 within 300 fs before it relaxes to the value of the liquid phase. This maximum value of the reflectivity in the subpicosecond time scale is exactly that of the crystalline phase and is the same for any fluence value above 19 mJ/cm^2. This can be clearly seen in Fig.1B by comparing the traces for Δt = 500 fs and Δt = ∞. The flat-top behavior at Δt = 500 fs discards any excitational effect (electron-hole plasma) and is a clear evidence for the formation of an intermediate phase of the material which is (optically) different from the normal molten state and identical to the crystalline one. Moreover it must be noticed, that this intermediate phase is observed exactly in that region of the spot which finally crystallizes.

These observations provide strong experimental evidence for a connection between the transient phase and the crystalline state. Our data can be thus interpreted in terms of a non-thermal, electronically induced, *amorphous-to-crystalline* transition occurring prior to melting. The structural similarities between Sb-rich GeSb and pure Sb provide several arguments for considering the former as a clear candidate for such a transition [5].

4. Conclusion

We have we have analyzed the dynamics of phase transformations induced by fs laser pulses in amorphous GeSb. The results indicate the formation of a non-equilibrium phase of the excited material for fluences above 19 mJ/cm^2. This non-thermal transient phase is observed for only a few hundred fs, is optically different than the normal molten state and shows a clear connection with the crystalline one.

Acknowledgments. This work has been partially supported by CICYT (Spain) under project TIC93-0125 and by the University of Essen (*Forschungspool*). K.Sokolowski-Tinten gratefully acknowledges the funding support by *Flughafen Frankfurt Main* foundation. J.Siegel gratefully acknowledges the funding of the *European Community* through a grant (ERB 40001GT954352) within the program *Training and Mobility of Researchers*.

References

1 See for instance: C.V. Shank, R. Yen, and C. Hirlimann; Phys. Rev. Lett. 51, 454 (1983), K.H. Benneman and P. Stampfli; Phys. Rev. **B49**, 7299 (1994), K. Sokolowski-Tinten, J. Bialkowski, D. von der Linde; Phys. Rev. **B51**, 14186 (1995).
2 J. Solis, C.N. Afonso, S.C.W. Hyde,N.P. Barry, and P.M.W. French, Phys. Rev. Lett. **76**, 2519 (1996).
3 M.C. Downer, R.L. Fork, and C.V. Shank, J. Opt. Soc. Am. **B4**, 595 (1985).
4 J. Solis, C.N.Afonso, J.F. Trull, and M.C. Morilla, J. Appl. Phys. **75**, 7788 (1995).
5 K. Sokolowski-Tinten, J. Solis, J. Bialkowski, J. Siegel, C.N. Afonso, and D. von der Linde (unpublished).

Time of Flight Mass Spectroscopy of Femtosecond Laser Ablation of Solid Surfaces

A. Cavalleri, K. Sokolowski-Tinten, J. Bialkowski, D. von der Linde

Institut für Laser und Plasmaphysik, Universität-GHS-Essen, Germany

Abstract. Velocity distributions of evaporated particles from laser irradiated GaAs and Silicon give a temperatures in excess of 3000 K at the ablation threshold. The results suggest that explosive boiling causes material removal.

After excitation with femtosecond laser pulses, ablation can be achieved at the surface of an absorbing solid. While for lower fluences normal evaporation/sublimation causes particle desorption from the surface, above a fluence threshold F_a macroscopic removal of material takes place and craters are left on the surface.

By means of time-of-flight (TOF) mass spectroscopy, we measure the velocity distributions of gallium and silicon particles from GaAs and Si surfaces, respectively. A quadrupole mass spectrometer (QMS) provides the TOF distribution of a given particle mass after excitation with a 100 fs, 620 nm laser pulse. In order to provide information on single shot ablation, the sample is translated after every exposure. The ablation threshold F_a is measured independently by means of *post mortem* analysis.

Figure 1 shows the total number of particles as a function of laser fluence. Below F_a, the number of particles depends on the specific surface, consistent with what expected for normal desorption from a hot semiconductor surface.

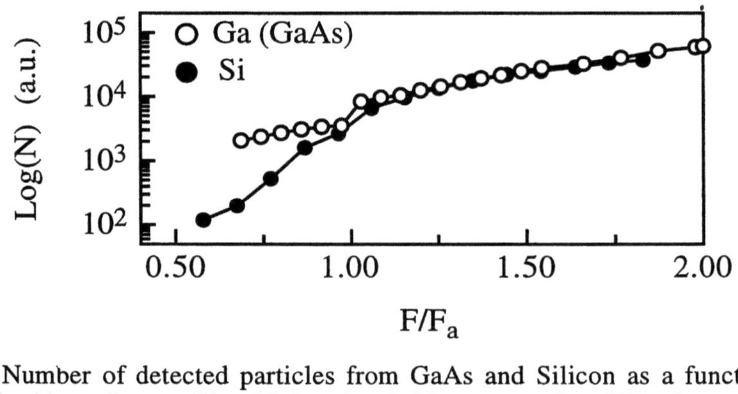

Fig. 1 Number of detected particles from GaAs and Silicon as a function of normalised laser fluence. The ablation threshold corresponds to $F/F_a=1$

Above threshold the total number of detected particles shows instead a common behavior for both Si and GaAs, suggesting that a process of general nature is taking place. The step observed at F_a in the total number of particles, together with the abrupt creation of a 20 nm crater, suggest that ablation is a bulk effect.

The measured velocity distributions have been quantitatively analysed taking into account the relevance of mutual interaction of the released particles. If the density of desorbed particles is low, no interaction occurs and a free flight model of the TOF signal can be applied (equation 1). Since the desorption probability is an exponential function of the surface temperature, the measured T_0 provides a good approximation of the maximum surface temperature $T_s^{(max)}$ [1].

$$f_{free}(t) = A / t^4 \exp[-(m / 2k_b T_0) L^2 / t^2] \qquad (1)$$

Figure 2 shows the result of our fits below the threshold for ablation. In both cases the measured maximum surface temperature increases monotonically from the respective melting temperatures at the corresponding fluences F_m to a value in excess of 3000 K at F_a.

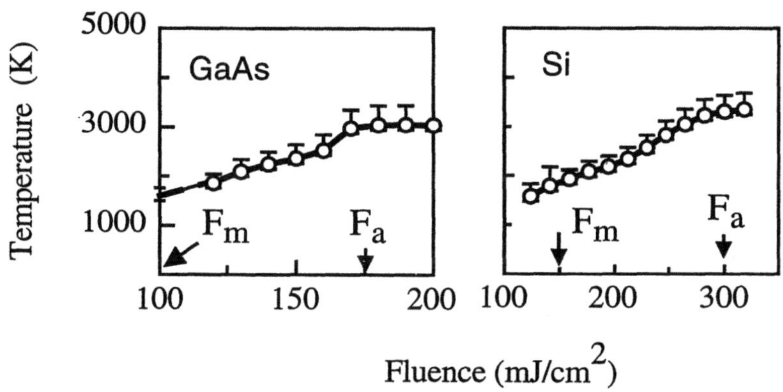

Fig. 2 Fitted maximum surface temperature for laser fluences below the ablation threshold

For higher laser fluences ($F>F_a$) the density in front of the sample increases. Mutual interaction is responsible for the developement of a "full range" Maxwellian distribution [2,3]. Two flow parameters are then required to describe TOF distributions, temperature T_d and drift velocity v_d.

$$f_{full}(t) = A / t^4 \exp[-(m / 2k_b T_d)(L - v_d t)^2 / t^2] \qquad (2)$$

L is the flight distance before detection, m for the particle mass and k_b is the Boltzmann constant. Figure 3 shows the corresponding fitting results fo gallium atoms, T_d decreases towards approximately 1800 K, whereas v_d reaches values of the order of 600 m/sec. For $F>F_a$ we interpret only the results on Ga (GaAs), where only monomers are detected.

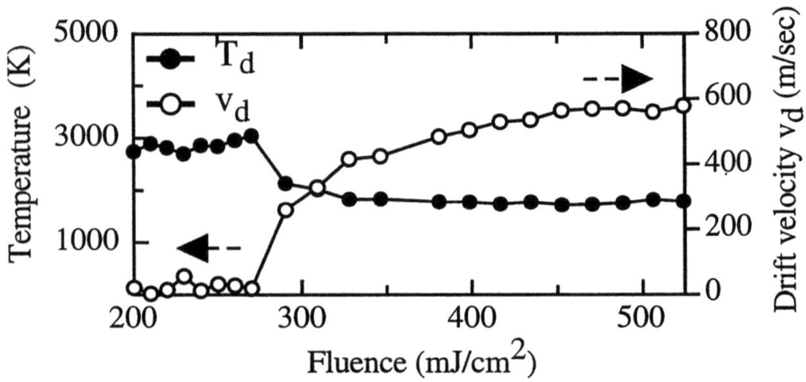

Fig. 3 Temperature and drift velocity resulting from a Full-range-Maxwellian fit of the velocity distributions of gallium particles for $F>F_a$

Ablation starts from a hot liquid surface whose temperature is below the crititical temperature (approx 5000 K in silicon). Optical data [4] suggest that rapid phase decomposition (explosive boiling) occurs after creation of a hot, liquid layer of about 20 nm. The initial phase of the expansion, where formation of the gas phase occurs, is to be described within a hydrodynamic picture [4]. Upon dilution however, the system goes through a phase where the description as a continuum is not appropriate any more and single collisions between the particles occur on a timescale which is comparable with that of the expansion. A Knudsen laye, corresponding to that region in front of a sample where collisions are relevant, is formed. After further dilution, the particles become ballistic and fly towards the mass spectrometer where they are detected. According to Knudsen layer theory [2] the relationship between the two fitted parameters should be $v_d=(\gamma k_b T_d/m)^{1/2}$, where $\gamma=C_p/C_v$ in the Knudsen layer. Above $2.5*F_a$, the theoretical relationship between v_d and T_d holds within less than 5% with $\gamma=5/3$.

In conclusion we have measured the maximum temperature of GaAs and silicon surfaces irradiated with visible femtosecond pulses for several fluences below the ablation threshold. We find peak temperatures at F_a of 3200-3500 K. Phase decomposition occurring at the beginning of the expansion process results eventually into a dense vapor in which a Knudsen-layer is formed.

References

[1] B. Stritzker et al. *Phys. Rev. Lett.* **47**, 356, (1981)
[2] R. Kelley, R.W. Dreyfus, *Surf. Sci.* **198**, 263, (1988)
[3] A. Cavalleri et al. *Appl. Phys. Lett* **72**, 2385, (1998)
[4] K. Sokolowski-Tinten, et al.*Phys. Rev. Lett.* **81**, 224, (1998)

Nanometer precision material ablation and optical breakdown with sub-10 fs pulses

Matthias Lenzner[1], Sasa Sartania[1], Christian Spielmann[1], Ferenc Krausz[1], Jörg Krüger[2] and Wolfgang Kautek[2]

[1] Department of Quantum Electronics and Laser Technology, Vienna University of Technology, Gusshausstr. 27-29, A-1040 Wien, Austria
E-mail: lenzner@iaee.tuwien.ac.at
[2] Lab. for Thin Film Technology, Federal Institute for Materials Research and Testing Unter den Eichen 87, D-12205 Berlin, Germany
E-mail: joerg.krueger@bam.de

Abstract. We report measurements of the optical breakdown threshold and ablation depth in dielectrics for laser pulse durations between 5 ps and 5 fs at the Ti:sapphire wavelength. The application of the shortest pulses leads to sub-μm ablation rates, smooth morpholgies of the generated structures and reduces the statistical behaviour of material removal. With 5-fs pulses a light intensity of 10^{14} W/cm^2 can be applied to fused silica without causing damage

1. Introduction

Laser-induced optical breakdown has been the subject of extensive experimental and theoretical investigations since powerful lasers have been found to cause damage even in transparent materials [1-5]. We carried out experiments with laser pulses between 5 ps and 5 fs in duration in order to determine the optical damage threshold of several dielectric materials and to gain detailed insight into the physical processes responsible for ablation. Our measurements revealed that damage intensities continue to increase down to the sub-10 fs regime in dielectrics. As a result, a light intensity of >10^{14} W/cm^2 can be applied to fused silica without causing damage. This opens up the possibility to investigate non-perturbative nonlinear optical processes in solids in a reversible manner for the first time.

2. Experimental results

For the damage experiments we used a laser system consisting of oscillator, multipass amplifier and hollow fiber compressor delivering 5 fs pulses with an energy of 0.5 mJ at a repetition rate of 1 kHz [6]. Bandwidth-limited pulses with durations between 5 fs and 20 fs can be generated by varying the gas pressure in the hollow fiber. Pulse durations of $\tau > 20$ fs are realized by evacuating the fiber and temporally stretching the pulses in blocks of glass in front of the fiber, which acts as an efficient spatial filter for all pulse durations.

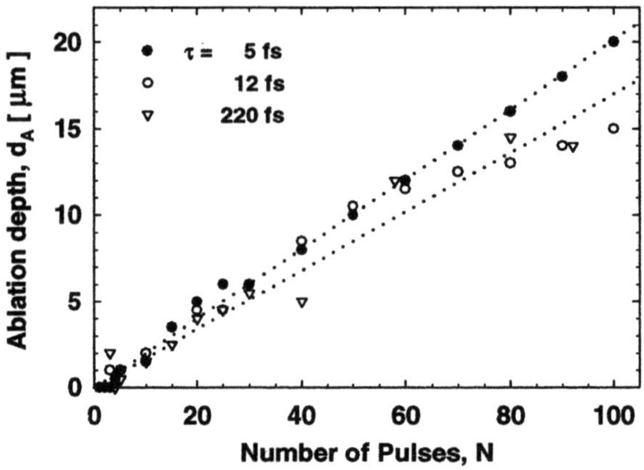

Fig. 1. Ablation depth of fused silica in dependence on the number of laser pulses applied to one spot for three different pulse durations.

The damage threshold has been determined by above-threshold measurements of the volume of ablated material V_a in dependence on the incident pulse energy and extrapolating to V_a=0. The dimensions of the ablation craters have been recorded for single- and multi-shot irradiation. Ablation with the shortest utilized pulses turned out to be advantageous in several ways. For longer pulses and fluence levels close to the damage threshold we found the onset of incubation effects (i. e. ablation starts only after several pulses have been applied to the surface without damage). This phenomenon is completely absent for pulses shorter than 100 fs. With these short pulses the ablation rate per pulse is also much more precise and reproducible than in the sub-picosecond range. We calculated an ablation rate per pulse of (200±10)nm for 5-fs pulses and (170±35)nm for 220-fs pulses, as shown in Fig.1. The given errors represent a 99% confidence interval.

Scanning electron microscope pictures for ablation craters obtained with the shortest pulses and, for comparison, with 3-ps pulses are shown in Fig.2. Both are generated with 60 pulses at an energy of about 5 times the damage threshold, the beam size of the 3-ps pulses was even 20% smaller than that for the 5-fs pulses. It can be clearly seen, that in the case of the 5-fs pulses the interaction region between laser pulses and matter is more confined and exhibits a much more regular morphology.

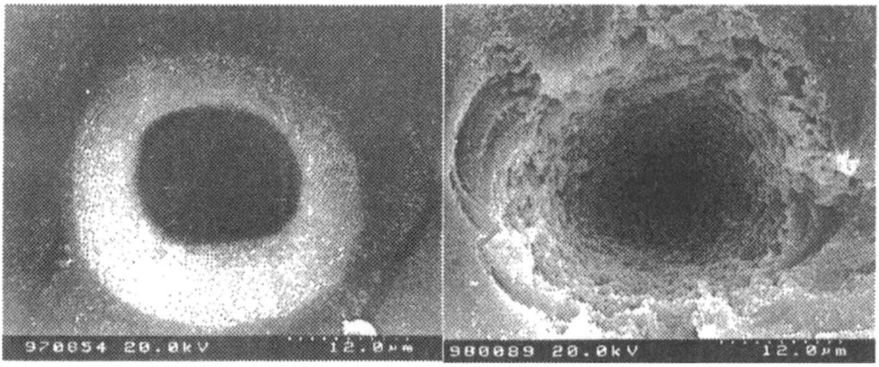

Fig. 2. SEM pictures for ablation craters generated with 60 laser pulses of 5 fs duration at 7 J/cm^2 (left) and 60 pulses of 3 ps duration at 20 J/cm^2 (right).

3. Conclusion

Sub-10 fs laser pulses open up the way ot ultrahigh-precision material ablation with submicron lateral and nanometer depth resolution.

Acknowledgements. We would like to thank Zhao Cheng for help with the experimental work. This work was supported by the Austrian Science Foundation (FWF) under grant Y44-PHY.

References

1 R. M. Wood, *Laser Damage in Optical Materials,* Hilger, Boston, 1986
2 D. Du, X. Liu, G. Korn, J. Squier, G. Mourou, Appl. Phys. Lett. **64**, 3071, 1994
3 B. C. Stuart, M. D. Feit, A. M. Rubenchik, B. W. Shore, M. D. Perry, Phys. Rev. Lett. **74**, 2248, 1995
4 H. Varel, D. Ashkenasi, A. Rosenfeld, R. Herrmann, F. Noack, E. E. B. Campbell, Appl. Phys. A **62**, 293, 1996
5 W. Kautek, J. Krüger, M. Lenzner, S. Sartania, Ch. Spielmann, F. Krausz, Appl. Phys. Lett. **69**, 3146, 1996
6 S. Sartania, Z. Cheng, M. Lenzner, G. Tempea, Ch. Spielmann, F. Krausz, Opt. Lett., **22**, 1562, 1997

Time Resolved Studies of Femtosecond Laser Induced Ablation from Solid Surfaces

K. Sokolowski-Tinten[1], J. Bialkowski[1], A. Cavalleri[1,2], and D. von der Linde[1]

[1]Institut für Laser- und Plasmaphysik, Universität-GHS-Essen, D-45117 Essen, Fed. Rep. of Germany, e-mail: kst@ilp.physik.uni-essen.de
[2]INFM, Dipartimento di Elettronica, Universitá di Pavia, Via Ferrata 1, 27100 Pavia, Italy

Abstract: Ultrafast time resolved microscopy has been used to study the dynamics of femtosecond laser induced ablation from the surfaces of various semiconductors and metals. Upon ablation initially metallic matter transforms into a transparent state with a high index of refraction.

The surface structure of solids can be modified by laser irradiation, and these modifications are often connected with ablation, the removal of macroscopic amounts of material from the surface. In the case of ablation with ultrashort (pico- or femtosecond) laser pulses the energy can be deposited *before* the actual removal of material occurs, and no interaction of the laser with ablated material takes place. Recent experimental work [1] indicates that this may offer some advantages in material processing applications, although a complete understanding of short pulse laser ablation is still missing.

We have studied the dynamics of femtosecond laser induced ablation by means of ultrafast time-resolved optical microscopy [2]. Ablation is induced by a first laser pulse of 100 fs duration at a wavelength of 620 nm. The evolution of reflectivity and transmission of the irradiated sample is monitored with high-resolution optical microscopes using a second, delayed ultrashort pulse for illumination. This method provides 100 fs snapshot pictures of the surface with a spatial resolution of a few micrometers over the entire time span from the deposition of the laser energy up to the appearance of the final structural modifications.

Measurements were performed on a number of different semiconductors (C, Si, Ge, GaAs, GeSb) and metals (Al, Au, Cr, Hg, Mg, Pt, Ti). In these absorbing materials ablation can take place without the formation of a plasma. In a relatively narrow range of excitation in the vicinity of the ablation threshold the laser pulse simply acts as a fast heat supply. Ablation occurs on a ps to ns time scale and is brought about by thermophysical processes, in particular hydrodynamic expansion of hot, pressurized material. The expansion is accompanied by striking changes of the optical properties indicating transitions from a metallic to an insulating state.

Characteristic snapshots during the ablation of different semiconductors and metals are shown in Fig. 1. The principal feature of these images is the observation of a transient ring-like structure which is common to all absorb-

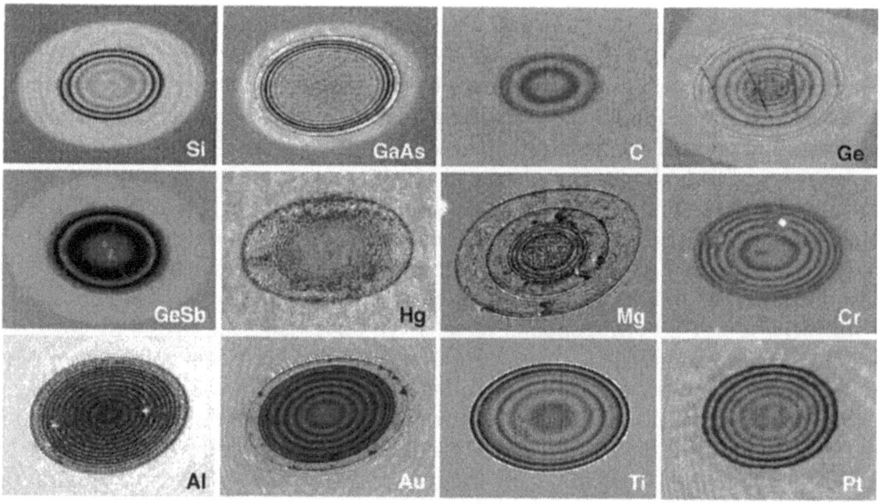

Fig. 1. Pictures of the transient ring structure on different semiconductors and metals at typical delay times between pump and probe of 1 - 5 ns.

ing materials investigated so far. The region covered by the ring structure is precisely congruent with the final ablated area. Thus the mechanism responsible for the formation of the rings is of rather general nature and directly related to the ablation process. By changing the wavelength of the probe pulse it can be shown that the ring structure is an interference pattern similar to Newton rings. From the change of the fringe spacing with time one can determine the (optical) thickness of the ablating layer and its expansion velocity (up to 1.000 m/s).

Important conclusions concerning the physical properties of the ablated material can be drawn from the existence of high contrast interference fringes: (i) The ablated layer must be confined between two sharp interfaces (scale-length $\leq 10\,\mathrm{nm}$); (ii) The material must be transparent ($\alpha \approx 10^3\,\mathrm{cm}^{-1}$) and possess a high refractive index ($n \approx 2$). Thus, initially metallic matter has transformed into a highly refractive transparent medium.

The interpretation of our observations proceeds along the following line [3]: The optically deposited energy thermalizes within a few picoseconds. This time is too short for significant material expansion to occur. Therefore the initial stage of excitation and rapid thermalization can be regarded as isochorical heating at solid density. The (metallic) material is left in a fluid state of high temperature and high pressure. At the ablation threshold maximum surface temperatures of 3.000 - 4.000 K have been measured on silicon and gallium arsenide [4], which correspond to pressures of a few tens of GPa. The pressure gradient drives then a hydrodynamic expansion of the heated layer into vacuum. The initial stage of the expansion can be described as an isentropic process. Upon expansion the temperature is decreased by adiabatic cooling and the diluted fluid reaches the boundary of the liquid-gas coexistence regime. The material can no longer exist as a homogeneous phase and rapid nucleation of the gas phase (*phase explosion* [5]) sets in. One of the characteristic properties of inhomogeneous two-phase mixtures is an extremely low sound velocity. It can be shown that a self-similar rarefaction wave [6] develops a sharp front when the discontinous drop of the sound velocity at the two-phase boundary is taken into account.

As pointed out above, the observed interference requires not only a steep ablation front, but also a high refractive index and low absorption in the ablated layer. Neither a Drude-model of an expanded metal nor a Clausius-Mosotti description of a dense atomic gas are capable of accountiung for these anomalous optical properties, giving further indication that the ablated material is not a homogeneous phase. We have used the Maxwell-Garnett model [7] to describe the optical properties of the heterogeneous two-phase mixture. This model describes a system of spherical particles (small compared to the optical wavelength) immersed in a homogenous background medium. Assuming that the spheres are droplets (clusters) of liquid material and the background is formed by the gas such a Maxwell-Garnett medium would become transparent and highly refractive when the metallic droplets fill approximately half of the available volume.

More work is necessary to clarify other aspects of the ablation process. For example, our observations reveal the existence of a well defined ablation threshold. The occurance of this threshold might be related to the strong dependence of the gas phase nucleation rate on temperature [5].

Acknowledgements. The authors are greatly indebted to S.I. Anisimov, J. Meyer-ter-Vehn, A. Oparin, and B. Rethfeld for important suggestions and stimulating discussions. KST acknowledges support by the *Flughafen Frankfurt Main* foundation. AC acknowledges support by the Europen Community (*Human Capital and Mobility*).

References

1 for example: X. Liu, D. Du, and G. Mourou, IEEE J. Quant. Electr. **QE-33**, 1706 (1997).
2 M.C. Downer, R.L. Fork, and C.V. Shank, J. Opt. Soc. Am. **B 4**, 595 (1985).
3 K. Sokolowski-Tinten, J. Bialkowski, A. Cavalleri, D. von der Linde, A. Oparin, J. Meyer-ter-Vehn, and S.I. Anisimov, Phys. Rev. Lett. **81**, 224 (1998).
4 A. Cavalleri, K. Sokolowski-Tinten, J. Bialkowski, and D. von der Linde, Appl. Phys. Lett. **72**, 2385 (1998).
5 A. Miotello and R. Kelly, Appl. Phys. Lett. **67**, 3535 (1995).
6 L.D. Landau and E.M. Lifshits, *Course of Theoretical Physics VI: Fluid Mechanics* (Pergamon Press, Oxford, 1982).
7 J.C. Maxwell Garnett, Phil. Trans. Roy. Soc. **A203**, 385 (1904).

Surface Modifications in Transparent Dielectrics Induced by a Mid-Infrared Free-Electron Laser

R. F. Haglund, Jr., H. K. Park, O. Yavas, D. R. Ermer, M. R. Papantonakis and A. H. Lines
Department of Physics and Astronomy and W. M. Keck Foundation Free-Electron Laser Center, Vanderbilt University, Nashville TN 37235

Abstract. Irradiation of wide-bandgap dielectrics using tunable picosecond infrared pulses at high repetition rate is shown to induce unusually efficient ablation, phase transformations and the evolution of a Kelvin-Helmholtz hydrodynamic instability.

High-repetition-rate free-electron lasers (FELs), with micropulse durations of 0.5-2 ps, permit ultrafast surface excitation and relaxation processes to evolve unmasked by long-term thermal changes. At the same time, the tunability of FELs makes it possible to target specific vibrational modes or to control energy deposition by adjusting the absorption depth relative to the thermal diffusion length. We have studied the phenomenology of these interactions using two model materials, crystalline calcite and glassy fused silica, irradiated at various wavelengths in air.

The FEL used in our experiments[1] produces a 4-μs macropulse at a repetition frequency of up to 30 Hz. Each macropulse consists of a train of 0.7-ps micropulses spaced at intervals of 350 ps. Typical micropulse peak powers are 10-20 MW. The FEL fluence is determined by the macropulse duration, which is varied from 50-4,000 ns by a broadband HgCdTe electro-optic switch (Pockels cell).

Laser ablation of natural calcite ($CaCO_3$) at wavelengths where the crystal is relatively transparent produces a surface morphology suggestive of subsurface boiling, melting and thermomechanical damage. Emission spectra of the ablation plume typically shows highly-excited atoms and ions. When the FEL is tuned to the ν_2- ν_4 stretch of CO_3 near 7 μm, however, the ablation threshold drops significantly, and the energy going into the ablation shock wave is much reduced. Figure 1 shows distance *vs* time curves $R(t)$ for blast waves generated from calcite at various wavelengths, measured by a plasmon-resonance spectrometer located a variable distance from the ablated surface. The functional

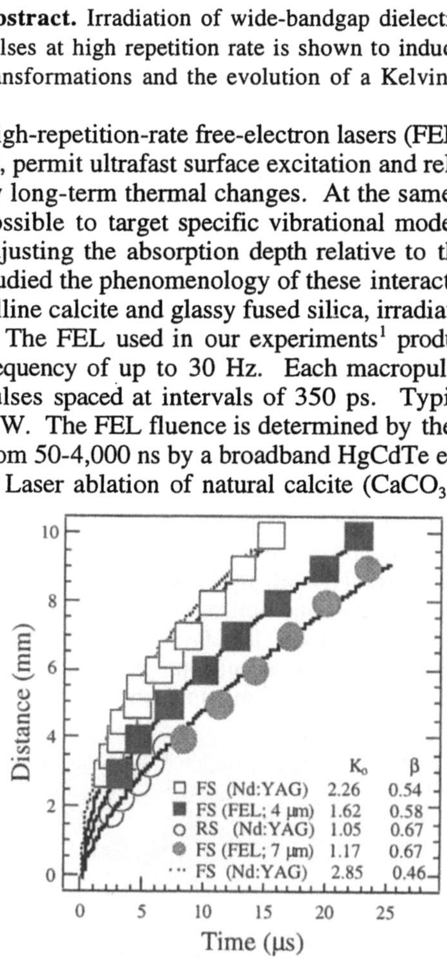

Figure 1. Distance *vs* time data for blast waves from calcite, for front- (FS) and rear-side (RS) ablation. The solid curves are fits to Eq. (1) in the text, with parameters K_0 and β.

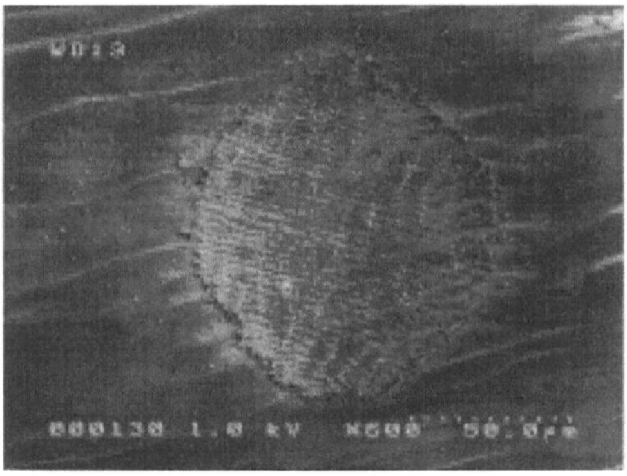

Figure 2. Morphology of natural calcite near ablation threshold fluence with the FEL tuned to 7.0 μm, the peak of the carbonate asymmetric stretch. The crater is 100 μm wide.

form of this relation is given, according to theory, by

$$R(t) = \xi_o\left(E_o/\rho_o\right)^{0.2} t^{-\beta} \qquad (1)$$

where the prefactor ξ_o is of order 1, $K_o = E_o/\rho_o$ is the ratio of explosive energy to the density of the wave medium, and the exponent β varies with the geometry of the shock front. Exponents nearer 0.4 indicate spherical shock waves, while those closer to 0.67 indicate nearly planar shocks.[2] The less explosive, nearly planar material removal inferred from the blast-wave measurement is confirmed by electron microscopy (Figure 2), as well as by plume-emission and time-of-flight ion spectra (not shown).

Figure 2 also demonstrates the possibility of laser-induced *structural phase transformations* in calcite, also at 7.0 μm and 100-ns macropulse duration.[3] The regular striations in the calcite form at intensities just below ablation threshold, independent of the polarization or angle of incidence of the FEL beam. The striations, around 11 μm apart, appear to be generated when the aragonite polymorph of calcite is formed under the heat and pressure loading of the laser pulse. Since aragonite is about 9% denser than calcite, the irradiated material shrinks upon cooling and the resultant strain is relieved by crack formation along cleavage planes. Small fragments of calcite are visible around the edges of the FEL-irradiated spot where they were vaporized and then refrozen upon ejection from the surface.

A contrasting model material is fused silica, a glassy material of immense technological importance in both microelectronics and optical systems. We irradiated a-SiO_2 flats (ultraviolet-grade fused silica from ESCO, Inc.) which were cleaned with methanol but were otherwise given no special preparation. These samples were irradiated in air at wavelengths ranging from 3 to 8 μm.

In this wavelength range, the optical properties of water-free a-SiO_2 are dominated by the Si-O absorption band which peaks at 9.6 μm. At wavelengths far from the Si-O absorption peak, changes in the surface morphology were dominated by fracture, subsurface boiling and melting, resulting in explosive ejection of material. A typical ablation crater in this wavelength range, approximately twice the size of the laser spot, is shown in Figure 3. The phenomenology is consistent with the fact that for weak absorption, thermalization results from the slow transfer of energy from the high-

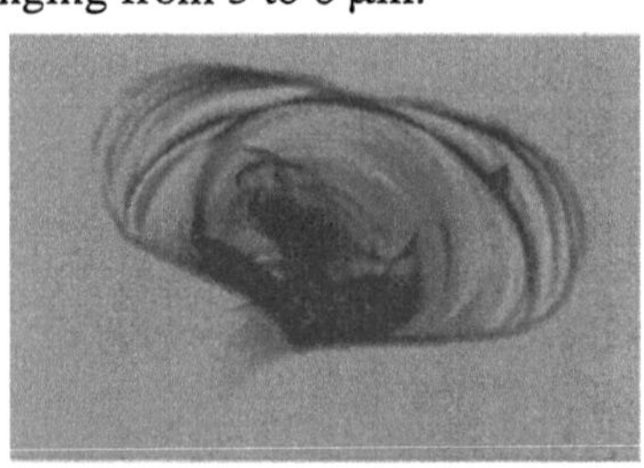

Figure 3. SiO_2 surface after absorption of a single FEL macropulse at a wavelength of 4 μm. Fluence approximately 10 J/cm^2.

frequency (>1000 cm^{-1}) anharmonic modes excited by the FEL to low-frequency (<200 cm^{-1}) harmonic modes of the phonon bath.

Ultrafast melting and quenching of insulators can be induced by adjusting the total energy deposited by the FEL macropulse while changing wavelength to control depth of penetration. Figure 4 shows a micrograph of fused silica irradiated at 8 μm at a fluence of 14 $J{\cdot}cm^{-2}$, using the full FEL macropulse duration of 4 μs. Under these conditions, the thermal diffusion length is nearly equal to the absorption depth. Heat-transfer calculations based on the method of moments[4] show that, under these conditions, the ablation depth of the silica is approximately 2 μm while the depth of the underlying liquid layer is about 8 μm. The radial expansion of the ablation plume over the hot liquid generates a Kelvin-Helmholtz instability, a transition to mesoscopic surface self-organization over a length scale of many microns, whose most probable wavelength depends on both laser intensity and surface tension. This instability is similar to that observed when the wind blows over a shallow expanse of water.[5]

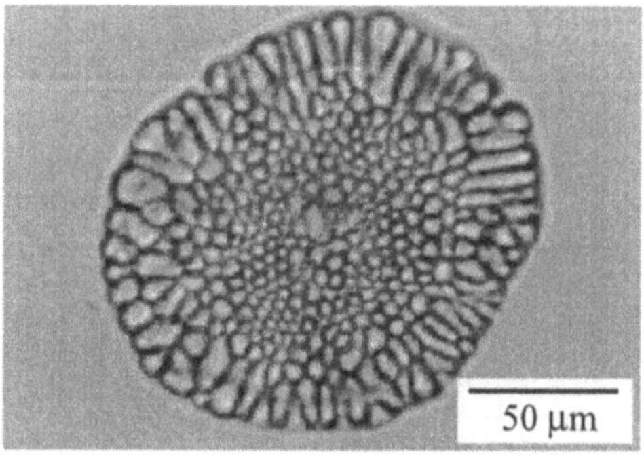

Figure 4. Visible micrograph of a picosecond FEL-irradiated SiO_2 surface showing the formation of a Kelvin-Helmholtz surface instability. FEL wavelength, 8.0 μm, fluence 14 J/cm^2.

Much work remains to develop a clear picture of the potential of ultrashort infrared pulses in materials processing. In particular, the microscopic mechanisms of vibrational energy transfer on an ultrafast time scale are only now being elucidated as tunable, femtosecond and picosecond sources in this wavelength region become available. Nevertheless, the fact that this form of high-intensity laser irradiation works on the vibrational, rather than the electronic, degrees of freedom of the modified material clearly opens new possibilities for texturing, etching and forming insulator surfaces with high precision.

Acknowledgements. This work was supported by the Office of Naval Research Medical Free-Electron Laser program, contract number N00014-94-1-0123.

REFERENCES

1. G. S. Edwards *et al.*, IEEE J. Sel. Topics Quantum Electron. **2**, 810 (1996).
2. O. Yavas *et al.*, Appl. Phys. Lett. **71**, 1287 (1997).
3. O. Yavas *et al.*, Appl. Surf. Sci., **127-129**, 26 (1998).
4 N. Arnold, B. Luk'yanchuk, N. Bityurin, Appl. Surf. Sci. **127-129**, 184 (1998).
5 D. Bäuerle, *Laser Processing and Chemistry,* 2nd Edition (Berlin: Springer, 1996),

Part VIII

Surfaces, Metals and Superconductors

Direct Measurement of the Valence Band Decoherence in Copper

Hrvoje Petek, Hisashi Nagano, and Susumu Ogawa

Advanced Research Laboratory, Hitachi, Ltd., Hatoyama, Saitama, 350-0395 Japan
E-mail: petek@harl.hitachi.co.jp

Abstract. Interferometric time-resolved photoemission is used to measure the decoherence of *d*-bands in copper. Temperature dependence of the decoherence times gives the hole lifetime of 24 ± 3 fs for the X_5 point (top-of-the *d*-bands) due to the Auger decay, and an electron-phonon mass enhancement factor $\lambda = 0.20 \pm 0.01$.

1. Introduction

The response of a metal surface to an external field is essentially a coherent, yet very little is known regarding the decoherence times of charge carriers in metals. This is in part due to a perception that optical response is nearly instantaneous and the strong Coulomb interaction between the carrier leads to exceedingly rapid decoherence. However, a survey of Drude relaxation times shows that the scattering times are ≤20 fs for alkali and noble metals for DC fields [1]. Furthermore, decoherence and control of coherent electron dynamics at surface states of metals, where the carriers are only weakly coupled to the bulk, have already been demonstrated [2-6]. Thus, with state-of-art laser technology it also may be possible to measure the decoherence times of bulk bands in metals [7].

2. Experimental Methods

The interferometric time-resolved two-photon photoemission (ITR-2PP) technique and data evaluation are described elsewhere [5, 8]. Briefly, frequency-doubled output of a Ti:sapphire laser ($h\nu$ - 3.08 eV; energy/pulse - 0.6 nJ; width of the pulse intensity envelope - 13 fs; 80 MHz repetition rate) is split into equal pump and probe pulses in a Mach-Zehnder (MZ) interferometer. The pump-probe delay is scanned under feedback control by ±150 fs with an accuracy and reproducibility of <50 as. The pulse pair is focused collinearly onto the Cu(100) surface, inside a UHV chamber. The photoemission current is induced with *s*- or *p*-polarized light incident at 30° from the surface normal, and measured as a function of energy, with a hemispherical electron energy analyzer (5° angular, and <30 meV energy resolution). The work function of Cu(100) is reduced by ~1 eV by depositing ~0.10 monolayer of Cs from a SAES getter, which makes it possible to observe 2PP from the *d*-bands. Cs reduces the vacuum potential, but it does not appear to affect the bulk carrier dynamics [9]. The sample temperature is controlled between 33 and 300 K by a closed-cycle He refrigerator/heater combination.

3. Results and Discussion

Figure 1 shows representative 2PP spectra of the *d*-band region with *s*- and *p*-polarized excitation. The s-polarized spectrum is dominated by single band, which makes it possible to analyze the shoulder feature in the p-polarized spectrum into two weaker bands. Observation of three peaks is consistent with the calculated relativistic band structure of copper, which predicts three bands with a binding energy of ~-2 eV [10]: assignment of the shoulder is to the top-of-the *d*-bands, X_5 point, which splits by 0.1 - 0.14 eV due to the spin-orbit coupling, and that of the main peak is to resonant ($\Delta_2 \rightarrow \Delta_1$) photoemission from Δ_2.

Band structure in Fig. 1, shows that two-photon absorption from the *d*-bands creates a d-hole and electron in the X band gap of Cu(100), where the electron cannot propagate in the bulk. The electron can only exist in the vacuum and has exponentially damped probability density inside the bulk. Therefore, two-photon absorption creates an e-h pair, which decoheres mainly through the scattering of the holes. The decoherence times are determined from interferometric two-pulse correlations (I2PC), where the pump-probe delay is scanned interferometrically while measuring the 2PP signal for specific electron energy and momentum [5].

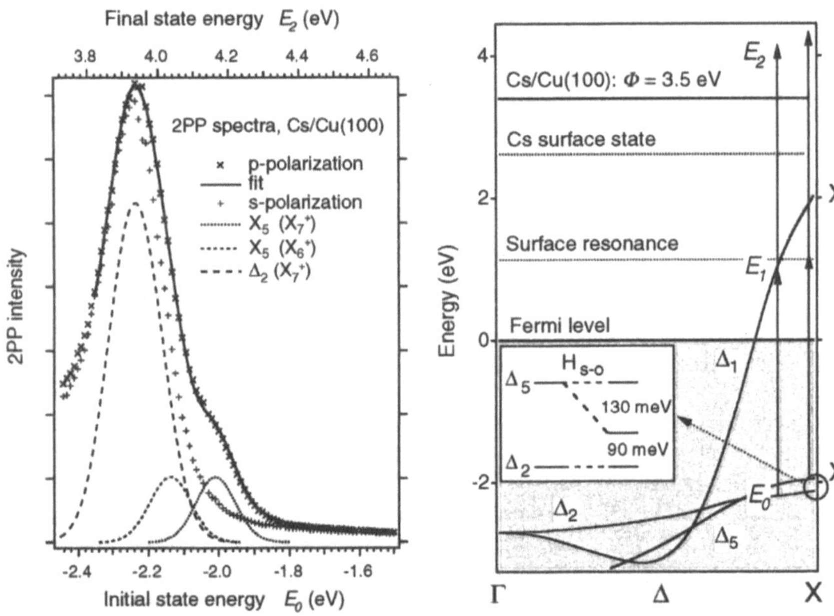

Fig. 1. 2PP spectra of Cs/Cu(100) showing the top-of-the *d*-bands, and the calculated band structure for $k_{||} = 0$. The energy splittings are determined from the spectrum.

A typical I2PC scan for excitation from X_5 at 33 K in Fig. 2a has oscillations with a period of approximately ω the excitation laser frequency and its second-harmonic. There is also a slowly varying signal which is due to photoemission of incoherent hot-electron population that is excited by the pump pulse. The origin of the ω and 2ω oscillations is optical interference in the MZ interferometer, and more importantly quantum interference between the coherent polarization that is created by the pump pulse, and the electric field of the probe pulse. The time scale on which the interference is determined by the decoherence of linear and non-linear polarizations. The decoherence times are extracted by determining the envelopes of the ω and 2ω oscillations, and fitting them to a convolution of and exponential with the autocorrelation of the laser pulse [5, 8]. With the precise knowledge of decoherence times, the hot-electron population decay can be extracted from the phase average of the I2PC signal. These reduced data for the I2PC signal for X_5 also are shown in Fig. 2a. The comparison of the ω and 2ω envelopes shows that the non-linear polarization decoheres much more slowly than the linear polarization. In case of the linear polarization, one photon absorption creates an electron in the *s,p*-band, and therefore, both *e-e* and *e-p* scattering contribute

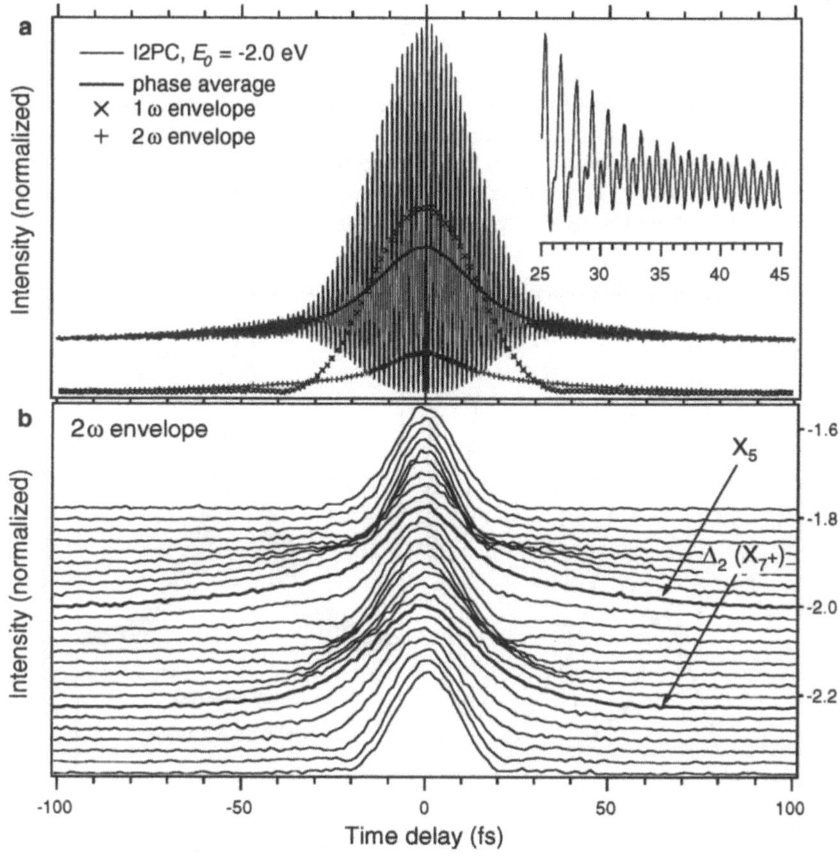

Fig. 2. a) I2PC scan measured at the X_5 point, and its decomposition into coherent and incoherent signal components. Inset shows an enlargement of ω and 2ω interference fringes. b) 2ω envelopes for different observation energies. Slow decays indicate long hole decoherence times.

to the e-h pair decoherence [5]. The following discussion will focus on the interpretation of the decoherence time of the nonlinear polarization T_2^{02}.

Figure 2b, gives the energy dependence of 2ω envelopes for excitation across the *d*-bands in Fig. 1. The data show that the T_2^{02} time has two maxima corresponding to the excitation from X_5 and Δ_2. More quantitative values of the decoherence times from fits of the envelopes are given in Fig. 3. It should be pointed out that the reported values are based on the assumption of exponential kinetics; however, at certain energies the signal decays in a distinctly non-exponential fashion, exhibiting clear quantum beat like oscillations. The quantum beats cannot be interpreted simply as a consequence of coherent excitation of the *d*-bands, but also probably are influenced by the band structure at the intermediate level for the 2PP process [7]. Nevertheless, at the X_5 and Δ_2 maxima, data are well represented by exponential kinetics and yield accurate values for the hole decoherence times. The results of fits in Fig. 3 show that the decoherence is slowest for X_5 and approximately 2.4 times faster at Δ_2.

Two main mechanisms contribute to the hole decoherence, namely, *h-h* (Auger decay) and *h-p* scattering. The contribution of the *h-h* scattering is deduced from the temperature dependence of the decoherence times. Fig. 3. gives several 2ω envelopes at X_5 for different temperatures. The temperature dependence gives the *e-p* mass enhancement factor $\lambda = 0.20 \pm 0.01$, and the intrinsic *h-h* scattering time at the top-of-the *d*-bands $\tau_{h-h} = 24 \pm 3$ fs. The *h-p* scattering is expected to be independent of the hole binding energy when the binding energy is much larger

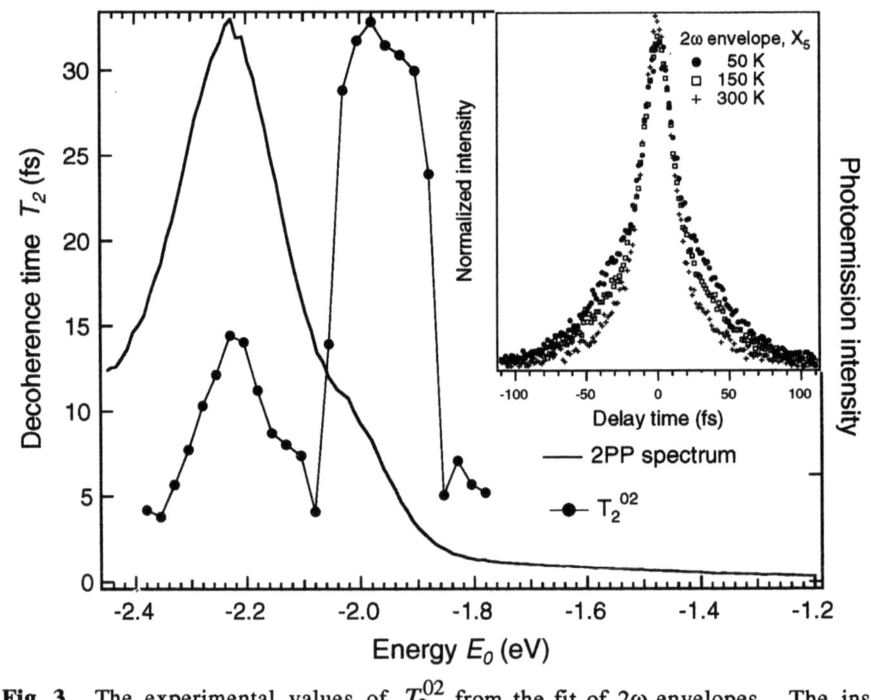

Fig. 3. The experimental values of T_2^{02} from the fit of 2ω envelopes. The inset shows the temperature dependence of the 2ω envelopes due to the *h-p* scattering at X_5.

than the Debye frequency. Thus the difference of decoherence times between for X_5 and Δ_2 must be due to differences in the *h-h* scattering rates. The holes at X_5 can only decay to the *s,p*-band through Auger decay that involves different shells (M→N). However, *d*-holes with larger binding energy can decay through both M→M and M→N processes. Since there is a larger spatial overlap between bands that belong to the same shell, the decay within *d*-bands can compete with slower decay form the valance to the conduction band. Thus the present work suggests that the energetic *d*-holes will preferentially float up to the top-of-the valance band before decaying into the conduction band.

4. Conclusions

In summary, the decoherence of *d*-band holes in Cu(100) due to the *h-h* and *h-ϕ* scattering is measured directly by the interferometric time-resolved two-photon photoemission technique. Coherent polarization induced by two-photon excitation of the spin-orbit split X_5 and nearby Δ_2 bands decays in a complex manner showing quantum beats between the coherently excited bands. The strong temperature dependence of the decoherence rates makes it possible to determine the *e-p* mass enhancement factor $\lambda = 0.20 \pm 0.01$ and an upper limit for the hole lifetime due to Auger decay at the top-of-the-*d*-bands of 24 ± 3 fs. The implications of such long lifetimes and slow decoherence is that in optical excitation with >2 eV photons, most of the energy is deposited in long lived *d*-holes, which could play an important part in surface photochemistry and in laser processing of copper surfaces. Finally, the discovery of long decoherence times of the *d*-bands of copper demonstrates that the decoherence of valance bands in metals can be measured by state-of-art laser techniques, and it opens the way for exploring coherent control of electron dynamics in bulk metals.

Acknowledgments. The authors thank N. Moriya, N. Matsunami, and S. Saito for technical support and NEDO International Joint Research Grant for partial funding of this project.

References

[1] N. W. Aschroft and N. D. Mermin, "Solid State Physics", Holt, Rinehart, Winston, 1987.
[2] E. Knoesel, A. Hotzel, and M. Wolf, J. Elect. Spect. Rel. Phen. (in press).
[3] S. Ogawa, *et al.*, Phys. Rev. Lett. **78**, 1339 (1997).
[4] H. Petek, *et al.*, Phys. Rev. Lett. **79**, 4649 (1997).
[5] H. Petek and S. Ogawa, Prog. Surf. Sci. **56**, 239 (1998).
[6] U. Höfer, *et al.*, Science **277**, 1480 (1997).
[7] H. Petek, H. Nagano, and S. Ogawa, Phys. Rev. Lett. (submitted).
[8] W. Nessler, *et al.*, J. Elect. Spect. Rel. Phen. (in press).
[9] S. Pawlik, M. Bauer, and M. Aeschlimann, Surf. Sci. **377-379**, 206 (1997).
[10] R. Courths and S. Hüfner, Phys. Reports **112**, 53 (1984).

Lifetime Effects of Xenon Adsorption on the Image-Potential States of Ru(0001)

W. Berthold[1], I.L. Shumay[1,2], P. Feulner[3], and U. Höfer[1,3]

[1] Max-Planck-Institut für Quantenoptik, Hans-Kopfermann-Str. 1, D-85748 Garching, Germany
[2] Lehrstuhl für Festkörperphysik, Universität Erlangen-Nürnberg, Staudtstr. 7, D-91058 Erlangen, Germany
[3] Physik Department, Technische Universität München, James-Franck-Str., D-85747 Garching, Germany

Abstract. The electron dynamics of the image-potential states of clean and Xe-covered Ru(0001) surfaces was investigated by means of time-resolved two-photon photoemission. Xe adsorption is observed to reduce the coupling of the first image-potential state to the metal and leads to lifetimes that increase from 11 fs in case of the clean surface to 125 fs for a bilayer of Xe. The results are discussed within the framework of a tunneling model wherein the Xe 6s-derived electron affinity level determines the barrier height.

The quantum yield of most electronically induced adsorbate reactions on metal surfaces is exceedingly small, because the primary excitation delocalizes into the bulk within a few femtoseconds or even on a sub-femtosecond timescale. Within the commonly employed Menzel-Gomer-Readhead model, these small yields depend exponentially on the excited state lifetime [1]. In order to be able to exploit some of the unique properties of two-dimensional adsorbate layers for time-domain experiments [2] it will thus be necessary to reduce the coupling between adsorbate and bulk electronic states. This may be accomplished by the use of rare-gas spacer layers [3]. As a model system to study related coupling issues we have investigated the relatively long-lived image-potential states [4,5] that are mainly located in the vacuum above a Ru(0001) surface and report here how their relaxation behavior depends on xenon adsorption.

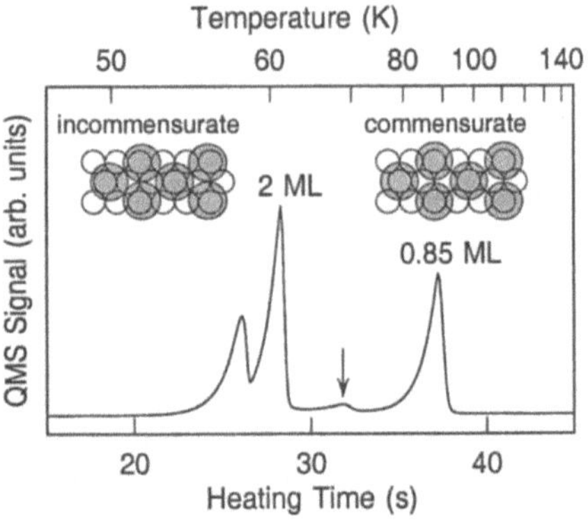

Fig. 1. Temperature programmed desorption of Xe from Ru(0001)

The experiments were performed by time-resolved two-photon photoemission (2PPE). Electrons were excited into the normally unoccupied image-potential states by 80-fs UV pulses ($\hbar\omega_a = 4.95\,\text{eV}$). They were photoemitted by subsequent 40-fs IR pulses ($\hbar\omega_b = 1.65\,\text{eV}$) and detected energetically resolved in a hemispherical analyzer. The Xe layers were prepared with temperature programmed desorption (TPD). The TPD spectra (Fig. 1) show the desorption of the second and

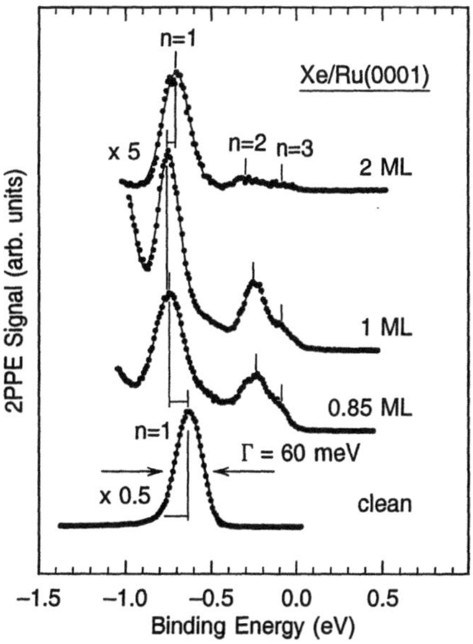

Fig. 2. 2PPE Spectra from clean and Xe covered Ru(0001) recorded at normal emission and an analyzer resolution of 40 meV.

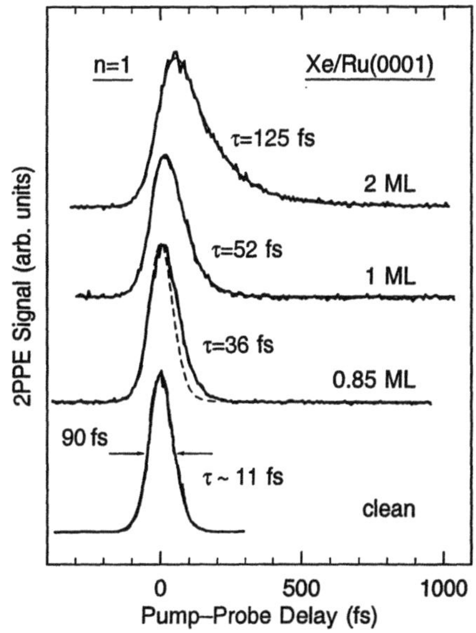

Fig. 3. 2PPE cross-correlation traces from the $n = 1$ state for different Xe coverages. At positive delays the UV pump pulses arrive before the IR probe pulses

the first monolayer (ML) as two prominent peaks; the small maximum in between arises from a phase transition from the incommensurate close-packed monolayer to the commensurate $\sqrt{3}\times\sqrt{3}$-superstructure with a coverage of 0.85 ML [6].

The 2PPE spectra recorded at zero delay between the UV and IR pulses are plotted in Fig. 2. The binding energy of the lowest image-potential state ($n = 1$) increases from -0.64 eV to -0.74 eV after adsorption of the commensurate and incommensurate monolayers; it decreases again slightly to -0.71 eV after adsorption of the bilayer. Image-potential states with higher quantum numbers ($n = 2, 3$) can be seen from the Xe-covered surfaces as well, because the decrease of the workfunction by 0.6 eV upon Xe adsorption enables their excitation with the UV pump photons. The lifetime of the $n = 1$ state on clean ruthenium was determined from a peak shape analysis to be 11 fs and served as a reference for the time-domain experiments. The 2PPE cross-correlation curves (Fig. 3) exhibit a strong increase in lifetime upon Xe adsorption. 36 fs, 52 fs and 125 fs were obtained for the commensurate and the incommensurate monolayer and the bilayer, respectively.

The observed behavior can be interpreted in a picture where the Xe layers form a repulsive barrier for electrons in the $n = 1$ state [7]. For all Xe layers we investigated, the binding energy of this state lies below the electron affinity of bulk Xe, which is 0.5–0.55 eV below the vacuum level. Thus, the wavefunction of the $n = 1$ state is mainly located in the vacuum above the Xe layer and gets exponentially damped when transversing this barrier. The measured increase in lifetime upon completing the second monolayer can be explained by a thicker tunneling barrier.

An estimate from plane waves tunneling through a rectangular barrier, with a Xe/Xe layer spacing of 3.6 $Å$, is in good agreement with our experimental results.

In contrast to the Ag(111) and Cu(111) surfaces [7,8], the $n = 1$ energy on clean Ru(0001) lies close to the center of the band gap. The effect that the lowering of the work function upon Xe adsorption shifts the image-potential states towards the gap center and thereby increases the bulk barrier [8] can thus be neglected in the case of Ru(0001). The significant longer lifetime of the incommensurate as compared to the commensurate monolayer can be understood qualitatively from the fact that the interaction of an excess electron with the closed Xe valence shell is repulsive. As the Xe coverage is 15% higher in the incommensurate phase, the enhanced interatomic overlap is likely to cause an upshift of the electron affinity and thus to increase the barrier height.

Valuable discussions with and support from K. L. Kompa and D. Menzel are gratefully acknowledged. Funded by the Deutsche Forschungsgemeinschaft through Sonderforschungsbereich 338.

References

1. P. Feulner and D. Menzel in: *Laser Spectroscopy and Photochemistry on Metal Surfaces*, H.-L. Dai and W. Ho (eds.), (World Scientific, Singapore, 1995), pp. 627.
2. E. W. Plummer, Science **277**, 1447 (1997); U. Höfer, Science **279**, 190 (1998).
3. U. Höfer, M. J. Breitschafter, and E. Umbach, Phys. Rev. Lett. **64**, 3050 (1990).
4. P. M. Echenique and J. B. Pendry, Prog. Surf. Sci. **32**, 111 (1990).
5. Th. Fauster and W. Steinmann in: *Photonic Probes of Surfaces*, P. Halevi (ed.), Electromagnetic Waves: Recent Developments in Research Vol. 2 (North-Holland, Amsterdam, 1995), pp. 347.
6. H. Schlichting and D. Menzel, Surf. Sci. **272**, 27 (1992); B. Narloch and D. Menzel, Chem. Phys. Lett. **270**, 163 (1997)
7. J. D. McNeill, R. L. Lingle, R. E. Jordan, D. F. Padowitz, and C. B. Harris, J. Chem. Phys. **105**, 3883 (1996).
8. M. Wolf, E. Knoesel, and T. Hertel, Phys. Rev. B **54**, R5295 (1996).

Dephasing of Image-Potential States on Cu(100) upon CO Adsorption Measured by Femtosecond Two-Photon Photoemission Spectroscopy

Christian Reuß, Michael Kutschera, Igor L. Shumay, Ulrich Thomann, Martin Weinelt, and Thomas Fauster

Lehrstuhl für Festkörperphysik, Universität Erlangen-Nürnberg, Staudtstraße 7, D-91058 Erlangen, Germany; Tel. +(49)9131/858408; Email: c.reuss@fkp.physik.uni-erlangen.de

Ulrich Höfer

Max-Planck-Institut für Quantenoptik, D-85740 Garching, Germany, Tel. +(49)89/32905/232, Email: Hoefer@mpq.mpg.de

Abstract. The relaxation dynamics of the image-potential states of Cu(100) under the influence of CO adsorption is investigated by time-resolved Two-Photon Photoemission Spectroscopy (2PPE). Pure dephasing due to elastic scattering leads to a fast decay of quantum-beat patterns of energetically close-lying image states at already very low CO coverages. For higher CO exposure an increase in structural order is correlated with a decrease in pure dephasing rates.

1. Introduction

An electron located at distance z in front of a metal surface is attracted by its image charge located at -z. The resulting image potential in front of the surface gives rise to a series of Rydberg-type unoccupied bound states converging towards the vacuum energy E_{vac} ($E_n \approx E_{vac}$ - 0.85eV / n^2) if there are no available bulk states (see Ref. 1). For increasing quantum number n the center of the wavefunction shifts more and more away from the surface, leading to a reduced overlap of the wavefunction with unoccupied bulk states which act as energy-decay channels (electron-electron scattering). For Cu(100) the lifetimes T_1 of the image-state population have been measured for $n \leq 6$ with time-resolved 2PPE (Ref. 2).

In this paper we focus on quasielastic scattering, which does not change the population of the image-potential states but destroys the coherence between the involved states (pure dephasing time T_2^*). We investigate the influence of disorder on the quasielastic-scattering channel and minimize a possible effect of quasielastic electron-phonon scattering (Ref. 3) by reducing the sample temperature down to 90 K. Adsorption of CO on Cu(100) allows to control the surface order (revealed by linewidth in spot-profile LEED analysis).

2. Experimental Methods

The method of choice for investigating the dynamics of image-potential states is the time-resolved Two-Photon-Photoemission Spectroscopy (Refs. 1-4). In our case the states are populated by the frequency-tripled light of a Ti-sapphire-Laser (Coherent, MIRA 900) (4.5 eV, 70 fs) and ionized by the fundamental (1.5 eV, 45 fs after optical compression). The number of electrons photoemitted normal to the surface is measured as a function of kinetic energy with a hemispherical analyzer (energy resolution = 40 meV, acceptance angle = 0.6°). Time-resolved pump-probe experiments are accomplished by varying the delay between excitation and ionization pulses.

From the exponential decay of the photoemission signal, the population relaxation time T_1 can be deduced. Within the framework of optical Bloch-equations population decay time T_1 and pure dephasing time T_2^* both contribute to the intrinsic linewidth $\Gamma = \hbar(1/T_1+2/T_2^*)$. In order to suppress short-lived offresonant excitations, which broaden the linewidth, one has to delay the probe pulse with respect to the pump pulse by about 100 fs.

3. Results and Discussion

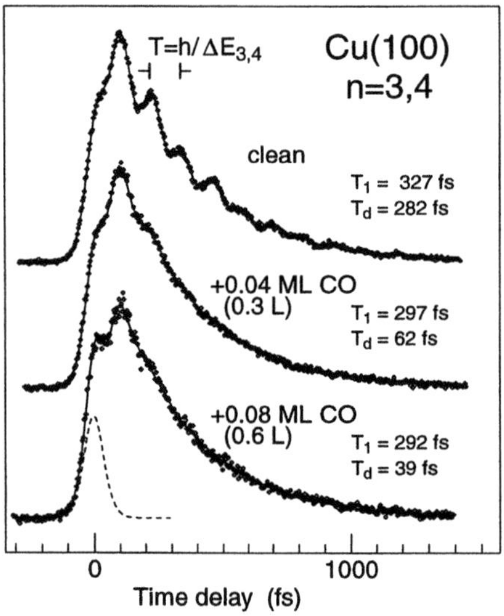

Fig. 1. Quantum-beat pattern showing a coherent superposition of the n = 3 and n = 4 states. Adsorption of CO leads to a rapid decay of the quantum-beat pattern (T_d), whereas the overall exponential decay (T_1) remains almost unchanged.

The effect of pure dephasing on the relaxation dynamics of image-potential states can be directly seen in Fig. 1 when one excites a coherent superposition of several states (Ref. 2). This is achieved, when the energy difference of the excited states gets comparable to the linewidth of the excitation pulse (30 meV). For the clean sample the 2PPE signal of the n = 3 state shows an exponential decay (T_1 = 327 fs) as a function of pump-probe delay. Superimposed on this decay is a quantum-beat pattern between the n = 3 and n = 4 states. The beat period T corresponds to the energy difference ΔE = h/T = 35 meV of these states.

Adsorption of 0.04 monolayers (ML) CO only slightly alters the exponential decay time (T_1 = 297 fs), but dramatically increases the damping of the quantum-beat structure, which can be explained in terms of rapid pure dephasing of the involved states: CO molecules initially adsorb randomly on the Cu(100) surface. Thus disorder is produced leading to elastic scattering, which destroys the coherence between the involved states.

Figure 2 shows a series of energy-resolved spectra of the states n = 1, 2, 3 as a function of CO exposure. The linewidth of the n = 1 state varies non-monotonically with CO exposure. Part of the overall linewidth increase can be explained by the increase of the inverse lifetime ($\hbar/T_1$). The lifetime T_1 decreases from 40 fs (clean surface) to 10 fs at 3.5 L and stays constant at higher coverages. The CO molecules offer additional unoccupied states ($2\pi^*$) acting as decay channels for the image-state electron.

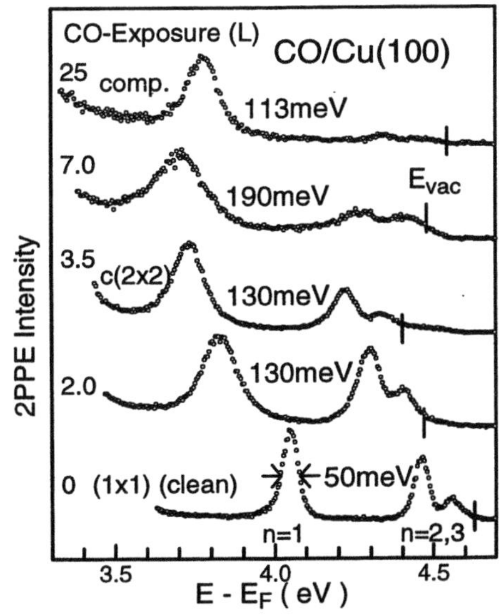

Fig. 2. Series of energy-resolved spectra as a function of CO exposure (1L = 10^{-6}Torr s). The spectra at 0 L, 3.5 L and 25 L correspond to the well-ordered (1x1), c(2x2) and compressed surface structures, respectively.

We interpret the difference between Γ and $\hbar/T_1$ in terms of pure dephasing rate, which shows pronounced minima at 0 L, 3.5 L and saturation-exposure. LEED analysis reveals that these exposures correspond to the well-ordered surface structures (1x1) (clean sample), c(2x2) and the c($7\sqrt{2} \times \sqrt{2}$)R45° compressed structure.

In conclusion, we have shown, that one can distinguish between energy relaxation and pure phase relaxation (pure dephasing) of image-potential states and that the structural order correlates with pure dephasing. We attribute the latter to quasielastic scattering due to the disorder of the surface.

References

1 Th. Fauster and W. Steinmann, in *Photonic Probes of Surfaces,* in Vol. 2 of *Electromagnetic Waves: Recent Developments in Research,* edited by P. Halevi (North-Holland, Amsterdam 1995), Chap. 8, p.347.

2 U. Höfer, I. L. Shumay, Ch. Reuß, U. Thomann, W. Wallauer, and Th. Fauster, *Science* **277**, 1480 (1997).

3 E. Knoesel, A. Hotzel, and M. Wolf, *J. Electron Spec. Relat. Phenom.* **88-91**, 577 (1998).

4 R. W. Schoenlein, J. G. Fujimoto, G. L. Eesley, and T. W. Capehart, Phys. Rev. Lett. **61**, 2596 (1988).

Excited Electron Dynamics at Benzene/Ag(111)

C.B. Harris, J.D. McNeill, N.-H. Ge, C.M. Wong, K.J. Gaffney, S.H. Liu, and A.D. Miller

Department of Chemistry, University of California, Berkeley, California 94720 and Chemical Sciences Division, Ernest Orlando Lawrence Berkeley National Laboratory, Berkeley, California 94720

Abstract. The electronic structure and excited electron dynamics of the benzene/Ag(111) interface was studied with angle- and femtosecond time-resolved two-photon photoemission.

A detailed understanding of charge transport across interfaces requires a thorough knowledge of the interfacial electronic structure. Angle- and femtosecond time-resolved two photon photoemission (TPPE) provides a means for studying interfacial electronic structure and the dynamics of electron transport across interfaces. The measurement of the lifetime, binding energy, and effective mass of excited electronic states at the benzene/Ag(111) interface extends our study of dielectric/metal interfaces to include unsaturated hydrocarbon systems.

Previous studies of dielectric/metal interfaces show that the electron affinity of the adlayer influences the spatial overlap of the excited electron with the metal substrate. Excited interfacial electrons decay via electron hole excitations in the metal, leading to lifetimes which decrease as the electron probability density in the metal increases. An overlayer with a negative electron affinity, such as n-heptane, will exclude the image state electrons from the layer, placing a majority of the electron density at the adlayer vacuum interface. As the layer width increases the electron lifetime increases exponentially because the electron must tunnel through the adlayer potential barrier [1]. Adlayers with positive electron affinity and wide conduction bands, such as xenon, form quantum well states within the layer and result in electron lifetimes that do not increase exponentially with layer thickness [2]. Table 1 shows these general trends in image state lifetimes for both n-heptane and xenon overlayers on Ag(111).

Table 1: Lifetime of Image States for Various Adsorbates

	n=1 (fs)	n=2 (fs)	n=3 (fs)
xenon monolayer	210	210	180
xenon trilayer	520	-	-
n-heptane monolayer	150	260	700
n-heptane trilayer	17 600	-	-
benzene monolayer	45	70	110
benzene 2-4 layers	100	-	-
Clean Ag(111)	30	<20	70

Angle-resolved TPPE allows the determination of the band structure parallel to the interface. Image state electrons usually exhibit free electron-like dispersion parallel to the surface, whereas localized electrons possess angle independent binding energies. Dynamic localization of initially delocalized image state electrons occurs at the n-heptane/Ag(111) interface via small polaron formation

[3]. Benzene on Ag(111) provides an opportunity to further our study of dielectric/metal interfaces because solid benzene possesses a positive electron affinity like xenon, but a narrow conduction band like n-heptane.

The coverage dependent evolution of the benzene/Ag(111) interfacial electronic structure and electron dynamics have been studied with angle and femtosecond time-resolved TPPE. Image states for a monolayer and roughly 2-4 layers of benzene on Ag(111) exhibit shorter lifetimes than those for monolayers and trilayers of xenon and n-heptane as shown in Table 1. In addition to image states, the benzene/Ag(111) interface supports benzene negative ion states as shown in Figure 1. The negative ion state labeled (A) occurs at a monolayer coverage and persists for multilayer coverages, while at higher coverages a second negative ion state, labeled (B), develops. The higher coverage negative ion has an angle independent binding energy of -0.57 eV, indicating the electron is spatially localized. Studies of the electronic structure of aromatic crystals suggest benzene solids should have an attractive electron affinity of roughly 0.5 eV and that excess electrons in crystalline benzene should form spatially localized negative ions [4]. These studies in conjunction with the results of an inverse photoemission study of 2-3 layers of benzene on a Ag(111) surface [5], concur with our assignment of the localized high coverage state as a benzene negative ion. The multilayer negative ion has a lifetime of 130 fs for 2-4 layers of benzene on Ag(111).

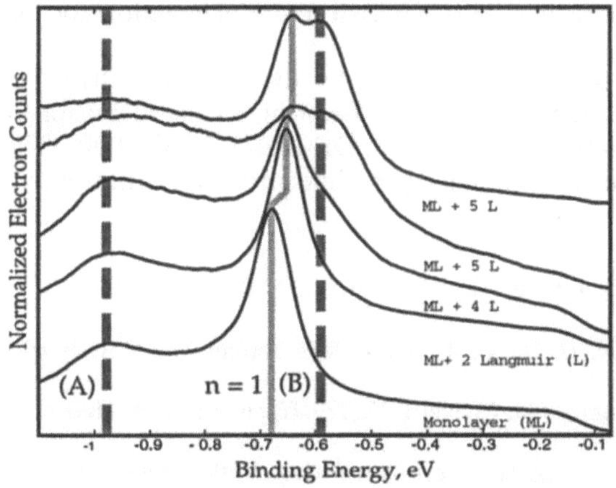

Figure 1. Coverage dependent evolution of the interfacial benzene/Ag(111) band structure from a monolayer to 2-4 layers. State (A) is the dispersive monolayer negative ion, while state (B) is the non-dispersive multilayer negative ion.

Two factors lead to assigning the feature (A) in Figure 1 as the benzene monolayer negative ion despite the state's free electron-like effective mass of 0.6, indicative of a free electron-like state. Firstly, the binding energy of the state lies within the Ag(111) surface projected band gap, eliminating the possibility of a Ag(111) derived state. Secondly, the purportedly weak interaction between benzene and silver lessens the likelihood that the state results from a mixing of benzene and silver electronic states [6]. The assignment does not account for the dispersive nature of the excitation, but that could result from the unique structure of the benzene monolayer. In the monolayer, the molecular planes of nearest neighbor benzene molecules are parallel to one another [6], while in the benzene crystal the molecular planes are perpendicular to one another [7]. This change in nearest neighbor orientation will increase the electronic wavefunction overlap between adjacent molecules and increase the band width parallel to the surface. Whether the extent of this increase accounts for the effective mass of the monolayer negative ion remains unclear. The multilayer structure will be dominated by benzene-benzene interactions and resemble the structure found in the benzene crystal.

The monolayer negative ion lifetime could not be measured directly because of its short duration, but linewidth analysis gives a lifetime of 4-10 fs. This lifetime is an order of magnitude shorter than the lifetime of the higher coverage negative ion. This difference in lifetime supports the conclusion that the monolayer negative ion is closer to the metal than the higher coverage negative ion. The presence of an unoccupied state below the image states provides a decay channel for image state electron relaxation not present for xenon or n-heptane. This could explain the shorter lifetime for image states at the benzene/Ag(111) interface.

The electronic structure and dynamics of excited electronic states at the benzene/Ag(111) interface have been studied. The image state lifetimes are significantly shorter than those for n-heptane/Ag(111) and xenon/Ag(111). The benzene/Ag(111) interface supports two states which have been assigned as benzene negative ions, one for monolayer coverage and one for higher benzene coverages. The unusual electronic structure of the monolayer negative ion could result from the unique crystal structure of the monolayer. Multilayer benzene supports a negative ion state with a binding energy and dispersion consistent with the results of previous studies. More conclusive assignment of the monolayer negative ion state would appear to require a band structure calculation for the benzene/Ag(111) interface.

Acknowledgments. This work was supported by the Director, Office of Energy Research, Office of Basic Energy Sciences, Chemical Sciences Division of the U.S. Department of Energy, under Contract No. DE-AC03-76SF00098. The NSF supported specialized equipment used in the experiments described herein.

References

[1] R.L. Lingle, Jr., N.-H. Ge, R.E. Jordan, J.D. McNeill, and C.B. Harris, 1996. *Chem. Phys.* **205**:191, **208**:297.

[2] J.D. McNeill, R.L. Lingle, Jr., N.-H. Ge, C.M. Wong, R.E. Jordan, and C.B. Harris, 1997. *Phys. Rev. Lett.* **79**:4645.

[3] N.-H. Ge, C.M. Wong, R.L. Lingle, J.D. McNeill, K.J. Gaffney, and C.B. Harris, 1998. *Science* **279**: 202.

[4] E.A. Silinsh, Organic Molecular Crystals , (Springer-Verlag, Berlin Heidelberg 1980), p.47.

[5] K.H. Frank, P. Yannoulis, R. Dudde, and E.E. Koch, 1988. *J. Chem. Phys.* **89**:7569.

[6] Ph. Avouris and J.E. Demuth, 1981. *J. Chem. Phys.* **75**:4783.

[7] D.E. Williams and Y. Xiao, 1993. *Acta Cryst.* **A49**:1.

New Directions in Nonlinear Surface Spectroscopy Enabled by Ultrafast Lasers

P. T. Wilson, Y. -S. Lee, Y. Jiang, R. Kempf, M. C. Downer

Physics Department, University of Texas at Austin, Austin, TX 78712
Tel: (512) 471-6054, Fax: (512) 471-9637
Email: downer@physics.utexas.edu

Abstract. We demonstrate Fourth Harmonic Spectroscopy below the damage threshold and real-time parallel measurement of Second Harmonic Spectra from semiconductor surfaces using amplified and broad bandwidth femtosecond pulses.

The contribution of femtosecond lasers to Second Harmonic Surface Spectroscopy has been recognized and exploited by many investigators. We demonstrate two new directions in nonlinear surface spectroscopy, each enabled by a unique feature of femtosecond pulses.

1. Fourth Harmonic Generation

The high peak intensity (~10^{11} W/cm^2) of Ti:Sapphire laser pulses amplified to μJ energy enables the first demonstration of Fourth Harmonic Generation (FHG) in reflection from crystalline surfaces below the damage threshold.[1] Our results[1] show that FHG possesses superior surface specificity and sensitivity to ≥ 4-fold surface anisotropies compared to SHG. Analysis of azimuthal FHG anisotropy from the surface of a noncentrosymmetric crystal (GaAs(001)) for 4 polarization configurations (Fig. 1a) reveals that the FH signal originates from a mixture of bulk and surface dipole-allowed nonlinear polarizations. For example, for P-in/P-out, the 8-fold-anisotropic FH intensity component originates purely from the 4mm surface (which responds isotropically in SHG), while the 2-fold-anisotropic component originates from the cubic bulk. Surprisingly, the former is nearly as strong as the latter, in marked contrast to much weaker surface/bulk ratios typical of SHG from GaAs.[2] This raised the intriguing possibility that FHG from a centrosymmetric crystal , where the bulk contribution is dipole-forbidden, might possess unprecedented surface-specificity compared to SHG. FHG from Si (001) (Fig. 1b) bears out this possibility.[1b] Although a bulk quadrupole contribution exists in theory, the absence of S-in/S-out FHG shows it is at least 10x weaker relative to surface dipole contributions than in SHG. Moreover, FHG for other polarization configurations, unlike SHG, is consistent with purely surface contributions. Extreme sensitivity to surface roughness has also been demonstrated. Thus FHG from Si(001) is the most purely surface-specific optical probe ever reported.

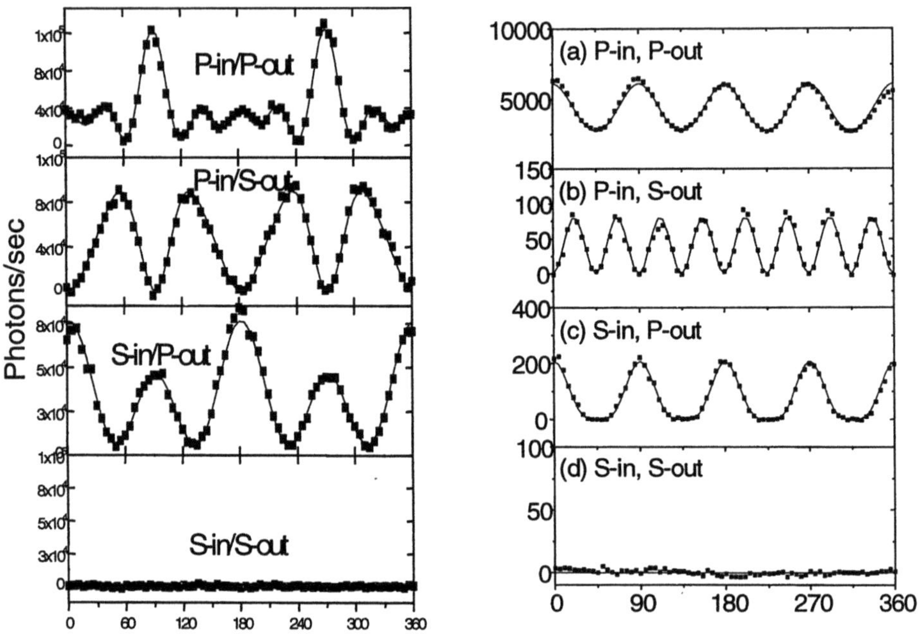

Figure 1 Azimuthal anisotropy of Fourth Harmonic Generation from **a)** GaAs. and **b)** Si

2. Broad Bandwidth Second Harmonic Generation

Because the spectral bandwidth of sub-20 fs lasers exceeds that of major surface critical point features such as the E_1 resonance, the reflection SHG spectra of these features can now be acquired in parallel, using a spectrometer on the detection side, rather than in serial fashion by slow manual tuning of a narrow bandwidth laser and single-element SHG detection. Alterations to the E_1 SH resonance in Si, acquired in the latter fashion,. have been used to diagnose interface roughness, strain, chemical composition, band bending and other key characteristics, but required >30 minutes data acquisition time.[3] Figure 2 demonstrates the E_1 SH spectra of $Si_{1-x}Ge_x$ alloys acquired in ~ 1 second, using unamplified 15 fs fundamental pulses from a Ti:S laser, and clearly show the redshift of E_1 with increasing Ge content. With this new capability, SHG spectroscopy for the first time acquires real-time capability equivalent to that of Spectroscopic Ellipsometry. The large spectral bandwidth also allows the measurement of the phase difference of two SH signals through frequency domain interference. Figure 3 shows the measured phase difference of SH from Si, relative to a GaAs reference, across the spectral range of the E_1 critical point.

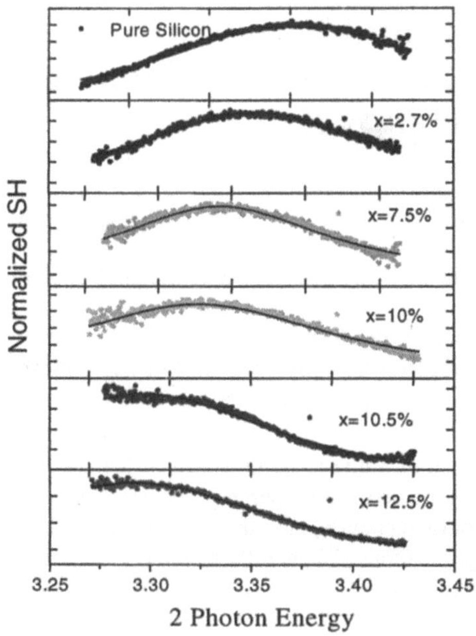

Figure 2. SH spectra of $Si_{1-x}Ge_x$ acquired in parallel with unamplified15 fs Ti:Sapphire laser

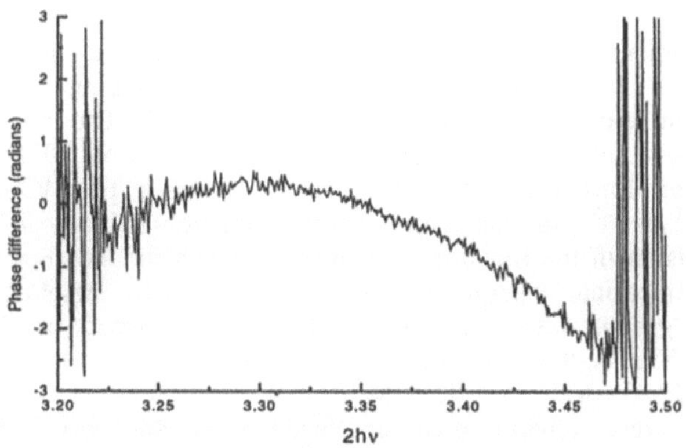

Figure 3. Phase difference of SH from Si and GaAs measured by SH spectral interferometry

References

1. Y. –S. Lee and M.C. Downer, *Opt. Lett.*, **22**, 973 (1997), Lee and Downer submitted to publication
2. C. Yamada and T. Kimura, *Phys. Rev. B*, **49**, 14,372 (1994)
3. C. Meyer et al., *Phys. Rev. Lett.*, **74**, 3001 (1995), G. Lupke et al., Opt. Lett., **20**, 1997 (1995)

Ultrafast broadband (0.5-2.5eV) absorption spectroscopy of a one-dimensional mixed-valence metal complex

Atsushi Sugita[1], Masahiro Yamashita[2], and Takayoshi Kobayashi[1]

[1] Department of Physics, University of Tokyo, 7-3-1, Hongo, Bunkyo-ku, Tokyo, 113-0033, JAPAN
E-mail: sugita@femto.phys.s.u-tokyo.ac.jp
[2] Division of Informatics for Sciences, Graduate School of Human Informatics, University of Nagoya, Furo-cho, Chikusa-ku, Nagoya, 464-0814, JAPAN

Abstract. Ultrafast dynamics of a quasi-one-dimensional halogen bridged mixed valence metal complex (HMMC) is studied by the pump-probe spectroscopy extending over 0.5-2.5 eV. Just after the photo-excitation, non-thermal self-trapped excitions (STE's) are generated as well as charged solitons. The non-thermal STE's have two relaxation channels to thermal STE's and spin solitons.

1. Introduction

Quasi-one-dimensional halogen-bridged mixed valence metal complexes (HMMC's) have been attracting an interest as compounds which are well described by an extended Peierls-Hubbard model [1]. The one dimensional structure consists of alternating transition metal ions ($M^{3+} = Ni^{3+}$, Pd^{3+}, or Pt^{3+}) and bridged halogen ions ($X^- = Cl^-$, Br^-, or I^-). Among HMMC's, complexes containing Pd^{3+} or Pt^{3+}, the electron-phonon interaction is larger than the on-site Coulomb repulsion in the transition metal ions, which induces Peierls instability resulting in fractional charge transfers between two neighboring metals. Therefore, the ground state of the materials shows an aspect of charge density wave (CDW). That is, the structure is described as

$$—M^{3+\rho}—X^- \cdots\cdots M^{3-\rho} \cdots\cdots X^-—M^{3+\rho}—X^- \cdots\cdots.$$

Here, M and X are a transition metal and bridged halogen and ρ is the amount of the charge transfer. When photo-excited, charge transfer (CT) free excitons or free electron-hole pairs are generated depending on pump photon energy and they relax to various transient species, like self-trapped excitions (STE's), solitons, and polarons. In this paper, the formation and relaxation dynamics of these species are discussed in the ultrafast time region.

2. Experimental Methods

Femotosecond time-resolved photoinduced absorption (PA) spectra are measured by the pump-probe method in the visible to near-infrared region.

Femtosecond light pulses are generated by a ML Ti:sapphire laser and the pulse energy is amplified by the Ti:sapphire regenerative amplifier system (0.8 μm, 200 fs, and 0.4 mJ). The one part of the pulse is frequency doubled (λ =0.4 μm (3.1 eV)) in a LiB_3O_5(LBO) crystal to photoexcite a sample. For measurement in 0.5 - 1.6 μm (2.4 - 0.8 eV), probe and reference pulses are the white light continuum in carbon tetrachloride (CCl_4) by the self-phase modulation effect. The broadband difference-frequency-generation (DFG) in a $KTiOPO_4$ (KTP) crystal is exploited for wavelength λ > 1.6 μm [2]. The fundamental pulse and the white light continuum pulse are collinearly focused into the KTP crystal to produce the light with their difference frequency. In this experimental configuration, the wavelength of the DFG can be tuned between 1.6 and 2.9 μm by rotating the crystal and its bandwidth is about 700cm^{-1}. The probe and reference spectra are measured by a platinum silicon multichannel infrared detector in λ > 1.2 μm and by a MCPD detector in λ = 0.5 - 1.2 μm. The sample is an oriented thin film of $[Pt(en)_2][Pt(en)_2Cl_2]\ (ClO_4)_4$ (en=ethylenediamine), which is one of the most typical HMMC's, and it is thereafter abbreviated as Pt-Cl.

3. Results and discussion

Figure 1 shows the time dependence of the PA, ΔA , at 0.60, 1.20, 2.10 eV. The ΔA is found to consist of four components as in equation (1).

$$\Delta A(t,v) = A(v)\cdot \mathrm{erf}\left(\sqrt{\sigma}\cdot t\right) + B(v)\cdot\left(1-\exp(-t/\tau_1)\right) \\ + C(v)\cdot\exp(-t/\tau_2) + D(v)\cdot\exp(-t/\tau_3) \qquad (1)$$

where parameters are evaluated as σ^{-1}=1.3±0.2 ps, τ_1=0.9±0.3 ps, τ_2=1.0± 0.1 ps, and τ_3=90 ± 20 ps. In the equation (1), two exponential decay components, $C(\nu)$ and $D(\nu)$, are attributed to the relaxation of the non-thermal and thermal self-trapped excitons respectively, in the reference to the two fluorescence lifetimes of Pt-Cl previously studied; that is, the fluorescence of the non-thermal STE's with lifetime of 270 fs - 1.2 ps [3] and that of thermal STE's with lifetime of 70 ps [4] at room temperature.

The spectrum $B(\nu)$ is isolated and found to have two characteristic peaks around 1.6 and 2.0 eV, which are similar to the PA due to the spin-soliton in a previous work [5]. In the work, it is reported that the lifetime of spin-soliton is very long over 1 sec, which is consistent to our experimental result. On the other hand, the formation process of the component is seen with time constant of 0.9 ps, and it is nearly the same with the decay time of the non-thermal STE's. Hence, the component $B(\nu)$ due to the spin-solitons are generated from the non-thermal STE's.

In the last, the component $A(\nu)$ is discussed. The first term in equation (1) corresponds to geminate recombination of a pair of photo-induced species the process that pair of photo-induced species geminatively recombine after the random walk on a one-dimensional chain. The component $A(\nu)$ is attributed to the absorption of another type of solitons, that is, charged solitons. The photo-induced absorption of the charged solitons have no formation time and this means that the solitons are generated directly from free electron-hole pairs.

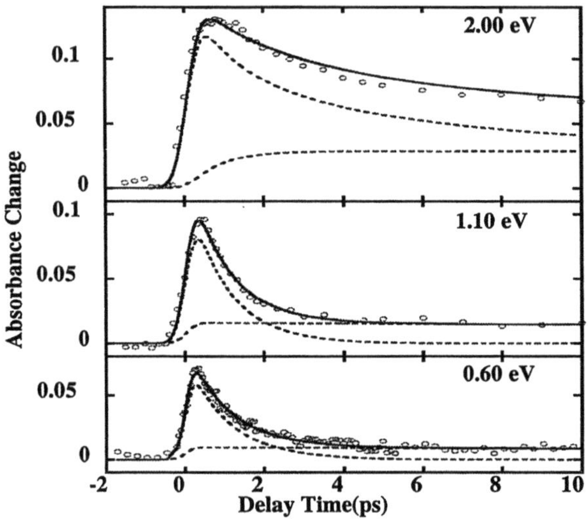

Fig. 1. Time dependence of the photo-induced absorption of Pt-Cl at three probe energy.

4. Conclusions

Ultrafast dynamics of Pt-Cl, one of the HMMC's compounds, are studied by pump-probe spectroscopy extending over 0.5 - 2.5 eV. The four transient states are found, that is, non-thermal and thermal STE's, spin-solitons and charged-solitons. The non-thermal STE's have two relaxation channels both to the thermal STE's and the spin-solitons, and the charged-solitons are generated directly from free electron-hole pairs.

References

1 K. Nasu and A. Mishima, Rev. Solid State Science, **2,** 539, 1988.
2 S. Takeuchi and T. Kobayashi, J. Appl. Phys. **75**, 2757, 1994.
3 T. Kobayashi, T. Sekikawa and M. Yamashita, Cehm. Letters 1029, 1997.
4 Y. Wada, U. Lemmer, E. O. Göbel, M. Yamashita and K. Toriumi, Phys. Rev. B **52**, 8276, 1995.
5 N. Kuroda, Y. Wakabayashi, M. Nishida, N. Wakabayashi, M. Yamashita and N. Matsushita, **79**, 2510, 1997.

Ultrafast Surface Plasmon Resonance Nonlinearities in Noble Metal Nanoparticles

Natalia Del Fatti, Marc Achermann, Fabrice Vallée, Christos Flytzanis

Laboratoire d'Optique Quantique du CNRS, Ecole Polytechnique,
91128 Palaiseau cedex, France
E-mail: delfatti@leonardo.polytechnique.fr

Abstract. Modification of the surface plasmon resonance properties induced by creation of an off-equilibrium electron distribution is investigated in silver nanoparticles. Experiments performed using a two-color 25fs pump-probe technique show that the observed short time delay (<50fs) surface plasmon resonance shift and broadening are dominated by interband absorption changes, in agreement with our theoretical model.

The optical properties of nanostructured metallic systems have been extensively investigated both for elucidation of the fundamental processes and for their technological applications. A large interest has been devoted to noble metal nanocrystals in a glass matrix whose optical properties are strongly modified by the dielectric confinement. In the visible spectral range the main feature is the appearance of a morphological resonance, the surface plasmon resonance (SPR), that can be associated to a collective electron oscillation in a nanoparticle [1]. This resonance dominates the linear and nonlinear responses of these materials. An important aspect here is the physical origin of the observed nonlinearities and the related influence of the free electron motion on the SPR.

The SPR lineshape is determined by the complex dielectric constant, $\varepsilon(\omega) = \varepsilon_1(\omega) + i\varepsilon_2(\omega)$, of the metallic crystallites. This can be written as the sum of a bound (interband) and free (intraband) electron contribution, ε^b and ε^f respectively. Using a modified Drude model for ε^f and assuming a low volume fraction of small metal nanoparticles in a transparent matrix with dielectric constant ε_m, the collective electron motion leads to a quasi-Lorentzian resonance in the composite material absorption spectrum [1] (Fig.1):

$$\alpha(\omega) = \frac{A\omega^2\Omega_R^4\gamma}{(\omega^2 - \Omega_R^2)^2 + (\Omega_R^2\gamma/\omega)^2} \tag{1}$$

where A is a constant. The SPR frequency Ω_R is related to the real part of ε^b :

$$\Omega_R = \frac{\omega_p}{\sqrt{\varepsilon_1^b(\Omega_R) + 2\varepsilon_m}}, \tag{2}$$

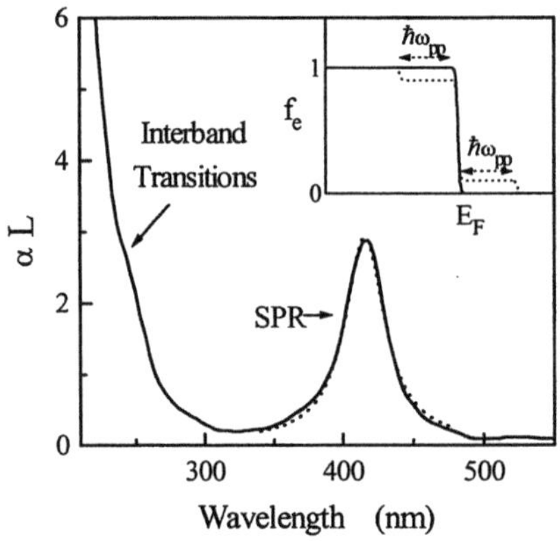

Fig. 1. Absorption spectrum of R~13nm Ag nanoparticles in a glass matrix. The inset shows the equilibrium and perturbed electron distribution.

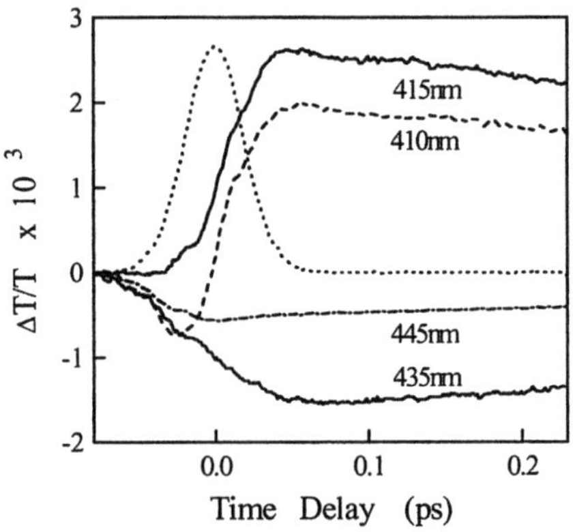

Fig. 2. Transmission change $\Delta T/T$ for four probe wavelengths around the SPR. The dotted line is the pump-probe cross correlation.

where ω_p is the bulk plasmon frequency. The SPR width, γ, is proportional to the imaginary part of the dielectric constant $\varepsilon_2(\Omega_R)$ and is, a priori, the sum of an intraband and interband electron contribution, γ_f and γ_b, respectively, with [1]:

$$\gamma_b = \frac{\Omega_R^3}{\omega_p^2} \varepsilon_2^b(\Omega_R). \tag{3}$$

γ_f is the inverse of the electron optical scattering time, $\tau(\Omega_R)$, in a nanoparticle [1]:

$$\gamma_f = \frac{1}{\tau(\Omega_R)} \approx \frac{1}{\tau_0(\Omega_R)} + \frac{v_F}{R} g(\Omega_R). \tag{4}$$

It is the sum of an intrinsic bulk-like electron scattering rate (including electron-phonon and umklapp electron-electron scattering) and a term corresponding to electron scattering off the nanoparticle surfaces (proportional to the electron Fermi velocity v_F with a proportionality factor g). Change of the electron distribution alters the metal dielectric constant and thus both the frequency and width of the SPR.

Using a two-color pump-probe technique with 25fs resolution, we have investigated the transient SPR property changes induced by selective perturbation of the free electron gas distribution in silver nanoparticles. In this system, the SPR is well separated from the interband transitions (Fig.1) and its behavior can thus be selectively studied and related to the free electron gas dynamics. Measurements were performed in the low perturbation regime as a function of the probe wavelength for nanoparticles with radius R ~ 13nm. The two pulses are created from the pulse train delivered by a 25fs Ti:sapphire oscillator tunable in

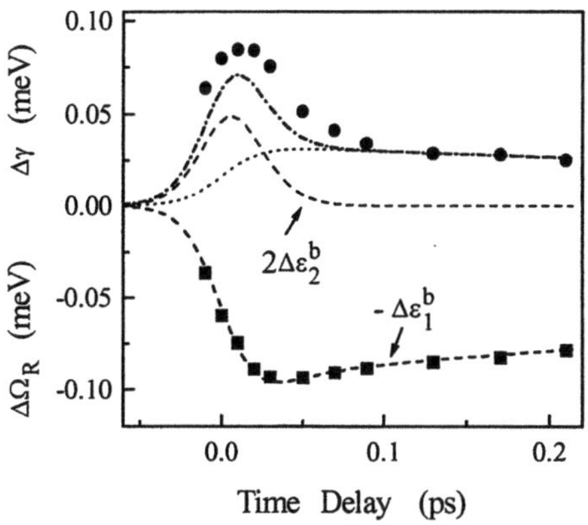

Fig. 3. Measured SPR frequency shift, $\Delta\Omega_R$ (squares) and broadening, $\Delta\gamma$ (circles). The dash-dotted line is the estimated broadening (see text).

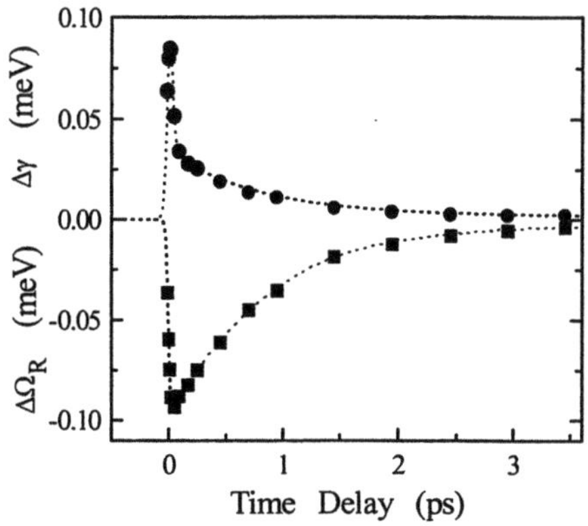

Fig. 4. Measured SPR frequency shift, $\Delta\Omega_R$ (squares) and broadening, $\Delta\gamma$ (circles), on a picosecond time scale.

the 780-920nm range. The electron distribution is driven out of equilibrium by intraband absorption of an infrared pump pulse $\hbar\omega_{pp}$. Electrons with energies down to E_F-$\hbar\omega_{pp}$ are excited from below to above the Fermi energy, creating a strongly athermal electron distribution (Fig.1, inset). This has been shown to subsequently internally thermalize by electron-electron collisions with a full thermalization time of few hundred femtoseconds [2]. The induced changes of the sample transmission, $\Delta T/T$, are probed in the vicinity of the SPR using the 28fs blue pulses created by frequency doubling part of the pulse train. The two beams are sent into a standard pump-probe setup and focused into the sample over focal spots of 40μm in diameter.

The measured transient differential transmission, $\Delta T/T$, is shown in Fig.2 for different probe photon energies $\hbar\omega_{pr}$ around the SPR frequency ($\hbar\Omega_R = 2.93$eV ≡ 425nm). The pump fluence is 100μJ/cm^2 corresponding to a maximum electron temperature increase ΔT_e ~ 160K. Similar results were obtained over the investigated pump fluence range 20-180μJ/cm^2, corresponding to ΔT_e ranging from 40K to 260K. Reflectivity changes being negligible, $\Delta T/T$ is only determined by the sample absorption change: $\Delta T/T(\omega_{pr})=-\Delta\alpha(\omega_{pr})L$, where L is the sample thickness. For very short times, absorption increases on both the high and low energy sides of $\hbar\Omega_R$, indicating a strong broadening ($\Delta\gamma$) of the SPR line. Subsequently, the absorption decreases for the short probe wavelengths and increases for the large ones, yielding evidence for an additional contribution due to a red shift ($\Delta\Omega_R$) of the SPR.

For positive probe delays, the $\hbar\omega_{pr}$ dependence of $\Delta T/T$ can be very well reproduced assuming that the SPR line stays quasi-Lorentzian (Eq.1) and is frequency shifted and broadened. Fitting is actually only sensitive to the ratio $\Delta\gamma/\Delta\Omega_R$ and, as the linear sample absorption has been independently measured (Fig.1), the absolute $\Delta\Omega_R$ and $\Delta\gamma$ values are determined by comparing the computed and measured signal amplitudes (Fig.3). The deduced red shift rises instantaneously while, in contrast, $\Delta\gamma$ exhibits a transient overshoot for probe time delays < 50fs.

On a longer time scale (> 100fs) both $\Delta\Omega_R$ and $\Delta\gamma$ decay exponentially with a time constant τ_{th} reflecting electron gas energy losses to the lattice (Fig.4). This electron-phonon effective coupling time is comparable to the one measured in silver films [3] for the largest nanoparticles ($\tau_{th} \sim 900$fs for $R \geq 5$nm) and is found to significantly decrease for the smallest one ($\tau_{th} \sim 500$fs for $R = 2$nm) [4]. As the long time scale response is discussed elsewhere [4], we will focus here on the short time scale behavior dominated by induced interband absorption changes.

As the free electron density is unchanged, the observed frequency shift is ascribed to the change of the real part of the interband contribution to the metal dielectric function, concomitant with alteration of the interband absorption spectrum (optical Kerr effect). Using Eq.2 for the small changes induced here, $\Delta\Omega_R$ can be written:

$$\Delta\Omega_R \approx -\frac{1}{2}\frac{\Omega_R^3}{\omega_p^2}\Delta\varepsilon_1^b(\Omega_R)\,. \tag{5}$$

Quantum confinement being negligible in our samples, the lowest energy interband absorption is expected to be dominated by transitions from the full d-bands to empty states around the Fermi surface, as in bulk systems [5]. Energy injection into the electron gas results in a smearing of the Fermi surface and thus to an induced (reduced) absorption for interband transitions from the d-bands to electron states below (above) the Fermi energy. Using the Kramers-Kronig relationship and taking into account that Ω_R is smaller than the interband transition threshold ($\hbar\Omega_{ib} \sim 4.1$eV in Ag (Fig.1)) one can show that the induced $\Delta\varepsilon_1^b(\Omega_R)$ is positive, resulting in a red shift of the SPR (Eq.5) in agreement with our experimental results.

For a thermal electron distribution with temperature T_e, the difference between the interband transition threshold and the SPR energy is much larger than the thermal energy, $\hbar\Omega_{ib} - \hbar\Omega_R >> k_B T_e$. Consequently, the interband absorption at the frequency Ω_R, and thus the interband induced width γ_b (Eq.3), plays a minor role as it essentially involves electron states well below E_F, that are fully occupied. The situation is different during and shortly after the excitation process when the electron distribution is strongly athermal. The occupation numbers of the electron states with energies down to $E_F - \hbar\omega_{pp}$ are then reduced (Fig.1, inset) leading to the opening of interband transitions for frequencies down to $\Omega_{ib} - \omega_{pp}$. This is smaller than Ω_R and a strong transient induced broadening of the SPR is thus expected, in agreement with the experimental results (Fig.3). This additional optically-activated broadening corresponds to SPR damping with excitation of a bound d-electron into the conduction band. As the involved final electron states have an energy much smaller than E_F, they return to equilibrium much faster than the full thermalization time of the electron gas [6] resulting in a fast decay of the interband induced broadening on a few femtosecond time scale, as observed experimentally (Fig.3).

The transient shift and broadening due to interband processes are directly related to the electron distribution. Neglecting quantum confinement effects, this can be calculated as in bulk noble metals by numerically solving the electron Boltzmann equation for free electrons [2]. The transient change of the interband term $\Delta\varepsilon_1^b$ is then calculated from the computed electron distribution using the band structure model of Rosei et al [5]. Using this approach, the computed time behavior of $\Delta\varepsilon_1^b$ is in very good agreement with that of $\Delta\Omega_R$ (Fig.3, note that the only parameter here is the normalization factor, β, for comparing the $\Delta\varepsilon_1^b$ and $\Delta\Omega_R$ amplitudes). In contrast to $\Delta\varepsilon_1^b(\Omega_R)$, $\Delta\varepsilon_2^b(\Omega_R)$ is sensitive to the local change of the electron state occupation number and exhibits a fast rise and decay (Fig.3). The observed broadening can be fairly well reproduced as the sum of contributions proportional to respectively the electron gas excess energy and $2\Delta\varepsilon_2^b(\Omega_R)$ (this latter term has been multiplied by the same normalization factor, β, as for comparing $\Delta\varepsilon_1^b$ and $\Delta\Omega_R$, see Eq.3 and Eq.5). This good agreement confirms the common interband origin of the SPR frequency shift and short time broadening.

Femtosecond investigations of the ultrafast electron dynamics have been recently reported in copper and gold nanoparticle systems, where the SPR partly overlaps the interband transition spectral region [7-10]. In these systems, for the investigated strong perturbation regime, a large interband induced broadening is thus expected. Furthermore, its temporal evolution is expected to exhibit a non instantaneous rise and a slow decay reflecting the occupation number dynamics of the conduction band electron states close to the Fermi energy that are involved in the interband transitions at the SPR frequency in these systems [2]. This is consistent with the experimental results, showing that electron distribution perturbation induces a very large broadening of the SPR with a complicated time behavior, and, comparatively a negligible frequency shift [8-10].

Acknowledgments. The authors acknowledge Y.Hamanaka and A.Nakamura for helpful discussions and S.Omi for providing the silver nanoparticle samples.

References

1. U.Kreibig, in Handbook of Optical Properties, edited by R.E.Hummel and P.Wissmann, CRC Press, New York, Vol. 2, p.145 (1997).
2. C.K.Sun, F.Vallée, L.Acioli, E.P.Ippen, and J.G.Fujimoto, Phys. Rev. B **50**, 15337 (1994).
3. R.Groeneveld, R.Sprik and A.Lagendijk, Phys. Rev. B **51**, 11433 (1995).
4. N.Del Fatti, F.Vallée, C.Flytzanis, to be published.
5. R.Rosei, Phys. Rev. B **10**, 474 (1974).
6. M.Aeschlimann, M.Bauer and S.Pawlik, Chem. Phys. **205**, 127 (1996).
7. T.Tokizaki, A.Nakamura, S.Kavelo, K.Uchida, S.Omi, H.Tanji, and Y.Asahara, Appl. Phys. Lett. **65**, 941 (1994).
8. J.Y.Bigot, J.C.Merle, O.Cregut, and A.Dannois, Phys. Rev. Lett. **75**, 4702 (1995).
9. M.Perner, P.Bost, U.Lemmer, G.von Plessen, J.Feldmann, U.Becker, M.Mennig, M.Schmitt, and H.Schmidt, Phys. Rev. Lett. **78**, 2192 (1997).
10. H.Inouye, K.Tanaka, I.Tanahashi, K.Hirao, Phys. Rev. B **57**, 11334 (1998).

Nonlinear Optical Interplay Between Different Surface Plasmons in Silver Nano-Ellipsoids

M. Perner[1], J. März[1], G. von Plessen[1], J. Feldmann[1], J. Porstendorfer[2], K.-J. Berg[2], and G. Berg[2]

[1] Lehrstuhl für Photonik und Optoelektronik, Sektion Physik, Ludwig-Maximilians-Universität, Amalienstr. 54, 80799 München, Germany
E-Mail: martin.perner@physik.uni-muenchen.de
[2] Fachbereich Physik, Martin-Luther-Universität Halle-Wittenberg, Friedemann-Bach-Platz 6, 06108 Halle, Germany

Abstract. Using femtosecond pump-probe experiments we measure the interplay between the two, perpendicularly polarized and spectrally separate, surface plasmons of silver nano-ellipsoids. Pulsed optical excitation leads to two normal-mode vibrations of the ellipsoidal particles.

1. Introduction

Noble-metal nanoparticles show large optical resonances associated with surface plasmons (SP's). After ultrashort excitation of the SP or interband transitions, the electron temperature rises, changing the linewidth and spectral position of the SP resonance [1,2]. Ellipsoidal nanoparticles with axis dimensions $a=b<c$ (so-called prolate spheroids) show two different SP's which lie at different wavelengths for light polarized parallel to the short and long axes, respectively. In the present contribution, we show that these two collective excitations influence each other via hot-electron processes. In addition, we observe two normal-mode vibrations of the ellipsoidal particles.

2. Experimental Methods

The ellipsoidal silver nanoparticles were produced by embedding spherical particles in a glass matrix and subjecting the sample to tensile deformation [3]. They have axis dimensions of $a=b\approx 42$ nm and $c\approx 102$ nm, resulting in two different SP resonances that can be excited with light polarized parallel to the short and long axes, respectively. The short-axis SP resonance (SAP) peaks at a wavelength of 405 nm, while the long-axis SP (LAP) peaks at 660 nm (Fig. 1a). Femtosecond pump and probe experiments are performed with 150 fs laser pulses that are generated from regeneratively amplified Ti:Sapphire pulses via optical parametric amplification (OPA), frequency doubling (SHG), or white-light continuum generation.

3. Results and Discussion

In the first experiment, the SAP is pumped and a white-light pulse probes the response of the LAP. This response consists mainly in a transient red-shift of the LAP, as shown by the differential transmission spectrum in Fig. 1b taken at a delay time of 2 ps between the pump and probe pulses. The red-shift of the LAP is explained as follows: The decay of the SAP into electron-hole pairs causes a transient electron heating, which results in changes in the interband and intraband susceptibility that shift the SP energy. Similar temperature-induced red-shifts (not shown here) are observed when pumping the LAP and probing the SAP, or when pumping and probing the same SP resonance.

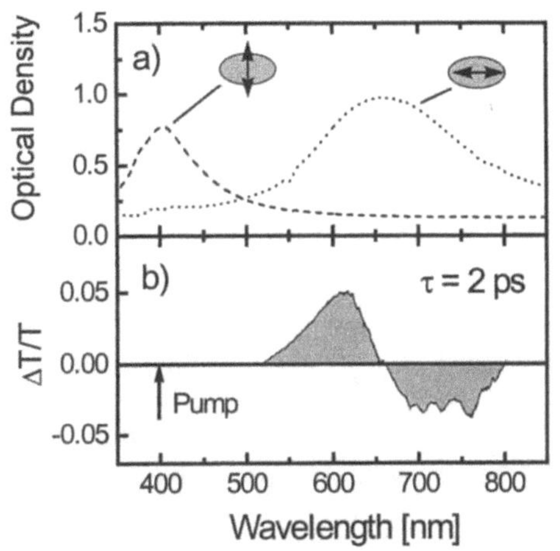

Fig. 1: (a) Extinction spectra taken with perpendicular light polarizations as indicated by the arrows. (b) Differential transmission spectrum. The arrow indicates the pump wavelength.

The red-shift allows us to monitor the time dependence of the electron temperature after excitation. Fig. 2 shows two differential transmission transients, taken for excitation of the SAP; the upper trace reflects the transient red-shift of the SAP when probed at 550 nm, while the lower one shows that of the LAP probed at 700 nm. During the first ps after excitation, an initial rapid increase of the red-shift is observed, which reflects the thermalization of the non-equilibrium electron-hole distribution resulting from the decay of the optically excited SAP. The red-shift subsequently undergoes a two-step decay. During the first 5-10 ps the hot electrons cool down by emitting phonons, thereby heating the lattice rapidly. In a second step, which takes hundreds of ps, the nanoparticle cools to room temperature, via heat transfer to the surrounding matrix.

An interesting observation is that the signal in the second decay step exhibits periodic modulations, similar to the signal beats recently observed on Ga and Sn nanoparticles [4]. A striking difference, however, is that here the time period of the modulation depends on the nanoparticle axis probed. A time period of $\approx$23 ps is observed when probing the SAP (upper trace in Fig. 2), while a period of $\approx$47 ps is observed when probing the LAP (lower trace). We argue that these anisotropic modulations are associated with particle vibrations, for the following reasons: (i) The lattice heating during the first $\approx$5 ps after the electron thermalization leads to an expansion of the silver lattice and thus to an increase of the particle size. The spectral position of each SP resonance depends sensitively on the particle length along the corresponding axis, hence the temperature-induced length increase results in red-shifts of the SP's. For the present excitation

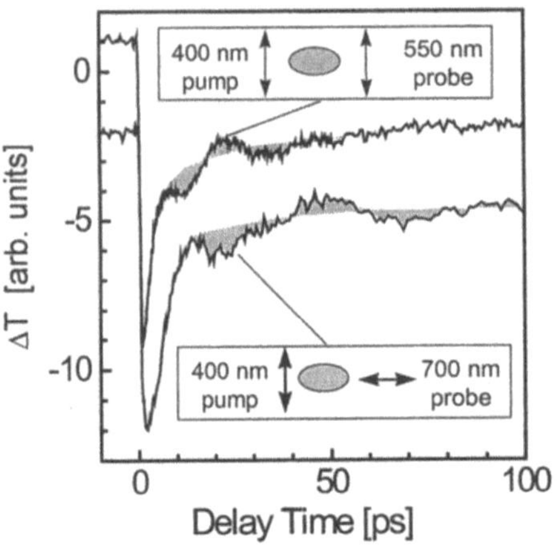

Fig. 2: Time dependence of the pump-probe signal for two perpendicular probe polarizations as indicated by the arrows. Transients are vertically offset for clarity. Shaded areas underline the signal modulation.

conditions, we calculate a length increase of ≈0.5%, resulting in a transmission change of ≈2.5%; this value is in agreement with the observed modulation amplitudes. (ii) Any temperature-induced changes of the particle size are based on lattice vibrations. Here the vibrational modes are energetically discrete due to the finite particle size; a simple analysis in terms of standing acoustic waves yields time periods of 22 ps and 55 ps for vibrations along the short and long particle axes, respectively. These time periods are much longer than the time taken by the heating process (≈5 ps). This means that the particle lengths cannot immediately follow the increase of the lattice temperature, and initially will be smaller than the new equilibrium lengths at elevated temperature. When the lengths increase, they will overshoot the new equilibrium values, thus initiating particle vibrations around these values, with time periods as estimated above. In fact, a comparison shows that the estimated time periods are in good agreement with the experimentally observed modulation periods.

4. Conclusion

In conclusion, we show in femtosecond pump-probe experiments the nonlinear interplay between the two perpendicularly polarized and spectrally separate SP's of ellipsoidal silver particles. The generation of a hot-electron distribution due to SP decay into electron-hole pairs causes a red-shift of the SP resonances. Furthermore, we observe polarization-dependent modulation periods in pump-probe transients caused by particle vibrations along their principal axes.

References

1 M. Perner et al., *Phys. Rev. Lett.* **78**, 2192, 1997
2 Y. Hamonoka, N. Hayashi, A. Nakamura, S. Omi, *J. Lumin.* **76&77**, 221, 1997
3 R. Borek, T. Rainer, K.-J. Berg, DE patent 19502321, 1996
4 M. Nisoli et al., Phys. Rev. B **55**, 13424 (1997).

Femtosecond spectroscopy of silver nanoparticles in different dielectric matrixes

Valérie Halté[1], Bruno Palpant[2], B. Prevel[3], Jean-Claude Merle[1], Michel Broyer[2], Alain Perez[3], and Jean-Yves Bigot

[1] Institut de Physique et Chimie des Matériaux de Strasbourg
Groupe d'Optique Non-Linéaire et d'Optoélectronique
23 rue du Loess, 67037 Strasbourg Cedex, FRANCE
E-mail: bigot@valholl-u.strasbg.fr
[2] Laboratoire de Spectrométrie Ionique et Moléculaire,
[3] Département de Physique des Matériaux
Université C; Bernard Lyon I, 69622 Villeurbanne Cedex, FRANCE

Abstract. The dynamics of electrons heated with femtosecond optical pulses is investigated in silver nanoparticles. The electron-lattice relaxation is compared for particles in various surrounding matrixes.

1. Introduction

The relaxation of energy in metal clusters excited by femtosecond optical pulses displays several features that are not present in bulk material. It has been shown for instance that the collective modes of the electron gas are damped by the electron–electron scattering. This mechanism manifests as a time dependent spectral broadening of the surface plasmon (SP) mode in the case of Cu and Au nanoparticles [1, 2] and as a shift of the SP resonance in the case of Ag [3]. The transfer of energy to the lattice may also be size dependent through electron and surface modes coupling and is expected to rise with decreasing size. The above processes are intrinsic to the metallic nanoparticle and do not take into account the heat transfer to the surrounding medium. Here we investigate this mechanism in silver nanoparticles embedded either in a dielectric matrix made of glass or porous Al_2O_3. In addition, the electron relaxation in particles of different sizes is reported.

2. Experiments

Time resolved measurements have been performed using the degenerate pump-probe technique. A tunable, amplified and frequency doubled Ti : Sapphire laser provides pulses of 150 fs duration. The pump and probe are crossed polarized to avoid their coherent coupling in the material and the maximum energy density of the pump pulse is ~2mJ/cm^2. Two different types of silver nanoparticles have been studied. The first ones, which are prepared by a nucleation and growth process, are embedded in a silicate glass matrix (M1) with an average diameter of 5 nm and display a SP resonance at 434 nm. The second ones, which are produced by

the low-energy cluster-beam deposition technique (LECBD) [4], are encapsulated in an Al_2O_3 matrix (M2), have a diameter, which ranges from 3 to 7 nm, and SP resonance is situated around 425 nm. We have measured time resolved differential transmission ΔT/T as a function of size particle for a given matrix and also for two different matrices with a given size. Each graph in figure 1 shows ΔT/T for particles of ~5 nm diameter embedded in M1 and M2 and for identical pump excitation conditions (wavelength and incident pump power). These curves exhibit a relaxation time τ_1, corresponding to the cooling of the electrons to the lattice. It is slower for the matrix M1 (1.7 ps) than for the matrix M2 (0.77 ps). One can observe a similar behavior near the SP resonance as seen in Fig. 1 (b).
These results clearly indicate that the electronic energy relaxation is faster when the nanoparticle is surrounded by Al_2O_3 matrix. We attribute this effect to an energy transfer to the environment. The higher thermal conductivity of Al_2O_3 (κglass $\approx$ 0.01 and $\kappa Al_20_3 \approx 0.3$ W.cm^{-1}K^{-1}) provides an efficient channel for the heat dissipation out of the metal particle. This mechanism is responsible for the residual signal observed at long time delays ($\geq$ 5 ps) in the glass matrix. For such delays the electronic and lattice temperatures are in equilibrium and the dynamics is essentially determined by the heat propagation in the environment.

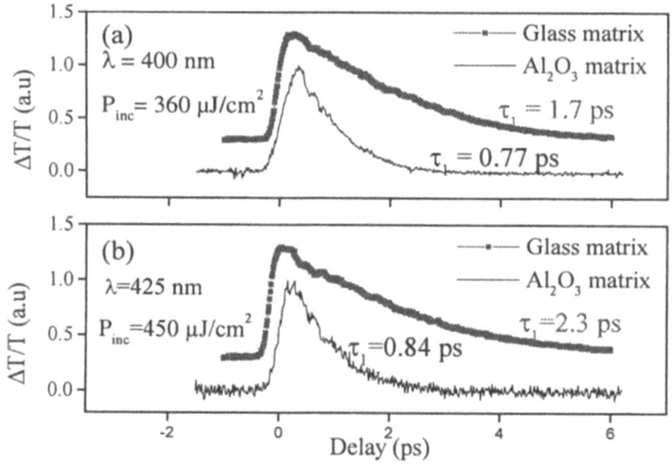

Fig. 1. ΔT/T for the ~5 nm nanoparticles embedded in the glass (plotted with a vertical offset) and Al_20_3 matrixes and for different conditions of excitation : (a) λ=400 nm, P_{inc}=360 μJ/cm^2 and (b) λ=425 nm, P_{inc}=450 μJ/cm^2

It is important to mention that this mechanism is not hindered by the density dependent heat capacity of the electrons. In figure 2 (a), the exponential relaxation times τ_1 are plotted as a function of pump energy density for the two matrixes. It shows that τ_1 increases with increasing pump fluence (as expected from the temperature dependent electronic heat capacity) but it is higher for the glass embedded particles over the two orders of magnitude of pump fluence that we explored.

Figure 2 (b) shows the variation of the relaxation time for different particle sizes embedded in the alumina matrix. τ_1 decreases with decreasing size. Two mechanisms can be attributed to this observation. The first one corresponds to a coupling of the electrons with surface phonons. This effect becomes more important as the particle size decreases as suggested by Nisoli *et al.* in the case of gallium [5]. The second mechanism is a heat transfer from the nanoparticle to the matrix, an effect that is also more efficient as the ratio of the surface to volume increases.

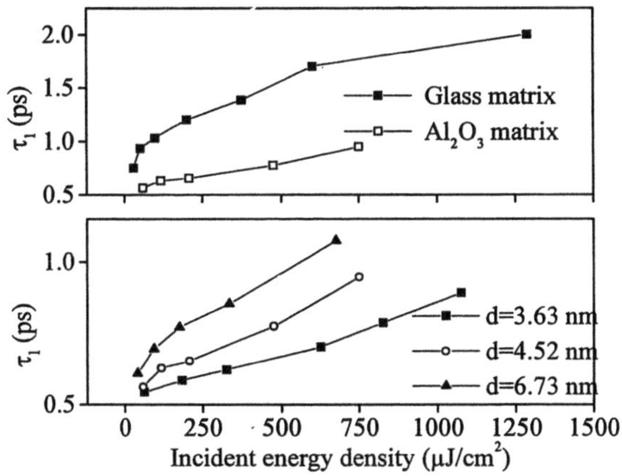

Figure 2 : Relaxation time τ_1 as a function of the energy density of excitation for the matrixes M1 and M2 (a) and for different sizes in matrix M2 (b)

3. Conclusions

In these experiments, we have studied two remarkable effects in the dynamics of silver nanoparticles. First, in addition to the intrinsic mechanisms involved in the electronic energy relaxation in the subpicosecond time scale, we have stressed the importance of an extrinsic mechanism associated to energy transfer to the matrix. Secondly, the relaxation is faster in particles with smaller size. It shows that, for small enough clusters (with diameters typically less than a few nanometers), an appropriate engineering of the environment allows to slow down or accelerate the energy relaxation on the picosecond time scale.

References

1 J.-Y. Bigot *et al*, Phys. Rev. Lett. **75**, 4702 (1995).
2 M. Perner *et al.* Phys. Rev. Lett. **78**, 2192 (1997).
3 V. Halté *et al.*, OSA Technical Digest Series (1998).
4 B. Palpant *et al.*, Phys. Rev. B **57** (3), 1963 (1998).
5 M. Nisoli *et al.*, Phys. Rev. Lett. **78**, 3575 (1997).

Ultrafast Relaxation Processes in Au Quantum Dots at High Pulse Repetition Frequencies

R. F. Haglund Jr.,[1] G. Lüpke,[1] D. H. Osborne,[1] H. Chen,[1] R. H. Magruder III,[2] and R. A. Zuhr[3]
[1]Department of Physics and Astronomy, Vanderbilt University, Nashville TN 37235
[2]Department of Physics and Chemistry, Belmont University, Nashville, TN 37235
[3]Solid-State Division, Oak Ridge National Laboratory, Oak Ridge, TN 37831

Abstract. We employed pump-probe, transient-grating and Z-scan measurements to distinguish electronic from thermal relaxation processes in metal quantum dots in a glass matrix as functions of dot size and pulse-repetition frequency. Novel observations include a negative-time probe-pump transient and an extremely long-lived relaxation which produces a thermal background above a few MHz pulse repetition frequency.

The electronic and vibrational relaxation mechanisms of metal quantum-dots embedded in dielectric matrices following ultrashort-pulse laser excitation is a topic of intense current interest.[1] The electronic dephasing process in metal quantum dots occurs on a time scale of 10 fs,[2] while the electron-phonon (e-p) coupling in the quantum dot, and the phonon-phonon (p-p) coupling between the dot and the matrix, are typically picoseconds and tens to hundreds of picoseconds, respectively. Among the critical issues are variations in these time constants with quantum-dot size and pulse-repetition frequency (prf).

We have measured these time constants for Au nanocrystals embedded in fused silica, using pump-probe (PP), transient-grating (TG) and Z-scan (ZS) techniques. A cw mode-locked, cavity-dumped dye laser, tunable from 565-610 nm, provided 0.7-1.0 ps pulses at prfs from 0.24 to 38 MHz. Samples were prepared by multi-energy ion implantation at beam energies of up to 3.75 MeV. This process[3] yields thin (200-300 nm) layers of randomly-distributed quantum dots with relatively uniform spatial and dot-size distributions, depending on annealing protocol, at volumetric concentrations of a few *per cent.*

Figure 1 shows PP traces for Au dots in SiO_2 (average diameter 12-15 nm) as a function of prf at 575 nm, to the red side of the Au surface-plasmon resonance (526 nm) and the interband transition (533 nm). The e-p relaxation time varies from about 9 ps at 3.8 MHz prf to 8 ps at 230 kHz, consistent with the pulse energies and electron temperatures in our experiment. The p-p coupling time is much longer, on the other hand, and we observe PP signal more than 200 ps after the initial excitation. Both time constants are substantially longer than those

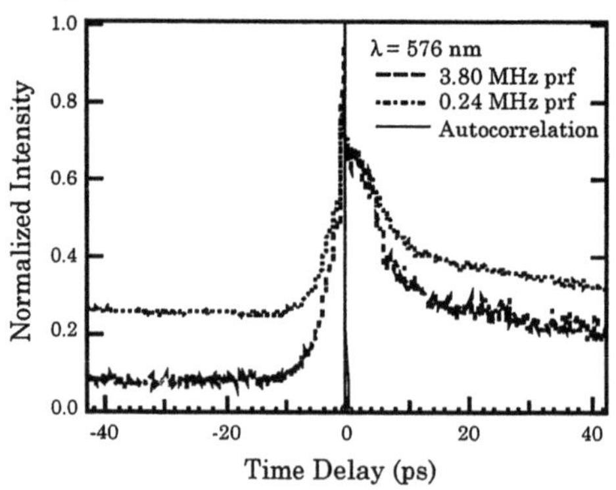

Figure 1. PP and autocorrelation traces of Au quantum dots in SiO_2, at two repetition rates.

observed in 1 kHz PP experiments, indicating that long-term heating of the matrix affects the e-p coupling by changing the temperature of the quantum dot prior to the arrival of the next pump pulse.

TG measurements, on the other hand, show a size-dependent e-p relaxation time with little sign of a p-p relaxation tail (Figure 2) or the rising thermal background seen in Fig. 1. When the pump wavelength was tuned further to the red of the plasmon resonance, the e-p relaxation time was reduced somewhat for a given dot size. The appearance of an e-p relaxation signal in our measurements contrasts with previous TG studies of Au nanoparticles in liquids at prfs in the MHz range[4] in which e-p coupling was masked by the long 5-ps pulses. On the other hand, those p-p relaxation times are similar to our measurements, consistent with the thermal conductivities of the matrix. This suggests that dramatic reductions in the pp-relaxation time might be achieved by embedding quantum dots in matrix materials with higher thermal conductivities, such as Al_2O_3.

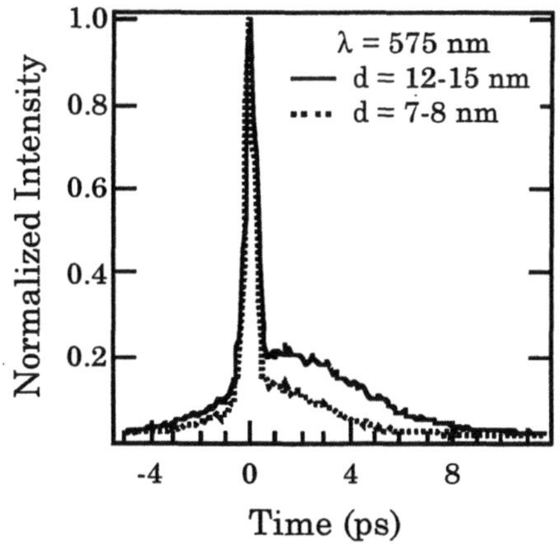

Figure 2. TG signal from Au nanocrystals of differing sizes embedded in SiO_2.

Both in Figure 1 and Figure 2 we observe negative-time-delay signals which are much broader than the measured autocorrelation widths. This phenomenon has been observed by others,[5] and arises from the thermal decay of the probe signal as it interacts with the pump pulses.

The inhibited phonon-phonon coupling between the quantum dots and the host matrix is a critical barrier to the application of metal-quantum-dot materials in all-optical switching schemes, because long-lived thermal relaxation can mask high-speed electronic switching. While model calculations of this effect[6] show reduced thermal conductivity with diminishing nanocrystal size, the calculated effects are smaller than those measured in semiconductor-doped glasses.[7]

In addition to the PP measurements shown in Fig. 1, we have studied this effect using the Z-scan , a single-beam technique which monitors changes in the radial intensity profile of a laser beam transmitted through a nonlinear material as it is scanned through the focal plane (z=0) of a fixed lens.[8] Measuring the variation in the center of the transmitted beam with a small-aperture detector in the far field (FF) gives the combined effects of nonlinear absorption and refraction, while measuring the entire beam in the near field (NF) yields the nonlinear absorption alone. In our experiments, Au nanocrystals were irradiated at varying prfs, and the measured NF and FF profiles were fitted to extract values for the nonlinear absorption and the electronic and thermal components of the nonlinear refractive index.

Figure 3 clearly shows that the thermal component of the nonlinear index of refraction n_2(th) increases gradually with average power, whereas the electronic component n_2(el) remains unchanged. This is consistent with the PP data shown in

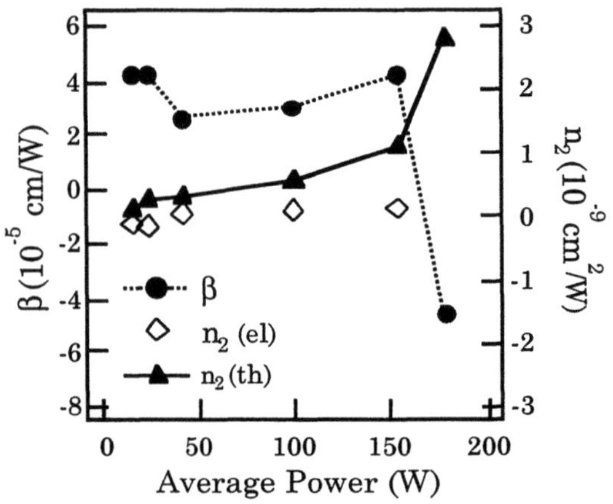

Figure 3. Nonlinear refractive index and absorption as a function of average laser power for Au in SiO_2.

Figure 1. The nonlinear absorption β shows a sharp change above a threshold of about 150 mW indicating the onset of strong thermal distortion in the sample above a prf of 7.6 MHz The sign change in β marks an alteration in the optical physics, from nonlinear absorption to nonlinear saturation.

In conclusion, we have observed picosecond e-p relaxation times for Au quantum dots irradiated at high pulse repetition frequencies. The times are consistent with measurements by other workers, but show some evidence for a modest size dependence in the e-p coupling. The potential for applying quantum-dot materials in optical switching and sensing, however, is likely to be dominated by pulse distortion and long-term thermal relaxation processes observed at high prfs. These have yet to be studied in detail in these materials, and will be the subject of future experiments.

Acknowledgement. We thank the Army Research Office (DAAH04-93-G-0123) and the Vanderbilt University Research Council for support, and the Alexander von Humboldt Foundation for a Fyodor Lynen Fellowship (GL). Oak Ridge National Laboratory is partially supported by the U. S. Department of Energy under contract DE-AC05-96OR22464 with Lockheed-Martin Energy Systems, Inc.

REFERENCES

1. T. Tokizaki *et al.*, Appl. Phys. Lett. **69**, (1994). J.-Y. Bigot *et al.*, Phys. Rev. Lett. **75**, 4702 (1995). M. Perner *et al*, Phys. Rev. Lett. **78**, 2192 (1997). M. Nisoli *et al.*, Phys. Rev. Lett. **78**, 3575 (1997).
2. T. Klar *et al.*, Phys. Rev. Lett. **80**, 4249 (1998).
3. R. H. Magruder, III *et al.*, J. Appl. Phys. **76**, 708 (1994).
4. K. Puech *et al.*, Opt. Lett. **20**, 1613 (1995).
5. H. J. Bakker and H. Kurz, Phys. Rev. B **50**, 7805 (1994). A. Harten *et al.*, Phys. Rev. Lett. **69**, 852 (1992).
6. G. Chen, J. Heat Transfer **118**, 527 (1996).
7. P. A. M. Rodrigues *et al.*, J. Appl. Phys. **80**, 5963 (1996).
8. M. Sheik-Bahae *et al.*, IEEE J. Quantum Electron. **QE-26**, 760 (1990)

Femtosecond demagnetization of ferromagnetic $CoPt_3$ thin films and Ni nanoparticles

J.-C. Merle, E. Beaurepaire, V. Halté, G. Garreau, A. Daunois and J.-Y. Bigot

Institut de Physique et Chimie des Matériaux de Strasbourg
Groupe d'Optique Nonlinéaire
23 rue du Loess, 67037 Strasbourg Cedex, France,
E-mail : bigot@valholl-u.strasbg.fr

Abstract. We have studied the spin dynamics in $CoPt_3$ thin films and Ni nanoparticles. It is shown that a ferro-paramagnetic transition can be induced in a few hundreds of femtoseconds. In addition, the spin relaxation in the nanoparticles displays a super-paramagnetic behavior.

1. Introduction

The understanding of the electron dynamics in metals has benefited from the techniques used in ultrafast spectroscopy, where either the optical properties [1] or the electron photoemission [2] can be time resolved. Until now, most of the experiments concern effects associated with the occupation densities of electrons without considering their spin relaxation. Such spin dynamics is of major importance in magnetic materials. For instance, it has been shown recently that thin films made of a transition metal can be partially demagnetized by an incident femtosecond pump pulse [3]. This process occurs in the femtosecond time scale, that is much shorter than the characteristic time of the spin-lattice interaction. It raises challenging questions for the understanding of the dynamics of magnetism in metals since, on this time scale, the Coulomb exchange, the spin-orbit interaction and the photon polarization must be considered. In the present paper we report new results on the spin dynamics in $CoPt_3$ thin films. A complete demagnetization is obtained with a dynamics comparable to the pulse duration. In addition, the demagnetization dynamics of nickel nanoparticles embedded in a glass matrix is shown to occur in two steps. The analysis of this dynamics suggests a super-paramagnetic behavior of the nanoparticles.

2. Experiments

The experiments are performed in the time resolved magneto-optical Kerr configuration using degenerate pump and probe pulses of 120 fs duration from a Ti : sapphire cavity amplified at 5 kHz. The time dependent reflectivity signal is measured in the polar geometry at an angle of 50° with an applied magnetic field varying in the range ±3 kOe. The thin $Co_{0.25}Pt_{0.75}$ films (48.5 nm thick) are grown by epitaxy on a Ru layer deposited on a mica substrate. Figure 1 (a) shows the normalized time dependent magnetization defined as the ratio $I_K(t)/I_{K\infty}$ where I_K and $I_{K\infty}$ are respectively the time dependant and static Kerr signals measured with opposite applied magnetic field H ($I_K(t) = I_K(H,t) - I_K(-H,t)$). It shows that the characteristic time of the demagnetization process is close to the pulse duration. The curve in full line is a fit obtained with a convolution between the two incident pulses and an exponential decay of 100 fs, the complete process (minimum of magnetization) being accomplished in ~500 fs. It is interesting to compare the magnetization dynamics with the change of reflectivity obtained in the sample with a conventional pump-probe measurement. As seen in figure 1 (b), the differential reflectivity $\Delta R/R$ decreases with a dynamics comparable to the magnetization except for an initial weak coherent contribution. The comparison of figures 1a) and 1b) shows that the demagnetization process associated with the spin dynamics is simultaneous with the electronic relaxation. As it is well known, in metal films the initial process in the electronic relaxation is the thermalization of the electrons to a hot Fermi distribution, a process which occurs within a few hundreds of femtoseconds [2]. Therefore, the present experiment shows that it is possible to induce a phase transition from a ferromagnetic to a paramagnetic phase in the time scale when the electrons reach an equilibrium distribution.

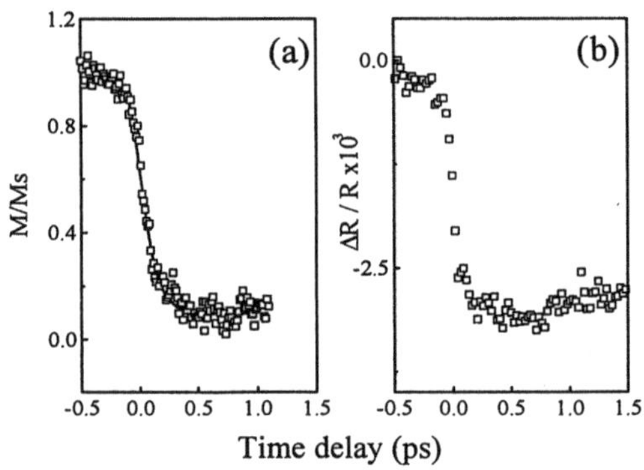

Figure 1 : Time resolved polar Kerr (a) and reflectivity measurements (b) obtained with p polarized pump and probe and a pump energy of 4 μJ.

Figure 2 shows both the reflectivity and magnetization dynamics using different probe wavelengths (800 and 400 nm), the pump being set at 800 nm. No significant change shows up. Similar results are obtained when pumping at 400 nm. These results suggest that the temporal behavior of the demagnetization is independent of the laser wavelength. Since the ultrafast demagnetization reported here occurs in a time scale when the lattice plays a minor role, it can be attributed to scattering between electrons in the d and conduction bands of the transition metal. A many body theory including such dynamical processes is currently not available.

The spin dynamics of nickel nanoparticles (mean diameter 10 nm), embedded in a glass matrix, is shown in figure 3. It is compared to the one obtained with Ni thin films. In both cases a fast demagnetization occurs during the pump pulse. The dynamics at longer time delays is different. For the films, the demagnetization comes back to its original polarization under the applied H field as previously reported [3]. For the nanoparticles, it still decreases. We attribute this slow component of the demagnetization to an increasing lattice temperature which provides enough energy to the spins in each nanoparticle to reach a super-paramagnetic state above a critical temperature.

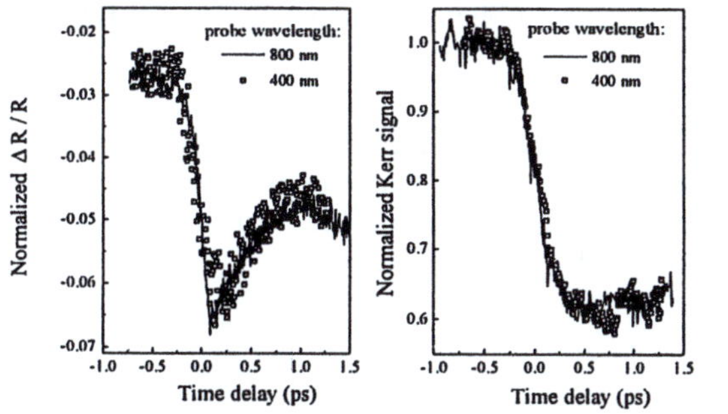

Figure 2 : Time resolved reflectivity and magnetization for a probe wavelength of 800 nm (full curve) and 400 nm (squares). The pump wavelength is 800 nm and the signals have been normalized.

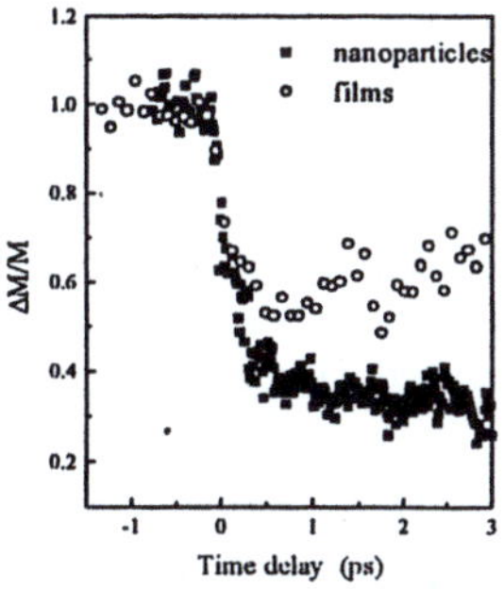

Figure 3 : Time resolved magnetization of nickel thin films (circles) and nanoparticles (squares).

3. Conclusions

Time resolved Kerr magneto-optical measurements of ferromagnetic materials reveal an ultrafast complete demagnetization of the samples. It shows that the collective properties of a correlated spin system can be altered without heating the lattice. The underlying microscopic mechanism may be attributed to the modification of the spin populations via the Coulomb exchange interaction. This type of technique also allows to investigate the dynamics of the super-paramagnetism which occurs when the dimensionality of the spin system is reduced like in ferromagnetic nanoparticles.

References

1 Eesley, Phys. Rev. B, **33**, 2144 (1986)
2 Fann, Storz, Tom, Bokor, Phys. Rev. Lett., **68**, 2834 (1992)
3 Beaurepaire, Merle, Daunois, Bigot, Phys. Rev. Lett. **76**, 4250 (1996)

Generation of Coherent and Squeezed Phonon Fields Using Femtosecond Laser Pulses

G. A. Garrett[1,2], J. F. Whitaker[2], and R. Merlin[1,2]

[1] Department of Physics, University of Michigan, Ann Arbor, MI 48109-1120, USA

[2] Center for Ultrafast Optical Science, University of Michigan, Ann Arbor, MI 48109-2099, USA

Abstract. Squeezed phonon fields in $KTaO_3$ and $SrTiO_3$ are generated using femtosecond laser techniques. Coherent, first-order, phonon oscillations were observed in $SrTiO_3$ at temperatures below its cubic-tetragonal phase transition at ≈ 110K. Time-domain data are compared with measurements of spontaneous Raman scattering.

1. Introduction

Femtosecond light pulses have been used to create first-order coherent phonon fields in a variety of solids [1]. Recently, this technique has been used to generate second-order squeezed phonon states in $KTaO_3$ [2] and $SrTiO_3$ [3]. Here, we present recent results and compare spontaneous with time-resolved stimulated Raman scattering, focusing on $SrTiO_3$.

2. Theory

To understand the generation of coherent and squeezed phonon fields, we consider the following Hamiltonian for a single phonon branch [1-3],

$$H = \sum_{\mathbf{q}} \frac{1}{2}\left(P_{\mathbf{q}}^2 + \Omega_{\mathbf{q}}^2 Q_{\mathbf{q}}^2\right) - \frac{1}{2}\left(\frac{\partial \chi}{\partial Q_0}\right) Q_0 |E(t)|^2 - \sum_{\mathbf{q}} \frac{1}{4}\left(\frac{\partial^2 \chi}{\partial Q_{\mathbf{q}}^2}\right) Q_{\mathbf{q}}^2 |E(t)|^2 \qquad (1)$$

where $P_{\mathbf{q}}$ and $Q_{\mathbf{q}}$ are the momentum and amplitude of a phonon with frequency $\Omega_{\mathbf{q}}$ and wave vector $\mathbf{q}$. The first term is the harmonic contribution to the lattice energy while the second and third terms correspond to the interaction between the incident electric field, $E(t)$, and the lattice. The coupling reflects the modulation of the electronic susceptibility, χ, by the ion motion. These susceptibility derivatives are intimately related to the Raman scattering cross-section. This establishes a relationship between time-domain data and spontaneous Raman spectroscopy [1,2]. With a laser pulse much shorter than the period of the phonons, we can approximate $E^2(t)$ as a delta function. Thus, the equation of motion for $Q_{\mathbf{q}}$ follows that of a harmonic oscillator with the second term in (1) acting as an impulsive driving force [4]. The third term behaves as a time-dependent frequency or, equivalently, a force proportional to $Q_{\mathbf{q}}$. Solving for the expectation values of $Q_{\mathbf{q}}$ and its variance, we find [2,3]

$$\langle Q_0(t)\rangle = \frac{E^2}{2\Omega_0}\left(\frac{\partial\chi}{\partial Q_0}\right)\sin(\Omega_0 t) \tag{2}$$

$$\langle Q_{\mathbf{q}}^2(t)\rangle \approx \langle Q_{\mathbf{q}}^2(0)\rangle\left[1+\frac{E^2}{2\Omega_{\mathbf{q}}}\left(\frac{\partial^2\chi}{\partial Q_{\mathbf{q}}^2}\right)\sin(2\Omega_{\mathbf{q}} t)\right]. \tag{3}$$

Equations (2) and (3) represent, respectively, a coherent phonon field and a two-phonon or squeezed field. While there is a single, or more generally, a discrete set of coherent phonons at $\mathbf{q}\approx 0$, there is a continuum of two-phonon states resulting from the creation of phonon pairs at $\pm\mathbf{q}$. In $KTaO_3$ and $SrTiO_3$, this continuum exhibits a dominant frequency due to a sharp peak in the density of states at large $\mathbf{q}$. Quantum *squeezing* occurs when the time varying variance of a particular mode falls below that of the standard quantum limit, $\hbar/(2\Omega_{\mathbf{q}})$ [3].

3. Experiment

Time-resolved data on $SrTiO_3$ at 5 K, acquired using a basic Ti:Sapphire oscillator and pump-probe system [2], is shown in Fig. 1 along with spontaneous Raman scattering results. Both spectra show A_{1g}- and E_g-symmetry phonons at 1.3 and 4.3 THz. The presence of the latter mode indicates that the tetragonal axis has developed along a direction perpendicular to the propagation of the laser [3]. $KTaO_3$ data (not shown) was also collected and previous line-width discrepancies between the spontaneous and time-domain results, [2], are now understood as being due to poor thermal contacts. Temperature dependence of the time-domain $SrTiO_3$ data shows softening and disappearance of the first-order coherent signal as the sample passes through the phase transition at 110 K. This phenomenon was expected and found to be unaffected by the presence of the squeezed phonon

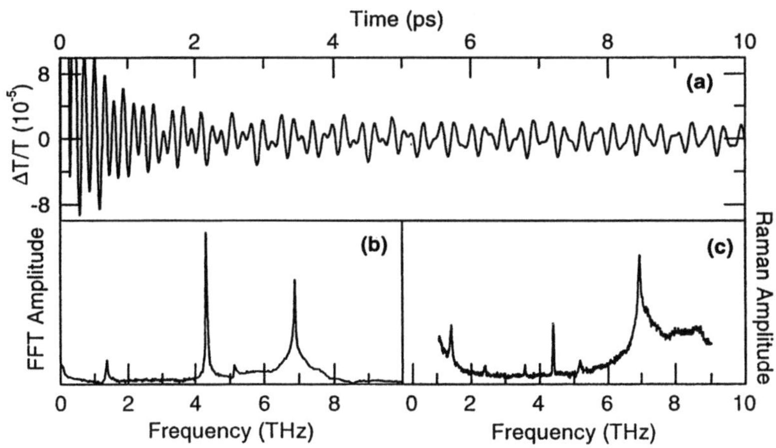

Fig. 1. (a) Time-resolved differential transmission in $SrTiO_3$ at 5 K. The Fourier transform (b) compares well with the spontaneous Raman spectrum (c).

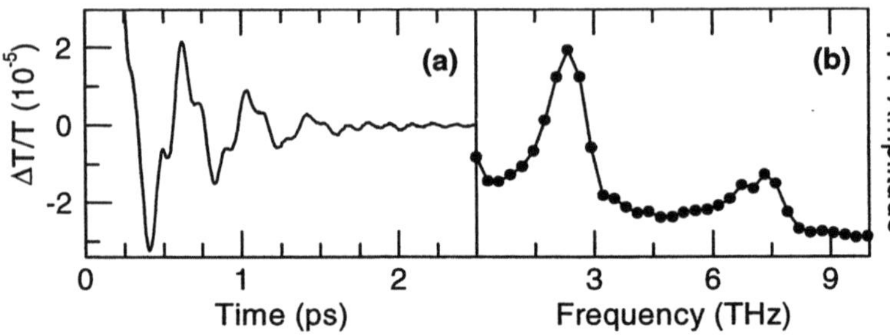

Fig. 2. (a) Time-resolved differential transmission in $SrTiO_3$ at 298 K. (b) Fourier transform indicates the presence of two second-order phonon fields, one at 2.3 THz due to difference scattering and an oscillation at 7.3 THz due to sum scattering.

field. Above the transition temperature, only second-order scattering is allowed. Fig. 2 shows room temperature time-domain data revealing two distinct oscillations. The fast one at 7.3 THz, which evolves into the narrow feature at 6.9 THz in Fig. 1, represents thermal squeezing. The slower oscillation at 2.3 THz is also seen in spontaneous Raman scattering [5]. It is associated with difference scattering involving a transverse optical and a transverse acoustic mode, an effect that can only occur in the presence of thermal phonons at higher temperatures.

4. Conclusions

Spontaneous Raman scattering in both $KTaO_3$ and $SrTiO_3$ has shown no significant difference with time-resolved stimulated Raman results at temperatures from 5 to 300 K. Room temperature data on $SrTiO_3$ shows the interesting effect of a coherent second-order difference field.

References

1 R. Merlin, Solid State Commun. **102**, 207-220, 1997.
2 G. A. Garrett, A. G. Rojo, A. K. Sood, J. F. Whitaker and R. Merlin, Science **275**, 1638-1640, 1997.
3 G. A. Garrett, A. K. Sood, J. F. Whitaker and R. Merin, Optics Express **1**, 385-389, 1997.
4 L. Dhar, J. A. Rogers and K. A. Nelson, Chem. Rev. **94**, 157, 1994.
5 W. G. Nilsen and J. G. Skinner, J. Chem. Phys. 48, 2240-2248, 1968.

Nonequilibrium Superconductivity and Quasiparticle Dynamics in $YBa_2Cu_3O_7$

Jennifer L.W. Siders, R. N. Jacobs, C. W. Siders, S. A. Trugman, and A. J. Taylor

Materials Science and Technology Division, Los Alamos National Laboratory, Los Alamos, New Mexico, 87545, USA. E-mail: sidersj@ibm.net

Abstract. We use optical pump, coherent terahertz probe spectroscopy to transiently excite nonequilibrium populations of quasiparticles in $YBa_2Cu_3O_{7-\delta}$ and monitor, with picosecond resolution, the superfluid and quasiparticle dynamics.

We use an optical (1.5 eV, 150 fs) pump to transiently excite nonequilibrium populations of quasiparticles in superconducting $YBa_2Cu_3O_{7-\delta}$ (YBCO) and coherent terahertz spectroscopy [1] to monitor, with picosecond resolution, the subsequent interplay between the superfluid and the quasiparticle dynamics with the ultimate goal of understanding the character of the excited quasiparticles, their decay channels and relaxation times.

Previously, quasiparticle dynamics in YBCO have been studied using optical pump, optical probe spectroscopy.[2] However, the coherent detection of a terahertz probe permits a more direct picture of both the superfluid and the quasiparticle dynamics, yielding the real, $\sigma_R(\omega)$, and imaginary, $\sigma_I(\omega)$, parts of the conductivity over the frequency range from 0.2 to 0.8 THz, as a function of pump-probe delay. Using a two-fluid model,[3] the conductivity below T_c, consists of a Drude-like component proportional to the number of quasiparticles and an imaginary component proportional to the superfluid population with an ω^{-1} dependence. Figure 1 reveals such transient spectra, $\sigma_R(\omega)$ and $\sigma_I(\omega)$, for $YBa_2Cu_3O_7$ at 4 K with an initially photoexcited quasiparticle density of $5\times10^{19}cm^{-3}$. Before the pump arrives, $\sigma_R(\omega)$ is very small, indicating the

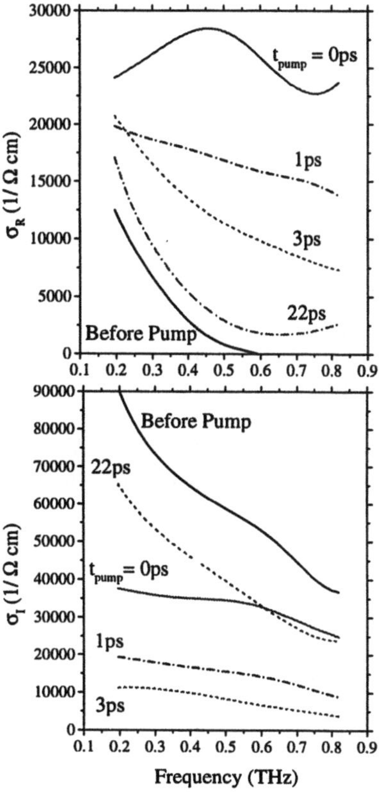

Figure 1: σ_R and σ_I, are plotted for different pump delays at 4 K for $YBa_2Cu_3O_7$ and n is 5×10^{19} cm^{-3}

presence of very few quasiparticles and $\sigma_I(\omega)$ is large and exhibits a ω^{-1} dependence due to the superfluid, consistent with a two-fluid picture. Immediately after excitation (0 ps) $\sigma_R(\omega)$ becomes large due to the creation of quasiparticles, while $\sigma_I(\omega)$ decreases and becomes roughly constant in ω because of both destruction of the superfluid component and a significant increase in a Drude component with a relaxation time of 0.2-0.4 ps.

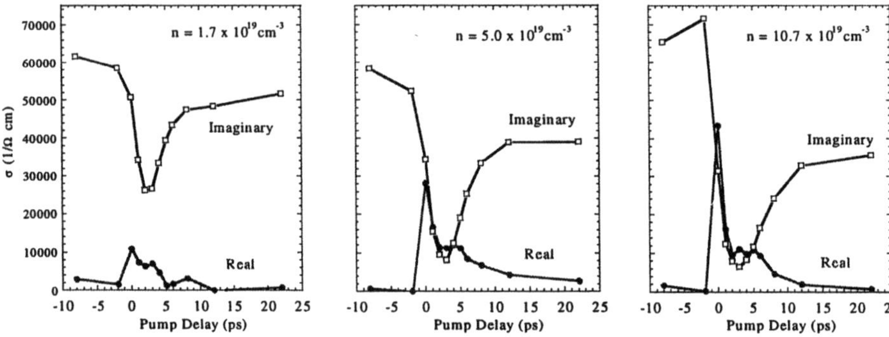

Figure 2: σ_R and σ_I are plotted for f=0.50 THz versus pump delay for $YBa_2Cu_3O_7$ at 4 K for three different excitation densities.

To reveal the dynamics of the conductivity more clearly, we plot, in Figures 2 and 3, σ_R and σ_I at 0.5 THz and 4 K, as a function of pump delay for $YBa_2Cu_3O_7$ (T_c = 85 K) and $YBa_2Cu_3O_{6.5}$ (T_c = 55 K) for three different excitation densities. For $YBa_2Cu_3O_7$ below T_c, the 1.5 eV pump pulse creates a highly excited quasiparticle distribution by absorbing $5x10^{19}$ cm^{-3} photons. The resulting large decrease in σ_I corresponds to nearly total destruction of the condensate (> 10^{21} cm^{-3} Cooper pairs), whereby, given a superconducting gap, Δ, of 0.03 eV, a 1.5 eV quasiparticle destroys 1.5 eV/0.030 eV = 50 superfluid carriers, creating as many quasiparticles. This destruction of the superfluid occurs in 1-2 ps. A temperature depend-

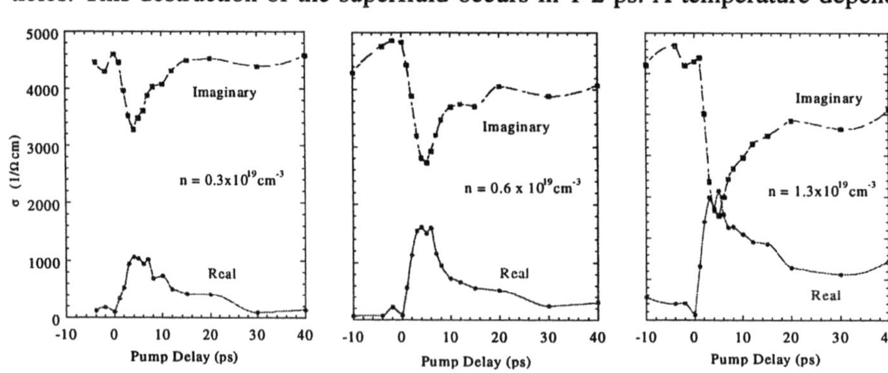

Figure 3: σ_R and σ_I are plotted for f = 0.50 THz versus pump delay for $YBa_2Cu_3O_{6.5}$ at 4 K for three different excitation densities.

ent bottleneck in the condensate recovery occurs around 2-5 ps, resulting from an excess population of high energy (2Δ) phonons released during the quasiparticle recombination process which become locally trapped and can rebreak Cooper pairs. The initial recovery dynamics of the superfluid is both temperature and density dependent, occurring in 5-10 ps (10-20 ps) for $YBa_2Cu_3O_7$ ($YBa_2Cu_3O_{6.5}$). As the excitation density is increased the number of phonons increases, and, as can be seen in Figures 2 and 3, delays the total recovery of the superfluid. In Figure 2, the rapid decrease in σ_R at high excitation densities, without the complementary increase in σ_I indicates that the quasiparticles are relaxing in such a way that they no longer contribute to the conductivity but are not yet reforming Cooper pairs. This behavior may be caused either by Anderson localization of the quasiparticles as they relax to the gap or an increase in mass due to a dressing of the quasiparticles by phonons. This relaxation mechanism is not seen in $YBa_2Cu_3O_{6.5}$ (Figure 3).

In Figure 4, σ_R and σ_I for $YBa_2Cu_3O_7$ at a moderate excitation density of 2.5 x 10^{19} cm^{-3}, are plotted for three temperatures near T_c(85K). In all cases a significant fraction(> 20%) of the superfluid has not recovered after 20 ps indicative of a critical slowing down of the order parameter near a second order phase transition. Above T_c, for $YBa_2Cu_3O_7$, σ_R decreases by about 16%, corresponding to an increase in the scattering time, τ, which does not recover within 20 ps.

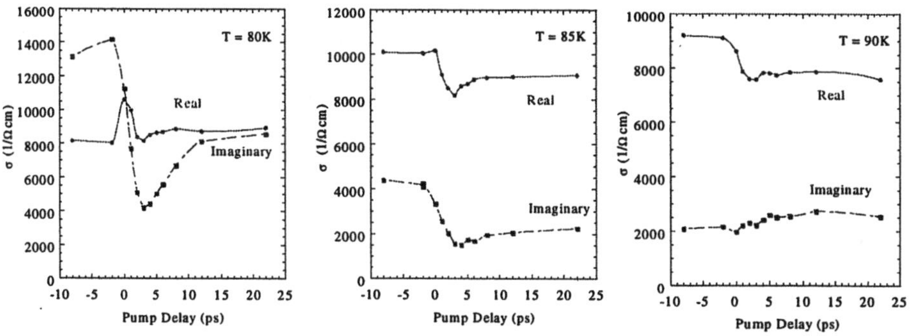

Figure 4: σ_R and σ_I are plotted for f=0.50 THz versus pump delay for $YBa_2Cu_3O_7$ for three temperatures near T_c (85 K). The excitation density, *n*, is 2.5 x 10^{19} cm^{-3}.

References

1 J. F. Frederici, *et al.*, *Phys. Rev. B* **46**, 11153 (1992).
2 W. Albrecht, *et al.*, *Appl. Phys. A* **57**, 203 (1993).
3 A. Berlinsky, *et al.*, *Phys. Rev. B* **48**, 4074 (1993)

Femtosecond Mid-Infrared Spectroscopy of Nonequilibrium Excitations in $YBa_2Cu_3O_{7-\delta}$

R. A. Kaindl[1], D. C. Smith[2], M. Woerner[1], C. J. Stevens[3], G. A. Wagner[4], J. E. Evetts[4], J. F. Ryan[2], and T. Elsaesser[1]

[1] Max-Born-Institut für Nichtlineare Optik und Kurzzeitspektroskopie, Rudower Chaussee 6, D-12489 Berlin, Germany
[2] Department of Physics, Clarendon Laboratory, Oxford University, Oxford OX1 3PU, UK
[3] Department of Engineering Science, Oxford University, Parks Road, OX1 3PJ, UK
[4] Superconductivity Materials IRC, Cambridge University, Cambridge, UK

Abstract. We report on the first time-resolved experiment on high-temperature superconductors probing the ultrafast nonlinear (ab)-plane reflectivity in the mid-infrared spectral range (10 to 20 μm).

The understanding of the mid-infrared response of high-temperature superconductors is closely linked with the underlying physics of the strongly correlated quasiparticles. In a limited range of the infrared spectrum, i.e. 10 to 20 μm in $YBa_2Cu_3O_{7-\delta}$ (YBCO), the (ab)-plane conductivity of these materials decreases sharply for temperatures below the superconducting transition temperature T_C, exhibiting a broad maximum at higher photon energies [1]. Different models have been proposed to explain this behavior. One picture suggests an infrared response composed of a Drude-type contribution at low photon energies and a broad mid-infrared absorption peak at larger energies which is due to boson-mediated quasiparticle absorption [1]. In a different approach, a strongly energy dependent scattering rate and quasiparticle mass is proposed to account for the observed features [2].

Femtosecond spectroscopy in the near-infrared has shown that above T_C, YBCO exhibits very fast transients corresponding to phonon scattering, similar to the behaviour observed in metals [3]. For temperatures below T_C, a distinctly different dynamics has been found, which has been linked to quasiparticles created from excitation-induced Cooper pair breakup [3,4]. This picture suggests that a pronounced quasiparticle dynamics should be seen in the mid-infrared range, close to the superconducting gap. So far, this range has not been investigated in ultrafast experiments. In this paper, we present the first time-resolved study of the ultrafast mid-infrared response in YBCO after femtosecond near-infrared excitation providing new insight into the underlying mechanisms.

The investigated $YBa_2Cu_3O_{7-\delta}$ samples are optimally doped c-axis oriented films of 100 nm thickness sputtered onto a MgO substrate [4]. The superconducting transition occurs at $T_C = 86$ K with a transition width of 1.5 K. The samples are mounted in a cryostat equipped with KBr windows and held at temperatures between 14 and 300 K. Carriers are excited using 20 fs Ti:sapphire pulses at $\lambda = 830$ nm with a repetition rate of 88 MHz. Sub-200 fs probe pulses in the mid-infrared spectral range (10 to 20 μm) are generated by phasematched difference frequency mixing in GaSe [5]. Average powers in the experiment are

30 mW for the pump, and ≤ 1 μW for the probe. The beams of parallel polarization are focused onto the sample near normal incidence, leading to focus diameters of 200 μm for the pump and ≈ 80 μm for the probe beams.

In the experiments, the time-resolved reflection change is monitored as a function of delay between the near-infrared pump and mid-infrared probe beams. Fig. 1 shows the observed signals at low temperatures below T_C at a probe energy of 88 meV (λ = 14 μm). The mid-infrared reflection sharply increases on a sub-picosecond timescale by about $\Delta R/R_0 \approx 10^{-3}$ and subsequently exhibits a fast picosecond decay. The signal clearly cannot be described with a single-exponential decay and shows additional, much slower relaxation behaviour. The temperature dependence of the signal amplitude is similar to that observed at 1.5 eV [4], decreasing strongly as the transition temperature is approached.

Scans with higher time-resolution, shown in Fig. 1 (b), reveal a finite risetime of ≈ 400 fs (solid circles), resulting in a delayed peak. A transient from a Ge sample (open squares) reflecting the time-integrated cross-correlation of pump and probe [5] is shown for comparison. This demonstrates that the finite rise time of the YBCO signal is due to the ultrafast dynamics in the sample. For temperatures above T_C, the signals show a distinctly different behaviour. A reflection *decrease* is observed with a much faster decay.

The energy dispersion of the signals was studied over a wide spectral range (10 - 18 μm) and is shown in Fig. 2. The amplitudes peak around 90 meV, falling off sharply towards lower and higher energies.

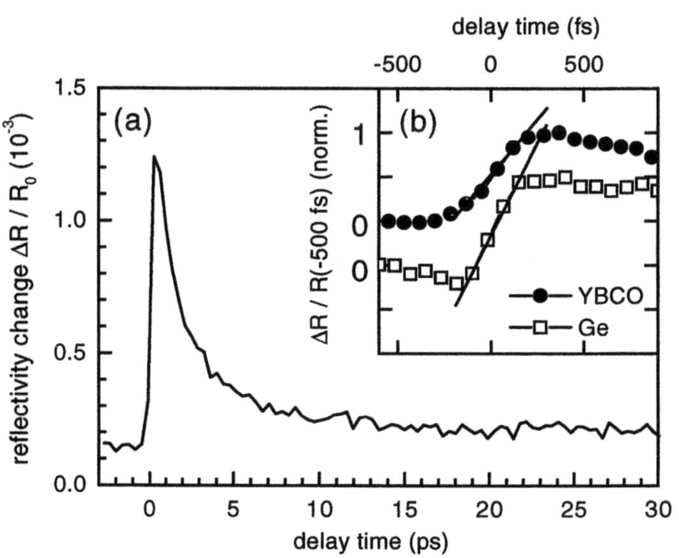

Fig. 1 (a) Transient change of reflectivity $\Delta R / R_0$ at a probe wavelength of λ = 14 μm after ultrafast excitation at λ = 830 nm of optimally doped YBCO at T = 14 K. (b) Solid circles: smaller delay range of YBCO signal in comparison with a Ge sample (open squares) measured with the same experimental setup. The straight lines point to the different slopes of rise times.

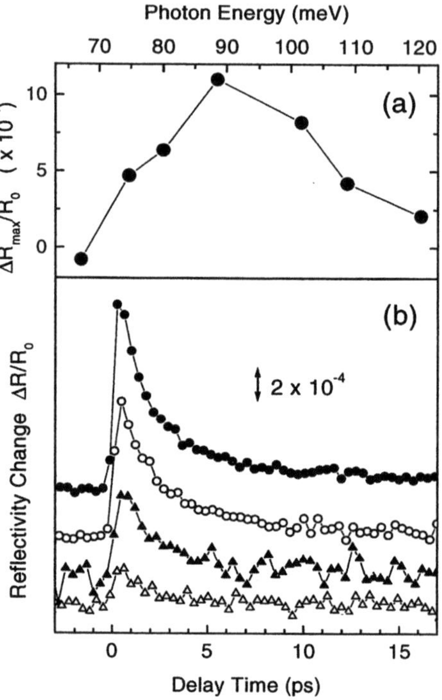

Fig. 2. (a) Spectral dependence of the maximum signal amplitudes. (b) Transients at probe wavelengths of 14 μm (solid circles), 12.2 μm (open circles), 10.3 μm (open triangles), and 16.6 μm (closed triangles). The curves are displaced vertically for clarification.

The observed finite risetime of the signals is in accordance with a break-up of Cooper pairs by the injected high-energy carrier, leading to a cascade of scattering processes [6]. A hot distribution of quasiparticles remains, which will subsequently recombine into Cooper pairs on a longer timescale. The observed sign of the reflection change (reflectivity increase) in this spectral range is expected when looking at reflectivity spectra at different temperatures (cf. Fig. 2 (a) in Ref. [7]). From a more detailed analysis one would expect that our transient spectra should exhibit a change in sign at left and right wings. Here, further extension of the probe wavelength in our setup will clarify this open question.

The spectral dependence of the signal suggests that it will not be possible to explain it using only the temperature dependence of the reflectivity [7] which shows a node in the thermal differential of the cw reflectivity. Although this has yet to be confirmed it would be interesting to speculate that the peak at 90 meV could be associated with four times the superconducting gap energy which is important in theories of the cw MIR reflectivity [8].

References

1 see e.g. J. R. Waldram, *Superconductivity of Metals and Cuprates*, Institute of Physics Publishing, Bristol and Philadelphia, 1996.
2 D.N. Basov, et al, Phys. Rev. Lett. **77**, 4090, 1996.
3 S. D. Brorson et al, Solid State Commun. **74**, 1305, 1990. J. M. Chwalek et al, Appl. Phys. Lett. **57**, 1696, 1990.
4 C. J. Stevens et al, Phys. Rev. Lett. **78**, 2212, 1998.
5 R. A. Kaindl et al., Opt. Lett **23**, 861, 1998.
6 A. Frenkel, Phys. Rev. B **48**, 9717, 1993.
7 R. T. Collins, et al, Phys. Rev. Lett. **63**, 422, 1989.
8 S. M. Quinlan et al, Phys. Rev. B **53**, 8575, 1996.

Part IX

Higher Harmonics, X-Ray Generation and Diffraction, High Intensity Physics

Phase-matched Generation of Short Wavelength, Ultrashort-pulse Light in Capillary Waveguides

C. G. Durfee III, A. Rundquist, S. Backus, Z. Chang, C. Herne, H. C. Kapteyn and M. M. Murnane

Center for Ultrafast Optical Science, U. of Michigan, 2200 Bonisteel Blvd., Ann Arbor, MI 48109-2099 USA.
E-mail: charlesd@eecs.umich.edu

Abstract. We demonstrate that a balance of gas and waveguide dispersion can be used to phase-match the upconversion of ultrafast pulses into the deep-ultraviolet and the extreme-ultraviolet regions of the spectrum. Ultrashort pulses tunable in the range 200-305 nm are obtained with a peak efficiency >18% by mixing the Ti:sapphire second harmonic with either the fundamental or the output of an optical parametric amplifier in a gas-filled capillary waveguide. Since neutral gases are anomalously dispersive to XUV high-order harmonics, these may be directly phase-matched without the need to use difference-frequency mixing. For high-harmonic generation, we have obtained an increase of conversion efficiency of 100-1000x over that obtained using a gas jet.

1. Introduction

Nonlinear frequency-conversion techniques have almost exclusively relied on crystalline solids as nonlinear media. Phase matching has generally been achieved by taking advantage of the birefringence of the nonlinear material. This reliance on solid materials has generally limited the application of nonlinear optical techniques to wavelengths longer than 200 nm[1]. Throughout the deep-UV, group velocity walkoff limits the output pulse duration to >150 fs. In contrast, many gases are transparent well into the vacuum ultraviolet (VUV), and even shorter wavelengths can propagate with moderate absorption through low-pressure gases. However, a focused-beam geometry results in a strong propagation phase mismatch that must be compensated. Previous work in VUV frequency conversion in gasses, making use of anomalous dispersion for light at wavelengths just shorter than a resonant absorption line,[2] has limited applicability for ultrashort pulses because of the narrow bandwidth.

Recently we demonstrated a novel approach to phase-matching of harmonic generation and parametric mixing in gases[3] in which the frequency conversion takes place in a gas-filled hollow waveguide. Not only can the interaction length at high intensity be increased, but the focusing phase shift is eliminated. For beams propagating as spatial modes of a waveguide, the phase of the wavefronts evolves linearly with propagation distance and can be used to balance the dispersion of the gas to achieve phase matching. The propagation constant for a beam (wavelength λ) propagating in a hollow waveguide of radius a, filled with a partially ionized gas (atomic density N_a, electron density N_e) is given in Eq. 1. The gas refractive index is written as $1+N_a\delta(\lambda)$, r_e is the classical electron radius, and u_{nm} is the modal constant for a spatial mode of radial and azimuthal indices n, m. The respective contributions to the propagation constant

$$k(\lambda)=\frac{2\pi}{\lambda}+\frac{2\pi N_a\delta(\lambda)}{\lambda}-N_e r_e\lambda-\frac{u_{nm}{}^2\lambda}{4\pi a^2} \qquad (1)$$

result from (1) vacuum, (2) the gas, (3) free-electron; and (4) waveguide dispersion. The first term cancels in any evaluation of phase mismatch of a frequency conversion process. In this work, we demonstrate that the remaining terms can balance to make phase matching possible in the UV and XUV spectral regions.

2. Phase-matched Deep-UV Generation

Frequency conversion in gases exhibits a dramatically lower group velocity walkoff than in nonlinear crystals, making it possible to generate much shorter UV pulses[4, 5]. However, due to poor phase matching in a simple focused beam configuration, the conversion efficiency is relatively low (0.1%). In our novel phase-matching technique[3], we dramatically increase the efficiency, while retaining the advantages of low group velocity walkoff. Gases are normally dispersive in this wavelength region ($\delta(\lambda)$ increases with decreasing λ). Phase-matching of direct third-harmonic generation in the waveguide is possible with the harmonic in a higher spatial mode[6], but better signal/nonlinear polarization overlap is possible with all beams in the lowest mode. Parametric amplification ($\omega_{signal}= 2\omega_{pump}-\omega_{idler}$) may be phase-matched under these conditions, resulting in near-Gaussian beam quality with a high degree of focusability.

Experimental Method and Results. Our kilohertz dual amplifier stage multipass Ti:sapphire amplifier system (<20 fs, 4.5 mJ)[7] was used for these experiments. The output pulses (0.5-1 mJ) are down-collimated and frequency-doubled in a nonlinear crystal. After delay and alignment of the polarizations, the blue pump and near-IR idler pulse are recombined with a dichroic splitter. The collinear beams are then focused together with a curved Al-coated mirror to the entrance aperture of a 140 µm ID capillary held in a cell. The cell has thin (250 µm) sapphire windows, and is evacuated and filled with gas to a pressure that gives the optimum conversion efficiency.

Our most recent work has improved over our previously-reported results. By using a better quality and larger diameter, we have increased the conversion efficiency to >18% of the blue pump light, obtaining > 10 µJ per pulse at 267 nm. Using self-diffraction frequency resolved optical gating[4], we measured a pulse duration of the 267 nm pulse at <40fs, with a linear chirp that would allow compression to < 25fs. With further optimization expect to obtain sub 10-fs pulses by optimizing the chirp, duration and energy of the input pulses.

In other recent work, we show that by using "idler" light obtained from an optical parametric amplifier (OPA, Spectra Physics Inc.) , we can generate ultrashort light pulses throughout the deep-UV. Using both the signal and the idler light, and the second harmonics of these, a tuning range of at least 590-2600 nm can be obtained. Combining this light with 400 nm pump light in the waveguide allows tuning the output from 217-305 nm. For each wavelength (see Fig. 1), the pressure of the gas (argon) was readjusted to maintain the proper phase-matching condition. The signal pulse energy in each case was microjoules or greater. Improvements in coupling the beams into the capillary will further increase the output energy.

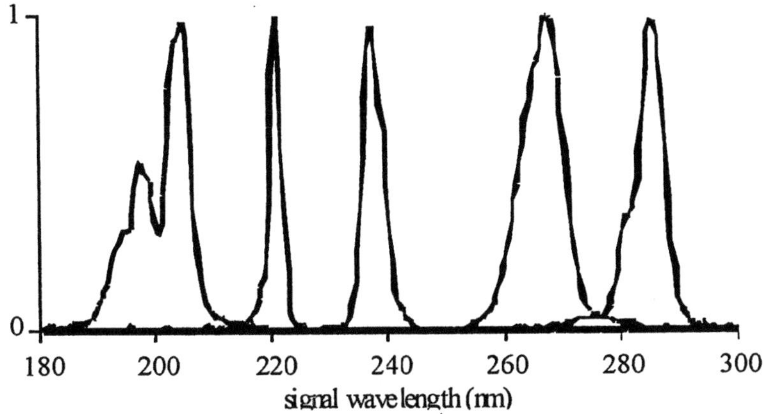

Fig. 1. Output spectra for high-order mixing in a capillary waveguide filled with argon. First peak: six-wave mixing (3 ω(400 nm) – 2 ω(800 nm)); right four peaks: four wave mixing (2ω(400 nm) – ω_{idler}), with idler wavelengths 2200 nm, 1280 nm, 800 nm and 670 nm respectively.

Guided-wave phase matching may be extended to shorter wavelengths in the VUV range through higher-order mixing. The spectrum centered at 200 nm in Fig. 1 resulted from a six-wave mixing process: $\omega_{signal}= 3\omega_{pump}-2\omega_{idler}$ (where second harmonic and the fundamental were used as the pump and idler, respectively). The broadening of the spectrum in this case resulted from self-phase modulation of the 800 nm light.

3. Phase-matched High-order Harmonic Generation

In recent years the technique of high-harmonic generation has proven to be a useful source of XUV and soft x-ray light[8, 9]. In this process, atoms in a dilute gas radiate harmonics of the incident light in the process of being ionized at the focus of an intense ultrashort-pulse laser. The use of very short laser pulses has been shown to produce coherent light to wavelengths below 3 nm[10, 11] and may be a source of sub-femtosecond pulses[12-14]. Despite the low efficiency of this process ($< 10^{-8}$ conversion of laser light into one harmonic peak), the photon flux generated is still sufficient to be useful for some applications in ultrafast x-ray spectroscopy. However, significantly more applications would be possible if the process could be phase-matched. Most experiments in high harmonic generation to date have been done in the tight focusing geometry, where the phase shift due to the laser focus dominates and limits phase-matching. Although proper choice of focusing conditions can improve the coherence length, the useful interaction length is still fundamentally limited by the laser beam divergence.

For photon energies well above the ionization potential, the refractive index of neutral gas is slightly less than unity ($\delta(\lambda)<0$). This anomalous dispersion allows direct phase matching of high-order harmonics without frequency mixing simply through pressure tuning to balance the normal dispersion of the waveguide. Free electrons share the same dispersive qualities as the waveguide so ionization tends to increase the optimum phase-matched pressure. This balance can be maintained for ≈10% ionization in argon; the shorter the pump pulse, the higher the intensity that can be reached at a given level of ionization, and the higher the harmonic order that can be phase-matched. The interaction length is limited by the absorption of the harmonic light by the neutral gas. This can be a long as 1-3 cm in 30 Torr of argon for harmonics >27.

Experimental Method and Results. For these experiments[15], pulses from the kilohertz system (25 fs, 100-300 μJ) were focused with a fused silica lens into a differentially pumped capillary cell such that the beam was carefully coupled to the lowest order spatial mode of the waveguide. In the novel cell design a constant gas pressure is held in the main capillary by using two short capillary sections at either end to restrict the gas flow out into the vacuum system. This arrangement greatly reduces the system gas load compared with a jet operating a high repetition rate. The sections are held in alignment by a V-groove with <1mm spacing. The inner diameters were 150 μm; the central section was 3 cm long. An aluminum filter rejected the pump light while passing harmonic 11-45.

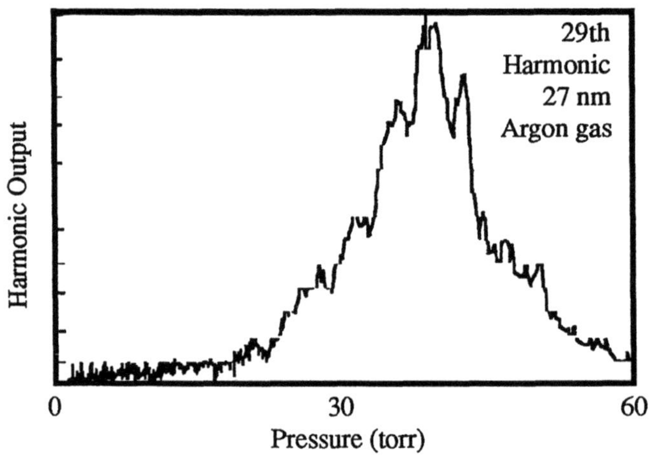

Fig. 2. Yield of the 29th harmonic order versus argon pressure in the capillary.

If phase-matching of conversion of laser light into x-rays is achieved, we should observe a dramatic increase in x-ray output at the optimum pressure for a given harmonic order. The yield of the 29th harmonic, measured using a grazing incidence spectrometer and a microchannel plate detector, is shown in Fig. 2. At a pressure of 35 torr, phase-matching is predicted and observed to occur and causes a dramatic increase in harmonic output. We see no evidence of absorption or spatial distortion of the fundamental at high pressure; the decrease in harmonic signal at high pressures is due to phase mismatch. By measurement of the signal with an x-ray diode we estimate a conversion efficiency of 10^{-5}–10^{-6}, with approximately 0.5 nJ energy in each of 5 harmonics centered on the 29th order. The optimum phase-matching pressure for each harmonic is slightly different, because the higher harmonics are generated at higher values of ionization. The signal for harmonic orders below the 23rd was greatly attenuated because of strong absorption in the argon gas. We see improved yield at harmonics above order 33, though the phase matching is not as optimal because these harmonics are generated at higher levels of ionization.

4. Conclusions

The use of capillary waveguides in nonlinear frequency conversion is a powerful and useful new technique for generating short-wavelength light. The capillary has two principle functions: it increases interaction length with the nonlinear medium, and it facilitates phase matching of the process. We are currently working to further increase the conversion efficiency, and broaden the wavelength into the VUV and XUV. This technique should enable the extension of ultrafast experiments to short wavelengths while using only small-scale lasers.

Acknowledgements. The authors gratefully acknowledge support for this work from the National Science Foundation. H. Kapteyn acknowledges support from an Alfred P. Sloan Foundation Fellowship.

References

[1] J. Ringling, *et al.*, Opt. Lett. **18**, 2035 (1993).
[2] A. H. Kung, J. F. Young, and S. E. Harris, Appl. Phys. Lett. **22**, 301 (1973).
[3] C. G. Durfee, *et al.*, Opt. Lett. **22**, 1565 (1997).
[4] S. Backus, *et al.*, Opt. Lett. **21**, 665 (1996).
[5] C. W. Siders, *et al.*, J. Opt. Soc. Am. B **12**, 330 (1996).
[6] C. G. Durfee, *et al.*, in *Applications of High Field and Short wavelength Sources*, edited by L. DiMauro, *et al* (Plenum, New York, 1998).
[7] S. Backus, *et al.*, Opt. Lett. **22**, 1256 (1997).
[8] A. L'Huillier and P. Balcou, Phys. Rev. Lett. **70**, 774 (1993).
[9] J. Zhou, *et al.*, Phys. Rev. Lett. **76**, 752 (1996).
[10] Z. Chang, *et al.*, Phys. Rev. Lett. **79**, 2967 (1997).
[11] C. Spielmann, *et al*, Science **278**, 661 (1997).
[12] P. B. Corkum, N. H. Burnett, and M. Y. Ivanov, Opt. Lett. **19**, 1870 (1994).
[13] I. P. Christov, M. Murnane, and H. Kapteyn, Phys. Rev. Lett. **78**, 1251 (1997).
[14] P. Antoine, *et al*, Phys. Rev. A **56**, 4960 (1997).
[15] A. Rundquist, *et al.*, Science **280**, 1412 (1998).

Generation of Near-keV High Harmonic Radiation With a Sub-10fs Laser

M. Schnürer, S. Sartania, Z. Cheng, G. Tempea, M. Lenzner, Ch. Spielmann, and F. Krausz

Abteilung Quantenelektronik und Lasertechnik, Technische Universität Wien, Gusshausstr. 27, A-1040 Wien, Austria
E-mail: schnuer@ps1.iaee.tuwien.ac.at

Abstract. Intense few-cycle near-infrared light pulses generate spatially coherent xuv continuum from helium with photon energies extending beyond 0.5 keV at 1 kHz repetition rate.

1. Introduction

High harmonic generation from ultrashort laser pulses has been a promising candidate for laboratory size coherent xuv sources. Due to the physics of the generation process the characteristics of this source are strongly coupled to the laser driver. Recently 0.1 TW, 5 fs pulses at a repetition rate of 1 kHz became available in the near infrared [1], where the oscillation period of the radiation field is about 2.7 fs.

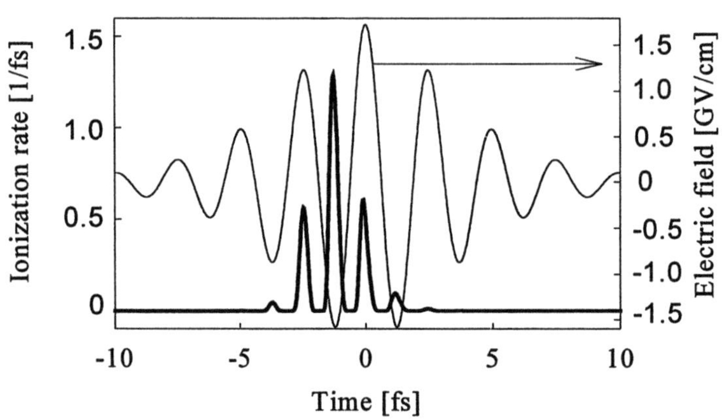

Fig. 1. Evolution of the electric field in a linearly-polarized 5 fs laser pulse of a peak intensity of $4*10^{15}$ W/cm^2 carried at a wavelength of 750 nm (thin line) and the ionization rate in He gas induced by this pulse (thick line), the barrier suppression field is about 10^9 V/cm for helium

With these light "transients" a new regime of strong-field atomic processes becomes accessible. This is primarily due to the fact that the laser field can exceed the barrier-suppression limit well before the ground state is depleted and the emission of the highest-energy photons can be confined to a fraction of a single oscillation cycle, because the field amplitude is significantly changing during the laser period. These features are evident from Fig.1.

2. Experiment and Results

These extremely short pulses allowed the generation of coherent water window (2.3 nm - 4.4 nm) radiation [2] for the first time at sub-mJ pulse energy levels and a kHz repetition rate [3]. The collimated xuv beam emerges from a thin (<0.2 mm) He jet ($p \approx 2$ bar) illuminated by the infrared laser pulses with parameters close to those used in Fig.1. In this work the short wavelength end of the emitted spectrum has been characterized with an energy dispersive spectrometer based on a Si(Li) detector [4]. The observed spatially coherent xuv beam with photon energies extending to above 0.5 keV is characterized in its spectrum by a rapid decrease of the intensity with increasing photon energy

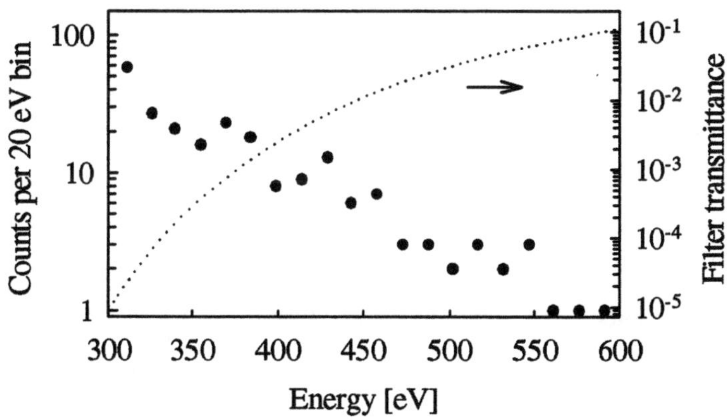

Fig. 2. The high energy end of the xuv spectrum transmitted through a 400 nm boron and 200 nm aluminium filter together with the transmittance of the filters

which has been predicted theoretically [4]. Owing to the high repetition rate of the driver laser the energy dispersive spectrometer allows xuv spectral characterization over a considerable dynamic range. Careful suppression of the high flux low energy photons is necessary to ensure safe pile-up rejection, which is a prerequisite for proper spectral analysis. With our detector (Z-MAX, Noran Instr.) we could perform spectral measurements for photon energies above 200 eV. The spatial distribution of the xuv beam displayed in fig.3 has been explored with a scanning knife egde and binning of energy channels of the EDX-spectrometer.

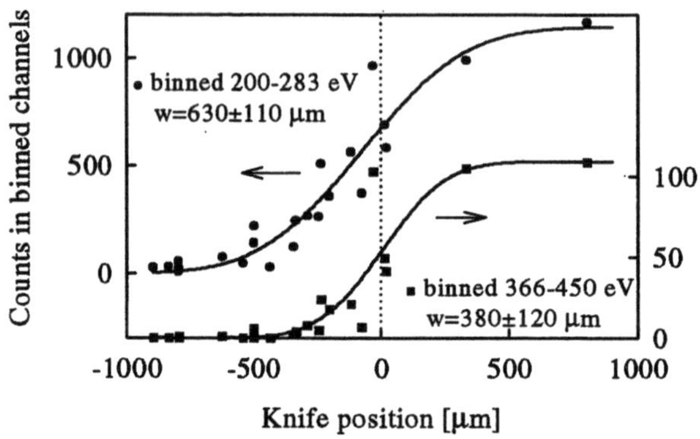

Fig.3. Measurement of the radial xuv beam distribution at a distance of about 30cm from the source with a knife edge aperturing the beam at different positions (lines are gaussian fits of the profile)

The energy of the optical laser driver pulses of about 0.4 mJ requires a sub-10 fs pulse width to achieve necessary intensities in an appropriate focusing geometry for high harmonic production in the water window region. More importantly, due to the ultrafast ionization dynamics of the atoms sub-10 fs driver pulses are required to minimize phase mismatch between the fundamental and the highest harmonics in the ionizing gas and thereby ensure efficient soft x-ray generation which has been verified with different pulse length [4].

One of the great challenges of the sub 10-fs laser driven coherent xuv source is the pulsewidth predicted to be in the attosecond range for wavelength below 10 nm. Such a source may open up new prospects for atomic spectroscopy and nonlinear optics in the x-ray regime.

Acknowledgements. This work has been sponsored by the National Science Fund in Austria in form of a Lise Meitner fellowship M445 (M.S.) and research grants Y44-PHY and P12361-PHY.

References

1 S. Sartania, Z. Cheng, M. Lenzner, G. Tempea, Ch. Spielmann, F. Krausz, Opt. Lett. **22**, 1562, 1997.

2 Z. Chang, A. Rundquist, H. Wang, M.M. Murnane, H.C. Kapteyn, Phys. Rev. Lett. **79**, 2967, 1997.

3 Ch. Spielmann, N.H. Burnett, S. Sartania, R. Koppitsch, M. Schnürer, C. Kan, M. Lenzner, P. Wobrauschek, F. Krausz, Science **278**, 661, 1997.

4 M. Schnürer, Ch. Spielmann, P. Wobrauschek, C. Streli, N.H. Burnett, C. Kan, K. Ferenc, R. Koppitsch, Z. Cheng, T. Brabec, F. Krausz, Phys. Rev. Lett. **80**, 3236, 1998.

Temporal Broadening Of Femtosecond High Harmonic Radiation

T. E. Glover[1], A.H. Chin[2], R. W. Schoenlein[1], and C.V. Shank[1,2]

[1] E. O. Lawrence Berkeley National Laboratory, Berkeley, California 94720, USA
Email : teglover@lbl.gov
[2] University of California Berkeley, Berkeley, California 94720, USA

Abstract. Measurement of femtosecond high harmonic radiation indicates that significant pulse shortening can be accomplished by limiting the transverse intensity variation of the generating laser pulse.

High-order harmonic generation has attracted considerable attention as a source of short-pulse, short-wavelength radiation. While a number of authors have established that ultrashort pulses are expected from the single-atom dipole response [1], comparatively few studies have accounted for the macroscopic nature of harmonic generation and associated effects on the harmonic temporal envelope. Lewenstein and co-workers have accounted for both the atomic dipole response and the full (three-dimensional) phase-matching integral and calculated harmonic durations in a regime where ionization effects could be neglected [2]. Harmonic durations (~ 60 - 150 fs) were calculated to be on order of the generating laser duration (150 fs) [2]. Subsequently, Kan and co-workers predicted that harmonics significantly shorter (~ 5 fs) than the generating laser pulse (150 fs) could be obtained by working at laser intensities sufficiently high to strongly ionize the harmonic generating (gas) medium [3]. These calculations accounted for one-dimensional phase-matching and pulse shortening resulted from two factors. First, atomic dipole emission is rapidly terminated due to ionization. Second, the presence of free electrons born (via ionization) preferentially in the temporal tail of the harmonic pulse inhibits phase-matching and leads to pulse shortening.

We have measured harmonic durations in laser intensity regimes corresponding to both weakly and strongly ionized gas media. In contrast to the one-dimensional results of Ref. [3], we find that harmonic pulses produced in a strongly-ionized medium are significantly longer than those generated in a weakly-ionized medium. Our results indicate that the macroscopic nature of harmonic generation permits pulse shaping effects which can significantly broaden the harmonic pulse. Experiments indicate that significant pulse shortening can be achieved by limiting the transverse intensity variation of the generating laser pulse.

Harmonics generated (in argon) by a 110 fs, 800 nm, laser pulse were measured using a laser and harmonic cross correlation technique [4]. Cross-correlation curves are shown in Figs. 1 and 2 (diamonds) for the 25th harmonic generated by a spatially gaussian laser pulse at intensities of $2.4x10^{14}$ W/cm^2 (weakly ionized medium) and $1.1x10^{15}$W/cm^2 (strongly-ionized medium).

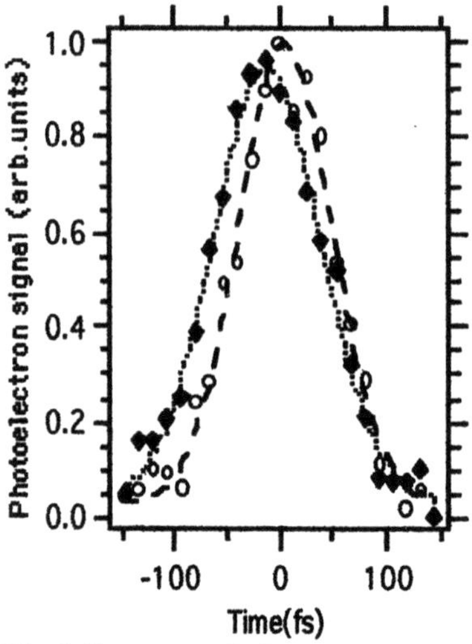

Fig.1 Cross correlation measurements obtained at $2.4x10^{14}$ W/cm^2. Deconvolved harmonic durations are 50 fs (diamonds, spatially gaussian laser) and <40fs (circles, spatially clipped gaussian).

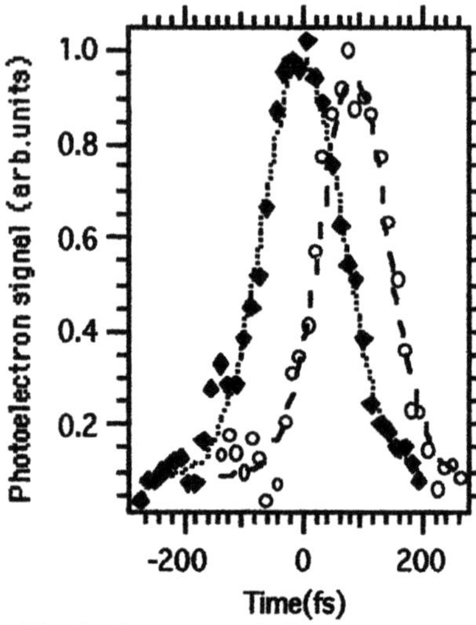

Fig.2. Cross correlation measurements obtained at $1.1x10^{15}$ W/cm^2. Deconvolved harmonic durations are 105 fs (diamonds, spatially gaussian laser) and 70fs (circles, spatially clipped gaussian).

The data establish that significantly longer harmonics are observed in the strongly ionized medium (105 fs vs 50 fs). This result stands in contrast to expectations based on the duration of single atom dipole emission (calculated using the model of Ref. [5]). The calculated single-atom durations are significantly shorter than the measurements of Figs. 1 and 2 and *decrease* with increased laser intensity (15 fs at $1.1x10^{15}$W/cm^2, 30 fs at $2.4x10^{14}$ W/cm^2).We note that the trends in the single-atom calculation are consistent with the one-dimensional phasematching calculations of [3].

The long (compared to single-atom expectations) pulses shown in Figs. 1 and 2 (diamonds) can be understood as arising from a combination of two factors : 1) the (gaussian) transverse variation of the generating laser beam and 2) ionization of the medium. We calculate harmonic generation using the model of Ref. [5] for the atomic dipole response and by using a fast hankel transform algorithm to calculate the three-dimensional phase-matching integral. The calculation indicates that the transverse spatial variation of the generating laser beam causes a temporal delay in when harmonics are produced across the transverse profile of the laser beam. This temporal delay would not occur for a spatially flattop laser beam. Additionally, in the absence of ionization variation of the transverse laser profile does not lead to harmonic pulse broadening since harmonic generation would (for all transverse points in the medium) peak at the temporal peak of the laser pulse. We calculate that broadening associated with these two effects can be eliminated by reducing transverse variations in the generating laser beam to < 10%.

To demonstrate temporal broadening due to time-delayed harmonic generation in the spatial wings of the laser pulse, the spatial wings of the laser pulse were reduced and the harmonic durations re-measured. The transverse laser profile was clipped by an aperture ($0.5*\omega_0$) and imaged into the harmonic gas medium. Two effects are observed (Figs. 1 and 2 circles). First, significant pulse shortening was observed at both the lower (<40 fs, resolution limited) and higher (70 fs) laser intensities. Second, the correlations are shifted in time in a direction which indicates that aperturing the fundamental has eliminated harmonics produced at late time (i.e. closer to the center of the laser in time). Significantly, the harmonic pulse measured at the higher laser intensity (70 fs) is appreciably longer than expected. We calculate a 20 fs harmonic duration for generation by the apertured laser beam. The discrepancy suggests that propagation of the intense laser pulse through the gas medium (a plasma) may produce non-uniformities across the spatial profile of the laser which act to broaden the harmonic pulse.

In conclusion, our results indicate that while single atom calculations predict that high laser intensity (rapid atomic ionization) is an efficient pulse shortening mechanism, variations in the transverse profile of the laser beam significantly broaden the harmonic pulse. We demonstrate that this temporal broadening can be controlled by limiting the transverse intensity variation of the generating laser pulse. Discrepancy between measurement and theory suggests that propagation of the laser through the harmonic generating plasma may lead to additional intensity variations which broaden the harmonic pulse. This mechanism would represent an obstacle to generating very short harmonic pulses.

Acknowledgements. The authors would like to thank G. Cerullo for fruitful discussions. This work was supported by the U.S. Department of Energy under Contract No. AC0376SF00098 and by the National Science Foundation under grant PHY-9512693.

References

1. P.B. Corkum, *et al.*, Opt. Lett. 19, 1870 (1994); I.P. Christov, *et al.*, Phys. Rev. Lett. **78**, 1251 (1997); P. Antoine, *et al.*, Phys. Rev. A **56**,4960 (1997).
2. Pascal Salieres, *et al.*, Phys. Rev. Lett. **74**, 3776 (1995).
3. C. Kan, C.E. Capjack, R. Rankin, N.H. Burnett, Phys. Rev. A **52**, R4336 (1995).
4. T.E.Glover, R.W.Schoenlein, A.H. Chin, C.V. Shank, Phys. Rev. Lett. **76**, 2468 (1996).
5. M. Lewenstein, Ph. Balcou, M. Yu. Ivanov, Anne L'Huiller, P.B. Corkum, Phys. Rev. A **49**, 2117 (1994).

High-Order Harmonic Generation and Quasi-Phase-Matching With Self-Guided Femtosecond Laser Pulses

H. R. Lange[1], J. -F. Ripoche[1], A. A. Chiron[2], B. S. Prade[1], P. Breger[1], P. Agostini[1] and A. Mysyrowicz[1]

[1] Laboratoire d'Optique Appliquée, École Polytechnique, Chemin de la Hunière, F - 91761 Palaiseau CEDEX, France
phone: +33-1-69 31 98 27, fax: +33-1-60 10 60 85,
mysy@enstay.ensta.fr

[2] Dipartimento di Fisica, Universita di Pisa, Piazza Torricelli 2, I - 56100 Pisa, Italy

[3] Service des Photons, Atomes et Molecules, Centre d'Etudes de Saclay, F- 91191 Gif sur Yvette

Abstract. We generate high order harmonics with self-guided femtosecond laser pulses and demonstrate quasi phase matching for the 3rd harmonic.

Odd harmonic generation in gases using intense sub-picosecond pulses provides coherent ultrashort tuneable emission up to the XUV [1]. However, the conversion efficiencies of this process is small. One of the limitations is imposed by the phase-mismatch due to the focused beam geometry.

In this contribution we present an experiment overcoming this limitation by using a self-guided beam as a source [2]. Self-guided beams exhibit constant intensities of typically 4 10^{13} W/cm^2 over several meters in air with quasi-planar wavefronts. We have coupled such a filament into a gas cell of variable length filled with Xenon and observed up to the fifteenth harmonic [3].

The appearance of Maker fringes due to the group velocity dispersion between the fundamental and third harmonic (TH) wave demonstrate the absence of geometric dephasing (Figure 1a)).

The dispersive dephasing can be compensated by a thin silica plate introduced in the cell at the position corresponding to one coherence length L_{coh}. The thickness of the plate (about 1.3 μm) was chosen so that it adds an additional phase-shift of π between fundamental and third harmonic (TH) wave. The length of the gas cell is adjusted such that the optical length is equal to 2 L_{coh} at 100 mbar. Figure 1b) shows the TH intensity of the Maker fringes as a function of pressure - without (filled dots) and with the silica plate (opaque square). No detectable TH was generated in the silica plate alone. The quasi-phase-matched TH intensity is a factor 1.5 higher than the maximum signal observed in absence of the plate at any pressure (see dashed lines in Figure 1b) for comparison).

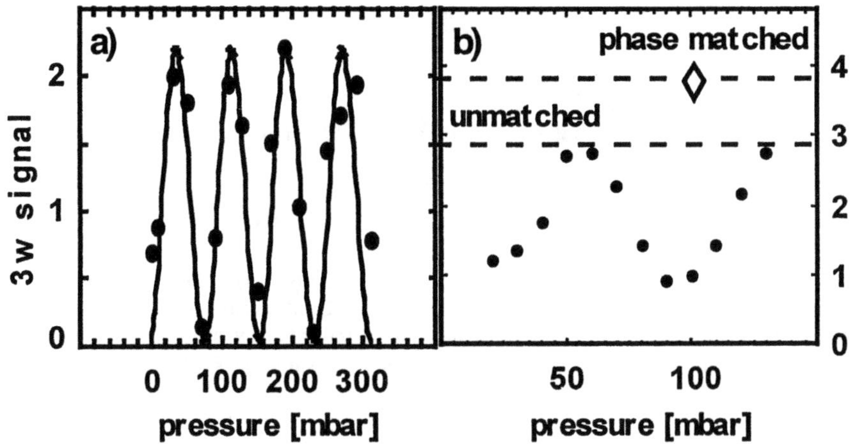

Figure 1. a) Observed (dots) and calculated third harmonic Maker fringes in Xe. b) Quasi-phase-matching of the 3rd harmonic in Xenon for a cell of 35 mm.

A plane wave and perturbative calculation of the TH intensity generated in Xenon as a function of the cell length in the unmatched and quasi-phase-matched case is shown in Figure 2. Curve a) shows Maker fringes in the absence of the phase plate. Curve b) gives the result for perfect QPM: a gain factor of 4 in intensity is expected at twice the coherence length. The difference between the observed and expected enhancement factor is probably due to inhomogeneities in the plate thickness. Curve c) illustrates the consequence of a 15 % deviation from optimal thickness.

In conclusion, we have demonstrated 1) the absence of a geometric phase mismatch in the filament within the gas cell 2) quasi-phase-matching of TH in Xenon. At harmonics of higher order than the third order, the use of a glass plate is excluded. Nevertheless, this experiment opens the perspective of QPM in higher harmonics through modulation of optical properties of the gas as discussed in [4].

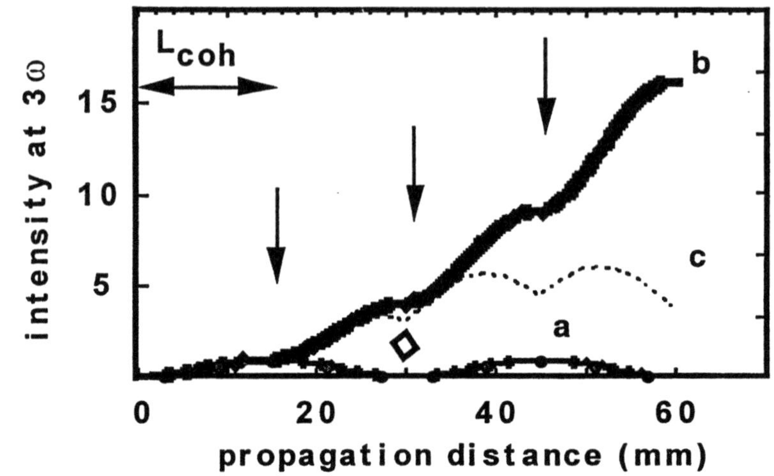

Figure 2. Theoretical Maker fringes and quasi-phase-matching (QPM) of the 3rd harmonic: a - Maker fringes in absence of dephasing plates, b - calculated QPM signal and c - calculated sub-optimal QPM signal with positions of rephasing plates indicated by arrows. The dispersive coherence length is indicated top left, the experimentally measured quasi-phase-matched value by a square.

References

1 Ch. Spielmann *et al.*, Science 278, 661, 1997.
2 Zenghu Chang *et al.*, Phys. Rev. Lett. 79, 2923, 1997.
3 H. R. Lange *et al.*, Opt. Lett. **23**, 120, 1998; and references therewithin..
4 H. R. Lange *et al.*, Application of High Fields and Short Wavelength Sources VII Santa Fe, March 19-22, 1997.
5 P.L. Shkolnikov *et al.*, Phys. Rev. A **50**, R4461 1994.

Kilohertz High Harmonic Generation in Hollow Core Fibers.

D. Garzella[1], P. Breger[1], and P. Agostini[1].

E. Constant[2], E. Mevel[2], and F. Salin[2].

C. Dorrer[3], and C. Le Blanc[3].

[1]SPAM, CEA Saclay, 91191 Gif sur Yvette, France (Tel : 33 1 69 08 69 01).
[2]ELIA, CPMOH Université Bordeaux 1, 351 cours de la libération, 33405 Talence, France
[3]LOA, chemin de la hunière, 91761 Palaiseau, France.

Abstract. High harmonics generation in hollow-core glass fibers by a 40 fs Ti :Sa pulses at a KHz repetition rate is obtained with a high conversion efficiency ($3x10^{-5}$ for the 13th harmonic i. e. ~ 10^{10} photons per shot). Hollow-core fibers reduce significantly the gas consumption.

Introduction

Using very short pulses to generate high order harmonics is a way to increase the saturation intensity [1] and to generate shorter wavelength [2]. Short pulses in principle also result in an higher conversion efficiency [3]. However this efficiency is limited by the maximum gaz pressure that can be used without inducing defocusing of the fundamental [4]. At a 1 KHz repetition rate this maximum pressure can even be reduced by the pumping speed of the vacuum system since the gas flow is important. Using hollow core glass fibers filled with noble gases to generate high harmonics allowed us to overcome partly these limitations : the defocusing is reduced since the fundamental is confined inside the fiber and the gas consumption is reduced (by ~ 35 for a given number of photons as compared to the 0.8 mm wide jet that we used) by the small conductance of the fiber.

Experimental Set-up

We used 1.5 mJ, 40 fs pulses and hollow core fibers filled with Ar or Xe to generate high harmonics.

The gas was injected in the fibers through a 500 μm diameter hole drilled in the fiber's wall and exited the fibers by flowing through their cores. The fundamental was focused with a 1 m lens to a 250 μm focus (W_0 ~ 125 μm) at the entrance of the fibers. Harmonics were detected with a magnetic-bottle time-of-flight electron spectrometer filled with argon at a pressure of $3x10^{-6}$ mbar. The sensitive volume of the spectrometer was overfilled by the harmonic beam whose dimensions are not precisely known but are likely to be larger than the sensitive zone (300 μm wide).

The number of XUV photons deduced from the measured number of detected photoelectrons, the estimated 10% detection efficiency and the photoionization cross section is therefore underestimated. The fibers we used had an inner diameter of 200 μm and lengths L = 1, 2, 4 and 8 cm.

Results

With Xe energy conversion efficiency as high as ~ $3x10^{-5}$ for the 13th harmonic was obtained (~$1.4x10^{10}$ photons per shot) similar to the efficiency that we obtained by using a standard atomic jet. This efficiency is significantly higher than usual efficiencies of few 10^{-6} obtained for longer pulses [5, 6] and correspond to an average power of 50 μW for q= 13.

The maximum orders that we observed were 27 with Ar and 19 with Xe, identical to those observed with the jet showing that the coupling was good (and quasi-identical for all fibers). For all fibers the efficiency peaked at an optimum pressure (Fig. 1).

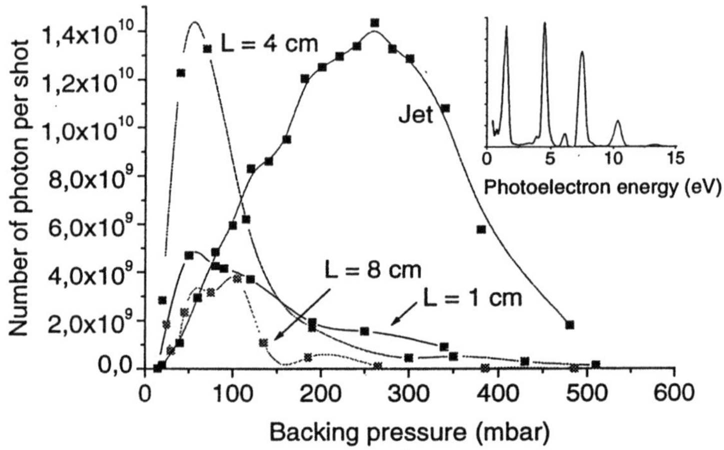

Fig. 1: Number of detected photons of the 13th harmonic of a Ti:Sa laser generated in Xe. The inset shows the 11th, 13th, 15th, 17th and 19th harmonics generated in Xe for an optimised run where we obtained $2.5x10^{10}$ photons per shot for q = 15

In an hollow-core fiber phase mismatch between the high harmonics and the fundamental results primarily from the phase-shift of the fundamental due to the guiding [7] while high harmonics travel freely and experience no phase shift (if we neglect the dispersion of the atomic gas). This dephasing, given by $\Delta\psi = -2.3x10^6 L q (\lambda_0/\phi)^2$ (L is the propagation length, ϕ inner diameter of the fiber, λ_0 fundamental wavelength and q harmonic order), is larger than the Gouy phase shift that occurs at the focus of a Gaussian beam having the same size. The efficiency of high harmonic on-axis emission is therefore expected to be lower with fibers, at variance with the present experiment which furthermore showed that the efficiency increases with the fiber length for L = 1, 2 and 4 cm, even if these lengths are longer than the coherence length for these harmonics (0.6 cm for q = 13). However, this efficiency dropped for L =8 cm.

These results could agree with the predictions of a simple model [8] showing that when the fundamental is guided in a cylindrical fiber, good phase-matching conditions are obtained *off-axis*. The efficiency of this conical emission increases *linearly* with the length of the emitting medium (Cerenkov-like emission). However the angle of emission is ~ $3x10^{-3}$ rad and after a propagation length of ~ 4 cm harmonics are reflected on the wall of the fiber with a π shift. Harmonics emitted in the first and second half of the 8 cm long fiber are therefore out-of phase and the efficiency is reduced.

Conclusions

In this experiment we found that high order harmonics can be efficiently generated at KHz enhancing significantly the average brightness as compared with 10 Hz systems. Furthermore, using hollow core fibers reduces the gas consumption. Intense field wave guiding has other interesting features for high harmonic generation. For instance when the fundamental is guided, its intensity can be shaped along the propagation by carefully choosing the diameter of the tapered wave guide. In this way on-axis phase matching lengths of several cm can be obtained even for high harmonics. This should further increase the high harmonic efficiency.

References

[1] J. Zhou *et al*, Phys. Rev. Lett **76**, 752 (1996).
[2] Ch. Spielmann *et al*, Science **278**, 661 (1997).
[3] K. J. Schaffer and K. C. Kulander, Phys. Rev. Lett. **78**, 638 (1997).
[4] C. Altucci *et al.* J. Opt. Soc. Am. B. **13**, 148 (1996).
[5] T. Ditmire *et al*, Phys. Rev. A **51**, R902 (1995).
[6] Anne L'Huillier and Ph. Balcou, Phys. Rev. Lett. **70**, 774 (1993).
[7] M. Nisoli *et al.* Appl. Phys. Lett. **68**, 2793 (1996).
[8] E. Constant *et al.*,to be published.

Generation of "Dark" Femtosecond Synchrotron Pulses from the Advanced Light Source

R. W. Schoenlein[1], P. Balling[2], H.H.W. Chong[1,3], T. E. Glover[1], P.A. Heimann[1], C.V. Shank[1,3], A. Zholents[1], M. Zolotorev[1]

[1] E. O. Lawrence Berkeley National Laboratory, Berkeley, California 94720, USA
[2] University of Aarhus, Ny Munkegade, DK-8000 Aarhus C, Denmark
[3] University of California Berkeley, Berkeley, California 94720, USA

Abstract. Femtosecond optical pulses extract ~100 fs slices of electrons from the electron beam at the ALS. We directly measure the resulting ~500-fs "dark" pulses in the synchrotron radiation from a bend-magnet.

An important new area of scientific research in chemistry, physics, and biology is the investigation of ultrafast structural dynamics in materials using x-rays. While high-brightness synchrotrons are powerful tools for probing the "static" structure of materials, their time resolution is limited to >30 ps. We report our first results using a novel scheme[1] to generate femtosecond pulses of synchrotron radiation. We directly measure "dark" femtosecond pulses in the visible synchrotron emission from a bend-magnet at the Advanced Light Source (ALS) beamline 6.3.2. The measured femtosecond time structure should be invariant over the entire bend-magnet radiation spectrum (near infrared to hard x-ray). This work is a first step toward developing a synchrotron-based source for ultrafast x-ray spectroscopy.

The time structure of synchrotron radiation is determined by the stored electron bunches, each typically longer than 30 ps in duration. We modify this time structure by using a femtosecond laser pulse to extract ~100 fs slice from an electron bunch[1]. The radiation from the core of a sliced bunch consists of a "dark" femtosecond pulse within the 30 ps synchrotron pulse. The generated pulse is absolutely synchronized to the femtosecond laser and is thus ideal for visible/x-ray pump-probe spectroscopy on the ultrafast time scale. Slicing of the long electron bunch is accomplished through the interaction of a femtosecond laser pulse with an electron bunch as they propagate through a wiggler (Fig. 1A). The high electric field in the optical pulse produces an energy modulation in the electrons as they traverse the wiggler. Electrons that are in phase with the optical field are accelerated and electrons that are out of phase are decelerated. The optimal interaction occurs when the wiggler period λ_W satisfies the resonance condition $\lambda_L = (\lambda_W / 2\gamma^2)(1 + K^2/2)$ (where γ is the Lorentz factor, K is the wiggler deflection parameter and λ_L is the laser wavelength) and when the far-field laser radiation overlaps (in both the transverse and longitudinal phase space) with the far-field spontaneous emission from the wiggler. Under these conditions, the energy absorbed by the electrons from the laser field (or transferred to the laser field) is calculated by considering the far-field superposition of the laser radiation and the spontaneous radiation from an electron in the wiggler[1]. Under proper conditions, a 50 fs laser pulse at 800 nm with a pulse energy of only 100 μJ will produce an ~10 MeV energy modulation on the stored 1.5 GeV electron beam at the ALS. The energy modulation is several times larger than the rms beam energy spread at the ALS (1.2 MeV), and only a thin slice of the electron bunch (the portion which is temporally overlapped with the laser pulse) experiences this modulation. The accelerated and decelerated electrons are then spatially separated from the original electron bunch by a transverse distance which is several times larger than the rms transverse size of the electron beam. This separation occurs in a dispersive section (bend magnet) of the storage ring, which follows the wiggler (Fig. 1B). The interaction with an ultrashort slice of the electron beam leaves a femtosecond hole or dark pulse in the electron bunch which directly maps to the synchrotron emission (Fig. 1C) originating from a bend magnet.

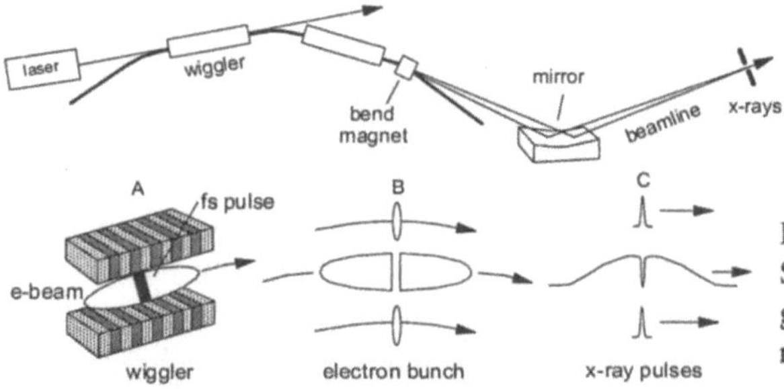

Figure 1.
Schematic of the technique for generating femtosecond synchrotron pulses

Such a dark x-ray pulse can be used for ultrafast pump-probe spectroscopy in the same way that "bright" femtosecond pulses are used, since the dark pulse consists essentially of a constant flux minus a femtosecond pulse. In addition to the dark pulse, complementary bright femtosecond x-rays from the displaced electron beam slices can be isolated for ultrafast spectroscopy using a radially offset aperture to block synchrotron radiation originating of the e-beam core (Fig. 1C).

The experiments are conducted at the ALS at the Berkeley Lab. We inject ~500 μJ, 70 fs pulses from a 1 kHz Ti:sapphire laser into the main storage ring where they co-propagate with the 1.5 GeV (γ~3000) electron bunch through a 19 period (λ_W=16 cm) adjustable-gap wiggler. Timing between the laser pulse and the electron bunch is synchronized using laser-cavity phase locking. The near and far-field spatial modes of the fundamental wiggler emission and the laser beam are matched. Optimum interaction of the laser with the electron beam is verified by measuring the gain experienced by the laser propagating with the electron beam through the wiggler (Fig. 2). Electrons that are accelerated by the laser radiate at slightly shorter wavelengths, and electrons that are decelerated by the laser radiate at slightly longer wavelengths as they transit the wiggler. Thus, the electron beam preferentially absorbs shorter wavelengths from the laser pulse (loss) and emits longer wavelengths to the laser pulse (gain). The agreement of the measured gain with the theoretical prediction shows that the laser interaction with the electron-beam is nearly ideal.

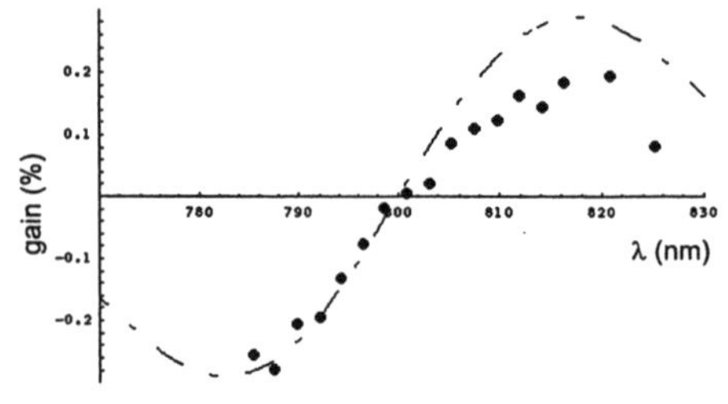

Figure 2. Laser gain from FEL action in wiggler. Dashed line is theoretical prediction based on beam parameters.

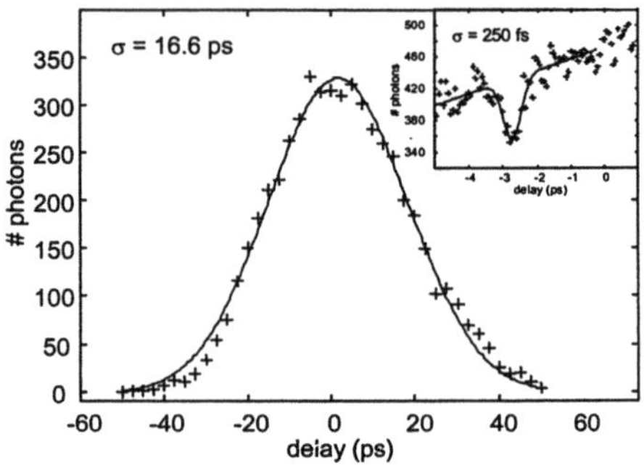

Figure 3. Cross-correlation of visible synchrotron radiation with laser pulse. Inset: dark pulse from e-beam slicing.

The time structure of the synchrotron pulse is measured by cross-correlating the visible synchrotron emission from a bend-magnet (ALS beamline 6.3.2) with a delayed pulse from the laser system in a BBO crystal. Figure 3 shows a laser/synchrotron radiation cross-correlation on a long-time scale. The measured pulse duration, σ=16 ps, corresponds to the electron bunch duration. Measurement with higher time resolution (inset Fig. 3) shows the femtosecond "dark" pulse which appears as a narrow hole in the main pulse. Fig. 4 shows the femtosecond dark pulse in more detail, along with a gaussian fit (data is inverted and background is subtracted). The measured pulse duration is ~500 fs FWHM.

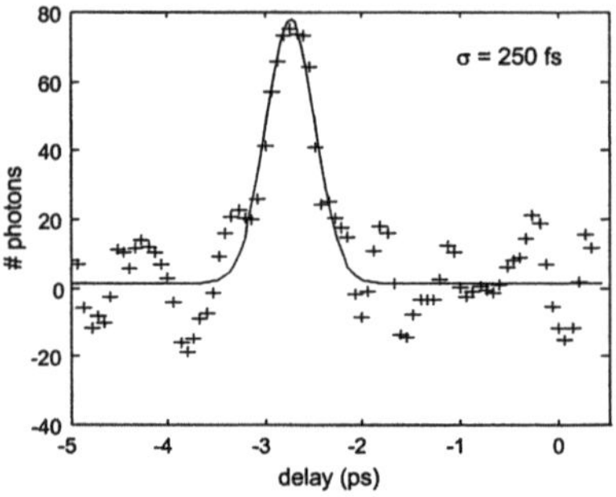

Figure 4. Femtosecond dark pulse of visible synchrotron radiation. Solid line is a gaussian fit. Note, background is subtracted and data is inverted (re fig. 3).

In conclusion, femtosecond "dark' x-ray pulses are generated from a synchrotron storage ring using a novel laser slicing technique. A pulse duration of ~500 fs is measured in the visible using cross-correlation techniques, and the limit on the pulse duration is less than 100 fs. This technique also generates complementary bright femtosecond x-ray pulses, and is the first step toward development of a high-brightness femtosecond source for ultrafast x-ray spectroscopy.

References

[1] A. A. Zholents, M. S. Zolotorev, *Phys. Rev. Lett.*, **76**, 916-918 (1996).

Work supported by the U.S. Department of Energy under Contract No. AC03-76SF00098

Disorder in Solids Studied on an Ultrashort Timescale using Femtosecond X-Ray Diffraction

A. Rousse[1], C. Rischel[1*], I. Uschmann[2], P.A. Albouy[4], J.P. Geindre[3], P. Audebert[3], J.C. Gauthier[3], E. Förster[2], J.L. Martin[1], A. Antonetti[1]

[1] Laboratoire d'Optique Appliquée - ENSTA, CNRS UMR7639, INSERM U451, Chemin de la Hunière, 91761 Palaiseau, France.
* Permanent address: Department of Mathematics and Physics, The Royal Veterinary and Agricultural University Thorvaldsensvej 40, DK-1871 Frederiksberg C, Denmark.
E.mail: rousse@enstay.ensta.fr

[2] X-ray Optics Group, Institute of Optics and Quantum Electronics, Friedrich Schiller University Jena, Max-Wien-Platz 1, 07743 Jena, Germany.
E.mail: ingo@roentgen.physik.uni-jena.de

[3] Laboratoire pour l'Utilisation des Lasers Intenses, Unité Mixte N°7605 CNRS-Ecole Polytechnique-CEA-Université Paris VI, Ecole Polytechnique, 91128 Palaiseau, France.
E.mail: gauthier@greco2.polytechnique.fr

[4] Laboratoire de Physique des Solides, Bâtiment 510, Centre Universitaire, 91405 Orsay, France. E.mail: albouy@hprx.lps.u-psud.fr

Abstract. We've studied the sub-picosecond response of an organometallic Langmuir-Blodgett film to an ultrashort laser excitation using ultrafast time-resolved x-ray diffraction. The ultrafast and monochromatic K_α x-ray source is produced from the interaction of an intense 100 fs laser pulse with a silicon solid target. Disordering in the film has been found to rise in less than 1 ps at pumping fluences larger than 20 J/cm^2.

1. Introduction

Ultrafast phenomena have been studied previously by using IR, visible and UV light, but spectroscopy based on light at those wavelengths only probes the sample's electronic and vibration states. On the other hand, X-ray diffraction studies provide direct information on atomic positions and molecular structure. The field of femtosecond x-ray science is thus important to biologists, chemists and physicists because it allows them to view fast atomic motion, which ultimately drives processes like (bio-) chemical reactions and phase transitions, on the time-scale at which they occur.

We have developed [1,2,3,4], at LOA-LULI laboratories, a source of very short and bright x-ray pulses that allows us to perform time-resolved visible laser pump - x-ray probe experiments on a sub-picosecond time-scale. We present here the study of the response of a Langmuir-Blodgett film of Cd arachidate to the excitation by a visible laser pulse, by observing the Bragg reflection for different delays between x-rays and exciting pulse [5].

2. Experimental Methods

Two techniques could provide sub-picosecond time resolution with x-rays. For both techniques one must excite the sample with a pumping pulse lasting less than the process to probe. First, you may use a long x-ray pulse ($\approx$ 50 ps) associated with an ultrafast detector like a streak camera [6,7,8]. These detectors must be synchronized with the visible pump pulse and with the x-ray probe pulse. Being poorly efficient, accumulation over many shots is required which induces a temporal jitter around 1ps. Second, you may use short pumping and probing pulses associated with slow detectors. These pulses are produced from the same laser system, which prevents any temporal jitter. The detector can be large and very efficient. Such ultrafast x-ray sources may be produced by the interaction of high intensity 100 fs laser pulses focused onto electron beams (linear accelerator, Synchrotron) [9] or solid materials [1,2,3,4,10,11].
In our laboratory, the x-ray source is produced from the interaction of a 10 Hz, 120 fs, 20 mJ Ti:Sa laser beam (λ = 800 nm) focused on a silicon solid target. For typical laser intensities of few 10^{16} W/cm^2, 10 % of the laser energy is converted into energetic electrons with a temperature of 10 keV. These hot electrons loose their energy deeper inside the solid target producing monochromatic K_α x-ray radiation [1]. 10^8 photons per steradian are produced in one laser shot. The efficiency of this K_α x-ray source is improved by focusing a pre-pulse onto the silicon target 6 ps before the main pulse. The electronic density gradient (L) in front of the target is then optimized ($L/\lambda \approx 0.23$) to increase the absorption of the incident laser energy through the process producing the energetic electrons. We have measured that the x-ray source becomes seven times more efficient [2].
The emitted 7.12 Å x-ray radiation is then collected by a toroïdal quartz crystal and focussed onto the sample. The sample is a 2000 Å thick organometallic Langmuir-Blodgett film which consists on multilayer of cadmium atoms separated by 55 Å. A second laser pulse from the same laser system is used to excite the film at fluences from few J/cm^2 to 30 J/cm^2. The transient Bragg x-ray diffracted signal (first order of diffraction) is monitored with a cooled x-ray CCD camera as a function of the delay between the visible exciting pulse and the x-ray probing pulse. The experimental setup is shown in figure 1.
In one laser shot, we estimated the number of x-ray photons detected by the CCD camera to be 50. To increase the signal-to-noise ratio of each measurement, one experimental run consists on 5 reference shots without the pumping beam, 1 shot with the pumping beam and 5 shots onto the damaged area without the pumping beam.

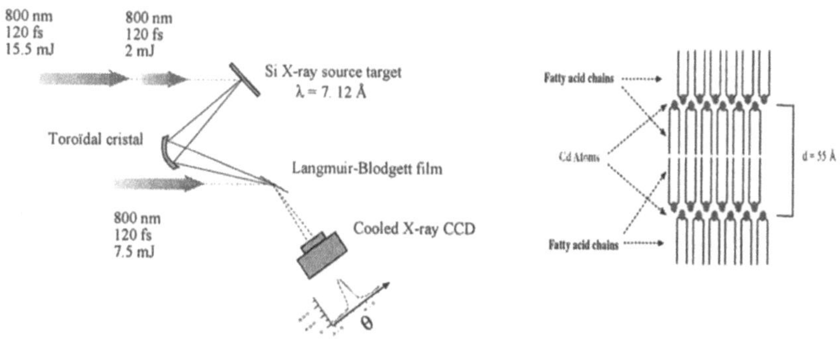

Fig. 1. Experimental setup for the ultrafast time resolved x-ray diffraction experiment with a laser-produced plasma x-ray source; and structure of the Langmuir-Blodgett film.

3. Results and Discussion

For all the pumping fluences, we observed a strong decrease of the x-ray diffracted intensity by 75 % in a timescale from few hundreds of femtoseconds to few tens of picoseconds depending on the energy of the exciting pulse. Results are shown in figure 2 for 4 different pumping fluences. The time response of the film is found to be 15 ps at a pumping fluence of 1.8 J/cm^2. It drops down to 1 ps at 11.8 J/cm^2 and is lower than 600 fs at 27.6 J/cm^2. This loss in diffracted intensity can be understood as a disordering of the diffracting atoms produced inside each cadmium layer before any thermal expansion of the lattice has time to occur. Using the Debye-Waller theory [12], a broadening of 8 Å of the electronic density in each layer of the film must be induced to provide a 75 % loss of the x-ray diffracted intensity [5]. For pumping fluences higher than 27 J/cm^2, the signal-to-noise ratio on the detector becomes very poor due to the x-ray emission produced from the sample itself, which hides the transient x-ray diffracted signal. Such large atomic displacements occurring in less than 1 ps probably comes from the very specific film structure. Actually, each layer of the film contains 2 layers of cadmium atoms separated by 3 Å and the distance between 2 atoms is about 5 Å. At high pumping intensities (10^{14} W/cm^2), bonds are broken and atoms are ionized. Ions then experience strong electrostatic repulsion between each other's. Rough numerical simulations without taking into account the influence of the organic chains are consistent with the observed sub-picosecond film time response. In the case of the Langmuir-Blodgett film, cadmium atoms experience weak ionic bonding with carboxyl group of the organic chain. We made experiments on Thallium Acid Phtalat crystal in which the thallium atoms are strongly linked to the crystal structure by 4 covalent bonds. At high pumping fluence (21 J/cm^2), the time response is found to be 3 ps which is much slower than for the film at the same fluence. This confirms the ability of the Langmuir-Blodgett film to experience high disordering.

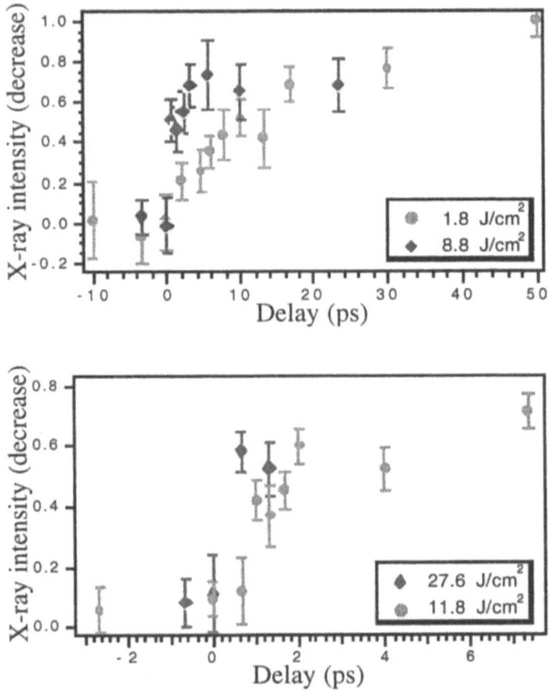

Fig. 2. Decrease of the Bragg x-ray diffracted signal as a function of the delay between the visible pumping pulse and the x-ray probing pulse. Negative delays correspond to a probe of the sample before excitation.

4. Conclusions

This experiment demonstrates that ultrafast dynamics in simple systems can be studied in the x-ray regime with a laser-produced plasma x-ray source at an atomic spatial resolution level. It also confirms experimentally for the first time the very short pulse duration of the keV x-ray source (at least less than 600 fs).

Acknowledgements. This work is supported by the Centre National de la Recherche Scientifique and by the European community under the large scale facility program under the contract RTD: ERB FMGE CT950019.

References

1 Rousse A., Audebert P., Geindre J.P., Falliès F., Gauthier J.C., Mysyrowicz A., Grillon G., Antonetti A., "An Efficient Kα X-ray Source from Femtosecond Laser-Produced Plasmas," Phys. Rev. E 50 (3), 2200 (1994).

2 Bastiani S., Rousse A., Geindre J.P., Audebert P., Quoix C., Hamoniaux G., Antonetti A., Gauthier J.C., "Experimental Study of the Interaction of Subpicosecond Laser Pulses with Solid Targets of Varying Initial Scale Lengths," Phys. Rev. E 56, 7179 (1997).

3 Geindre J.P. et al, "X-rays Generated by Femtosecond Laser-Produced Plasmas," presented at the Ultrafast Phenomena VIII, Antibes - France, 1993, Edited by Martin J.L., Migus A., Mourou G.A., and Zewail A.H., Series in Chemical Physics, Vol. 55, p. 272, Springer-Verlag.

4 Gauthier J.C., Bastiani S., Audebert P., Geindre J.P., Rousse A., Quoix C., Grillon G., Mysyrowicz A., Antonetti A., Mancini R., Shlyaptseva A., "Characterization of a femtosecond laser-produced plasma X-ray source by electronic, optical and X-ray diagnostic techniques," presented at the SPIE, San Diego, 1997, Edited by Kyrala G.A. and Gauthier J.C., Vol. 3157, p. 52, SPIE - International Society of Optical Engineering.

5 Rischel C., Rousse A., Uschmann I., Albouy P.A., Geindre J.P., Audebert P., Gauthier J.C., Foerster E., Martin J.L., Antonetti A., "Femtosecond time-resolved X-ray diffraction from laser-heated organic films," Nature 390 (6659), 490 (1997).

6 Helliwell J.R.Rentzepis P.M., in *Time-resolved Diffraction*, Oxford University Press ed. (Clarendon - Press, Oxford, 1997).

7 Larsson J., Heimann P.A., Lindenberg A.M., Schuck P.J., Bucksbaum P.H., Lee R.W., Padmore H.A., Wark J.S., Falcone R.W., "Ultrafast structural changes measured by time-resolved X-ray diffraction," Applied Physics A 66, 587 (1998).

8 Larsson J., Chang Z., Judd E., Schuck P.J., Falcone R.W., Heimann P.A., Padmore H.A., Kapteyn H.C., Bucksbaum P.H., Murnane M.M., Lee R.W., Machacek A., Wark J.S., Liu X., Shan B., "Ultrafast X-ray diffraction using a streak-camera detector in averaging mode," Optics Letters 22 (13), 1012 (1997).

9 Schoenlein R.W., "Femtosecond X-ray pulses at 0.4 Å generated by 90° Thomson scattering: a tool for probing the structural dynamics of materials," Science 274, 236-238 (1996).

10 Chen P., Tomov V., Rentzepis P.M., "Time resolved heat propagation in a gold crystal by means of picosecond x-ray diffraction," J. Chem. Phys. 104, 1001-1007 (1996).

11 Rose-Petruck C., Guo T., Jimenez R., Ràski F., Squier J.A., Walker B.C., Wilson K., Barty C.P.J., "Picosecond-milliangstrom resolution dynamics by ultrafast x-ray diffraction," submitted to Nature (1997 (July)).

12 H.P. Klug and L.E. Alexander, in *X-Ray Diffraction Procedure*, John Wiley and Sons Inc., New York, 1959.

Vacuum Heating vs. Resonance Absorption in the Solid to Plasma Transition of fs-Laser-Irradiated Iron and Aluminum

Mikal K. Grimes[1], Yun-Shik Lee[2], and Michael C. Downer[1]

[1] Department of Physics, The University of Texas at Austin, Austin, Texas 78712, U.S.A.
E-mail: downer@physics.utexas.edu
[2] Current address: The Center for Ultrafast Optical Science, the University of Michigan, 2200 Bonisteel Blvd., Ann Arbor, MI 48109

Abstract. Femtosecond reflectivity measurements on Fe and Al targets provide evidence for the vacuum heating absorption mechanism. The effect occurs for obliquely incident laser pulses polarized in the plane of incidence at intensities above 10^{14} W/cm^2. The electron oscillation normal to the surface leads to this widely predicted absorption effect, whose first experimental observation is reported in the current work.

1. Introduction

When an intense, p-polarized fs light pulse is incident obliquely on a sharply bounded overdense plasma, strong nonlinear absorption from electrons that are pulled into the vacuum then sent back into the plasma with quiver velocity v_{osc} has been repeatedly predicted theoretically [1-3]. However, experimental confirmation of this "vacuum heating" (VH) remains sparse and indirect. We report fs reflectivity and interferometry measurements in a regime in which VH absorption is dominant, and is clearly distinguished from the major competing mechanism: resonance absorption (RA) in the surface density gradient.

We first measured self-reflectivity on Fe and Al targets with high-contrast, s- and p-polarized, $\lambda = 620$ nm, 120 fs pulses with intensity $10^{11} < I < 10^{15}$ W/cm^2, at angles of incidence $20° < \theta < 70°$ in 5° increments. As exemplified by the data in the Figure 1 we cannot reproduce the high-intensity reflectivity changes with Fresnel formulae, which include only laser-induced changes to the linear dielectric constant and assume an atomically abrupt surface. This indicates the presence of VH and/or RA. By invoking RA with density gradient scale length as large as $L \sim .02\lambda$ we can fit the Al data, but not the Fe data, for which p-polarized reflectivity R_p drops much too strongly with increasing I. A simple VH model [2] (solid curves), however, agrees well with both Al and Fe data, a preliminary indication of its dominance. Vacuum heating is present for Al, but the large increase in collisional absorption obscures the much smaller VH contribution to the drop in R_p. The rising reflectivity in Fe facilitates identification of VH, but the effect is present for both targets.

Femtosecond-time-resolved reflectivity and frequency domain interferometry experiments [4] then distinguish VH from RA in the time domain based on two criteria. First, although electron oscillations driven by $E_\perp$ cause both RA and VH,

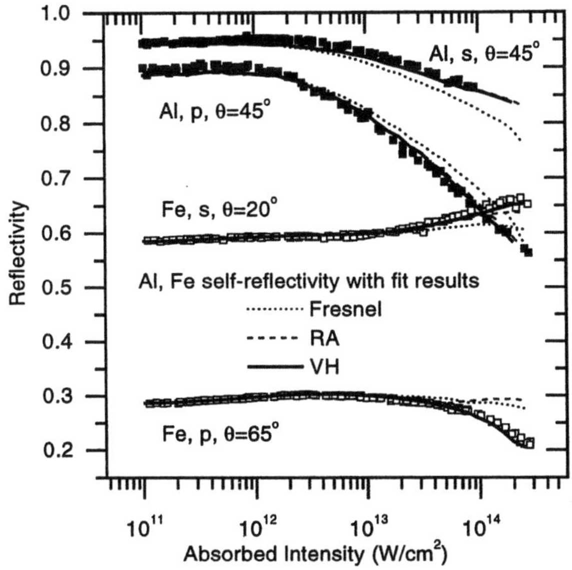

Fig 1. Reflectivity of 120 fs pulses at λ =620 nm vs. laser intensity (points). The samples used are Fe and Al. The fit results for the three models described in the text are shown (lines).

RA requires quiver distance $x_{osc} << L$, while VH dominates when $x_{osc} \geq L$. Second, RA is linear while VH scales as $(I\lambda^2)^{1/2}$. The time-resolved data in Figure 2 reveal that the weak probe sees *rising* R_p at $t = 0$ where the self-reflectivity experiment saw *falling* R_p, consistent with the non-linear nature of VH. Moreover, time-resolved interferometry (not shown) reveals $L = .0035\lambda$ at $t = 0$ (much smaller than the value $.02\lambda$ noted above) for $I \sim 10^{15}$ W/cm^2, for which $x_{osc} = .003\lambda$. Thus the self-reflectivity experiment indeed satisfied $L \sim x_{osc}$, placing it in the regime in which VH dominates. Finally, the p-polarized probe reflectivity pattern for $.02 < t < 2$ ps is explained quantitatively by the emergence of RA as L grows to $L >> x_{osc}$.

In summary, fs self- and time-resolved-reflectivity show that VH dominates for $t \sim 0$, $I > 10^{14}$ W/cm^2 (*i.e.* $L \sim x_{osc}$), that RA dominates for $.2 < t < 2$ ps and is nearly I-independent (*i.e.* $x_{osc} << L$). Femtosecond interferometry verified those regimes by direct measurement of L.

References

1 F. Brunel, Phys. Rev. Lett. 59, 52 (1987).
2 V. F. D'yachenko and V. S. Imshennik, Sov. J. Plasma Phys. 5, 413 (1979).
3 P. Gibbon and A. Bell, Phys. Rev. Lett. 68, 1535 (1992).
4 J. P. Geindre, *et al.*, Optics Letters **19**, 1997 (1994).

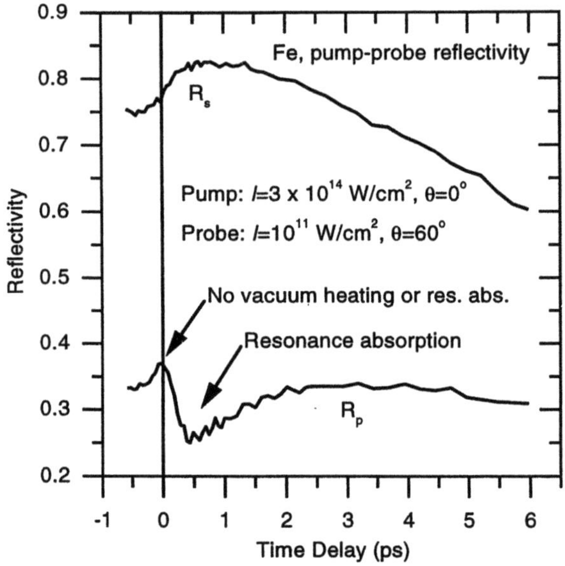

Fig. 2. Reflectivity vs. pump-probe time-delay for an Fe target.

Ultrafast Structural Dynamics in InSb Probed by Time-Resolved X-ray Diffraction

A. H. Chin[1,2], R. W. Schoenlein[2], T. E. Glover[2], P. Balling[3], W. P. Leemans[2], and C. V. Shank[1,2]

[1] University of California, Department of Physics, Berkeley, California 94720, USA
[2] E. O. Lawrence Berkeley National Laboratory, Berkeley, California 94720, USA
[3] Institute of Physics and Astronomy, University of Aarhus, Ny Munkegade, DK-8000 Aarhus C, Denmark

Abstract. Ultrafast structural dynamics in InSb are studied using time-resolved x-ray diffraction. A delayed onset of significant lattice expansion and an indication of subpicosecond disordering are observed.

Time-resolved x-ray diffraction is a useful tool for studying structural dynamics in condensed matter. Time-resolved x-ray diffraction can be performed with femtosecond resolution by using x-ray pulses generated via 90° Thomson scattering [1, 2]. This source of ultrashort x-ray pulses was used to study the structural dynamics of laser perturbed InSb.

The experiments were carried out in the Center for Beam Physics Beam Test Facility at the Advanced Light Source. We performed time-resolved x-ray diffraction on InSb by using the ultrashort x-ray pulses (~ 300 fs, 0.4 Å) generated via Thomson scattering along with the synchronized femtosecond laser pulses involved in the x-ray generation [1, 2]. These intense laser pulses (~100 fs, 800 nm, ~20 mJ/cm^2) perturbed the InSb crystal around the disordering threshold, accumulating ~ 10^4 shots per spot before moving the sample. The InSb crystal used for lattice expansion studies was cut 3° off the (111) orientation to allow x-ray diffraction in the asymmetric Bragg geometry, with the x-ray beam incident ~ 0.3° from the crystal surface. This geometry better matches the x-ray photoelectric absorption depth (~ 500 nm) and attenuation depth (~ 300 nm) to the laser (~ 100 nm) penetration depths (perpendicular to the surface). For the ultrafast disordering studies, the photon energy of the x-rays generated via Thomson scattering was reduced by lowering the electron beam energy. The resulting x-ray pulses (~ 600 fs, 1.7 Å) along with a different asymmetrically cut InSb crystal (12.8° off the (111) orientation) provided more sensitivity to the surface (~ 40 nm probing depth at an incident angle of ~ 0.5° from the surface).

Representative 0.4 Å x-ray diffraction spectra for –20 ps and +100 ps time delays are shown in Figure 1. The overall number of diffracted photons increases with laser perturbation, and the peak of the diffraction spectrum at +100 ps is slightly shifted towards lower energy relative to the diffraction spectrum at –20 ps. This is consistent with previous observations of laser-induced lattice expansion in time-resolved x-ray diffraction experiments on Si [3] and in GaAs [4].

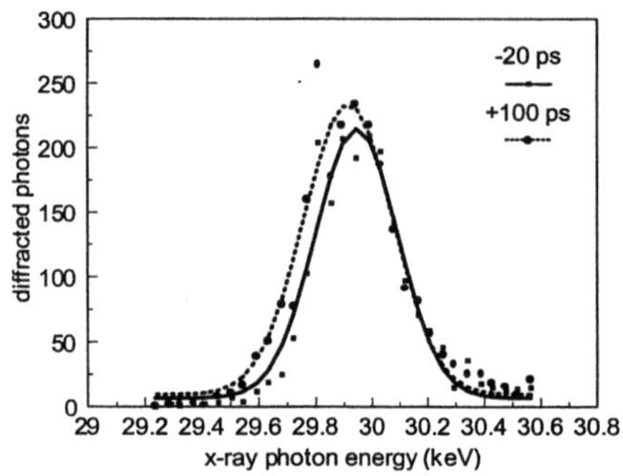

Figure 1. X-ray diffraction spectra from laser perturbed InSb at time delays of –20 ps (squares) and +100 ps (circles). Solid and dashed lines are Gaussian fits.

The x-ray diffraction spectra integrated over the Bragg peak as a function of time delay is shown in Figure 2. We call attention to three features of the data. First, a delay (~ 10 ps) in the onset of the rise is observed. Second, there is a rise in the integrated counts for long time delays. Finally, the peak of the diffracted spectrum shifts towards lower energy.

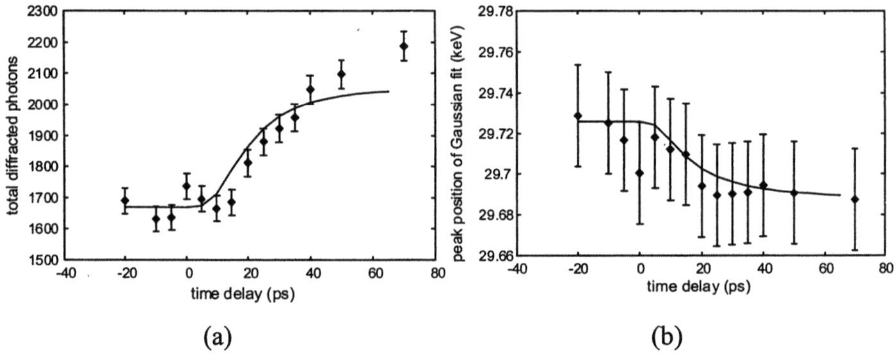

Figure 2. (a) Total diffracted x-ray photons (~ 30 keV) as a function of time delay. (b) The peak of the Gaussian fit to the x-ray diffraction spectra as a function of time delay. Solid lines are from the laser heating simulation.

The delay in the onset of lattice expansion can be attributed to energy relaxation from the electron-hole plasma eventually to acoustic phonons. The absorbed laser pulse energy, initially deposited into the electron-hole plasma, is subsequently transferred to the lattice via longitudinal optical (LO) phonon emission. The time scale for this energy relaxation has been recently observed by optical experiments in GaAs [5]. Lattice expansion (nucleating at the surface) starts to occur as LO phonons decay into acoustic phonons. This lattice expansion is induced by the generation and propagation of acoustic phonons.

A simulation, based on that of Lietoila and Gibbons [6], taking into account the strained crystal x-ray diffraction response [3] qualitatively fits the data (see Figure 2). This simulation accounts for Auger recombination and LO phonon emission leads to the transfer of energy from the carriers to the lattice. Also included in the simulation are the LO phonons to acoustic phonon relaxation time, and the expanded layer propagation at the speed of sound (5×10^5 cm/s) into the sample.

Figure 3 shows data taken using ~ 7.3 keV x-ray photons. There are indications of a rapid decrease in diffracted photons after 0 ps and a shift towards lower energy beginning around +10 ps. The rapid decrease in diffracted photons would be consistent with x-ray absorption due to the formation of a disordered layer on a subpicosecond time scale [5, 7, 8]. The spectral shift at +10 ps is consistent with the onset of lattice expansion described above.

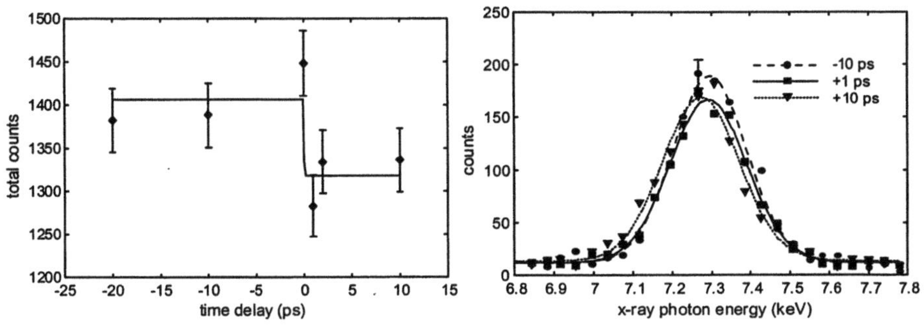

(a) (b)

Figure 3. (a) Total diffracted x-ray photons (~ 7.3 keV) as a function of time delay. (b) X-ray diffraction spectra for different time delays. Solid and dashed lines are Gaussian fits.

Time-resolved x-ray diffraction was used to study ultrafast structural dynamics in laser perturbed InSb. A delay in the onset of lattice expansion is observed, due to the time for energy relaxation from electron-hole pairs to acoustic phonons. An indication of subpicosecond disordering of a thin surface layer is also observed. This work was supported by the U.S. Department of Energy under Contract No. AC03-76SF00098 and by the National Science Foundation under grant PHY-9512693.

References

[1] R. W. Schoenlein, et al., *Science* **274**, 236 (1996).

[2] W. P. Leemans, et al., *Phys. Rev. Lett.* **77**, 4182 (1996).

[3] B. C. Larson, et al., *J. Mater. Res.* **1**, 144 (1986).

[4] T. Guo, et al., *SPIE* **3157**, 84 (1997).

[5] L. Huang, et al., *Phys. Rev. Lett.* **80**, 185 (1998).

[6] A. Lietoila and J. F. Gibbons, *Appl. Phys. Lett.* **40**, 624 (1982).

[7] C. V. Shank, et al., *Phys. Rev. Lett.* **50**, 454 (1983).

[8] I. L. Shumay and U. H:ofer, *Phys. Rev. B* **53**, 15878-15884 (1996).

Time-resolved x-ray diffraction of GaAs with a 20-fs, laser-driven plasma source

Ralph Jimenez, Christoph Rose-Petruck, Ting Guo, Kent R. Wilson, and Christopher P.J. Barty

Department of Chemistry and Biochemistry, University of California, San Diego, La Jolla CA 92093-0339, USA
E-mail: rxj@chem.ucsd.edu

Abstract. Laboratory-scale, optical pump/x-ray probe measurements are used to measure GaAs lattice dynamics following excitation with a 25-fs, 800-nm pulse.

1. Introduction

X-ray techniques provide scientists with one of the most direct means of measuring the positions of atoms in molecular, inorganic, and protein crystals. For some time, it has been hoped that the emergence of time-resolved x-ray methods will allow the the positions of the atoms to be charted on ultrafast time scales during chemical, physical and biological processes. We discuss a laboratory-based, ultrafast x-ray experiment which is used to follow changes in the structure of a GaAs crystal following short pulse optical excitation. Our ultrafast x-ray source[1] is based on an ultrafast laser-produced plasma. Since laser-produced plasma x-rays can simultaneously be of subpicosecond duration [2] and reasonably high flux, this method is a practical means of performing time-resolved x-ray measurements with a laboratory scale system.

2. Experimental Methods

The 800-nm laser pulses are generated at 20 Hz by an ultrafast Ti:sapphire chirped pulse amplification system [3]. The output of the laser system is split for use in sample excitation and x-ray generation. This scheme provides jitter-free synchronization between optical and x-ray pulses. The x-ray pulses are generated by focusing 20-fs, 50-mJ, laser pulses onto a moving Cu wire in vacuum. The x-ray spectrum features a Cu K_α line at 1.54 Å (8.04 keV) (which consists of two spin orbit split lines, $K_{\alpha 1}$ (1.5407 Å) and $K_{\alpha 2}$ (1.5439 Å)) and a weaker Cu K_β line at 1.39 Å (8.96 keV). Our laser-generated x-ray yield is 10^{10} Cu K_α photons$(4\pi$ steradians s$)^{-1}$. The K_α radiation diffracted by the optically pumped GaAs is detected by a Princeton Instruments x-ray CCD camera. The GaAs sample, a 500-μm thick single crystal wafer cut along the (111) plane, is located 25 cm from the Cu wire x-ray source and the CCD image plane is placed 57 cm from the sample. We excite the crystal with ~25-fs pulses at fluences ranging from 25 to 100 mJ/cm^2. The pump beam spot size is set to about 4 mm^2, such that it spatially encompasses the Bragg angles of both K_α lines and the optically induced diffraction lines.

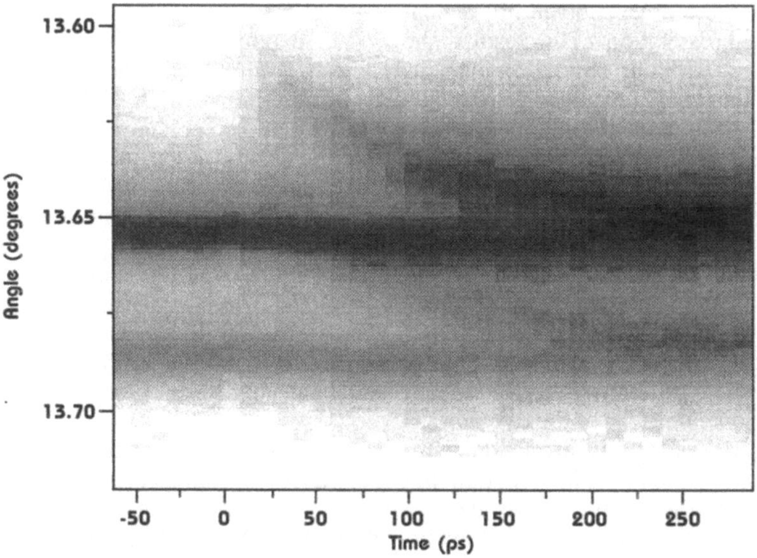

Figure 1. Image of time-resolved x-ray diffraction intensity from GaAs pumped with an excitation fluence of 59 mJ/cm^2.

3. Results and Discussion

The data show the static diffraction of Cu $K_{\alpha 1}$ and $K_{\alpha 2}$ radiation from the (111) planes, as well as new diffraction lines which appear upon laser excitation. In Figure 1, we show an image of time-resolved diffraction. The line at smaller angles (larger lattice spacing) from $K_{\alpha 1}$ increases in intensity and moves towards the static $K_{\alpha 1}$ line as a function of delay. As the dynamics proceed, the total diffraction intensity from the optically pumped region increases by almost a factor of two from the static value. The new diffraction peaks merge with the static peaks as the lattice returns to equilibrium on a timescale of hundreds of picoseconds; the dynamics are fully reversible. The diffraction pattern at each delay time was measured in two minutes, yielding data collection rates comparable to many optical pump-probe experiments.

It is important to note that femtosecond optical excitation causes the appearance of new diffraction lines rather than the broad shoulders typically observed for nanosecond optical excitation [4]. The nature of the diffraction pattern arises because of the mismatch between the optical excitation depth and the x-ray probe depth. The x-rays probe a ~1 μm, non-uniformly excited depth of the GaAs crystal. Modelling with dynamic diffraction theory explicitly considers the strain profile (lattice spacing vs depth) induced by the optical excitation [4]. We have performed preliminary dynamic diffraction modelling, as well as separate optical transmission measurements which show that the strain profile is not exponential, such as those found in nanosecond studies [4, 5]. The model, which fits the intensity and position of the time-resolved diffraction peaks, neglects lateral inhomogeneity of the optical excitation. The results show that exponential strain profiles are inconsistent with our diffraction data, and indicate strain profiles which are flatter over depths of a few hundred nanometers. Furthermore, both diffraction and separate optical transmission measurements show that the excitation pulsewidth is important for this effect. When the duration of the excitation pulse is increased to >500 fs, the optically induced diffraction peak disappears, and the optical transmission drops. However, the absorbed energy, as measured by monitoring the reflected light, remains constant, clearly indicating that the initial dynamics are non-thermal. The measured lattice expansion and contraction therefore result from excitation of electrons into the conduction band and subsequent carrier relaxation [6]. In principle, the time-resolved strain profiles, which are purely a measure of nuclear dynamics, may be used to infer the spatial distribution of the carriers. In this way, the ultrafast x-ray diffraction measurement can be used to separate nuclear and electronic portions of the dynamics.

Time resolved x-ray diffraction with femtosecond laser produced plasmas is beginning to yield novel information on photoinduced dynamics in crystals. These table-top ultrafast x-ray sources have evolved to the point that data acquisition times are comparable to those of many "all-optical" ultrafast techniques. In the near future, we expect these methods will be used to resolve directly the molecular dynamics of chemical systems.

References

1 T. Guo, C. Rose-Petruck, R. Jimenez, F. Råksi, J. Squier, B. Walker, K.R. Wilson and C.P.J. Barty, in *Applications of X-Rays Generated from Lasers and Other Bright Sources* SPIE, **3157**, 84-92 (1997).

2 C. Rischel, A. Rousse, I. Uschmann, P.-A. Albouy, J.-P. Geindre, P. Audebert, J.-C. Gauthier, E. Förster, J.-L. Martin and A. Antonetti, *Nature* **390**, 490-492 (1997).

3 C.P.J. Barty, T. Guo, C. Le Blanc, F. Raksi, C. Rose-Petruck, J. Squier, K.R. Wilson, V.V. Yakovlev and K. Yamakawa, *Opt. Lett.* **21**, 668-670. (1996).

4 B.C. Larson, J.Z. Tischler and D.M. Mills, *J. Mater. Res.* **1**, 144-154 (1986).

5 J.Z. Tischler, B.C. Larson and D.M. Mills, *App. Phys. Lett.* **52**, 1785-1787 (1988).

6 P. Stampfli and K.H. Bennemann, *Phys. Rev. B* **49**, 7299-7305 (1994).

Interaction of Superintense Femtosecond Laser Pulses with Cluster-like Solids

A.B.Savel'ev, D.M.Golishnikov, V.M.Gordienko, P.M.Mikheev, A.A.Shashkov, T.M.Vlasov, and R.V.Volkov

International Laser Centre & Physics Faculty M.V.Lomonosov Moscow State Universit, Vorobyevy gory, Moscow, 119899, Russia, Phone: +7-095-9395318; e-mail:savelev@femto.ilc.msu.su

Abstract. We present results on plasma formation in porous silicon (cluster size 3-5 nm, mean density 0.1-0.5 of crystalline silicon) by femtosecond laser pulses at intensity above 10 TW/cm^2.

Introdiction.

Recently the new type of target consisting of nanoclusters (cluster-like solids or CLS) was introduced in femtosecond laser plasma interaction. Namely, colloidal metal films of Au and Al, jet of noble atoms clusters, porous silicon all revealed very specific behavior under femtosecond laser plasma interaction [1-7]. The porous silicon here is only one representative of wide class of materials (porous Al and glass, carbon foam, etc). This class has one distinct feature, near solid state density, which differs it from other cluster materials and causes more efficient laser absorption, hot electron production, and hard x-ray generation.

In this paper we report on our recent progress in both experimental and theoretical investigation of interaction of superintense femtosecond laser pulses with CLS - highly porous silicon.

Interaction of ultrashort laser pulse with CLS differs in some chief aspects from that with bulk solids (BS), badly depending on porosity P (ratio of BS to CLS densities). Firstly, initial reflectivity of CLS is lower than that for the BS - it looks almost black. Secondly, huge surface area of CLS (for P=6 ratio of the CLS surface to the BS one can reach the value of 600) allows effective "vacuum heating" and generation of hot electrons. Thermal conductivity λ of CLS could be estimated neglecting ablation as $\lambda_{CLS}/\lambda_{BS}\sim P^2$, that gives the ratio for the specific absorbed heat per atom of $\varepsilon_{CLS}/\varepsilon_{BS}\sim P^{1/2}$.

Finally, hydrodynamic and Coulomb explosion of clusters fills the intercluster void on the time scale of 100-300 fs. This leads to the two main consequence - efficient resonant absorption and ion jet collisions. The first one again forces hot electron generation, whereas due to the second process mean ion temperature in 100-200 fs becomes of the order of 20-30 keV, much higher than electron temperature [7]. On the possible application for such non-equilibrium plasma will be discussed in the end of the paper.

In performing experiments on femtosecond laser-plasma interaction with CLS we used femtosecond laser system described elsewhere (see, for instance, [5]), which provides intensity in excess of 10^{16} W/cm^2 with intensity contrast ratio better than 10^5. Soft X-ray emission from plasma was measured by p-i-n diodes. Hard X-rays were detected after 100 μm Be output window (transparency 0.1 for quanta energy E~2.3 keV) by means of NaI(Tl) crystal along with photomultiplier tube. A set of filters was used to distinguish between different spectral bands from 2.5 up to 40 keV. For the porous silicon (por-Si) samples preparation procedures see [6].

To estimate from our hard X-ray data electron temperature we used double-channel registration scheme. For each channel with transparency function $H_i(E)$ ($E = \hbar\omega$ - X-ray quantum energy) the output signal could be written as:

$$s_{1,2} = f(T_e)\int H_{1,2}(E)\cdot e^{-\frac{E}{T_e}}\cdot dE$$

(assuming that quantum energy is much higher than photorecombination threshold E_{th}~2.5 keV for deeply ionised Si). To get the value for T_e directly from single channel data one needs to know function $f(T_e) = N_e \cdot Z \cdot T_e^{1/2}$ (N_e- electron concentration and Z - mean ion charge). At the same time for the ratio r of two channels with different filters one cam easily obtain the following:

$$r = \frac{\int H_1(E)\cdot e^{-\frac{E}{T_e}} dE}{\int H_2(E)\cdot e^{-\frac{E}{T_e}} dE}. \quad (1)$$

Hence measuring in one shot signals from two channels one can get electron temperature, making use of Eq.(1).

Fig.1. shows the result of this procedure for two samples (c-Si and por-Si). It should be noticed here that this Figure collects results for different filter pairs. The first filter was the same - 100 μm Be, while the second one was one of the following - 180 μm Al (E>9keV), 13 μm Ta (E>11 keV), 1 μm Al (E>17 keV), 26 μm Ta (E>20 keV).

For all the intensities we obtain for the electron temperature the value far above 1 keV - typical value for mean electron temperature in femtosecond plasma. This should be attributed as the consequence of hot electron production in plasma. Fitting data from the Fig.1 we obtained $T_e \sim I^{0.6}$ for c-Si sample. This appears to be in good agreement with different theories of hot electron production on sharp boundary (so-called "vacuum heating") or on the smooth one ("resonant absorption"). For the latter case the scaling should be slightly lower - $T_e \sim I^{0.3}$. For the porous sample the scaling is even higher - $T_e \sim I^1$.

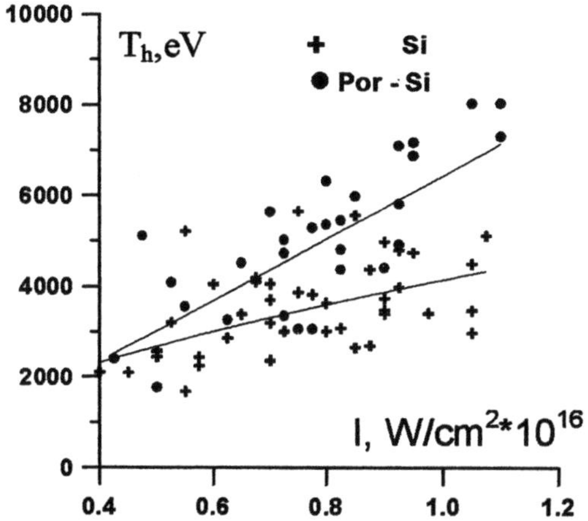

Figure 1 Hot Electron temperature T_h dependency on laser intensity I for c-Si (crosses) and por-Si (full circles).

Besides silicon itself, large amount of hydrogen atoms and OH groups are absorbed on the surface of silicon cluster. This could strongly affect the interaction process or either cause new physical phenomena. For instance, hydrogen atoms could be substituted by deuterium ones. Ion kinetic energy emerging from cluster explosion should transform to thermal energy through ion-ion collisions, thus giving rise to direct D-D fusion: D+D → *^{3}He(0,82 MeV) + n (2,45 MeV).* The cross-section for this reaction reaches $4*10^{-28}$ cm^2 for D ion mean energy E_D~30 keV [8]. Hence, the number of D+D reactions and consequently, the number of neutrons in one laser shot, would be 10^4 (D ions concentration $\sim10^{21}$cm^{-3}, plasma volume $\sim10^{-10}$ cm^3, plasma cooling time 10 ps).

Acknowledgments. Authors greatly acknowledge support from Russian Foundation for Basic Research (grant #96-02-19146a) and State Scientific Programs "Fundamental Metrology" and "Russian Universities".

References.

1. C.Wulker, W.Theobald, D.R.Gnass, et. al., Appl.Phys.Lett., 68 (10) 1338 (1996).
2. A.McPherson, K.Boyer, C.K.Rhodes, J.Phys.B: At.Mol.Opt.Phys., 27 L637 (1994).
3. Y.L.Shao, T.Ditmire, J.W.G.Tisch, et.al., Phys.Rev.Lett., 77 (16) 3343 (1996).
4. T.Ditmire, T.Donelly, A.M.Rubenchik, et al., Phys.Rev., A53(5) 3379 (1996).
5. R.V.Volkov, V.M.Gordienko, M.S.Dzhidzhoev, et al., Quantum electronics, 27 #12, 1081-1093 (1997).
6. R.V.Volkov, V.M.Gordienko, M.S.Dzhidzhoev, et al., Quantum electronics, 28 #1, 1-2 (1998).
7. A.B.Savel'ev, V.G.Babaev, M.S.Dzhidzhoev, et.al., Laser Physics, 8 (3), 1 (1998).
8. K.Brakner, S.Jorna, Laser driven fusion, KMS Fusion, Inc, 1973.

Probing Black Hole Physics in the Laboratory Using High Intensity Femtosecond Lasers

W. Theobald[1], R. Häßner[1], R. Kingham[1], T. Feurer[1], H. Schillinger[1], G. Schäfer[2], and R. Sauerbrey[1]

[1]Institut für Optik und Quantenelektronik, Friedrich-Schiller-Universität Jena, Max-Wien-Platz 1, D-07743 Jena, GERMANY, Phone: +49 3641 947200, Fax: +49 3641 947202, E-mail: sauerbrey@qe.physik.uni-jena.de
[2]Theoretisch-Physikalisches Institut, Friedrich-Schiller-Universität Jena, Max-Wien-Platz 1, D-07743 Jena, GERMANY

Abstract. Mass acceleration in femtosecond laser plasmas exceeding 10^{18} g is measured using the frequency resolved optical gating technique. Accelerations of this magnitude appear in the stretched horizon regime of black holes. Detection of chirped light from black hole candidates would allow for unique identification.

1. Introduction

One of the major unresolved problems in astrophysics is the unique identification of black holes. Due to the limited spatial resolution of astronomical observations a large mass concentration alone is not sufficient. Processes showing the existence of an event horizon are, however, clear signatures of a black hole. Here we describe a scenario where the investigation of light emission from stretched horizons of black holes can be simulated in intense femtosecond laser-plasma interaction experiments and simultaneously leads to a clear signature for a stellar mass black hole.

Femtosecond pulse generated laser-plasmas are created using high power table-top laser systems providing intensities exceeding 10^{18} W/cm^2. At these intensities rapid ionization of a solid during the first few optical cycles leads to the formation of a high density plasma with electron densities $n_e \geq 10^{23}$ cm^{-3} exceeding that of solids [1,2]. The femtosecond laser pulse rapidly heats the electrons in the plasma and a pressure gradient is built-up.

Internal pressures approaching hundreds of Megabars accelerate the plasma away from the target. Similar accelerations in the opposite direction are caused by the radiation pressure which reaches about 1 Gigabar at an intensity of $I \approx 10^{18}$ W/cm^2. These accelerations actually dominate the motion of plasma material during the laser pulse duration of $\tau_L \approx 100 fs$. For an expanding plasma the final velocity reached by the plasma corresponds to the ion acoustic velocity of $v_i \approx 10^7 \ldots 10^8$ cm/s and is small compared to the speed of light. Due to the short laser-pulse duration and the rapid heating of the plasma, however, this velocity is reached in a very short time and thus the acceleration may be estimated to $b \approx v_i / \tau_L \geq 10^{18}$ m/s^2 [3,4]. We will show that for a direct measurement of the acceleration of the plasma surface (more precisely, the location of the critical

electron density), amplitude and phase of the reflected laser pulse have to be measured.
For an evaluation of such phase sensitive experiments we consider laser pulses of pulse duration $\tau_L \approx 100\,\text{fs}$ and intensities ranging from 10^{16} to 10^{19} W/cm^2. For typical plasma parameters we obtain for intensities in the 10^{16} W/cm^2 range and a sufficiently high contrast ratio $L \leq v_i\tau_L \approx 20\,\text{nm}$ as an upper limit for the plasma scale length, using a value of $v_i \approx 2\times10^7$ cm/s for the ion acoustic velocity. $L/\lambda << 1$ holds for typical laser wavelengths of $\lambda = 800$ nm and the kinematics of the plasma surface may be described by a moving mirror of temporally changing reflectivity. The electric field of frequency ω_0 impinging on the plasma mirror has the form $\exp\{i(\omega_0/c)[ct - x]\}$. The solid surface is located at $x = x_0$ before arrival of the laser pulse and the location of the critical surface as a function of time is given by $x(t) = x_0 + v_0t + bt^2/2$, where v_0 is an arbitrary initial expansion velocity and b is the acceleration of the plasma front. For perpendicular incidence the electric field $E_R(t_r)$ reflected from the accelerated mirror has for $|v_0| << c$, $|b_0|\tau_L << c$ the following form in the retarded time $t_r = t - (x - x_0) / c$:

$$E_R(t_r) = E(t_r) \exp\{ i(\omega_0/c) [2x_0 + (c + 2v_0)t_r + bt_r^2] \} \qquad (1)$$

where $E(t_r)$ is the reflected pulse envelope [3,4]. It is evident from this equation that light emitted from a uniformly accelerated source shows a linear chirp in frequency. The observation of chirp from kinematically or gravitationally accelerated mirrors can therefore be used to measure acceleration.

2. Experiments

The experimental set-up is shown in Figure 1. The experiment was performed with a Chirped-Pulse-Amplification Ti:sapphire laser system delivering unchirped laser pulses of 110 fs pulse duration with a repetition rate of 10 Hz at a wavelength of 793 nm. The available pulse energy was 105 mJ on target with a contrast ratio of peak pulse power to that of the pedestal on the order of 10^4. The whole experiment took place inside a target chamber evacuated to a pressure of 1 mbar. The laser beam was focused with an angle of incidence of 12° by an off-axis parabolic mirror with an effective focal length of 17.9 cm onto a polished glassy carbon or a glass target. The target was mounted on a xyz-positioning table to provide a fresh surface for each shot. The focus diameter was measured with a microscope objective to be less than 10 µm and hence an intensity of up to 2×10^{18} W/cm^2 was reached in the spot.
A collimating lens with a focal length of f = 250 mm is used to collect the reflected light into a parallel beam and send it to a Second-Harmonic-Generation (SHG) FROG-autocorrelator. The central part of the laser beam with a diameter of 5 mm is taken out and is split into two replicas which are focused by cylindrical lenses to two spatially overlapping lines on a BBO crystal. Provided that the intensity distribution of the fundamental is constant along the line focus the generated SHG signal is symmetric in time and consequently the FROG trace obtained by spectrally resolving the autocorrelation trace is symmetric about the $\tau = 0$ axis. By using a FROG retrieval algorithm [5] it is possible to reconstruct

the temporal amplitude of the electric field and the absolute value of the phase of the laser pulse, respectively, out of the FROG traces in an iterative minimization process. The sign of the phase can not be determined with a SHG FROG.
In addition, the spectrum of the reflected laser pulse has been measured separately by a spectrograph to investigate the expected broadening and the shift in the spectrum and to test the FROG trace quality.

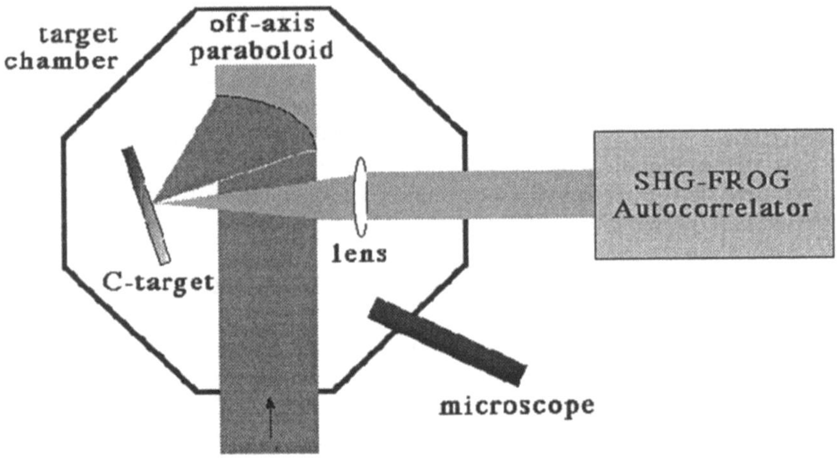

Fig. 1. Experimental set-up

3. Results and Discussion

Figure 2 shows the result of a FROG retrieval algorithm for two single shot measurements of the incident (gray) and the reflected (black) waveform. The solid curves represent the electric field amplitudes and the dotted curves the phases.
The pulse duration of the reflected electrical field is increased by a factor of 1.7 compared to the incident laser beam and the phase exhibits a strongly parabolic curved shape. This is attributed to an accelerated plasma mirror which leads to the parabolic phase modulation of the reflected pulse.
By fitting the phase to a second order polynomial function $\varphi = C\tau^2$ a value of $\varphi = (3.2 \pm 0.5)\times 10^{26}\,(\text{rad}/\text{s}^2)\times \tau^2$ is obtained. The measurement yields for the 793 nm laser radiation an acceleration of $b = (2.0 \pm 0.3)\times 10^{19}\,\text{m}/\text{s}^2$ for the plasma, which is in good agreement with theoretical estimates. An acceleration on the same order is also obtained by analyzing the spectra of the reflected pulse with a time integrating spectrograph. A red shift of 6 nm is observed which is interpreted as the result of the reflection from a plasma that is pushed into the target by the light pressure. From the spectral broadening [3] of the reflected light an acceleration on the order of $2\times 10^{19}\,\text{m}/\text{s}^2$ is estimated in agreement with the results of the FROG measurement. From the spectral shift, however, an

acceleration of only about 1×10^{19} m / s^2 is obtained indicating again that spectral shifts are less reliable in determining the plasma motion than spectral shapes and phase sensitive analysis [3].

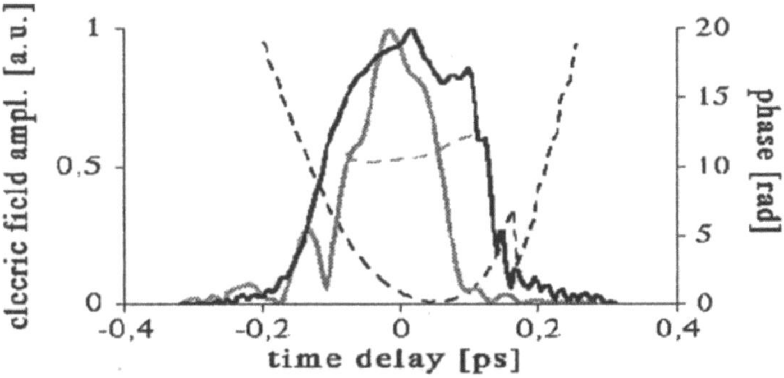

Fig. 2. Results of the FROG retrieval algorithm: Electric field amplitudes (solid lines) and phases (dashed lines) of single shot incident laser pulse (gray) and laser pulse after reflection on a plasma (black).

Accelerations of the order of 10^{18} g exceed even the gravitational acceleration on the surface of neutron stars by a factor of 10^6. Accelerations of this magnitude occur as gravitational accelerations only close to the event horizon of a black hole in the stretched horizon regime. Applying the Einstein equivalence principle, the high-acceleration laboratory experiments can directly be transferred to stretched horizons of black holes. The mirror, accelerated in Minkowski space, corresponds to a mirror located at a black hole stretched horizon and the laser source, at rest in the Minkowski space, corresponds to a radially freely falling source near the stretched horizon, having zero velocity at the time t = 0. To know what an observer measures far away from the stretched horizon we have to transform the phase of the wave to Schwarzschild coordinates and obtain [6]: $\exp(i\Omega u_0)\exp\{i(\Omega u_0 c / 2r_s)[t_r + ct_r^2 / 4r_s]\}$. In this expression, $\Omega u_0 c / 2r_s$ and $b_F \equiv c^2 / 2r_s = c^4 / 4GM$ are respectively the angular frequency and the phase acceleration measured with respect to the observer's proper time t far away from the horizon. G is the gravitational constant, M is the black hole mass, Ω is the angular frequency in Kruskal time, u_0 corresponds to the position of the stretched horizon in Schwarzschild coordinates, and $r_s = 2GM / c^2$ is the Schwarzschild radius. It is important to note that the measured acceleration is independent from the chosen stretched horizon. It is also worthwhile to mention that direct light from freely back falling sources in the stretched horizon regime, i.e. light not being reflected at a stretched horizon, reveals an acceleration pattern in its phase with just the opposite sign for the acceleration.

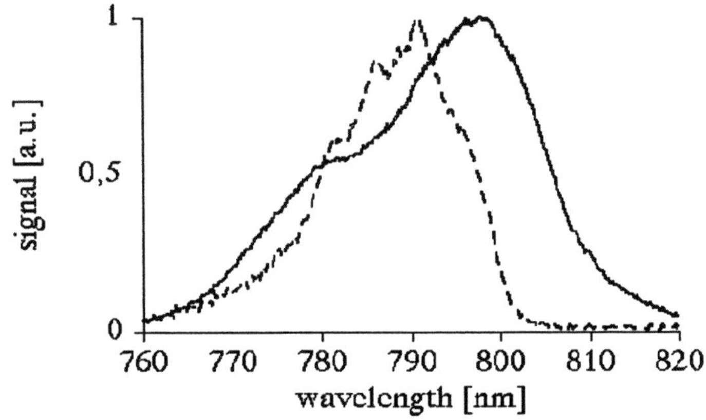

Fig. 3. Spectrum of the laser pulse before (dashed line) and after (solid line) reflection on the accelerated plasma.

For a neutron star with mass M the gravitational acceleration at its surface is smaller than $(8/9)^2\, b_F$ so, using phase measurements, a discrimination between black holes and neutron stars is possible if the mass of the object is known with sufficient precision by other means. For a neutron star the effect should also be observable and the reflecting mirror is made by its surface and the source for light are near-surface eruptive plasma processes.

In summary we have measured with phase sensitive techniques the acceleration of a fs-laser produced plasma and we find that the application of the equivalence principle and the concept of plasma loaded stretched horizons to such experiments lead to a new experimental method to identify stellar mass black holes.

References

1. J. C. Kieffer, Z. Jiang, A. Ikhlef, C. Y. Cote, and O. Peyrusse, J. Opt. Soc. Am. B **13**, 132 (1996).
2. W. Theobald, R. Häßner, C. Wülker, and R. Sauerbrey, Phys. Rev. Lett. **77**, 298 (1996).
3. R. Sauerbrey, Phys. Plasmas **3**, 4712 (1996).
4. R. Häßner, W. Theobald, S. Niedermeier, K. Michelmann, H. Schillinger, T. Feurer, G. Schäfer, and R. Sauerbrey, Superstrong Fields in Plasmas, ed. L. Lontano, American Institute of Physics 1998, p. 213.
5. R. Trebino and D. J. Krane JOSA A **10**, 1101 (1993).
6. N. D. Birell and P. C. W. Davies, Quantum Fields in Curved Space Time (Cambridge University Press, Cambridge 1982).

Experimental observation of photon acceleration of femtosecond laser pulses in relativistic under-dense ionization fronts.

C. Stenz[1], F. Blasco[1], J. Dias, N. Lopes[2], L. Oliveira e Silva[2], T. Mendonça[2], A. Dos Santos[3], A. Mysyrowicz[3], and A. Antonetti[3]

[1] ELIA/CPMOH, Université Bordeaux 1,33405 Talence, France
E-mail: cstenz@celia.u-bordeaux.fr
[2] GoLP, Centro de Fisica dos Plasmas, Instituto Superior Técnico, 1096 Lisboa Codex, Portugal
[3] LOA, ENSTA/Ecole Polytechnique, 91120 Palaiseau, France

Abstract. Frequency up-shifts from photon acceleration ands flash ionization of ultrashort laser pulses counter-propagating and co-propagating at a given angle with respect to a laser produced relativistic under-dense ionization are studied.

1. Introduction

The 'photon acceleration' concept was initially proposed by Wilks *et al* [1] to label the frequency shift produced by a plasma wave wake field. The Hamiltonian formulation of the ray-tracing trajectories [2] has been used to describe the various mechanisms of the photon acceleration. This leads to a straightforward derivation of the frequency shifts. When a laser pulse (probe pulse) encounters a relativistic ionization front (a gas-plasma interface) it will be up-shifted in frequency [1-4] and can be compressed in time [5]. Such a front can be created by optical-field ionization of a gas with an intense ultra-short laser (ionization pulse). Significant up-shifts will occur even for under-dense ionization fronts [2,6]. The frequency up-shift of the probe pulse is connected to a fast variation of the phase due to :
-the variation in time of the refractive-index during the leading edge of the ionization pulse: the 'flash ionization blue-shift effect' [6].
-the variation in space of the refractive-index when the probe pulse crosses the moving ionization front at velocity v_f : the 'ionization front blue-shift effect' [2,7]. Three types of phenomena have been looked at : the frequency shift of the transmitted photons for a counter-propagating probe pulse, the Doppler shift of reflected photons for a co-propagating probe pulse, witch correspond to the co-propagating relativistic mirror [7] and the flash ionization blue-shift effect.

We consider for both cases (co-propagation and counter-propagation) an oblique incidence of the probe pulse with respect to the ionization front. Using the ray-tracing and the phase continuity equations at the plasma front, the dispersion equation of the em wave respectively in gas (vacuum) and in a cold collision-less plasma we obtain the following equations :

-for the transmitted wave and for $\omega_o > \omega_c$

$\omega_0^2 = k_0^2c^2$; $\omega_1^2 = k_1^2c^2 + \omega_p^2$; $\omega_0 - k_{0x} v_f = \omega_1 - k_{1x} v_f$; $k_{0y} = k_{1y}$

where ω_p is the plasma frequency and ω_c the cutoff frequency given by

$\omega_c = \omega_p (1- \beta^2)^{1/2} / \{(1 - \beta \cos \theta_0)^2 - (1- \beta^2) \sin^2 \theta_0\}^{1/2}$ where $\beta = v_f/c$

-for a reflected probe pulse when the front overtakes it and for $\omega_o > \omega_c$:

$\omega_0^2 = k_0^2c^2$; $\omega_1^2 = k_1^2c^2$; $\omega_0 - k_{0x} v_f = \omega_1 - k_{1x} v_f$; $k_{0y} = k_{1y}$

It is straightforward to derive the frequency up-shift, the refraction angle as well the group velocity in the moving front or in the laboratory frame for configurations of interest. The frequency up-shifts for counter and co-propagating probe pulses are given for the transmitted waves by

$$\omega_1/\omega_0 =(1-\beta\cos\theta_0)/(1- \beta^2) \pm [\beta^2 (1-\beta\cos\theta_0)^2 - \omega_p^2 (1-\beta^2)/\omega_0^2]^{1/2} / (1-\beta^2) \qquad (1)$$

$$\omega_1/\omega_0 =(1-\beta\cos\theta_0) / (1- \beta^2) \pm \beta\left| \beta - \cos\theta_0 \right| / (1-\beta^2) \qquad (2)$$

For the flash ionization the frequency shift is simply given by [7]
$\omega_2/\omega_0 = (1 + \omega_p^2/ \omega_0^2)^{1/2}$

2. Experimental Results

The experimental set-up is presented elsewhere [7].The ionization front is produced in an argon gas jet by a 5mJ energy, 65 fs pulse duration, 620 nm wavelength and 10 Hz repetition rate laser pulse. The probe laser pulse correspond to a fraction (0.05%) of the ionization laser pulse. Time resolved plasma density (Moiré interferometry) and front velocity (shadowgraphy) measurements have been realized by using a 35 fs , 560 nm wavelength and 10 Hz repetition rate laser pulse. Typical spectra of the blue-shift of the probe beam at 400 mm away front the focal region (a) and at the focal region (b) are shown on fig 1 for counter and co-propagation. In counter-propagation the blue-shift of the transmitted probe pulse increases from 1.9 mm to 2.3 nm at focal region where the electron density is maximum. This shift is due to photon acceleration process in the ionization front as predicted by equation (1). A second peak is also obtained at 4.8 nm. The same blue-shift of 4.8 nm is observed in both counter and co-propagation experiments and can be clearly associate to the flash ionization process. For co-propagation we observe a blue-shift of 8.6 nm, as expected from the photon acceleration equation (2). In this case the co-propagating relativistic mirror effect is involved. The difference between the observed up-shifts in both cases is a clear evidence of the photon acceleration process in the relativistic moving ionization front.

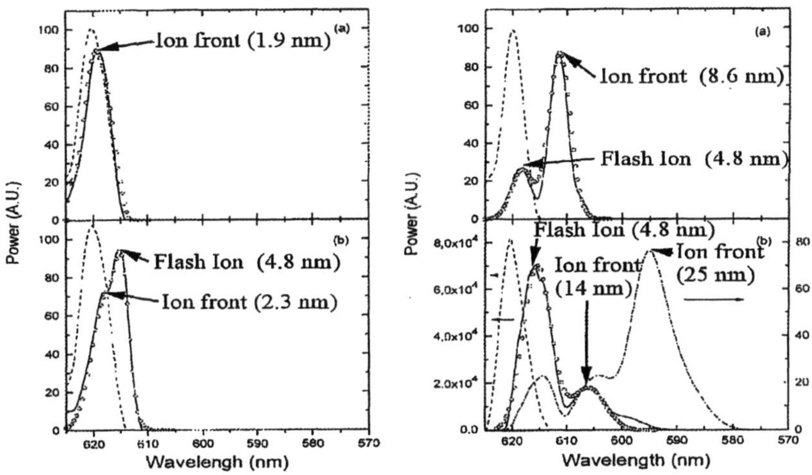

Fig. 1. Counter-propagating spectra : θ_0=160° (left) and co-propagating spectra θ_0=20° (right). Dashed line : original probe pulse spectrum, solid line : up-shifted spectrum, dash-dotted line : filtered up-shifted spectrum, circles : numerical simulation.

3. Conclusions

We have observed the frequency up-shifts for an ultra-short laser pulse with an ionization front in opposite directions. Efficiencies of frequency shift as high as 75% have been obtained. We have also observed, for the first time, the co-propagating relativistic mirror as predicted by the ray-tracing theory. Blue-shifts due to photon acceleration in the ionization front and due to flash ionization processes have been clearly identified..

Acknowledgements. This work was supported by EC Access to Large Scale Facilities Program under the contract No ER-BFMGECT950019..

References

1 S.C Wilks, *et al*, Phys. Rev. Lett., vol 62 , 2600, 1989
2 J.T Mendonça, and L. Oliveira e Silva , Phys. Rev. Vol E49, 3520, 1994
3 V.I. Semenova Sov. Radiophys., vol 10, 199, 1967
M. Lampe, E. Ott, and J.H. Walker, Phys. Fluids, vol 21,42, 1978 ;J.T Mendonça, J. Plasm Phys., vol 22, 15, 1979 ; P. Sprangle, E. Easery, and A. Ting Phys. Rev., vol A41, 4463, 1990
4 H.C. Kapteyn, and M.M. Murname, J. Opt. Soc. Am., vol B8,1657, 1991
5 W. B. Mori, Phys. Rev, vol A44, 5118, 1991
6 N. Bloembergen, Opt. Comm, vol 8, 285, 1973; E. Yablonovitch, Phys. Rev. Lett. vol 31, 877, 1973 ; S.C. Rae, and K. Burnett, Phys. Rev. Vol A46, 1084, 1992 ; J.T Mendonça, *et al.*, IEEE Trans. Plasma Sci. Vol 24,147, 1996
7 J.M. Dias *et al*, Phys. Rev. Lett., vol 78, 4773, 1997

Generation of Ultrashort Electron Bunches Using Table-Top Laser-Plasma-Based Electron Accelerators

Szu-yuan Chen, Robert Wagner, Anatoly Maksimchuk, and Donald Umstadter

Center for Ultrafast Optical Science, University of Michigan,
Ann Arbor, MI 48109-2099, USA

Abstract. As an ultrashort high-intensity laser pulse propagates in a plasma, an electron plasma wave with relativistic phase velocity is excited by stimulated Raman forward scattering instability. A low emittance, relativistic electron beam is produced when the electron plasma wave traps and accelerates hot background electrons. Such an electron beam itself, or an ultrashort x-ray pulse generated from it, can be used for studying ultrafast science.

In the focus of an ultrashort, high peak power laser pulse, electrons oscillate at nearly the speed of light, giving rise to novel effects in a new intensity regime. Above the relativistic self-focusing threshold, a laser is observed to self-guide over many Rayleigh ranges, boring a hole in the plasma in which it propagates [1], and generating an ultrashort pulse of relativistic electrons in a well-collimated beam. Besides of being of basic interest, these interactions have applications in nonlinear optics, astrophysics, advanced fusion energy, x-ray lasers and ultrahigh-gradient electron accelerators.

When the laser intensity exceeds 10^{18} W/cm^2, a laser pulse with a pulse duration larger than several plasma periods can excite a large-amplitude, relativistic-phase-velocity electron plasma wave as a result of stimulated Raman forward scattering instability. Such an electron plasma wave can trap hot background electrons in the plasma and accelerate them to relativistic energies. In this paper, we show that the field gradient of a plasma wave can exceed that of an RF linac by four orders of magnitude ($E \geq 200$ GV/m) and accelerate a picosecond bunch of electrons with over 0.1-nC of charge per bunch in a low-emittance beam (0.1 mm-mrad).

In this experiment, we used a Ti:sapphire-Nd:glass laser system based on chirped-pulse amplification that produces 2 J, 400 fs pulses at 1.053 μm. The 50 mm diameter beam was focused with an f/3.3 off-axis parabolic mirror to $r_0 = 7$ μm ($1/e^2$), corresponding to vacuum intensities exceeding 3×10^{18} W/cm^2. This pulse was focused onto a supersonic helium gas jet with a sharp gradient (250 μm) and a long flat-topped interaction region (750 μm).

Under these conditions, a large amplitude plasma wave was driven through the self-modulated laser wakefield instability (Raman forward scattering instability). This was inferred by observing the appearance of up to the fourth anti-Stokes Raman sideband in the forward scattered spectra [2]. The amplitude of the plasma wave and its temporal duration were measured using collinear collective Thomson scattering [3]. It is found that the plasma wave lasts about 1.5 ps FWHM and the amplitude varies from $\delta n/n = 0.1$ to 0.3 for laser powers of 1–3

TW at 3×10^{19}-cm^{-3} plasma density. The growth rate is observed to be consistent with the theory of Raman forward scattering. The decay of the plasma wave is believed to be a result of both the effect of electron beam loading and the modulational instability. Electrons in the plasma can be trapped and accelerated by the electron plasma wave, when their thermal velocities are higher than the trapping threshold. As a result, an electron beam was observed to be generated in the direction of the laser propagation.

The electron energy spectrum was measured using a dipole magnet and detected with a LANEX scintillating screen coupled with a CCD camera for the low energy range or scintillating-fiber-array coupled with PMTs for the high energy range. As shown in Fig. 1, the electron energy spectrum is a continuum which can be fitted by two Maxwellian distributions with different temperatures. The temperature of the low energy electrons (1 - 4 MeV) is about 750 keV and the temperature of the high energy electrons (5 - 45 MeV) is about 9 MeV, for the case of 3-TW laser power and 3×10^{19}-cm^{-3} plasma density.

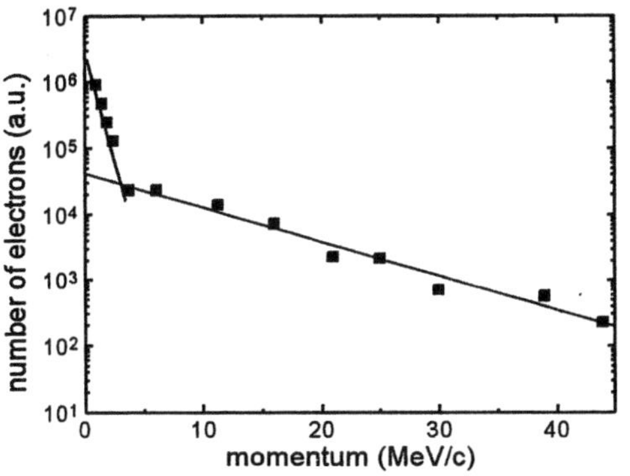

Fig. 1. Electron momentum distribution at 3-TW laser power and 3×10^{19}-cm^{-3} plasma density. The solid lines represent exponential fitting curves.

The electron beam profile was measured using a LANEX scintillating screen imaged by a CCD camera. The LANEX is placed behind an aluminum sheet which blocks the laser light and allows only the electrons with energy higher than 100 keV to be imaged. Fig. 2 shows the angular pattern of the electron beam measured under the same condition as in Fig. 1. Usually, the angular pattern of the electron beam contains two concentric components [4], both have a Gaussian profile but with different divergence (15° and 7.5° FWHM). Under certain conditions, a third component can show up, which has a beam divergence of 1.2° - 2.5° FWHM. Detailed measurements shows that the energy distributions of these components have the same temperature at least in the low energy range. The total number of electrons in the beam was measured to be 5×10^9 in this case using a Faraday cup. The size of the source is estimated to be ~ 10 μm in diameter, as determined by the size of the plasma wave. The temporal duration of the electron beam is estimated to be around 1 ps, determined by the duration of the plasma wave.

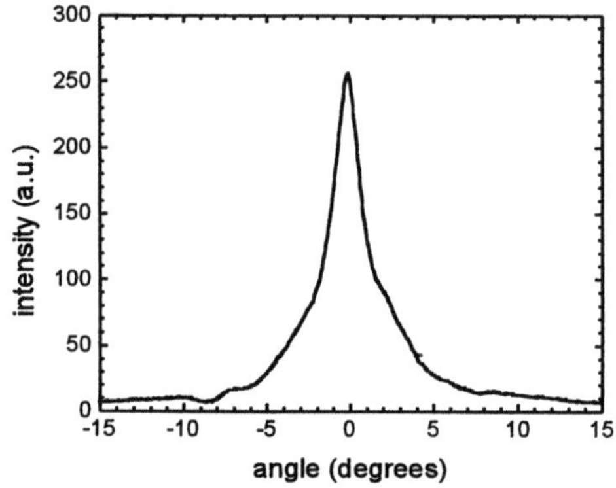

Fig. 2. Angular distribution of the electrons in the generated electron beam at 3-TW laser power and 3×10^{19}-cm^{-3} plasma density.

In this experiment, we demonstrated, using self-modulated laser wakefield, the generation of an electron beam with 1 MeV average kinetic energy, 10^9 electron number, 0.05 π-mm-mrad transverse emittance, and ~ 1 ps pulse duration. These parameters are comparable or better than the state-of-the-art radio-frequency photo-injector linear accelerator, except the huge spread in the energy spectrum. This emittance is better than any RF gun for the same number of electrons, due to the much higher field gradient, which mitigates emittance growth. We have proposed a new laser-plasma-based accelerator concept, in which a short laser pulse is used to drive a plasma wave resonantly and another laser pulse is used to inject electrons into the plasma wave. From the particle-in-cell simulations [5], we expect to produce an electron beam with 30 - 100 MeV energy (without guiding), 10^8 electron number, < 10 % energy spread, < 0.3 π-mm-mrad transverse emittance, and, most important, < 10 fs duration.

Such a femtosecond electron pulse can be of great use in ultrafast science. The short electron pulse can be used to probe a target directly, or to generate a femtosecond highly-coherent x-ray pulse (via Compton scattering of a laser pulse or a magnetic undulator), which can be used as a probe. Experimental demonstration of such sources is currently underway.

Acknowledgements. This project is supported by the U.S. Department of Energy and the U.S. National Science Foundation.

References

1 S.-Y. Chen *et al.*, in *Phys. Rev. Lett.*, Vol. 80, 2610, 1998.
2 D. Umstadter *et al.*, in *Science*, Vol. 273, 472, 1996.
3 S. P. Le Blanc *et al.*, in *Phys. Rev. Lett.*, Vol. 77, 5381, 1996.
4 R. Wagner *et al.*, in *Phys. Rev. Lett.*, Vol. 78, 3125, 1997.
5 D. Umstadter *et al.*, in *Phys. Rev. Lett.*, Vol. 76, 2073, 1996.

Ultrafast optical diagnostics for laser driven wakefields and plasma fibers

S. P. Le Blanc[1], E. W. Gaul[1], M. C. Downer[1], H. Langhoff[2], A. Babine[3], and A. Stepanov[3]

[1] Department of Physics, University of Texas at Austin, Austin, Tx 78712, USA
E-mail: spl@physics.utexas.edu
[2] Physikalisches Institut, Wurzburg University, Germany
[3] Institute of Applied Physics, Nizhny Novgorad, Russia

Abstract. Ultrafast frequency domain holography and chirped pulse interferometry are developed as techniques for characterizing phase modulation caused by laser induced plasma wakefields and plasma fibers.

1. Introduction

Laser induced plasma wakefields can excite large amplitude electrostatic fields (E >100 GV/m) which are potentially suitable for compact accelerators and advanced high energy colliders. Just as cavity geometry is important for conventional RF accelerators, plasma wave micro-structure is similarly critical for determining the accelerating properties of the laser wakefield accelerator (LWFA). In previous work, we have demonstrated that ultrafast optical techniques based on frequency domain interferometry [1,2] and collective Thomson scattering [2,3] can be used to diagnose critical wakefield parameters such as amplitude, decay time, and radial extension. Optical diagnostics with femtosecond resolution are important for measuring wakefield growth and decay, establishing feedback control on the wakefield, and synchronizing the injection of particles at the appropriate wakefield phase. In this paper, we report the development of a 2D frequency domain holographic technique capable of measuring the radial and temporal properties of the wakefield in a single laser shot. For practical high energy accelerators based on LWFA, the high intensity laser pulse must propagate though plasma fibers over many Rayleigh lengths. In part 2, we discuss a single shot method based on chirped pulse interferometry for resolving the temporal and spatial evolution of a plasma fiber created by hydrodynamic expansion of an axicon focused laser pulse.

2. Frequency Domain Holography

To record several wakefield oscillation cycles in a single measurement, we introduce a new technique based on the extension of frequency domain interferometry to the regime of frequency domain holography (FDH). In single shot frequency domain holography (SS-FDH), 1 short and 1 long probe pulse are used to measure the wakefield excited by an intense pump pulse. The broad probe pulse is made sufficiently long so that it samples several cycles of the wakefield. Spectral interference of the two probe pulses produces a frequency domain interferogram sensitive to temporal phase changes (caused by the wakefield or any other type of phase modulator) over the duration of the long pulse. The temporal resolution is approximately equal to the short probe pulse duration.

In a proof of principle experiment, we have used SS-FDH to measure phase changes caused by the non-linear susceptibility $\chi^{(3)}$ of a 150 μm thick fused silica plate in which 1 or 2 strong pump pulses co-propagate with and cross-phase modulate a 1 ps probe pulse. In this experiment, the short pulse was generated by type I phase matching in a thin (100 μm) $LiIO_3$ crystal, while the long probe pulse was generated in a 2 mm thick $LiIO_3$ crystal. Such a thick crystal has a small phase matching bandwidth ($\Delta\lambda = 0.45$ nm), resulting in a temporally long second harmonic pulse. Fig 1a shows a 1D lineout from a single shot 2D frequency domain hologram recorded by a CCD camera at the output of the spectrometer. The broad pedestal from the short (50 fs) probe, the sharp central feature from the long (1ps) probe pulse, and their frequency domain interference fringes are evident. In this example, the short and long probe pulses are separated by 3 ps. Fig. 1b shows the temporal phase changes extracted from such a hologram during a single shot experiment in which a sequence of two 50 fs pump pulses ($I=10^{11}$ W/cm^2) separated by 333 fs overlapped the 1 ps probe pulse as they co-propagated through the fused silica plate. Two 0.1 rad phase-modulation "peaks" separated by 333 fs (corresponding to the two 50 fs pump pulses) are recovered. On each laser shot, the phase shift on the second probe pulse is probed over a range of 1 ps with a resolution of ~50 fs. As the temporal separation ΔT1 between the pump pulse sequence and the probe pulse sequence is varied, Fig 1b shows the expected shifts in the phase modulation peaks. The magnitude of the cross phase modulation peak (0.1 rad) is in good agreement with the expected value based on the nonlinear index of refraction of fused silica [5]. When the phase modulation on the second probe pulse is caused by wakefield plasma oscillations, SS-FDH will enable measurement of the wakefield structure over several plasma oscillations.

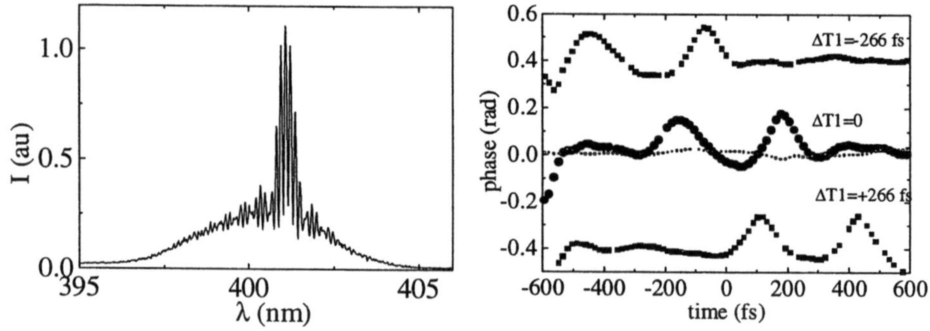

Fig. 1. a) Frequency domain hologram produced by the long and short probe pulses. b) Single shot temporal phase change measured during the 1 ps probe for different delays ΔT1 between the pump and probe pulse sequences.

3. Chirped pulse interferometry

Plasma channels can be used to guide high intensity laser pulses over distances much longer than the diffraction limit, which is crucial for future laser wakefield accelerators. In this section, we demonstrate a single shot optical diagnostic of the temporal and spatial evolution of a pre-formed plasma channel created by an axicon focused laser pulse [6] (λ=1 μm, E=250 mJ, τ=100 ps). A 30 fs pulse linearly chirped to 1 ns is used to probe the axicon plasma in a Mach-Zehnder interferometer arrangement (Fig. 2a). The interferogram is imaged to a spectrometer equipped with a CCD camera detector. The temporal behavior of the interferogram is mapped onto the horizontal (wavelength) axis of the CCD camera because of the linear chirp imparted to the probe pulse, while the spatial structure of the plasma channel is mapped to the CCD vertical axis. Fig. 3b shows a spectrally resolved interferogram 250 ps after the arrival of the Nd:YAG pulse focused in 400 Torr Ar. Radial expansion of the plasma is indicated by the ⊃-shaped intensity contour since frequency maps to time for the positively chirped probe pulse. The evolution of the plasma shock front versus radius agrees with previous results [6] and is in qualitative agreement with the expression for the expansion of a cylindrical explosion $r(t) = \xi(E/\rho)^{1/4}t^{1/2}$ [6], where r is the radius of the shock front, ρ is the mass density, ξ is a factor of order unity (ξ=1), and E is the absorbed energy per unit length (~10 mJ/cm). A wider bandwidth and more strongly chirped probe will enable single shot probing over several nanoseconds.

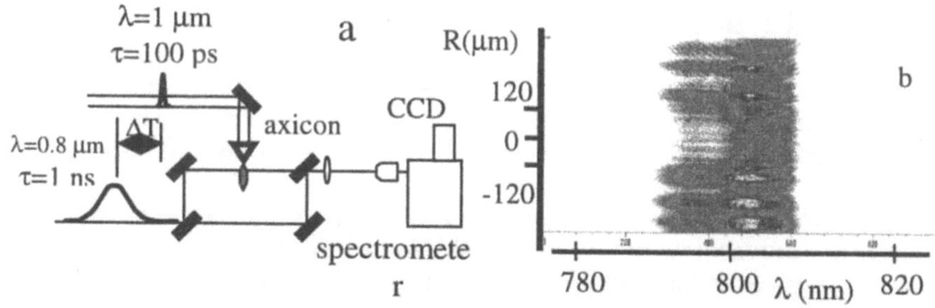

Fig. 2. a) Experimental setup for single shot, chirped pulse interferometry. b) Interferogram of the shock front evolution in 400 Torr argon recorded over a 1 ns time interval by the positively chirped probe pulse

Acknowledgements. This work is supported by the US Dept. of Energy.

References

1 C. W. Siders *et al*, Phys. Rev. Lett. **76**, 3570 (1996).
2 J. R. Marques *et al.*, Phys. Rev. Lett. **76**, 3566 (1996).
3 S. P. Le Blanc *et al.*, Phys. Rev. Lett. **77**, 5381 (1996).
4 A. Ting *et al.*, Phys. Rev. Lett. **77**, 5377 (1996).
5 G. Rodriguez and A. J. Taylor, Opt. Lett. **23**, 858 (1998).
6 T. H. Clark and H. M. Milchberg, Phys. Rev. Lett. **78**, 2383 (1997).

Ultrafast Plasma Studies by Phase Measurements with Femtosecond Spectral Interferometry

J.-C. Gauthier[1], C. Quoix[2], J.P. Geindre[1], P. Audebert[1], G. Hamoniaux[2], and A. Antonetti[2]

[1] Laboratoire pour l'Utilisation des Lasers Intenses (LULI), UMR 7605 CNRS-Ecole Polytechnique-CEA-Université Paris VI, Ecole Polytechnique, 91128 Palaiseau, (France)
E-mail: gauthier@greco2.polytechnique.fr
[2] Laboratoire d'Optique Appliquée - ENSTA, CNRS UMR7639, INSERM U451, Chemin de la Hunière, 91761 Palaiseau, (France)
E-mail: antonett@enstay.ensta.fr

Abstract. Dual quadrature femtosecond spectral interferometry is used to measure the gradient length, the peak temperature, and collision frequencies around critical density of an ultra-short pulse laser-produced plasma.

1. Introduction

In short pulse laser-matter interaction, it is crucial to obtain a time- and space-resolved picture of the very steep electron density gradient around the critical density region to study the laser absorption mechanisms. This is of interest for the production of short pulse X-ray emission [1,2].

In our previous studies [3], we have shown that the use of an ultra short (~100 fs) laser at relatively low irradiances ($I<10^{16}$ W/cm^2) contributes to disentangle the various laser-matter interaction mechanisms (hydrodynamic expansion, thermal conduction to the target bulk, ionization dynamics, and laser absorption) which play a role in X-ray conversion efficiency. With this end in view, we have recently demonstrated [4] a frequency-domain interferometer which enables to measure both the amplitude and the phase of the complex reflection coefficient of a plasma with simultaneously high spatial and temporal resolution. We have shown that the phase shift difference between a P- and S-polarized probe pulse is related to the electron density gradient scale length while the S-polarized phase shift alone gives the position of the probe laser turning point, and thus the hydrodynamic expansion velocity [5].

2. Experimental Methods

We have extended the frequency-domain interferometry technique by recording in a single shot the spectra of the P- and S-polarized light of the probe beam. For this, we have used the dual quadrature setup originally proposed by Joffre and his co-workers [6]. Present experiments have been performed at the Laboratoire d'Optique Appliquée with a CPA Titane:sapphire laser at 800 nm which can deliver 50 mJ in 120 fs on target. The contrast ratio, as measured by third order autocorrelation, is better than 10^7. The probe pulse of 150 fs duration is frequency-doubled before entering the target chamber.

3. Results and Discussion

Figure 1 shows the results obtained for the phase difference between S- and P-polarized light on an aluminum target irradiated at 4×10^{14} and 4×10^{15} W/cm^2. Hydrodynamic simulations with the MULTI hydrocode [7] are in rather good agreement with experiment. To get a better insight in the physical situation, model calculations assuming isothermal expansion of the plasma show that the time variation of the scale length around critical density ($N_e=1.6\times10^{21}$ cm^{-3}) is governed by the amount of energy deposited by the laser during the laser pulse, and by the ablation rate. As a result, assuming that the matter which is at critical density at any given time t_2 (late after the pump pulse) has been ablated at the time t_1, the relation between t_1 and t_2 is given by: $t_2=t_1(\rho_0/\rho_c)$ where ρ_0 is the matter density at the ablation time t_1, ρ_c is the density at the critical position for the probe pulse.

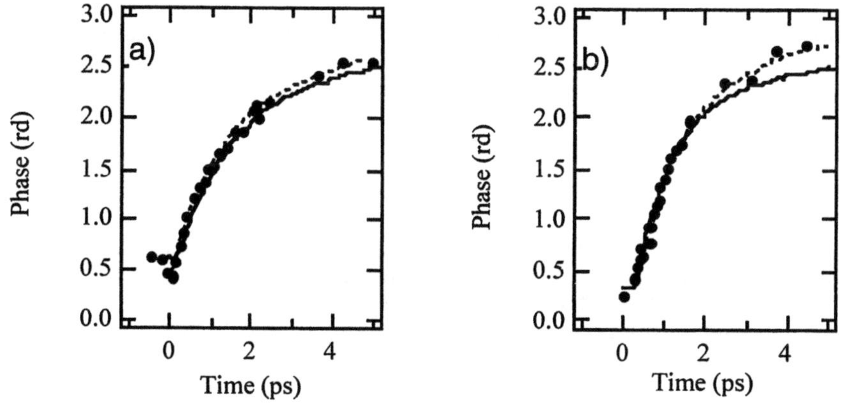

Fig. 1. Measured phase difference (dots) of the S- and P-polarized 400nm, 150 fs probe pulse as a function of time after the peak of the pump pulse for two laser irradiances: a) 4×10^{14} W/cm^2, b) 4×10^{15} W/cm^2. Experimental results are compared to MULTI simulations (solid lines) and model calculations (dashed lines)

In our experimental conditions, $t_2 \sim 70\ t_1$; the measurement of the gradient scale length around critical density between 1 and 6 ps is thus quite representative of the matter ablated during the 100 fs laser pulse. We show that the scale length around the critical density at t_2 is in fact only dependent of the temperature at the ablation time t_1 whatever the plasma cooling rate after t_1 may be. For the two laser irradiances used in Fig.1, the simple model gives 70 and 110 eV in rather good agreement with MULTI estimates of 45 and 120 eV. Knowing the gradient scale length, collision frequencies at critical density have also been determined through time-resolved reflectivity measurements of a S-polarized probe beam. Using the measured collision frequencies, we have calculated the P-polarized reflectivity with the use of a Drude model for the dielectric constant of the plasma. We have found an excellent agreement with the independently measured P-polarized probe reflectivity. This shows the overal consistency of our experimental procedure.

4. Conclusions

In summary, dual quadrature frequency-domain interferometry with twin femtosecond probe pulses has been used to infer sub-100nm density scale lengths in laser-produced plasmas with high time and space resolution. Using both hydrodynamic simulations and a simple model for isothermal expansion, we have measured the electron temperature at the peak of the pulse and we have determined collisional frequencies from time-resolved reflectivity measurements. For electron temperatures around 20 eV and densities around 10^{21} cm^{-3}, collision frequencies are found to be larger (by a factor of 2) than theoretical predictions [8] in aluminum and fused silica. Work is in progress to improve the dielectric constant modeling of these slightly-correlated plasmas.

Acknowledgements. We gratefully acknowledge the support of the laser staff at LOA where the experiments were carried out. This work was partially supported by the European Community grants CHRXCT930346 and FMRXCT960080.

References

1 J.-C. Gauthier, S. Bastiani, P. Audebert, J.P. Geindre, A. Rousse, C. Quoix, G. Grillon, A. Mysyrowicz, A. Antonetti, R. Mancini and A. Shlyaptseva, SPIE Proceedings **3157**, 52 (1997).
2 C. Rischel, A. Rousse, I. Uschmann, P.-A. Albouy, J.-P. Geindre, P. Audebert, J.-C. Gauthier, E. Förster, J.-L. Martin, and A. Antonetti, Nature, **390**, 497 (1997).
3 J.-C. Gauthier, J.P. Geindre, P. Audebert, S. Bastiani, C. Quoix, G. Grillon, A. Mysyrowicz, A. Antonetti, and R.C. Mancini, Phys. Plasmas **4**, 1811 (1997).
4 J.P. Geindre, P. Audebert, A. Rousse, F. Falliès, J.C. Gauthier, A Mysyrowicz, A. Dos Santos, G. Hamoniaux, et A. Antonetti, Optics Letters **19**, 1997 (1994).
5 P. Blanc, P. Audebert, F. Falliès, J.P. Geindre, J.C. Gauthier, A. Dos Santos, A. Mysyrowicz, et A. Antonetti, JOSA **B13**, 118 (1996).
6 L. Lepetit, G. Chériaux, and M. Joffre, JOSA **B12**, 2467 (1995).
7 R. Ramis, R. F. Schmalz, and J. Meyer-ter-Vehn, Comput. Phys. Commun. **49**, 475 (1988).
8 Y.T. Lee and R.M. More, Phys. Fluids **27**, 1273 (1984).

Part X

Ultrafast Dynamics and Control of Atoms, Molecules and Clusters in the Gas Phase

Complete Measurement and Control of Rydberg Wave Packets

T. C. Weinacht, J. Ahn, P. H. Bucksbaum

Physics Department, University of Michigan, Ann Arbor, MI 48109-1120
e-mail:tcw@gomez.physics.lsa.umich.edu

Abstract: We have used spectrally resolved Hanbury-Brown and Twiss[1] interferometry to measure the phase of atomic Rydberg wave packets [2], allowing us to fully reconstruct the Rydberg wavefunction. The wave packets were created with shaped optical pulses so that the measured wavefunction could be reshaped with feedback to match any predetermined target shape.

Introduction

Coherent control has been used to maximize fluorescence or ionization yields[3], manipulate currents in semiconductors[4], and determine pathways in simple chemical reactions[5]. These experiments have all had control goals which were defined by integrals or other functions of the quantum states interacting with the external fields. We have performed a coherent control experiment where the goal was expressed in terms of the wavefunction *itself.* Complete measurements of Rydberg wave packets were used as feedback to create predetermined target states.

Experiment

Sculpted Rydberg wave packets were created in atomic cesium with ultrafast shaped laser pulses[6]. The electronic wavefunctions excited by the shaped laser pulses were tomographically reconstructed from measured projections[7,8]. State selective field ionization (SSFI)[9] allowed us to make the spectrally resolved measurement of the wave packets by projecting them onto the Hamiltonian eigenstate basis. Amplitude measurements of the projections came directly from the SSFI spectra. Phase measurements came from analysis of fluctuations[10] in the SSFI spectra over several laser shots after the shaped wave packet had interfered with a well-characterized reference[11]. The phase measurement was time resolved. The arrival of the reference pulse defined the measurement time. This was essentially a spectrally and temporally resolved Hanbury-Brown and Twiss measurement.

The tomographic reconstruction identified the position of the initial wave packet in the configuration space spanned by the eigenstates of the Hamiltonian. Then, the vector from the current position to the target position of the wave packet in the configuration space was calculated and used to reprogram the shaped laser pulse in order to reach the target position. The calculation involved subtracting the phases of the measured wavefunction from that of the target. This procedure could be iterated until the target was reached.

Results and Discussion

We found that by using the phase and amplitude information for the shaped wave packets, we could reach the target within two iterations of the feedback loop. The experiment was repeated for four randomly selected measurement times. The laser pulses required to excite wave packets that would match the target at these four times were very different. The figure below shows the Wigner function for the shaped wave packet being controlled to match the target for one of the four delays.

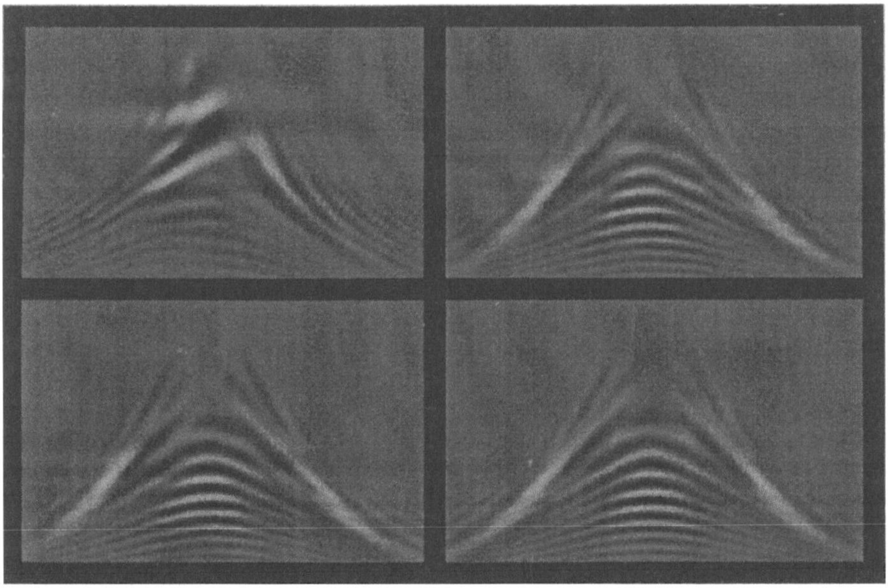

Fig. 1. Measured and target Wigner functions. The first three Wigner functions are measured and the final one (bottom right) is the target. The x axis is the radial momentum with a range of -0.1 to 0.1 atomic units. The y axis is radius with a range of 0 to 1800 atomic units.

The initial Wigner function (top left) represents a wave packet consisting of randomly phased eigenstates. After the first iteration of the loop, the reconstructed Wigner function looks similar to the target. After the second iteration (bottom left), the agreement is even better. The nonlinear response of the pulse shaper and other technical control problems were automatically corrected for in the feedback loop.

Conclusions

We have used tomographic reconstructions of shaped Rydberg wave packets as feedback in order to create target electronic wavefunctions. The use of tomography obviated the need for a sophisticated algorithm to implement the feedback, and resulted in rapid convergence to any target wavefunction spanned by the eigenstates within the excitation laser bandwidth.

References

1 Hanbury-Brown and R. Q. Twiss, *Nature* vol. 175, 1046 (1956).
2 T. C. Weinacht, J. Ahn, P. H. Bucksbaum *Phys. Rev. Lett.* vol 79 (1998).
3 C. J. Bardeen, V. V. Yakovlev, K. R. Wilson, S. D. Carpenter, P. M. Weber, and W. S. Warren, *Chem. Phys. Lett.* Vol. 280, 151 (1997).
4 R. Atanasov, A. Hache, J. L. P. Hughes, H. M. van Driel and J. E. Sipe *Phys. Rev. Lett.* vol. 76,1703 (1996).
5 L. Zhu, V. Kleiman, X. Li, K. Trentelman and R. J. Gordon, *Science* vol. 77, 270 (1995).
6 J. X. Tull, M. A. Dugan and W. S. Warren, *Adv. Mag. Resonance* vol. 20, 1 (1997).
7 I. A. Walmsley and L. Waxer *J. Phys. B* vol. 31, 1825 (1997).
8 U. Leonhardt, *Measuring the Quantum State of Light,* Cambridge University Press, Cambridge (1997).
9 T. F. Gallagher, L. M. Humphrey, W. E. Cooke, R. M. Hill and S. A. Edelstein, *Phys. Rev. A* vol. 16, 1098 (1977).
10 O. Kinrot, I. S. Averbukh and Y. Prior, Phys. Rev. Lett. vol. 75, 3822 (1995).
11 D. N. Fittinghof, J. L. Bowie, J. N. Sweetser, R. T. Jennings, M. A. Krumbugel, K. W. DeLong, R Trebino and I. A. Walmsley, *Opt. Lett.* vol. 56, 2316 (1996),
C. Xin and J. A. Yeazell, *Phys. Rev. A* vol. 56, 2316 (1997),
C. Leichtle, W. P. Schleich, I. Sh. Averbukh, and M. Shapiro, *Phys. Rev. Lett.* vol. 80, 1418 (1998).

Collective Multi-Electron Tunneling in State- and Charge-Selected Atoms in Strong Fields

H. Maeda, M. Dammasch, U. Eichmann, and W. Sandner

Max Born Institut, Rudower Chaussee 6, 12489 Berlin, Germany
E-mail: eichmann@mbi-berlin.de

Abstract. Ionization yields versus laser intensity have been measured for He-like In^+-Ions, metastable Kr^* and ground-state Kr, using a 50-fs Ti:Sapphire laser at $\lambda = 800$ nm. A clear signature of non-sequential ionization has been found for In^+, but the result is qualitatively different from previous helium data. A dramatic influence of ionization rate due to target excitation was found in metastable Kr^*. The multiple ionization yields in Kr (up to Kr^{4+}) and in Indium are excellently reproduced by a novel collective tunnel ionization formula with zero free parameters, which has been developed in our laboratory.

1. Introduction

The unexpected strength of simultaneous multi-electron ionization in strong fields, several orders of magnitude higher than expected from a sequential tunneling picture, has initiated numerous theoretical and experimental studies on the role of electron correlation [1-3]. Quantitative agreement between theory and experiment has been reported for some selected processes by elaborate and accurate ab-initio calculations [2]. In view of the importance of multiple ionization in plasma formation and dynamics a more comprehensive treatment appears, however, necessary. This requires a systematic experimental variation of the correlation strength in multi-electron atoms, as well as simple, but powerful analytical models to predict multiple ionization.

We have recently started a series of experiments on state- and charge-selected atoms to study multiple ionization as a function of electron correlation. Here we present experimental results of multiple ionization of In^+ (a helium-like system), Kr, and metastable Kr^*, using strong Ti:Sapphire laser pulses. Furthermore, we have developed a formula for collective multi-electron ionization in strong fields. It predicts, for the first time, accurate ion yields for arbitrary atoms and charge states from a surprisingly simple analytical tunneling model.

2. Experimental and theoretical results

Theory: In contrast to most other approaches our model is based on the idea of multi-electron tunneling. If it exists, this mechanism can only be non-sequential if the electrons move in a strictly collective fashion during the tunneling process (for a detailed discussion see [5]). Any asymmetry in the electron motion would immediately lead to recapture and, thus, to a sequential process. Starting from this ansatz we found the following result: The rate $W_{collective}(N, E_{total})$ for collective N-electron tunneling in a strong field (where E_{total} is the sum of binding energies of all active electrons) is represented by the one-electron ADK tunneling rate W_{ADK} [4], if the one-electron binding energy is simply replaced by mean binding energy $E_{eff} = (E_{total}/N)$ of each of the active electrons,

$$W_{collective}(N, E_{total}) \cong W_{ADK}(E_{eff}).$$

Comparison with available experimental ion yield data [1,3] gives generally satisfactory (within a factor of three), in many cases even excellent agreement over at least five orders of magnitude below saturation [5]. Further examples are given below as solid curves. We stress that there is no adjustable parameter in the theoretical calculation, although comparison with un-normalized experiments requires one single overall normalization factor per atom.

Experiment: The measurements were made with 50-fs-pulse duration, 800-nm-wavelength, linearly polarized light from Ti:Sapphire laser operated at a 10-Hz repetition rate. The laser was focused by a 160-mm focal length lens into an ultrahigh vacuum chamber crossing the atomic beam produced, in case of indium, by a resistively heated stainless-steel tube. The laser intensity was derived from the energy, pulse duration, and focus diameter of the laser pulse. The intensity was changed using a combination of a rotatable $\lambda/2$ plate and a polarizer. Produced ions were detected by a microchannel plate through a time-of-flight tube.

The experiments on ground state Kr extend previous Kr-ion yields to higher charge states, up to 4+. Figure 1 shows the results; the dotted curves are ion yields predicted by the sequential ADK theory. We could clearly recognize a large enhancement in the ion yield curves for Kr^{2+} - Kr^{4+}, which were attributed to non-sequential ionization by comparison with our collective tunneling ionization formula. These data complete the universal observation found in all rare gases, that collective ionization of electrons from the same shell seems to be dominant over sequential processes up to rather high charge states.

Next, by running a discharge through the atomic beam we obtained about 1% of metastable Kr-atoms in the beam. Figure 2 gives first, yet preliminary results on the strong field ionization mechanisms of excited atoms: "A" denotes the ion yield stemming from the ionization of Kr beam containing both metastable and ground state atoms, while "B" shows the previously known single ionization out of the ground state alone. The dotted curves are again ADK calculations. The dramatic influence of target excitation is clearly visible, although well reproduced by simple one-electron tunneling models. Hence, this technique will, in a next step, serve as a reliable basis to investigate how changes in the electron configuration influence the multiple ionization yield.

The experiments on In^+ are first results on a most basic correlated system other than ground state helium. Singly charged In^+ has two s-electrons outside a closed shell and thus resembles helium or alkaline-earth elements, which are common models for three-body Coulomb systems. In contrast to alkaline earths the ionization potentials of In^+ and In^{2+} are rather high, 18.86 eV and 28.03 eV, respectively, and require relatively high intensity for ionization. On the other hand, they are sufficiently different from helium (24.6 eV and 54.4 eV) to serve as an independent test system for ionization dynamics in strong fields.

Figure 3 shows the In^{2+} and In^{3+} ion yield as a function of the laser intensity. A clear deviation from the sequential ADK curve (dotted line) can again be seen in the In^{3+} ion yield, indicating an occurrence of non-sequential double ionization of In^{+}. This is confirmed by comparison with our two-electron tunneling formula (solid curve). It is surprising, however, that the enhancement due to non-sequential ionization is, both in theory and in experiment, much less dramatic than in helium [1,2]. Since In^{+} is the first quasi-two-electron system where collective ionization in strong fields has been studied, it remains to be systematically investigated whether this is an effect of the total binding energy, the structure of the $5s^2$ configuration, or of the influence of the core electrons. In any case, the results seem to be excellently reproduced by our new collective tunneling model.

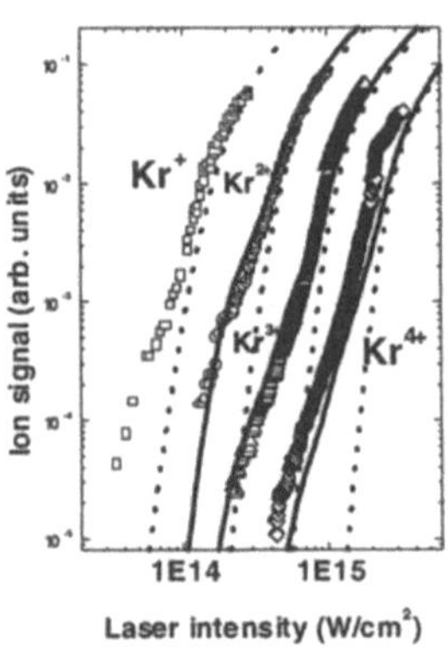

Fig.1 Ion yield of Kr^{+} - Kr^{4+} vs laser intensity

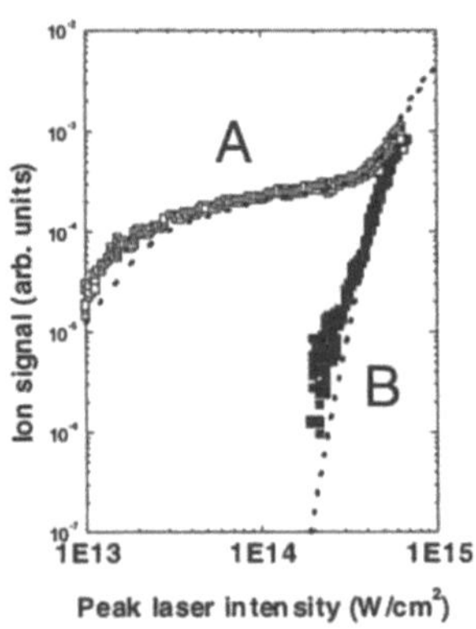

Fig.2 Ion yield of Kr^{+} vs laser intensity

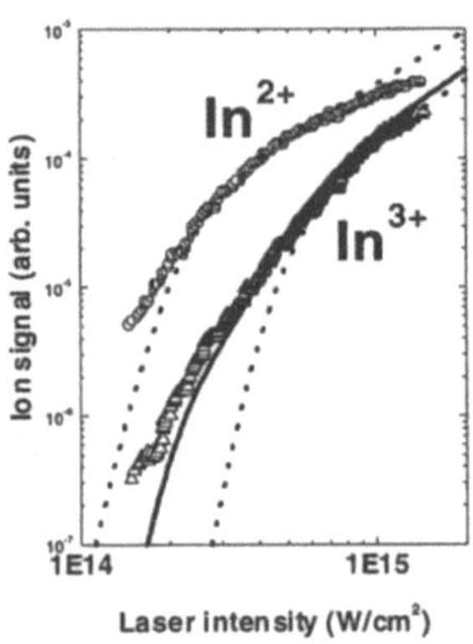

Fig.3 Ion yield of In^{2+} and In^{3+} vs laser intensity

Acknowledgement: One of us (HM) gratefully acknowledges fellowship support from the Alexander v. Humboldt Stiftung. The work is supported by DFG.

References

1. see, for instance, L.F. DiMauro and P. Agostini, Adv. At. Mol. Phys. **35**, 79 (1995) and references therein
2. A. Becker and F.H.M. Faisal, J. Phys. B **29**, L197 (1996).
3. S. Larochelle, A. Talebpour, and S.L. Chin, J. Phys. B **31**, 1201 (1998)
4. M.V. Ammosov, N.B. Delone, and V.P. Krainov, Zh. Eksp. Teor. Fiz. **91**, 2008 (1986) [Sov. Phys. JETP **64**, 1191 (1986)].
5. U. Eichmann, W. Sandner, and H. Maeda, submitted for publication.

Femtosecond time-resolved experiments in atomic calcium in weak and strong laser fields

M. Strehle, C. Völker, U. Weichmann, and G. Gerber

Physikal. Institut, Universität Würzburg, Am Hubland, D-97074 Würzburg, Germany
E-mail: gerber@physik.uni-wuerzburg.de

Abstract. We have investigated the two-valence electron atom calcium by employing femtosecond ion and photoelectron pump-probe spectroscopy. The radial Rydberg wave packet dynamics as well as multiphoton multiple ionization involving autoionizing states have been observed.

1. Introduction

The study of atomic systems with two valence electrons has attracted considerable interest since electronic excitations allow for time-resolved observations of pure one-electron dynamics as well as of phenomena arising from two-electron interactions. Both types of experiments may contribute to an understanding of the correlation between weakly bound valence electrons. The advantage to study atomic calcium compared to one-electron atoms is the existence of doubly excited states leading to autoionization and perturbation of Rydberg series. By employing different laser intensities of ultrashort laser pulses we studied the atomic dynamics both in the weak field limit (Rydberg wave packet dynamics) and in strong laser fields (time-resolved multiphoton multiple ionization).

The time-resolved measurements on atomic calcium were performed by combining femtosecond laser pulses and time-of-flight spectroscopy with atomic beam technology. Conventional pump-probe spectroscopy as well as phase sensitive detection were employed to reveal the temporal evolution of the studied processes.

2. Radial Rydberg wave packet dynamics

When employing moderate laser intensities we observed the real-time evolution of an electronic radial Rydberg wave packet on its orbit around the nucleus [1]. With the phase sensitive measurement technique applied to the detection of photoelectrons we monitored both the classical period of the Kepler electron and the quantum mechanical behavior of the Rydberg wave packet, illustrated by the (fractional) revival structures (Fig.1). Since the excitation occurs with spectrally broad fs-laser pulses several low lying electronic eigenstates ($n \approx 20$) are coherently coupled. This facilitates the detection of revival structures at early delay times. The inset in Fig.1 shows the Fast Fourier Transform (FFT) of the

Femtosecond Coherent Spectroscopy at 800nm: MI-FROG Measures High-Field Ionization Rates in Gases

Craig W. Siders[1,2], Jennifer L.W. Siders[1], and Antoinette J. Taylor[1]

[1] Materials Science and Technology Division, Los Alamos National Laboratory, Los Alamos, New Mexico, 87545, USA. E-mail: sidersc@ibm.net
[2] Present address: The Wilson Group, Department of Chemistry and Biochemistry, The University of California, San Diego, La Jolla, California 92093-0339, USA.

Abstract. We report the first quantitative phase-sensitive measurement of ultrafast ionization rates in gases. Ultrafast probe depletion via frequency mixing in the ionization front is observed.

The detailed dynamics of an atom in a strong laser field is rich in both interesting physics and potential applications. The source of both high-harmonic UV-VUV-XUV radiation [1] and ionization-based sources of tunable radiation [2,3], successful "mode-locking" of ultrafast ionization may yield the first attosecond pulses [4]. Previous studies [2,3] of ultrafast ionization have relied on spectral power density measurements using 100 fs pulses. As the detailed structure of the ionization-front blue-shifted spectra depends on a complicated interplay of the input pulse intensity/phase structure and the ionization dynamics [5], the conclusions made from such experiments are at best well-thought-out and limited inferences.

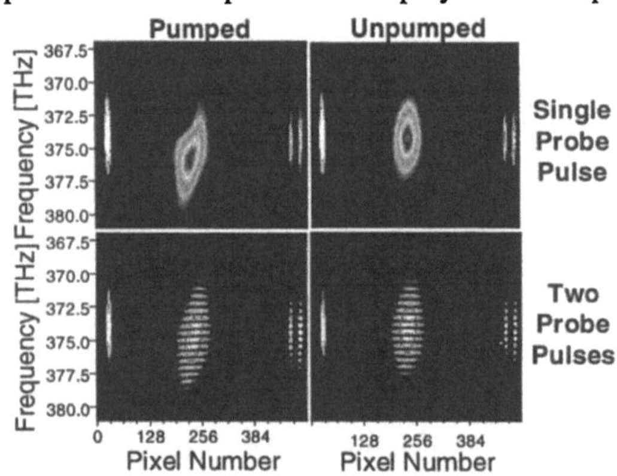

Figure 1: Example data.

To achieve a quantitative measure of the ionization rate independent of probe structure, we used an experiment configuration which, by utilizing three power spectra channels and one optically gated frequency resolved channel per each of four CCD images (pumped or unpumped cases with one or two probe pulses, see figure 1), provides simultaneous and self-consistent dual-channel power spectra and frequency domain interferograms, Multi-pulse Interferometric Frequency Resolved Optical Gating (MI-FROG), [6], and FROG [7] data. Using an experimental setup similar to that previously described [6], data was taken with various gas densities, N_{gas}, of He, Ne, Ar, Kr, Xe, H_2, N_2, O_2, and Air at pump-probe increments of 12 fs. The polarization of the 124 fs, 400 nm pump was ori-

play a role as intermediate states in the ionization process [4]. The autoionizing states have to decay before the subsequent transfer into the doubly charged continuum can take place. Additionally, since we coherently excite a manifold of doubly excited states, the existence of the multiple peak structure may be explained by an electronic Rydberg wave packet dynamics (T_{Kepler}= 200 fs) in this autoionizing manifold.

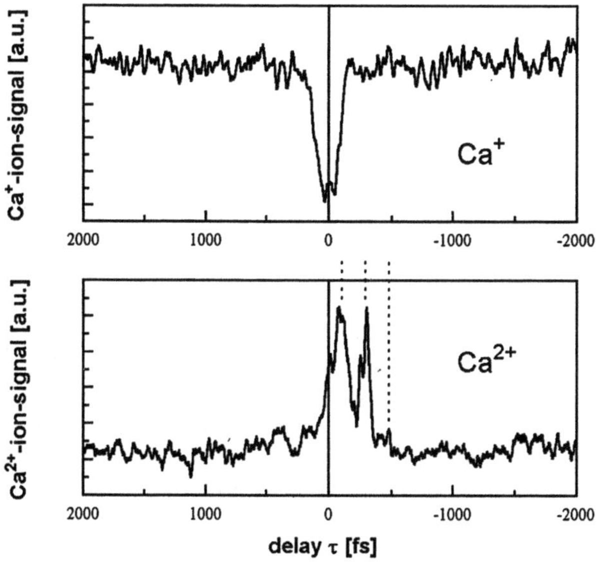

Fig. 2. Time-resolved multiphoton multiple ionization. The observed delay between the singly and doubly charged signal extremes may indicate the finite lifetime of an autoionizing state. The dotted lines point out the $T_{Kepler} \approx 200$ fs dynamics. It should be noted, that the ion signal axis cannot be compared.

4. Conclusions

In conclusion, we have reported on time-resolved experiments in a two-valence electron atom. The classical periodic motion of the Kepler electron as well as the quantum mechanical revival structure of the Rydberg wave packet were observed. By increasing the laser intensities we studied the process of multiple ionization with femtosecond time resolution. Under the given experimental conditions our results show no evidence for a non-sequential process. Instead, we observed a time delay in the formation of Ca^{2+} which may be attributed to the lifetimes of autoionizing states.

References

1 M. Strehle, U. Weichmann, and G. Gerber, *Pysical Review A*, Vol. **58**, 450, 1998.

2 L. G. Hanson and P. Lambropoulos, *Physical Review Letters*, Vol. **74**, 5009, 1995.

3 B. Walker, B. Sheehy, L. F. DiMauro, P. Agostini, K. J. Schafer, and K. C. Kulander, *Physical Review Letters*, Vol. **73**, 1227, 1994.

4 P. Lambropoulos and P. Zoller, *Physical Review A* , Vol. **24**, 379, 1981.

no pump or time-resolved probe absorption and by n_2I-like features in the frequency-shifts.), we find good agreement between the measured ionization rates with those calculated with a 1D plasma fluid model and Ammosov ionization rates [8]. Using Keldysh rates yielded ten times smaller predicted rates.

In the case of the homonuclear diatomic gases H_2 and N_2 (figure 4), we find similar good agreement for the model, using published molecular ionization potentials. However, in the case of O_2, we find a two order of magnitude disagreement, with tunneling theory *overpredicting* the ionization rate. Similar disagreement has been observed by Gibson [9], who has proposed an alternative calculation method for the tunneling rate.

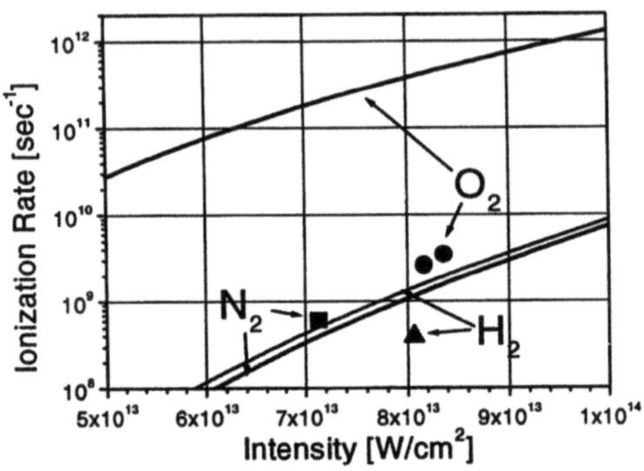

Figure 4: Measured and modeled ionization rates for three homonuclear diatomics.

Finally, modeling of the Figure 2 and similar cases reveal frequency mixing via the transverse plasma current [10] on such a large scale that ultrafast depletion of the probe pulse occurs within the ionization front. Though this mixing is clearly accompanied by 2D effects not included in the model, the observed ultrafast depletion, as it occurs only *within* the ionization front and for *both* polarizations, cannot be due to defocusing or neutral gas effects. We conclude, then, that this work represents the first direct, ultrafast-time-resolved observation of this phenomenon.

References

1 Atoms in Intense Laser Fields, M. Gavrila, Academic Press, Boston, 1992.

2 C.W. Siders, *et al.*, *JOSA B* **13**, 330 (1996); S. Backus, *et al.*, *Opt. Lett.* **21**, 665 (1996).

3 W.M. Wood, *et al.*, *Phys. Rev. Lett.* **67**, 3523 (1991); S.P. Le Blanc, *et al.*, *JOSA B* **10**, 1801 (1993).

4 I.P. Christov, *et al.*, *Phys. Rev. Lett.* **78**, 1251 (1997).

5 B. Rau, *et al.*, *JOSA B* **14**, 643 (1997).

6 C.W. Siders, A.J. Taylor, and M.C. Downer, *Opt. Lett.* **22**, 624 (1997).

7 D.J. Kane and R. Trebino, *Opt. Lett.*, **18**, 823 (1993).

8 B.M. Penetrante, *et al.*, *JOSA B* **9**, 2032 (1992).

9 G.N. Gibson, *et al.*, CLEO '98, JTuB7; *Ibid*, private communication.

10 F. Brunel, *JOSA B* **7**, 521 (1990).

ented either parallel or orthogonal to that of the weak 800 nm, 175 fs, multi-pulse probe.

The power-spectra channels allow both centroid-based power spectrum and interferogram analyses, which provide pulse-width averaged measures of frequency and phase shift, respectively. They can depend sensitively [5,3] on the detailed temporal and frequency-domain structure of the probe pulse. As MI-FROG recovers, to all orders, the phase difference between the pumped and unpumped probe pulses, we can utilize it to a) verify the intensity and phase structure of the probe pulse and b) recover the sub-pulsewidth time-resolved phase and frequency shifts impressed upon the probe by the ionization front, *independently* of the probe pulse structure.

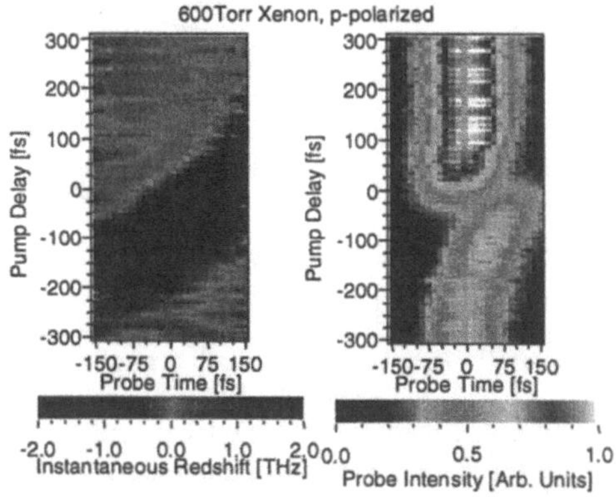

Figure 3: a) Doubly time-resolved frequency shift. b) Probe intensity profile.

Figure 2a illustrates the doubly time-resolved probe pulse frequency shift in the case of above-threshold ionization of xenon. The upper, red, diagonal corresponds to $n_2 I$ cross-phase modulation via the neutral gas and is strongest for early pump delays. This asymmetry suggests defocusing is significant, a hypothesis supported by independent probe spectral data. The lower, blue, diagonal is predominately due to blueshifting in the ultrafast ionization front, though comparison of the *p* and *s*-polarized data indicates that neutral gas contributions are not always negligible. Figure 2b similarly illustrates the probe pulse intensity profile versus time. It should be stressed that by using MI-FROG a zero of time is well defined and the apparent motion of the pulse profile is genuine.

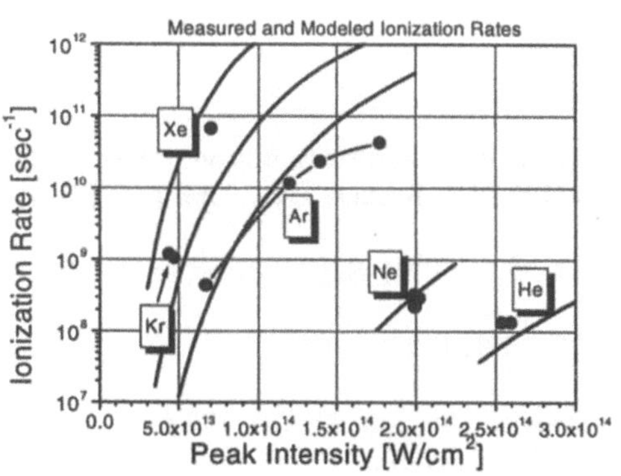

Figure 2 : Measured and modeled ionization rates for the Nobel gases, obtained assuming an interation length of two Rayleigh lengths.

To compare the measured ionization-front blueshifts with existing models of atomic ionization, we used a reduced frequency-shift parameter δ_ν given by $\Delta\nu / r_e \lambda N_{gas}$, where $\Delta\nu$ is the measured shift, $r_e = 2.8 \times 10^{-13}$ cm, and λ the laser wavelength. Thus, δ_ν is determined solely by the product of the single-atom ionization rate and an interaction length. Figure 3 presents the measured δ_ν divided by the confocal parameter. For those cases where the confocal parameter well approximates the plasma length (evidenced by no pump or time-resolved probe absorption and by n_2I-like features in the frequency-shifts.), we find good agreement between the measured ionization rates with those calculated with a 1D plasma fluid model and Ammosov ionization rates [8]. Using Keldysh rates yielded ten times smaller predicted rates.

In the case of the homonuclear diatomic gases H_2 and N_2 (figure 4), we find similar good agreement for the model, using published molecular ionization potentials. However, in the case of O_2, we find a two order of magnitude disagreement, with tunneling theory *overpredicting* the ionization rate. Similar disagreement has been observed by Gibson [9], who has proposed an alternative calculation method for the tunneling rate.

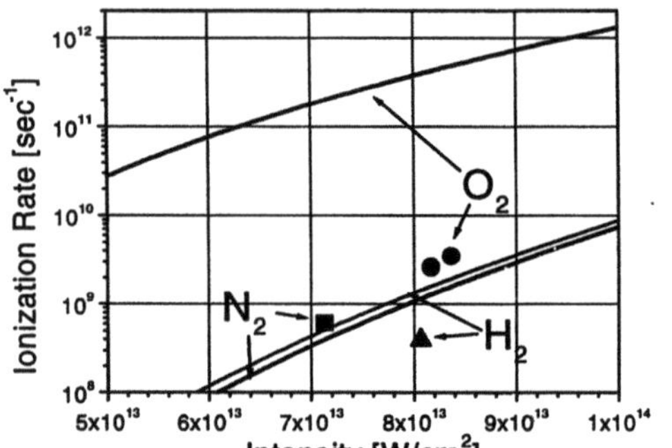

Figure 4: Measured and modeled ionization rates for three homonuclear diatomics.

Finally, modeling of the Figure 2 and similar cases reveal frequency mixing via the transverse plasma current [10] on such a large scale that ultrafast depletion of the probe pulse occurs within the ionization front. Though this mixing is clearly accompanied by 2D effects not included in the model, the observed ultrafast depletion, as it occurs only *within* the ionization front and for *both* polarizations, cannot be due to defocusing or neutral gas effects. We conclude, then, that this work represents the first direct, ultrafast-time-resolved observation of this phenomenon.

References

1 Atoms in Intense Laser Fields, M. Gavrila, Academic Press, Boston, 1992.
2 C.W. Siders, *et al.*, *JOSA B* **13**, 330 (1996); S. Backus, *et al.*, *Opt. Lett.* **21**, 665 (1996).
3 W.M. Wood, *et al.*, *Phys. Rev. Lett.* **67**, 3523 (1991); S.P. Le Blanc, *et al.*, *JOSA B* **10**, 1801 (1993).
4 I.P. Christov, *et al.*, *Phys. Rev. Lett.* **78**, 1251 (1997).
5 B. Rau, *et al.*, *JOSA B* **14**, 643 (1997).
6 C.W. Siders, A.J. Taylor, and M.C. Downer, *Opt. Lett.* **22**, 624 (1997).
7 D.J. Kane and R. Trebino, *Opt. Lett.*, **18**, 823 (1993).
8 B.M. Penetrante, *et al., JOSA B* **9**, 2032 (1992).
9 G.N. Gibson, *et al.*, CLEO '98, JTuB7; *Ibid*, private communication.
10 F. Brunel, *JOSA B* **7**, 521 (1990).

Resonant Effects in Above Threshold Ionization

Marc Hertlein[1], Phil Bucksbaum[1], Mark Walker[2], and Linn van Woerkom[2]

[1] Physics Department, University of Michigan, Ann Arbor, MI 48109-1120 USA
E-mail: hertlein@umich.edu
[2] Physics Department, Ohio State University, Columbus, OH USA
E-mail: lvw@mps.ohio-state.edu

Abstract. We have examined resonances in the high energy (plateau) part of the ATI spectrum of argon attributed to rescattering. As the laser wavelength is changed, the peaks shift in energy by a multiple of the photon energy. We have also observed the characteristics of photons emitted from the focal region perpendicular to the laser propagation direction.

The above-threshold ionization (ATI) electron spectrum of noble gases[1] consists of characteristic groups of peaks spaced by one photon energy, which are created by ionization enhancement when the atom levels Stark shift into resonance with a multiple of the photon energy. At higher energies these peaks do not continue to decrease monotonically with energy, but exhibit plateaus[2] or even enhancements. Classical rescattering theory[3] can account for the general nonresonant features at electron energies up to $10U_p$, however the process leading to the observed resonant substructure in the enhancement regions[4] has not yet been explained. Our examination of these high order resonant peaks and the photons emitted from the focus gives further insight into the dynamics.

We examined the substructure in the high order electron peaks of argon between 15 and 35eV, visible in the narrow intensity window between about $6.4*10^{13}W/cm^2$ and $6.9*10^{13}W/cm^2$. Figure 1 shows one typical set of peaks at an electron energy corresponding to about 13 times the photon energy. As the center wavelength of the laser pulse is varied, the peaks in this set shift in energy by an amount corresponding to 13 times the change of the photon energy. Similarly, the other peaks in the enhancement region at energies around $nh\nu$ shift by an amount of $nh\Delta\nu$, similar to peaks in the low order region which do not involve rescattering, where the energy of the produced electron is $E = nh\nu - E_{level}(I)$.

If the peaks were resulting from an Auger decay peak dressed by the laser field[5], this behavior would not be expected; its central peak should be fixed with a change in wavelength. A resonantly driven core transition[6,7] on the other hand would show this behavior, since the electrons are only produced at very definite light intensities at which the resonant condition for the transition is met, and at which the transition energy has a constant value. An electron temporarily excited to a Rydberg level will experience the fields produced by the ion core and also the light, the superposition of which contains harmonics capable of resonantly exciting electron and produce an energetic emerging electron. Another possible scenario is the dynamic stabilization of an outer electron temporarily trapped in a Rydberg state, which could lead to its resonantly-enhanced rescattering from the core[8].

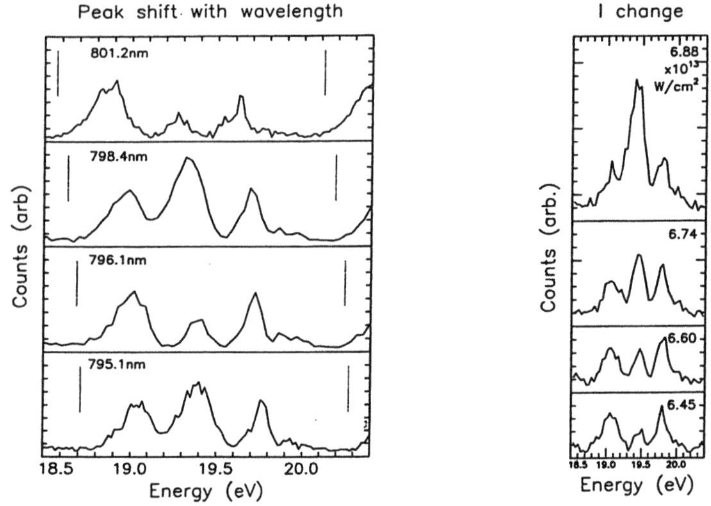

Figure 1: Left: ATI electron energy spectrum for Argon at an intensity of about $6.7*10^{13}$W/cm^2 (about 5% lower for 1st and 3rd spectrum) at the indicated wavelengths. The dotted lines on each graph indicate the position of the ionization threshold shifted by 12 and 13 times the photon energy. Spectra represent averages over the focal volume. Right: ATI spectrum taken at a constant wavelength of 790nm at varying intensities.

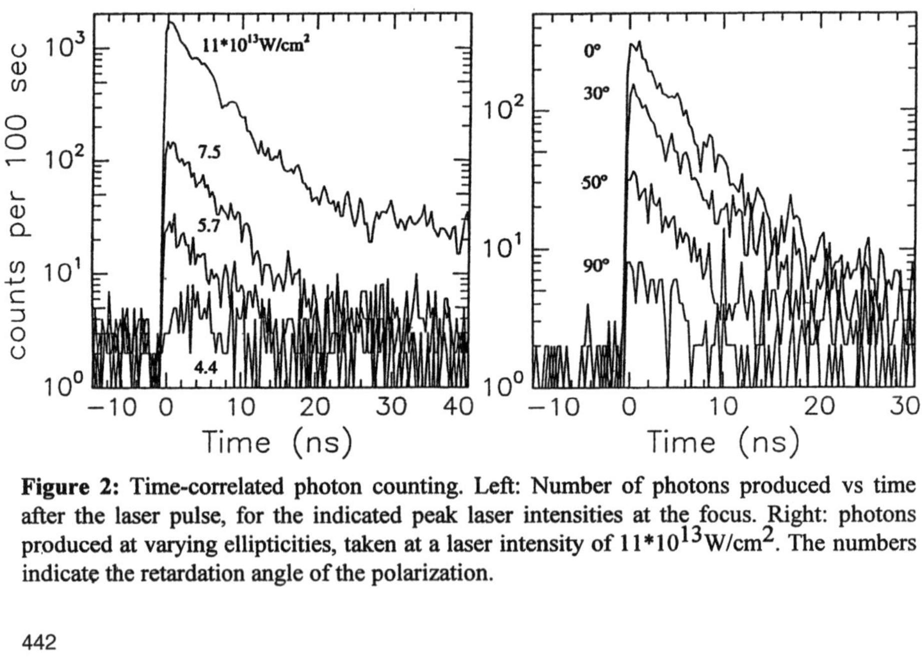

Figure 2: Time-correlated photon counting. Left: Number of photons produced vs time after the laser pulse, for the indicated peak laser intensities at the focus. Right: photons produced at varying ellipticities, taken at a laser intensity of $11*10^{13}$W/cm^2. The numbers indicate the retardation angle of the polarization.

We have also recorded the arrival time of photons emitted from the laser focus at right angles to the laser propagation direction. Our detector was a microchannel plate, located at a distance of 2cm from the focal region. The time correlated photon spectrum exhibits an exponential decay with a time constant of 5.3ns. The photons have an intensity appearance onset around $4*10^{13}$W/cm^2 (Figure 2), close to the intensity at which the high order (plateau) electrons start to appear. The number of photons emitted decreases strongly with increasing ellipticity of the laser light, however it does not go to zero even at circular laser polarization, where ionization enhancement due to resonances is negligible. Our data also shows an anisotropy of emission with respect to the angle to the laser polarization direction.

The decay time present in the spectrum shows that some electrons remain in an excited level in the atom after the laser pulse[9], emitting a photon upon decay. The microchannel plate is only sensitive to photons shorter than about 150nm. The only thus detectable transition with a lifetime close to the measured one is the 3d-3p transition in neutral argon. Interestingly, the appearance onset of the low order 3d electron peak coincides with the appearance of the first resonance visible in the enhancement region, and of the photons. However, no decay times corresponding to other resonant levels were detectable in our photon data.

References

1 H.B. van Linden van den Heuvell and H.G. Muller, in *Cambridge Studies in Modern Optics 8: Multiphoton processes*, eds. S.J. Smith and P.L. Knight (Cambridge, NY, 1988), p.25;

2 G.G. Paulus, W. Nicklich, Xu Huale,P. Lambropoulos and H. Walther, *Phys. Rev. Lett.* 72, 2851 (1994).

3 B. Yang, K.J. Schafer, B. Walker, K.C. Kulander, L.F. DiMauro, P. Agostini, *Acta Physica Polonica* A 86, no.1-2, 41-50 (1994).

4 M.P. Hertlein, P.H. Bucksbaum, H.G. Muller, *J. Phys.* B 30 L197 (1997).

5 J.M. Schins, P.Breger, P. Agostini, R.C. Constantinescu, H.G. Muller, G. Grillon, A.Antonetti, A.Mysyrowicz, *Phys. Rev. Lett.* 73, 2180 (1994).

6 L.G. Hanson, J. Zhang, P. Lambropoulos, *Europhys. Lett.* 30, 81 (1995).

7 P.H. Bucksbaum, A. Sanpera, M. Lewenstein, *J. Phys.* B 30, L843 (1997).

8 H.G. Muller, private communications.

9 N.J. van Druten, R.C. Constantinescu, J.M. Schins, H. Nieuwenhuize, H.G. Muller, *Phys. Rev.* A 55, 622 (1997).

Imaging of ions produced by femtosecond laser induced Coulomb explosion of molecules

Jakob Juul Larsen[1], Nis Bjerre[1], Nils Jakob Mørkbak[2], Johannes Olesen[2], Søren Rud Keiding[2], and Henrik Stapelfeldt[2].

[1] Institute of Physics and Astronomy, Ny Munkegade, Århus University, DK-8000 Århus C, Denmark
E-mail: juul@dfi.aau.dk
[2] Department of Chemistry, Ny Munkegade, Århus University, DK-8000 Århus C, Denmark
E-mail: henriks@kemi.aau.dk

Abstract. We use two dimensional ion imaging to analyze the fragments from femtosecond laser induced Coulomb explosion of molecules. The ion imaging technique enables measurements of the energy and angular distribution of highly charged ions produced in the Coulomb explosion process. Also, the dynamics of photodissociation of I_2 is studied by recording 2-D images as a function of the delay between a pump pulse, initiating the dissociation, and an intense probe pulse, Coulomb exploding the dissociating molecules.

1. Introduction

Coulomb explosion of molecules induced by intense femtosecond laser pulses is a new approach to study molecular structure and dynamics [1,2]. A typical experiment consists of initiating a molecular event of interest by a femtosecond pump pulse and probing the evolution of the event by removing several electrons with an intense delayed probe pulse. Measurement of the distribution of the velocities of the ionic fragments, resulting from the Coulomb fragmentation of the highly charged molecular ions, enables us to determine the molecular structure using Coulomb's law.

In our experimental setup the ions produced are extracted towards and detected by a detector system consisting of a microsphere plate and a phosphor screen. Each ion detected causes the phosphor screen to emit a localized flash of light which is recorded by a CCD camera. Hereby, we directly image the 2-dimensional distribution of ion velocities parallel to the detector plane.

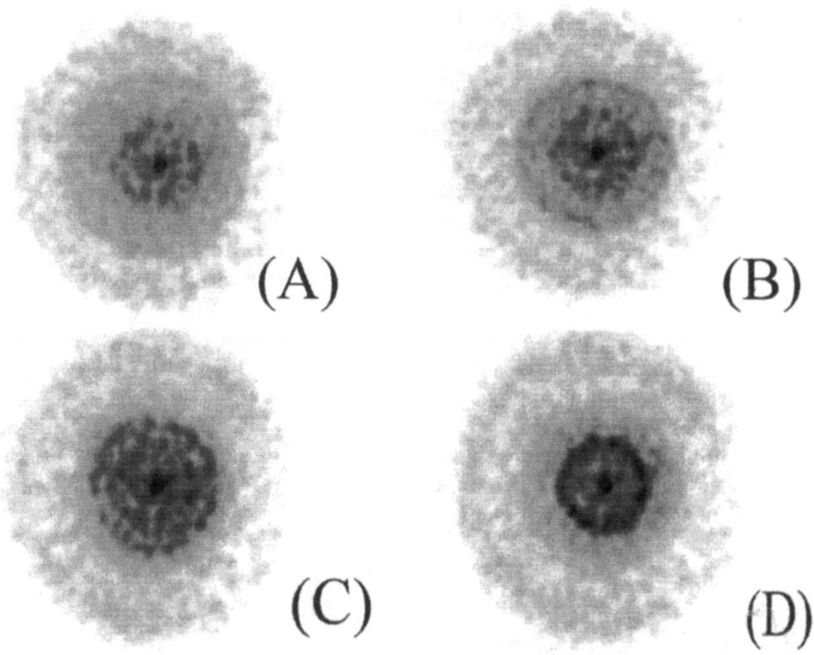

Fig. 1 Ion images of the pump-probe experiment described in the text for delays (A) : 500 fs, (B) : 1 ps, (C) : 2 ps, (D) : 20 ps. Both the pump and the probe laser are polarized perpendicular to the detector plane.

Figure 1 shows the 2-D images recorded when a 100 fs pump pulse (λ = 670 nm) photodissociates a sample of iodine molecules, via the A $^3\Pi_u$, and an intense delayed 100 fs (λ = 800 nm) pulse probes the evolution of the dissociation. The ion images are shown at four different pump-probe delays. The detection of I^+ ions originating from the Coulomb explosion of the dissociating molecules gives rise to rings on the 2-D detector since the molecules are preferentially dissociating in the plane of the detector due to molecular selection rules. At a fixed delay the diameter of a ring is determined by the kinetic energy of the I^+ ion. If only one electron is removed from the molecule by the probe pulse the dissociative wave packet is projected onto an I^+-I state. For delays larger than 300 fs both the initial and the final potential curves are flat and the fragments gain no additional energy. Thus, an I^+ ion created in this way emerges with half of the dissociation energy and gives rise to the innermost ring on the images.

The second ring with a larger diameter results from the Coulomb explosion channel, i. e. when the dissociative wave packet is projected onto an I^+-I^+ state. In addition to the energy from the dissociation energy the I^+ ion emerges with the Coulomb energy corresponding to the internuclear distance at which the probe pulse promotes the wave packet to the I^+-I^+ state. At small delays (Fig. 1(A)) the molecular bond is just broken and the I^+ ions obtain a large velocity due the large

Coulomb energy associated with the small internuclear separation. At larger delays the internuclear distance increases and the Coulomb energy decreases. Therefore, the diameter of the ring decreases as observed in the series of images in Fig. 1. A radial analysis of the pictures reveals that the asymptotic dissociation velocity is ~9.7 Å/ps in good agreement with the velocity expected at an excitation wavelength of 670 nm [2].

We have also studied the Coulomb fragmentation of multiply charged molecular ions, I_2^{n+}, formed when an intense 100 fs pulse irradiates I_2 molecules in the electronic ground state. As an example we show the results from these measurements when I^{2+} ions are recorded. Each half ring corresponds to one of the fragmentation channels, $I_2^{n+} \rightarrow I^{(n-2)+} + I^{2+}$. The large diameter rings results from the highest kinetic energy ions, i. e. from fragmentation of the highest charge states of the molecular ions. The ion images also reveal the angular distribution of the various fragmentation channels.

Currently, these studies are being extended to include 15 fs pulses. Using such short pulses the nuclei will be completely frozen during the pulse and the resulting ion images should provide spectroscopic information about the highly charged molecular ions at small internuclear distances.

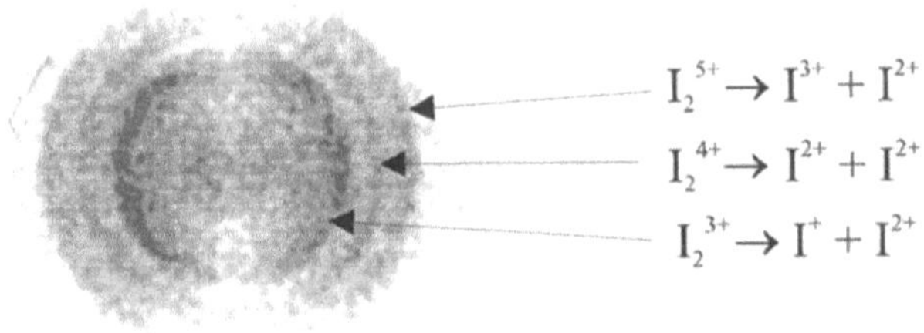

Fig. 2 Ion image of the I^{2+} fragments produced when an intense 100 fs laser pulse irradiates a gaseous sample of I_2 molecules. The laser is polarized perpendicular to the detection axis. The different rings correspond to fragmentation of highly charged molecular ions as indicated on the figure.

References

1 H. Stapelfeldt, H. Sakai, E. Constant, and P. B. Corkum, *Phys. Rev A.* **55**, R3319 (1997).

2 J. Juul Larsen et al, submitted to *J. Chem. Phys.*

Coulomb Explosion Imaging of H_2^+

S. Chelkowski[(1)], A. D. Bandrauk[(1)] and P. B. Corkum[(2)]

[(1)]Université de Sherbrooke, Sherbrooke, P.Q. Canada
[(2)]National Research council of Canada,Ottawa, Ont. Canada

Abstract. A molecular electron that ionizes in a strong optical field is removed from the vicinity of the molecule in a small fraction of a period. For H_2^+ this exposes the protons to the full force of Coulomb repulsion. We solve the time dependent Schroedinger equations for realistic laser parameters and accurately reconstruct the initial vibrational wave function $|\Psi(R)|^2$ by associating the fragment kinetic energy with an internuclear separation through Coulomb's law.

1. Introduction: Chemistry is often understood in terms of molecular structure and changes in molecular structure. For most molecules the equilibrium structure is well known having been indirectly determined by spectroscopic means. However, currently available high power laser pulses are short enough that all motion of even the lightest molecule can be frozen on this time scale, opening the path to directly imaging the structure of any small molecule using optically initiated Coulomb Explosion imaging techniques.

We solve the one dimensional Schroedinger equation, following both the electron and nuclear motion[1]. We concentrate on the simplest molecules where theoretical treatments are well established. Technically we determine the momentum spectrum of the dissociating protons long after the molecule has dissociated and use this distribution, together with Coulomb's law to reconstruct a distribution of the internuclear separations $|\Psi(R)|^2$. This paper should be viewed in the context of the extensive research on Coulomb explosion imaging experiments using ion-foil collisions [2]. Optical Coulomb explosion imaging replaces the molecule-foil collision with strong field ionization of the molecule.

2. Ionization Probability. Ideally, for optically triggered Coulomb explosion imaging the ionization rate should be independent of R. Uniform ionization rates can be achieved with very high intensities, such that the electron is well over the barrier (in the region of R's in which the imaging is performed).

Since our aim for Fig. 1 is to clarify the physical phenomena, we have used an artificial ½ period sine wave pulse and calculate the ionization probability by determining the electron probability that leaves a box 8 Bohr beyond the position of each ion. The dependence of strong field ionization rate on the instantaneous electric field and the relatively large excursion of the electron during the first half period following ionization justify this procedure. Figure 1 shows that the probability is nearly independent of the internuclear separation in the 3-5 atomic unit region above the laser intensity 2x10^{15} W/cm^2.

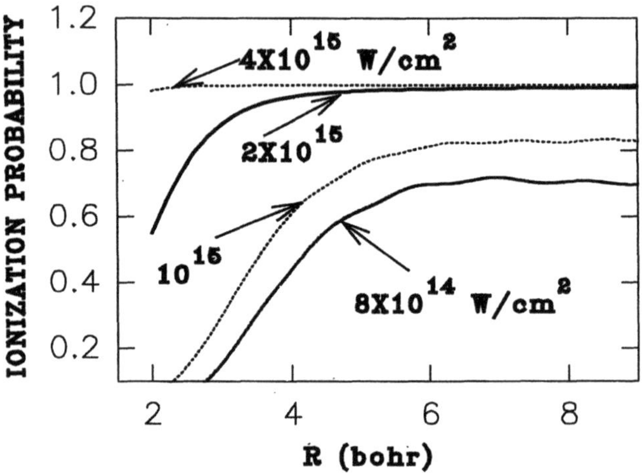

Fig. 1 is a plot of the ionization rate for H_2^+, for different intensities determined as a function of R in the 2-5 Bohr region.

3. Imaging. To obtain the results in Fig. 2 we arrange the peak intensity to be 4x10^{15}, ensuring that full ionization is reached during the portion of the pulse where the intensity is raising rapidly. Figure 2 was obtained for a 4.5 fs (full width at half maximum) 800 nm pulse, with the Gaussian pulse shape.

The initial and reconstructed vibrational (v=2) wave-functions $|\Psi(R)|^2$ of H_2^+ in Fig. 2 are very similar. In this short pulse limit, the primary limitation to the accuracy of the image reconstruction is the initial momentum distribution in the vibrational state of interest. This initial momentum can be neglected for all low vibrational states but noticeably affects the reconstruction for higher lying levels. However, this error is correctable even without knowledge of the H_2^+ potential curve.

By comparing results obtained with 3.2 and 6.4 fs pulses, we can show that, nuclear motion that occurs on the time-dependent Born-Oppenheimer surfaces between the turn-on of the laser pulse and when ionization reaches completion becomes the dominant source of error for longer pulses.

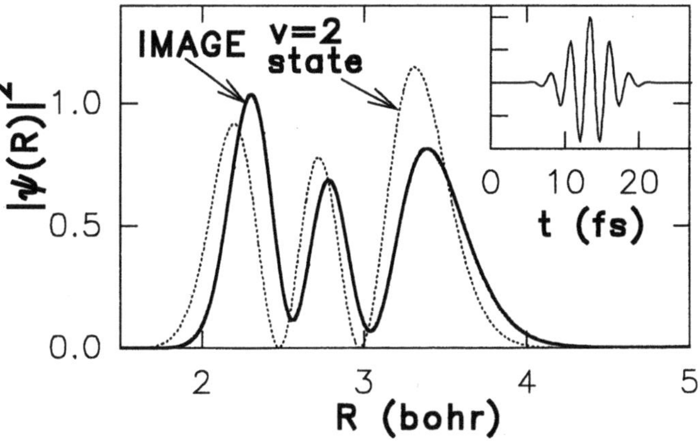

Fig. 2 is a plot of $|\Psi(R)|^2$ for the initial and reconstructed wavefunction.

4. Implications. Accurately imaging $|\Psi(r)|^2$ can be extended to other diatomic molecules provided they have closed-shell exploding fragments (eg Na_2^{2+}, Ca_2^{4+}). Coulomb explosion imaging will be a relatively simple diagnostic of the vibrational excitation of this class of molecules. Bonding electrons play a role if fragments do not have closed shells since then Coulomb's law does not describe the fragment interaction. However, this error can be minimized by requiring that the Coulomb exploding fragments originate from highly charged molecules[3]. Drawing on the experience of accelerator based Coulomb explosion imaging experiments, even without correction, triatomics and small polyatomics can be adequately imaged.

Our results, however, gain their main importance because of the promise of timed-Coulomb explosion imaging. In timed-Coulomb explosion imaging [4] the probe pulse in a femtosecond pump-probe experiment triggers a Coulomb explosion. Pump-probe techniques introduce dynamics and the image is a snapshot of a changing structure. Our results show that Coulomb explosion imaging has the potential to observe even the fastest nuclear motion.

References:

[1] S. Chelkowski, C. Foisy, A. D. Bandrauk, Phys Rev **A57**, 1176 (1998).
[2] Z. Vager, R. Naaman and E.P. Kanter, Science, **244**, 426 (1989).
[3] P. B. Corkum, M. Yu. Ivanov and J. S. Wright, Ann Rev. Phys. Chem. **48**, 387 (1997).
[4] H. Stapelfeldt, E. Constant, and P. B. Corkum, Phys. Rev. Lett. **74**, 3780 (1995); C. Ellert et al, Phil Trans. R. Soc. Lond. A **356** 329, (1998).

High Resolution Investigation of H_2^+ Exploding in a High Intensity Ultrashort Laser Pulse

Christoph Trump, Horst Rottke, and Wolfgang Sandner

Max-Born-Institut, Rudower Chaussee 6, 12489 Berlin, Germany,
E-mail: rottke@mbi-berlin.de

Abstract. We investigate for the first time high resolution H^+, D^+ ion kinetic energy distributions from dissociation of $H_2^+$and D_2^+ ions. Dissociation is initiated by ultrashort light pulses (~ 70 fsec) at 790 nm and light intensities up to 10^{15} W/cm^2. High resolution is achieved in this experiment by using a translationally cold beam of H_2/D_2 produced in a supersonic jet expansion.

1. Introduction

The dissociation and ionization mechanism of molecules in a high intensity non-perturbative optical radiation field has attracted much attention. Basic dissociation mechanisms which have been identified in diatomic molecules are bond softening two photon dissociation [1] and Coulomb explosion if higher charge states are created in the high intensity light field [2,3,4].

Of special importance is the molecular hydrogen ion. It shows all the dissociation features also found in more complicated diatomic molecules and has the advantage that it presently becomes tractable theoretically [5]. The kinetic energy distribution of ionic dissociation products is expected to give the most detailed insight into the strongly non-pertubative dissociation mechanisms at high light intensities. But the experimental investigations [1,3,4] up to now suffer from a poor resolution in H^+ ion kinetic energy distributions. The resolution is limited by the thermal energy distribution of the target gas [4]. In this experiment we eliminate the thermal energy distribution as resolution limiting factor by using a translationally cold molecular beam.

2. Experimental Method

A cold beam is formed in a supersonic jet expansion of H_2 or D_2 gas through a pulsed valve. The center part of this beam passes though a skimmer and a slit orifice before it reaches the interaction region with the laser pulses approximately 150 cm downstream from the valve. The terminal speed ratio in the molecular beam is larger than 20. This gives rise to a measured kinetic energy resolution in ion kinetic energy distributions which is better than 20 meV. Energy distributions are measured with an electric field free time of flight technique. We detect only ions which are emitted into a small solid angle along the molecular beam axis.

The pulses of a Ti:Sapphire laser system are used to excite the hydrogen molecules. It delivers pulses with a pulse width of 65 fsec (FWHM), and pulse energies of up to 4 mJ at 790 nm. The temporal shape of the pulses we used for this experiment consists of two peaks. This structure is designed by a suitable alignment of the gratings for pulse compression. An influence of the frequency chirp associated with the double pulses on H^+, D^+ kinetic energy distributions was not observed. The time delay between the two peaks was approximately 80 fsec.

The laser beam is focused onto the molecular beam to reach peak intensities in the range between 10^{14} and 10^{16} W/cm^2. Special care was taken to stabilize the output of the laser during experimental runs by monitoring the pulse energy on a single shot basis.

3. Results and Discussion

Fig. 1 shows first representative H^+ (a) and D^+ (b) ion kinetic energy distributions taken at a light pulse peak intensity of ~ 10^{15} W/cm^2. The overall shape of the spectra is independent of the isotopomere. But concerning the details distinct differences exist.

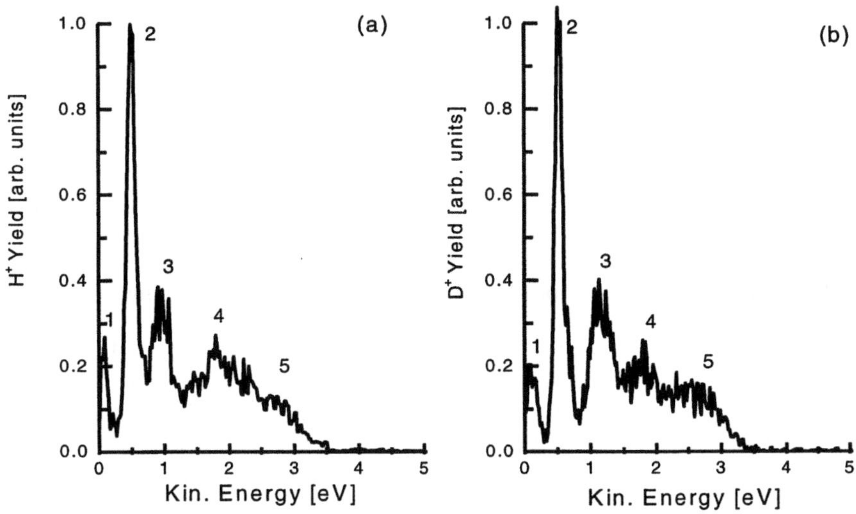

Fig. 1. H^+(a) and D^+ (b) kinetic energy distributions taken at a light pulse peak intensity of ~1 x 10^{15} W/cm^2. For more details see the text.

Different nuclear dynamics in H_2/D_2 multiphoton excitation gives rise to differing shapes of the broad structure consisting of parts (3), (4), and (5) in Fig. 1. Especially the position of peak (3) in the spectra sensibly depends on the

isotopomere. For H_2 and D_2 the narrow features (2) appear at the same kinetic energy. The same holds for the low energy structure (1). Based on low resolution experiments peaks (1) and (2) can be attributed to dissociation of H_2^+ into H^+ and H(1s) by bond softening (1) and two photon dissociation (2) [5,6]. The width of the two photon dissociation peak (2) is ~150 meV. This is significantly smaller compared to earlier experiments which are limited by the thermal motion of H_2^+ and D_2^+ in the gas phase [3]. In the present experiment the bandwidth of the light pulses gives a main contribution to the width. The observed narrow line width indicates that only one or at most two H_2^+ vibrational states are dissociated by two photon absorption.
The broad structures (4), (5) result from Coulomb explosion of H_2^+ and D_2^+ after photoionization at a critical internuclear separation near 8 a. u. [5]. The structure (3) which is strongly dependent on nuclear dynamics (i.e. depends on the isotopomere) can be attributed to Coulomb explosion of H_2^+/D_2^+ at large internuclear separations (15-30 a. u.). In this double pulse experiment, the first pulse initiates dissociation of H_2^+/D_2^+ and the second one (delay time ~ 80 fsec) then ionizes the dissociating molecule. The fixed time delay between the pulses is responsible for the different explosion energies found for H^+ and D^+ since H_2^+ and D_2^+ dissociate with different velocities and therefore have reached different internuclear separations when the second pulse arrives. The Coulomb repulsion of the bare nuclei at these large distances still results in a significant contribution to the final kinetic energy of the nuclei. We identified the H_2^+ and D_2^+ molecules which are photoionized through a pump-probe experiment with variable time delay and calculations to be those which dissociate after 2-photon absorption (peak (2) in Fig. 1).

References

1 A. Zavriyev, P. H. Bucksbaum, H. G. Muller, and D. W. Schumacher, Phys. Rev. A 42, 5500 (1990); B. Yang, M. Saeed, L. F. DiMauro, A. Zavriyev, and P. H. Bucksbaum, Phys. Rev. A 44, R1458 ,1991

2 J. H. Posthumus, L. J. Frasinski, A. J. Giles, and K. Codling, J. Phys. B 28, L349, 1995

3 T. D. G. Walsh, F. A. Ilkov, and S. L. Chin, J. Phys. B 30, 2167, 1997

4 J. Ludwig, H. Rottke, and W. Sandner, Phys. Rev. A 56, 2168, 1997

5 A. Giusti Suzor and F. H. Mies, Phys. Rev. Lett. 68, 3869 (1992); M. Ivanov, T. Seideman, P. Corkum, F. Ilkov, and P. Dietrich, Phys. Rev. A 54, 1541, 1996; S. Chelkowski, T. Zuo, O. Atabek, and A. D. Bandrauk, Phys. Rev. A 52, 2977, 1995; S. Chelkowski, A. Conjusteau, T. Zuo, and A. D. Bandrauk, Phys. Rev. A 54, 3235, 1996

6 C. Trump, H. Rottke, W Sandner to be published

Photodissociation of Na_2^+ in intense femtosecond laser fields

A. Assion[1], T. Baumert[2], V. Seyfried[1], U. Weichmann[1], and G. Gerber[1]

[1]Physikal. Institut, Universität Würzburg, Am Hubland, D-97074 Würzburg, Germany
E-mail: gerber@physik.uni-wuerzburg.de
[2]DLR, Institut für Optoelektronik, Postfach 1116, D-82230 Weßling, Germany

Abstract. Photodissociation of Na_2^+ in a well defined vibrational level is studied for laser intensities ranging from 10^{11} W/cm^2 up to 10^{14} W/cm^2. Two intensity dependent competing processes are observed: photodissociation by laser-induced potentials and by Coulomb explosion.

1. Introduction

High intensity multiphoton physics is now well established and a great deal of work has concentrated to understand processes within and beyond the regime of perturbation theory. Recently observed features of small molecules subjected to intense laser fields are: above-threshold dissociation, molecular bond softening, vibrational population trapping, Coulomb explosion, etc. [1, 2]. The model system H_2^+ and more recently Ar_2^+ in an intense laser field have been intensively studied. However, a detailed comparison between experiment and theory is very difficult because due to the experimental preparation of these molecular ions a number of different vibrational states contribute to the same spectrum of photofragments. If the interest is to study the influence of the laser pulse duration on the photodissociation dynamics, H_2^+ is again not a perfect candidate because the time scale of the vibrational period is about 15 fs and intense laser fields of such time durations are not yet generally available. Therefore we have chosen the molecular ion Na_2^+ to study the different processes arising from the interaction of the molecule with the intense and ultrashort laser field.

The molecular ion Na_2^+ can be prepared in a well defined vibrational level by nanosecond stepwise ionization in a molecular beam. Laser pulses with a duration of 90 fs, much shorter than the vibrational period of the Na_2^+-molecule (300 fs) are well suited for studying the photodissociation dynamics. This diatomic molecule is an ideal candidate to study photodissociation processes with ultrashort and intense laser pulses and in particular to investigate the influence of the laser pulse duration on the dissociation dynamics [3].

2. Experimental setup

The experimental setup is shown in Figure 1. The neutral diatomic molecule Na_2 was produced vibrationally cold in its electronic ground state by expanding pure sodium from an oven operated at 1000 K. With a ns-laser Na_2^+ was prepared predominantly in the vibrational state $v'' = 0$ of the ionic ground state by resonant two photon ionization. These ions were extracted from the molecular beam by a static electric field and illuminated in a second interaction region with the intense femtosecond laser pulses. Due to the separation of the two interaction regions space charge effects play no role. The resulting charged photodissociation products were detected in a linear time-of-flight (TOF) spectrometer.

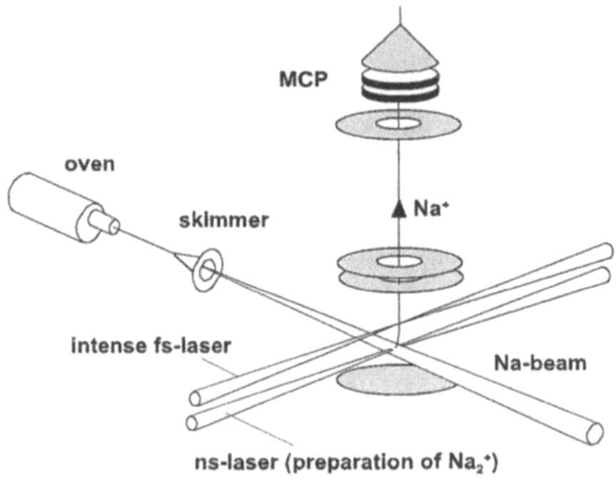

Fig. 1. Experimental setup. Two separate interaction regions are used. In the first region the Na_2^+-ions are prepared by resonant two photon ionisation. The ions are transferred by a small electric field into the second interaction region where the Na_2^+ ($v''= 0$) interact with intense femtosecond laser pulses.

3. Results and Discussion

The measured TOF-spectrum is shown in Figure 2(a) for different intensities of the fs-laser. With the laser polarization parallel to the TOF-axis we observe four peaks in the TOF-spectrum which we assign to different dissociation channels. To determine the released kinetic energy of the fragments and the total signal of the different dissociation channels we deconvoluted the measured TOF-spectrum into an energy spectrum. In Figure 2(b) the fragment distribution between the four channels and the corresponding energy release is displayed for several laser intensities.

The photodissociation processes can be understood in the picture of "adiabatically field-dressed state potentials" [4]. The molecule dissociates by moving from the field-free equilibrium distance towards larger internuclear distances on the laser-field induced repulsive potentials. Comparison of the results with theoretical calculations [3] show that the fragments with a released energy of 1.3 eV and of 0.5 eV originate from resonant 1- and 2-photon absorption in the field-dressed potential curves, respectively. At the lowest laser intensities (10^{11} W/cm^2) only these two channels are observed. At intermediate laser intensities (10^{12}-10^{13} W/cm^2) a third channel with a total released energy of 2.9 eV starts to emerge. We assign this channel to a resonant 3-photon absorption with subsequent ionization and formation of Na_2^{++}, which undergoes Coulomb explosion. At this intensity the molecule moves on the light-induced potential curves to larger internuclear separations and reaches a distance where the ionization probability is close to unity [4], still within the duration of the laser pulse. From the measured fragment energy we conclude that this Coulomb explosion takes place at an internuclear separation of about 4.8 Å. At the highest laser intensity (10^{14} W/cm^2) a strong contribution from a fourth channel with a released energy of 4.7 eV dominates the fragment energy distribution. From the measured energy we assign this channel to multiphoton ionization to the Na_2^{++}-Coulomb curve taking place close to the Na_2^+ internuclear equilibrium distance of 3.6 Å.

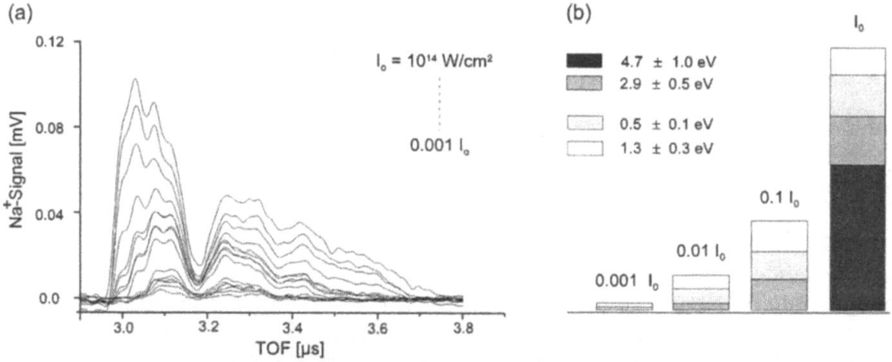

Fig. 2. (a) TOF-spectra of Na_2^+ obtained with an intense 790 nm, 90 fs laser and for different laser intensities. (b) Total kinetic energy of the fragments as obtained from the deconvolution of the TOF-spectra and distribution between the fragmentation channels for several laser intensities.

From these results we conclude that depending on the laser intensity, dissociation of Na_2^+ competes with ionization to Na_2^{++}. For low laser intensities dissociation clearly dominates. At moderate intensity of 10^{12} W/cm^2 ionization comes into play so that dissociation after 3-photon absorption cannot proceed entirely. At the highest laser intensities direct ionization to Na_2^{++} and fragmentation via the Coulomb curve dominates.

4. Conclusions

In conclusion the photodissociation of Na_2^+ in a given vibrational level and interacting with an intense femtosecond laser field has been studied. A comparison with recent theoretical results enabled us to assign each observed fragmentation channel. Additionally, the interplay between the two competing processes of dissociation via laser-induced potentials and ionization with subsequent Coulomb explosion was observed.

References

1 A. Zavrijev, P. H. Bucksbaum, H. G. Muller and D. W. Schumacher, *Phys. Rev. A*, Vol. **42**, 5500, 1990.

2 A. Giusti-Suzor, F. H. Mies, L. F. DiMauro, E. Charron and B. Yang, *J. Phys. B*, Vol. **28**, 309, 1995.

3 M. Machholm and A. Suzor-Weiner, *J. Chem. Phys.*, Vol. **105**, 971, 1996.

4 A. Giusti-Suzor, X. He and O. Atabek, *Phys. Rev. Lett.*, Vol. **64**, 515, 1990.

5 M. Machholm, *Dissertation,* Departement of Chemistry, University of Copenhagen, 1997.

Time-Resolved Configuration Interaction

Valerie Blanchet and Albert Stolow

Steacie Institute for Molecular Sciences, National Research Council of Canada, 100 Sussex Drive, Ottawa, Ontario, Canada K1A 0R6
E-mail: albert.stolow@nrc.ca , val@yosemite.ups-tlse.fr

Abstract. A new, general method for separating electronic from vibrational interactions within molecules uses Time-Resolved Photoelectron Spectroscopy into different electronic states of the ionization continuum.

The non-adiabatic coupling of electronic and vibrational motions, e.g. internal conversion, is a central yet difficult problem in the dynamics of excited polyatomic molecules. It has been suggested that one method to study this is via Time-resolved Photoelectron Spectroscopy (TRPES) [1,2]. The basic idea is to use the electronic and vibrational structure of the ionization continuum as a template for decomposing complex dynamics. The vibrational structure of the ion, via Franck-Condon factors, is sensitive to the nuclear dynamics. We now extend the method so as to include the excited electronic states of the ion. The partial ionization cross-sections into these different electronic continua depend on the electronic symmetry of the state being ionized. As the non-adiabatic coupling proceeds, the changing electronic symmetry alters ionization cross-sections into the ground vs. electronically excited states of the ion [3]. By monitoring the photoelectron spectra into these continua, we are able to separate the nuclear dynamics (IVR) from the purely electronic population dynamics, representing the time-resolved configuration interaction.

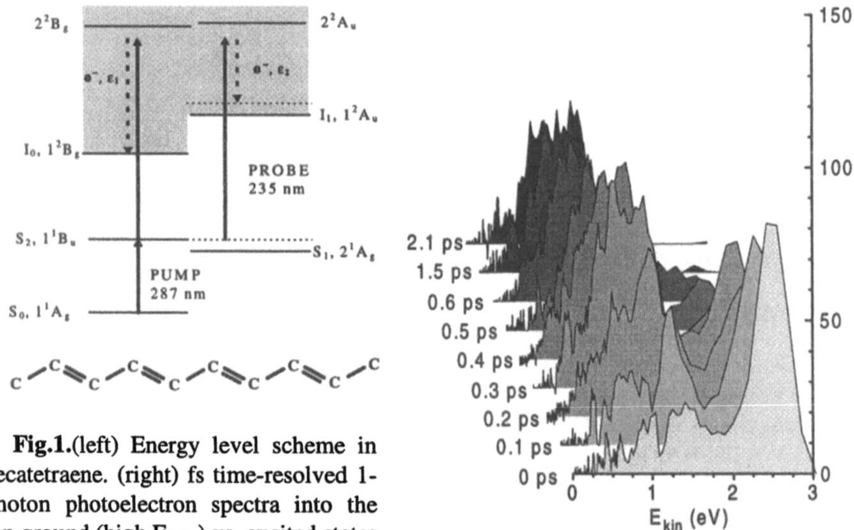

Fig.1.(left) Energy level scheme in decatetraene. (right) fs time-resolved 1-photon photoelectron spectra into the ion ground (high E_{kin}) vs. excited states (low E_{kin}).

We demonstrate the method with a linear $C_{10}H_{14}$ polyene: all-*trans* decatetraene (DT).
In Fig.1, we show the level scheme for DT excited with a 287 nm fs pump pulse to the S_2, 1^1B_u state origin. This is coupled via ultrafast internal conversion to the S_1, 2^1A_g state, forming a vibrationally hot molecule. A weak 235 nm fs probe pulse ionizes the excited state, yielding the TRPES shown on the right. In Table 1, the molecular orbital configurations are given for the S_0 ground state, the S_2 optically coupled state and the S_1 state to which it is non-adiabatically coupled. The S_0 and S_2 states are single configurations whereas the S_1 state is composed of three configurations. The right-hand column gives the electronic state of the cation formed upon removal of the outermost electron. These are the states expected to be formed upon one-photon, one active electron ionization (Koopman's approximation). It can be seen that the singly excited S_2 state correlates with the ion ground state, I_0, whereas the S_1 state, containing doubly excited character, correlates in large part with the first excited electronic state of ion I_1.

Table 1 Configuration Interaction in all-*trans* Decatetraene.

Electronic state	Energy (in eV)	Molecular Orbital Occupancy $1a_u$	$1b_g$	$2a_u$	$2b_g$	$3a_u$	$3b_g$	Weight	Correlated Ion State
Neutral									
S_0, 1^1A_g	0	2	2	2	2			100%	1^2B_g
S_1, 2^1A_g	3.6	2	2	2	0	2		33%	1^2A_u
		2	2	1	2	1		33%	1^2A_u
		2	2	2	1	0	1	33%	1^2B_g
S_2, 1^1B_u	4.3	2	2	2	1	1		99%	1^2B_g
Cation									
I_0, 1^2B_g	IP = 7.3	2	2	2	1			97%	
I_1, 1^2A_u	8.5	2	2	1	2			50%	
		2	2	2	0	1		50%	

The TRPES in Fig.1 is characterized by a rapid shift (400 fs) of electrons from a fast component ($\varepsilon_1 \approx 2.5$ eV) to broad, structured, slow component. This shift is a *direct* signature of the changing electronic configuration induced by the non-adiabatic coupling. Indeed, the fast component is due to ionization of the S_2 state into I_0. Therefore the disappearance of this peak represents the S_2-S_1 internal conversion time. The slower component results from photoionization of the S_1 state, which correlates to I_1, as discussed above. Its appearance is due to population of the S_1 state by internal conversion.

We confirm the above results by using a 352 nm probe pulse, (Fig.2 left) which by one-photon accesses only I_0. As in Fig.1, the initial S_2 state ionizes into I_0, here producing a slow, sharp electron peak ($\varepsilon_1 \approx 0.4$ eV). This peak decays in 400 fs, confirming the result of Fig.1. Upon internal conversion to the S_1 state, this one-photon ionization into I_0 becomes unfavourable for electronic symmetry reasons (Table 1). In contrast to Fig.1, the I_1 state is not energetically accessible via one photon. However, at 352 nm, the probe intensity is now high enough to allow above threshold ionization (i.e. absorption of a second photon in the continuum).

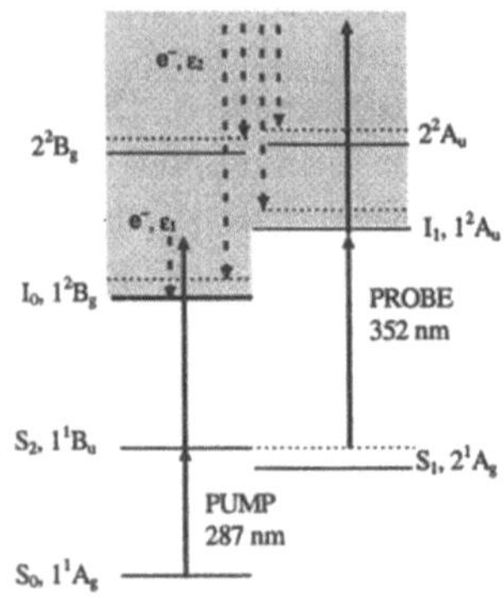

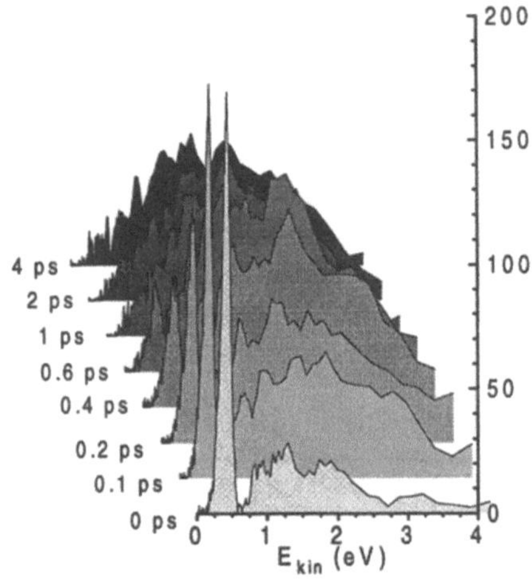

Fig.2.(left) Energy level scheme in decatetraene. (right) Time-resolved 2-photon photoelectron spectra into the ion ground (low E_{kin}) vs. excited states (high E_{kin}).

A two photon-ionization occurs from S_1, leading to the very fast electron component seen on the right. Due to the two-photon operator symmetry, the ionization from the S_1 state may now also occur to a 2^2B_g excited state and I_0, yielding a very broad electron distribution.

These observations show the separation of vibrational from electronic degrees of freedom in polyatomic molecules: the vibrational dynamics can be seen in the Franck-Condon structure within each component whereas the electronic dynamics is reflected by the overall changing intensity in the two components. In conclusion, the time-evolving configuration interaction in non-adiabatically coupled neutral excited states is revealed by TRPES into different ion electronic states.

References

1 I.Fischer, D. M. Villeneuve, M.J.J. Vrakking & A. Stolow, *J.Chem.Phys.*, 102, 5566, 1995
2 M.Seel & W.Domcke, *J.Chem.Phys.* 95, 7806, 1991
3 V. Blanchet & A. Stolow, *J.Chem.Phys.*, 108, 4371, 1998

Ultrafast Dissociation Dynamics of Polynuclear Organometallic Compounds in the Gas Phase

M. Bergt, B. Kiefer and G. Gerber

Physikalisches Institut, Universität Würzburg, D-97074 Würzburg, Germany
E-mail: gerber@physik.uni-wuerzburg.de

Abstract. The different cleavage dynamics of the metal-metal and two kinds of metal-ligand bonds are observed in femtosecond real time experiments.

1. Introduction

The photochemistry of transition-metal carbonyls is of great importance in the understanding of physical and chemical proporties of these compounds [1]. Metal carbonyls are widely used as photocatalysts in many organic reactions [2]. It has been shown that highly reactive unsaturated species resulting from the photolysis of transition-metal carbonyls are responsible for this catalytic activity. Hence, the characterisation of photochemical reaction pathways of saturated organometallic species is of great interest for applications.

The real time observation of the photodissociation dynamics of isolated metal carbonyls in femtosecond laser experiments has attracted considerable interest in recent years. Ultrashort laser pulses in combination with molecular beam techniques and time-of-flight (TOF) mass spectrometry allow the observation of the parent molecular ion as well as the different fragment ions. Ladder switching [3], as observed with nanosecond laser excitation is efficiently supressed. For the monometal carbonyl $Fe(CO)_5$ [4] it has been shown that the fragmentation patterns strongly depend on the laser wavelength, the laser intensity and the laser pulse duration. Femtosecond pump-probe experiments reveal that for the $Fe(CO)_5$ molecule the metal-CO bond breaks in a few hundred femtoseconds [4]. In a similar experiment on $Mn_2(CO)_{10}$, it has been shown that two different types of bonds, namely the metal-metal and metal-ligand bond, show different cleavage dynamics [5]. Femtosecond experiments on the three-bond organometallic compound $[CpFe(CO)_2]_2$ in solution show that either the metal-metal or the metal-CO bond is broken upon laser irradiation [6].

Here we report a study of the elementary photodissociation processes of isolated $[CpFe(CO)_2]_2$ molecules employing femtosecond pump-probe techniques. The dynamics of the competing bond cleavages for three different kinds of bonds are studied. We find different time-scales for the dissociation dynamics of the metal-metal and the two different metal-ligand bonds.

The experimental setup consists of an amplified Ti:Sapphire laser system, providing pulsed laser radiation centered at 800 nm with a pulse duration of 70 fs and a pulse energy of 1 mJ at a repetition rate of 1 kHz. The appropriately attenuated pump (400 nm) and time-delayed probe (800 nm) laser pulses are focused onto a molecular beam. The molecular beam apparatus consists of two differentially pumped chambers. One chamber contains the heated sample source and the other a reflectron TOF spectrometer. The $[CpFe(CO)_2]_2$ sample, used without further purification (98%; Strem Chemicals), is heated to about 100 °C in a home built evaporation cell.

2. Results

The multi-photon ionization (MPI) TOF mass spectrum obtained with 70 fs laser pulses at 800 nm and a pulse energy of 300 μJ is presented in Fig. 1a The spectrum is dominated by the parent ion $[CpFe(CO)_2]_2$ and a variety of observed fragment ions. Also present in the mass spectrum are the ionic fragments ferrocen $Fe(Cp)_2$, FeCp and atomic Fe. We believe the origin of $Fe(Cp)_2$ is thermal decomposition in the heated reservoir rather than photodissociation of the parent. Therefore the FeCp and Fe ion yields consist of photoproducts from $[CpFe(CO)_2]_2$ and from $Fe(Cp)_2$ which are difficult to distinguish. The smaller peaks indicated in the figure with '- CO' can be assigned to the CO-loss products $(Cp)_2Fe_2(CO)_{3\text{-}0}$ and the homolytic fragments $FeCp(CO)_2$ and $CpFe(CO)_{1,0}$. The asymmetric peak shape of the CO loss products $(Cp)_2Fe_2(CO)_{3\text{-}0}$ are an indication that at least two different dissociation channels lead to the formation of these ionic fragments: a fast dissociation in the interaction region of the reflectron spectrometer and a slower (metastable) fragmentation in the field free drift region of the spectrometer.

The mass spectrum displayed in Fig. 1a clearly shows the different dissociation channels of this large organometallic molecule. The mass peaks assigned to $(Cp)_2Fe_2(CO)_{3\text{-}0}$ indicate the CO loss channels, while the mass $FeCp(CO)_2$ is prove of the metal-metal bond breaking channel and the Cp loss results in the observation of Fe_2 and Fe. Note that no Cp-loss product with remaining CO ligands is observed. This is a clear indication that the CO loss occurs prior to the Cp loss.

Transient ionization spectra are recorded by employing the femtosecond pump-probe technique. One and two color pump-probe experiments are performed using the fundamental wavelength (800 nm) and the second harmonic (400 nm) of the Ti:Sapphire femtosecond laser system.

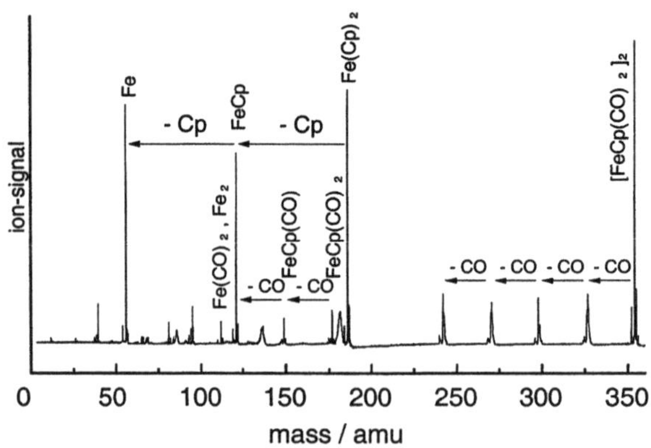

Fig. 1 Femtosecond MPI-TOF mass spectrum of $[CpFe(CO)_2]_2$ at 800 nm.

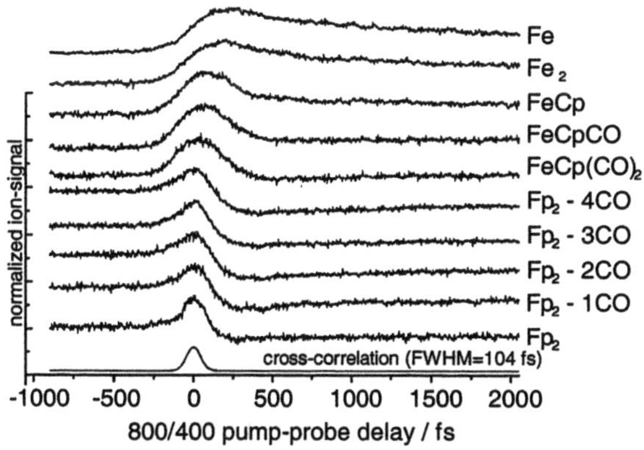

Fig. 2 Femtosecond transient ionization spectra of $[CpFe(CO)_2]_2$ (abbreviated Fp_2) and of all the fragments obtained with 400 nm (pump) and 800 nm (probe) laser radiation.

Figure 1b shows the measured femtosecond transient ionization spectra of $[CpFe(CO)_2]_2$ and of the fragments. For positive delay times, the 400 nm laser is the pump and the 800 nm is the probe. All the transients are measured with attenuated pump laser intensities (E_{pump}= 0.7 μJ ≙ $1\cdot10^{11}$ Wcm^{-2}, E_{probe}= 150 μJ ≙ $5\cdot10^{12}$ Wcm^{-2}), so that the pump laser alone produces less than 1% of the ion signal shown in Fig. 1a. Under these experimental conditions the transient ionization dynamics measured by the time delayed probe laser is only determined by the time evolution of the intermediate resonance in the neutral molecule. Since we need about four photons of 400 nm to ionize the parent we cannot exclude the possibility to probe different unknown intermediate states in the 400 nm/ 800 nm pump-probe experiments. The cross correlation of the pump and probe laser pulses is about 100 fs and is also included in Fig. 1b for comparison. Three different types of ionization transients can be distinguished when probing the various fragmentation channels. In the analysis we assume that all transients of similar time dependence show the dynamics of the same intermediate resonance, which after ionization by the probe laser dissociates into the observed ionic fragments.

Acknowledgements. This work has been supported by the Deutsche Forschungsgemeinschaft through SFB 347 and by the Fonds der Chemischen Industrie.

References

1 G. L. Geoffroy and M. S. Wrighton, in *Organometallic Photochemistry,* Academic Press, New York, 1979.

2 I. Wender and P. Pino, editors, *Organic Synthesis via Metal Carbonyls,* Vol. 1,2, Wiley, New York, 1977.

3 K. F. Willey, C. L. Brummel, and N. Winograd, *Chem. Phys. Lett.* **267**, 359, 1997.

4 L. Bañares, T. Baumert, M. Bergt, B. Kiefer, and G. Gerber, *J. Chem. Phys.* **108**, 5799, 1998, *Chem. Phys. Lett.* **267**, 141, 1997.

5 S. K. Kim, S. Pedersen, and A. H. Zewail, *Chem. Phys. Lett.* **233** 500, 1995.

6 P. A. Anfinrud, C.-H. Han, T. Lian, and R. M. Hochstrasser, *J. Phys. Chem.* **95**, 574, 1991.

Vibrational ladder climbing with chirped pulses

L. D. Noordam, D. J. Maas, and W. J. van der Zande

FOM-Institute AMOLF, Kruislaan 407, 1098 SJ Amsterdam, The Netherlands
E-mail: noordam@amolf.nl, www.amolf.nl/external/wwwlab/atoms/femtophys

Abstract. Enhanced vibrational ladder climbing has been demonstrated in NO using picosecond, chirped, free-electron laser pulses. Three steps of the vibrational ladder are climbed within a single infrared laser pulse. Quantum interference in the total yield of rotational pathways is observed.

1. Quantum ladder climbing in atoms and molecules

Laser excitation of molecules into a specific highly-excited vibrational state is a controllable way to enhance their chemical activity. Direct population of such a highly-excited state by absorption of a single photon is very inefficient. Step-wise excitation of a sequence of vibrational states is a more promising route. However, this method is frustrated by the anharmonicity of the molecular potential: an infrared laser which has the appropriate frequency to drive a molecule from the vibrational ground state v=0 to v=1 is non-resonant with consecutive vibrational transitions (v=1 to v=2, v=2 to v=3 .....). To overcome this bottleneck, the use of broadband short laser pulses that contain more frequencies in their spectrum has been suggested [1]. Such broadband pulses can be ``chirped", that is, their frequency can be made to change in time to follow the consecutive vibrational transitions. This chirping should strongly enhance the transfer to the higher states. Moreover, ultrafast excitation of a specific bond in a poly-atomic molecule by chirped short pulses can outpace intramolecular relaxation processes and, hence, enables bond-selective excitation.

In a series of experiments [2-4] we demonstrated efficient quantum ladder climbing of *electronic* transitions in *atomic* rubidium (5s -> 5p -> 5d). Using chirped 780 nm pulses of a large range of energies we found the following two regimes: (I) At low fluence the chirp enhances the population transfer if the pulses are chirped such that the laser frequencies match the consecutive transitions. The total population in the upper state oscillates as a function of chirp due to interfering pathways: sequential photoabsorption (5s -> 5p -> 5d) and direct two-photon absorption (5s -> -> 5d). (II) At high fluence the chirped pulse transfers all population to the upper state following the sequential pathway (the frequencies in the pulse are ordered in the intuitive way), moreover even for counter-intuitive chirp all population is transferred in the direct two-photon process. Remarkbly only unchirped pulses, having the highest intensity, are unable to inverse the population!

Up till now, laser experiments have not been very successful in driving *vibrational* ladder climbing in *molecules* for two reasons. Firstly, the required radiation to drive vibrational transitions is in the infrared (IR). In that wavelength regime, lasers that are both intense and tunable are scarce. Secondly, the bandwidth of most high power IR lasers is too small to resonantly drive consecutive transitions in the anharmonic potential. Nowadays, free-electron lasers provide tunable ultrashort infrared pulses with a sufficiently broad frequency spectrum to drive several consecutive transitions. We report on population transfer to high vibrational states of NO using broad band, chirped, IR laser pulses from the free-electron laser FELIX.

We studied [5,6] the enhancement of population transfer through the vibrational ladder of NO (see Fig.1) as a function of the chirp of an ultrashort IR laser pulse. The experiment was organized as follows. Short (0.37 ps) bandwidth-limited IR pulses from a free-electron laser drove the vibrational ladder climbing process in the diatomic molecule NO. The laser was tuned to the frequency matching the vibrational transitions in NO with sufficient bandwidth to support the v=0 to v=1 transition up the v=2 to v=3 transition, as indicated by the short arrows in fig. 1.

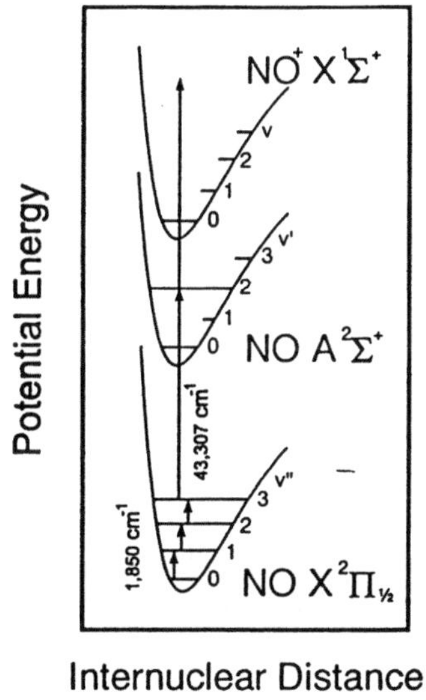

Fig. 1: Vibrational ladder climbing in NO. The step size of the vibrational ladder of the electronic ground state is 0.23 eV (short arrows). The anharmonicity of the NO ladder is 1.5% per step. Transfer to vibrationally excited states is monitored by 1+1 REMPI: a UV photon excites the vibrationally excited NO molecule from the electronic ground state to the first Rydberg *state (the gamma* band) and a second photon ionizes the molecule to the NO-ionic ground state (long arrows).

2. Interfering rotational pathways in the vibrational ladder

As an example we show the population in v=3 as a function of the chirp (α) of the infrared laser pulses. Negative values of α correspond to a decreasing laser frequency as a function of time. Hence for negative α, the laser frequency follows the decreasing vibrational transition frequencies. Figure 2 shows the three main observations reported here: (I) The broad bandwidth of the IR laser pulse enables population of the v=3 vibrational state by three consecutively resonant photons. (II) A strong enhancement of this population is observed when the chirp follows the consecutive steps up the vibrational ladder. (III) The population in excited vibrational states shows an oscillatory behaviour as a function of the chirp of the IR laser pulse.

The interferences originate from sequential transfer to the same final ro-vibrational states via different intermediate states. The interference is due to the accumulated phase difference of pathways involving different intermediate rotational states, as confirmed in a multistate time dependent quantum calculation (curve in figure 2).

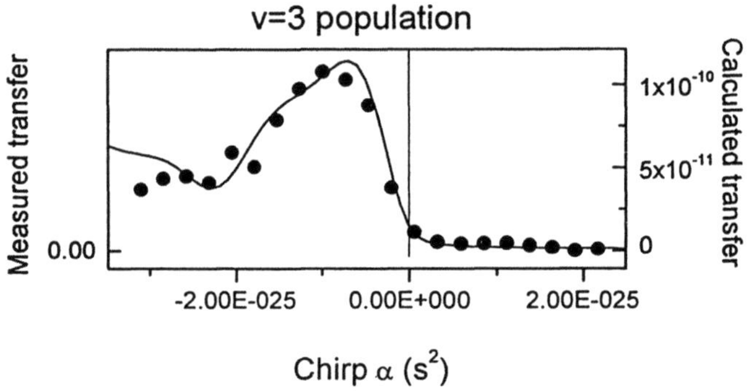

Fig. 2: Population of the v=3 vibrational mode of NO as a function of the chirp. For negative chirp the swept laser frequency follows the frequency of consecutive vibrational transition. The experiment (dots) shows indeed enhanced population transfer for negative α, as well as the destructive interference of the rotational pathways for $\alpha \sim -2\ 10^{-25}$.

Acknowledgements. The work described in this article is supported by the Stichting FOM which is financially supported by NWO.

References

1. S. Chelkowski, A. D. Bandrauk, and P. B. Corkum, Phys. Rev. Lett. **65**, 2355 (1990)
2. B. Broers, H. B. van Linden van den Heuvell, and L. D. Noordam, Phys. Rev. Lett. **69**, 2062 (1992)
3. P. Balling, D. J. Maas, and L. D. Noordam, Phys. Rev. A **50**, 4276 (1994)
4. D. J. Maas, C. W. Rella, P. Antoine, E. S. Toma, and L. D. Noordam, submitted to Phys. Rev. A (1998)
5. D. J. Maas, D. I. Duncan, A. F. G. van der Meer, W. J. van der Zande and L. D. Noordam, Chem. Phys. Lett. **270**, 45 (1997)
6. D. J. Maas, D. I. Duncan, R. B. Vrijen, W. J. van der Zande and L. D. Noordam Chem. Phys. Lett. **290**, 75 (1998).

Optimal laser control for photoisomerization

Regina de Vivie-Riedle[1], Jörn Manz[2], and Karsten Sundermann[2]

[1] MPI für Quantenoptik, Hans-Kopfermannstr. 1, D-85748 Garching, Germany
E-mail: rdv@mpq.mpg.de
[2] FU-Berlin, Institut für Physikalische und Theoretische Chemie, Takustr. 3,
D-14195 Berlin, Germany
E-mail: karstens@chemie.fu-berlin.de

Abstract. We present an extended theory of optimal control for the analytical design of feasible laser pulses for photoisomerization. The new algorithms incorporate feedback from the laser field and a subsequent pulse analysis. The method is applied for the photoisomerization of Li_2Na by means of wave packet propagation on ab initio potential surfaces. High product state specifity together with high quantum yield is achieved.

1. Introduction

Selective control of chemical reactions using ultrashort laser pulses recently advanced to a new field of great interest for theoretical [1-4] and experimental research [2,5-7]. A family of new iteration methods incorporating feedback control from the electric field was only recently presented by the group of Rabitz[4]. First results for this concept have been verified experimentally using ultrafast pulse shaping techniques in combination with genetic algorithms [6,7]. We implemented and modified the new method to deal with more complex quantum systems [8]. In order to increase the experimental feasibility of the resulting optimal laser pulse the algorithm is extended to allow for smooth switch on and off behavior of the designed fields. Specificly, our new method will be applied to Li_2Na as a model system for photoisomerization from the acute groundstate to the obtuse isomer via an electronically excited state. Up to now, it was not possible to separate the two different isomers by means of traditional nanosecond or cw-spectroscopy. The selective preparation of vibrational target states of the obtuse isomer by means of femtosecond laser pulses is therefore a demanding challenge.

2. Model and Theory

The system is treated in a two-dimensional model using Jacobi coordinates to represent the Li_2-Na and Li-Li distances. The potential energy surfaces (PES) involved have been calculated using the quantum chemistry program MOLPRO. The vibrational eigenfunctions for initial and target state are evaluated using filter diagonalization methods [9]. For the analytical determination of the optimal laser field, which drives the system from an initial state $\Psi_i(0)$ at t=0 to the selected target state $\Psi_f(T)$ at time t=T the subsequent set of coupled differential equations has to be solved.

$$\alpha_0\varepsilon(t) = -s(t) \cdot \text{Im}\,(\langle\Psi_i(t)|\,\Psi_f(t)\rangle\langle\Psi_f(t)|\mu|\,\Psi_i(t)\rangle)$$

$$i\hbar\,\partial/\partial t\,\Psi_i(t) = [H_0 - \mu\cdot\varepsilon(t)]\,\Psi_i(t)\,;\;\; i\hbar\,\partial/\partial t\,\Psi_f(t) = [H_0 - \mu\cdot\varepsilon(t)]\,\Psi_f(t) \quad (1)$$

The first equation represents the optimal electric field in terms of the evolving wavepackets Ψ_i and Ψ_f. The two subsequent equations guarantee the compliance of the Schrödinger equation for Ψ_i and Ψ_f under the influence of the laser field $\varepsilon(t)$. The shape function s(t) is the additional constraints for the laser field to satisfy the demand of smooth switch on and off behavior. The parameter α_0 limits the time averaged laser intensity and T denotes the overall pulse duration.

3. Results and Discussion

The dynamical calculations are based on ab initio PES and transition dipole moments for the electronic ground and excited states. For the realization of the population transfer we selected the $4^2A'$ state, shown in Fig. 1 together with the groundstate X^2A' and an intuitive pump-dump process indicated by the arrows. Exemplary, the optimal laser field to drive the system from the acute groundstate towards the vibrational target state Ψ_f=(1,0) of the obtuse isomer was computed. The time T of the overall process was set to 850 fs and the differential equations (1) were solved for the shape function $s(t)=\sin^2(\pi\cdot t/T)$. Increasing of the penalty factor α_0 leads to the smooth oscillations in the optimized laser field (Fig. 2 top, right) with two distinct frequencies in the corresponding power spectrum (Fig. 2 bottom, right). Accordingly, the electronic population dynamics during the pump-dump cycle (Fig.2 top,left) varies smoothly. In a subsequently performed windowed Fourier transform the frequency trace (Fig.2 bottom, left) of the optimized field is analyzed. With the obtained Fourier coefficients an efficient driving field composed of two pulses centered around the main frequencies shown in Fig. 2 (bottom, right) can be reconstructed. The decomposition of the optimized field into dominant contributions and the subsequent reconstruction signifies a further step towards experimental realization. The selected target state could be populated to about 67 %

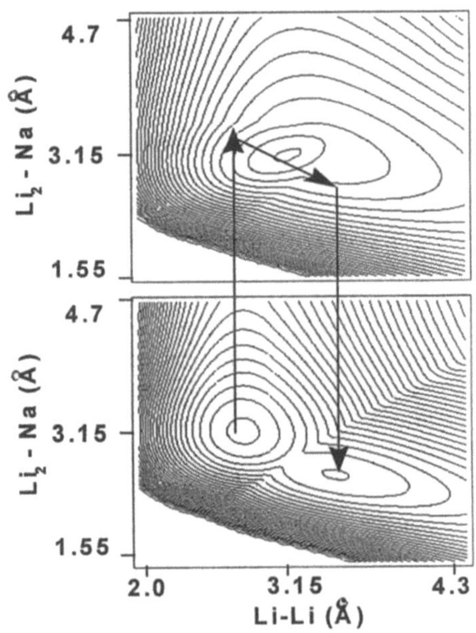

Fig. 1. Pump-dump scheme with electronic ground and excited state

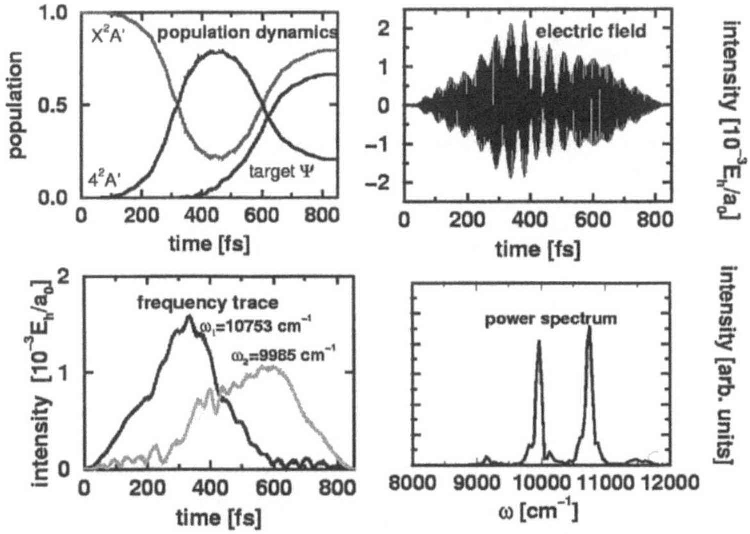

Fig. 2. Population dynamics, optimized field, frequency trace and power spectrum of the optimized laser pulse for the photoisomerization of Li_2Na.

with the optimized laser field and to about 63% with the reconstructed field, irrespective of the phase. In conclusion, our theoretical investigations indicate that optimal control theory is able to achieve photoisomerization of more complex systems with high efficiency and selectivity. Additional constraints for the electric field lead to more realistic laser pulses.

Acknowledgements We would like to thank the DFG for financial support through project No. Sfb 337 and the 'Fond der chemischen Industrie'.

References

1 S. A. Rice, Adv. Chem. Phys., **101**, 213, 1997.
2 R.J. Gordon and S. A. Rice, Ann. Rev. Phys. Chem., **48**, 601, 1997.
3 J. Manz, in: V. Sundström, Ed, Femtochemistry and Femtobiology (Singapore, 1998), p. 80.
4 W. Zhu, Jair Botina and H. Rabitz, J. Chem. Phys., **108**, 1953., 1998.
5 A.H. Zewail, in: V. Sundström, Ed, Femtochemistry and Femtobiology (Singapore, 1998), p. 1.
6 Ch. J. Bardeen , V.V. Yakovlev, K.R. Wilson, S.D. Carpenter, P.M. Weber, W.S. Warren, Chem. Phys. Lett., **280**, 151, 1997.
7 T. Baumert, J. Helbing, and G. Gerber. Adv. Chem. Phys., **101**, 47, 1997.
8 J. Manz, K. Sundermann, R. de Vivie-Riedle, Chem. Phys.Lett ,**299**, 415, 1998.
9 D. Neuhauser, J. Chem. Phys., **100**, 5076, 1994.

Quantum Control of Femtosecond Multiphoton Absorption via Chirp

Vladislav V. Yakovlev[1], Christopher J. Bardeen[2], Jianshu Cao[3], Jianwei Che[4], and Kent R. Wilson[5]

[1] Department of Physics, University of Wisconsin, Milwaukee, Physics Building, PO Box 413, Milwaukee, WI 53201, USA, E-mail: vyakovlev@physics.uwm.edu
[2] Department of Chemistry, University of Illinois, Urbana, Urbana, IL 61801, USA
[3] Department of Chemistry, MIT, Cambridge, MA 02139, USA
[4] Department of Chemistry, California Institute of Technology, Pasadena, CA 91125, USA
[5] Department of Chemistry and Biochemistry, University of California, San Diego, La Jolla, CA 92093-0339, USA

Abstract. We show both experimentally and theoretically how multiphoton absorption in a molecular system can be significantly affected by the phase structure of a single femtosecond pulse.

1. Introduction

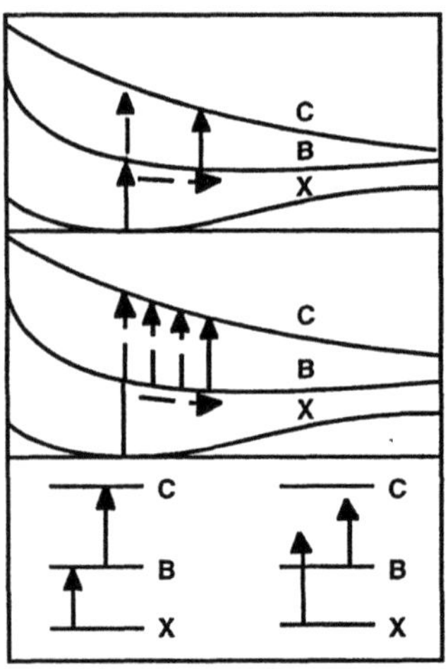

Fig. 1. Schematic of molecular multiphoton processes (see text)

Multiphoton processes play an important role in many areas of science, including photochemistry, fluorescence imaging, photoionization, and the generation of plasma and X-rays. To understand multiphoton molecular processes driven by ultrafast light pulses, one must consider the wavepacket dynamics of matter which occur during the light pulse [1, 2]. Furthermore, we show that the absorption of light and the dynamics of matter can be quantum controlled by playing off the color dynamics in the light pulse (by varying its chirp) against the color dynamics of receptivity to light in the molecules due to their wavepacket dynamics during the light pulse [1, 2].

We show theoretically, for the first time that these processes can be understood in terms of three physical effects: time-delayed resonance, wave-packet following and sequential resonance [1] as illustrated from top to bottom in the panels of Figure 1.

2. Experimental Methods

We have tested these ideas experimentally. A femtosecond kHz Ti:sapphire amplifier system is used in conjunction with an optical parametric amplifier of white light continuum and second harmonic generation to produce tunable 30-40 fs pulses in the range of 550-630 nm [3]. When these pulses, with energies of about 1 μJ in the visible, are focused inside a quartz bulb containing molecular iodine, a strong laser induced fluorescence (LIF) is observed around 340 nm. By varying the pump energy, we find that this LIF signal depends on the third power of the pump intensity, i.e. it is a result of three photon absorption.

By changing the chirp of the pump pulses we observe the variation of LIF intensity. One might expect that a transform limited pulse, i.e. a zero chirped pulse which has the highest intensity, would give the strongest absorption and fluorescence for this multiphoton process. Instead, we find the opposite (see Figure 2), for example that, for

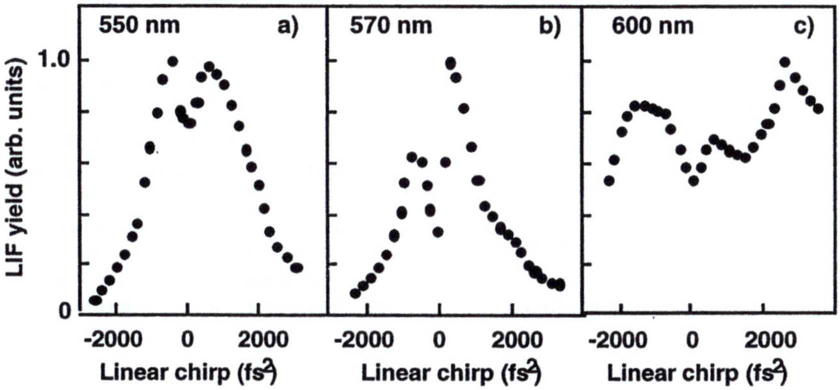

Fig. 2. Experimentally measured LIF yield as a function of chirp for different center wavelengths of the pump pulse

the pump wavelength centered at 570 nm, there is a local minimum of the LIF intensity with respect to the chirp for a nearly transform limited pulse, while for a positively chirped pulse, whose pulse duration is about two times longer than the nearly transform limited one, the LIF signal increases by a factor of three. We vary the pump wavelength in the range of 555 to 620 nm, and observe similar behavior, but the relative change of the signal varies.

We also perform a pump-probe experiment using the same excitation and probe wavelengths. The oscillating part of the LIF signal corresponds to the wavepacket oscillations on the B-state of iodine.

3. Discussion

Several mechanisms, as listed above, are responsible for this dependence of the multiphoton absorption on the chirp of the pulse [1, 2]. We believe that the major effect here is due to the dynamical evolution during the light pulse of the wavepacket on the first excited state (B-state) and the position-dependent energy detuning of the electronic state and is thus different in origin from that observed for Na_2 [4].

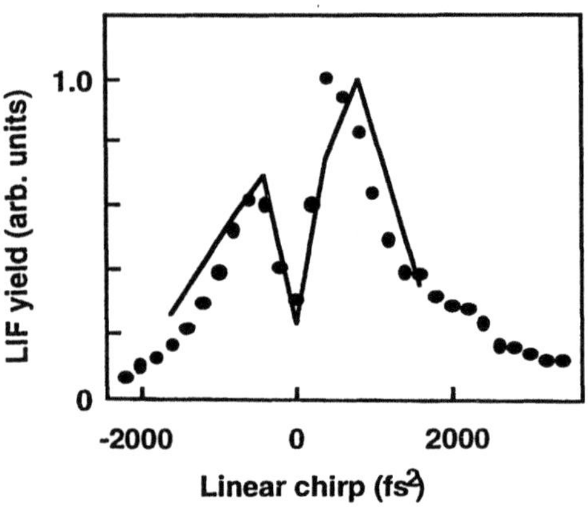

Fig. 3. Comparison of experimentally measured three photon LIF signal (points) and the calculated signal (solid line), calculated using measured *E(t)* of the light pulses

A comparison of theory (which involves full quantum dynamical calculation and includes ground state X, intermediate states B and C and a final ion-pair state D) with experiment at 570 nm is shown in Fig. 3. The qualitative agreement is quite good at this wavelength, as it also is at 550 nm (although models with other potential parameters might also fit the data). The calculation predicts a much stronger chirp enhancement for 600 nm than is experimentally observed, and this may be due to other states which are not taken into account by our calculation.

4. Conclusion

In summary, we have shown theoretically and experimentally how the phase structure of femtosecond pulses may be used to control multiphoton processes.

References

1 J. Cao, J. Che, and K. R. Wilson, J. Phys. Chem. **A102**, 4284 (1998).
2 V. V. Yakovlev, C. J. Bardeen, J. Che, J. Cao, and K. R. Wilson, J. Chem. Phys. **108**, 2309 (1998).
3 K. R. Wilson and V. V. Yakovlev, JOSA **B14**, 444 (1997).
4 A. Assion, T. Baumert, J. Helbing, V. Seyfried, and G. Gerber, Chem. Phys. Lett. **259**, 494 (1996).

Automated Coherent Control of Chemical Reactions and Pulse Compression by an Evolutionary Algorithm with Feedback

A. Assion[1], T. Baumert[2], M. Bergt[1], T. Brixner[1], B. Kiefer[1], V. Seyfried[1], M. Strehle[1], and G. Gerber[1]

[1] Physikal. Institut, Universität Würzburg, Am Hubland, D-97074 Würzburg, Germany
[2] Deutsches Zentrum für Luft- und Raumfahrt, D-82230 Weßling, Germany
E-mail: gerber@physik.uni-wuerzburg.de

Abstract. We experimentally demonstrate automated coherent control of a photodissociation reaction as well as femtosecond pulse compression using a pulse shaper and an evolutionary algorithm with feedback.

1. Introduction

The electric field of femtosecond laser pulses can be shaped nearly arbitrarily by modifying the relative phase of the spectral components. We combine a programmable pulse shaper and an evolutionary algorithm with feedback in order to automatically form femtosecond laser pulses. The phase function is improved iteratively by directly including a specific experimental output in an optimization procedure which enables totally automated search for the best adapted laser pulse shape [1]. An evolutionary computer algorithm addresses the programmable femtosecond pulse shaper, uses the desired experimental output as feedback, and iteratively improves the applied laser field (see Fig. 1). This opens up technological as well as molecular physical applications. We use the method to compress complex phase-shaped laser pulses from a chirped pulse amplification (CPA) Ti:Sapphire femtosecond laser system, and to optimize different reaction channels of the photodissociation processes of $Fe(CO)_5$.

The pulse shaper is based on the design of Weiner et al. [2]. It is set up as a zero dispersion compressor with a liquid crystal spatial light modulator (SLM) in its Fourier plane. With this device different optical path lengths can be introduced to the spatially separated spectral components of the laser pulse,

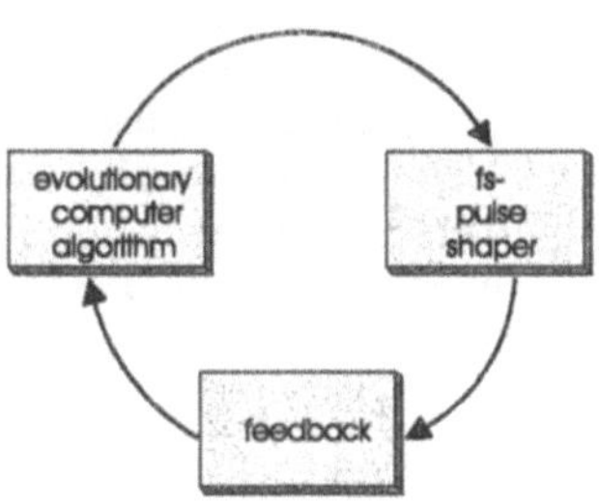

Fig. 1. Femtosecond laser pulses are modified by a computer-controlled pulse shaper. An evolutionary computer algorithm addresses the programmable femtosecond pulse shaper, uses the desired experimental output (signal of a 2-photon photodiode or the amount of ionic fragments recorded by a time-of-flight mass spectrometer) as feedback, and iteratively improves the applied laser field.

resulting in a shift of the relative phases. A detailed description of the experimental setup and the algorithm is given in Ref. [3].

2. Results and Discussion

First we demonstrate the applicability of this method on the automated compression of modulated femtosecond laser pulses from a CPA Ti:Sapphire laser system. The CPA technique implies a lot of optical components which introduce phase distortions. Since the amplified laser pulses are usually required to be bandwidth-limited, careful alignment of the laser system is necessary. Higher order dispersion terms can be compensated for only with extensive experimental effort, if at all. To optimize the output of our CPA system we employ a 2-photon photodiode which is irradiated upon by the modified laser pulses. The shorter the laser pulses, the higher will be the signal of the nonlinear photodiode. The evolutionary algorithm improves iteratively the 2-photon signal and thus the phase function until an optimum is found. Interferometric autocorrelations of the unshaped input pulse and the optimized output pulse are shown in Fig. 2. The wings in the autocorrelation (Fig. 2A) could not be removed by a simple compressor alignment, but the algorithm yields a bandwidth-limited pulse (Fig. 2B) without any prior knowledge of the input phase function. Therefore pulses from a misaligned laser system can be improved automatically, without the user having to worry about complex phase functions.

Another application of the feedback-controlled femtosecond pulse shaper is found in femtochemistry. Microscopic control of chemical reactions on a molecular level is an old dream in chemistry. We experimentally demonstrate how automated coherent control of the photodissociation of $Fe(CO)_5$ can be achieved. Interaction of 800 nm photons with a $Fe(CO)_5$ molecular beam leads to different multi-photon processes. One possible process is direct ionization of the parent molecule, another possibility is a combination of ionization and fragmentation processes in which the $Fe(CO)_5$ molecules loses an electron and also from one up to all five of its carbonyl ligands. The reaction products are

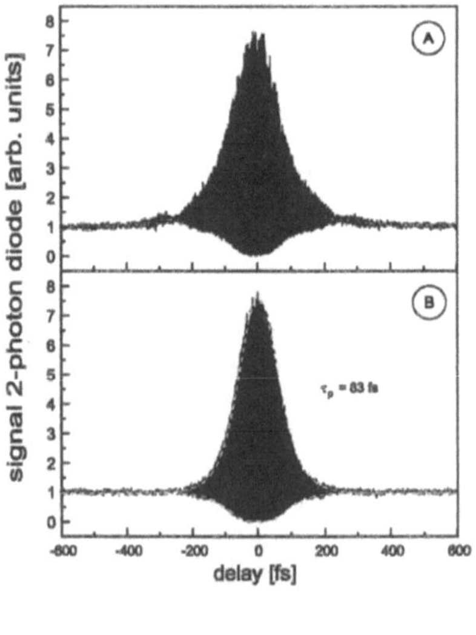

Fig. 2. Interferometric autocorrelations. The input pulse (A) clearly exhibits wings which cannot be removed by simple compressor alignment. They indicate a complex phase function which is automatically compensated for by the evolutionary algorithm, resulting in a bandwidth-limited pulse (B).

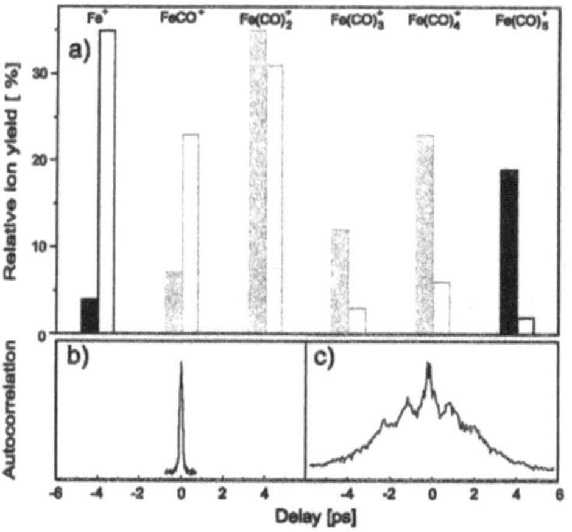

Fig. 3 a) The ratio of $Fe(CO)_5^+/Fe^+$ product yields is maximized (filled blocks), and minimized (empty blocks) by the optimization algorithm, yielding significantly different abundances of Fe^+ and $Fe(CO)_5^+$. The autocorrelations of the corresponding laser pulses are also included. A nearly bandwidth-limited pulse (b) leads to the production of mainly $Fe(CO)_5^+$ and a long laser pulse (c) to the production of mainly Fe^+.

detected in a time-of-flight (TOF) mass spectrometer and relative ion yields of the products are used as feedback. We maximized and minimized the branching ratio $Fe(CO)_5^+/Fe^+$, yielding significantly different abundances of reaction products, as illustrated in Fig. 3. The maximization procedure yields the product distribution of Fig. 3a (solid blocks). The result of this optimization is a very short (bandwidth-limited) laser pulse (Fig. 3b). If, on the other hand, we choose to minimize the $Fe(CO)_5^+/Fe^+$ ratio, that is, to maximize the inverse, an optimum is found as well (Fig. 3a, open blocks). This optimization leads to a long laser pulse of ps duration (Fig. 3c). The optimized branching ratios could be changed by a factor of 70. The results are in complete agreement with previous experiments of pure pulse length variation [4].

3. Conclusion

The results demonstrate totally automated coherent control of photodissociation reactions as well as automated pulse compression. For this purpose we employ a femtosecond pulse shaper which is feedback-controlled by an evolutionary computer algorithm. It should be noted that the applied method needs no calibration and no information about the system for a successful optimization.

References

1 R. S. Judson, H. Rabitz, *Phys. Rev. Lett.* **68**, 1500 (1992).
2 A. M. Weiner, D. E. Leaird, J. S. Patel, J. R. Wullert, *IEEE J. Quantum Electron.* **28**, 908 (1992).
3 T. Baumert, T. Brixner, V. Seyfried, M. Strehle, G. Gerber, *Appl. Phys. B* **65**, 779 (1997).
4 L. Bañares, T. Baumert, M. Bergt, B. Kiefer und G. Gerber, J. Chem. Phys **108**, 5799 (1998).

Dynamics of Charge-Transfer-to-Solvent Precursor States in $I^-(D_2O)_n$ Clusters

Martin T. Zanni,[1] Leo Lehr,[2] B. Jefferys Greenblatt,[1] Rainer Weinkauf,[2] and Daniel M. Neumark[1]

[1]Department of Chemistry, University of California, Berkeley, CA 94720, USA
[2]Institut für Physikalische und Theoretische Chemie, Technische Universität München, Lichtenbergstr. 4, 85747 Garching, Germany

Abstract. The dynamics of charge-transfer-to-solvent (CTTS) precursor states in $I^-(D_2O)_n$ clusters are studied using anion femtosecond photoelectron spectroscopy. For the n=4 cluster, excitation of a dipole-bound state results in a simple decay by vibrational autodetachment. In the n=5 and n=6 clusters, the initially excited dipole-bound state undergoes solvent reorganization in order to stabilize the excess electron.

1. Introduction

In 1928, Franck and Scheibe [1] discovered that aqueous solutions of the I^- anion exhibit two diffuse absorption bands in the ultraviolet. These bands, found at 5.5 and 6.4 eV, are attributed to a transition in which an electron is excited from the atomic iodide anion into the pre-oriented solvent cavity and are therefore termed "charge-transfer-to-solvent" (CTTS) bands. Since their discovery, the CTTS bands have proved to be fertile ground for physical chemists. While early work focused on understanding the frequency-domain spectrum of the CTTS bands,[2] more recent experimental [3] and theoretical [4] studies have concentrated on understanding the condensed phase dynamics resulting from their excitation.

In the gas phase I^- has no excited electronic states. However, Johnson [5] has recently shown that $I^-(H_2O)_n$ clusters (n=1-4) exhibit a diffuse absorption band at the detachment threshold. and that this band which shifts to higher energy and increases in intensity *n* increases. Similar effects were seen by Cheshnovsky [6] in $I^-(Xe)_n$ clusters. These bands were assigned to the cluster analog of the CTTS bands, termed CTTS precursor states, based on the reasonable expectation that they would converge to the lower energy bulk CTTS band for sufficiently large clusters. In order to better understand the nature of the CTTS precursor states in finite clusters, we have performed time-resolved experiments on a series on mass-selected $I^-(D_2O)_n$ clusters using the method of anion femtosecond photoelectron spectroscopy (FPES).

In anion FPES[7], a size-selected anion is electronically excited with a femtosecond pump pulse. The resulting non-stationary state is photodetached by a femtosecond probe pulse, and the photoelectron spectrum is measured. At each delay time, the photoelectron spectrum represents a complete "snapshot" of the evolving state. Previously, we have used anion FPES to study valence electronic excitation in bare and clustered anions. For example, in studies of $I_2^-(Ar)_n$ and

$I_2^-(CO_2)_n$,[8,9] the I_2^- chromophore was excited to a repulsive state, and anion FPES was used to monitor how the number and composition of solvent species affects the resulting dissociation and recombination dynamics.

The experiments on $I^-(D_2O)_n$ are conceptually quite different as the excited electronic state represents excitation of the excess electron into a highly delocalized state. The time-resolved photoelectron spectra then provide a detailed probe of the subsequent solvent response. In this sense our experiments are related to the transient absorption experiments that have been performed on solvated electrons in condensed media [10,11]. We discuss results for the n=4, 5, and 6 clusters. Our results suggest that for the n=4 cluster, the electron is photoexcited into a "dipole-bound" state of the four water molecules and decays by vibrational autodetachment. However, the n=5 and 6 clusters undergo solvent rearrangement that stabilizes the excess electron. These are therefore the smallest systems in which the cluster analog of electron solvation occurs.

2. Experimental

The anion FPES apparatus at Berkeley has been described previously. [7,9] Briefly, a mass-selected beam of $I^-(D_2O)_n$ anions is formed by expansion of a pulsed molecular beam of $Ne/CH_3I/D_2O$ into a vacuum chamber. The molecular beam is crossed by an electron beam just downstream of the valve orifice. The resulting anions are injected into a time-of-flight mass spectrometer. Mass-selected anions interact with the pump and probe pulses, and the resulting photoelectrons are collected with high efficiency (>50%) using a "magnetic bottle" time-of-flight photoelectron spectrometer. The pump and probe pulses are generated from a commercial Ti:sapphire oscillator/regenerative amplifier system operating at a repetition rate of 500 Hz. The fundamental wavelength at 780 nm (1.59 eV) is used as the probe pulse; the pump pulse at 260 nm (4.77 eV) is produced by frequency-tripling of the fundamental. The pump and probe are 120 and 80 fs in duration, respectively. The pump pulse is resonant with CTTS band seen by Johnson [5] for $I^-(H_2O)_4$; our results show that clusters with as many as ten water molecules can be excited at 260 nm.

3. Results and Discussion

Fig. 1 shows femtosecond photoelectron spectra for $I^-(D_2O)_{n=4,6}$. Results for the n=5 cluster (not shown) are similar to those for n=6. The spectra are represented as two-dimensional contour plots, with electron kinetic energy (eKE) as the vertical axis. Hence, vertical cuts through these plots represent photoelectron spectra at specific delay times.

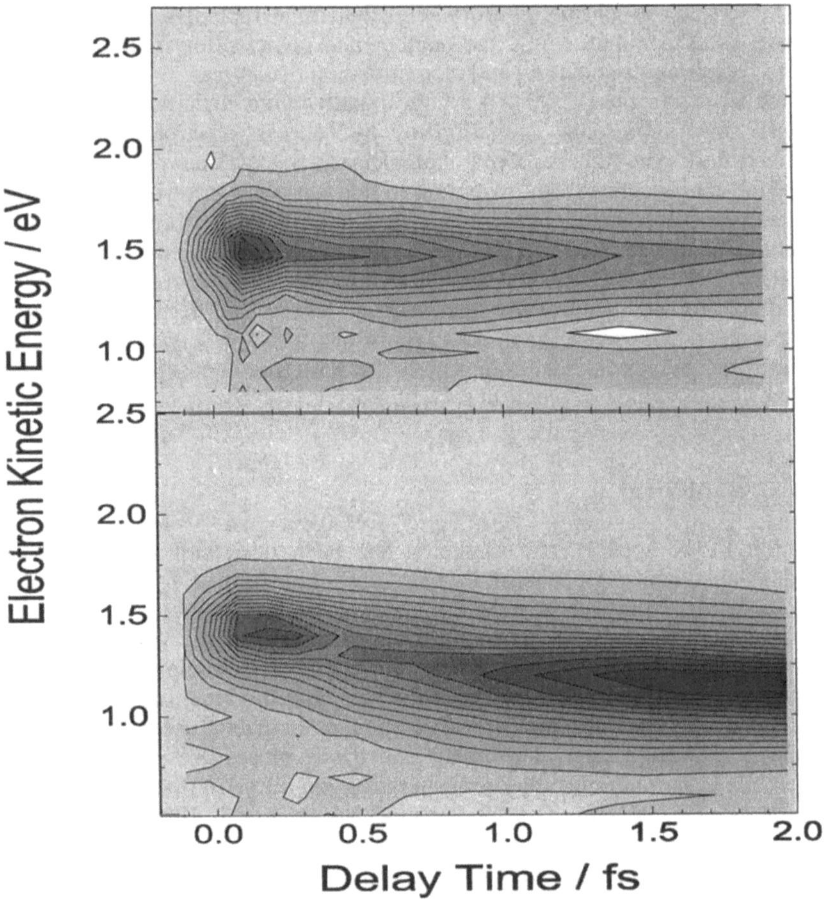

Fig. 1. Top to bottom: Femtosecond photoelectron spectra of $I^-(D_2O)_4$, and $I^-(D_2O)_6$

The spectrum for the n=4 cluster is centered at eKE≅1.6 eV for all delay times. The electron signal decays with a time constant of 190 fs. Along the energy axis, the spectrum is only about 0.33 eV wide at short delay times and narrows further at longer times. The n=6 spectrum is quite different; the average electron energy drops by 0.3 eV during the first picosecond. In addition, the integrated intensities clearly do not decrease monotonically with time, in contrast to the n=4 cluster. Instead, the intensity drops rapidly for a few hundred femtoseconds, rises again beyond the initial maximum, and then decays much more slowly with time constants of 37 and 96 ps for the n=5 and 6 clusters, respectively.

These spectra can be interpreted with the aid of *ab initio* calculations by Jortner [12], and Xantheas [13] on $X^-(H_2O)_n$ clusters. The ground state structure for $I^-(D_2O)_4$ is shown in Fig. 2. In this conformer, the free H atoms in the solvent network are oriented toward the iodide. Similar ground state structures with "surface" I^- anions are found for n=5 and 6. This arrangement of water molecules results in a large dipole moment, estimated to be 4.4 D in $I^-(H_2O)_4$.[5] This is sufficiently large to trap an excess electron in a "dipole-bound" state (DBS). suggesting that the CTTS precursor states of small $I^-(H_2O)_n$ clusters have dipole-bound character.

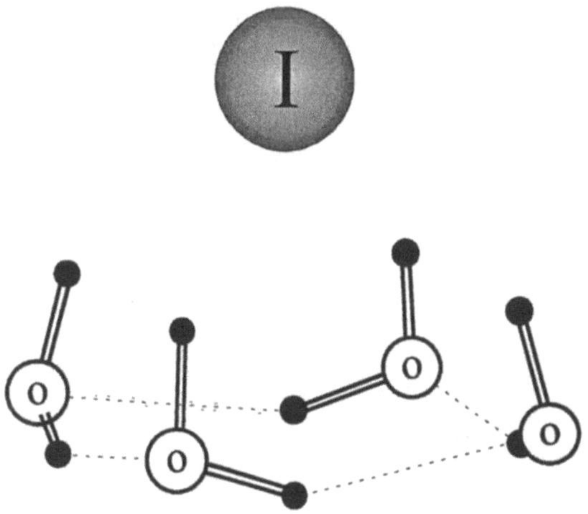

Fig. 2. Ground state structure of $I^-(D_2O)_4$, adapted from. Ref. 12

Bowen [14] has measured many photoelectron spectra of anions for which the *ground* state is a dipole-bound state (DBS), finding that the excess electron is typically bound by < 100 meV. In our results for $I^-(D_2O)_4$, the peak electron energy lies very close to the probe photon energy, consistent with excitation into a DBS by the pump pulse. However, the pump photon energy (4.77 eV) is larger by 0.45 eV than the adiabatic electron affinity of 4.3 eV for $I(D_2O)$,[15] indicating that the DBS has about 0.50 - 0.55 eV of vibrational excitation. We therefore attribute the 190 fs decay of the n=4 spectrum to vibrational autodetachment of the DBS.

The n= 5 and 6 spectra indicate significantly different dynamics subsequent to CTTS precursor state excitation. First, the shifting of the spectrum toward lower electron kinetic energy indicates that at longer times, it takes more energy to detach the excess electron than immediately after excitation. This implies that the solvent molecules re-arrange to stabilize the excess electron.

Secondly, the non-monotonic variation of integrated intensity with pump-probe delay time suggests that a transition between two conformer structures takes place which have different photodetachment cross sections. If only a single conformer were involved, the increase of integrated intensity with time cannot be explained. Johnson [16] has shown that there are two conformers of $(H_2O)^-_6$, and that the photodetachment cross section is larger (by a factor of three) for the conformer in which the electron is more strongly bound. It is therefore reasonable to attribute the intensity increase in our FPES spectra to isomerization to a conformer with a higher degree of electron solvation. Hence, we interpret the FPES intensities as a superposition of two effects: (i) a population decay by vibrational autodetachment and (ii) an increase of the transition moment for the photodetachment process due to isomerization.

The explanation most consistent with these results is that the CTTS precursor state initially accessed is a DBS, in which the water molecules are in a similar geometry as in the $I^-(D_2O)_n$ cluster. The water molecules then rearrange to form a more stable conformer. *Ab initio* calculations by Kim [17] on $(H_2O)_6^-$ suggest one such structure in which the water molecules form a "bowl", around the excess electron, resulting in less delocalization than is characteristic for a DBS.

Thus, in the n=5 and 6 clusters the initial state of the electron undergoes "solvation" by the water molecules in the cluster, but not in the n=4 cluster. The two larger clusters are therefore the first to show the type of dynamics that occur in bulk water, in which excitation of a CTTS band results in formation of a solvated electron.[3,4] A more complete account of these studies, involving larger clusters and different isotopes, will be presented in the near future.

Acknowledgements: This research is supported by the Air Force Office of Scientific Research under Grant No. F49620-97-1-0018. R. Weinkauf thanks the Deutsche Forschungsgemeinschaft for financial support. L. Lehr acknowledges support from E. W. Schlag.

References

1 J. Franck and G. Scheibe, in *Z. Phys. Chem.*, Vol. A139, 22, 1928.
2 M. J. Blandamer and M. F. Fox, in *Chem. Rev.*, Vol. 70, 59, 1970.
3 F. H. Long, X. Shi, H. Lu, and K. B. Eisenthal, in *J. Phys. Chem.*, Vol. 98, 7252, 1994.
4 W. S. Sheu and P. J. Rossky, in *Chem. Phys. Lett.*, Vol. 213, 233, 1993.
5 D. Serxner, C. E. Dessent, and M. A. Johnson, in *J. Chem. Phys.*, Vol. 105, 7231, 1996.
6 I. Becker, G. Marcovich, and O. Cheshnovsky, in *Phys. Rev. Lett.*, Vol. 79, 3391, 1997.
7 B. J. Greenblatt, M. T. Zanni, and D. M. Neumark, in *Chem. Phys. Lett.*, Vol. 258, 523, 1996.
8 B. J. Greenblatt, M. T. Zanni, and D. M. Neumark, in *Science*, Vol. 276, 1675, 1990.
9 B. J. Greenblatt, M. T. Zanni, and D. M. Neumark, in *Disc. Faraday. Soc.*, Vol. 108, 101, 1997.
10 A. Migus, Y. Gaudel, J. L. Martin, and A. Antonetti, in *Phys. Rev. Lett.*, Vol. 58, 1559, 1987.
11 Y. Kimura, J. C. Alfano, P. K. Walhout, and P. F. Barbara, in *J. Phys. Chem.*, Vol. 98, 3450, 1994.
12 J. E. Combariza, N. R. Kestner, and J. Jortner, in *J. Chem. Phys.*, Vol. 100, 2851, 1994.
13 S. S. Xantheas, in *J. Phys. Chem.*, Vol. 100, 9703, 1996.
14 J. H. Hendricks, H. L. deClerq, S. A. Lyupustina, and K. H. Bowen, in *J. Chem. Phys.*, Vol. 107, 2962, 1997.
15 G. Marcovich, S. Pollack, R. Giniger, and O. Cheshnovsky, in *Z. Phys. D*, Vol. 26, 98, 1993.
16 C.G. Bailey, and M.A. Johnson, in *Chem. Phys. Lett.*, Vol. 265, 185, 1997.
17 K. S. Kim, S. Lee, J. Kim, and J. Y. Lee, in *J. Am. Chem. Soc.*, Vol. 119, 9329, 1997.

Analysis of the femtosecond time-resolved photoelectron spectra of the ammonia dimer excited to the $\tilde{A}$ state

V. Stert, W. Radloff, P. Farmanara, H.-H. Ritze, I.V. Hertel

Max-Born-Institut für Optik und Kurzzeitspektroskopie,
Rudower Chaussee 6, D-12489 Berlin,
E-mail: radloff@mbi-berlin.de

Abstract. The H atom transfer state $NH_4...NH_2$ occupied about 150fs after the $\tilde{A}$ state excitation of $(NH_3)_2$ was analyzed by femtosecond time-resolved photoelectron spectroscopy.

In recent years the dynamics of ammonia clusters excited to their lowest electronic state ($\tilde{A}$) has been studied by pump-probe experiments with femtosecond laser pulses (e.g. [1]). This state decays very fast within a few 100fs by fragmentation and internal rearrangement. The rearrangement is suggested to lead to an internal H atom transfer state $(NH_3)_{n-2}NH_4NH_2$ which can not be excited directly by a Franck-Condon transition from the electronic ground state. Ist lifetime has been found to be 3-4ps [1]. The role of this H transfer state which is of general interest for the understanding of the dynamics in hydrogen bonded cluster systems has been studied in detail for the ammonia dimer.

In order to follow the energy flow in the excited neutral cluster states in real time we have analyzed the femtosecond time-resolved photoelectron spectra of the ammonia dimer. We used the novel developed method of combining the photoelectron-photoion coincidence measurements with the fs pump-probe technique [2]. The ammonia clusters are formed by adiabatic expansion through a pulsed nozzle. In order to avoid significant contributions of larger cluster fragments to the dimer signal we use a concentration of only 0.8% ammonia in the He seed gas of 1 bar. The clusters are excited into the $\tilde{A}$ state by a pump pulse at λ_1=200nm and ionized by a time-delayed probe pulse at λ_2=267nm. The wavelengths are obtained as the fourth and third harmonic, respectively, of a commercial Ti:sapphire laser and amplifier system (Clark MXR) tuned to 800nm. The width of the laser pulses is about 130fs. The ions are detected by a time-of-flight mass spectrometer while the coincident photoelectrons are analysed by a time-of-flight „magnetic bottle" electron spectrometer. To suppress accidental coincidences the laser fluences have to be strongly reduced to ensure ionization rates significantly below 1 per laser pulse.

In Fig. 1 the electron spectra are shown for the cluster ions $(NH_3)_2^+$ and NH_4^+, respectively, for different delay times τ between the pump and the probe pulse. The spectra are normalized to the maximum electron signal measured at τ =100fs. The drastical variations of the electron spectra especially for $(NH_3)_2^+$ reflect the changes of the geometry and the internal energy on the fs time scale.

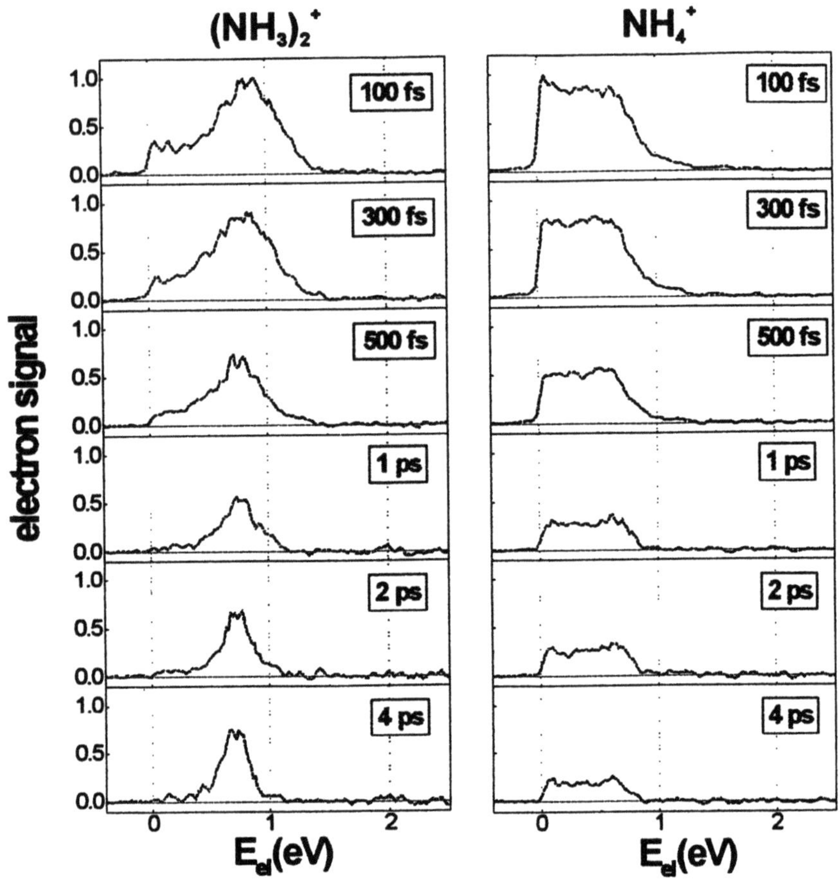

Fig. 1 Electron spectra for different delay times between the pump pulse at 200nm and the probe pulse at 267nm.

The electron spectra at τ =100fs reach from E_{el}=0 up to the maximum electron energy E_{el}^{max}= $h\nu_1$+ $h\nu_2$ −AP where AP represents the appearance potentials (9.19eV for $(NH_3)_2^+$ and 9.54eV for NH_4^+). The large width of the photoelectron spectra can be understood as due to the large geometry difference between the involved neutral and the ionized states. For larger delay times τ ≥ 500fs the electron spectra of $(NH_3)_2^+$ are characterized by a much narrower distribution down to a width of about 0.3eV. The position of the maximum at E_{el}=0.7eV is attributed to the energy of the H transfer state NH_4NH_2 populated about 150fs after the primary excitation. As our ab initio calculations show the geometry of the H transfer state is quite similar to that of the proton transfer ion state NH_4^+ ...NH_2 resulting in a relatively narrow Franck-Condon distribution as confirmed by our experiments. The calculated energy of the H transfer state agrees well with the measured value E_{el}=0.7eV. The H transfer state is stabilized by the superposition with a charge transfer state NH_4^+ ...NH_2^- . The crossing of the two potential curves at large N-N distances causes the lifetime of 3-4ps for this state.

Acknowledgement

Financial support be the Deutsche Forschungsgemeinschaft through Sonderforschungsbereich 337 is gratefully acknowledged.

References

1 Th. Freudenberg, W. Radloff, H.-H. Ritze, V. Stert, K. Weyers, F. Noack, I.V. Hertel, Z. Phys. D 36 (1996) 349

2 V. Stert, W. Radloff, Th. Freudenberg, F. Noack, I.V. Hertel, C. Jouvet, C. Dedonder-Lardeux, D. Solgadi, Europhys. Lett. 40 (1997) 515

Time-Resolved Observation of Geometrical Reorientations of Metal Clusters

Štefan Vajda, Sebastian Wolf, Ulrike Busolt, Henry Hess, Thomas Leisner, and Ludger Wöste

Institut für Experimentalphysik, Freie Universität Berlin, Arnimallee 14, D-14195 Berlin
E-mail: vajda@physik.fu-berlin.de, woeste@physik.fu-berlin.de

Abstract. Ultrafast geometrical relaxation of Ag_3 clusters in the gas phase was studied by two-color femtosecond NeNePo spectroscopy. The first fs laser puls created a vibrational wave packet in Ag_3 by photodetachment from Ag_3^-. The subsequent geometrical relaxation in the neutral Ag_3 was monitored by resonant two-photon ionization with a second laser pulse (Negative to Neutral to Positive, „NeNePo"). Fast bending of the initially linear Ag_3 was observed. Under given experimental conditions, no wave packet dynamics was resolved. This finding can be explained by very fast intracluster vibrational relaxation.

1. Introduction

NeNePo pump-probe femtosecond spectroscopy is a powerful tool to study the dynamics of clusters in the ground state[1,2]. This method is based on the preparation of a vibrational wave packet in the neutral cluster by one-photon photodetachment of the excess electron from the corresponding negatively charged cluster with a fs laser pulse. The temporal evolution of the induced wave packet is probed by a delayed pulse via two-photon ionization. This method can provide valuable dynamical information about clusters whose anionic and neutral geometries are different.

The NeNePo scheme has been successfully tested on Ag_3^-, using nonresonant two-photon ionization [1]. In these experiments, the same laser was used for photodetachment and for photoionization. A recent theoretical study has shown [3], that much of the dynamical information is lost in nonresonant two-photon ionization experiments due to the unknown energy of the free electron. In order to minimize the energy ambiguity in our experiment, we have employed two-photon ionization via an intermediate state. Since the cross-section for a bound-bound transition involved in the ionization step is very large in comparison with the nonresonant process, the resonant NeNePo should sensitively monitor the geometrical relaxation in real time. Our experimental results will be compared with the theoretical calculations of Bonacic-Koutecký *et al.* [3].

2. Experimental Methods

We used an ion beam apparatus similar to the one developed by Wöste *et al.* in Lausanne [4]. The cluster source chosen was a sputtering arrangement consisting of four silver targets, allowing us to produce Ag_n^- clusters. The scheme of the experimental setup is shown in Figure 1. The clusters are produced by bombardment of silver targets with energetic xenon cations. The emerging Ag_n^- clusters are reflected into the first quadrupole ion guide. Here, the Ag_n^- clusters are slowed down and cooled approximately to room temperature via collisions with He atoms. The Ag_3^- clusters are then selected by the first quadrupole mass filter and enter the quadrupole ion trap filled with He gas. Subsequently, the Ag_3^- clusters interact with the laser beams in the ion trap. The produced cluster cations leave the ion trap via the exit lens and are analyzed with the second quadrupole mass filter. A detailed description of the ion source is given in [5].

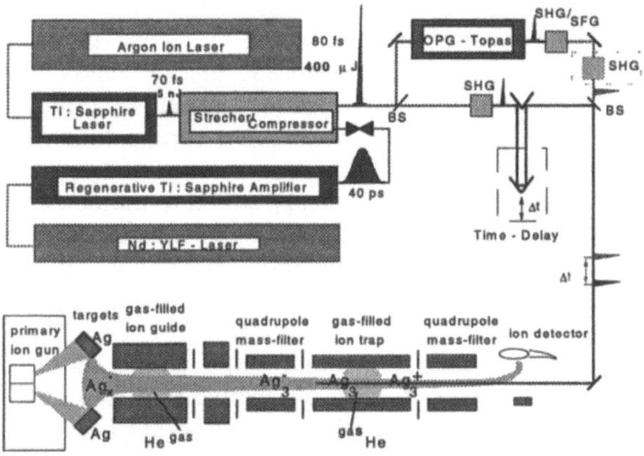

Figure 1. Scheme of the experimental setup.

The laser system consists of a Spectra Physics Tsunami titanium-saphire laser pumped with a Spectra Physics argon-ion laser. The pulses of the Tsunami laser are used to seed a Quantronix titanium-saphire regenerative amplifier (RegA) pumped with a Quantronix Nd:YLF laser. The amplifier running at 1 kHz produces pulses of 80 - 100 fs duration and about 400 µJ puls energy at 800 nm. Ninety percent of this power is used to pump an Light Conversion Topas optical parametrical generator (OPG). The infrared output light of the OPG is frequency doubled two times to provide femtosecond ultraviolet pulses in the wavelength region between 345 and 385 nm. Ten percent of the RegA's power is used for second harmonic generation of 400 nm. Photodetachment is performed with pulses of a central wavelength of 400 nm, the ionization of the neutral Ag_3 is performed by UV pulses of variable wavelength.

3. Results and Discussion

In Figure 2., the time-dependent Ag_3^+ signals obtained at different photoionization wavelengths are shown. In all cases, the signal shows a minimum at delay time Δt=0 and rises rapidly to a maximum after about Δt=600 fs. At shorter ionization wavelengths, shorter rise times of the signal are observed. This behavior can be explained by time-dependent absorption of Ag_3 due to geometrical relaxation. The modulation of the signal with time arises from the time dependent change of potential energy difference between the ground state and the $^2E''$ excited state of Ag_3 involved in resonant photoionization step during geometrical relaxation of the

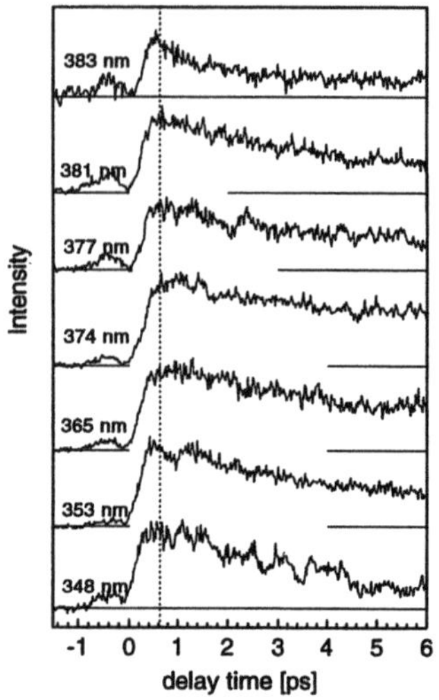

Fig. 2. Time-dependent Ag_3^+ signals

cluster [3,5]. No cations were detected at ionizing wavelengths shorter than 345 nm. In addition, no oscillatory signal was resolved in these experiments, indicating that the coherence is destroyed within the first vibrational period. This conclusion is supported by recent *ab initio* calculations of Bonacic-Koutecký and her collegues [3]. The neutral cluster is initially linear and starts to bend. After certain time, the terminal atoms collide. The analysis shows, that much of the kinetic energy stored in the bending motion is transferred to the symmetric stretching mode. The asymmetric stretching mode participates in the relaxation process only at later times. The relaxation is strongly temperature dependent. At the initial anion temperature of 300 K a very efficient dissipative IVR leading to the loss of vibrational coherence takes place [3].

4. Conclusions

We presented two-color NeNePo measurements on Ag_3 using resonant two-photon ionization. The fast rise of the ion signal at very short times indicates a large change of the differential ionization potential in the region close to the linear geometry of Ag_3. The lack of oscillations in the cation signals is explained by very fast IVR taking place after intracluster collision of the terminal atoms, demonstrating the important role of IVR processes in clusters as small as a trimer.

Acknowledgements. The financial support provided by the Deutsche Forschungsgemeinschaft within the SFB 337 project is gratefully acknowledged.

References

1 S. Wolf, G. Sommerer, S. Rutz, E. Schreiber, T.Leisner, L. Wöste, and R. S. Berry, Phys. Rev. Lett., 74, 4177, 1995.

2 D.W. Boo, Y. Ozaki, L.H. Andersen, and W.C. Lineberger: J. Phys. Chem. A, 101, 6688, 1997.

3 M. Hartmann, A. Heidenreich, J. Pittner, V. Bonacic-Koutecký, and J. Jortner: Phys. Chem. A, 102, 4069, 1997.

4 M. Fayet, M.J. McGlinchey, and L.Wöste, J. Am. Chem. Soc. 109, 1733, 1987

5 T.Leisner, S. Vajda, S. Wolf, L. Wöste, and R. S. Berry, J. Phys. Chem., subm., 1998.

Experimental Observation of Resonance Effects in Intensely Irradiated Atomic Clusters

J. Zweiback, T. Ditmire, A. M. Rubenchik, A. Komashko, M. D. Perry

Lawrence Livermore National Laboratory
PO Box 808, L-477
Livermore, CA 94550, USA
Email: zweiback1@llnl.gov

Abstract. We have resolved the expansion of intensely irradiated atomic clusters on a femtosecond time scale. These data show evidence for resonant heating, similar to resonance absorption, in spherical cluster plasmas.

1. Introduction

There have been many studies of the irradiation of atomic clusters with short pulse high intensity laser radiation [1-4]. The high local density of cluster targets greatly increases the coupling of the laser energy to the atoms. Of particular interest are the cluster heating and expansion mechanisms, along with the time for disassembly. Recent experiments [1,4] have started to examine these dynamics. Theory [3] predicts resonance effects should play a major role in these interactions. These will occur when the laser frequency equals the natural frequency of the plasma. Similar to resonance absorption, this resonance should greatly enhance the absorption of laser energy by the clusters. Due to the spherical geometry of the clusters, the resonance occurs when the electron density is three times the critical density. In this paper we report on a series of experiments investigating these resonance effects.

2. Experimental Methods

The laser used for these experiments was a Ti:Sapphire CPA laser system operating at a wavelength of 810 nm, capable of producing about 50 mJ in a 50 fs pulse [5]. We could vary the laser pulse length by changing the grating spacing in the compressor. The beam could also be directed through a beamsplitter, with one beam going into a variable delay leg. The beam was then focused with an f/3 off axis paraboloid onto a cluster target. The clusters were produced by expansion in a supersonic Laval nozzle backed with up to 200 psi of xenon. From a calculation of the Hagena parameter we estimate the average number of atoms per cluster, N, to be up to 6×10^5 [6,7]. The cluster radii ranged from 85 Å to 205 Å. The average density produced by the nozzle is estimated to vary from approximately 1.5×10^{17} to 6×10^{17} atoms/cm^3 [8].

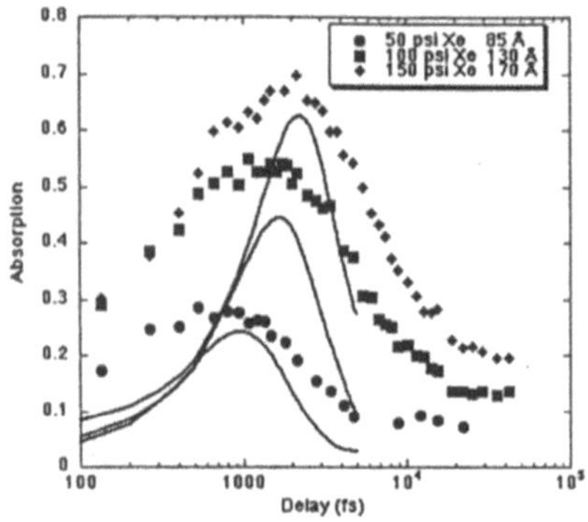

Fig. 1. Data from the pump probe experiments. Solid lines are model calculations.

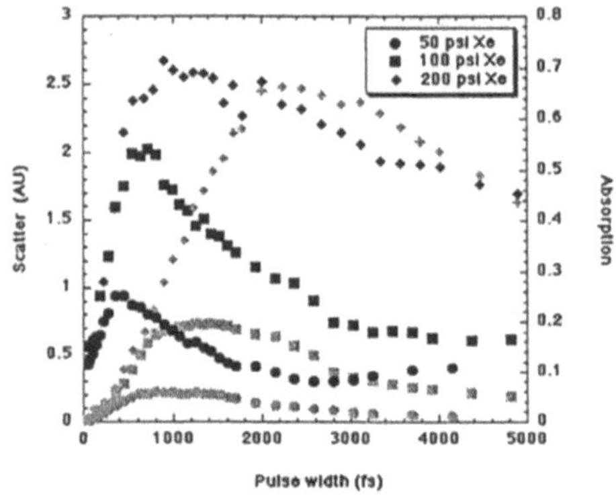

Fig. 2. Absorption (black) and Mie scattering (gray) measurements. Note that they peak at different pulse widths.

3. Results and Discussion

In the pump-probe experiments a small (10%) pump pulse was sent into the target to expand the clusters before the probe pulse arrives. This allows us to probe the disassembly of the clusters as they expand into a bulk plasma. Both pulses were about 50 fs in length. The probe pulse had a peak intensity in vacuum of 1.6 x 10^{17} W/cm^2. Figure 1 shows absorption as a function of probe delay for different backing pressures of xenon. The absorption peak in Fig. 1 indicates the presence of resonant heating as the clusters expand. At very small delays, absorption is low; the cluster has little time to expand and the second pulse arrives long before the electron density reaches the resonance condition ($n_e/n_{crit}=3$). However, as we increase the delay, absorption increases. The longer delay allows the electron density to be near the resonance condition when the pulse arrives, greatly enhancing the absorption of laser light. For longer delays, the cluster has continued to expand and the probe pulse is poorly absorbed by the, now underdense, plasma. Eventually, we reach a condition in which the clusters are fully disassembled and the absorption approaches a constant depending only on the average gas density. Our model, similar to Ref. [3], is shown as solid lines in Fig 1. The delay for peak absorption is in good agreement with the data.

Figure 2 shows the results of a simultaneous measurement of absorption and 90° Mie scattering of the laser radiation, for different pulse lengths. Clearly the scattering signal reaches a maximum at a longer pulse length than the absorption. As a cluster expands, not only does the electron density decrease, but the radius increase as well. For spheres of the relevant size, the ratio of scattering cross section to absorption cross section increases as radius increases [9]. This should cause the observed increase in scattering after the absorption has reached it peak.

4. Conclusions

In conclusion, we have resolved the expansion of atomic clusters on a femtosecond time scale. These data show that spherical resonance absorption plays an important role in cluster heating. The effect of resonance is also seen in Mie scattering measurements; with the maximum scattering signal occuring at longer pulse widths than the maximum absorption.

Acknowledgments. This work was performed under the auspices of the U.S. Department of Energy by the Lawrence Livermore National Laboratory under contract W-7405-ENG-48.

References

1. T. Ditmire, E. Springate, J. W. G. Tisch, Y. L. Shao, M. B. Mason, N. Hay, J. P. Marangos, M. H. R. Hutchinson, Phys. Rev. A 57, 369 (1998).
2. J. Purnell, E. M. Snyder, S. Wei, A. W. Castleman Jr., Chem. Phys. Lett. 229, 333 (1994).
3. T. Ditmire, T. Donnelly, A. M. Rubenchik, R. W. Falcone, M. D. Perry, Phys. Rev. A 53, 3379 (1996).
4. M. Lezius, S. Dobosz, D. Normand, M. Schmidt, Phys. Rev. Lett. 80, 261 (1998).
5. P. Banks, J. Zweiback, S.N. Fochs, B.C. Stuart, M.D. Perry, in Conference on Lasers and Electro-Optics, Vol 9 1996 OSA Technical Digest Series (Optical Society of America, Washington, DC, 1996), paper CWI3.
6. O. F. Hagena and W. Obert, J. Chem. Phys. 56, 1793 (1972).
7. J. Wörmer, V. Guzielski, J. Strapelfeldt, T. Möller, Chem. Phys. Lett. 159, 321 (1989).
8. M. D. Perry, C. Darrow, C. Coverdale, J. K. Crane, Opt. Lett. 17, 523 (1992).
9. M. Kerker, The Scattering of Light and other Electromagnetic Radiation, (Academic Press, New York, 1969).

Part XI

Vibronic Excitations and Radiationless Processes

Conical Intersections and Femtosecond Dynamics

Wolfgang Domcke[1], Luis Seidner[1], and Gerhard Stock[2]

[1] Institute of Theoretical Chemistry, Heinrich-Heine-University, D-40225 Düsseldorf, Germany
[2] Department of Physics, University of Freiburg, D-79104 Freiburg, Germany

Abstract. Conical intersections of potential-energy surfaces have emerged as a central mechanistic concept in photochemistry. This paper discusses characteristic features of the ultrafast non-Born-Oppenheimer dynamics mediated by conical intersections as well as the real-time detection of these processes.

1. Introduction

Although the possibility of exact degeneracies of potential-energy (PE) surfaces of polyatomic molecules had been known for decades [1], it has been assumed for a long time that true surface crossings are rare phenomena, being primarily associated with high symmetries of the nuclear frame. Recent extensive *ab initio* computational studies have revealed, however, that so-called conical intersections of adiabatic PE surfaces, even between states of the same symmetry, are a very common phenomenon in polyatomic systems [2]. For triatomic systems, the topology of the surfaces and the seams of intersections have been analyzed in detail. For larger polyatomic systems, the minima of the hypersurfaces of intersection have been located with the help of geometry optimization methods. The ubiquity of conical intersections confirmed by these studies strongly supports the conjecture that conical intersections play an universal mechanistic role in photochemistry [3,4].

A generic model of PE surfaces with a conical intersection is shown in Fig. 1. The PE energies are displayed as a function of a torsional coordinate φ, which plays the role of the reaction coordinate for *cis-trans* isomerization, and a vibronically active mode (the so-called coupling mode) which couples the electronic states.

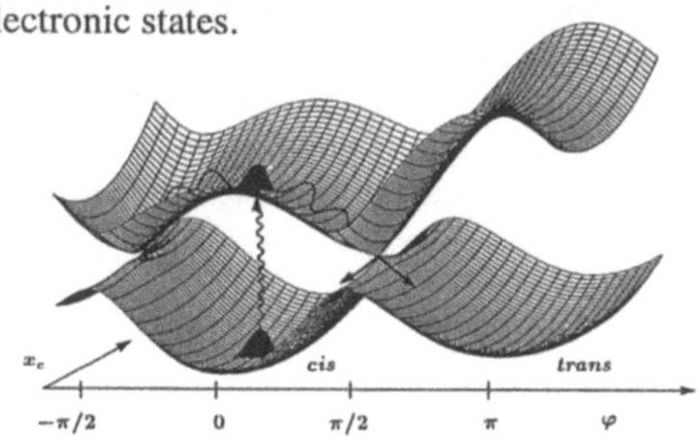

Fig. 1. Schematic view of conically intersecting potential-energy surfaces associated with a *cis-trans* photoisomerization process.

2. Ultrafast dynamics at conical intersections

Early attempts to treat the dynamics at conical intersections were based on the Landau-Zener approach [5]. While the Landau-Zener model provides a transparent picture for onedimensional avoided-crossing situations, its generalization to multi-dimensional nuclear motion is nontrivial. No quantitative results have been obtained in these early studies.

In an alternative approach, exact (numerical) time-dependent quantum wave-packet methods have been employed since the early eighties to explore the dynamics for realistic models of conical intersections, see [6-8] for reviews. These calculations have revealed that the fundamental dissipative processes of population and phase relaxation are clearly expressed already in few-mode systems, if a conical intersection of the PE surfaces is involved. The results strongly support the idea that conical intersections provide the microscopic mechanism for ultrafast nonadiabatic relaxation processes. More recently, these calculations have been extended to describe photodissociation and photoisomerization processes associated with conical intersections. The latter are particularly relevant for our understanding of the elementary mechanisms of photochemistry.

The time-dependent wave-packet calculations are generally performed in a so-called diabatic electronic representation. By expansion of the wave function in a vibrational basis set, the time-dependent Schrödinger equation is converted into a large set of coupled first-order differential equations

$$i\hbar d/dt\mathbf{C}(t)=\mathbf{H}\mathbf{C}(t)$$

where $\mathbf{C}(t)$ denotes the vector of expansion coefficients and $\mathbf{H}$ is the matrix representation of the molecular Hamiltonian. Efficient numerical techniques are available to solve Eq. (1) as an initial value problem for systems including up to about seven vibrational modes [7,8].

A characteristic example of the dynamics at a conical intersection is shown in Fig. 2. $P_1(t)$ represents the total population of the upper adiabatic surface, assuming the initial condition $P_1(0)=1$. $P_{cis}(t)$ represents the population of the initial chemical conformer, c. f. Fig. 1. The time evolution of these two populations reflects internal-conversion and photoisomerization processes, respectively. The model includes, in addition to the torsional coordinate and the coupling coordinate displayed in Fig. 1, a Condon-active so-called accepting mode [9]. Fig. 2 illustrates the ultrafast and irreversible character of the dynamics triggered by a conical intersection. The phenomena illustrated by Fig. 2 appear to be fairly general: very similar results for the electronic population dynamics have been found for a variety of systems, e. g. $C_2H_4^+$, $C_6H_6^+$, NO_2, pyrazine. In all cases that have been studied in detail, it has been found that a minimum of three strongly coupled nuclear degrees of freedom is required to obtain ultrafast relaxational behavior of electronic populations [6-8] .

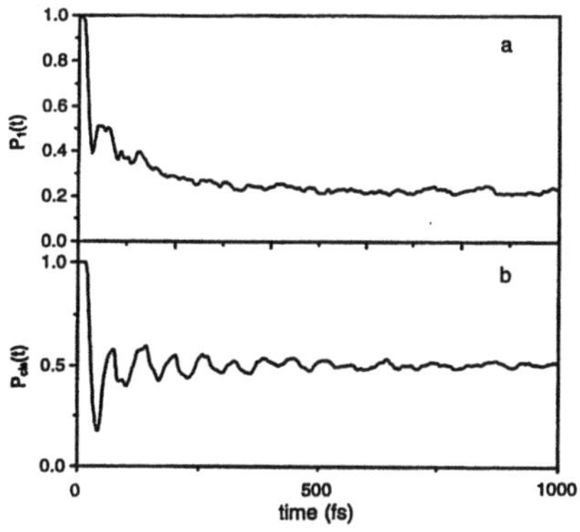

Fig. 2. Time evolution of the population $P_1(t)$ of the upper adiabatic surface (a) and of the population $P_{cis}(t)$ of the *cis* conformer (b).

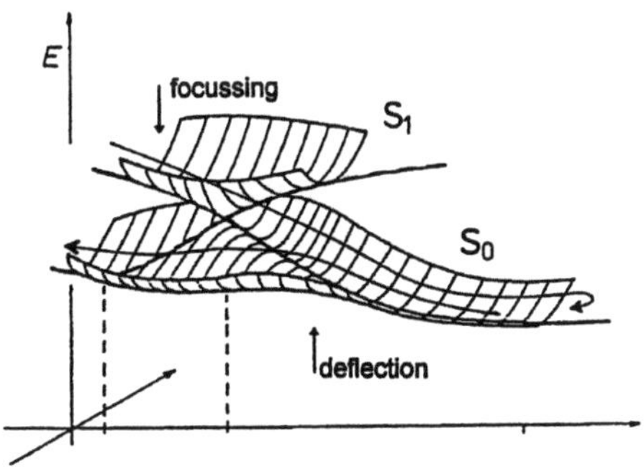

Fig. 3. Mechanism of a conical intersection.

A detailed picture of the dynamics at a conical intersection can be obtained by analyzing the time-dependent wave packet during its transition through the conical intersection. From such investigations, which have been reported in [9], the following qualitative picture of the mechanism of a conical intersection emerges. As schematically shown in Fig. 3, the relevant dynamical phenomena are (i) focussing of the initial wave packet towards the apex of the cone due to the convex shape of the upper (S_1) adiabatic surface, (ii) first transition through the intersection which leads to the lower (S_0) surface with high probability, (iii) deflection of the recurring wave packet by the concave shape of lower (S_0) adiabatic surface, which prevents return of the wave packet to the upper surface, and (iv) vibrational energy relaxation involving one ore more additional (accepting) modes, which lowers the energy content in the active modes below the energy of the point of intersection and ensures the irreversibility of the process.

A significant aspect of this picture of ultrafast relaxation dynamics via conical intersections is the fact that only few degrees of freedom are directly involved in the process, even in large molecules. In contrast to the traditional picture of radiationless transitions in polyatomic molecules, the dynamics is rationalized in terms of the local topology of PE surfaces, rather than the overall density of states of the molecule. This point of view, which emphasizes local properties of the multidimensional PE surfaces such as slopes, barriers and photochemical funnels, becomes now widely accepted in photochemistry [3,4] .

3. Real-time detection of the dynamics at conical intersections

We consider as an example the integral transient transmission signal

$$I = \int_{-\infty}^{\infty} dt\, \dot{\mathbf{E}}_2(t)\mathbf{P}_{\mathbf{k}_2}(t)$$

which implicitly depends on the carrier frequencies of the pump and probe fields, the pulse durations, and the pulse delay time Δt. $\mathbf{P}_{\mathbf{k}_2}(t)$ denotes the nonlinear polarization propagating in the direction of the wave vector $\mathbf{k}_2$ of the probe field [7,10]. The standard approach in nonlinear optical spectroscopy is the evaluation of the polarization in the lowest nonvanishing order of perturbation theory. In this formulation, the relevant features of the material dynamics are compactly contained in so-called nonlinear dipole response functions [10]. The perturbative approach is most suitable when the multitime dipole response functions can be evaluated analytically, e. g., for few-level systems or the harmonic oscillator.

For the ultrafast dynamics at conical intersections no simple analytic solutions exist and the evaluation of the nonlinear polarization via numerically computed dipole response functions would be extremely tedious. It appears preferable to adopt the nonperturbative approach, in which the time-dependent field-matter interaction is directly included in the solution of the time-dependent Schrödinger equation. The overall electric polarization is then given by the expectation value of the transition dipole operator μ. For a pump field with wave vector $\mathbf{k}_1$ and a probe field with wave vector $\mathbf{k}_2$, the macroscopic polarization $\mathbf{P}(\mathbf{x},t)$ has the form

$$\mathbf{P}(\mathbf{x},t) = 2\,Re \sum_{n,m} \mathbf{P}_{n,m}(t) e^{i(n\mathbf{k}_1 + m\mathbf{k}_2)\mathbf{x}}.$$

A general and simple prescription for the determination of the polarization $\mathbf{P}_{n,m}(t)$ corresponding to a specific spectroscopic process has been developed in [11]. Within the rotating-wave approximation (RWA), four wave packet propagations with different phases of pump and probe pulses are sufficient to extract the signals in $\mathbf{k}_1$ (pump absorption), $\mathbf{k}_2$ (probe absorption), and $2\mathbf{k}_1$-$\mathbf{k}_2$, $2\mathbf{k}_2$-$\mathbf{k}_1$ (photon echo) directions [11].

Fig. 4 shows, as an example, the excited-state contribution of the integral pump-probe signal for the conical intersection of Fig. 1, calculated with pulses of 6 fs duration (see [11] for details). There is a direct correspondence between the wave-packet dynamics indicated in Fig. 1 and the spectral features observed in Fig. 4. The initial decay of the signal reflects the rapid delocalization of the wave-packet. At time $\Delta t \approx 100$ fs the wave packet has reached the conical intersection, which is reflected in a splitting of the spectrum into two components, corresponding to the electronic gap of the system in the *cis* and *trans* configurations, respectively. The first component describes the stimulated emission ($S_1 \rightarrow S_0$) of the reactants, while the latter describes the absorption ($S_0 \rightarrow S_1$) of the photoproducts.

This calculation demonstrates that the theoretical tools are available for a detailed simulation of the real-time spectra associated with the ultrafast dynamics of nontrivial systems. To extend the dynamical treatment to truly multi-dimensional systems, the reduced density-matrix formulation as well as mixed quantum-classical schemes appear promising (see Refs. in [7]). As representative examples, we mention the density-matrix formulation [12] that allows for the inclusion of

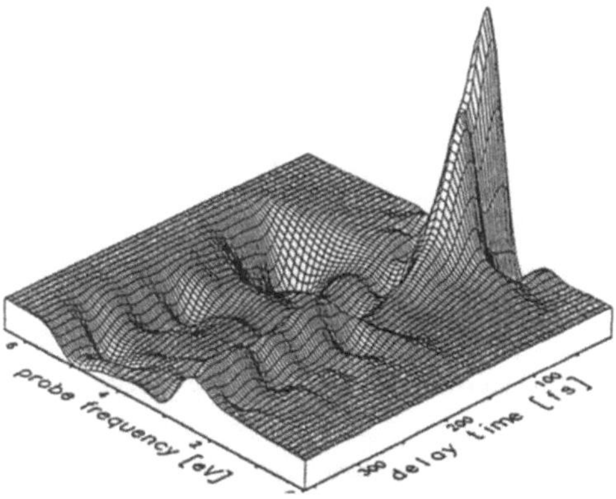

Fig. 4. Integral stimulated-emission pump-probe spectrum reflecting the transition through a conical intersection.

environmental effects in the simulation of femtosecond spectra, as well as the successful classical modeling of the dynamics and the real-time spectroscopy of the photoisomerization system studied above [13].

References

1 J. von Neumann and E. Wigner, Physik. Z., **30**, 467, 1929; E. Teller, J. Phys. Chem., **41**, 109, 1937.

2 S. S. Xantheas, G. J. Atchity, S. T. Elbert, and K. Ruedenberg, J. Chem. Phys., **94**, 8054, 1991; D. R. Yarkony, J. Chem. Phys., **100**, 3639, 1994; F. Bernardi, S. De, M. Olivucci, and M. A. Robb, J. Am. Chem. Soc., **112**, 1737, 1990; A. L. Sobolewski, C. Woywod, and W. Domcke, J. Chem. Phys., **98**, 5627, 1993.

3 M. A. Robb, F. Bernardi, and M. Olivucci, Pure & Appl. Chem., **67**, 783, 1995.

4 M. Klessinger and J. Michl, *Excited States and Photochemistry of Organic Molecules*, VCH, New York, 1995.

5 E. E. Nikitin, in *Chemische Elementarprozesse*, Ed. H. Hartmann, Springer, Berlin, p. 70, 1968.

6 H. Köppel, W. Domcke, and L. S. Cederbaum, Adv. Chem. Phys., **57**, 59, 1984.

7 W. Domcke and G. Stock, Adv. Chem. Phys., **100**, 1, 1997.

8 H. Köppel and W. Domcke, in *Encyclopedia of Computational Chemistry*, Ed. P. v. R. Schleyer, Wiley, New York, in press.

9 L. Seidner and W. Domcke, Chem. Phys., **186**, 27, 1994; R. Schneider, W. Domcke, and H. Köppel, J. Chem. Phys., **92**, 1045, 1990; U. Manthe and H. Köppel, J. Chem. Phys., **93**, 345, 1658, 1990.

10 S. Mukamel, *Principles of Nonlinear Optical Spectroscopy*, Oxford UP, New York, 1995.

11 L. Seidner, G. Stock, and W. Domcke, J. Chem. Phys., **103**, 3998, 1995.

12 B. Wolfseder, L. Seidner, G. Stock und W. Domcke, Chem. Phys., **217**, 275, 1997.

13 G. Stock, J. Chem. Phys., **103**, 10015, 1995.

Vibrational Coherence in Excited State of an Electron Donor-Acceptor Complex.

Igor V. Rubtsov[1,2] and Keitaro Yoshihara[1]

[1] Japan Advanced Institute for Science and Technology, Hokuriku, Tatsunokuchi 923-1292, Japan, tel.: +81-761-51-1523
E-mail: ivr@jaist.ac.jp, yosihara@jaist.ac.jp
[2] Institute for Chemical Physics Research, Chernogolovka, Moscow Region, Russia

Abstract. The oscillatory fluorescence decay of an electron donor-acceptor complexes was observed including one with hexamethylbenzene (HMB) and tetracyanoethylene (TCNE). A fast fluorescence decay component was assigned to vibrational relaxation of intramolecular modes. The 140 fs characteristic time was obtained from reconstructed shift of the fluorescence spectrum (C(t)). The underdamped oscillation (212 ± 4 fs) superimposed with the fast decay corresponds to a frequency of 157 ± 3 cm^{-1}. It is shown that the fluorescence spectrum is oscillating in time with the same frequency as a fluorescence observed at a single wavelength. The fluorescence anisotropy data show a fast decay component of *ca.* 500 fs which was assigned to conversion from CT2 to CT1 - too closely located excited charge-transfer states. The angle between transition moments of CT1 and CT2 states was estimated to be *ca.* 30 degrees

The study on dynamics of electron donor-acceptor (EDA) complex in the excited state gives basic information on molecular interactions and relation between different vibrational modes and electron transfer. The optical excitation of the EDA complex to a charge transfer band leads to formation of the ionic excited state. As the charge distribution after photoexcitation changes significantly, different vibrations (intra- and inter-molecular) could be excited. In the present conditions low frequency vibrations up to several hundred wavenumbers can be excited coherently and observed in time resolved experiment.

We present investigation of dynamics of various EDA complexes including one with hexamethylbenzene (HMB) and tetracyanoethylene (TCNE) in CCl_4 which has been studied rather intensively in the last few years [1-5].

Experimental details. The second harmonic of a chromium-forsterite femtosecond laser (pulse width of 40 fs at 630 nm) was used to excite the complex. The spontaneous fluorescence was measured by a fluorescence up-conversion setup [4]. The instrument response function reaches 70 fs (FWHM of crosscorrelation).

Results and Discussion

The fluorescence from the excited state of an EDA complex (recombination fluorescence) was observed in the spectral range from 700 to 970 nm. Under-damped oscillations in fluorescence are clearly observed (Figure 1) for wavelengths up to ~900 nm. The period of oscillations is same for all wavelengths.

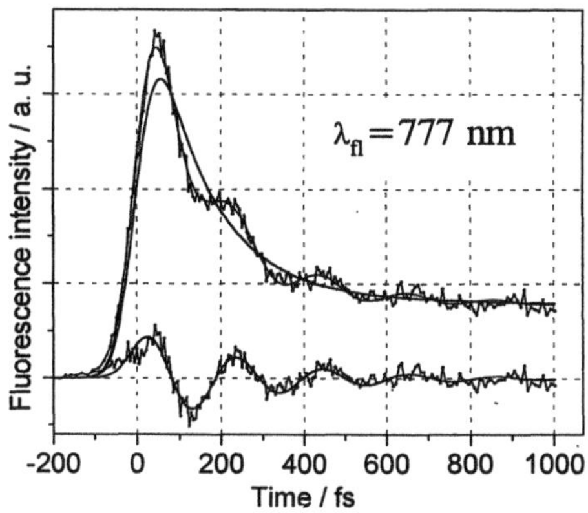

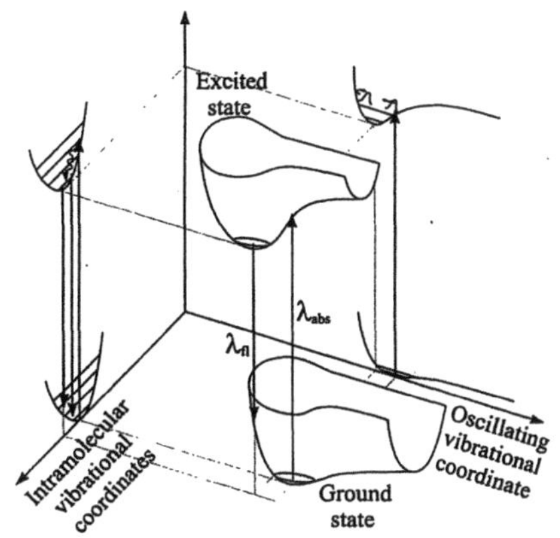

Figure 1. Fluorescence decay and the best fit of convolution with instrument response function. The oscillatory component is also shown separately.

Figure 2. The schematic potential energy diagram.

It is 212 ± 4 fs that corresponds to a frequency of 157 ± 3 cm^{-1}. At 907 nm the oscillation is still observable, but another frequency is involved. At 970 nm a twice higher frequency is expected, as a wave-packet passes the maximum of fluorescence spectrum twice in a period. This was unable to be observed, due probably to a lack of time resolution and a limited S/N ratio in the near IR region. A fast exponential component (which has a characteristic time of about 100 fs at the blue edge of fluorescence spectrum) is superimposed with oscillations. The fast decay component has a character of spectral diffusion. The amplitude of this component decreases and the characteristic time increases from 700 to 970 nm. This component is assigned to vibrational relaxation of intramolecular vibrations prepared by laser excitation at 630 nm (the excess of photon energy above 0-0 transition is about 1800 cm^{-1}). The vibrational relaxation is schematically shown in Figure 2 in a projection of multidimensional energy surface to the intramolecular vibrational coordinates.

The reconstruction of the fluorescence spectrum time behavior was made. The time dependence of the fluorescence maximum is shown in Figure 3. The initial component decays with a time constant of *ca.* 140 fs, which is the upper limit for vibrational relaxation time as the row data without deconvolution was used for spectral reconstruction. From Figure 3 one can see that fluorescence maximum not only moves to the relaxed excited state of complex, but also oscillates with the same frequency - 157 cm^{-1} as observed by single wavelength measurements. This analysis demonstrates that there is a wave-packet motion in the excited state, rather than synchronous fluctuations of fluorescence efficiency in all wavelength simultaneously. Determination of the phase of oscillations at each wavelength of observation is not that easy, as it requires proper and complete analysis. An the same time, oscillations of fluorescence maximum (Figure 3) prove the wave-packet character of the motion. The origin of the oscillating component can be either intermolecular or intramolecular. The ground state intermolecular frequency was calculated to be about 59 cm^{-1} [3]. In the excited state it could be expected to be 2-3 times larger because of Coulombic attraction in the excited

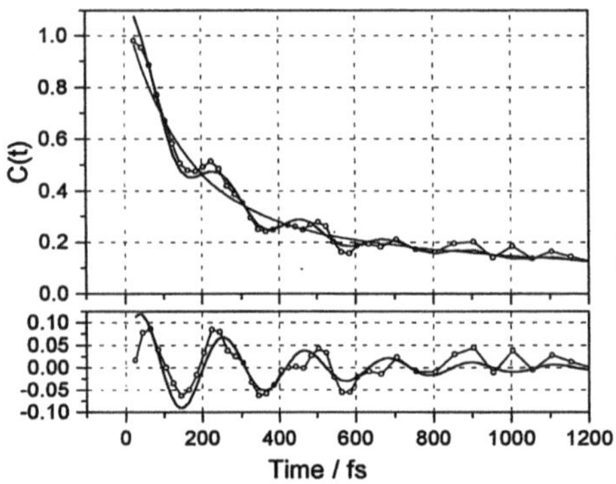

Figure 3. Dynamics of fluorescence maximum: $C(t) = (\nu(t) - \nu(\infty)) / (\nu(0) - \nu(\infty))$.

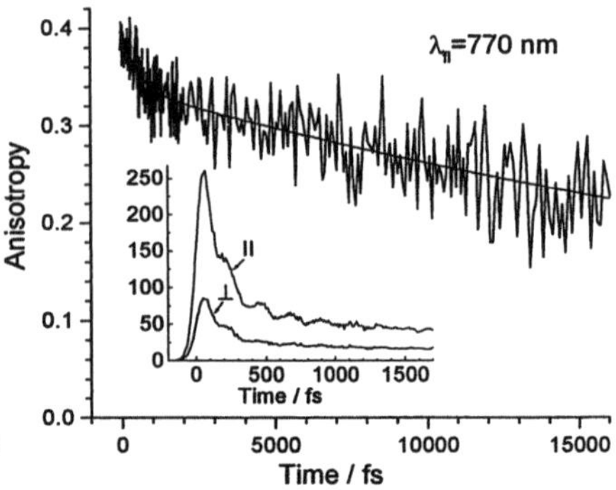

Figure 4. Anisotropy decay, λ_{fl}=770 nm. Thick line is a biexponential fit with τ_1= 470 ± 200 fs (10%), τ_2=30 ± 3 ps (90%).

state and increased force constant. There are several motivations to assign oscillating component to the intermolecular mode. Generally, to have a large modulation of the transition energy between the excited and ground states during vibration, the displacement of these two energy surfaces should be large along particular vibrational coordinate. The intermolecular vibrational mode changes the dipole moment of the complex and should substantially modulate the transition moment. Also, the intermolecular mode should be less coupled to intramolecular modes and it should be less affected by vibrational relaxation of intramolecular modes (in our case the damping time of oscillations is more than twice longer than vibrational relaxation time). At present, more studies are necessary to assign confidently about the origin of the observed oscillations.

The presence of two closely located charge transfer states (CT1 and CT2) has recently been predicted and quantum mechanically calculated [3, 5]. The fluorescence anisotropy measurements show a fast decay component of *ca.* 500 fs (Figure 4) which is assigned to conversion from CT2 to CT1. Based on the absorption spectra calculated for CT1 and CT2 states [3] the angle between transition moments of the two excited states is estimated to be *ca.* 30 ± 10 degrees.

Acknowledgments. The assistance by Inoue Foundation for Science is greatly acknowledged by IVR.

References

1 K. Kulinowski, I. R. Gould, A. B. Myers, *J. Phys. Chem.*, **99** (1995) 9017
2 L. Wynne, C. Gali, R. M. Hochstrasser, *J. Chem. Phys.*, **100** (1994) 4797
3 M. Hayashi, T. -S. Yang, J. Yu, A. Mebel, S. H. Lin, *J. Phys. Chem. A* ,**101** (1997) 4156
4 I. V. Rubtsov and K. Yoshihara, *J. Phys. Chem. A*, **101** (1997) 6138
5 M. Hayashi, T. -S. Yang, J. Yu, A. Mebel, R. Chang, S. H. Lin, I. V. Rubtsov, K. Yoshihara, *J. Phys. Chem. A* , **102** (1998) 4256

Ultrafast Vibrational Energy Transfer Studied by a Novel Spectrometer

I. Hartl and W. Zinth

Institut für Medizinische Optik, Universität München, Oettingenstr. 67, D-80538 München, Germany

Abstract. A new experimental system for the study of vibrational energy relaxation is presented. Infrared excitation of vibrational modes and probing via anti-Stokes Raman scattering allows the direct observation of vibrational energy transfer between and out of CH-stretching modes with a temporal resolution of approx. 160 fs.

Detailed information on the transfer of vibrational energy within a molecule (intramolecular vibrational redistribution, IVR) is of major importance for the understanding of many photochemical processes. For femtochemistry (the manipulation of chemical reactions by suitably tailored light pulses) a detailed knowledge of IVR-processes is a key information. Unfortunately, most methods for the study of IVR in the condensed phase with sufficient temporal resolution are only indirect and do not yield the necessary mode specific information. Details can be obtained only when the modes are directly excited by resonant IR absorption and when mode selective probing, e.g. via the observation of anti-Stokes Raman scattering is used. While this method has been applied since the early days of picosecond spectroscopy [1], it is only via the recent progress in pulse generation and sensitive light detection that broadband vibrational spectroscopy can be performed with time resolution in the picosecond [2] and femtosecond time range. Until that date only few experimental studies had been performed. This infrequent application was directly related to the very small values of typical Raman cross-sections in combination with the strong disturbance by background light caused by nonlinear optical processes [3]. In the present experimental system we optimized the anti-Stokes Raman signal via different improvements:

(i) High repetition rate to obtain sufficient signal at low light intensities.
(ii) Optics with high aperture.
(iii) Highly efficient low-noise multichannel detection.
(iv) Temporal and spectral resolution adjusted to the molecular transition.

The experimental system is based upon a 1 kHz Ti-sapphire laser system. Excitation pulses (3μJ, t_p = 130fs, 2.5μm < λ < 4μm) are generated by parametric generators. The probing proceeds via the light pulses from the Ti-sapphire laser system. Suitable shaping of the pulses leads to probing pulse lengths adjustable between approx. 100fs and 1ps. By this way spectral and temporal resolution can be adapted to the molecular system and nonlinear background is minimized by reduction of light intensity. A highly sensitive, liquid

N_2 cooled CCD array is used in conjunction with notch filters and a spectrograph for the recording of the anti-Stokes Raman spectra. The optimum temporal resolution (cross-correlation function) of the system is shown in Fig. 1, left. A width (FWHM) of 160 fs could be obtained. This temporal resolution should be sufficient to study even the fastest vibrational relaxation processes in the electronic ground state.

Examples for experiments on the energy relaxation out of CH-stretching modes are given in Fig. 1, right. Here time constants of 14ps (1,1,1-trichloroethane) and 1.7ps (cyclopentane) are obtained. The experiments show that the energy transfer out of the CH-stretching modes strongly depends on the type of molecule, the existence of other CH-modes and the anharmonic coupling to overtone and combination levels [4,5]

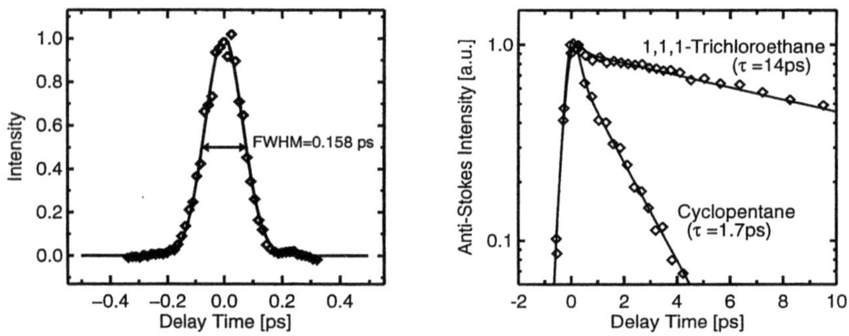

Fig. 1: Cross-correlation function for the experimental system measured in the probing setup via two photon fluorescence in oxazine 1 (left). Measured decay of the population of CH-stretching modes in 1,1,1 trichloroethane and cyclopentane (right).

The necessity for a technique with high temporal resolution is evident when the transfer of vibrational energy between different modes has to be studied. As an example, Fig. 2 shows the results for the excitation transfer between the antisymmetric ν_7 mode (3129cm^{-1}) and the symmetric ν_1 CH-stretching mode (3034cm^{-1}) of 1,1-dichloroethylene and for the relaxation out of these modes. Experiments were performed with all combinations of excitation and detection wavelengths. Fig. 2 shows the population (measured via the anti-Stokes Raman signal) of the two CH-stretching modes after excitation of the antisymmetric mode at 3129cm^{-1}. The experimental data reveal a more rapid decay of the initially populated mode and the delayed rise of the population of the symmetric CH-stretching mode. The complete set of experimental data (time constants, amplitudes) allows to analyze the energy transfer process in detail (see Fig. 3): The fastest relaxation process (τ = 2.8ps) is energy relaxation out of the symmetric stretching mode to lower lying modes. The transfer time between the two CH-stretching modes is relatively slow (5.3ps and 8.6ps). The slowest process found in 1,1 dichloroethylene is the energy relaxation out of the antisymmetric mode to low lying modes which takes about 10.7ps. The combination of these microscopic reaction times leads to the experimentally observed time constants of approx. 1.7 ps and 5.5 ps.

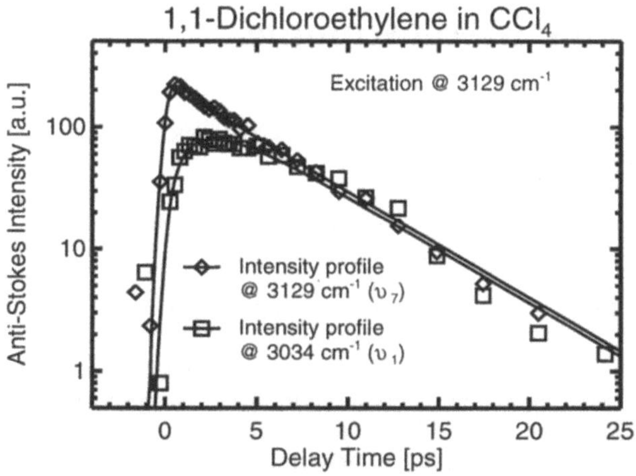

Fig. 2: Anti-Stokes Raman signal recorded for the CH-stretching modes of 1,1-dichloroethylene after excitation of the antisymmetric CH-stretching mode at 3129 cm^{-1}.

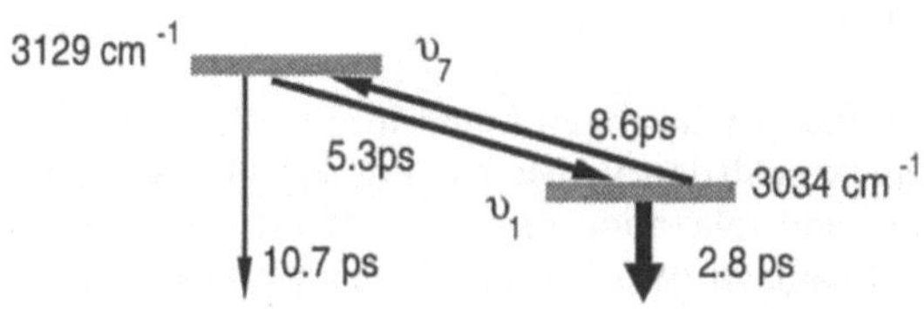

Fig. 3: Microscopic model for vibrational energy redistribution between and out of CH-stretching modes of 1,1-dichloroethylene.

In conclusion, the examples given in this paper show that the new experimental system allows time resolved spectroscopy of vibrational levels with subpicosecond resolution. Due to the weak scattering cross-section of Raman transitions and a strong nonlinear background a careful optimization of the experimental system is required to obtain high temporal resolution and sufficient signal to noise ratio.

References

1 A. Laubereau, W. Kaiser, Rev. Mod. Phys., Vol. 50, 607, 1978.
2 H. Graener, R. Zürl, M. Hofmann, J. Phys. Chem. B, Vol. 101, 1745, 1997.
3 R. Zürl, H.Graener, Apl. Phys. B, Vol. 66, 213, 1998.
4 A. Fendt, S.F. Fischer, W. Kaiser, Chem. Phys. , Vol. 82, 350, 1981.
5 C. Kolmeder, W. Zinth, W. Kaiser, Chem. Phys. Lett., Vol. 91, 323, 1982.

Ultrafast Relaxation Processes of Metal-to-Ligand Charge Transfer Excited State in One-Dimensional Metal Complex

Shinichiro Iwai[1], Toshihide Kamata[1], Kaoru Yamamoto[2], Toshiaki Ohta[2], Toshio Fukaya[1], Shigeo Murata[1]

[1] Department of Physical Chemistry, National Institute of Materials and Chemical Research, 1-1 Higashi, Tsukuba 305-8565, Japan
E-mail: iwai@nimc.go.jp
[2] University of Tokyo, Bunkyo-ku, Tokyo 113-0033, Japan

Abstract. We have investigated the optical response of $Ni(dmg)_2$ thin film by pump-probe spectroscopy under excitation of the metal-to-ligand charge transfer(MLCT) absorption band. Almost the same photo-induced spectral change was observed as in the case of the resonant excitation to 1-D exciton absorption in $Pt(dmg)_2$. It rises up within 200fs. This shows that the MLCT 'molecular ' excited state relaxes to 1-D intermolecular exciton state in such a short time.

1. Introduction

Large optical nonlinearity has been found in one-dimensional(1-D) dimethylglyoxime metal complexes($M(dmg)_2$, M: Ni, Pd, Pt) [1]. We have investigated the electronic structures and relaxation processes of the complexes, and revealed that the 1-D exciton delocalized over the metal chain plays an important role for the appearance of their nonlinear optical properties[2, 3]. On the other hand, optical transition due to the metal-to-ligand charge transfer (MLCT) transition can occur in it[4]. Charge transfer excitation can change the charge distribution of the above complexes. In order to investigate the effect of a change in charge distribution in the complex on the electronic structure of the 1-D exciton state, we investigated the optical response of the 1-D exciton under excitation in the MLCT absorption band by using femtosecond pump-probe method.

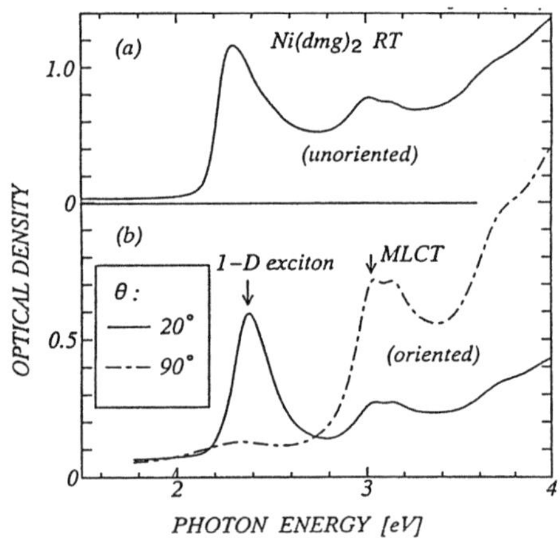

Fig.1 Polarized absorption spectra in unoriented (a) and in oriented (b) films of $Ni(dmg)_2$ at room temperature. θ is defined as the angle between the incident beam and the chain direction.

2. Experimental

Vacuum evaporated thin films of $Ni(dmg)_2$ were prepared on fused quartz substrates of 1mm thickness. We also prepared oriented thin film on a KBr substrate of 1mm thickness in which the 1-D chain is perpendicular to the substrate. Pump-probe experiment was performed by the amplified pulse of a mode locked $Ti:Al_2O_3$ laser(800nm) and its second harmonic pulse. The time resolution of the whole instrument was ~200fs .

3. Results and Discussion

Fig. 1 shows the absorption spectrum of an unoriented film (a) and polarized absorption spectra of an oriented films (b) of $Ni(dmg)_2$. The solid and dashed curves in (b) indicate the spectra for θ =20 and 90° , respectively. An absorption band due to the 1-D exciton is observed at ~2.3eV. The MLCT absorption band is located around ~3eV. The transition energy of the MLCT band of the oriented film is roughly the same as that of the unoriented film, although a small change less than 20meV is noticed. We see the anti-correlated relation between the 1-D exciton(2.3eV) and MLCT(~3eV) absorptions in (b). The MLCT band in a single molecule is polarized along the molecular plain. The fact that the 3eV band is anticorrelated with the 1-D exciton band in Fig. 1 suggests that the MLCT band retains its intramolecular character also in the solid state.

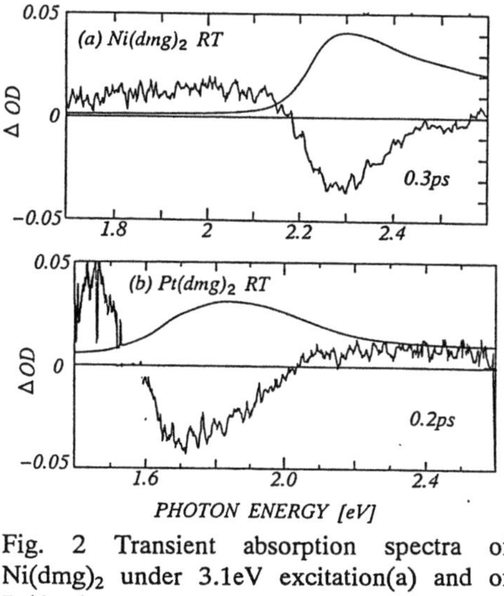

Fig. 2 Transient absorption spectra of $Ni(dmg)_2$ under 3.1eV excitation(a) and of $Pt(dmg)_2$ under 1.7eV(b) excitation, respectively

Fig. 2(a) shows the transient absorption spectrum at room temperature of a thin unoriented film of $Ni(dmg)_2$ under 3.1eV excitation. Bleaching of the exciton absorption around ~2.3eV and a very broad induced absorption at lower energies were observed. The spectral shape of the transient absorption is similar to that of $Pt(dmg)_2$ under resonant excitation of 1-D exciton shown in Fig.2(b). Excitation density was 4 and 0.2mJ/cm^2 in (a) and (b), respectively. These photo-induced spectral changes rise up instantaneously within the time resolution as seen in Fig. 3(a). The recovery of the bleaching and the decay of the induced absorption can be described by a superposition of 2 exponential functions (solid curve: τ =25ps, >200ps in Fig3(b)). The ratio of decay of the induced absorption is the same as that of recovery of the bleaching of the exciton band as shown in Fig.3(b). This shows that the initial state of the induced absorption is the lowest exciton state. At a lower temperature (80K), the bleaching of the exciton band first decays in a time region of several picoseconds and then grows again (not shown). The grow is considered to be due to a thermal process as described in a previous paper for Pt(dmg_2)[2]. However, we do not mention about this more in detail in this paper. The excited state absorption from the lowest exciton was found to rise up within the time resolution as shown in Fig.3(a). This rise-up is as fast as in the case where the 1-D exciton state is resonantly excited in Pt$(dmg)_2$. This suggests that the intramolecular "MLCT" excited state relaxes to the intermolecular lowest exciton state in such a short time.

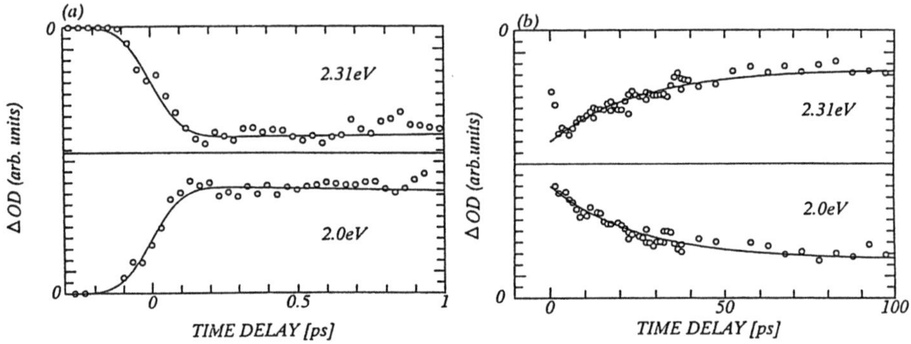

Fig. 3 Time evolution of the transient absorption of Ni$(dmg)_2$ observed at 2.31 and 2.0eV, respectively, which correspond to the energy region of the bleaching and the induced absorption. (a) -0.5~1.5ps, (b) -10~100ps.

One of the important problems is whether the absorption around 3eV is due solely to the MLCT transition; transition to the free electron-hole pair state may also occur in this region in Ni$(dmg)_2$, although the absorption due to free e-h pair cannot be identified clearly in our spectrum. If the excited e-h pair state relaxes to the lowest exciton state within 200fs, the transient absorption change as shown in Fig.2(a)would also be observed. To clarify this point, further measurements should be made on Pt$(dmg)_2$ films and on oriented Ni$(dmg)_2$ films using polarized light.

In summary, we measured the relaxation processes of Ni$(dmg)_2$ by pump-probe spectroscopy under excitation at 3.1eV which corresponds to the transition energy of the MLCT absorption band. The lowest exciton formed within the time resolution (200fs) after excitation. The 'molecular' MLCT excited state possibly relaxes into the lowest intermolecular exciton state within 200fs.

References

[1] T. Kamata et al. *Chem. Phys. Lett.* **221**, 194 (1994).
[2] K.Yamamoto et al. to be published.
[3] S.Iwai et al. *J. Lumin.* **76**,226(1998).
[4] B. G. Anex, F. K. Krist J. Am. Chem. Soc. 22, 6114(1967)

Sub-picosecond Interfacial Electron Injection in Dye Sensitized Titanium Dioxide Films

Todd A. Heimer and Edwin J. Heilweil

Optical Technology Division, Room B208, Building 221,
National Institute of Standards and Technology, Gaithersburg, MD 20899

Abstract. Sub-picosecond infrared (IR) spectroscopy in the 5 μm region was employed to study the properties of Ru(4,4'-$(COOH)_2$-2,2'-bipyridine$)_2(NCS)_2$,**1**, and Ru(5,5'-$(COOH)_2$-2,2'-bipyridine$)_2(NCS)_2$,**2**, anchored to nanostructured thin films of TiO_2 and ZrO_2. For these molecules attached to TiO_2 semiconductor films, a transient absorbance appears which is attributed to electrons injected into the TiO_2 semiconductor. This absorbance appears within the instrumental time resolution yielding an approximate 350 fs upper limit time constant for electron transfer from the sensitizer excited state to TiO_2. This implies that sensitizers with lifetimes as short as a few picoseconds can effectively transfer electrons to TiO_2 with high quantum yield.

Introduction

Sensitization of TiO_2 electrodes with ruthenium polypyridyl chromophores has proven to be an effective method for the conversion of solar to electrical energy [1]. Monochromatic efficiencies on the order of 80% have been reported. However under white light irradiation the overall conversion efficiency drops to 10% due to less than optimal overlap of the sensitizer absorbance spectrum with the solar spectrum [1a]. In order to improve this overlap, sensitizers with enhanced absorbance at long wavelengths are required. For the metal-to-ligand charge transfer (MLCT) dyes typically used, this presents a problem: As the energy gap between the metal and ligand orbitals decreases, the non-radiative decay rate increases exponentially, decreasing the lifetime of the sensitizer excited state [2]. Excited state decay competes with electron transfer to TiO_2 in the overall energy production scheme. For these reasons, it is important to measure the rate of interfacial electron transfer as well as to determine how the energetics at the interface affect it.

Near IR and mid-IR measurements have recently established the usefulness of time-resolved infrared spectroscopy as a tool to measure the appearance time of electrons in semiconductors such as TiO_2 [3,4]. A previous report established a lower limit of 5×10^{10} s^{-1} for electron injection from Ru(4,4'-$(COCH_2CH_3)_2$-2,2'-bipyridine)(2,2'-bipyridine$)_2$ [4]. We have now examined two additional sensitizers using ~250 fs probe and pump pulses. Species **1** is the most effective sensitizer reported to date, giving 10% efficiency under solar irradiation [1a]. Species **2** is a related dye with the carboxylated ligand modified, pictured in Figure 1. This dye was developed to have enhanced absorbance at long wavelengths. The carboxylated ligand in **2** has a 350 mV lower electrochemical reduction potential than in **1**, which results in a significant red shift in the absorbance spectrum [5]. The ruthenium based redox potential is identical for the two species (Figure 1).

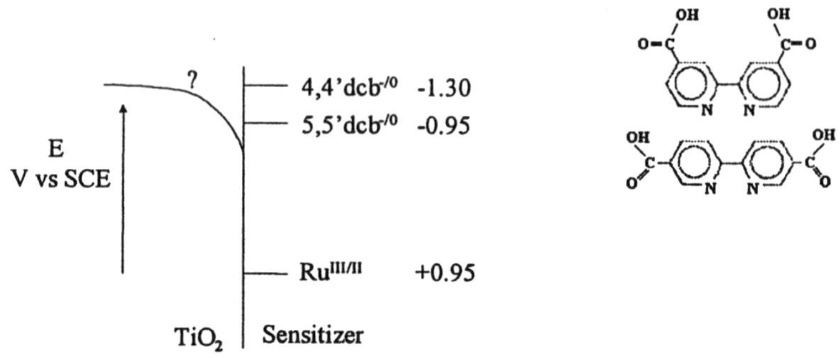

Figure 1. Energetics at the semiconductor/sensitizer interface for $Ru(4,4'dcb)_2(NCS)_2$ and $Ru(5,5'dcb)_2(NCS)_2$. Voltages are relative to a standard calomel electrode.

Experimental Results

The probe pulse consists of a broadband (~ 100 cm^{-1}, centered at 1850 cm^{-1}), ~250 fs pulse generated by difference frequency mixing of two visible pulses in $LiIO_3$. The pump beam (590 nm, ~250fs, 8 μJ) and probe beams were focused to approximately 200 μm diameter at the sample. Broadband detection of the probe was achieved directly using a spectrograph coupled to an InSb array. The pump-probe cross correlation, measured in a Si wafer, indicates an ~350 fs instrument response function, see Figure 2. For sensitized TiO_2 samples following visible excitation, the absorbance by electrons in the TiO_2 is flat across the entire probed region, and the average value as a function of time is shown in Figure 2. For both **1** and **2** anchored to TiO_2 the adsorption rise tracks the cross correlation with no further increase in intensity, indicating that electron injection occurs within the excitation laser pulse duration.

The coincidence of the cross correlation and the electron absorbance caused us to suspect that two-photon excitation of the TiO_2 could give rise to the observed signal. Substitution of non-sensitized TiO_2 for the sample revealed a small absorbance change (Δ Abs = +0.005 at 8 μJ pump power) contribution by this effect, indicated by the arrow in Figure 2. However, the absorbance signal for **1** anchored to TiO_2 exceeds this value by a factor of four, and displays a linear dependence on pump power in the range of 3-12 μJ. The two-photon signal for non-sensitized TiO_2 is undetectable at 3 μJ. No transient absorbance was observed in the 1800-1900 cm^{-1} region for the dye molecules anchored to insulating ZrO_2, indicating that the observed transient absorbance in TiO_2 is not a thermal effect. The intensity of the absorbance signal is not likely an accurate estimate of the injection quantum yield, and the smaller signal in the case of **2** is probably a result of less than optimum pump-probe spatial overlap.

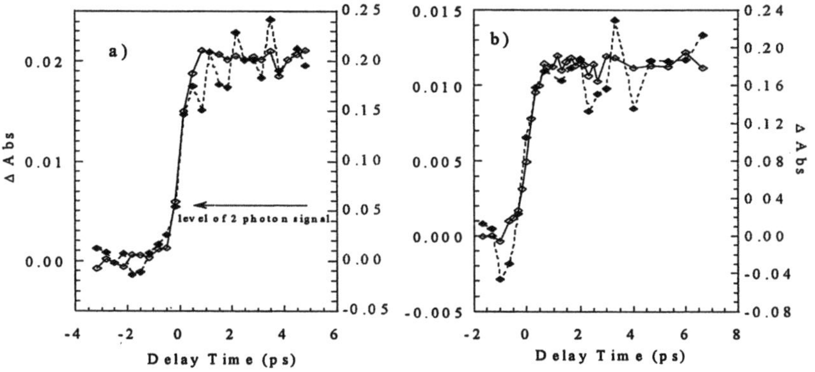

Figure 2. Pump-probe cross correlation (——, right scale) and injected electron absorbance rise-time (--------, left scale) for a) **1** and b) **2** on TiO_2.

It is interesting that it is possible to lower the ligand reduction potential (and presumably the excited state reduction potential) by 350mV and still have < 350 fs electron injection. Two possible explanations for this behavior are: 1) the acceptors in the TiO_2 form a broad continuous energy band, overlapping well with both sensitizer excited state reduction potentials. 2) ultrafast electron transfer at these interfaces differs completely from the Marcus-Gerischer "weak electronic interaction" electron transfer and does not involve redistribution of vibrational excitation energy. The latter concept has been suggested by Willig and co-workers, who observed < 25 fs electron transfer from **1** to TiO_2 using an 1100 nm probe [3a].

We have performed transient visible absorbance experiments on these systems to explore the cause of a 50% lower photon-to-current conversion efficiency for **2** [5]. Following electron injection to TiO_2, dye ground state recovery occurs roughly 2x faster for **2**, suggesting that recombination of injected electrons with the oxidized ruthenium center is a significant energy loss mechanism for **2**.

References

1. a) M. K. Nazeeruddin, A. Kay, I. Rodicio, R. Humphry-Baker, E. Müller, P. Liska, N. Vlachopoulos and M. Grätzel, *J. Am. Chem. Soc.* **115**, 6382, 1993. b) T. A. Heimer, C. A. Bignozzi, G. J. Meyer, *J. Phys. Chem.* **97**, 11987, 1993.
2. K. F. Freed and J. Jortner, *J. Chem. Phys.* **52**, 6272, 1970.
3. a) T. Hannappel, B. Burfeindt, W. Storck and F. Willig, *J. Phys. Chem. B*, **101**, 6799, 1997, and references therein. b) N. J. Cherepy, G. P. Smestad, M. Grätzel and J. Z. Zhang, *J. Phys. Chem. B*, **101**, 9342, 1997.
4. T. A. Heimer and E. J. Heilweil, *J. Phys. Chem. B*, **101**, 10990, 1997.
5. R. Argazzi, C. A. Bignozzi, T. A. Heimer, F. N. Castellano and G. J. Meyer, *Inorg. Chem.* **33**, 5741, 1994.

Local Density Effects in the Energy Relaxation of Vibrationally Highly Excited Molecules from the Gas to the Compressed Liquid Phase

D. Schwarzer[1], C. Hanisch, J. Troe, and M. Zerezke

Max-Planck-Institut für Biophysikalische Chemie, Abt. Spektroskopie und Photochemische Kinetik, D-37077 Göttingen, Germany
[1] E-mail: dschwar@gwdg.de

Abstract. Collisional deactivation rates of vibrationally highly excited model compounds and solvatochromic shifts of their electronic absorption spectra show similar density dependencies indicating that both quantities are determined by the local bath gas density.

1. Introduction

Collisional energy transfer of vibrationally highly excited molecules is of fundamental importance for many gas and liquid phase reactions. Therefore, the determination of energy relaxation rates has been the subject of many experimental studies [1]. In this paper we report on systematic investigations of the density and temperature dependence of vibrational energy transfer from highly excited small (CS_2) and large (azulene) molecules to a bath, respectively [2]. Energy relaxation rates are compared with solvatochromic shifts $\Delta\nu$ of electronic absorption bands of these molecules under identical conditions. Employing supercritical fluids, such as xenon, CO_2, and ethane, at pressures between 0.03 and 4000 bar and temperatures between 298 and 640 K, measurements covering the complete gas-liquid transition were performed.

2. Experimental Methods

Experiments were performed with an oscillator-amplifier laser system based on a colliding pulse mode locked dye laser. Activated azulene with an energy of ~20000 cm^{-1} was generated by light absorption into the S_1- and internal conversion to the S_0^*-ground state. The subsequent loss of vibrational energy was monitored by following the transient absorption at the red wing of the $S_3 \leftarrow S_0$ absorption band near 290 nm (Fig.1). Vibrationally highly excited CS_2 was generated with excess energies of 32000 cm^{-1}. The energy decays were monitored at 266 nm. Transient signals were converted into energy-time profiles using hot band absorption coefficients from shock wave experiments for calibration [3,4] and accounting for solvent-shifts of the spectra.

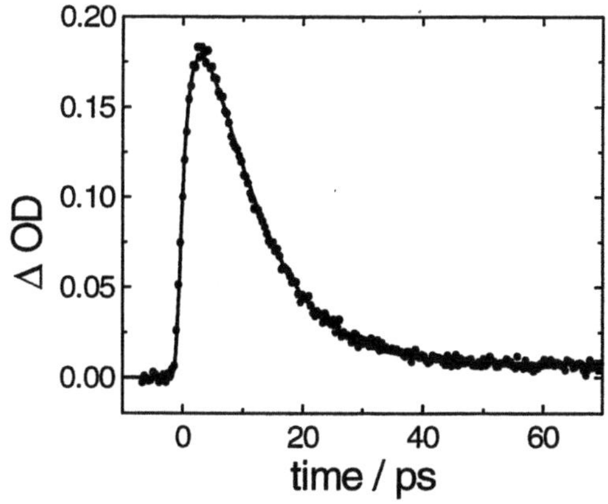

Fig. 1. Energy relaxation of azulene in ethane at 3470 bar, 296 K (full line: simulation assuming an exponential energy decay with a time constant of 8.0 ps).

3. Results and Discussion

Under all conditions, the energy decays of azulene were found to be exponential with phenomenological deactivation rate constants k_c. A comparison of k_c and spectral shifts $\Delta\nu$ showed quite similar density dependencies (see Fig. 2): the low pressure linear increase of both quantities at $\rho_r=\rho/\rho_c\approx0.1$ (ρ_c is the critical density of the solvent) starts to level off, before it finally becomes stronger again. These observations suggest that both quantities are determined by the same local bath gas density around the azulene molecule.

In contrast to azulene the energy-time profiles of CS_2 are highly non-exponential, with faster decays at early times. However, for a given collider the shapes of these energy decay curves are the same in the gas phase and in the liquid, if time axes are scaled properly. This allows for extracting mean deactivation rate constants which show similar density dependencies as spectral shifts of the CS_2 $^1B_2 \leftarrow {}^1\Sigma_g^+$ transition at 200 nm (Fig. 3).

The data were analyzed employing isolated binary collision (IBC) models and perturbation theory. Within the IBC approach we described $k_c(\rho)$ by a collision frequency Z which is proportional to the value $g(\sigma)$ of the radial distribution function at the surface of an attractive hard-sphere particle in a Lennard-Jones fluid [5]

$$Z = Z_0 \cdot g(\rho, T, \sigma) \tag{1}$$

where Z_0 is the dilute gas hard-sphere collision frequency. By means of Monte Carlo simulations excellent agreement between experimental data and eq. (1) was obtained assuming well depths for azulene-ethane and ethane-ethane interactions of ε/k_B = 720 K and 220 K, respectively (solid lines in Fig. 2a). Similar $k_c(\rho)$ dependencies were modeled applying weak coupling theory and relating the energy relaxation rate to the spectral density of the fluctuating force exerted by a Lennard-Jones solvent on a breathing Lennard-Jones sphere [6]. Radial distribution functions from these calculations were also used to model $\Delta\nu(\rho)$ (Fig. 2b). For CS_2 in ethane attractive hard-sphere and weak coupling model yield ε/k_B = 600 K for solute-solvent interactions (see Fig. 3).

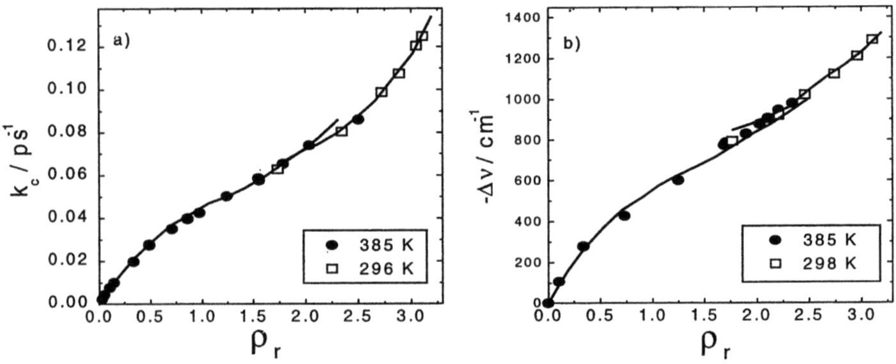

Fig. 2. Azulene in ethane: a) vibrational energy relaxation rate constants (full lines: simulation); b) shifts of the $S_3 \leftarrow S_0$ absorption band (full lines: simulation)

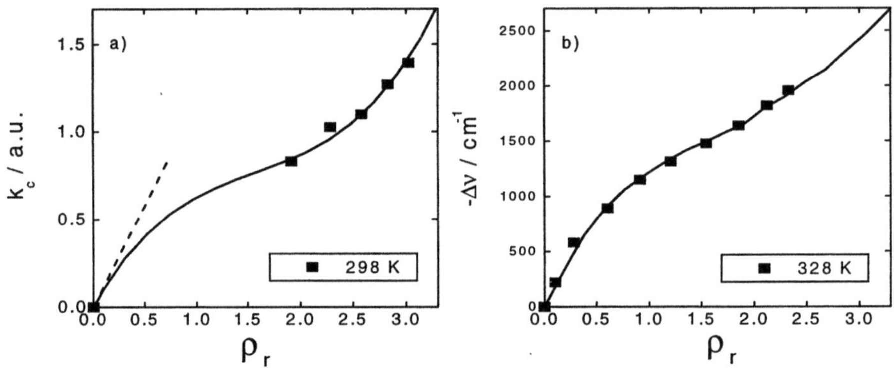

Fig. 3. CS_2 in ethane: a) vibrational energy relaxation rate constants (dashed line: extrapolation from gas phase experiments; full line: simulation); b) shifts of the $^1B_2 \leftarrow {}^1\Sigma_g^+$ absorption band (full line: simulation).

References

1 T. Elsaesser and W. Kaiser, Ann. Rev. Phys. Chem. **42**, 83 (1991).
K. E. Schultz, D. J. Russel, and C. B. Harris, J. Chem. Phys. **97**, 5431 (1992).

2 D. Schwarzer, J. Troe, M. Votsmeier, and M. Zerezke, J. Chem. Phys. **105**, 3121 (1996); D. Schwarzer, J. Troe, and M. Zerezke, *ibid.* **107**, 8380 (1997).

3 L. Brouwer, H. Hippler, L. Lindemann, and J. Troe, J. Phys. Chem. **89**, 4608 (1985).

4 J. E. Dove, H. Hippler, J. Plach, and J. Troe, J. Chem. Phys. **81**, 1209 (1984).

5 C. Delalande and G. M. Gale, J. Chem. Phys. **71**, 4804 (1979).

6 S. A. Egorov and J. L. Skinner, J. Chem. Phys. **105**, 7047 (1996)

The UV Band System of the Allyl Radical, Studied by Time-Resolved Photoelectron Spectroscopy

Thomas Schultz, and Ingo Fischer

Laboratorium für Organische Chemie, ETH Zürich, CH-8092 Zürich
Fax: 41-1/632 1280, email: fischer@org.chem.ethz.ch, schultz@org.chem.ethz.ch

Abstract: The nonradiative decay of the UV band system of the allyl radical is studied by picosecond time-resolved photoelectron spectroscopy. Lifetimes are determined for all vibronic bands previously assigned in ns-MPI experiments.

Hydrocarbon radicals constitute important intermediates in combustion processes. Being a relatively stable radical, allyl can accumulate in a reactive environment, making its photochemistry relevant for those systems. The UV bands between 250 and 238 nm were studied by MPI spectroscopy[1] allowing the identification of many vibronic states and their zero order assignments to the electronic B $^2A_1$, C $^2B_1$ and D $^2B_2$ states. In our experiments we investigated the nonradiative decay of those states by time-resolved photoelectron spectroscopy (TR-PES), a sensitive, state-specific and background-free detection technique. Recently we communicated the decay of low-lying vibrational levels of the B $^2A_1$ state within 20 ps, probably due to internal conversion to the ground state[2]. Meanwhile those studies were extended to the C- and D-state[3], as well as higher B-state levels.

We used a regeneratively amplified Ti:Sapphire laser with a pulse duration of 2 ps. To excite the UV bands to the blue of 250 nm, we Stokes-shifted the frequency doubled output of the Ti:Sa laser in a high-pressure hydrogen cell[4]. For 2-photon excitation of the low-lying B-state bands this light was employed directly, for 1-photon excitation of the C- and D-state it was frequency doubled in a BBO crystal. The excited molecules were ionized by the frequency-tripled output of the Ti:Sa laser. Mass and photoelectron spectra were recorded in a magnetic bottle time-of-flight spectrometer, allowing for a very efficient detection of photoelectrons.

A clean beam of allyl radicals was generated by supersonic jet flash pyrolysis[1] of 1,5 hexadiene. The mass spectrum recorded via the C 7_0^1 band at 245.8 nm is depicted in figure 1. It is dominated by a peak at m/e = 41, the mass of the allyl radical. A smaller peak at m/e = 39 due to fragmentation in the ion is also visible, but no precursor molecules appear at mass 82.

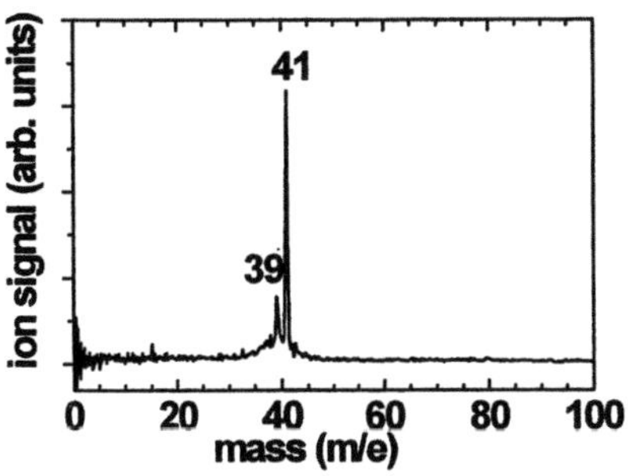

Fig. 1: *Mass spectrum recorded via the C 7_0^1 band of the allyl radical.*

In figure 2 we present the time-resolved photoelectron spectrum from the C 7_0^1 band, employing a probe wavelength of 267.3 nm. A time-independent signals at an electron kinetic energy of 1.89 eV due to a 2-photon process induced by the pump laser alone can be distinguished from the time-dependent [1+1'] signal at 1.39 eV excess energy. This separation between the time-dependent and the time-independent signal is the reason for considering photoelectron spectroscopy a background-free detection method. Note that the order of the processes can be deduced from the electron energies.

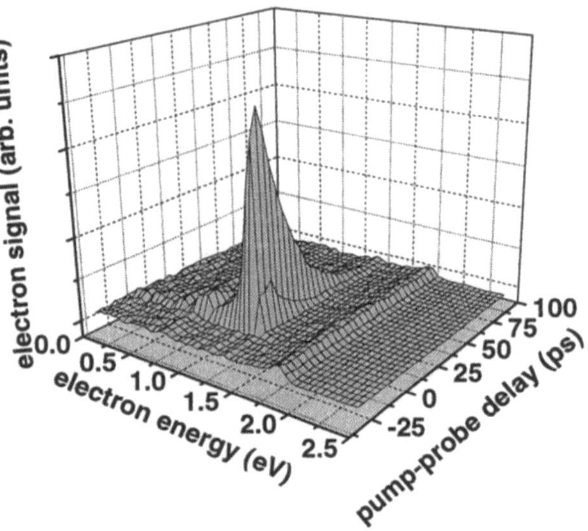

Fig. 2: *Time-resolved photoelectron spectrum from the C 7_0^1 state, showing a time-dependent [1+1'] as well as a time-independent [1+1] signal. Not all data points are shown for ease of viewing*

As the geometry of the C-state and the ionic ground state are very similar, only little vibrational structure is apparent in the spectrum. The ion is formed with an internal energy of 60 meV, corresponding to the excitation of one quantum in the CCC bending vibration ν_7.

Lifetimes for individual vibronic states are obtained by integrating over the major pump-probe peak and looking at its intensity as a function of the pump-probe delay. The resulting decay traces for as selected number of vibronic states are given in Fig. 3. The solid lines constitute a convolution of a monoexponential decay with an instrument response function of 4 ps. From the spectrum given in Fig. 2 a lifetime τ=20 ps is obtained for the C 7_0^1 state. The lifetimes decrease relatively smoothly with increasing excitation in the intermediate state, becoming as short as 9 ps for the D 7_0^1 state. Only the decay of the C-state origin is slightly faster than expected, which will be subject of further investigations.

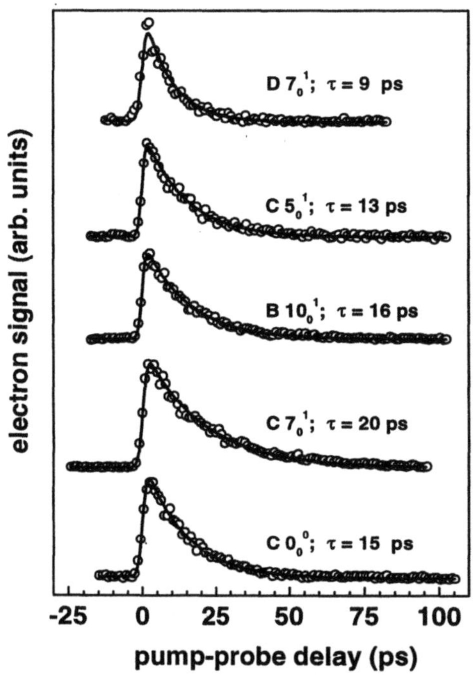

Fig. 3: Typical decay traces for several of the UV vibronic bands

At present the results are interpreted in the following way: The UV-bands of the allyl radical decay within 20 ps or less by internal conversion, forming hot ground state molecules. The hot allyl subsequently loses hydrogen atoms on a nanosecond time-scale, as reported elsewhere[5].

Due to the different time-scales of the initial decay of the UV bands and the appearance of the reaction products[2,5] we can rule out a dissociation or predissociation from the excited state. As allyl does not possess low-lying electronic states of higher multiplicity intersystem crossing will not occur.

Acknowledgments: We acknowledge the support of our work by Prof. P. Chen, the *ETH Zürich* and *the Schweizerische Nationalfonds*.

References

1 J. A. Blush, D. W. Minsek, and P. Chen, J. Phys. Chem. **96**, 10150 (1992)

2 T. Schultz, and I. Fischer, J. Chem. Phys. **107**, 8197 (1997)

3 T. Schultz, and I. Fischer, submitted to J. Chem. Phys.

4 I. Fischer, and T. Schultz, Appl. Phys. **B 64**, 15 (1997)

5 H.-J. Deyerl, T. Gilbert, I. Fischer, and P. Chen, J. Chem. Phys. **107**, 3329 (1997)

Vibrational relaxation and dephasing of small molecules strongly interacting with water

Peter Hamm, Manho Lim, and Robin M. Hochstrasser

Department of Chemistry, University of Pennsylvania, Philadelphia, PA 19104, USA

Abstract: Nonlinear infrared experiments were used to measure vibrational relaxation and spectral diffusion processes of small molecules (CN^-, N_3^-, CO_2) in water. Vibrational relaxation of CN^- is dominated by Coulomb interaction and intramolecular solvent modes are important to understand the mechanism. For N_3^- and CO_2 the relaxation appears to be directly into the solvent without involvement of lower lying intramolecular modes. For both N_3^- and CO_2, a structural inhomogeneity was observed, which decays on a ca. 1 ps time scale and which is attributed to the dynamics of the hydrogen bonds between water and the solute.

1. Vibrational relaxation

Vibrational energy dissipation considerably influences the pathways of chemical reactions in the condensed phase. Therefore, a molecular level description of energy relaxation and dephasing processes of polar molecules or ions in aqueous solutions is of great interest. In this paper, we present nonlinear IR experiments on small strongly interacting molecules (CN^-, CO_2, N_3^-) in water in order to provide new experimental data for a microscopic description of this important processes.

CN^- is a diatomic molecule and vibrational relaxation must be due to direct energy transfer to the solvent. Several isotopomers of CN^- have been investigated both in D_2O and H_2O. The dependence of the relaxation rate on the vibrational frequency of the isotopomers provides a small portion of the friction spectrum of the solvent. In D_2O, the vibrational relaxation rate is accelerated from 120 ps to 71 ps when increasing the vibrational frequency from 2004 cm^{-1} ($^{13}C^{15}N^-$) to 2079 cm^{-1} ($^{12}C^{14}N^-$) and in H_2O, the relaxation rate varies much less between 31 ps and 28 ps, respectively [1]. Interestingly, the vibrational relaxation rate of the solute is strongly correlated with the IR absorption cross section of the solvent. This suggests that vibrational relaxation in this spectral range is dominated by Coulomb interaction, but also shows that internal solvent modes are important to understand intermolecular vibrational relaxation.

In a next step, the asymmetric stretching mode of the tri-atomic molecules N_3^- and CO_2 has been investigated in order to address a possible contribution of intra-molecular energy pathways. The transient spectrum of N_3^- is shown in Fig. 1. In addition to the bleach of the ground state absorption $v=0 \rightarrow v=1$ and the dominating $v=1 \rightarrow v=2$ absorption, a population of higher vibrational levels can be observed ($v=2$ and $v=3$, see arrows). The data were fit to a global function which models the absorption contribution of each of the vibrational levels ($v=0$ to $v=3$) by a Lorentzian line and assumes a rate equation system for the population relaxation between adjacent levels. This procedure yields 2.3 ps for the $v=1 \rightarrow v=0$

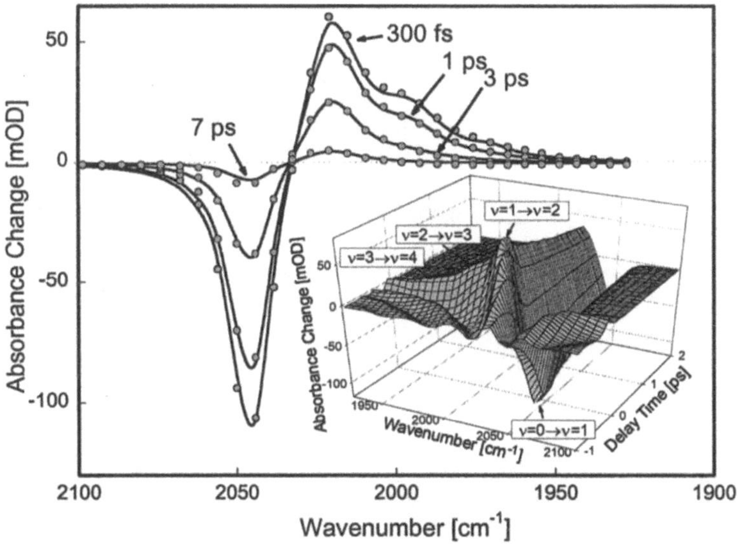

Fig. 1 Transient pump probe signal of N_3^- in D_2O together with a global fit (see text)

relaxation, 1.2 ps for the ν=2→ν=1 relaxation and 0.4 ps for the ν=3→ν=2 relaxation, respectively. Interestingly, this treatment, which takes into account only the directly excited mode, can perfectly reproduce the data without the need of pathways to coupled lower lying N_3^- modes, suggesting that N_3^- essentially behaves like a diatomic molecule. Vibrational relaxation of CO_2 in H_2O, on the other hand, is somewhat slower at 10 ps. Nevertheless, also in this case the decay is single exponential and there is no evidence for population relaxation into lower lying vibrational modes of the CO_2 molecule.

2. Vibrational dephasing

Fig. 2 shows the integrated photon echo signal of N_3^- in D_2O (at 2043 cm^{-1}, in $-k_1+k_2+k_3$ direction) as a function of the time separation between first and second pulse (τ) and second and third pulse (T) (see insert). The signal is not symmetric in τ for early times T. The asymmetry diminishes as time T increases and has mostly disappeared by ca. 4 ps. This is a clear indication of spectral diffusion. At early times T, an inhomogeneous broadening enables rephasing for $\tau>0$, yielding an integrated photon echo signal which is larger than that measured at the corresponding time $-\tau$. Spectral diffusion destroys the phase memory and eventually, the signal becomes symmetric in τ for large enough times T.

The system-bath interaction was modeled by a line shape function calculated from the correlation function $\langle \delta\omega_{10}(t)\delta\omega_{10}(0)\rangle$ of the fluctuations of transition frequencies. With the help of this line shape function one can describe the photon

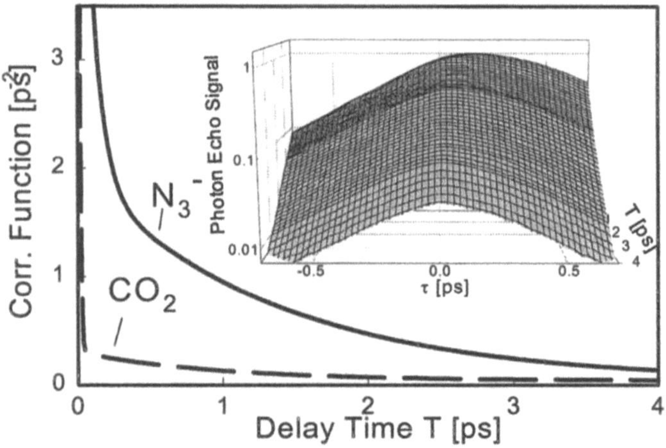

Fig. 2 Three pulse photon echo signal of N_3^- in D_2O at 2043 cm^{-1} (insert) and the correlation function obtained from a global fit of the data on N_3^- (solid line) and CO_2 in H_2O (dashed line).

echo signal in a similar way as was done for optical two level systems [2]. However, an additional term considering the multi-level character of a vibrator has to be taken into account [3]. With this formalism, the data were globally fit assuming a sum of two exponentials as a model correlation function. The resulting correlation function is shown in Fig. 2. The slowly decaying part (time constant 1.3 ps) is assigned to the dynamics of the hydrogen bonds between water molecules and the negatively charged terminal nitrogen atoms. MD simulations have revealed a hydrogen bond lifetime of ca. 1.3 ps [4], in excellent agreement with our observations. The initial fast component of the correlation function represents an essentially homogeneous contribution, possibly originating from long range interaction and the interaction between water molecules and the positively charged central nitrogen. Equivalent experiments were performed on CO_2 in H_2O (at 2340 cm^{-1}). The correlation function obtained from the fit is also shown in Fig. 2. Again, an inhomogeneity is observed which decays on a similar time scale but whose width is considerably smaller than that of N_3^-, presumably due to the smaller quadropole moment of the CO_2.

Acknowledgement: This research was supported by the NIH and NSF.

References

[1] P. Hamm, M. Lim, R. M. Hochstrasser, J. Chem. Phys. **107**, 10523, 1997.
[2] S. Mukamel, in *Principles of Nonlinear Optical Spectroscopy,* Oxford University, New York, 1995.
[3] P. Hamm, M. Lim, R. M. Hochstrasser, submitted to Phys. Rev. Lett.
[4] M. Ferrario, M. L. Klein, and I. R. McDonald, Chem. Phys. Lett. **213**, 537, 1993.

Anomalous 1-ps relaxation of excited exciton in J-aggregates of three-level molecules

Kazuhiko Misawa and Takayoshi Kobayashi

Department of Physics, University of Tokyo,
7-3-1 Hongo, Bunkyo-ku, Tokyo 113, Japan
email: takakoba@phys.s.u-tokyo.ac.jp

Abstract. We discovered an anomalous 1-ps relaxation channel of excited exciton in one-dimensional J-aggregates of three-level porphyrin molecules.

1. Introduction

Molecular J-aggregates are attracting interest of many scientists as ultrafast nonlinear optical material based on molecular excitons [1]. Especially, femtosecond nonlinear absorption due to the 1-exciton to 2-exciton transition in cyanine J-aggregates was discussed within a simple model described by a one-dimensional chain of two-level molecules [2]. The optical nonlinearlity of excited exciton (S_2 exciton) in three-level molecules was discussed theoretically [3], but no experimental study has been performed. In the present paper, we report an anomalous relaxation channel of excited exciton in one-dimensional J-aggregates of three-level molecules. The molecule studied is tetraphenyl porphine sulfonic acid (TPPS), of which information may offer an important clue to the light-harvesting mechanism in a system of antenna chlorophyll.

2. Experimental

The sample was aqueous solution of TPPS J-aggregates stabilized with KCl. The concentrations of TPPS and KCl were 5×10^{-4} and 5×10^{-2} mol/l, respectively. The excitation and probe pulses were both derived from a Ti:sapphire regenerative amplifier seeded by a mode-locked Er-doped fiber laser. The photon energy, pulse energy, and temporal duration of the amplifier output were 1.6 eV, 200 μJ, and 130 fs, respectively. The incident light was sub-resonant with the excitonic transition (Q-band) of the TPPS J-aggregates at 1.76 eV. Kerr-shutter configuration was used to improve signal-to-noise ratio. We also measured transient phase and absorbance changes using a Sagnac interferometer [4].

3. Results and Discussion

Figure 1 shows the time dependence of the absorbance change measured at several excitation power densities. The detected Kerr intensity is the squared sum of the real and imaginary parts of the nonlinear susceptibility. However, the imaginary contribution was found to dominate the nonlinear response by means of the femtosecond interferometry using the Sagnac interferometer as shown in Fig. 2. The profiles in Fig.1 correspond to the absorbance change converted from the detected Kerr signal.

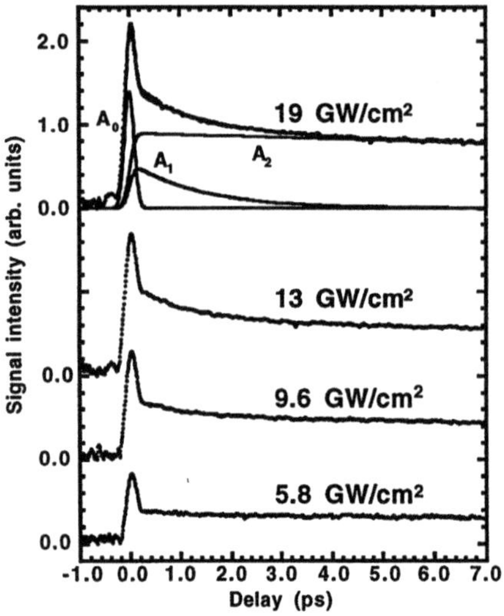

Fig. 1. Time dependence of the Kerr signal intensity at 19, 13, 9.6, and 5.8 GW/cm^2. The profiles corresponds to the absorbance change.

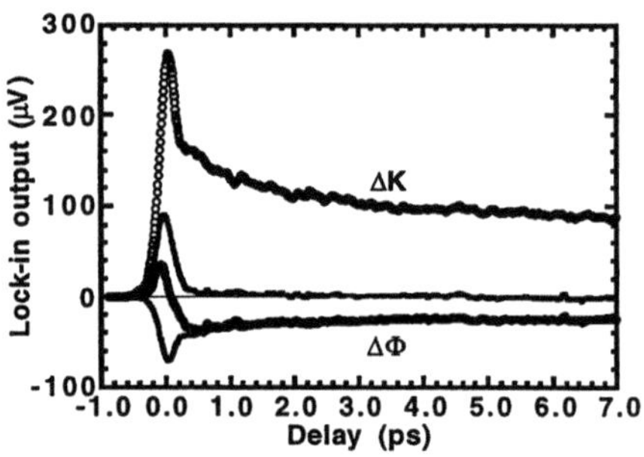

Fig. 2. Transient response corresponding to the real and imaginary parts of nonlinear optical susceptibility measured with a femtosecond Sagnac interferometer

At 19 GW/cm^2 and other intensity levels, the decay profile can be fitted to a convolution of the system response function with a sum of instantaneous response and two exponential decay components as,

$$A_0\delta(t) + A_1 \exp(-t/\tau_1) + A_2 \exp(-t/\tau_2).$$

The time constant of the slow component, τ_2, was determined to be about 50 ps, which agrees with the luminescence lifetime measured by the time-correlated single photon counting. The decay constant of the fast component τ_1 was obtained to be 1.2±0.1 ps. Below 5.8 GW/cm^2, the amount of the fast component was negligibly small. The instantaneous nonlinearity is considered to include the non-resonant contribution from the solvent. The real part changes its sign around zero delay as shown in Fig. 2, which indicates the different origin of the nonlinearity.

The main feature to be emphasized is that the contribution of the fast component was strongly dependent on the excitation power. The obtained amplitudes of the three contributions are plotted as a function of the excitation power density in Fig. 3. The signal amplitude of the fast component shows the remarkable quadratic dependence, while the amplitudes of the instantaneous response and slow component are propotional to the incident power. According to the hierarchical structure of J-aggregates proposed by the present author, these relaxation channels are associated with the coherent mesoaggregates [5]. The slow component is due to the population relaxation in the mesoaggregates, and the S_1 exciton lifetime is concluded to be 50 ps. The fast relaxation in the order of 1 ps is due to excited exciton (S_2 exciton) created with two photons.

The 1-ps decay constant is considered to be the lifetime of the two-photon accessible S_2-exciton. The 1-ps lifetime of excited exciton is much slower than the relaxation time of $S_2 \rightarrow S_1$ transition through internal conversion observed in other organic dyes. Because of the weak coupling between the excitons in J-aggregates and molecular vibrational modes, the efficiency of internal conversion becomes low. Bi-excitonic annihilation yields a non-exponential decay [6], but the observed temporal profile could not be reproduced by the model based on the bi-excitonic annihilation. The τ_1 lifetime depends only weekly on the excitation density, which is not expected from the bi-excitonic annihilation.

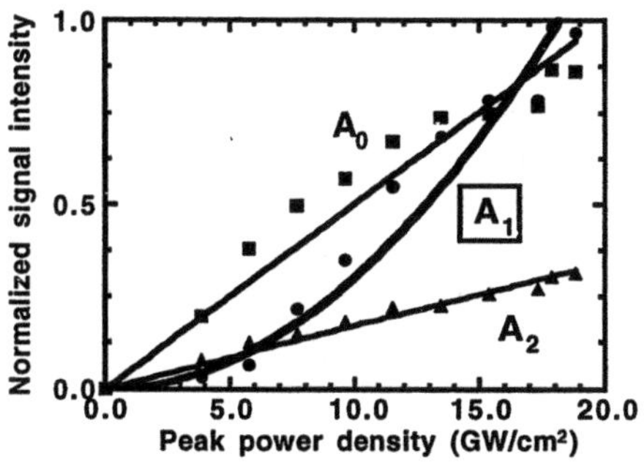

Fig. 3. Excitation power density dependence of the amplitudes of the instantaneous response, fast and slow components.

4. Conclusions

Femtosecond excitation of J-aggregates with a substantially detuned laser from the exciton peak induces simultaneously three different nonlinearity, (1) one-photon virtual S_1-exciton, (2) one-photon real S_1-exciton, and (3) two-photon real S_2-exciton. The S_2-exciton lifetime was found to be anomalously slow.

References

[1] J-aggregates, ed. T. Kobayashi (World Scientific, Singapore (1996)).

[2] K. Minoshima, M. Taiji, K. Misawa, and T. Kobayashi, Chem. Phys. Lett. 218, 67 (1994).

[3] J. Knoester and F.C. Spano, Phys. Rev. Lett. 74, 2780 (1995).

[4] K. Misawa and T. Kobayashi, Opt. Lett. 20, 1550 (1995).

[5] T. Kobayashi and K. Misawa, J. Lumin. 72, 38 (1997).

[6] V. Sundström, T. Gillbro, R.A. Gadonas, and A. Piskarskas, J. Chem. Phys. 89, 2754 (1988).

Coherence from Femtosecond Fluorescence Interference Noise-Terahertz Electronic Beatings in Pentacene

A. Tortschanoff, K. Brunner, Ch. Warmuth, and H. F. Kauffmann

Institut für Physikalische Chemie, Währingerstr. 42, A-1090 Wien, +43-1-31367/2512

Abstract: By measuring fluctuations of fluorescence correlations in Pentacene/p-Terphenyl novel fs-beats are observed. They arise from polarisation interferences due to descrete diagonal disorder between spatially different Pentacene sites.

1. Introduction

Femtosecond relaxation studies on the loss of coherence of optically prepared molecular excitations in nonlinear laser-spectroscopic configurations provide one of the most powerful tools to interrogate excited-state dephasing and thus, quantum-stochastic molecule-reservoir interaction. In the particular case of simultaneous coherent excitation of typically discrete, few-level systems, the superposition of the individual OFIDs can be resolved as an oscillatory modulation, generally referred as quantum interference phenomena. Beatings of different origin from electronic states of organic dyes in the condensed state were observed by photon-echo experiments[1], pump-probe measurements[2], the fluorescence up-conversion technique[3] and by transmission correlation techniques in interferometric pump-probe geometries[4].

Fluorescence quantum beats with fs-resolution have been demonstrated to occur in simple gas-phase systems, by measuring the correlated fluorescence signal generated by a pair of low-intensity pulses in interferometric set-ups[5][6]. An elegant approach to stabilizing the interferometric line is based on the use of *phase-randomized* pairs of pulses[6]. In this experiment the stochasticity of the relative phases between the pulses gives rise to fluctuating quantum interferences in the excited state population. Thus analyzing the *variance* of the (incoherent) interference fluorescence fluctuations yields information on the coherent transient, such as dephasing, wave-packet dynamics and static level-splitting (Coherence Observation by Interference Noise, COIN[6]).

2. Experimental Methods

In preliminary studies we have applied the COIN-technique to solid state systems to demonstrate its utility by measuring electronic coherence decays in large organic molecules. The guest/host system under consideration here is a high-dilute pentacene/p-terphenyl binary crystal which has been extensively studied in ps-photon-echo experiments[7] and in high-resolution, frequency-domain single-molecule spectroscopy[8]. From these investigations it is well known that the mixed crystal exhibits a special type of discrete, diagonal disorder at low phonon temperatures that splits the long-wavelength $S_1 \leftarrow S_0$ 0-0 transition ($^1L_a \leftarrow ^1A$) into four absorption lines at 16883 cm^{-1}, 16887 cm^1, 17006 cm^{-1}, and 17064 cm^{-1}, labeled as O1, O2, O3, and O4, respectively.

In our experiments a Ti:S laser/regenerative amplifier/OPA system generates excitation pulses (~80 fs) in the tuning range of the $^1L_a \leftarrow ^1A$ transition with a spectral width of ~10 nm (~200cm^{-1}). The relative phase of the pulse pair is randomized by heating the air in one of the two arms of a Michelson interferometer. Time-integrated fluorescence is detected at λ_{exc} = 640nm; by using Lock-In sum-frequency detection background and single-pulse contribution is eliminated.

3. Results and Discussion

Experimental COIN-results for T=3.8 K are shown in Fig.1 where the fluorescence variance is plotted as a function of the delay time. For delays τ>250 fs the data profiles depict significant fs-oscillations that strongly depend on the central excitation frequency of the pulse with respect to the spectral positions of the individual sites O_n. For example, different patterns occur, when setting the excitation wavelength either at λ_{exc}=585nm or at λ_{exc}=589 nm (profiles a and b in Fig.1).

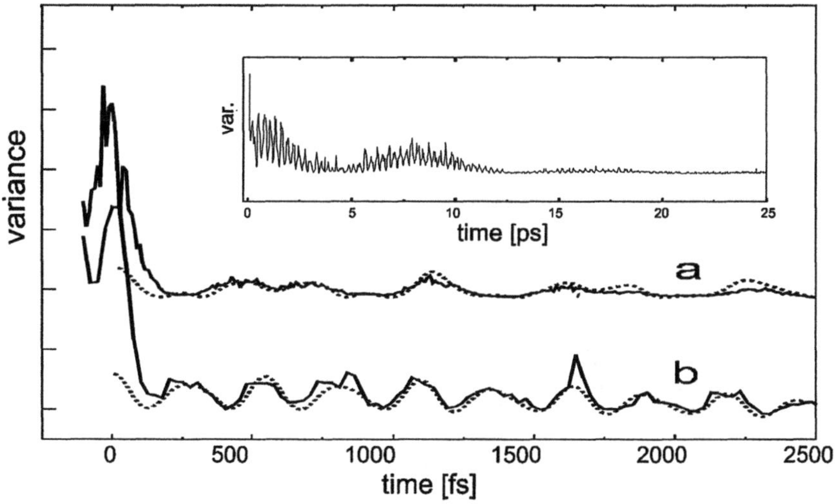

Fig. 1 COIN measurements with excitation at (a) 585 nm (b) 589 nm. Dotted curves show theoretical simulations. The inset shows the longtime behavior of curve (b)

While for the pattern a) the beat frequency corresponding to the O3-O4 interference (ω_{34}) is predominant in the Fourier analysis, the inversion of beat pattern b) reveals frequencies with main components arising from specific O1/O2-O4 and

O1/O2-O3 superpositions (ω_{14}=181cm^{-1}, ω_{24}=177cm^{-1}, ω_{13}=123cm^{-1},ω_{23}=119cm^{-1}) In the inset of Fig.1 a measurement to longer time scales is presented (λ_{exc}=589nm). The pattern displays the low frequency component due to the small O1-O2 site energy difference (4cm^{-1}, clearly concealed in pattern b) superimposed by the rapid oscillations. From the longer relaxation scale of the beat oscillation (insert) a phenomenological damping constant can be evaluated ($T_2^* \approx 10$ ps) which, quite obviously, reflects the time-domain analog of the residual inhomogeneous broadening in the spectral line profiles of the individual site $S_1 \leftarrow S_0$ transitions (Γ=0.5cm^{-1}). This value directly comes out from unbiased data fitting in preliminary simulations using (i) Gaussian shapes for the laser excitation pulse and absorption lines and (ii) the four site energies as inputs. The fluorescence variance profiles resulting from this treatment are drawn as dotted curves in Fig.1. The amplitudes and phases of the beats as well as the variation of the beat frequency pattern with laser excitation detuning are in satisfactory agreement with the experiments. However, the origin of the steep, ultrafast decay for τ < 250 fs remains obscure, for the time being.

4. Conclusions

In summary we have shown that the oscillating fluorescence fluctuations obtained by the COIN method originate from the initial coherent superposition of four energetically different uncoupled two-level ensembles which are, in addition, *spatially* separated. Thus the COIN-experiment reflects one of the very few typical polarization interferences from individual two-level, molecular constituents. These COIN-patterns are the first observation of fs-fluorescence beats for a large molecule in the solid state. We anticipate that this fluorescence technique will probe dynamic molecular interactions as well.

Acknowledgements: This work was supported by Österreichische Nationalbank (P-5512) and the Austrian Science Foundation (P-12566 PHY)

References

/1/ P. C. Becker, H.L. Fragnito, J.Y. Bigot, C.H. Brito Cruz, R.L. Fork, and C.V. Shank, *Phys. Rev. Lett.* **63**, 505 (1988);
/2/ H. L. Fragnito, J.Y. Bigot, P.C. Becker, and C.V.Shank, *Chem. Phys. Lett.* **160,** 101 (1989);
/3/ A. Mokhtari and J. Chesnoy, *IEEE, J. Quantum Electr.* **25**, 2528 (1989)
/4/ F. W. Wise, M.J. Rosker, and C.L. Tang, *J. Chem. Phys.* **86**, 2827 (1987)
/5/ N. F. Scherer, R. J. Carlson, A. Matro, M. Du, A. J. Ruggiero, V. Romero-Rochin, J. A. Cina, G. R. Fleming, and S. A. Rice, *J. Chem. Phys.* **95,** 1487 (1991)
/6/ O. Kinrot, I.Sh. Averbukh, and Y. Prior, *Phys. Rev. Lett.* **75**, 3822 (1995)
/7/ D.A. Wiersma, *Adv. Chem. Phys.* XLVII, 5421 (1981); F.G. Patterson, H.W.H. Lee, W.I. Wilson, and M. D. Fayer, *Chem. Phys.* **84**, 51 (1984)
/8/ L. Kador, *Phys.Stat.Sol.(b)* **189**, 11 (1995)

Fluorescence of Dye Molecules Embedded in a Photonic Crystal

E. P. Petrov,[1] V. N. Bogomolov,[2] I. I. Kalosha,[3] and S. V. Gaponenko[3]

[1] B. I. Stepanov Institute of Physics, National Academy of Sciences of Belarus, F.Skaryna Ave. 68, Minsk 220072, Belarus
E-mail: petrov@ifanbel.bas-net.by
[2] A. F. Ioffe Physico-technical Institute, Russian Academy of Sciences, Politekhnicheskaya Str. 26, St. Petersburg 194021, Russia
[3] Institute of Molecular and Atomic Physics, National Academy of Sciences of Belarus, F.Skaryna Ave. 70, Minsk 220072, Belarus

Abstract. Modification of fluorescence spectra and decay kinetics of dye molecules embedded in artificial opals is interpreted in terms of redistribution of the photon density of states.

1. Introduction

In is well known that the spontaneous emission can be modified in inhomogeneous optical environments such as interfaces, dielectric slabs, microcavities, etc., and can be strongly suppressed in periodic dielectric structures (photonic crystals) due to changes in the photon density of states (DOS) as compared to the free space (see, e.g., [1]). Despite a number of works devoted to modification of the spontaneous emission in 1D and 2D structures, effects of 3D photonic band structures were left almost unexplored, except for Ref. [2], where inhibition of the spontaneous emission of dye molecules in a 3D aqueous colloidal polystyrene crystal was observed. However, it has been found in [3] that the fluorescence lifetime modification observed in [2] is mainly due to intermolecular interactions of dye molecules with polystyrene and reabsorption of fluorescence.

We have recently shown that artificial opals consisting of submicrometer-sized amorphous silica globules arranged in an fcc lattice are promising structures for the photonic band engineering [4]: they exhibit a pronounced stop band in the visible range which manifests itself in optical transmission and reflection experiments and affects fluorescence spectra of embedded dye molecules. In the present paper, we report on the effect of 3D solid-state photonic crystals on the spontaneous emission of dye molecules.

2. Experimental Methods

Artificial opals were fabricated using a standard technology described elsewhere [5]. In the first set of experiments [4], Rhodamine 6G (R6G) dye molecules were embedded in opals by impregnating samples with an ethanol dye solution and

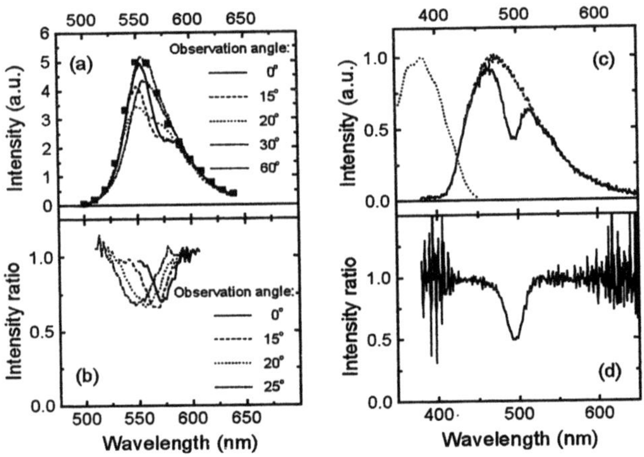

Fig. 1. (a) Fluorescence of R6G molecules embedded in opal sample measured at different observation angles and the "vacuum" spectrum (squares) recorded in an opal sample with the stop band located well to the blue from R6G fluorescences (a), and (b) their ratios; (c) fluorescence emission of NBIA in PMMA filling voids in opal (–) and fluorescence excitation (λ_{em}=500 nm) (···) and emission (---) spectra of the dye in a reference polymer film, and (d) ratio of fluorescence intensities of NBIA/PMMA solution in opal and free-standing film. λ_{ex}=337.1 nm.

subsequent evaporation of the solvent. In the second set of experiments [6], a photonic crystal was fabricated by filling voids in opal by a solid solution of 1,8-naphthoylene-1′,2′-benzimidazole (NBIA) dye in poly(methyl methacrylate) (PMMA). An optically thick polymer film was prepared from the same solution to perform reference measurements. Dissolution of the dye in the polymer eliminates uncontrollable dye–silica interactions and thus makes it possible to study the fluorescence of molecules in an identical chemical environment. In addition, NBIA fluorescence and absorption spectra overlap insignificantly, which makes it possible to avoid multiple reabsorption and reemission of fluorescence.

3. Results and Discussion

Steady-state fluorescence spectra of both R6G and NBIA dyes embedded in opal samples with stop bands overlapping the dye fluorescence show pronounced angular-dependent dips (Fig. 1) with positions corresponding to those of stop bands.

Time-resolved fluorescence investigations revealed a slowing down by a factor of ~1.5 of the fluorescence decay of R6G molecules in an opal sample with the photonic stop band overlapping the fluorescence spectrum compared to the sample with the stop band located well to the blue from the R6G fluorescence. We concluded, however, that the effect is most likely explained by an interplay of modification of the photon DOS, reabsorption and reemission of fluorescence due to the strong overlap of absorption and emission spectra of the R6G dye, and possible differences in intermolecular interactions of dye molecules with the silica surface in different opal samples.

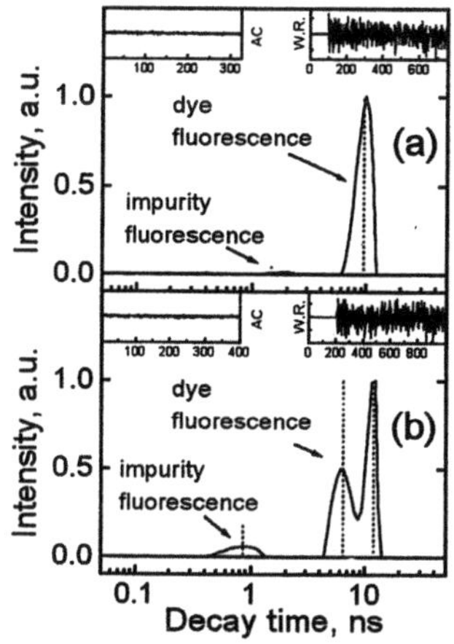

Fig. 2. Decay time distributions of NBIA fluorescence: (a) in free-standing polymer film and (b) in polymer network filling voids in opal. Results of two- (a) and three-exponential (b) analysis of the same decay curves are shown by dotted lines. Insets show plots of weighted residuals (WR) and their autocorrelation function (AC). λ_{ex}=337.1 nm, λ_{em}=510 nm.

An analysis of fluorescence kinetics of the solid solution of NBIA in PMMA filling voids in opal revealed both accelerated and inhibited components as compared to the dye fluorescence in the reference polymer film (Fig. 2). Taking into account that dye molecules are randomly oriented within opal and that opals show angular-dependent stop bands, one can expect that the spontaneous emission rate of a molecule will depend on its orientation with respect to the elementary crystal cell due to the angular dependence of the photon DOS, which is supported by numerical investigations [7].

4. Conclusions

Thus, we conclude that modification of the photon density of states can lead to both inhibition and acceleration of the spontaneous emission of dye molcules embedded in in photonic crystals.

Acknowledgements. The work was supported by the INTAS Project.

References

1 *Photonic Bandgaps and Localization*, Edited by C. M. Soukoulis, 1993.
2 J. Martorell and N. M. Lawandy, Phys. Rev. Lett. **65**, 1877, 1990.
3 B. Y. Tong, P. K. John, Y.-T. Zhu, Y. S. Liu, and W. R. Ware, J. Opt. Soc. Am. B **10**, 356, 1993.
4 V. N. Bogomolov, S. V.Gaponenko, I. N. Germanenko, A. M. Kapitonov, E. P. Petrov, N. V. Gaponenko, A. V. Prokofiev, A. N. Ponyavina, N. I. Silvanovich, and S. M. Samoilovich, Phys. Rev. E. **55**, 7619, 1997.
5 V. N. Bogomolov, D. A. Kurdyukov, A. V. Prokof'ev, and S. M. Samoilovich, JETP Lett. **63**, 520, 1996.
6 E. P. Petrov, V. N. Bogomolov, I. I. Kalosha, and S. V. Gaponenko, Phys. Rev. Lett. **81**, 77, 1998.
7 T. Suzuki and P. K. L. Yu, J. Opt. Soc. Am. B **12**, 570, 1995.

Part XII

Dynamics in Liquids and Solutions

Chirped wavepacket dynamics of solvating HgBr

Manho Lim, Matthew F. Wolford, S. Gnanakaran, Peter Hamm and Robin M. Hochstrasser

Department of Chemistry, University of Pennsylvania, Philadelphia, PA 19104-6323, USA.
E-mail: hochstra@sas.upenn.edu

Abstract. Ultrafast photodissociation of $HgBr_2$ in DMSO shows chirped wavepacket motion with nonexponential decay, providing microscopic insight into the structural and dynamical properties of solvation.

1. Introduction

A detailed understanding of chemical reactions in solution requires knowledge of the dynamics of solute-solvent structures. The dynamics of vibrational wavepackets of chemical reaction products has been utilized to resolve the reaction dynamics and the product state relaxation in detail [1]. The relaxation dynamics are most interesting when the solvent interaction is strong and the time scales of the reactive motion, the molecular vibrational motion and the solvent nuclear dynamics are similar. Here we report the photodissociation of $HgBr_2$ in DMSO which produces vibrational coherent states of HgBr. We have discovered that the HgBr potential energy surface is being modified on the time scale of the solvation, which is manifested as a chirped wavepacket motion with nonexponential decay.

X-ray diffraction experiments show that $HgBr_2$ becomes bent (165°) upon solvation by DMSO and is coordinated by the solvent giving rise to a well determined solvent structure, indicating strong interaction between $HgBr_2$ and DMSO [2].

2. Results and Discussion

Probe wavelength dependent transient absorption after photolysis of $HgBr_2$ in DMSO with a 290 nm pulse is shown in Fig. 1. All the transients observed show instrument time limited rises and have oscillatory signal components which decay to a plateau within a few picoseconds. The oscillatory component arises from the HgBr product molecules being formed in a vibrational superposition state.

The oscillatory part of transients is expanded in Fig. 2. It is clearly seen that there is an abrupt change in amplitude between the second and third peaks. Furthermore, the oscillation period between the first two peaks is longer than that of the remaining ones. The first two peaks oscillate with frequencies of ca. 155 cm^{-1} and decay with time constants of 150 - 200 fs. The recovered phase, 2.5 ±

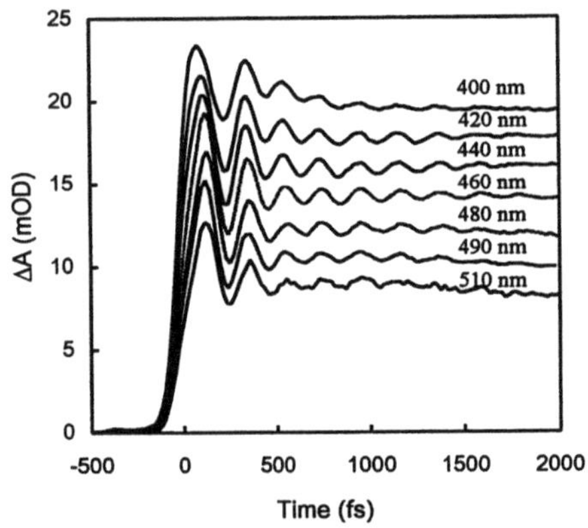

Fig. 1. Transient absorption after photolysis of $HgBr_2$ in DMSO.

0.3 radian is consistent with the reactive motion toward stretched HgBr taking place within 100 fs. The slow oscillation with abrupt damping is attributed to the breaking up of the solvent structure associated with the triatomic molecule. Since $HgBr_2$ has more charge displacement than HgBr, its Coulombic interaction with the solvent is stronger, leading the nascent photoproduct to be oversolvated. This solvent induced hyper-ionic character of HgBr results in a smaller vibrational frequency. As the solvent molecules dissociate from the nascent HgBr, its vibrational frequency becomes closer to the gas phase value of 189 cm^{-1} and the vibrational coherence decays abruptly due to inhomogeneous solvent motion. As shown in Fig. 2, the fast decay time of 150 - 200 fs is similar to the fast DMSO nuclear motion (214 fs) measured by dynamic Stokes shifts [3], corroborating the attribution to breaking the solvent structure.

After the third peak the period increases with delay time. This is caused by a linear chirp, i.e $\omega(t) = \omega_0(1 + \alpha\ t)$, not by two wavepackets [4]. Representative transients and their fit are shown in Fig. 3. To manifest the chirp in the oscillatory signal, a model function without chirp is overlapped in one of the transients. The oscillation frequency changes from ca. 171 cm^{-1} at 500 fs to ca. 160 cm^{-1} at 2 ps and the dephasing time constant is 600 - 800 fs. As expected, the vibrational frequency at early times is closer to the gas phase value. The chirped wavepacket is due to the construction of the equilibrium solvent structure around HgBr radical. As the solvation progresses, the potential energy surface of the HgBr is being modified.

Pair distribution functions of Hg and O or S of DMSO obtained by equilibrium

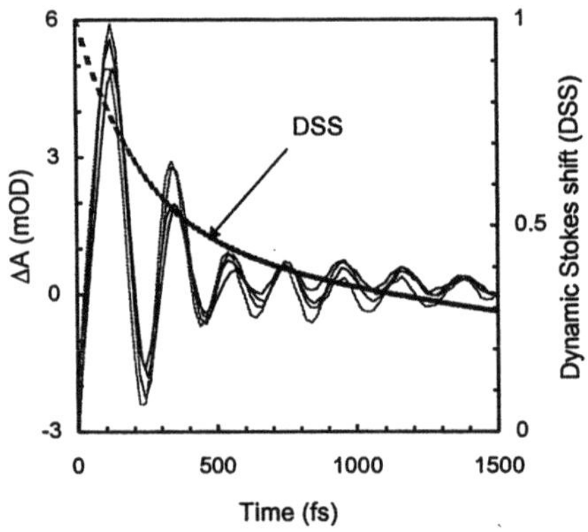

Fig. 2. The oscillatory part of transients.

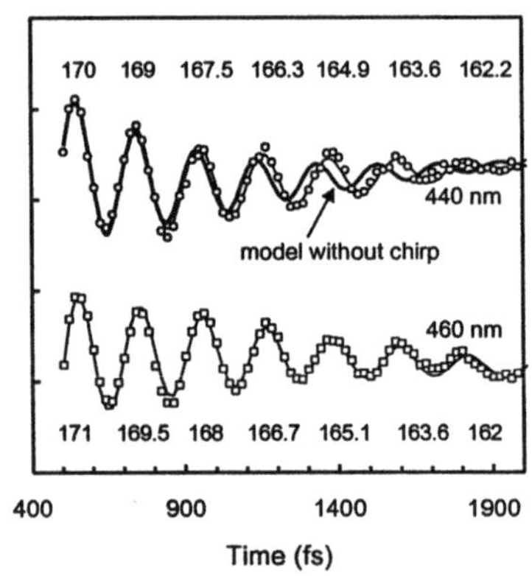

Fig. 3. Chirp in the oscillation frequency.

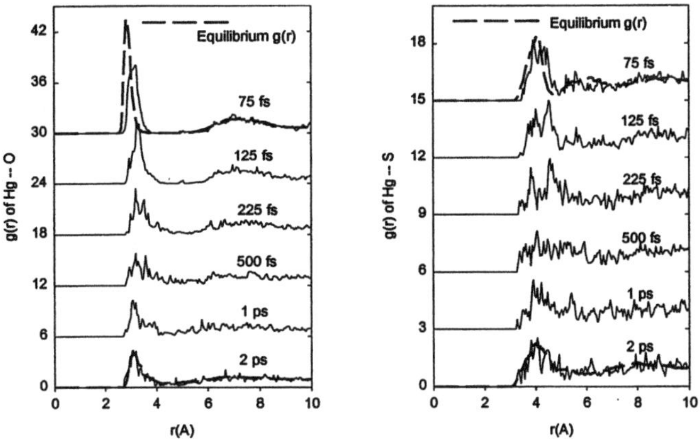

Fig. 4. Pair distribution functions of Hg--O and Hg--S of DMSO.

(thick dashed lines) and nonequilibrium (thin solid lines) MD simulations are shown in Fig. 4. $HgBr_2$ has a well-defined solvent structure coordinated by O's of four DMSO molecules, which is consistent with the X-ray diffraction experiment [2]. HgBr has a less well-defined solvent structure. As photodissociation progresses, the solvent rotates and moves away from the reacting molecule breaking the initial solvent structure. The most random solvent structure is seen at about 500 fs. At this point the HgBr potential is expected to be closest to that of the gas phase. A new solvent structure for HgBr is then fully established by 2 ps. The microscopic solvation processes obtained from the MD simulations are consistent with the foregoing interpretation that the chirped wavepacket dynamics are from the modification of HgBr potential during solvation.

In summary, the photolysis of $HgBr_2$ produces an oscillating signal component that arises from the HgBr molecule formed in a vibrational superposition state. A chirped wavepacket with a nonexponential decay is observed in the photolysis of $HgBr_2$ in DMSO, which has permitted experimental evaluation of dynamic solvation at the molecular level for the first time. Preliminary results from MD simulations provide a microscopic description of solvation consistent with the observed chirped wavepacket dynamics.

Acknowledgements. This research was supported by the NSF and NIH.

References

1 For example, N. Pugliano, A. Z. Szarka and R. M. Hochstrasser, J. Chem. Phys., **104**, 5062 (1996).

2 M. Sandstrom, Acta. Chem. Scand. **A32**, 627 (1978).

3 M. L. Horng, J. A. Gardecki, A. Papazyan and M. Maroncelli, J. Phys. Chem., **99**, 17311 (1995).

4. M. Lim, M. F. Wolford, P. Hamm and R. M. Hochstrasser, Chem. Phys. Lett., **290**, 355 (1998).

A Quantum Mechanical/Molecular Mechanical Approach to Solvation Dynamics Tested by Three Pulse Photon Echo Measurements

Ian P. Mercer[1], Stefan Abend, Ian R. Gould, David R. Klug,
Department of Chemistry,
Imperial College,
South Kensington,
London. SW7 2AY.
E-mail: d.klug@ic.ac.uk

[1]Current address:
L-440, Lawrence Livermore National Laboratory,
Livermore. California. 94550. US.
E-mail: ianmercer@llnl.gov

Abstract. Three pulse echo peak shift experiments are compared to mixed Quantum Mechanical/Molecular Mechanical (QM/MM) calculations of electronic-vibrational coupling for solvated Chlorophyll-a.

1. Introduction

The connection between optical observables and molecular dynamics calculations can be made via the auto-correlation function of the time-varying electronic energy gap [1]. We have previously used this approach to calculate the optical spectra of solvated Chlorophyll-a and Bacteriochlorophyll-a via a mixed QM/MM calculation [2]. The optical absorption and emission profiles are well reproduced, with the predicted Stokes shift (related to the reorganization energy) and spectral widths lying within 20% and 10% of the experimental values respectively.

The steady state spectra only sample the system dynamics on the timescale of the optical dephasing, with the dynamics that occur on a longer timescale giving an inhomogeneous contribution to the spectra. It is difficult to determine the degree of contribution to the steady state spectra from different time scales of dynamics. The three pulse echo peak shift technique (3PEPS) however can sample and resolve the contributions of dynamics from femtoseconds to nanoseconds [3,4]. The 3PEPS thus provides a stringent test of the performance of the molecular dynamics calculations over a broad range of time scales.

2. Methods

Measurements of the 3PEPS were performed using a 1kHz regenerative amplified Ti:sapphire system delivering pulses of 30fs duration at the millijoule level. This was used to seed and pump an optical parametric amplifier, delivering 29fs near transform limited pulses tunable in the visible. Measurements on solvated Chl-a were performed on the first singlet absorption transition centered at 659nm.

The protocol employed for the calculation of the energy gap dynamics involves a combination of classical molecular mechanics to calculate the time varying nuclear positions, and semi-empirical quantum mechanics to calculate the associated electronic transition energies.

The chromophore was placed in a cubic box of side 50 Angstroms, containing approximately 1500 methanol molecules. The system was energy minimized and an initial 80ps run was performed to equilibrate the system. The system was then propagated in the electronic ground state with 1fs steps, using Molecular Dynamics (MD) employing the AMBER [5] suite of programs. Quantum mechanical calculations of the transition energy were performed every 5fs using the semi-empirical package MOPAC93 [6], delivering the energy gap auto-correlation function. Applying a density matrix approach and with the application of a cumulant expansion, the nonlinear response functions can be generated [1]. Taking into account the finite pulse duration this yields the predicted 3PEPS signal.

3. Results and Discussion

The 3PEPS experimental and calculated results are shown in figure 1. The experimental data has been adjusted in peak shift and dwell time by –5.5 and –10fs respectively, to match the calculated profile over the first picosecond. The 5.5fs shift results predominantly from an effect of beam geometry while the 10fs shift is of unknown origin. Despite these offset problems, the dynamics measured by the 3PEPS agrees well with the dynamics calculated by the QM/MM approach for the first 1ps.

The initial decay of the 3PEPS is limited by the duration of our optical pulses. The QM/MM calculations show that there is an initial 50% loss of the system memory in 3fs. This is far too rapid to be well sampled by the echo however these dynamics do contribute significantly to the steady state optical spectra, and this is reflected in the good agreement between calculation and the measured spectra [2]. It is encouraging that there is significant agreement between the 3PEPS dynamics and the QM/MM calculations up to the picosecond timescale. This suggests that, at least for solvated Chl-a that classical dynamics explores the phase space of the system rather well, and that semi-empirical quantum mechanical methods are useful for evaluating the contribution of these parts of phase space to experimental observables. Beyond 1ps the predicted and experimental 3PEPS appear to diverge, the cause of which remains unexplained.

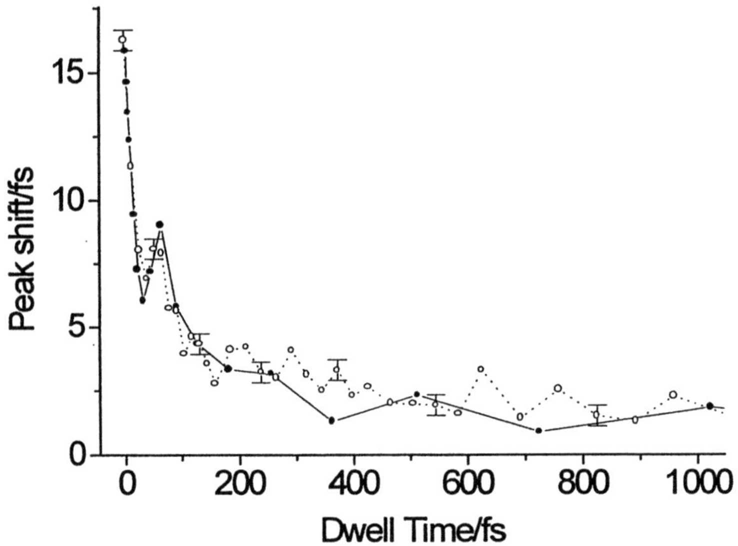

Fig. 1. Calculated (solid) and experimental (dotted) 3PEPS. The experimental profile has been shifted by –5.5 and –10fs in peak shift and dwell time respectively.

Acknowledgements. This work was supported by the EPSRC.

References

1 S. Mukamel, *Principles of Nonlinear Optical Spectroscopy*, Oxford University Press, 1995.

2 I.P.Mercer, I.R.Gould, D.R.Klug, *in Philosophical Transactions: Mathematical, Physical and Engineering Sciences* 1736, 1997.

3 S.A.Passino, Y.Nagasawa, T.Joo and G.R. Fleming, J. Phys. Chem. Vol. 101 (4), 725, 1997.

4 W.P. de Boeij, M.S. Pshenichnikov and D.A. Wiersma, Chem. Phys. Lett. Vol. 253, 53, 1996.

5 D.A. Pearlman, D.A. Case, J.W. Caldwell, W.S. Ross, T.E. Cheatham III, D.M. Ferguson, G.L. Seibel, U.C. Singh, P.K. Weiner, P.A. Kollman, AMBER 4.1, University of California, San Francisco, 1995.

6 MOPAC 93.00, J.J.P. Stewart, Fujitsu Limited, Tokyo, Japan, 1993.

Dynamics of Site-Specific Excited-State Solute-Solvent Interactions as Probed by Femtosecond Vibrational Spectroscopy

C. Chudoba, E. T. J. Nibbering, and T. Elsaesser

Max-Born Institut für Nichtlineare Optik und Kurzzeitspektroskopie, Berlin, Germany
E-mail: nibberin@mbi-berlin.de

Abstract. We study site-specific hydrogen-bonding interactions of coumarin 102 bonded to $CHCl_3$ or phenol with femtosecond vibrational spectroscopy. We demonstrate that upon electronic excitation of coumarin the intermolecular hydrogen bond is cleaved within 200 fs, followed by (sub)-picosecond reorganization of the hydrogen donor.

Study of ultrafast optical dynamics of vibronic excitations reveals the solute-solvent interactions responsible for chemical reaction dynamics in liquids. Optical dephasing and solvation dynamics have been widely studied via the transient Stokes shift of fluorescence from probe molecules [1] and by a variety of coherent techniques [2]. In many cases, solvation dynamics consist of ultrafast (100-200 fs) and slower picosecond components. Most of the experimental techniques applied so far give no detailed information on local solute-solvent geometries and - thus - provide only limited insight into the microscopic mechanisms. This is due to the fact that such experiments are sensitive only to an ensemble-averaged two-point time-correlation function for liquid motions [3]. Here we demonstrate that vibrational spectroscopy can serve as a sensitive probe for determination of local interactions, in particular intermolecular hydrogen bonding and its dynamics upon electronic excitation of a probe molecule. We study the probe coumarin 102 (C102) bonded to hydrogen donors $CHCl_3$ or phenol.

Vibrational spectra of ground-state C102 dissolved in nonpolar C_2Cl_4 and in polar $CHCl_3$ are shown in Figures 1 and 2, respectively (solid lines). In the nonpolar solvent the C=O stretching band occurs around 1735 cm^{-1}, a frequency characteristic of a free carbonyl group. In contrast, the C=O band in $CHCl_3$ is strongly down-shifted to 1700 cm^{-1}, indicating a strong hydrogen bond between solute and solvent. In the femtosecond measurements, C102 is excited to the S_1 state by a 100 fs pulse in the visible and the resulting change of the vibrational spectra is probed by 100 fs pulses tunable in the mid-infrared between 1500 and 1780 cm^{-1}.

For the case of C_2Cl_4 as solvent, the transient change of absorbance ΔA (Fig. 1b) shows a bleaching at the positions of the C=O stretching band and of the ring mode at 1610 cm^{-1}. The transient spectra of excited molecules are derived from such data (symbols in Fig. 1a) and demonstrate a decrease of the transition moments of these vibrations without significant spectral shifts. In the case of $CHCl_3$ as solvent, the ring mode also exhibits a transient bleach. However, a very different behavior of the C=O band is found: we observe a bleaching of the ground-state C=O-mode and the appearance of a new band upshifted to

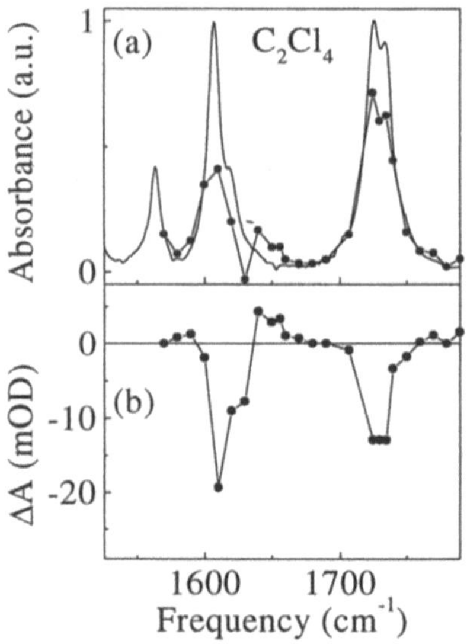

Fig.1. Ground-(solid) and excited-state (dots) IR-spectra (a) of C102 in C_2Cl_4. Time-resolved difference spectrum at 10 ps delay (b).

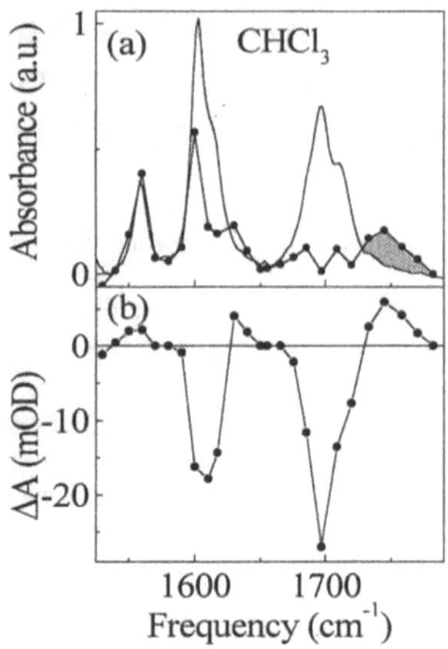

Fig. 2. Ground- (solid) and excited-state (dots) IR-spectra (a) of C102 in $CHCl_3$. Time-resolved difference spectrum at 10 ps delay (b). The grey-shaded area shows the shifted C=O band.

1740 cm^{-1} (see Fig. 2). Apparently, the spectral position of the C=O mode shifts upon electronic excitation to about the same position as that in C_2Cl_4. These results indicate a total diminishing of the site-specific hydrogen bonding interaction between the C=O-mode and $CHCl_3$ upon electronic excitation.

Inspection of the temporal dynamics show that the C=O band of C102 in $CHCl_3$ shifts about 38 cm^{-1} to the blue in less than 200 fs, with an additional slower component of 7 cm^{-1} on a time scale of 2.5 ± 0.3 ps (Fig. 3). We describe the fast initial shift to a cleavage of the hydrogen bond between solute and solvent. It can be shown [4] that the slower component can be directly correlated to the dielectric friction function of $CHCl_3$ using a van der Zwan-Hynes-type relation [5]:

$$\Theta(t) = [\nu_{el}(t) - \nu_{el}(0)]/[\nu_{el}(0) - \nu_{el}(\infty)] \quad (1)$$

where $\nu_{el}(t)$ is the time-dependent solvent-induced vibrational line shift.

In the case of C102-phenol complexes dissolved in C_2Cl_4 we study both the C=O mode of hydrogen acceptor C102 and the hydroxyl-moiety of the hydrogen donor phenol (see Fig. 4). The ground-state C=O-stretch is again strongly downshifted in the C102-phenol complex. The spectral region of the O-H stretch of the phenol- group shows two different bands: a narrow line at 3610 cm^{-1} corresponding to free phenol, and a broad feature corresponding to the hydroxyl band of the ground-state C102-phenol complex at 3250-3450 cm^{-1}.

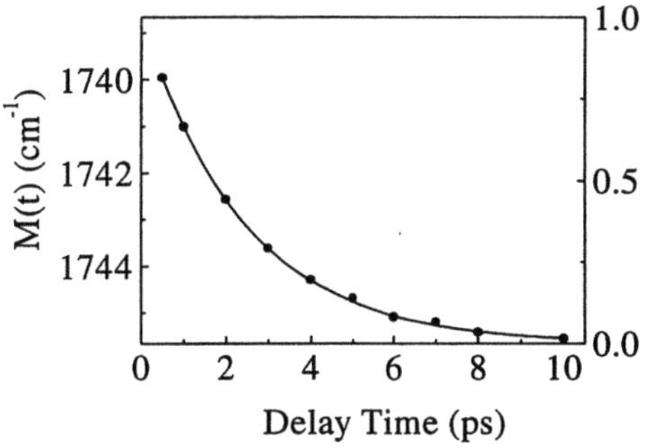

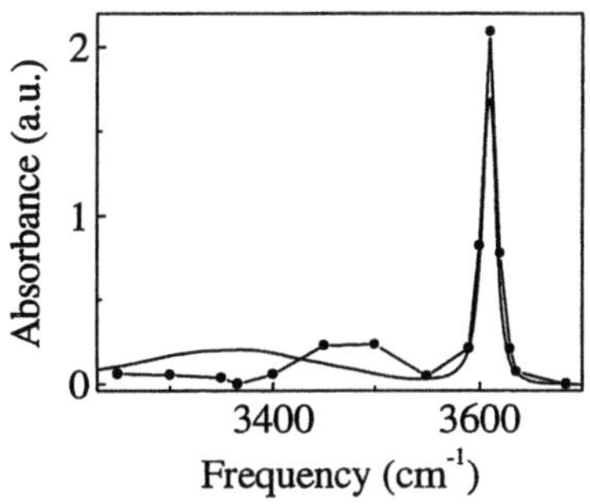

Fig. 3. Long-time behaviour of the normalized first-order moment M(t) and vibrational frequency shift Θ(t) of the C=O band (dots) and exponential fit with time constant 2.5 ± 0.3 ps (line).

Fig. 4. Ground- (solid) and excited-state (dots) IR-spectra (a) of the C102-phenol complex in C_2Cl_4. Time-resolved difference spectrum at 10 ps delay (b). The data depict the hydroxyl-stretch region of the phenol-moiety.

The excited-state dynamics of the hydrogen acceptor C102 is similar as in the case of $CHCl_3$ as hydrogen donor. The ground-state C=O mode exhibits a bleach within our temporal resolution of 200 fs, while the excited-state C=O appears at 1745 cm^{-1}. The O-H stretch of the C102-phenol complex bleaches while the absorption band of free O-H groups increases pointing to an increased number of free phenol molecules in the sample. An additional absorption increase around 3450-3550 cm^{-1} may probably be due to phenol-phenol complexes. From these observations we conclude that the hydrogen bonded C102-phenol complex breaks down within 200 fs after electronic excitation of C102. Additional slower dynamics of the hydroxyl bands with a time scale of 0.8 ps hint at reorientational dynamics of the phenol group relative to C102.

Acknowledgements. This work has been supported by the European Union (contract no. ERBCHBGCT940598).

References

1 R. M. Stratt and M. Maroncelli, J. Phys. Chem. **100**, 12981 (1996).
2 W. P. de Boeij, M. S. Pshenichnikov and D. A. Wiersma, J. Phys. Chem. **100**, 11806 (1996); S. A. Passino, Y. Nagasawa and G. R. Fleming, J. Chem. Phys. **107**, 6094 (1997).
3 S. Mukamel, *Principles of Nonlinear Optical Spectroscopy*, Oxford 1995.
4 C. Chudoba, E. T. J. Nibbering and T. Elsaesser, in preparation.
5 G. van der Zwan and J. T. Hynes, J. Phys. Chem. **89**, 4181 (1985).

Studies of Molecular Interactions and Collective Liquid Motions with Two-Dimensional Raman Spectroscopy

A. Tokmakoff[1,2], M.J. Lang[1,3], D.S. Larsen[1], X.J. Jordanides[1], and G.R. Fleming[1]

[1] Department of Chemistry, University of California, Berkeley, CA 94720, USA and Physical Biosciences Division, Lawrence Berkeley National Laboratory, Berkeley, CA 94720, USA
[2] Department of Chemistry, MIT, Cambridge, MA 02139, USA
[3] Department of Molecular Biology, Princeton University, Princeton NJ 08544, USA

Abstract. Two-dimensional Raman spectroscopy using femtosecond pulse sequences is used to better understand the interactions between molecules and collective dynamics in liquids.

Multidimensional time domain spectroscopy has been used for several years as a powerful structural and dynamic tool in the field of nuclear magnetic resonance (NMR). The power of multidimensional spectroscopies in studies of condensed phases is their ability to measure the correlation or interaction between different degrees of freedom by virtue of measuring higher correlation functions. Most traditional spectroscopies can be related to an ensemble-averaged two-point (or one-time-variable) correlation function, which is necessarily insensitive to the make-up of the ensemble. Two-dimensional (2-D) Raman spectroscopy [1-6] allows the principles of 2-D NMR to be extended to vibrational frequencies in order to probe the details of molecular dynamics in liquids and peer beneath the ensemble average of the Raman and light scattering spectrum.

Coherent 2-D Raman spectroscopy is a non-electronically-resonant technique that uses 45 fs pulse sequences in order to probe a fifth-order nonlinearity as a function of two time variables. By measuring the full 2-D Raman response, one measures an observable that is directly related to the interactions between Raman active vibrations. For low frequency overdamped motions at light scattering frequencies, the contours in the time domain are characteristic of the ensemble-makeup of the collective liquid dynamics. For higher frequency underdamped vibrational modes, the fifth-order Raman response can be 2-D Fourier transformed in order to obtain a 2-D Raman spectrum, in which interactions between vibration

The 1-D polarized Raman spectrum of liquid CCl_4 shows three modes that arise from C-Cl stretching and bending motions. These modes are seen in the 2-D spectrum (Fig. 1) as the three features along the frequency diagonal (ω_2=ω_4). What is also apparent are a number of peaks that lie off the diagonal, which are representative of interaction or couplings between these vibrational modes. Cross peaks between the features on the diagonal (for instance ω_2 = 460 cm^{-1}; ω_4 = 219 cm^{-1}) and difference frequency peaks (ω_2 = –219 cm^{-1}; ω_4 = (460–219) cm^{-1}) are direct indications of couplings between them [4].

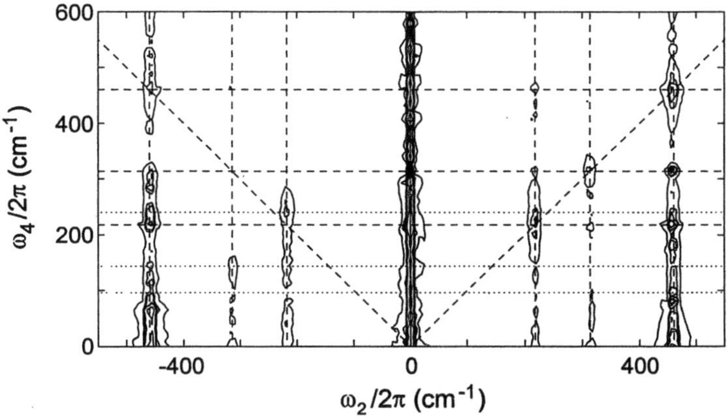

Fig. 1. Two-dimensional absolute value Raman spectrum of liquid CCl_4

Two possible coupling mechanisms have been shown to lead to cross peaks in the 2-D spectrum [4,5]. One mechanism is mechanical anharmonicity between vibrational modes. In particular, it has been shown that the amplitude of a cross peak is related to the magnitude of a particular cubic anharmonic expansion coefficient for the liquid potential. The other coupling mechanism is the nonlinear polarizability, which reflects the fact that the polarizability along a particular nuclear coordinate can be strongly influenced by the motion of another. A prime example of this is interaction-induced or collision-induced effects in liquids. If the liquid polarizability is expanded in a set of nuclear coordinates it can be shown that the magnitude of the nonlinear polarizability is related to the amplitude of cross peaks.

The same principles that are used to interpret 2-D Raman spectra can also be applied to understanding correlations and heterogeneity within collective motions in liquids. Figure 2 shows the intrinsic heterodyne 2-D Raman response $R^{(3)}(\tau_4)R^{(5)}(\tau_2,\tau_4)$ from liquid CS_2 at 165 K, just above its freezing point. The two-dimensional shape of this response forms a basis for understanding the time scales over which motions become averaged and diffusive, and allow a manner of quantifying the cross over from an inhomogeneous system at short-time to a diffusive homogeneous liquid at long times. This data set has been analyzed by breaking up the 1-D response of liquid CS_2 into characteristic time scales that are represented by a set of oscillators. The 2-D data can be fit to measure the correlation (or strength of interaction) between the collective liquid motions on these time scales. This analysis indicates a large degree of inhomogeneity on a 1 ps time scale in liquid CS_2 at 165 K that is not present at 295 K [6].

Acknowledgements. This work was supported by the National Science Foundation. A.T. thanks the National Science Foundation for a postdoctoral fellowship.

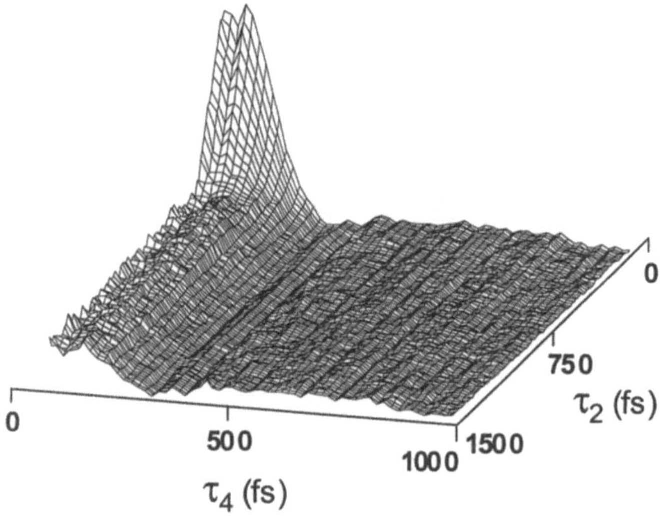

Fig. 2. Two-dimensional heterodyned Raman response $R^{(3)}(\tau_4)R^{(5)}(\tau_2,\tau_4)$ from liquid CS_2 at 165 K. The heterodyne technique is discussed in reference 7. The hyperpolarizablity contributions at delay times <80 fs have been truncated

References

1 Y. Tanimura and S. Mukamel, J. Chem. Phys. **99**, 9496, 1993.
2 K. Tominaga and K. Yoshihara, Phys. Rev. Lett. **171,** 179, 1995; J. Chem. Phys. **104**, 4419, 1996; J. Chem. Phys. **104**, 1159, 1996.
3 T. Steffen and K. Duppen, Phys. Rev. Lett. **76**, 1224, 1996; T. Steffen and K. Duppen, J. Chem. Phys. **106**, 3854, 1997.
4 A. Tokmakoff, M. J. Lang, D. S. Larsen, G. R. Fleming, V. Chernyak, and S. Mukamel, Phys. Rev. Lett. **79,** 2702, 1997.
5 K. Okumura and Y. Tanimura, J. Chem. Phys. **106**, 1687, 1997.
6 A. Tokmakoff, M. J. Lang, X. J. Jordanides and G. R. Fleming, Chem Phys., in press.
7 A. Tokmakoff, M. J. Lang, D. S. Larsen, and G. R. Fleming, Chem. Phys. Lett. **272**, 48, 1997.

Two-Dimensional Raman-Echo Spectroscopy; Femtosecond View of Vibrational Coherence

Andrei Piryatinski, Vladimir Chernyak, and Shaul Mukamel

Department of Chemistry, University of Rochester, Rochester NY 14627, USA
E-mail: mukamel.chem.rochester.edu

Abstract. Two semiclassical algorithms (high temperature and low temperature, weak nonlinearity) for computing 2D fifth-order Raman spectra are compared. In both cases the response can be described using phase-space wavepackets.

1. Introduction

Time-resolved vibrational spectroscopy is a powerful tool in the studies of intramolecular and intermolecular dynamics of condensed phase systems such as liquids, glasses, polymers, and proteins where nuclear dynamics spans a broad range of timescales [1,2]. The complete control of pulse shapes, durations, and phases allows the development of multidimensional Raman techniques whereby the spectrum is disentangled by spreading it onto several frequency or time axes [3-6]. Their modeling should take into account the quantum character of nuclear motions, the non-linear dependence of electronic polarizability on nuclear coordinates, and anharmonicities. Semiclassical expansions can be used to carry out these simulations.

Two different semiclassical expressions for the necessary nonlinear response functions are developed. The first, high temperature algorithm (HTA), holds when thermal energy is much greater than vibrational frequency ($k_B T >> \hbar\Omega$) [7]. The other, weak nonlinearity algorithm (WNA), applies to systems with weak nonlinearity and anharmonicity [8] and holds better at low temperatures. In both cases the response function is expanded in powers of $\hbar$. In the former this is done for a fixed temperature T whereas in the latter the thermal frequency $\omega_T = k_B T/\hbar$ is fixed.

2. Multiple Echoes

We applied the WNA and HTA to the fifth-order 2D response of a single mode coupled to the off-resonant radiation field via the electronic polarizability $\alpha(Q) = \alpha_0 \exp(Q/Q_0)$, where Q is the oscillator coordinate and Q_0 is characteristic length scale. Since all order terms in the power expansion of $\alpha(Q)$ are finite, the interaction with the radiation field excites coherences between all vibrational states. The fifth-order 2D signal for an underdamped mode is [9]

$$S(t_1,t_2) = \sum_m e^{-im\Omega_0 t_2} \sum_n \exp\left[-\frac{(m\Delta)^2}{2}\left(t_2 - \frac{nt_1}{m}\right)^2\right] e^{in\Omega_0 t_1} F_{nm}\left(e^{-\gamma_2} e^{-\gamma_1}\right) \quad (2.1)$$

where a Gaussian inhmogeneous distribution with central frequency Ω_0 and width Δ is assumed. Equation (2.1) shows multiple echoes at times $t_2 = (n/m)t_1$, m, n = 1,2,3.... The amplitudes of the echo peaks are determined by function F_{nm} which is different for both semiclassical expansions and varies on the homogeneous relaxation timescale.

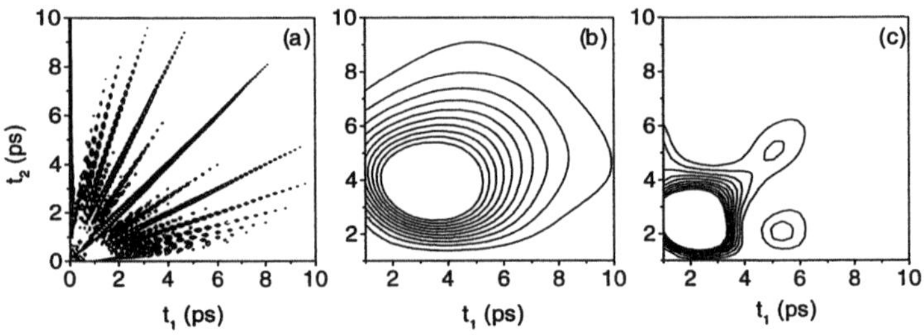

Fig. 1 2D Raman signal: (a) underdamped mode, (b)-(c) low frequency mode.

Figure 1 shows computer simulations of $|S(t_1,t_2)|$. Panel (a) presents an underdamped mode (Eq. (2.1)) with $\Omega_0 = 620\,cm^{-1}$, $\gamma = 0.7\,cm^{-1}$, and $\Delta = 60\,cm^{-1}$. Multiple echoes are clearly seen in the directions t_1 =1, 1/3, 1/2, 2/3, 1/4, 3/2, 2, 3, 4 t_2. The HTA is applied to a low frequency vibration $\Omega_0 = 20\,cm^{-1}$ with strong homogeneous damping. The signal in panel (b) where $\Delta > \Omega_0$ ($\Delta = 10\,cm^{-1}$), shows no echo whereas weak echo is clearly seen in panel (c) where $\Delta < \Omega_0$ ($\Delta = 40\,cm^{-1}$).

3. Phase-Space Wavepackets

The exact fifth-order off-resonance quantum response function is expressed as a sum of four Liouville space pathways each given by a three-point correlation function of polarizability $\alpha(Q)$. Each term is described by a Feynman diagram [10] and is represented by a phase-space wavepacket. For a harmonic system (with $\gamma = 0$) these wavepackets are Gaussian [9]:

$$\rho_\alpha(P,Q) = (-1)^\alpha \frac{\alpha_0^2}{(i\hbar)^2} \exp\left(\frac{\overline{Q}(0)+\overline{Q}_\alpha(t_1)}{Q_0}\right) \times \frac{\Omega}{2\pi\sigma} \exp\left[-\frac{(P-\overline{P}_\alpha(t_1+t_2)-\overline{P}_\alpha(t_2))^2}{4M\sigma} - \frac{M\Omega^2(Q-\overline{Q}_\alpha(t_1+t_2)-\overline{Q}_\alpha(t_2))^2}{4\sigma}\right] \tag{3.1}$$

where $\alpha = 1,\ldots,4$, $\sigma = (\hbar\Omega/2)\coth(\hbar\Omega/k_B T)$, and center coordinates:

$$\overline{Q}_{\alpha}(t_n) = \overline{Q}(0)\cos(\Omega t_n) + \varepsilon_{\alpha}^{n}\frac{\overline{P}(0)}{M\Omega}\sin(\Omega t_n), \quad n = 1,2 \tag{3.2}$$

$$\overline{Q}_{\alpha}(t_1 + t_2) = \overline{Q}(0)\cos\big(\Omega(t_1 + t_2)\big) + \varepsilon_{\alpha}^{1}\frac{\overline{P}(0)}{M\Omega}\sin\big(\Omega(t_1 + t_2)\big) \tag{3.3}$$

$$\overline{P}_{\alpha}(t) = M\Omega\dot{\overline{Q}}_{\alpha}(t) \tag{3.4}$$

with initial coordinate $\overline{Q}(0) = \sigma/M\Omega^2 Q_0$, initial momentum $\overline{P}(0) = i\hbar/2Q_0$, and the auxiliary variables $\varepsilon_1^1 = \varepsilon_2^1 = -\varepsilon_3^1 = -\varepsilon_4^1 = 1$, $\varepsilon_1^2 = -\varepsilon_2^2 = -\varepsilon_3^2 = \varepsilon_4^2 = -1$. According to Eqs. (3.1) – (3.4) only the center of each wavepacket evolves in time while its shape remains Gaussian. Different phase-space trajectories (Eqs. (3.2)–(3.4)) coincide if $\overline{P}(0) = 0$ and in both semiclassical expansions these four wavepackets are approximated by a single wavepacket which represents the total density matrix.

Acknowledgements. We gratefully acknowledge the support of the National Science Foundation and the United States Air Force Office of Scientific Research.

References

1. R. Laenen, C. Rauscher, and A. Laubereau, Phys. Rev. Lett., **80**, 2622, 1998.
2. D. T. Leeson and Wiersma, Phys. Rev. Lett., **74**, 2138, 1995.
3. Y. Tanimura and S. Mukamel, J. Chem. Phys., **99**, 9496, 1993.
4. K. Okamura and Y. Tanumura, J. Chem. Phys., **106**, 1687, 1997.
5. A. Tokmakoff, M. J. Lang, D. S. Larsen, G. R. Fleming, V. Chernyak, and S. Mukamel, Phys. Rev. Lett., **79**, 2702, 1997.
6. S. Mukamel, A. Piryatinski, and V. Chernyak, *Two-Dimensional Raman Echoes; Femtosecond View of Molecular Structure and Vibrational Coherence*, Acc. Chem. Res., (in press).
7. S. Mukamel, V. Khidekel, and V.Chernyak, Phys. Rev. E, **53**, 1, 1996.
8. V. Chernyak and S. Mukamel, J. Chem. Phys., **108**, 5812, 1998.
9. A. Piryatinski, V. Chernyak, and S. Mukamel, J. Chem. Phys., (submitted).
10. S. Mukamel, *Principles of Nonlinear Optical Spectroscopy*, Oxford, N. Y., 1995.

Ultrafast solvation dynamics and quantum beats

B. Zolotov[1], D. Huppert[1], and B. D. Fainberg[1,2]

[1] Raymond and Beverly Sackler Faculty of Exact Sciences, School of Chemistry, Tel Aviv University, Tel Aviv 69978, Israel
E-mail: huppert@tulip.tau.ac.il
[2] Center for Technological Education Holon, 52 Golomb St., Holon 58102, Israel
E-mail: fainberg@barley.cteh.ac.il

Abstract. We present new results concerning low frequency beats in several polymethine dyes. The important findings of our study is the coupling of the low frequency intramolecular vibrations, responsible for the beats, with the polar solvent.

1. Introduction

Quantum beats in transient four-photon spectroscopy of nondegenerate electronic states have been observed both on nonresonance [1] and also on resonance excitation [2,3].

We present here our new experimental and theoretical results concerning the low frequency beats in several polymethine dyes. Our studies show that the influence of the solvation dynamics on the beat behavior is not reduced only to the time evolution of the chromophore spectra (the increase in beats contrast due to the hole-burning effect in a solvation process). The important feature of our consideration is the coupling of the low frequency intramolecular vibrations, responsible for the beats, with the polar solvent. As a consequence, the beat attenuation has the fast non-Markovian component with a memory, corresponding to the solvation correlation function.

2. Experimental Methods

Using the heterodyne optical Kerr effect (HOKE) and pump-probe spectroscopy, we have observed low frequency (~150cm^{-1}) beats during the solvation process of a number of polymethine dyes in different solvents. Fig.1 shows our results concerning the HOKE signal of cryptocyanine in alcohols (dots) and their computer fit (solid lines), corresponding to the model, described in Ref.[4], which includes also the interaction between a low frequency intramolecular vibration z, responsible for the beats, and a polar solvent in the form $\upsilon_z xz$, where x is the solvation coordinate. One can see that the depth of the modulation of the beats is larger in the slow relaxing solvents. The beats are almost invisible in methanol, a solvent with a fast relaxing Gaussian component of appreciable amplitude.

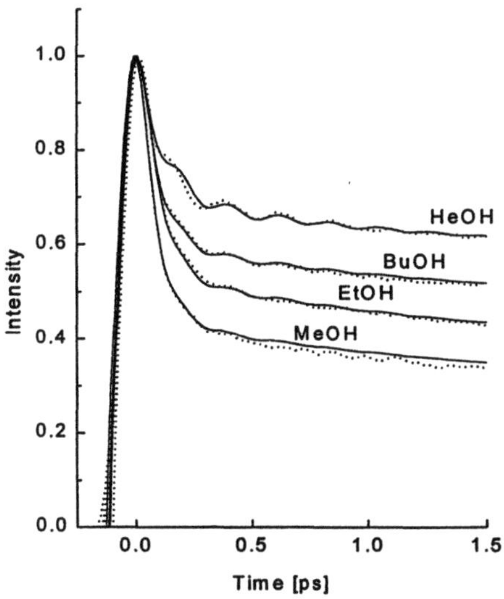

Fig. 1 The HOKE signal of alcohols

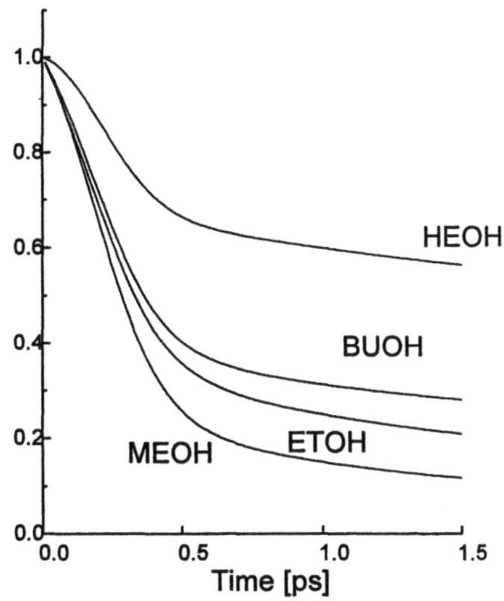

Fig. 2 Correlation functions of "pure solvation" in alcohols

3. Theory and Discussion

The beat frequencies for the molecules under investigation are also active in the infrared spectra of these molecules. Therefore, we consider first, that the beats under consideration correspond to low frequency intramolecular vibrations *z*, and second, that a component of the dipole moment of an IR transition $D'_z = (\partial D/\partial z)_{z_0}\, z$ exists.

Using the coupling between motions *z* and *x* in the form $\upsilon_z xz$, one can obtain that the attenuation of the mean value $\langle z\rangle$ of the coordinate *z* in the ground electronic state is described by the integral term of the form $\int_0^t \left(\upsilon_z^2/\Omega_x^2\right) f_{rs}(t-t')\langle \dot{z}(t')\rangle dt'$ where the "memory" $\left(\upsilon_z^2/\Omega_x^2\right) f_{rs}(t-t')$ is determined by the normalized relaxation function of "pure" solvation $f_{rs}(t)$, and its amplitude-by the quantity υ_z^2/Ω_x^2 (here Ω_x is the frequency corresponding to the coordinate *x*). The last value can be expressed by the following thermal average:
$\upsilon_z^2/\Omega_x^2 = \left\langle \overline{W}_{ns}'^2(0)\right\rangle/(kT)$ where $\overline{W}_{ns}'^2(t)$ denotes the derivative of the interaction energy between a solute in electronic state *n* and the nuclear degrees of freedom of a solvent with respect to coordinate *z*. The thermal average $\left\langle \overline{W}_{ns}'^2(0)\right\rangle$ can be

expressed by the amplitude of the orientational polarization correlation function of the solvent C(**k**,0):

$$\left\langle \overline{W}_{ns}^{\prime 2}(0) \right\rangle = (2\pi)^{-3} \int d\mathbf{k} \left| \widetilde{\mathbf{E}}^{\prime (n)}(\mathbf{k}, z_0) \right|^2 C(\mathbf{k},0) \qquad (1)$$

where $\widetilde{\mathbf{E}}^{\prime (n)}(\mathbf{k}, z_0)$ is the spatial Fourier transform of $\partial \mathbf{E}^{(n)}(\mathbf{r}, z(t))/\partial z$ at the point z_0, $\mathbf{E}^{(n)}(\mathbf{r}, z(t))$ is the electric field created by the solute molecule in electronic state n at point **r**, which depends on the vibrational coordinate *z(t)*. The quantity $\left\langle \overline{W}_{ns}^{\prime 2}(0) \right\rangle$ can be evaluated by the comparison with the characteristic value of the Stokes shift in the dipole solvent ω_{St} which is also expressed by C(**k**,0). In the dipole approximation, ω_{St} is proportional to $|\Delta\mathbf{D}|^2$,where $\Delta\mathbf{D} = \mathbf{D}_{11} - \mathbf{D}_{22}$ is the change of the dipole moment of a solute under electronic excitation, and $\left| \widetilde{\mathbf{E}}^{\prime (n)}(\mathbf{k}, z_0) \right|^2 \sim |\mathbf{D}'|^2$. A typical value of $|\Delta\mathbf{D}|$ is of the order of $|\mathbf{D}_0|$ -the constant part of the dipole moment of a solute in electronic state n *(~1Debye)*. The magnitude of $\mathbf{D}'$ can be evaluated as $|\mathbf{D}'|^2 \sim 0.01 |\mathbf{D}_0|^2 (2\Omega_z/\hbar)$ where $\hbar/(2\Omega_z)$ is the average value of the zero vibration amplitude squared of the vibration z. Thus, we obtain:

$$\left(\upsilon_z^2/\Omega_x^2\right)\left[\hbar/(2\Omega_z)\right]/(\hbar\omega_{St}) \sim |\mathbf{D}'|^2 \left[\hbar/(2\Omega_z)\right] |\mathbf{D}_0|^{-2} \sim 0.01 \qquad (2)$$

For $\Omega_z \approx 150 cm^{-1}$ and a typical value $\omega_{St} \sim 1000\ cm^{-1}$, we have $\upsilon_z^2/\Omega_x^2 \sim 3 \cdot 10^3 cm^{-2}$ that conforms to the values obtained by the computer fit of the experimental signal.

The fit of the experimental data by the model developed in this study is very good for all solvents. Fig.2 shows the normalized correlation functions of "pure" solvation in alcohols, corresponding to the computer fit shown in Fig.1, which can be obtained if we neglect the coupling between the vibration z and the solvent. These functions determine the "memory" of the relaxation of the vibration z.

References

1 Y. Yan and K. A. Nelson, *J. Chem. Phys.* **87**, 6257 (1987)
2 F. W. Wise, M. J. Rosker, and C. L. Tang, *J. Chem. Phys* **86**, 2827 (1987)
3 J. Chesnoy and A. Mokhtari, *Phys. Rev. A* **38**, 3556 (1988)
4 B. Zolotov, A. Gan, B. D. Fainberg, and D. Huppert. *Chem. Phys. Lett.* **265**,418 (1997)

Ultrafast Relaxation of the Dye LDS-750 in Solution: Dependence on Excitation Wavelength

N. Eilers-König, S.A. Kovalenko and N.P. Ernsting

Walter-Nernst-Insitut für Physikalische und Theoretische Chemie,
Humboldt Universität zu Berlin, Bunsenstr. 1, D-10117 Berlin, Germany
E-mail: ninak@npe01.chemie.hu-berlin.de

Abstract. CT-emission is achieved along three routes. Fs-excitation at 630 nm of the dye in chloroform is followed by the CT-reaction under solvent control. In more polar solvents, the primary excited state decays faster than solvent relaxation to a dark intermediate state prior to CT emission. This implies an internal reaction coordinate bypassing solvation. Following fs-excitation at 226 nm, the dark state is also accessed by internal conversion.

1. Introduction

The first observation of an inertial part of solvation predicted by MD simulations was reported for the styryl dye LDS-750 in acetonitrile [1]. Since then evidence accumulated that the photoexcitation of this dye involves also intramolecular charge transfer. Then one needs to clarify whether solvent dynamics alone controls the rate as in the strongly adiabatic barrierless regime, or whether a description demands an intramolecular coordinate as well. The purpose of the present work is therefore empirical: to characterize the different electronic states which are involved in the relaxation of LDS 750 after excitation in the first absorption band, and to correlate the observed state-to-state time constants with those from pure solvation dynamics. We report spectra of transient absorption and gain in polar solvents for excitation wavelengths of 630 nm and 226 nm.

2. Experiment and Results

The broadband transient absorption measurement has been described elsewhere [2]. The time resolution was approximately 110 fs for 630 nm excitation and 100 fs for the 226 nm excitation.

Figs. 1 and 2 show transient spectra for LDS 750 in acetonitrile. For red excitation (fig. 1), a broad stimulated emission band is observed after 100 fs with maxima at about 645 and 710 nm. Over the next 600 fs, the emission at 645 nm decays whereas the emission at 710 nm increases and shifts to a final peak wavelength of 722 nm, while an isosbestic point exists at 693 nm. - Directly after UV excitation (fig. 2), only the bleach of the ground-state absorption band around 570 nm and excited-state absorption (ESA) are observed.

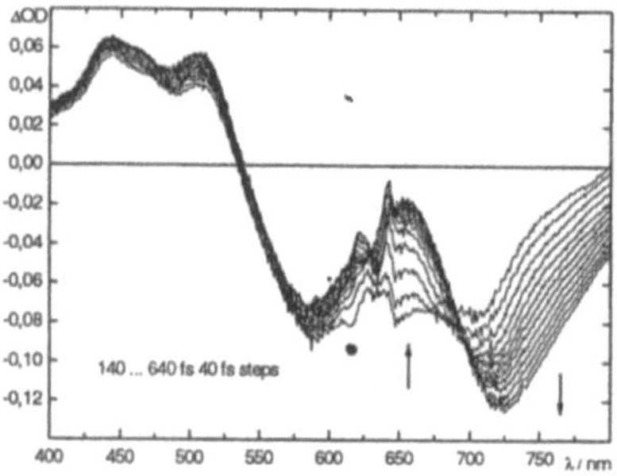

Fig. 1. Transient spectra in acetonitrile after excitation at 630 nm.

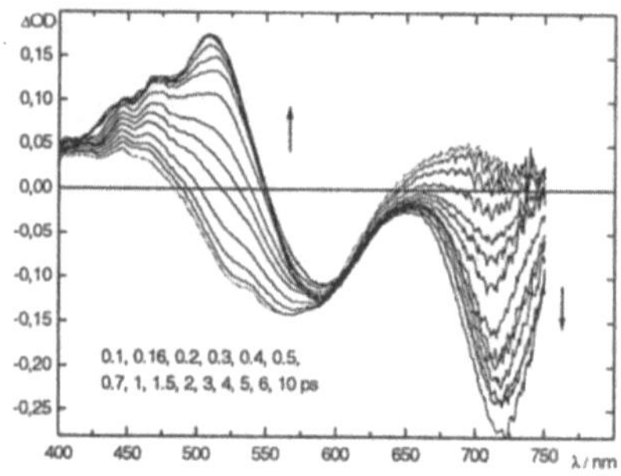

Fig. 2. Transient spectra in acetonitrile after excitation at 226 nm.

A stimulated emission band rises with time constants of 0.52 and 6.1 ps (with relative amplitudes 0.71 and 0.29), but even at earliest time it is already centered at the final peak wavelength of 722 nm. This is accompanied by a corresponding rise of absorption around 535 nm which develops into the ESA band of the previous figure. The spectral behaviour in propylene carbonate and methanol after red excitation is essentially similar to that in acetonitrile. For propylene carbonate, an additional slow (2 ps) part is found in the kinetics. In methanol, the isosbestic point is wiped out by a spectral shift on the same timescale as the rise of the 722 nm emission. In chloroform (fig. 3) the initial stimulated emission band at 655 nm is quite prominent. It decays on a longer timescale while a final emission band rises at 711 nm. The existence of state-to-state kinetics is thus clearly established.

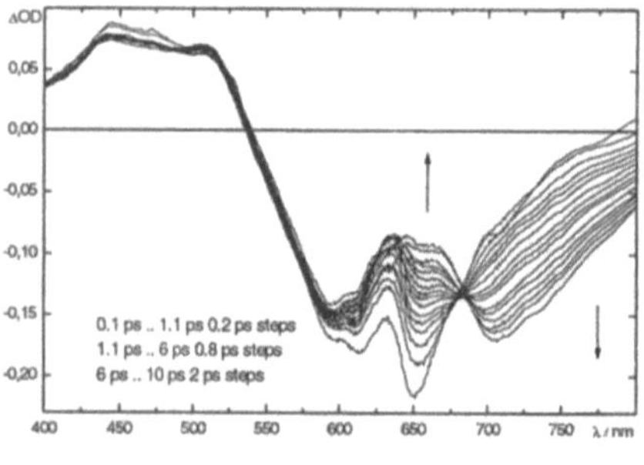

Fig. 3. Transient absorption spectra in chloroform after excitation at 630 nm.

Because of the strong spectral overlap, we do not attempt to separate the contributions from the ground and excited states and instead consider only the kinetic traces at 645 nm and 722 nm, representative for emission from the primary and final emissive states. Fits to a sum of exponentials and effectively instantaneous contributions convoluted with the apparatus function have been performed.

	decay 645 nm	solvent rel.	rise 722 nm
acetonitrile	0.08 (0.86)	0.089 (0.686)	inst. (0.35)
	0.56 (0.14)	0.63 (0.314)	0.53 (0.65)
methanol	inst. (0.6)	inst. (0.101)	inst. (0.47)
	0.99 (0.25)	0.28 (0.340)	0.66 (0.53)
	7.7 (0.15)	3.2 (0.298)	
		15.3 (0.261)	
propylen carbonate	0.13 (0.74)	inst. (0.116)	0.19 (0.45)
	1.82 (0.26)	0.180 (0.429)	0.99 (0.38)
		2.03 (0.237)	2.6 (0.17)
		6.57 (0.218)	
chloroform	0.48 (0.53)	0.289 (0.356)	inst. (0.3)
	4.4 (0.47)	4.15 (0.644)	4.6 (0.7)

Tab.1. Time constants (in ps) for fits of selected kinetic traces to sums of exponentials compared to data for solvent relaxation in [3].

Fig. 4. Kinetic scheme for the intramolecular relaxation and charge-transfer reaction A->C of LDS-750 in solution

3. Discussion

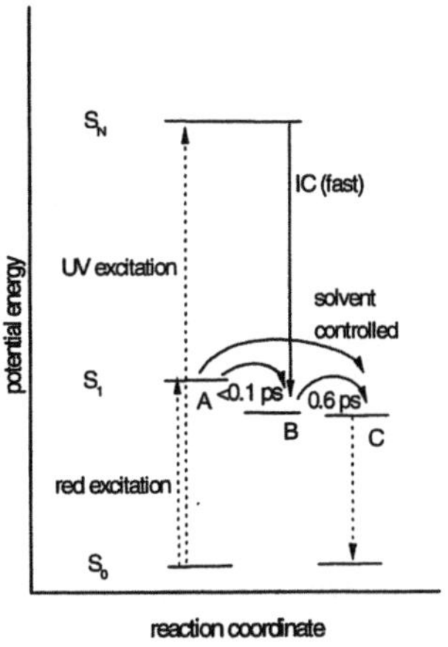

Large differences in width of the stationary absorption (4600 cm^{-1}) and emission (1490 cm^{-1}) spectra of LDS-750 in solution already indicate intramolecular changes upon photoexcitation. The decay of the primary emissive state A and the rise of the final emissive state C are compared in Table 1 to solvation dynamics [3]. In chloroform, the mean times for decay of the primary state and rise for the final state are both similar to the mean solvation time. This indicates that the CT-reaction A $\rightarrow$ C is solvent-controlled. In the other solvents, the mean lifetime for primary emission is essentially shorter than the mean solvation time and does not correspond to the mean risetime for the final emission. There are two conclusions for the more polar solvents: first, there must be a dark intermediate excited form B, second, the reaction A $\rightarrow$ B is faster than solvation. This implies a reaction route which bypasses solvation along an intramolecular coordinate (namely through B), possibly in the inverted regime for charge-transfer. By 226 nm photoexcitation higher electronic states (S_N) of the molecule are populated. From the fast increase of the stimulated emission in fig. 2 we conclude that the final state C is reached by subsequent internal conversion already at early time. The observed dynamics point also to B as precursor. We summarize the desactivation scheme in fig. 4.

References

1 S. J. Rosenthal, X. Xie, M. Du, G. R. Fleming, J. Chem. Phys. 95, 4715, 1991.
2 S. A. Kovalenko, N. P. Ernsting, J. Ruthmann, Chem. Phys. Lett. 258, 445, 1996.
3 M. L. Horng, J. A. Gardecki, A. Papazyan, M. Maroncelli, J. Phys. Chem. 99, 17311, 1995.

Optically Induced SHG as a Probe of Orientational Dynamics in Liquids

Shujie Lin, Ian D. Hands, David L. Andrews and Stephen R. Meech

School of Chemical Sciences, University of East Anglia, Norwich NR4 7TJ, UK

Abstract. We report the application of a resonant non-degenerate six wave mixing experiment to the measurement of ultrafast orientational dynamics in solution. Simultaneous excitation of the sample by a femtosecond pulse and its second harmonic leads to an anisotropic distribution of excited states, which supports SHG. It is shown that measurement of the SHG generated with a time delayed pulse yields a more detailed picture of orientational relaxation than is attainable from four wave mixing.

1. Introduction. The generation of second harmonic radiation from an isotropic medium is both intrinsically interesting and of some practical importance; the former because it demonstrates that higher order nonlinear optical interactions yield signals which are forbidden in lower order, and the latter because, as we will show, the time resolved signal contains unique information on the solution dynamics.

The generation of second harmonics through higher order interactions was predicted a number of years ago [1] and experimentally observed and characterized by Nunzi and co-workers.[2] Recently we extended these measurements to the ultrafast time domain [3] and presented a detailed quantum electrodynamical model of the six wave interaction.[4] In the following we report two further extensions of these studies. First, the original phase conjugate geometry is ill suited to ultrafast measurements, and a more practical geometry is described below. Secondly we will describe ultrafast time resolved measurements as a function of polarisation. It is shown that the orientational dynamics measured depend on the first and third moments of the orientational distribution function, i.e. the measurements truly probe orientational dynamics rather than the alignment dynamics obtained through four wave mixing or time resolved fluorescence.

2. Experimental Methods. The geometry used in our experiments is shown in figure 1. The output of a regeneratively amplified titanium sapphire laser was divided into three. Beams 1 (800 nm) and 2 (400 nm) were incident onto the sample simultaneously at time zero. These beams write the grating in the sample (see below). The subsequent orientational dynamics were monitored through the decay of the grating by recording the SHG signal (beam 4) generated from the delayed beam 3 (800 nm). The signal beam 4 was detected in the expected [5] quasi-phase matched direction, $\mathbf{k}_4(2\omega) = 2\mathbf{k}_3(\omega) + [\mathbf{k}_2(2\omega) - 2\mathbf{k}_1(\omega)]$.

All measurements were made on *ca* 10^{-3}M solutions of diethylamino nitrostilbene (DEANS) in non-polar solvents (the use of polar solvents complicates the excited state dynamics of this solute[3]). This dye has an

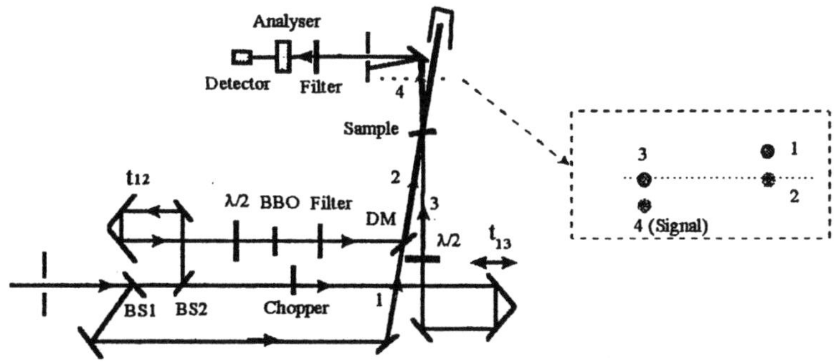

Fig. 1. Experimental layout for six wave mixing.

electronic transition at around 400 nm, and a large hyperpolarisablity, yielding a large signal. The samples were contained in a cell of pathlength 30 micron

3. **Results and Discussion.** Simultaneous irradiation of the sample at ω and 2ω leads to three contributions to the population of the excited state. The first two are simply the two and one photon transition probabilities arising from beams 1 and 2 respectively. These cannot support the generation of SHG since the orientational distribution of excited (or ground) states has inversion symmetry. The contribution of interest is that which results from excitation at both frequencies

$$P_{10} = C_3[\Delta\vec{\mu} \cdot \vec{E}_1(\omega)][\vec{\mu} \cdot \vec{E}_1(\omega)][\vec{\mu} \cdot \vec{E}_2(2\omega)] \cos[(\vec{k}_2(2\omega) - 2\vec{k}_1(\omega)) \cdot \vec{R} + \Delta\varphi_o] \quad (1)$$

In (1) μ is the transition dipole, Δμ the difference in permanent dipole moment on excitation and **R** the position. It is easy to see that (1) leads to an orientational grating. Clearly the excitation probability changes unsymmetrically as a function of molecular orientation. Thus the resultant distribution of excited (and ground) states does not have inversion symmetry, and so the medium has a net $\chi^{(2)}$. The medium $\chi^{(2)}$ is modulated by the grating term which leads to SHG in the phase match direction (Fig 1). We note that this is a modulation in the orientational distribution of the excited (and ground) states, rather than a population grating. The relationship between equation (1) and similar equations describing phase selective orientational excitation schemes[6] is discussed in detail elsewhere.[7]

The total second harmonic response of the medium can be calculated from

$$E_4(2\omega, t) \propto \int [\hat{\mu} \cdot \hat{e}_4(2\omega)][\hat{\mu} \cdot \hat{e}_3(\omega)]^2 \Delta\beta N_1 \rho(\theta, \phi, t) \sin\theta d\theta d\phi \quad (2)$$

where a time independent excited state population and hyperpolarisability have been assumed. The time dependence thus arises from the orientational distribution ρ(θ,φ,t), which has an initial condition defined by (1). This can be obtained from a solution to the Debye equation which, with the assumption of a single diffusion coefficient D for the rod-like DEANS molecule yields

$$\vec{E}_4'(2\omega,t) \propto \left[\frac{4}{7}e^{-12Dt} + 3e^{-2Dt}\right] \tag{3}$$

for the case of all four beams vertically polarised, and

$$\vec{E}_4''(2\omega,t) \propto \left[-\frac{2}{7}e^{-12Dt} + e^{-2Dt}\right] \tag{4}$$

for beams 1 and 2 vertically polarised, beam 3 at 45° and detection of the horizontal component of beam 4. The results of the experimental measurements under these conditions are displayed in figure 2 along with curves generated from equations (3) and (4). The only variable parameter is the value of 2D, which is accurately determined from exponential fits to data recorded over a 600 ps time window. The agreement between theory and experiment is very good, given the simplifying assumptions that have been made concerning the shape of DEANS.

Figure 2 contains the main result of this work. The two exponential terms in (3) and (4) which differ from each other by the factor of 6 have their origin in the first and third moments of the orientational distribution function. These moments are required for a description of the truly orientational dynamics of the solute molecules. They become particularly important in describing anisotropic media. These odd moments are not accessible through lower order, four wave mixing, transient grating experiments.

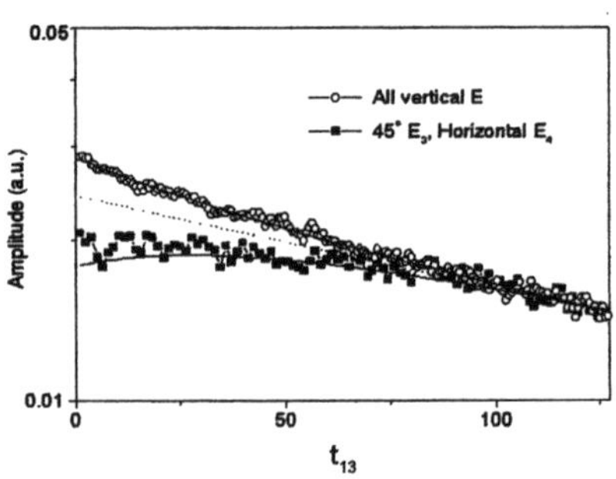

Fig.2. Polarisation dependence

4. Conclusions. The generation of second harmonic emission from an isotropic solution through six wave mixing has been described in terms of an orientational grating scattering mechanism. The results unambiguously show that the experiment is sensitive to the first and third moments of the orientational distribution function. Such odd higher moment are not available through lower order measurements.

Acknowledgements. We are grateful to EPSRC for financial support.

References

1 M. Ducloy, Applied Phys. Lett. , Vol. 46, 1020, 1985.
2 C. Fiorini, F. Charra, and J.-M. Nunzi, J. Opt. Soc. Am. Vol B11, 2347, 1994.
3 S. Lin, D.L Andrews, I. D. Hands, S. R. Meech, Chem. Phys. Lett. Vol. 285, 321, 1998.
4 I. D. Hands, S. Lin, D. L. Andrews, , S. R. Meech, J. Chem. Phys. in press.
5 S. Lin, I. D. Hands, D. L. Andrews, , S. R. Meech, Submitted to J. Phys. Chem.
6 M. Vrakking, S. Stolte, Chem. Phys. Lett., Vol. 271, 209, 1997.
7 I. D. Hands, et. al. in preparation.

Solvation Dynamics in Restricted Environments: Solvent Immobilization in Reverse Micelles

Nancy E. Levinger[1], Ruth E. Riter, Dale M. Willard, and Debi Pant

Chemistry Department, Colorado State University, Fort Collins, CO 80523-1872 USA
[1]E-mail: levinger@lamar.colostate.edu

Abstract. Polar solvation dynamics in various reverse micellar environments are investigated. The solvent is immobilized in the smallest micelles and becomes more mobile with increasing water content and micellar size.

1. Introduction

Ultrafast solvation dynamics have dramatically impacted our understanding of fast chemical reactions, such as electron transfer and isomerization[1]. Yet much chemistry occurs not in bulk homogeneous solution but in highly heterogeneous environments. The efficiency of biological reactions, such as electron and proton transfer are attributed, in part, to the highly-structured environments in which they occur. To learn how structured environments affect chemical reaction dynamics, we have launched a study of polar solvation dynamics in heterogeneous systems.

Reverse micelles, composed of polar solvent sequestered by surfactant in a nonpolar continuous phase, have been used to explore restricted environments. The molar ratio of polar solvent to surfactant, w_0 characterizes the reverse micelle size.

The goal of these studies is to investigate the solvation dynamics within reverse micelles and to determine the influence of the restricted environment. We have targeted two different systems employing the surfactants Aerosol OT (AOT) and lecithin. While many surfactants form reverse micelles, AOT is by far the most common[2, 3]. AOT reverse micelles are spherical and swell as they solubilize water. Lecithin reverse micelles are useful models for biological membranes[2]. Upon small additions of water, lecithin in cyclohexane forms reverse micelles that grow from discrete spherical droplets to an extended, entangled tubular structure, while lecithin reverse micelles in benzene are spherical.

2. Experimental Methods

Samples were prepared with high purity solvents, Milli-Q filtered, 18.2 $M\Omega/cm^2$ water, and HPLC grade formamide, isooctane (2,2,4-trimethylpentane), cyclohexane and benzene(Aldrich). Reverse micelles were created with AOT (sodium bis(2-ethylhexyl)sulfosuccinate, Aldrich) or lecithin (Avanti Polar Lipids). K^+ or Ca^{2+} counterion-exchanged AOT was prepared according to the literature[4]. Coumarin 343 (C343, Exciton) was the probe molecule.

The time-resolved fluorescence upconversion apparatus used for these studies has been previously described[5]. Briefly, the doubled output from an ultrafast mode-locked Ti:sapphire laser (410 nm) excites C343; the resulting fluorescence combines with the residual laser fundamental gate pulse in a BBO crystal a variable time later thus time-resolving the fluorescence. Pump and gate polarizations kept at the magic angle eliminate effects from rotational diffusion. The time resolution of the spectrometer is 100-200 fs precluding full resolution of the solvation dynamics inertial component. Time-resolved fluorescence spectra are reconstructed from fluorescence decays collected over a range of wavelengths across the emission spectrum. The time correlation function,

$$C(t) = \frac{\nu(t) - \nu(\infty)}{\nu(0) - \nu(\infty)} \tag{1}$$

is calculated from the peak position of the reconstructed fluorescence spectra.

3. Results and Discussion

Aerosol OT Reverse Micelles. We have measured solvation dynamics in various AOT reverse micellar samples as a function of hydration level, countercation, and intramicellar solvent, see Fig. 1. For the smallest reverse micelles, a ≥100ps decay component dominates the solvation dynamics, regardless of countercation or intramicellar solvent. This component is significantly longer than the dynamics of bulk water, which is complete within 2 ps[6]. Additionally, the fluorescence Stokes shift observed for the micellar solutions are an order of magnitude smaller than the ~1000cm^{-1} Stokes shift observed for C343 in bulk water. These results show that the intramicellar solvent is highly immobilized. As w_0 increases, an ultrafast relaxation component with a bulklike time constant is observed, the time scale for the second diffusive component decreases, and the overall Stokes shift increases, see Fig. 1, revealing some bulklike solvent in the micelles. We attribute the immobilization to solvent interactions with the surfactant head groups and the restrictive size of the micellar environment.

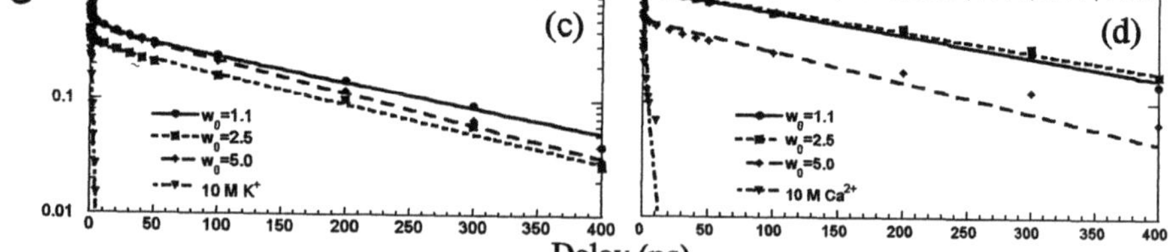

Fig. 1 C(t) for (a) water/Na-AOT; (b) formamide/Na-AOT; (c) water/K-AOT; and (d) water/Ca-AOT reverse micelles in isooctane.

Lecithin Reverse Micelles. We have probed water motion in lecithin reverse micelles, see Fig. 2. Small reverse micelles in cyclohexane, $w_0 \leq 4.8$ are spherical and bulklike water motion is absent. At $w_0 \geq 5.8$, the macroscopic phase changes from a nonviscous fluid to a viscous gel as the micellar tubules grow longer. While the dynamics are still dominated by long relaxation times, we also observe a small ultrafast component attributable to bulklike water.

The results for the tubular reverse micelles conflict with predictions for water at lecithin surfaces. Both IR and NMR spectroscopy show ≤3 water molecules interact strongly with the lecithin head group. Thus, a significant bulklike water population should exist at hydration levels $w \geq 4$. We attribute these results to significant lecithin head group hydration that effectively immobilizes water.

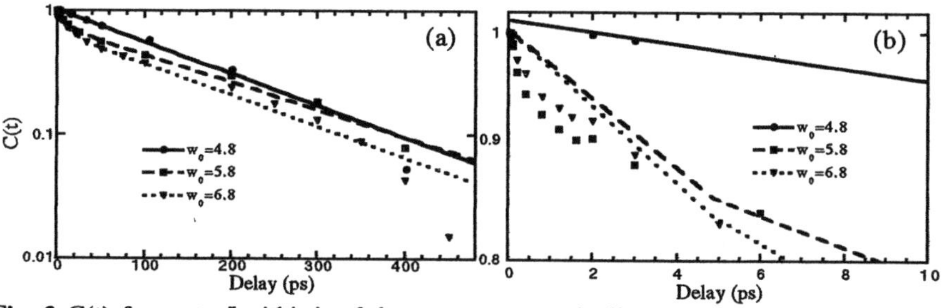

Fig. 2 C(t) for water/lecithin/cyclohexane reverse micelles: (a) long and (b) short time behavior.

Contrast of Reverse Micellar Systems. Since water should interact more strongly with the ionic AOT head group than with the zwitterionic lecithin head group, we anticipated that the water motion in lecithin would be faster than it is in AOT reverse micelles. However, our solvation dynamics experiments showed exactly the opposite! In AOT reverse micelles with $w_0=5$, we observed a substantial bulklike relaxation component, while lecithin reverse micelles with $w_0=4.8$ exhibited only dynamics on the ≥100ps time scale. We ascribe this result to notably higher lecithin head group hydration in the tubular micelle form.

Acknowledgements. This work was supported by the Petroleum Research Fund of the American Chemical Society and the US National Science Foundation.

References

1 RM Stratt and M Maroncelli, *J. Phys. Chem.*, **100**, 12981, 1996.
2 *Reverse Micelles: Biological and Technological Relevance of Amphiphilic Structures in Apolar Media*, edited by PL Luisi and BE Straub, Plenum Press, New York, 1984.
3 T De and A Maitra, *Adv. Colloid Interface Sci.*, **59**, 95, 1995.
4 J Eastoe, TF Towey, BH Robinson, J Williams and RK Heenan, *J. Phys. Chem.*, **97**, 1459, 1993.
5 RE Riter, DM Willard and NE Levinger, *J. Phys. Chem. B*, **102**, 2705, 1998.
6 R Jimenez, GR Fleming, PV Kumar and M Maroncelli, *Nature*, **369**, 471, 1994.

Photon Echo Studies of Solute-Solvent Interaction Dynamics in Polymer Glasses

Yutaka Nagasawa[1], Jae-Young Yu, and Graham R. Fleming[2]

Department of Chemistry and the James Franck Institute
The University of Chicago, Chicago, Illinois 60637

Present address:
[1] Department of Chemistry, Graduate School of Engineering Science, 1-3 Machikaneyama Toyonaka, Osaka 560, Japan
E-mail: nagasawa@chem.es.osaka-u.ac.jp
[2] Department of Chemistry, B84 Hildebrand MC1460, University of California, Berkeley, Berkeley, CA 94720-1460
E-mail: Fleming@CChem.Berkeley.EDU

Abstract The influence of coupling strength and temperature on the solute-solvent interaction of two chromophores with two different polymer glasses are reported.

The dynamical role of solvents in chemical and biological processes has been studied extensively in recent years. A universal characteristic appears to be the clear separation of the relaxation into ultrafast femtosecond inertial response and a slower diffusive relaxation occurring on the picosecond timescale. The three pulse photon echo peak shift (3PEPS) measurement has recently been developed into a powerful tool for the study of chromophore-solvent interactions [1]. It is known that photon echo exhibits drastic solute dependencies. For a dye, cresyl violet, in polymethylmethacrylate (PMMA), Ippen and coworkers reported that the peak shift increased from 0 to 30 fs when the temperature was decreased from 290 to 15 K while that of Nile blue and oxazine 720 always remained at zero [2]. Shank and coworkers also found a significant decrease in the photon echo decay rate of LD690 in PMMA at lower temperatures while that of LDS750 did not change [3]. Our own group used IR144 and found an ultrafast Gaussian component (~100 fs) with varying amplitude in all the studied solvents [1, 4] and it appears to be a severely damped oscillation in a polymer glass [5]. Wiersma and coworkers also found Gaussian component with DTTCI but time constants were 1.5-2 times longer than ours [6]. These studies give us the opportunity to investigate the extent to which the probe molecule contributes to this difference.

In our previous study of IR144 in PMMA we showed by simulation that probe molecules with different coupling strength can exhibit different temperature dependencies [5]. For example, the echo signal itself becomes wider and the long population time ("asymptotic") peak shift becomes larger as temperature is lowered and these effects are more significant for molecules with weaker coupling strength.

Here we confirm this prediction and clarify number of coupling strength dependent aspects of echo spectroscopy. The samples investigated are IR144 or DTTCI doped PMMA and polyvinylformal (PVF). The advantage of using polymers as solvents is that the diffusive portion of the response is so slow that it can be treated as static inhomogeneity. Thus it becomes easier to focus on the role of solvent inertial motion in the nonlinear spectroscopy.

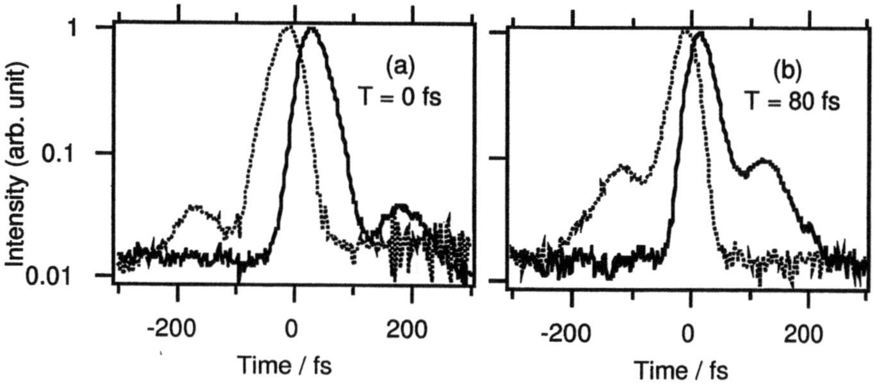

Figure 1. 3 pulse photon echo signal of DTTCI in PMMA at 31 K. Population time, T, is set to (a) 0 fs and (b) 80 fs.

Our main experimental results are: (1) The absorption spectrum of IR144 in PMMA is considerably wider than DTTCI, about 1900 and 960 cm^{-1} for IR144 and DTTCI, respectively. (2) In the 3 pulse photon echo signals of DTTCI at 31 K (Figure 1), an intramolecular vibrational quantum beat can be observed. On the other hand, echo signals of IR144 decay faster and no quantum beat can be found at any temperature [5]. (3) The 3PEPS signal of IR144 as a function of population time (Figure 2) exhibits an ultrafast decay while that of DTTCI is dominated by strong quantum beats. (4) The temperature dependencies of the echo peak shift with population time (the delay between second and third pulse) fixed to 120 ps are shown in Figure 3.

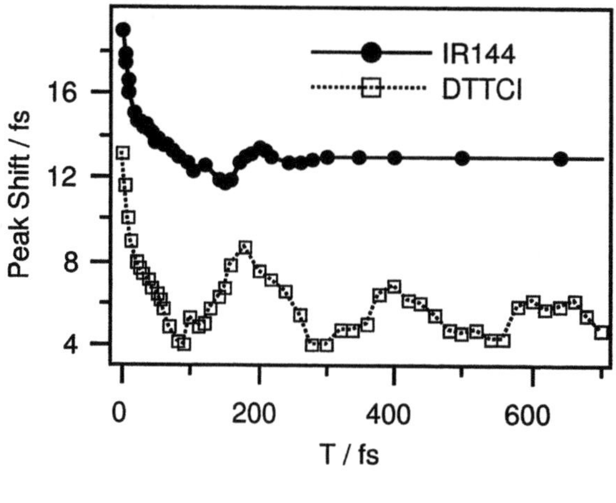

Figure 2. 3PEPS signals of IR144 and DTTCI in PMMA. IR144 signal is shifted 9 fs upward.

It can be seen that the slope of DTTCI/PMMA is much steeper than IR144/PVF. FWHM of the absorption spectrum of IR144/PVF (1790 cm^{-1}) is approximately the same to that of IR144/PMMA.

From these results we can say that: (1) Different width of absorption spectra indicates that IR144 is more sensitive to the inhomogeneity or strongly coupled to the solvent dynamics. (2) From the slower echo decay of DTTCI with quantum beats, it can be concluded that the electronic dephasing time is longer for DTTCI and IR144 is more strongly coupled to the system dynamics. (3) The appearance of large quantum beats in 3PEPS shows that electronic transition of DTTCI is mainly coupled to the intramolecular vibrations. (4) The steeper temperature dependent slope of the DTTCI/PMMA peak shift confirms the theoretical prediction that peak shift will show stronger temperature dependence for molecules with smaller coupling strength [5]. The values of reorganization energy and inhomogeneous bandwidth obtained from the analysis are 339 and 315 cm^{-1} for IR144/PVF and 138 and 190 cm^{-1} for DTTCI/PMMA, respectively. IR144 is more sensitive to the inhomogeneity or strongly coupled to the solvent dynamics.

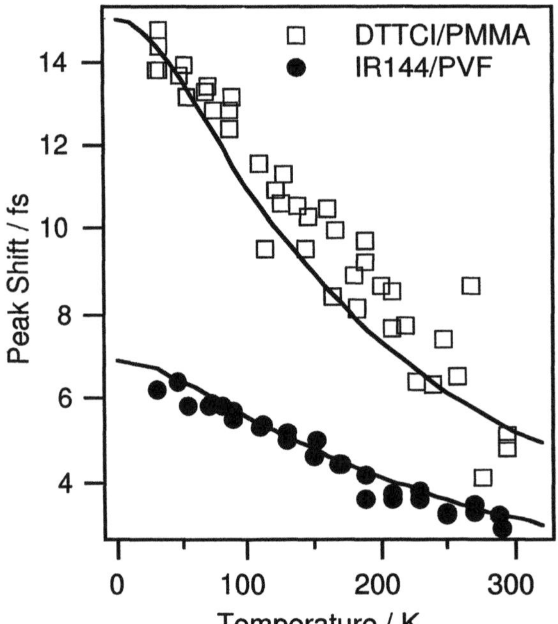

Figure 3. Temperature dependencies of peak shift with population time set to T = 120 ps.

[1] T. Joo, et al, J. Chem. Phys. **104**, 6089 (1996).
[2] A. M. Weiner, et al., J. Opt. Soc. Am. B **2**, 654 (1985).
[3] C. J. Bardeen, et al., Chem. Phys. Lett. **280**, 127 (1997).
[4] S. A. Passino, et at., J. Phys. Chem. **101**, 725 (1997).
[5] Y. Nagasawa, et al., J. Chem. Phys. **106**, 4840 (1997).
[6] W. P. de Boeij, et al., J. Phys. Chem. **100**, 11806 (1996).

Ultrafast measurement of microscopic chemical environments using high intensity chirped pulses

Jeff A. Squier[1], Chris Bardeen[1], Vladislav V. Yakovlev[1], Kent R. Wilson[1], Arjan Buist[2], G. J. Brakenhoff[2]

[1]University of California, San Diego, LaJolla, CA 92093-0339 USA
[2]BioCentrum Amsterdam, Institute for Molecular Cell Biology, Kruislaan 316, 1098 SM Amsterdam, The Netherlands.

Abstract. High intensity chirped pulses are used to detect the pH of the environment of fluorophores, with possible application to confocal microscopy of living cells.

In order to discover and understand the biological functioning of living systems in terms of the underlying biochemical and biological dynamics, one would like to measure the time evolution of the three dimensional microscopic chemical and physical nature of the sample. In this paper we discuss an example of a new kind of optical diagnostic which uses high intensity chirped pulses [1-4] to interrogate, using femtosecond time scale pulses, a reporter molecule with respect to the pH of its molecular surroundings. We use a molecule, SNAFL2, whose fluorescence is known to be pH dependent, in solutions of differing pH, pump it with intense broadband chirped pulses and measure the ratio of fluorescence intensity for positive and negative chirp. This ratio is seen to be strongly pH dependent as shown in Figure 1.

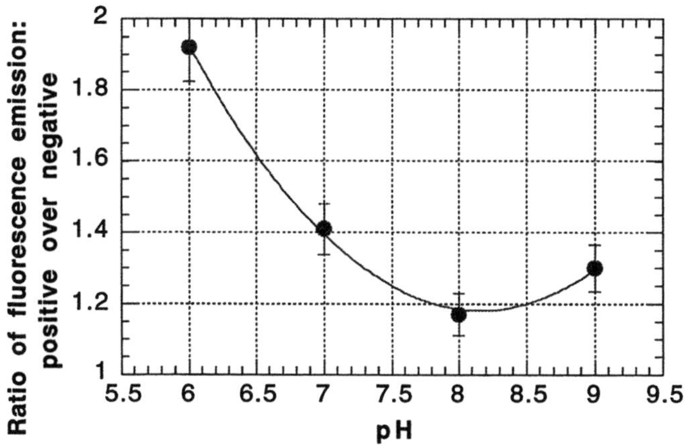

Figure 1. Ratio of fluorescence emission between positively and negatively chirped excitation pulses as function of pH.

These measurements are based on the recently developed general techniques which use high intensity chirped pulses [1-4] to produce an intra-pulse pump-dump effect which is dependent of the sign of chirp. Positively chirped pulses produce a higher fluorescence emission than negatively chirped pulses at equal excitation power. Notably, this effect decreases as the level of saturation decreases. Thus, the ratio plotted in Figure 1, will also diminish to 1 with a lower degree of saturation. The level of saturation could vary, in this case for instance, due to the change in absorption cross section as a function of pH. In this case however, the higher pH values were actually saturated more heavily than the lower pH values. This indicates that the functional dependence measured in Figure 1 is not simply due to a change in the degree of saturation resulting from a shift in the absorption cross section with pH.

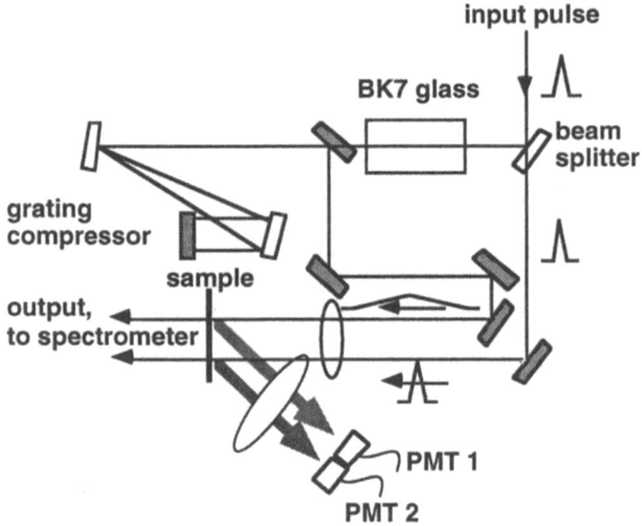

Figure 2. System layout

The apparatus used to measure the fluorescence emission as a function of pulse chirp is shown in Figure 2. For SNAFL2 a pump pulse centered at 555 nm is generated from a kilohertz optical parameteric amplifier (OPA) system. The pulses produced from the OPA are split: 50% travels directly to the sample and 50% of the light travels through a variable path length of BK7 glass. The pulses which are dispersed by the BK7 develop a significant positive chirp. These pulses are then directed through a 300 l/mm grating pair arranged in a parallel, double-pass geometry. This combination of glass and grating compressor enables the dispersion of the pump pulse to be conveniently varied from positive chirp to negative chirp. The chirped pulse and the unchirped pulse are focused onto separate spots within an isotropic sample. The fluorescence emission from the chirped pump pulse volume, and the transform limited pump pulse volume are separately imaged, and collected by PMT 1 and 2 respectively. The individual PMT signals are integrated, than ratioed and averaged. This detection scheme thus corrects for large shot-to-shot energy fluctuations in the excitation pulse, enabling an accurate measurement of the functional dependence of the fluorescence emission as the chirp is varied The excitation pulses are of sufficient energy to saturate the SNAFL2 at all pH values, but the intensity is carefully monitored (by measuring the spectrum of the transmitted excitation pulse) to ensure that nonlinear effects such as self phase modulation are negligible. When the SNAFL2 is excited with a low degree of saturation, the fluorescence emission is essentially independent of chirp. As the degree of saturation is increased, the situation changes quite drastically -- the fluorescence emission demonstrates a strong dependence of chirp (Figure 3). In fact, the fluorescence emission effectively doubles, in this example, as the dispersion is changed from -10,000 fs^2 to +10,000 fs^2. A series of these curves is then generated for SNAFL2 dissolved in solution with pH ranging from 6 to 9. This results in the pH dependence shown in Figure 1.

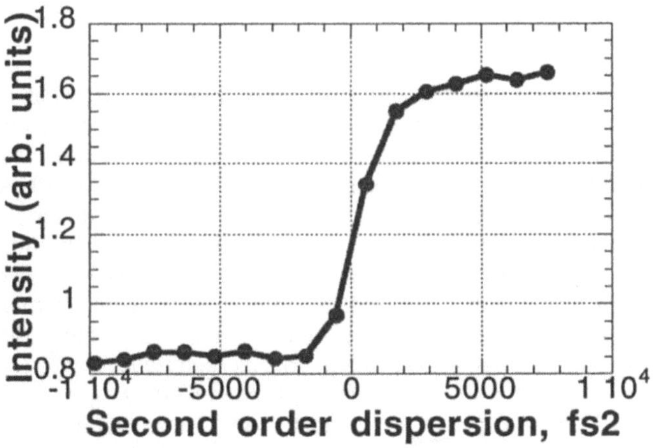

Figure 3. SNAFL2 fluorescence emission as function of chirp in pH 6 solution.

Due to the robust nature of this effect, its application to imaging in confocal microscopy is now being pursued. It is easy to envision, based on Figure 1, how ratio imaging techniques could be applied to make microscopic images which are sensitive to the chemical environment of the reporter molecule.

References

1. J. Cao, C. J. Bardeen and K. R. Wilson, Phys. Rev. Lett. [In Press]

2. V. V. Yakovlev, C. J. Bardeen, J. Che, J. Cao and K. R. Wilson, J. Chem. Phys. [In Press]

3. S. Ruhman and R. Kosloff, J. Opt. Soc. Am. **B7**, 1748 (1990)

4. G. Cerullo, C. J. Bardeen, Q. Wang and C. V. Shank, Chem. Phys. Lett. **262**, 363 (1996)

Femtosecond mid-infrared pump-probe spectroscopy of water and ice

Sander Woutersen, Uli Emmerichs and Huib J. Bakker

FOM-Institute Amolf, Kruislaan 407, 1098 SJ Amsterdam, The Netherlands
E-mail: s.woutersen@amolf.nl

Abstract. We have used femtosecond mid-infrared pump-probe spectroscopy on the OH-stretching mode of HDO dissolved in D_2O to investigate both the orientational and vibrational dynamics of water and ice. The orientational data provide evidence for a two-component structure of liquid water. The vibrational lifetime of the OH-stretch mode shows a remarkable temperature dependence: in water it is twice as long as in ice, and is found to *increase* with temperature.

Knowledge about the orientational and vibrational dynamics of water is essential for understanding the (bio)chemical and physical processes that take place in this liquid, notably chemical reactions and solvation. In previous studies, it was found that the orientational relaxation in water occurs both on a slow (τ_R~10 ps) and fast (τ_R~0.2-2 ps) time scale. However, all techniques employed to date probe the orientational dynamics averaged over all molecules in the liquid, and the interpretation of the results was difficult. We have studied the orientational dynamics of liquid water with a different technique: polarization-resolved pump-probe spectroscopy on the OH-stretching mode of HDO molecules dissolved in D_2O. In this technique, an IR pump pulse, which is tuned to the ν_{OH}=0→1 frequency, creates an anisotropic orientational distribution of vibrationally excited OH groups, with a maximum in the direction of the polarization of the pump pulse. Due to the random motion of the water molecules, the anisotropy rapidly decays. This decay, which directly mirrors the orientational motion of the OH groups, is monitored by a weak, delayed probing. Our laser setup generates intense fs pulses (200 fs, 30 μJ) in the 3000-3600 cm^{-1} frequency region[1]. With this setup, both one- and two-color pump-probe experiments have been performed.

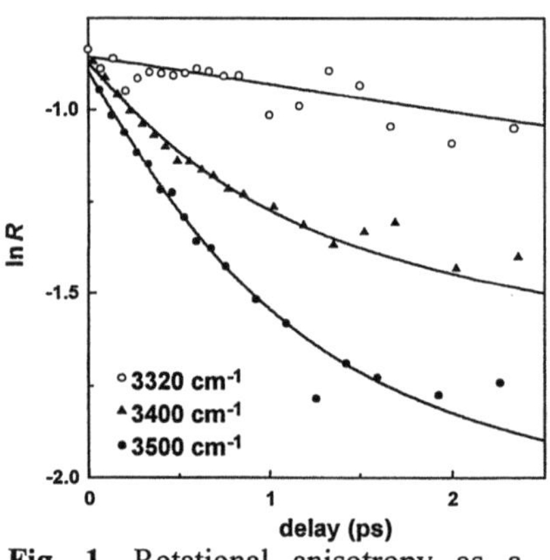

Fig. 1. Rotational anisotropy as a function of delay at three different excitation frequencies.

We have measured the decay of the orientational anisotropy at three excitation frequencies in the broad OH-stretch absorption band[2]. In these experiments, the probe was obtained by splitting off a small fraction of the pump pulse. The results are shown in Figure 1. Both at the center and at the high-frequency side of the absorption band, the decay occurs in a bi-exponential manner, with time constants of 13 ps (already observed at the low-frequency side) and 0.7 ps. There is a strong frequency dependence of the relative contribution of the two time scales. This means that the two time scales correspond to different species of water: strongly H-bonded molecules that can only reorient via the slow process and weakly H-bonded molecules that reorient via the fast process. This identification of the two time scales with two different species of water illustrates the advantage of polarization-resolved pump-probe spectroscopy. The frequency tunability of the IR pulses makes it possible to excite subensembles of the water molecules and to investigate their specific orientational dynamics.

Since the temperature dependence of vibrational lifetimes in the condensed phase can supply important information concerning the vibrational relaxation mechanism, we investigated the temperature dependence of the lifetime T_1(OH) of liquid water and ice using two-color pump-probe spectroscopy[3]. In these experiments, the probe is set to the $\nu_{OH}=1\rightarrow2$ frequency, and thereby directly monitors the excited state ($\nu_{OH}=1$) population. By measuring the probe absorption as a function of the delay between pump and probe, we determine the vibrational lifetime T_1(OH). The probe polarization is at the magic angle with respect to the pump, ensuring that the observed transients are not affected by orientational relaxation.

Fig. 2 presents typical pump-probe scans in ice and water. These scans show that in water the vibrational relaxation occurs significantly *slower* at high temperature than at low temperature. Fig. 2 also shows that the vibrational lifetime is approximately *twice as short* in ice as in water. The lifetimes were determined by fitting mono-exponential decays to the pump-probe scans. Since we observed that for small values of the delay the $\nu_{OH}=0\rightarrow1$ signal is influenced by spectral

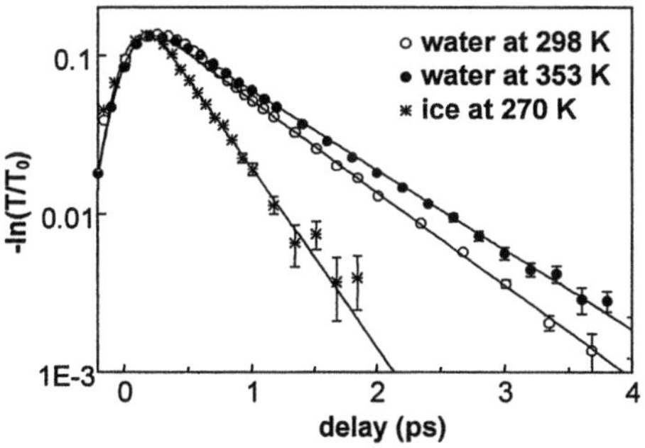

Fig. 2. Pump-probe scans recorded in dilute $HDO:D_2O$ in the liquid (ν_{pu}=3400, ν_{pr}=3150 cm^{-1}) and solid (ν_{pu}=3330, ν_{pr}=3090 cm^{-1}) phase, showing the absorption of the probe pulse as a function of delay between pump and probe.

diffusion, and since it cannot be excluded that the spectral diffusion influences the $\nu_{OH}=1\rightarrow2$ signal as well, only the data points for delay values larger than 1 ps were used in the fits on liquid water. Fig. 3 presents T_1 as a function of temperature. In ice, no significant temperature dependence is observed all the way from 30 K up to the melting point. Note that even for undercooled water at 270 K the value of T_1 is two times larger than for ice of the same temperature. In liquid water, the vibrational lifetime *increases* with increasing temperature. To our knowledge, up to now such anomalous temperature dependence of a vibrational T_1 has been observed in only very few other systems.

The decrease of the density with temperature in liquid water could in principle lead to an increasing T_1[4], but the density is clearly not an important parameter, since the T_1 of liquid water is much longer than that of the less dense ice. Rather, we think the anomalous temperature dependence is caused by the fact that in water the H-bond acts as an accepting mode. This has two important consequences. First, the H-bond potential is strongly anharmonic, and therefore the generally made assumption that the accepting modes are harmonic oscillators, does not hold. In fact, the H-bond potential is such that increased occupation of the H-bond mode is not expected to result in a change of T_1(OH)[3]. This would explain the absence of

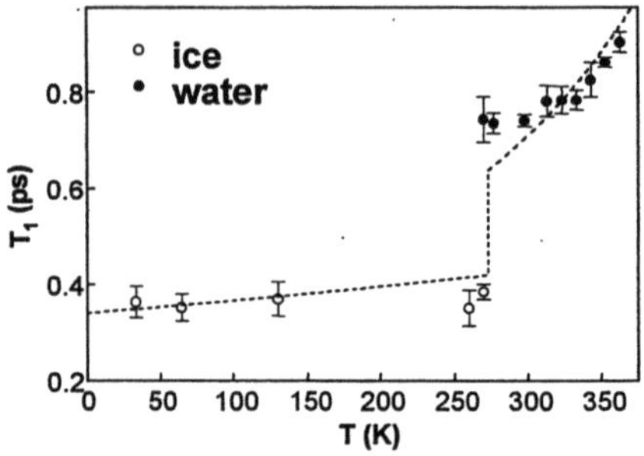

Fig. 3. Vibrational lifetime T_1(OH) in dilute HDO:D_2O as a function of temperature. The dashed curve has been calculated using a power law dependence of T_1 on the H-bond induced redshift of the OH-stretch frequency: $T_1 \propto (\delta\nu_{OH})^{-1.8}$[5].

temperature dependence observed in ice. Second, it is well-known that the average H-bond strength decreases with increasing temperature (especially in liquid water). This decrease will lead to a decrease in the anharmonic coupling between the OH-stretch mode and the H-bond mode, and hence to an increase of T_1(OH) with temperature. We found that the data can be well described with a model[5] which assumes that in the relaxation process all vibrational energy is transferred from the OH-stretch to the H-bond mode, and which gives T_1(OH) in terms of the H-bond induced redshift of the OH-stretch frequency. This redshift is known experimentally as a function of temperature, and hence T_1(OH) can be calculated as a function of temperature (apart from a scaling factor). We find good agreement between theory and data over the entire temperature range from 30 to 363 K.

Acknowledgements. The work described in this article is supported by the Stichting FOM, which is financially supported by NWO. The authors would like to thank Hinco Schoenmaker for technical support.

References

1 U. Emmerichs, S. Woutersen, and H. J. Bakker, J. Opt. Soc. Am. B **14**, 1480 (1997).
2 S. Woutersen, U. Emmerichs, and H. J. Bakker, Science **278**, 658 (1997).
3 S. Woutersen, U. Emmerichs, H.-K. Nienhuys, and H. J. Bakker, to appear in Phys. Rev. Lett. **81** (1998).
4 P. Moore, A. Tokmakoff, T. Keyes, and M. D. Fayer, J. Chem. Phys. **103**, 3325 (1995).
5 A. Staib and J. T. Hynes, Chem. Phys. Lett. **204**, 197 (1993).

Time-Resolved THz Studies of Liquid Dynamics

D. S. Venables and C. A. Schmuttenmaer

Yale University, Chemistry Department, Department, 225 Prospect St, New Haven, CT 06511, U.S.A.
E-mail: charles.schmuttenmaer@yale.edu

Abstract. Low frequency, intermolecular solvent dynamics in the vicinity of a photoexcited dye molecule have been probed directly with time-resolved THz pulse spectroscopy. We compare and contrast two distinct dye / solvent systems: Betaine-30 in chloroform and TBNC in toluene. The transient process for the former is dominated by an electronic response of the photoexcited dye molecule, whereas the latter does not display the polarization dependence expected for an electronic response.

1. Introduction

We report the low frequency intermolecular response of two polar solute / solvent systems upon photoexcitation of the solute molecule. Unlike typical time resolved laser studies, the spectrum of both the *solvent* and the solute is monitored. Two dye / solvent systems have been chosen for study: Betaine-30 [2,6-diphenyl-4-(triphenyl-1-pyridinio) phenolate] in chloroform, and TBNC (2,11,20,29-Tetra-*tert*-butyl-2,3-naphthalocyanine) in toluene.

The generation of FIR pulses with sub-picosecond duration is a relatively new development,[1] and the full capability of this technology has only begun to be tapped. The spectral content of these pulses extends from 3 to 200 cm^{-1} (0.09 to 6 THz), which allows the FIR spectrum of a sample to be measured with previously unattainable temporal resolution. Low frequency intermolecular dynamics and solvation in liquids can now be studied directly in the time domain.[2]

2. Experimental Methods

A 1 kHz regeneratively amplified Ti:Sapphire laser (Spectra Physics) is used to generate and detect the THz pulses, and also serves as the photoexcitation source. The THz pulses are generated by optical rectification of the visible pulse by a 1.0 mm thick ZnTe (110) single crystal,[3] and are detected *via* free space electro-optic sampling (FSEOS) in a separate 1.0 mm thick ZnTe (110) single crystal.[4] The THz pulse amplitude is 1000 times greater than the baseline noise, as measured with lockin detection with a 10 ms time constant. About 60 μJ/pulse is used to photoexcite the dye molecules. Typical transient absorption data are presented in parts (b) and (c) of Figure 1. These data are collected using a 30 ms lockin time constant, and requires 8 seconds to cover between 25 and 50 ps of

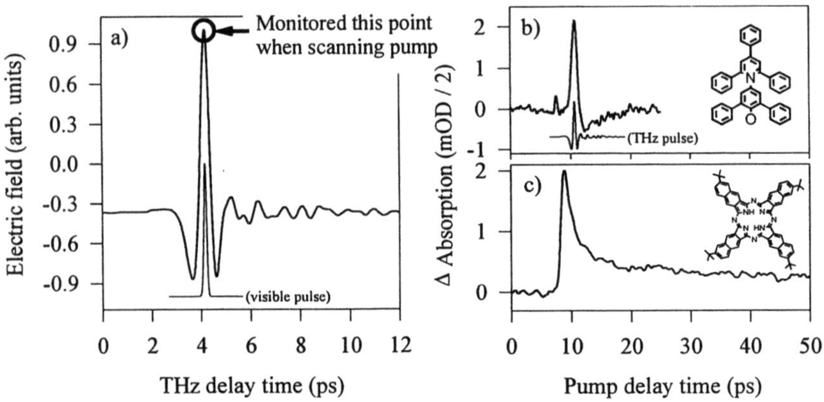

Figure 1. Part (a): Time-domain trace of THz probe pulse. The visible pulse used to generate it is shown for reference. Part (b): Photoinduced change in absorption of the Betaine-30 / chloroform system. The visible pump pulse arrives at a pump delay time of roughly 11 ps. The THz pulse from part (a) is shown for reference. Part (c): Same as (b), but for the TBNC / toluene system.

delay time in a 1024 point scan. The data presented here are the average of 100 to 150 scans. The dye concentration is about 0.03 moles/L.

3. Results and Discussion

The dye / solvent combinations chosen provide two distinct behaviors upon photoexcitation. The Betaine-30 molecule undergoes a 9 Debye change in dipole moment directed roughly along its symmetry axis, whereas the TBNC molecule changes from having excess negative charge located on the central porphyrin ring to a more delocalized charge distribution. In addition, the excited state lifetime of the Betaine-30 is on the order of 5 ps, leading to a convolution of lifetime effects with the change in the FIR absorption properties.

Part (a) of Figure 1 shows the THz probe pulse. The group velocity mismatch[5] between the visible and THz pulses in both the transmitter and receiver broadens the THz pulse relative to the 100 fs visible pulse(also shown the figure). The average transient FIR response shown in parts (b) and (c) is obtained by monitoring the highest amplitude point of the THz waveform. The magnitude of the transient absorption varies linearly with pump laser fluence. We have also recorded the full THz waveform at a variety of pump-probe delay times in order to describe the dynamics over the entire range of bandwidth contained in the FIR pulses.

In the Betaine-30 / chloroform system, the primary response is a narrow transient absorption (which is not pulsewidth-limited), and is followed by a slower and weaker bleach signal. In Figure 2, a polarization dependence study shows that the transient absorption feature is dominated by an instantaneous electronic response. The slower bleach component appears to have a weaker dependence on

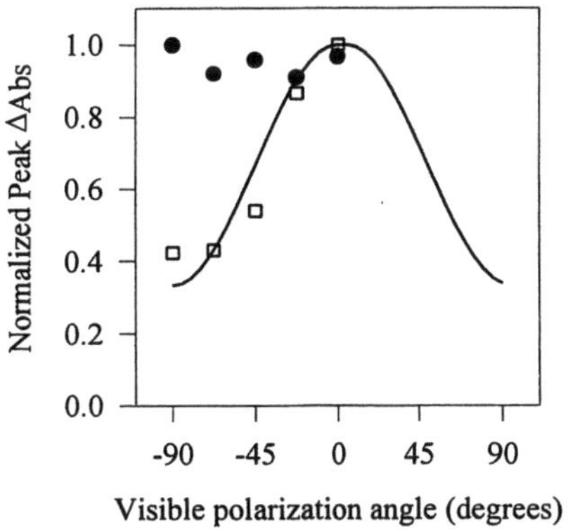

Figure 2. Dependence of the THz absorption on angle between THz and visible polarization. A calculation analogous to fluorescence depolarization ratio is shown with the solid line. The measured polarization dependence for the Betaine-30 system is shown with the hollow squares, and that for the TBNC system is shown with the solid circles.

polarization (not shown), indicating a thermal and / or reorientational response in addition to the electronic response, but a higher signal-to-noise ratio is needed before it can be interpreted quantitatively.

The TBNC / toluene system displays a sharp transient absorption feature with a 2 ps decay time constant, followed by a much longer bi-exponential recovery process. The fast component has a 50 ps time constant, and the slow component has a 700 ps time constant. In marked contrast to the Betaine-30 / chloroform system, the entire curve is independent of the relative polarization of the visible pump and THz probe beams, as shown in Figure 2. This indicates that instantaneous electronic processes do not contribute to the FIR absorption of this system in the same manner as in Betaine-30 / chloroform system.

4. Conclusions

We find that the FIR transient response of Betaine-30 / chloroform is dominated by an electronic process, whereas the response for TBNC / toluene has no polarization dependence. This could possibly arise from the absence of an electronic response. Future work will focus on the reasons for this difference.

Acknowledgments. This work was partially supported by the National Science Foundation, the Camille and Henry Dreyfus Foundation, and the Petroleum Research Fund, administered by the American Chemical Society.

References

1 J N. Katzenellenbogen and D. Grischkowsky, Appl. Phys. Lett. **58**, 222, 1991; X.-C. Zhang, B. B. Hu, J. T. Darrow, and D. H. Auston, Appl. Phys. Lett. **56**, 1011, 1990.

2 G. Haran, W. D. Sun, K. Wynne, and R. M. Hochstrasser, Chem. Phys. Lett. **274**, 365, 1997; R. McElroy and K.Wynne, Phys. Rev. Lett. **79**, 3078, 1997.

3 A. Rice, Y. Jin, F. Ma, and X.-C. Zhang, Appl. Phys. Lett. **64**, 1324, 1994.

4 Z. G. Lu, P. Campbell, and X. -C. Zhang, Appl. Phys. Lett. **71**, 593, 1997.

5 A. Nahata, A. S. Weling, and T. F. Heinz, Appl. Phys. Lett. **69**, 2321, 1996; Q. Wu and X.-C. Zhang, Appl. Phys. Lett. **70**, 1784, 1997.

THz Reflection Spectroscopy of $H_2O(l)$ and $D_2O(l)$

Cecilie Rønne and Søren R. Keiding

Chemistry Dept., Aarhus University, Langelandsgade 140, DK-8000 Århus C, Denmark
E-mail: keiding@kemi.aau.dk

Abstract. Using THz-spectroscopy we study the dielectric response of $H_2O(l)$ and $D_2O(l)$ between 0.1 and 2 THz from 270 to 363K. The response is modeled using a Debye model with a fast and slow decay time. Our results confirm that the temperature dependence of the slow decay time may be described by the recently reintroduced two-component model, with a critical point at 228K for H_2O and 235K for D_2O. As a consequence of this model, shifting the temperature scale for the slow relaxation time of D_2O by 7.2K, corresponding to the difference in temperature of maximum density, yields identical relaxation for D_2O and H_2O. This indicates a structural cause for the slow process.

1. Introduction

We present an experimental investigation on the temperature dependence of the dielectric properties of $H_2O(l)$ and $D_2O(l)$. Using THz time-domain spectroscopy (THz-TDS), we studied the decay of the induced polarization with an ultrashort far-infrared pulse, whose bandwidth enabled observation of a fast (τ_2<300fs) relaxation time in addition to the well-known Debye relaxation time (τ_D>2ps). The results were analyzed in view of the recently reintroduced two-component model for liquid water. In 1976 Speedy and Angell observed [1] that the temperature dependence of several anomalous thermodynamic properties of $H_2O(l)$ could be extrapolated from a critical point at 228K. This caused renewed interest in two-component models where water is considered to consist of a temperature-dependent equilibrium between two liquid components. Each component possesses distinct structural properties; one is a strongly hydrogen-bonded low density liquid (LDL); the other is a weakly hydrogen-bonded high density liquid (HDL). Many anomalous properties of liquid water, e.g. the density maximum occurring in H_2O (D_2O) at 277.15K (284.35K) can be reproduced by a two-component model. Recent experiments [2,3,4] and molecular dynamic simulations [5,6,7] provide evidence that water may be described as a two-component liquid. In the present work, we show that the effects of temperature and isotope substitution on the Debye relaxation of liquid water can be interpreted within the two-component model.

2. Experimental Method and Results

We use reflection mode THz-TDS, to obtain the complex dielectric constant of $D_2O(l)$ and $H_2O(l)$ in the region from 0.1 to 2 THz at temperatures ranging from

271 to 363K. Reflection mode THz-TDS [8] uses the reflection of ultrashort THz-pulses (Fig. 1a) from a Si-liquid interface to obtain the complex Fresnel reflection coefficient of the interface from which the linear dielectric function of the liquid can be extracted. The complex dielectric function is fit to a Debye model including two relaxation times [9], a slow (τ_D) and a fast (τ_2), $\hat{\varepsilon}(\omega) = \varepsilon_\infty + (\varepsilon_s - \varepsilon_1)/(1+i\omega\tau_D) + (\varepsilon_1 - \varepsilon_\infty)/(1+i\omega\tau_2)$. The amplitude of each of the relaxation modes is given by $\varepsilon_s - \varepsilon_1$ for the Debye process (τ_D) and $\varepsilon_1 - \varepsilon_\infty$ for the fast process (τ_2). The temperature dependence of τ_D for H_2O(l) and D_2O(l) is shown in Fig. 1b.

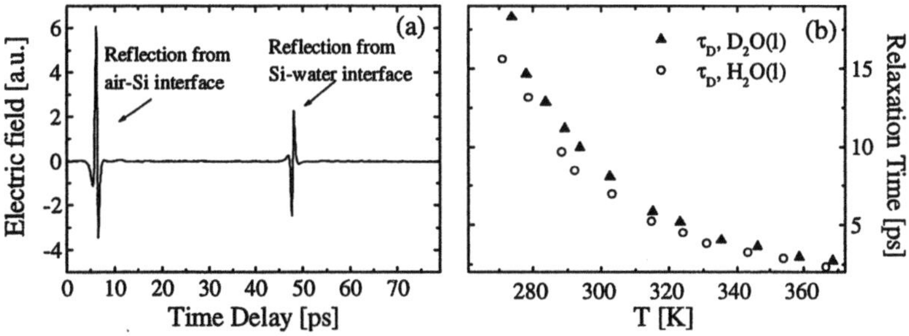

Fig. 1. a) A short THz-pulse reflected from the window enclosing the liquid is partitioned into a pulse reflected from the front of the Si-window and a delayed pulse reflected from the Si-liquid interface. Measurement of the frequency-dependent change in phase and amplitude of the second pulse with respect to the first pulse permits the dielectric constant of water to be obtained over the entire 0.1 to 2 THz bandwidth.
b) Comparison of the Debye relaxation time in H_2O(l) and D_2O(l) versus temperature.

3. Discussion

In the two-component model the temperature dependent mole fraction, f(T), describes the partition in liquid water between the LDL (f) and HDL (1-f) water structures. At the density maximum the mole fractions of H_2O(l) and D_2O(l) is assumed [10] to be equal, f(H_2O, 277.2K)=f(D_2O, 284.4K). Accordingly, if the density of H_2O(l) at temperature T are compared with those of D_2O(l) at temperature T-7.2K the intermolecular structure is identical for the two isotopes. Thus, all properties that solely depend on the liquid structure may be described with this temperature shift. Note, that this it not the case for a dynamic process that involve only one liquid component. In Fig. 2a the temperature shift is applied to the Debye relaxation times, and we find that the relaxation times are equal for the isotopes. We suggest that the Debye process involve both liquid components.

A similar comparison can be made between the fast relaxation times, but contrary to the Debye relaxation times, we do *not* find correlation between the fast relaxation times on the shifted temperature scale. Consequently, we suggest that the fast relaxation process involves only the weakly bonded HDL structure.

Assuming that the anomalous water behavior can be extrapolated from a critical point in the deep supercooled region, a given anomalous property, τ, of water can be fitted to a functional form [1], $\ln[\tau(T)]=\gamma\ln[\tau_0(T/T_S-1)]$, with a critical point at the temperature T_S. In Fig. 2b, is shown the Debye relaxation times on a log scale versus $-\ln(T/T_S-1)$, where T_S=228K (235.2K) for H_2O(l) (D_2O(l)). The observed linear dependence for both isotopes indicates that the temperature dependence of the Debye relaxation is determined by the fluctuations that lead to the critical point at T_S. Furthermore, a shift of 7.2K between the critical temperature for H_2O(l) and D_2O(l) makes the two curves coincide.

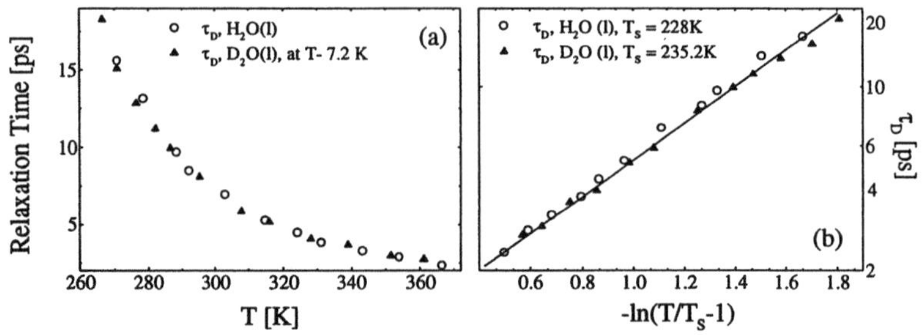

Fig. 2. a) Debye relaxation times compared where the D_2O(l) relaxation times measured at T are shown at temperature T-7.2K. b) The Debye relaxation times versus $-\ln(T/T_S-1)$, the temperature dependence that is expected if the water dynamics is influenced by the critical point in the supercooled region where the LDL and HDL phase boundary ends.

4. Conclusions

We found that the observed dielectric function for H_2O(l) and D_2O(l) was successfully represented by a biexponential Debye model. The effects of temperature and isotope substitution on the Debye relaxation of liquid water can be interpreted within a two-component model where the structural properties of H_2O(l) and D_2O(l) become equal when compared on a temperature scale shifted by 7.2K for D_2O(l). The Debye relaxation times are equal for the isotopes on the shifted temperature scale. Furthermore, the temperature dependence of the Debye relaxation may be modeled from a critical point in the supercooled region.

References

1 R. J. Speedy and C. A. Angell, *J. Chem. Phys.,* Vol. 65, 851, 1976.
2 G. E. Walrafen et al., *J. Chem. Phys.,* Vol. 85, 6970, 1986.
3 S. Woutersen et al., *Science,* Vol. 278, 658, 1997.
4 M.-C. Bellissent-Funel, *Europhysics Letters,* Vol. 42, 161, 1998.
5 P. H. Poole et al., *Nature,* Vol. 360, 324, 1992.
6 E. Shiratani and M. Sasai, *J. Chem. Phys.,* Vol. 108, 3264, 1998.
7 H. Tanaka, *Phys. Rev. Lett.,* Vol. 80, 113, 1998.
8 C. Rønne et al. *J. Chem. Phys.,* Vol. 107, 5319, 1997.
9 J. T. Kindt and C. A. Schmuttenmaer, *J. Phys. Chem.,* Vol. 100, 10373, 1996.
10 M. Vedamuthu et al., *J. Phys. Chem.,* Vol. 100, 3825, 1996.

Subpicosecond Holeburning Spectroscopy of Water in the Infrared: Dynamics of Local Structures

R. Laenen and C. Rauscher

Physik-Department E11, Technische Universität München,
James-Franck-Str., D-85748 Garching, Germany
E-mail: Robert.Laenen@Physik.TU-Muenchen.de

Abstract. Spectral holes are observed in the OH-band of HDO with 45 cm^{-1} width and $\approx$ 1 ps lifetime. Structural relaxation with 1.5 ± 0.5 ps (273 K) is measured for the first time.

Because of its outstanding role in nature, the investigation of the structural and dynamical properties of water[1] has a long tradition. This liquid consists of one of the smallest kind of molecules with specific intermolecular interaction via hydrogen-bonds. Changes of the transient OH-bandshape as a function of the frequency position of the excitation pulse provided evidence for the "inhomogeneous" character of the OH-band of HDO at 298 K.[2] Three major spectral components were inferred from the data and assigned to different environments while the time resolution of the investigation of 11 ps however did not allow to observe a spectral hole and determine the structural relaxation dynamics. For this reason we investigate this system with notably increased temporal resolution.
The IR-pulses are derived from parametric oscillator-amplifier devices with durations of 1 ps (2 ps) in the range 1600 to 3700 cm^{-1} (2800 to 3700 cm^{-1}; numbers in brackets refer to the pump). In part of the investigations the pulses were as short as 420 fs due to an improvement of the laser system. The relative transmission changes $\ln(T/T_0)_{\|,\perp}$ for variable probe frequency ν and delay time t_D represent the relevant quantities, from which an isotropic signal, $\ln(T/T_0)_{is} = (\ln(T/T_0)_{\|} + 2\ln(T/T_0)_{\perp})/3$ is determined delivering information on the vibrational population dynamics. A HDO concentration of 0.8 M is adjusted in heavy water.
An example for transient spectral hole-burning in the center of the OH-band is demonstrated in Fig. 1. The isotropic signal contribution is plotted versus probe frequency for $\nu_{Pu} = 3430$ cm^{-1} and $t_D = 0$ ps at room temperature. A narrow bleaching with width of 45 ± 10 cm^{-1} is indicated at the frequency position of the pump pulse with the dotted line, representing the spectral hole. A careful analysis of data with varying pump frequency and delay time shows that the transient spectra consists of three further bleaching components I-III of approximately Gaussian shape, depicted by the calculated curves in the figure.[3] In addition, a red-shifted excited-state absorption occurs around 3200 cm^{-1}, due to excess population of the v=1 state of the excited vibration.

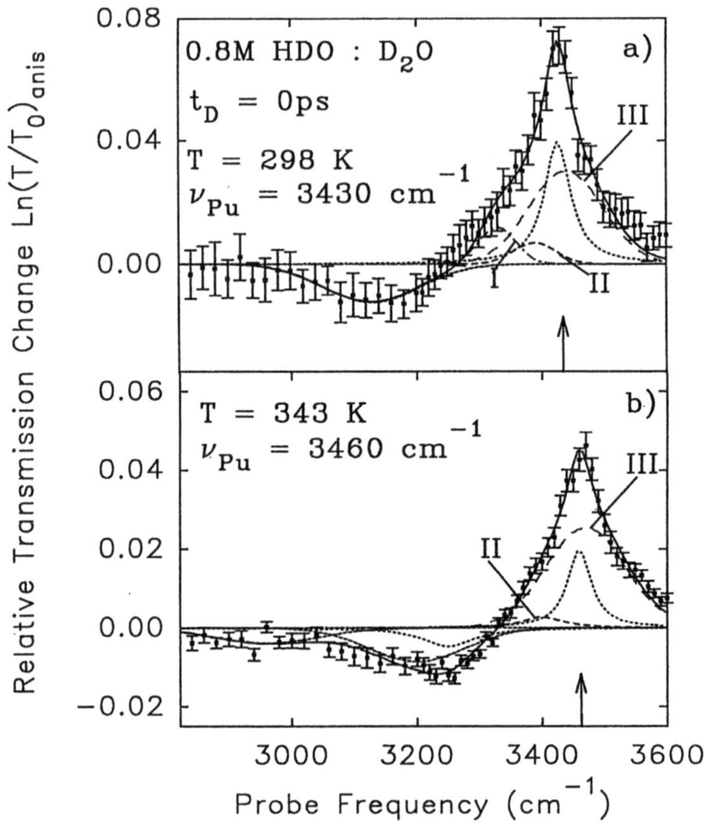

Fig. 1. Transient hole-burning within the OH-band of HDO is demonstrated. The spectral hole is denoted by the dotted line, the different structural components by different line styles. The spectral hole is only seen at higher temperatures of 298 K (a) and 343 K (b) where component III attributes to an high extend to the overall transient spectrum.

Components I, II and III are positioned respectively around 3340, 3410 and 3450 cm^{-1} with widths of approximately 60, 80, and 140 cm^{-1}. The relative amplitudes of the spectral constituents depend on temperature and excitation frequency while the relative contribution of component III is rising with temperature as can be seen for example from Fig. 1. Component I is believed to represent evidence for local hexagonal structures present even in liquid water as suggested from comparison with data from infrared spectroscopy.[4] Component III exhibits in turn an time constant for reorientation[5] of 3 ps at 343 K and this together with its spectral features gives some support to interprete this structure in terms of evidence for bifurcation[6] within the hydrogen-bonded network.

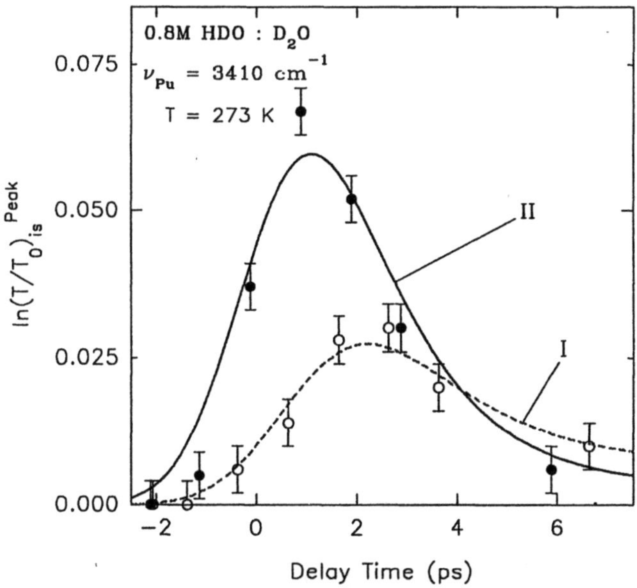

Fig. 2. The peak amplitudes of components I and II derived from fitting of the transient spectra taken at different delay times are depicted here. Direct excitation of component II results in a delayed increase of component I. A time constant of $\tau_s = 1.5 \pm 0.5$ ps is calculated at 273 K.

From the lifetime of the transient hole and the temporal evolution of the peak amplitudes of the three major components we infer a time constant for structural relaxation of $\tau_s = 0.8$ - 1.5 ps, depending on the temperature.

Acknowledgements. R.L. thanks the DFG for financial support.

References

1 *Water, A Comprehensive Treatise*, Edited by F. Franks, Plenum Press, New York, 1972.

2 H. Graener, G. Seifert, and A. Laubereau, Phys. Rev. Lett. **66**, 2092, (1991).

3 R. Laenen, C. Rauscher, and A. Laubereau, Phys. Rev. Lett. **80**, 2622, (1998).

4 T.A. Ford and M. Falk, Can. J. Chem. **46**, 3579, (1968).

5 R. Laenen, C. Rauscher, and A. Laubereau, submitted for publication.

6 P.A. Giguere, J. Chem. Phys. **87**, 4835 (1987).

Dynamics of the Hydrogen-Bond in Liquid Water

G. M. Gale, G. Gallot, F. Hache and N. Lascoux

Laboratoire d'Optique Quantique, Ecole Polytechnique
91128 Palaiseau, FRANCE
gale@leonardo.polytechnique.fr

Abstract. Intermolecular O-O movement in the fundamental electronic state is fully resolved for the first time for the hydrogen bond in liquid water and compared with a theoretical model. Strongly varying energy relaxation times are also measured and interpreted in terms of a V-V transfer model.

Liquid water is a ubiquitous and essential element in almost all biochemical reactions. Many of its properties stem from its ability to form extensive oriented hydrogen-bonded networks. Most of the available information on water has been obtained from various types of classical spectroscopy. However, until now, it has been very difficult to study the spectrally resolved dynamics of the hydrogen-bond in real time.

Employing a recently-developed 150 fs duration, Fourier-transform-limited, two channel, tunable mid-infrared source [1], we have been able to follow, on a time-scale of several hundred fs, the movement of the O-H..O bond when excited away from equilibrium. To our knowledge, this is the first time that hydrogen-bond dynamics have been fully resolved in real time in liquid water.

In the following we will consider a 0.5% solution of HDO in D_2O. This low concentration provides manageable transmission in our 100 μm thick flow cell and avoids problems due to sample heating. Our experimental technique relies on the well-known fact that the frequency of the O-H vibration in a hydrogen bond is a sensitive function of O-O distance. We may thus use the frequency of the O-H vibration as a probe of the hydrogen-bond dynamics. In an initial excitation stage, a 10 μJ, 150 fs mid-IR pulse excites a sub-ensemble of well-defined non-equilibrium O-O distances by choosing the frequency of excitation away from the central O-H vibration frequency of 3420 cm^{-1} (hole-burning). The return of the O-O bond to equilibrium is followed by time-resolved spectroscopy employing a second, independently tunable, 10 nJ, 150 fs probe pulse.

The result of one such experiment is shown in fig. 1. The excitation frequency was 3510 cm^{-1} and the first moments M_1 of the transient bleaching spectra are plotted as a function of probe delay time in fs. Spectral relaxation is initially non-exponential and subsequently decays exponentially with a time-constant of 800 fs. Spectral line-width (not shown) also increases with probe delay. The variation of the first moment is in quantitative agreement with the theory of Bratos and Leicknam [2].

$$M_1(\tau) = \frac{[1+\mathrm{erf}\left(\frac{\tau}{\tau_p} - \frac{\tau_p}{2\tau_\Omega} - \frac{\tau_p}{2\tau_\nu}\right)]\exp[-\frac{1}{\tau_\Omega}\left(\tau - \frac{\tau_p^2}{\tau_\Omega} - \frac{\tau_p^2}{2\tau_\nu}\right)]}{1+\mathrm{erf}\left(\frac{\tau}{\tau_p} - \frac{\tau_p}{2\tau_\nu}\right)} \cdot \Delta_1$$

where τ is the probe pulse delay, τ_P is the pulse half-width at 1/e height, τ_ν is the energy relaxation time , τ_Ω is the spectral relaxation time and Δ_1 is the frequency difference between the pump pulse and the band center.

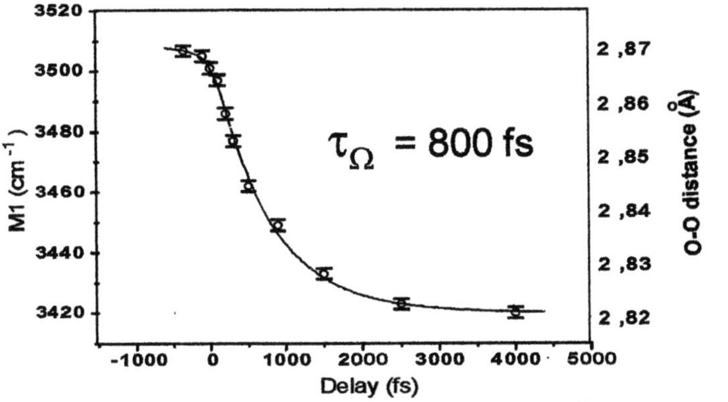

Figure 1

For a value of τ_Ω of 800 fs we obtain excellent agreement with experiment (full line in figure). Using the well-known relationship between O-H stretch frequency and O-O distance we can add the right-hand scale which shows that O-O distance varies from 2.87 to 2.82 Angstroms during the experiment. This is the first fully time-resolved measurement of intermolecular O-O movement in a liquid. Observations at other pump frequencies indicate that τ_Ω tends to increase with pump frequency. This may be due to the weakening of the hydrogen bond as O-H frequency increases. So far, no clear evidence for spectral substructures [3] has been seen in our experiments.

We have also measured the energy relaxation time T1 of the O-H stretching vibration of the same 0.5% solution of HDO in D_2O. Fig. 2 shows the measured T1 value in fs as a function of excitation and identical probe frequency. Note that T1 varies from 450 fs to 900 fs across the vibrational band-width (band center is 3420 cm^{-1} and band-width is 260 cm^{-1}). These values have been slightly corrected for the effects of spectral relaxation. These results can be explained quantitatively as an intramolecular $\nu_3 \rightarrow 2\nu_2$ process. The probability of this process has been measured as 11% per collision in water vapor [4]. One may correct for the collision frequency in the liquid [5] and calculate the frequency dependence using

$$1/\tau_\nu = \xi\, P_{SSH}\, \nu_{PL} + 1/T'_1$$

where

$$P_{SSH} = \exp\left\{\frac{-3}{2}\left(\frac{4\pi^2\mu\Delta\omega^2 L^2}{kT}\right)^{1/3}\right\}$$

ξ is a correction factor close to 1, ν_{PL} is the local phonon frequency (O-O stretch), T'_1 is a global frequency-independent relaxation tim, L is a potential range parameter which can be calculated from the result in the vapor and $\Delta\omega$ is the frequency difference between the ν_3 pump and probe frequency and $2\nu_2$ at 2960 cm^{-1}. Curve (a) in Fig. 2 is obtained using $\xi = 1$ and $T'_1 = \infty$. Curve (b) employs the parameters $\xi = 1.05$ and $T'_1 = 4$ ps and is in perfect agreement with experimental results.

In conclusion, for the first time, we have completely time-resolved spectral and population relaxation in liquid water and measured intermolecular O-O movement in a liquid. Undoubtedly, subsequent experiments will reveal further information on this extremely important subject.

This research was performed in the framework of the French Research Grouping 1017 "Elementary Chemical Processes in Liquids".

The Laboratoire d'Optique Quantique is a Unité mixte de Recherche between the CNRS and the Ecole Polytechnique.

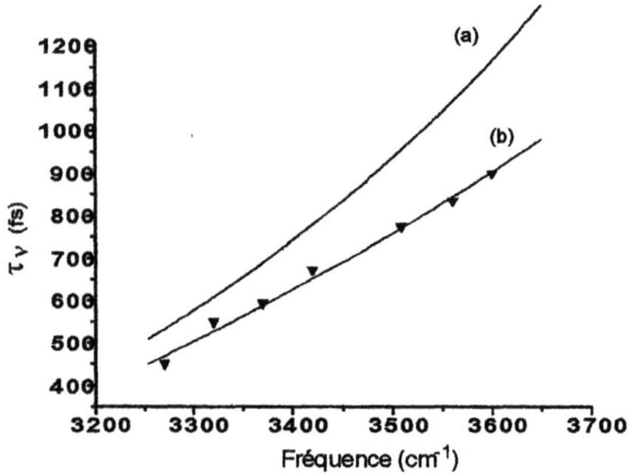

Figure 2

References

1. Generation of intense highly coherent femtosecond pulses in the mid infrared, G. M. Gale, G. Gallot, F. Hache and R. Sander, Opt. Lett **22**, 1253-1255 (1997)
2. Subpicosecond transient infrared spectroscopy of water. A theoretical description, S. Bratos and J-Cl. Leicknam, J. Chem. Phys.**103**, 4887-4893 (1995)
3. New spectroscopy of water using tunable picosecond pulses in the infrared, H. Graener, G. Seifert and A. Laubereau, Phys. Rev. Lett. **66**, 2092-2095 (1991)
4. Vibrational relaxation of water vapor, J. Finzi, F. E. Hovis, V. N. Panfilov, P. Hess and C. Bradley Moore, J. Chem. Phys. **67**, 4053-4061 (1977)
5. Vibrational relaxation in condensed phases, J. Chesnoy and G. M. Gale in Evolution of size effects in chemical dynamics (part 2); eds. I Prigogine and S. A Rice, Advances in Chemical Physics **LXX**, 297-355 (1988)

Mode Coupling Mechanisms in Liquids Studied by 2D Raman Scattering

Koos Duppen and Thomas Steffen

Ultrafast Laser and Spectroscopy Laboratory, Materials Science Center
University of Groningen, Nijenborgh 4, 9747 AG Groningen, The Netherlands
E-mail: duppen@chem.rug.nl

Abstract. Femtosecond temporally two-dimensional Raman scattering, that was originally designed to characterize the time scale(s) of intermolecular dynamics in liquids, also provides information on mode coupling mechanisms. Polarizability mode coupling suppresses the formation of motional echoes, that normally should occur when there are slow components in the time scales of local structural changes in the liquid.

1. Introduction

During many years, a large variety of one- and two-dimensional Raman techniques has been employed to study the complicated problems associated with liquid state dynamics. All 1D experiments, whether measured in time- or frequency domain, yield similar kind of information, that can be expressed in the form of a spectrum of intermolecular motion. The underlying dynamics that give rise to this spectral response is not elucidated, though. The width may indicate, for instance, a distribution of local molecular environments, that changes very slowly on the time scale of the inverse of the spectral width (inhomogeneous broadening). However, it may also result from ultrafast local molecular motions, that are expected to strongly influence electronic processes of molecules in liquids, such as solvation and (femto-)chemistry. In that case the spectrum may be called homogeneously broadened. The 1D techniques do not allow to discriminate these processes.

Multi-dimensional Raman response is required to separate the various contributions to the motional spectrum. Recently, the proposal of fifth-order 2D Raman scattering has received considerable attention, since this method promises to probe the structural dynamics in condensed phase systems on a microscopic level [1]. In the first of these experiments [2-4], that were performed on liquid CS_2 with five nonresonant laser pulses (Fig.1), it was found that the fifth-order signal along the second delay time T_2 depends only slightly on the value of the first delay time T_1. The absence of a clear echo signal

Fig. 1. Pulse sequence of the fifth-order Raman experiment. The two delay times T_1 and T_2 can be varied independently.

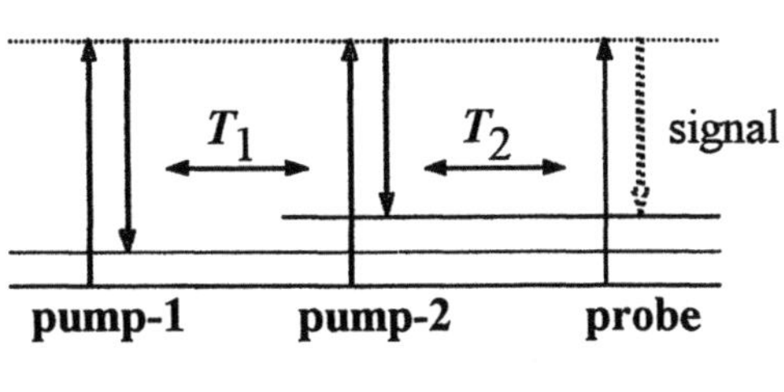

was interpreted in terms of small heterogeneity of local structures. Detailed analyses of these signal shapes revealed the inhomogeneous width of the motional spectrum and an estimate of the correlation time of the fluctuations [5].

One of the most surprising results of these experiments was that a distinct asymmetry was observed in the dynamics along the two time variables: the signal as function of T_1 is governed mainly by slow diffusive dynamics on a ps time scale, while it is dominated by ultrafast inertial motions along T_2. This interplay of vibrations and diffusion cannot be understood by the originally proposed models, in which the excitations of the liquid are treated as independent nuclear modes.

2. Mode coupling mechanisms

The most simple model for nuclear dynamics in liquids, that is still sufficient to describe the observed one-dimensional (third-order) response reasonably well, comprises two harmonic modes Q_V and Q_D for the ultrafast interia-limited vibrations and for the slow collective diffusion, respectively. As mentioned above, the calculation of the fifth-order response by simply adding the contribution of these two modes does not reproduce the data. When the modes are coupled by anharmonicities in the motions [6], the quality of the fits is not significantly better [7,8], so we will not treat this situation here.

Another coupling mechanism is provided by the coordinate dependent polarizability [7-9]. For the evaluation of the third- and fifth-order nonlinear optical response, the polarizability is usually expanded in powers of the nuclear coordinate:

$$\alpha(Q_V,Q_D) = \alpha^{(0)} + \sum_{i=V,D} \alpha_i^{(1)} Q_i + \sum_{i,j=V,D} \alpha_{ij}^{(2)} Q_i\, Q_j + \ldots$$

where V and D refer to vibration and diffusion, respectively. The linear and second-order polarizability coefficients $\alpha_i^{(1)}=\partial\alpha/\partial Q_i$ and $\alpha_{ij}^{(2)}=\partial^2\alpha/\partial Q_i\partial Q_j$ govern the 2D Raman effect. The diagonal elements $\alpha_{ii}^{(2)}$ give rise to two-quantum Raman transitions

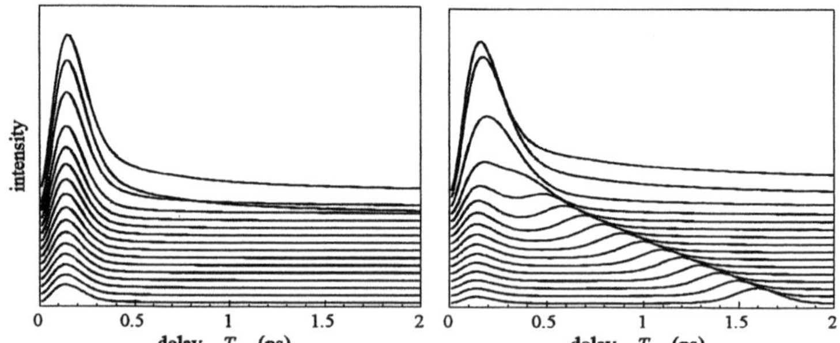

Fig. 2. The 2D Raman response due to polarizability coupling between a diffusive and a vibrational mode does not show any echo, despite the presence of inhomogeneity (left). With an additional contribution due to the vibrations only, the echo is clearly present (right). The value of T_1 is 100 fs for the top traces, and increases by another 100 fs for each subsequent lower trace.

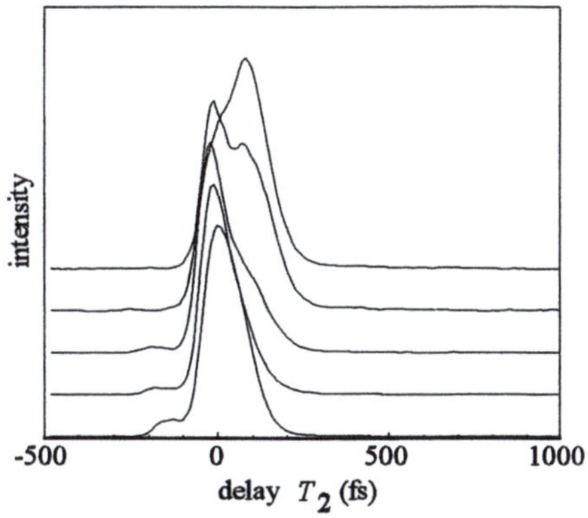

Fig. 3. The 2D Raman response of benzene as function of T_2. Two maxima can be recognized, that are interpreted as a free induction decay due to polarizability coupling, and an echo due to pure vibrational response. Delay T_1 is varied from 100 fs to 150, 200, 250, and 300 fs (bottom to top)

of the modes separately, while the off-diagonal element $\alpha_{ij}^{(2)}$ describes simultaneous excitation and/or deexcitation of the two modes. In the simulation of Fig. 2, it is shown how this so-called polarization-induced coupling gives a 2D Raman signal in which echo formation is totally suppressed, despite the presence of an inhomogeneously broadened vibrational mode. When there is an additional contribution to the signal due to the vibrations alone, the echo shows up as a signal that peaks at T_2=T_1.

With the polarization mode coupling model, the 2D Raman signals of CS_2 can be described very well, including the asymmetry along the two delay times [7,8]. In particular when the coefficients $\alpha_{VD}^{(2)}$ and $\alpha_{DV}^{(2)}$ are allowed to have different values (the vibrations couple to the diffusion but not vice versa), the full 2D response can be quite adequately simulated. However, no motional echo is observed.

In the case of benzene, shown in Fig. 3, a signal shape is seen along delay T_2 that resembles the simulation of Fig. 2. At small delays, a peak in the signal can be seen that does not depend much on delay T_1, while at larger delays a second maximum seems to develop that peaks near T_2=T_1. In the current model the interpretation is clear: the first maximum is due to the polarizability mode coupling term in the nonlinear response, while the second maximum reflects echo formation due to an inhomogeneously broadened set of intermolecular vibrations. Whether this can be considered solid proof of the presence of polarization coupling in this liquid remains to be seen, though. Further experiments are required to shed further light on this matter.

References

1 Y. Tanimura and S. Mukamel, J. Chem. Phys. **99**, 9496 (1993).
2 K. Tominaga and K. Yoshihara, Phys. Rev. Lett. **74**, 3061 (1995).
3 T. Steffen and K. Duppen, Phys. Rev. Lett. **76**, 1224 (1996).
4 A. Tokmakoff and G. R. Fleming, J. Chem. Phys. **106**, 2569 (1997).
5 T. Steffen and K. Duppen, J. Chem. Phys. **106**, 3854 (1997).
6 K. Okamura and Y. Tanimura, Chem. Phys. Lett. **278**, 175 (1997).
7 T. Steffen and K. Duppen, Chem. Phys. Lett. **290**, 229 (1998).
8 A. Tokmakoff, M. J. Lang, X. J. Jordanides and G. R. Fleming, Chem. Phys. accepted.
9 M. Cho, K. Okumura and Y. Tanimura, J. Chem. Phys. **108**, 1326 (1998).

Electronic "Bubble" Dynamics in Solid Argon

Catherine Jeannin, Marcia T. Portella-Oberli and Majed Chergui

Institut de Physique Expérimentale, Faculté des Sciences,
Université de Lausanne, CH-1015 Lausanne, Switzerland
Phone: (+41 21) 692 3678, Fax: (+41 21) 692 3635

Abstract: We follow the ultrafast dynamics of the medium response to impulsive Rydberg state excitation of an NO molecule trapped in solid argon using a femtosecond fluorescence depletion pump-probe technique. Our results show that the cage expands by undergoing collective coherent recurrencies for over ~3 ps, on its way to the fully relaxed electronic "bubble" conformation.

The detailed description of the medium dynamical response to photoexcitation of an impurity in many-body systems is of crucial importance in physics, chemistry and biology. We have investigated the dynamics of the lattice response upon excitation of the lowest Rydberg state A (3sσ) of the NO molecule embedded in solid argon. In this case, the electron orbital radius becomes comparable to the nearest neighbour distance in the solid and a repulsive overlap ensues with the closed shell matrix species. This induces an expansion of the matrix cage, the so-called electronic "bubble" formation, which corresponds to an increase by ~ 10% of the initial (ground state) cage radius and several hundreds of meV are dissipated in the process /1/. In this work, we have probed in real-time, the evolution of the cage on its way to the relaxed "bubble" state, by inducing transitions from the A-state to the next higher lying C(3pπ) and D(3pσ) Rydberg states.
The observables of the dynamics have been established on the basis of the A-state fluorescence spectra /1/ and of the ground state /2/ and A-state /3/ absorption spectra. From these data, intermolecular potential curves were derived /3/ which describe the molecule-matrix interaction in each state (ground, A, C, D) on the basis of the harmonic and the effective mode approximations (fig.1a). By tuning the probe laser wavelength to match the difference potentials A-C and/or A-D, we can probe the evolution of the system from t=0 (1.23 μm) to the fully relaxed cage (1.1 μm), along the cage radius which is the relevant deformation coordinate in the "bubble" formation.
The experimental procedure was described previously /4/. Briefly, impulsive excitation of the A-state is achieved by a 200 fs pulse at 196 nm while probing the cage evolution is done with a tunable 120 fs IR pulse (1.32 μm to 1.1 μm). The detected signal is the depletion of the A-state fluorescence to the ground state. Because of the difference potentials, our probe pulse can only induce A-C transitions, except for the 1.1 μm probe which opens two windows, on the D state at early times and on the C state at long time delays. In fig.1a, absorption lineshapes represent the approximate observation windows corresponding to the different probe wavelengths used in this work.
Fig.1b shows a set of fluorescence depletion signals as a function of pump-probe delay and for parallel pump and probe polarizations. The transients show an oscillatory behaviour which is lost in ~3 ps and is indicative of coherent cage

motion. Thereafter, the signal corresponds to a depletion level that remains constant. The signals were recorded at low probe laser intensities to avoid saturation of the main peaks. On the other hand, the saturation regime unravels details of the dynamics by enhancing signals in the wings at early times or the signal at longer times. The probe wavelength dependence of the signal at long times is consistent with the depletion level derived in nanosecond experiments /3/, indicating that the fully relaxed cage has been reached.

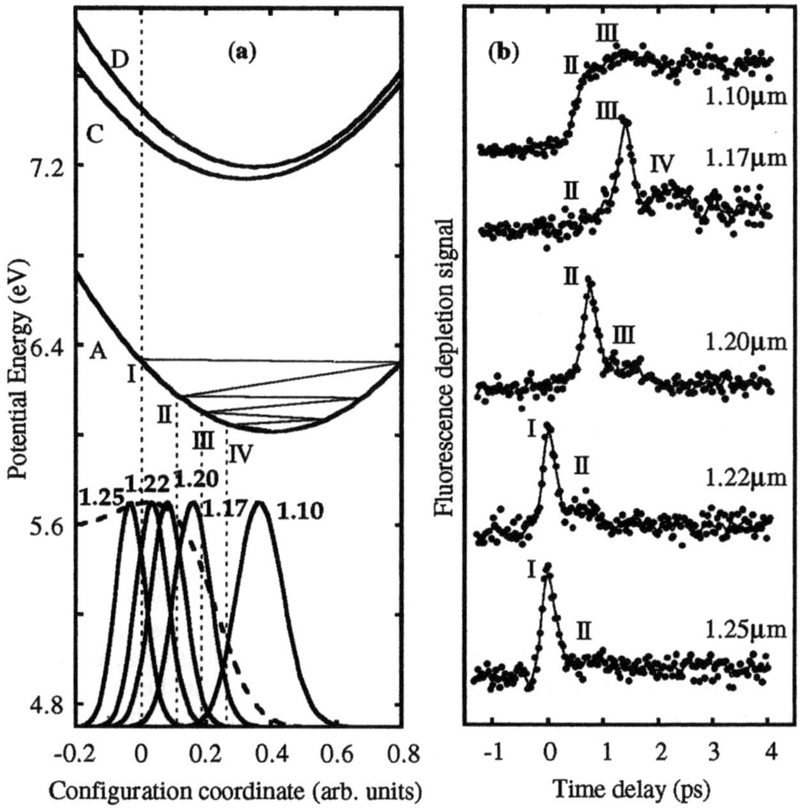

Figure 1: a) Intermolecular potentials and caculated A-C observation windows for different probe wavelengths (in μm); The broken line represents the 1.1 μm window on the D state. The trajectory of the wavepacket is shown in the A state potentiel well. b) Fluorescence depletion signals as a function of pump-probe delay.

The peaks in the transients correspond to the passage of the wavepacket of cage vibrational modes in the various observation windows (fig.1a) as described hereafter. The 1.25 μm and 1.22 μm probes sample the start of the dynamics and the transients show two peaks. Peak I follows the pump-probe cross-correlation. It reflects the formation of the wavepacket by the pump pulse at t=0. Peak II occurs ~700 fs later. Since the outer turning point is reached in ~250 fs, according to MD simulations /5/, we attribute peak II to the return of the wavepacket in the

observation window following an energy loss of ~50% after the first cage expansion. The 1.2 μm probe samples intermediate configurations of the dynamics and indeed, the corresponding transient is characterized by a wing at early times followed by peak II and a weaker peak (III) some 700 fs later. Because of the short dwell time of the wavepacket in the observation window at early times /5/, the signal is weak and appears as a wing at short times. Peak II appears prominently because the first recurrence is now well sampled by the 1.2 μm probe at the inner turning point. Peak III occurs some 700 fs later and is attributed to a second recurrence of the wavepacket at an inner turning point that lies lower in energy than before and is therefore less well sampled by the 1.2 μm probe window. At 1.17 μm, the observation window shifts further down the potential curve and the wing at early times becomes more pronounced than in the 1.2 μm transient. On the basis of the above, it would correspond to the first 3 passages, in the observation window, of the wavepacket at high velocity. The main peak (III) is now better sampled than at 1.2 μm and appears therefore more prominently. It is followed by a peak (IV) which is the fourth inner turning point recurrence of the wavepacket after an additional loss of energy so that it is more poorly sampled by the 1.17 μm probe. For the transient at 1.1 μm two windows occur in the C and D states (broken line, fig.1a). The signal shows a wing at short times followed by two structures and a high constant signal at long times. At this probe wavelength, the window on D samples well the formation (peak I in the other transients), the first recurrence (peak II) and reasonably well the following recurrencies of the wavepacket. The wing is due to the formation of the wavepacket (peak I) and the following structure to the first recurrence (II). The fact that peak I is weaker than peak II is due to the parallel pump-probe polarization and to depolarization effects /6/. The following structure corresponds to peak III probed on the C and D states. The last recurrence (IV) is lost in the high constant signal.
The scenario presented here shows that the lattice response to an impulsively generated bubble-like deformation is a coherent process. We observe up to four coherent recurrencies of the cage vibrational modes on the way to complete relaxation which occurs in ~3 ps. Most of the energy is lost after the first expansion of the cage. The results presented here are in qualitative agreement with recent MD simulations /5/ and are completed by polarization studies presented in a forthcoming publication /6/. This is, to our knowledge, the first direct visualization of coherent recurrent medium responses to a charge redistribution in a chromophore.

This work was supported by the Swiss National Science Foundation via grants 46019.96 and 50427.97.

1. M. Chergui, N. Schwentner and V. Chandrasekharan, J. Chem. Phys. 89 (1988) 1277.
2. M. Chergui, N. Schwentner and W. Böhmer, J. Chem. Phys. 85 (1986) 2472.
3. G. Zerza, F. Vigliotti, A. Sassara, M. Chergui and V. Stepanenko, Chem. Phys. Lett. 256 (1996) 63; F. Vigliotti and M. Chergui, J. Chem. Phys. (submitted).
4. M.T. Portella-Oberli, C. Jeannin and M. Chergui, Chem. Phys. Lett. 259 (1996) 475.
5. S. Jimenez, A. Pasquarello, R. Car and M. Chergui, Chem. Phys. (in press).
6. C. Jeannin, M.T. Portella-Oberli and M. Chergui, to be published.

Femtosecond Pump-Probe Spectroscopy on the Equilibrated Aqueous Solvated Electron: Isotope Effects and Saturation Studies

Kazushige Yokoyama, Dong Hee Son, Carlos Silva, and Paul F. Barbara

Department of Chemistry, University of Minnesota, 207 Pleasant St. S.E., Minneapolis, MN 55455, USA

Abstract. Femtosecond pump-probe spectroscopy on the hydrated electron was carried out with 35 fs time resolution over a broad probe wavelength and excitation intensity range.

Pump-probe experiments on the *equilibrated* hydrated electron were performed with 35-60 fs time resolution in the 600-920 nm probe wavelength range. The transient absorption data on the solvated electron in H_2O and D_2O were collected by pumping equilibrated s-state electrons at 800 nm with 25 fs pulses derived from an amplified multipass Ti:sapphire laser system [1], and probing at various wavelengths selected from a single-filament white-light continuum. Figure 1 presents representative absorption transients showing previously unresolved dynamics (<300 fs) monitored at various probe wavelengths. The data are well fit by a function consisting of one Gaussian and two exponential components convoluted with the instrument response function. The Gaussian component is found to be in the range of 30-80 fs depending on probe wavelength. The average time constant for the bleach recovery is in the range of 70-200 fs. The increased absorption observed around 780 nm to 920 nm decays on the ~1 ps timescale, as described in earlier publications [2]. The Gaussian and fast exponential kinetic components are assigned to inertial and diffusive solvation motion, respectively, as predicted by recent quantum molecular dynamics (MD) simulations of the hydrated electron [3-7].

The Gaussian component is observed to be especially sensitive to solvent deuteration [1]. The time constants are 45-65% longer in D_2O than in H_2O for the red wavelengths where the isotope effect was investigated. The isotope effect in the Gaussian component is consistent with simulations by Barnett et al. [5], where the inertial component of solvation was longer by $\sqrt{2}$ in D_2O due to solvent librations (hindered rotations). The presence of the isotope effect strongly suggests that the initial solvation dynamics are dominated by librational motion of the water molecules. *This is the first direct evidence for the librational character of the inertial solvent response of the solvated electron.* Analogous inertial solvation has been observed by time resolved fluorescence measurements for polar dyes in water but with significantly slower time resolution that was not able to directly time resolved the sub 50 fs dynamics [8]. Recent related experiments by Wiersma and co-workers have also found evidence for the librational character of the inertial response, and for coherent wavepacket dynamics in the pump-probe spectroscopy of the hydrated electron [9].

No significant isotope effect is observed for the slower exponential kinetic components (described above) within experimental error. This is in agreement with the nonadiabatic MD simulations by Rossky and co-workers [3], which predict a small net isotope effect for these process which have been assigned, by comparison to simulation, to diffusive solvation and p→s radiationless decay.

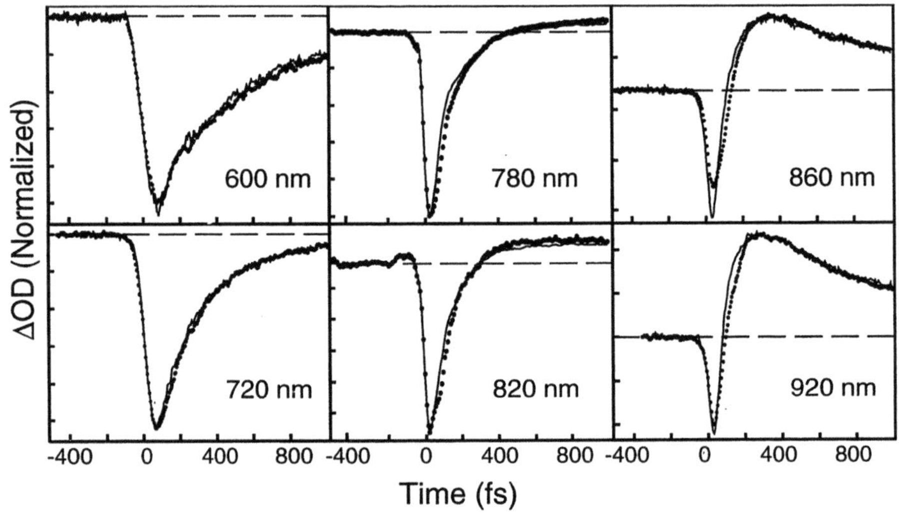

Fig. 1. A comparison of the absorption transients of the solvated electron in H_2O (solid lines) and D_2O (data points) in 10 mM $K_4Fe(CN)_6$ solutions. The probe wavelengths are indicated in each panel.

Since the observed spectral behavior can be sensitive to the extent of excitation, we performed a careful study of the pump-pulse energy dependence over a large range (200 nJ-100 μJ/pulse). Absorption transients collected with pulse energies of 96 μJ and 1.2 μJ are shown in Figure 2 for 720 nm and 920 nm probe wavelengths. Two different pump intensity regimes are exhibited for all wavelengths studied. In the low intensity regime the dynamics are not dependent on excitation intensity. At the high excitation intensity regime, however, the dynamics become significantly slower. This is considered to be the result of a sequential two-photon process which induces the solvated electron to be ejected from its original solvent site, and re-trap and relax at a site that is distinct from the original one. This two-photon channel is closely analogous to previous two-pulse electron photoinjection experiments [10-12], in which excess electrons are injected by direct multiphoton ionization of the solvent. This is supported by the role played by the slow kinetic components of the data. The timescales of the slow components in the high intensity regime are similar to those observed in photoinjection experiments, which play a major role in the data. On the other hand, the slow components play a lesser role in the one-photon pump-probe data.

An alternative model for the dynamics that also is consistent with the data assigns the initial dynamics on the sub 100 fs timescale to the p→s radiationless transition process. Thus, in this model the radiationless decay is controlled by inertial solvation. The slower components are accordingly assigned to relaxation of the "hot" ground-state electron. Since the spectral characteristics displayed in Figure 1 can be equally explained by either of these two distinct models, further experiments and simulations will be necessary to distinguish between the two possibilities.

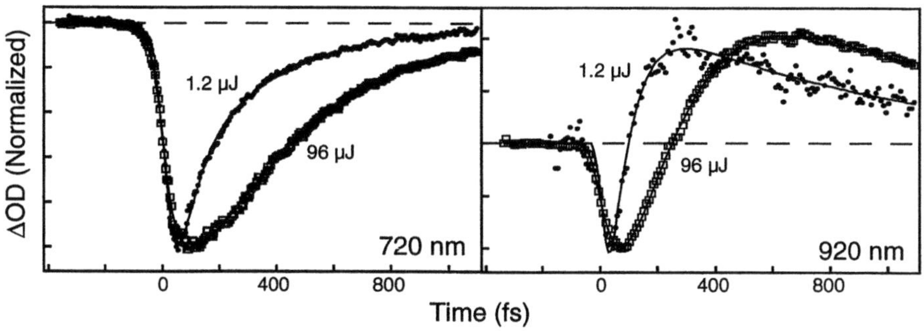

Fig. 2. Comparison of the transient ΔOD signals of the hydrated electron pumping at low and high intensity regimes. The pump and probe spot sizes were 300 and 40 μm (FWHM), respectively.

Acknowledgments. This work was supported by a grant from the Basic Energy Sciences Program of the Department of Energy.

References

1 C. Silva, P.K. Walhout, K. Yokoyama, and P.F. Barbara *Phys. Rev. Lett.* **80**, 1086 (1998); K. Yokoyama, C. Silva, D. Son, P.K. Walhout, and P.F. Barbara *J. Phys. Chem. A* (in press).

2 Y. Kimura, J.C. Alfano, P.K. Walhout, and P.F. Barbara *J. Phys. Chem.* **98**, 3450 (1994).

3 B.J. Schwartz and P.J. Rossky *J. Chem. Phys.* **101**, 6902 (1994); B.J. Schwartz and P.J. Rossky *J. Chem. Phys.* **101**, 6917 (1994); B.J. Schwartz and P.J. Rossky *J. Chem. Phys.* **105**, 6997 (1996); B.J. Schwartz, E.R. Bittner, O.V. Prezhdo, and P.J. Rossky *J. Chem. Phys.* **104**, 5942 (1996).

4 S. Bratos, J.-C. Leicknam, D. Borgis, and A. Staib *Phys. Rev. E.* **55**, 7217 (1997); D. Borgis and S. Bratos *J. Molec. Structure* **436-437**, 537 (1997).

5 R.B. Barnett, U. Landman, and A. Nitzan *J. Chem. Phys.* **90**, 4413 (1989).

6 P. Graf, A. Nitzan, G.H.F. Diercksen *J. Phys. Chem.* **100**, 18916 (1996).

7 A. Staib and D. Borgis *J. Chem. Phys.* **103**, 2642 (1995).

8 R. Jimenez, G.R. Fleming, P.V. Kumar, M. Maroncelli *Nature* **369**, 471 (1994).

9 M.F. Emde, A. Baltuska, A. Kummrow, M.S. Pshenichnikov, and D.A. Wiersma *Phys. Rev. Lett.* **80**, 4645 (1998).; A. Kummrow, M.F. Emde, A. Baltuska, M.S. Pshenichnikov, and D.A. Wiersma *J. Phys. Chem. A.* **102**, 4172 (1998).

10 X. Shi, F.H. Long, H. Lu, and K.B. Eisenthal *J. Phys. Chem.* **100**, 11903 (1996).

11 C. Pépin, T. Goulet, D. Houde, J.P. Jay-Gerin *J. Phys. Chem. A* **101**, 4351 (1997).

12 R.A. Crowell and D.M. Bartels *J. Phys. Chem.* **100**, 17940 (1996).

Hydrated Electron Dynamics on a 5-fs Timescale

Michel F. Emde[1], Andrius Baltuška[1], Andreas Kummrow[2], Maxim S. Pshenichnikov[1], and Douwe A. Wiersma[1]

[1] Ultrafast Laser and Spectroscopy Laboratory, Department of Chemistry, University of Groningen, Nijenborgh 4, 9747 AG Groningen, The Netherlands
E-mail: D.A.Wiersma@chem.rug.nl
[2] Max-Born-Institut, Rudower Chaussee 6, 12489 Berlin, Germany
E-mail: kummrow@mbi-berlin.de

Abstract. We report on frequency-resolved pump-probe and photon echo experiments on the equilibrated aqueous electron using 5-fs pulses. All experiments exhibit a signature of wave packet dynamics attributed to librational motions of water molecules.

1. Introduction

Since its discovery in the early sixties, the hydrated electron has received continued interest from chemists and physicists. Basically, the hydrated electron is an excess electron trapped in a potential well formed by surrounding water molecules, with an *s*-like ground state and three non-degenerate *p*-type excited states [1] as shown in Fig.1. The electron surrounded by oriented water molecules is a chemical reactant with unusually high electron donor capacity as its characteristic chemical feature. On the other hand, it seems one of the simplest physical systems to study solvation dynamics and to test mixed quantum-classical theories experimentally. Yet, even after decades of intensive experimentation and calculations on the hydrated electron, understanding of its solvation and (non-) adiabatic relaxation dynamics is far from being complete.

Pioneering optical studies on the hydrated electron started about ten years ago [2,3], when 100-fs pulses became available in the wavelength region where the equilibrated aqueous electron has an intense absorption band. However, the crucial part of the solvation process, predicted to proceed on a 10-fs time scale, was not resolved due to the limited time resolution. Clearly, much shorter pulses are needed [4,5]. In this contribution we present the first results of frequency-resolved pump-probe and photon-echo experiments on hydrated electron with 5-fs resolution.

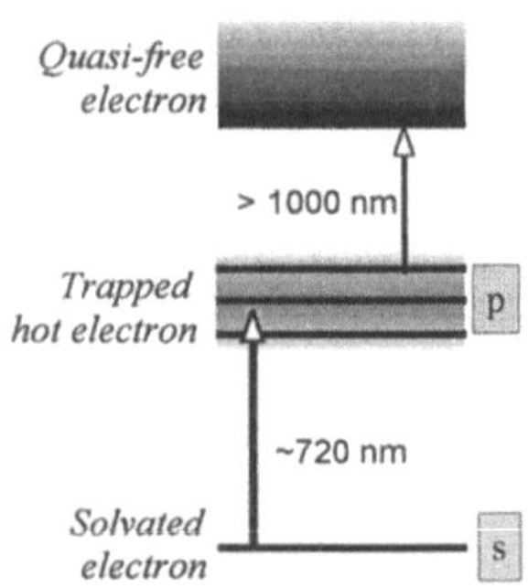

Fig. 1 Energy level scheme of solvated electron.

2. Experimental Methods

The hydrated electron is generated by photoionising potassium ferrocyanide dissolved in water using the quadrupled output of a Q-switched Nd:YAG laser [3,4]. The time-delayed excitation pulses are produced by compressing the white light continuum resulting from injection of the output of a cavity-dumped Ti:sapphire laser into a fibre, yielding pulses shorter than 5 fs [6]. The sample is flown in a jet through a 50-μm sapphire nozzle to avoid dispersive pulse broadening. Two-pulse and three-pulse photon echo signals, generated in different phase-matched directions, as well as pump-probe signals are detected simultaneously using photodiodes and lock-in amplifiers. The symmetric geometry ensures that the signals in the two conjugated echo directions are equivalent, but mirror-imaged, thus giving an intrinsic calibration of zero delays [5].

3. Results and Discussion

Figure 1 shows the results of the frequency-resolved pump-probe experiment on the hydrated electron. Next to the coherent artifact around zero delay, a distinct recurrence at ~40 fs manifests an underdamped solvent motion that is strongly coupled to the electron. On a longer time scale the induced transparency turns into induced absorption which begins to dominate at shorter delays for longer wavelengths.

The transient grating signal also

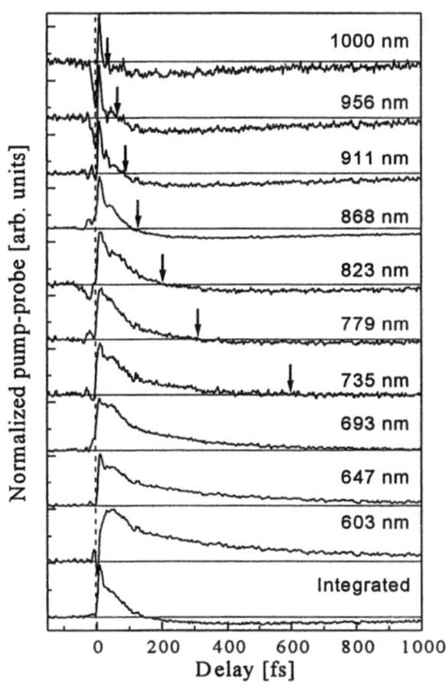

Fig. 1 Results of frequency-resolved pump-probe experiments with 5-fs pulses. Positive signals correspond to induced bleaching. The monochromator settings are shown along the traces. Arrows indicate delays at which excited-state absorption starts to dominate. The signals are arbitrarily normalized. Note that a 400-nm bandwidth is accessible at once.

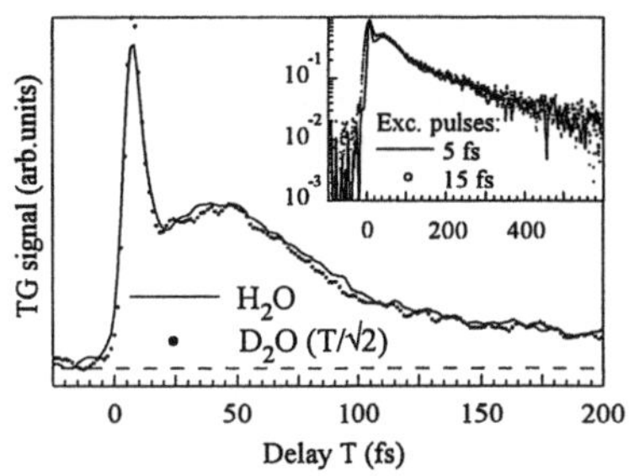

Fig. 2 The transient grating (TG) signals obtained with 5-fs pulses for electron in normal (solid curve) and deuterated (solid dots) water. The time scale for the D_2O signal is compressed by the factor of $\sqrt{2}$. Inset: TG signals obtained with 5-fs (solid curve) and 15-fs (open circles) pulses. Note the identity of the latter two signals.

exhibits a strong beat around ~40 fs (Fig.2, solid curve). This recurrence was recently predicted by dielectric-continuum model calculations and assigned to a libration of the water molecule [7]. It can be modeled by a strongly damped oscillatory mode with a frequency of ~670 cm^{-1}. This frequency as well as the required damping are in good agreement with the librational band observed in the Raman spectrum of neat water. Upon deuteration the recurrence occurs at the time $\sqrt{2}$ larger than that for normal water (Fig.2, circles), confirming the libration nature of the solvent motion. Further investigations are needed to elucidate whether the librations act through a modulation of the transition frequency (vibronic effect) or the transition dipole moment (non-Condon effect).

The TG signals obtained with 5-fs and 15-fs pulses turn out to be virtually identical (Fig.2, inset) which indicates that the fastest part of solvation dynamics is over after ~20 fs. Furthermore, neither pump-probe nor TG experiments exhibit quantum beats associated with coherent excitation of several *p*-states. From this we conclude that the correlation time between the *p*-states is shorter than the expected beat period (~20 fs). This can also explain the fact that the linear absorption spectrum of the hydrated electron is completely featureless.

4. Conclusions

Pump-probe and transient grating experiments on the hydrated electron have been performed with an unprecedented 5-fs time resolution. Both experiments show a distinct recurrence at ~40 fs. Experiments with D_2O indicate that the this beat is caused by an underdamped librational mode that is strongly coupled to the optical transition. We are currently working on a model of the hydrated electron that comprises 3 electronic states as well as possible anharmonicity and non-Condon effects.

Acknowledgements. We would like to thank my Prof. P. F. Barbara and Dr. P. Walhout for useful discussion, and the Netherlands Foundation for Chemical and Physical Research and the Netherlands Organization for the Advancement of Science for financial support.

References

1 J. Schnitker *et al*, Phys. Rev. Lett. **60**, 456, 1988
2 A. Migus *et al*, Phys. Rev. Lett. **58**, 1559, 1987; F. H. Long *et al*, Phys. Rev.Lett. **64**, 1469, 1990
3 J. C. Alfano *et al*, J. Chem. Phys. **98**, 5996, 1993; Y. Kimura *et al*, J. Phys. Chem. **98**, 3450, 1994
4 C. Silva *et al*, Phys. Rev. Lett. **80**, 1086, 1998
5 M. F. Emde *et al*, Phys. Rev. Lett. **80**, 4645, 1998; A. Kummrow *et al*, J. Phys. Chem. **A102**, 4172, 1998
6 A. Baltuska *et al*, Opt. Lett. **22**, 102, 1997; Appl. Phys. **B65**, 175, 1997
7 C. -P. Hsu *et al*, J. Phys Chem. **B101**, 2546, 1997

Femtosecond Pump-Probe-Spectroscopy of the Solvated Electron in Water and Aqueous Salt Solutions

Michael Assel, Robert Laenen, and Alfred Laubereau

Technische Universität München, Physik Department E11, James Franck Str., 85748 Garching, Germany
E-mail: assel@physik.tu-muenchen.de

Abstract. Transient pump-probe experiments on the equilibrated, solvated electron in pure water and aqueous NaCl solution are carried out with femtosecond time resolution. A deconvolution of the experimental data using a 5-level scheme results in the following features: A first absorption spectrum appears immediately after excitation which can be attributed to a frequency shifted excited p'-state. The first relaxation inbetween the p-states and/or the reorganisation of the solvation shell occures on a timescale <80 fs. After an internal conversion p' $\rightarrow$ s'' with 230 $\pm$ 40 fs the solvent surrounding cools down on a charactristic timescale of 1.35 $\pm$ 0.15 ps. The corresponding timeconstants in pure water are merkedly accelerated.

The solvated electron in aqueous systems is of common interest in physics and chemestry. It is a first reaction step in many chemical and photochemical reactions as well as this quantum mechanical easy to describe system (electron in a potential well) can serve as a test system for theoretical calculations in respect to the description of the solvent surrounding.

The long lifetime of the solvated electron of hundreds of nanoseconds depending on the concentration of quencher molecules or ions in the solution allows pump-probe experiments of the equilibrated electron. We have performed such measurements using 620 nm pump pulses (120 fs) and a white light continuum (180 fs) in the range 520 to 1040 nm for the probing. Our excitation wavelength differs from previous measurements [1] and prepares a new physical situation where the electron is promoted to the highest of the three predicted p-states. As a result, new insights in the subsequent rapid intra p-state relaxation, solvation and internal conversion

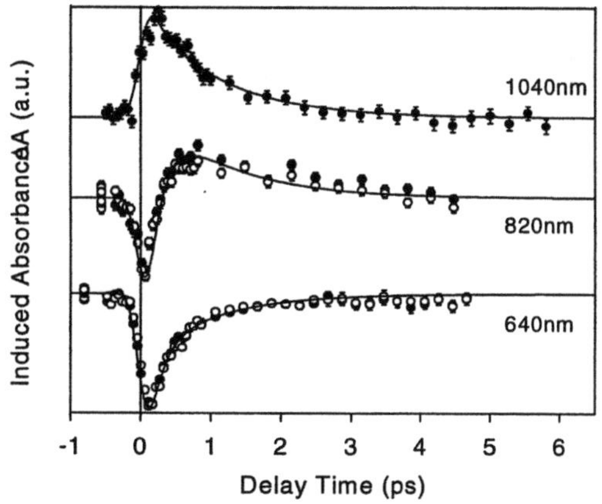

Fig 1: Induced probe absorbance at different wavelengths

dynamics are obtained. A special experimental setup allows simultaneous measurement of the parallel and perpendicular components of the probe transmission, yielding different results that reported previously [2]. An experimental example for a 5.9 M solution of NaCl is shown in Fig. 1 [3]. Three signal transients at a probe wavelength of 640 nm, 820 nm and 1040 nm are plotted, taken 1 ns after the generation of the solvated electrons by an intense UV- pulse with intensities in the range of TW/cm^2.

In the short wavelength region (< 750 nm) a bleaching is observed that disappears within a few picoseconds. At wavelengths > 900 nm induced absorbance occurs on a similar time scale. In the intermediate spectral region a rapid transition from an early bleaching to an induced absorbance indicates a rather complex dynamics. The parallel and perpendicular probe components (full points and open circles in Fig. 1) clearly indicate isotropic absorption of the solvated electrons in the solution and also for neat water (data not shown), quite different to previous observations and some theoretical predictions [4]. Our findings suggest a first fast solvent relaxation step below our experimental time resolution of 80 fs consistent with more recent predictions [5].

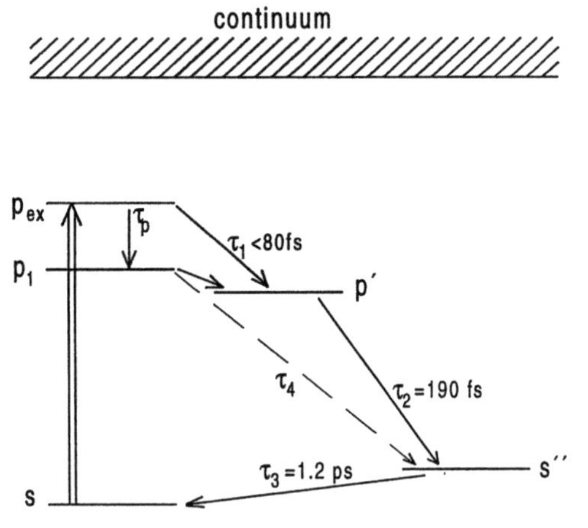

Fig 2: Novel energy level scheme of the solvated electron with intermediates p‘ and s“ .

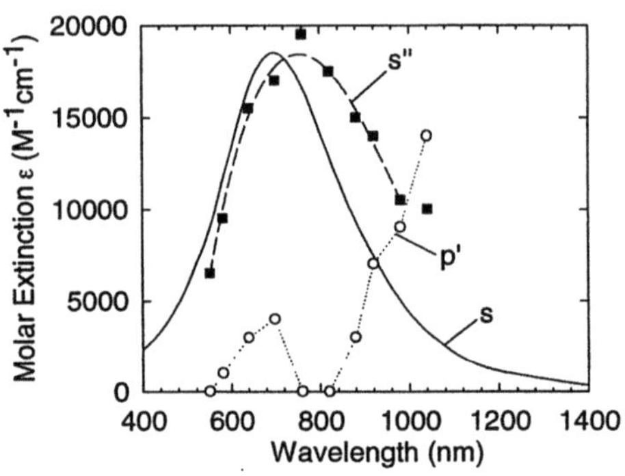

Fig 3: Molar extinction coefficients of the groundstate s (solid), first intermediate p' (dotted) and second intermediate state s'' (dashed), derived from the transient probe spectra.

The interpretation of our signal transients requires at least two intermediates populated after photoexcitation of a primary p-state p_{ex}. The level scheme accounting for our data is depicted in Fig. 2. The population relaxation of p_{ex} leads to the lowest p-state (p_1) accompanied by a rapid reorganization of the surrounding hydration layer involving the ionic part of the water molecules [2,4], leading to a modified p_1-state (p' in Fig. 2).

Subsequent internal conversion populates a modified ground state s'', the solvation character of which is proposed to differ from p' and finally relaxation back to the original ground state. Fitting rate equations for the model of Fig. 2 to the experimental data we obtain the probe absorption spectra of the intermediates p' and s'', in addition to the known absorption of the ground state s. The results are presented in Fig. 3. The data uniquely demonstrate that the first intermediate (p') is an excited electronic state (open circles) while the second intermediate (s'') is a modified ground state (full points). Strong evidence for the assignment of p' is the pronounced dip in the absorption at approximately 800 nm. The finding is explained as a superposition of excited state absorption p'→ continuum and of stimulated emission from the populated p'-state to the unoccupied modified ground state. The latter process is reported for solvated electrons for the first time. The energy change between p_1 and p' due to the rearranged solvent cavity is small so that the wavelength of maximum stimulated emission is only slightly red-shifted to the s→p_1 absorption band. The red-shift of the s''-absorption compared to the s-p band is in qualitative agreement with the assignment of s'' as a hot s-state.

For the dynamics we find the following: The first relaxation step p_{ex}→ p' is faster than our time resolution, τ_p, $\tau_1 < 80$ fs. Internal conversion from p' to s'' proceeds with $\tau_2 = 230 \pm 40$ fs and is uniquely identified. The electrons finally equilibrate from s'' to s with a time constant of $\tau_3 = 1.35 \pm 0.15$ps.

The results obtained for the solvated electron in pure water are qualitativly the same as in the aqueous salt solution. The absorption spectra of the excited states p' and s'' in water and the salt solution are undistinguishable within the measurement accuracy. Especially the dip in the p'-state absorption at 800 nm appears independent on the salt concentration. However the lifetime of the p'- and s''-state are noteably accelerated [6].

References

1 Y. Kimura, J. C. Alfano, P. K. Walhout and P. F. Barbara; J. Phys. Chem. **98**, 3450, 1994.

2 P. J. Reid, C. Silva, P. K. Walhout, P. F. Barbara, Chem. Phys. Lett. 228, 658, 1994

3 M. Assel, R. Laenen and A. Laubereau; J. Phys. Chem. A **102**, 2256, 1998.

4 B. J. Schwartz, P. J. Rossky; Phys. Rev. Lett. **72**, 3282, 1994.

5 S. Bratos, J.-Cl. Leicknam, D. Borgis, and A. Staib; Phys. Rev. E **55**, 7217, 1997.

6 M. Assel, R. Laenen, and A. Laubereau, to be published.

Excitation Energy Dependence of the H/D Isotope Effect in the Geminate Recombination of Hydrated Electrons

Martin U. Sander[1,2] *Mark S. Gudiksen,*[1] *and Klaus Luther*[1,2]

[1]Institut für Physikalische Chemie, Universität Göttingen,
Tammannstr. 6, D-37077 Göttingen, Germany.
☎ +49-551-393129, e-mail: msander@gwdg.de

[2]Max-Planck-Institut für Biophysikalische Chemie,
Am Fassberg, D-37077 Göttingen, Germany

Abstract. The probability for the geminate recombination of hydrated electrons was measured after two-photon excitation of liquid H_2O and D_2O at 10 eV two-photon energy. As compared to earlier results with an excitation energy of 8 eV, the H/D isotope effect on the recombination probability is reversed.

The details of the mechanism of the photo-ionization of liquid water have been under investigation for many years and are still rather controversial [1-6]. The prinicipal observables in these investigations are the rates and yields of geminate electron-ion pair recombination in normal and heavy water, obtained under varying experimental conditions. The key reactions are

$$(1)\quad (H_2O)_n \xrightarrow{2h\nu} (\bullet OH, H_3O^+)_{aq} + e^-_{aq}$$

$$(2)\quad e^-_{aq} + H_3O^+ \rightarrow H_{aq} + H_2O$$

$$(3)\quad e^-_{aq} + \bullet OH \rightarrow OH^-$$

Comparing the 8 eV and 10 eV excitations of H_2O, our previous study [3] showed that different mechanisms for the formation of the solvated electrons must be at work at these two excitation energies. In the present study, we further substantiate this finding by reporting a strong excitation energy dependence of the H/D isotope effect on the geminate recombination probability.

Experimental

Pure liquid H_2O and D_2O were pumped through a 1 mm stainless-steel cell equipped with very thin (0.2 mm) Suprasil windows. An excimer-pumped dye laser system with a Distributed-Feedback Dye Laser at its center provided 0.5 ps pulses at 497 nm. One part of this green light was used for the generation of a white-light continuum, while the other was frequency-doubled with subsequent amplification of the UV pulses in a KrF excimer amplifier. The samples were excited by the resulting 0.5 ps, 1-2 mJ pulses at 248.5 nm, delivering a two-photon energy of 10 eV. The decay of the hydrated electrons thus generated was followed by transient absorption measurements at detection wavelengths between 450 nm and 650 nm. Typical absorption-time profiles are shown in Fig. 1.

Results

As can be seen in Fig. 1, the initial absorption decays to a lower level in D_2O than in H_2O. Since hydrated electrons are the only absorbing species at the probe wavelength chosen, this means that the fraction of hydrated electrons diffusing away from their parents rather than undergoing geminate recombination is higher in H_2O than in D_2O. This result came as a surprise, as previous work by Gauduel *et al.* at excitation energies in the 8 eV range had established a marked isotope effect in the opposite direction [7], and Crowell and Bartels [5] had found this effect to be constant from 7.8 to 9.0 eV excitation (see Fig. 2).

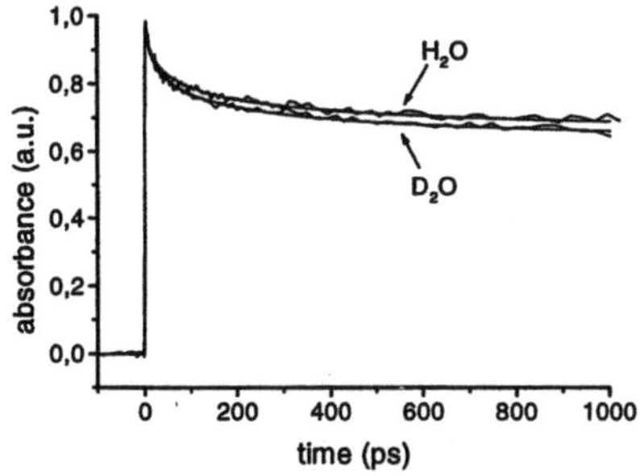

Fig. 1. Absorbance vs. time profiles of H_2O and D_2O at 600 nm upon two-photon excitation at 248.5 nm. Smooth lines: fit to the independent pairs model of ref. 8. The curves were scaled by the amplitude factors obtained from the fits.

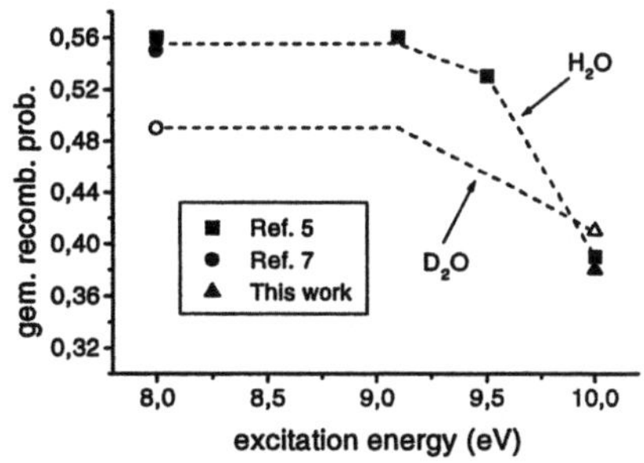

Fig. 2. Geminate recombination probabilities at different excitation energies for H_2O and D_2O. The dotted lines emphasize the inversion of the isotope effect, but do not correspond to any theoretical model.

We fitted our experimental curves to the independent pairs model, put forward by Pimblott [8]. A Gaussian distribution was used to approximate the spatial distribution of solvated electrons about their geminate partners. Extrapolating to infinite times, we found a geminate recombination probability p_{gr} of 0.38 for H_2O and 0.41 for D_2O. The value for H_2O is in good agreement with ref. 5, reporting $p_{gr}(H_2O)=0.39$. (Our previous value of $p_{gr}(H_2O)=0.30$ [3] was based on the parametrization of the experimental signal to a one-dimensional random-walk formalism, which was found to under-estimate recombination at longer times.)

These values should be compared with published results at 8 eV excitation: p_{gr} was found to be 0.55 [7] or 0.56 [5] for H_2O and 0.48 for D_2O [7]. As a general trend, going to the higher excitation energy leads to a smaller geminate recombination probability both for H_2O and D_2O. This lowering of p_{gr} is so much more pronounced for H_2O than for D_2O, however, that the isotope effect is inverted somewhere between 9.5 and 10 eV.

Discussion

The isotope effect at 8 eV excitation has been explained through [7,5]

a) the isotope effect of the recombination reaction (2) which requires considerable rearrangement of the reactants and is only partially diffusion-controlled, in contrast to the more important reaction (3);

b) the more orderly structure of liquid D_2O, in which the lower concentration of electron trap states supposedly leads to a larger average distance between ejected or transferred electrons and their parent molecules than in H_2O;

In order to account for their data at 7.8 to 9 eV excitation energy, Crowell and Bartels [5] needed a marginally narrower distribution for H_2O than for D_2O, with most of the measured isotope effect ascribed to reason a).

Naturally, these effects must also be the same for hydrated electrons generated by 10 eV excitation. Consequently, there must be some strong effect which is able to over-compensate for them. Indeed, our simulations show that the spatial distribution of the hydrated electrons is about 15% broader for H_2O than for D_2O (r.m.s. distances of 1.34 nm for H_2O and 1.16 nm for D_2O).

Apparently, the movement of the electron away from its parent molecule is correlated to some type of O-H bond breakage at 10 eV, but not at 8 eV excitation. In the context of a photo-induced electron transfer (PIET) mechanism [3], one excited H_2O^* molecule would transfer an electron to an acceptor site, gaining some of the necessary energy by simultaneously transferring a proton to a neighbouring water molecule. The proton transfer should be slowed down in D_2O, allowing the positive core of the D_2O^* a longer time to polarize its immediate neighbours, before the actual electron transfer occurs. This would lead to an enhanced formation of acceptor sites closer to the parent.

In an autoionization mechanism, on the other hand, one has a competition between the proton transfer and the Internal Conversion of X_2O^* into the ionization continuum. The slower proton transfer from D_2O^* could impart less kinetic energy to the leaving electron than in the case of H_2O^*.

Finally, the occurrence of neutral intermediates like the $H_3O\bullet$ radical cannot be ruled out (curiously, the theoretical absorption spectrum of $H_3O\bullet$ closely resembles the measured one of e^-_{aq} [9]!). The diffusion of this radical should be slower in D_2O, leading to a narrower spatial distribution.

Again: each of these explanations would be the same for 8 eV as for 10 eV and, in conclusion, only a change of the the dominant mechanism could explain the experimentally observed inversion of the H/D isotope effect.

References

[1] A. Migus, Y. Gauduel, J.L. Martin, A. Antonetti, *Phys.Rev.Lett.* **1987**, *58*, 1559

[2] H. Lu, F.H. Long, R.M. Bowman, K.B. Eisenthal, *J.Phys.Chem.* **1989**, *93*, 27

[3] M.U. Sander, K. Luther, J. Troe, *J.Phys.Chem.* **1993**, *97*, 11489-11492, and *Ber.Bunsenges.Phys.Chem.* **1993**, *97*, 953

[4] J.L. McGowen, H.M. Ajo, J.Z. Zhang, B.J. Schwartz, *Chem.Phys.Lett.* **1994**, *231*, 504

[5] R.A. Crowell, D.M. Bartels, *J.Phys.Chem.* **1996**, *100*, 17713

[6] A. Hertwig, H. Hippler, A.N. Unterreiner, P. Vöhringer, *Ber.Bunsenges.Phys.Chem.* **1998**, *102*, 805

[7] Y. Gauduel, S. Pommeret, A. Migus, A. Antonetti, *J.Phys.Chem.* **1991**, *95*, 533

[8] S. Pimblott, *J.Phys.Chem.* **1991**, *95*, 6949

[9] F.F. Muguet, H. Gelabert, Y. Gauduel, *J.Chim.Phys.* **1996**, *93*, 1808

Part XIII

Ultrafast Photochemistry

Investigating the Photochemical Dynamics of Chlorine Dioxide in Water and Acetonitrile using Ultrafast Pump-Probe Spectroscopy

Matthew J. Philpott, Sophia C. Hayes, and Philip J. Reid

Box 351700, Department of Chemistry, University of Washington, Seattle, WA 98195, USA. Email: preid@chem.washington.edu

Abstract. The photochemical reaction dynamics of chlorine dioxide (OClO) are investigated using femtosecond pump-probe spectroscopy. It is found that the majority of the dynamics can be attributed to the formation and vibrational relaxation of OClO from geminate recombination of the primary ClO and O photofragments.

1. Introduction

The condensed-phase reaction dynamics of chlorine dioxide (OClO) are of interest due to the importance of this compound in stratospheric chemistry.[1] Photoexcitation of OClO results in either the formation of ClO and O, or Cl and O_2:

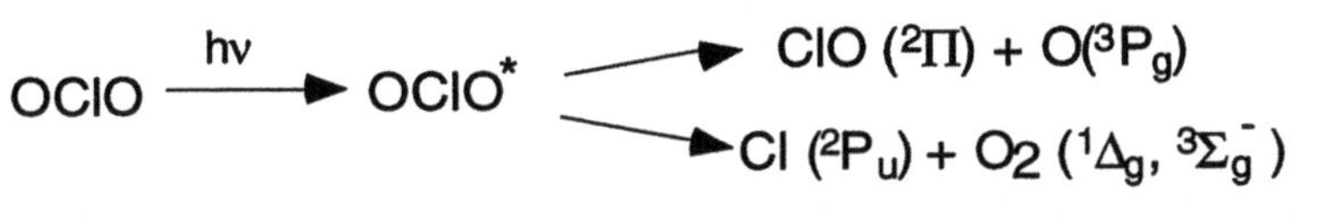

The environmental impact of OClO originates from its ability to photochemically produce atomic chlorine. Interestingly, the quantum yield for Cl formation is phase dependent ranging from 0.04 in the gas phase to unity in low-temperature matrices. [1-3] The quantum yield in solution is intermediate with values of 0.1-0.2 in water and methanol. Current interest involves understanding the origin of this phase dependence, especially the role of solvent in defining the photochemical reaction dynamics. Recently, ultrafast pump-probe studies have been employed to investigate the appearance kinetics and ground-state reactivity of the OClO photoproducts.[4-7] Early studies interpreted the pump-probe dynamics as originating from a precursor to atomic chlorine, ClOO, which is formed following photoexcitation of OClO.[4] However, it has recently been suggested that geminate recombination of ClO and O to form vibrationally hot OClO is responsible for the observed dynamics.[5] Therefore, the events that occur following OClO photoexcitation remain unclear. In this paper, we present the results of our recent subpicosecond, pump-probe studies of OClO dissolved in water and acetonitrile that were performed to ascertain the importance of geminate recombination in the condensed-phase reaction dynamics of OClO.[6-8]

2. Experimental Methods

Photoexcitation of OClO at 400 nm is accomplished using the frequency doubled output of a regeneratively amplified Ti:Sapphire laser. The details of the laser spectrometer have been presented elsewhere.[6,7] The change in optical density following photoexcitation was measured for probe wavelengths from 267 nm to 1 μm using either the fundamental and various harmonics of the amplified laser, or by spectrally filtering a white-light continuum generated in water. The change in optical density was measured on a shot-to-shot basis, with 200 laser shots averaged per time point. The temporal resolution of the spectrometer as determined by the optical Kerr effect in water is 700 fs.

3. Results and Discussion

Figure 1 presents a subset of the pump-probe data obtained for OClO dissolved in water and acetonitrile.[6,7] Comparison of the dynamics observed in acetonitrile to those in water provides information on the contribution of geminate recombination to the pump-probe dynamics. The 400 nm data provide information on the depletion and recovery of ground-state OClO, and the increase in long-time optical density depletion observed for acetonitrile relative to water at this wavelength demonstrates that the quantum yield for ClO/O cage-escape increases by a factor of five in acetonitrile. This increase should result in a corresponding reduction in the amount of vibrationally excited OClO present in acetonitrile relative to water. This expectation is confirmed by the smaller optical density changes observed between 450 and 600 nm for acetonitrile. However, for probe wavelengths greater than 650 nm, the optical density changes in both solvents are roughly equivalent. This observation suggests that vibrationally hot OClO is not the only contributor to the observed dynamics in this spectral region, and that ground-state ClOO may be present. This conclusion is supported by calculations employing the description of the optically prepared excited state derived from analysis of the absolute resonance Raman cross sections of OClO.[9]

A second result of this study is that the observed dynamics in acetonitrile occur on a longer timescale relative to water. Specifically, the optical density decay times in water range from 0.5 at 900 nm to 3 ps at 550 nm whereas the respective times in acetonitrile are 4 and 11 ps. This result demonstrates that the timescale for intermolecular vibrational relaxation is dependent on the structure of the solvent. In particular, the vibrational relaxation time of OClO and/or the decomposition of ClOO may be accelerated in solvents capable of intermolecular hydrogen bonding. This hypothesis is currently being tested by investigating OClO dissolved in halogenated alcohols were the strength of intermolecular hydrogen bonding can be systematically adjusted.[10]

Acknowledgments. This work was supported by the NSF CAREER Program (CHE-9701717). Acknowledgment is also made to the donors of the Petroleum Research Fund, administered by the American Chemical Society. P.J.R. is the recipient of a Camille and Henry Dreyfus New Faculty Award, and is a Cottrell Scholar of the Research Corporation.

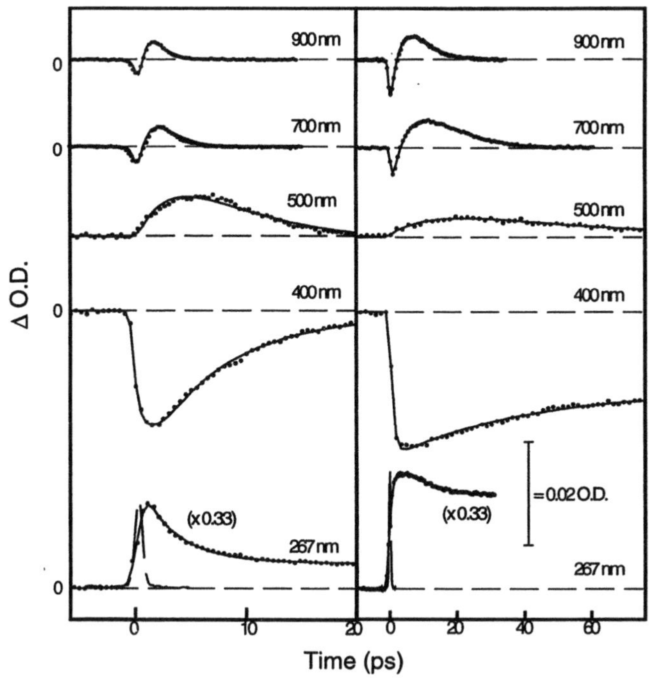

Fig. 1. Pump-probe data of OClO dissolved in water (left) and acetonitrile (right).

References

1 V. Vaida and J. D. Simon, **268,** 1443, 1995.
2 H. F. Davis and Y. T. Lee, J. Chem. Phys. **105,** 8142, 1996.
3 M. M. Rochkind and G. C. Pimentel, J. Chem. Phys. **46,** 4481, 1967.
4 Y. J. Chang and J. D. Simon, J. Phys. Chem. **100,** 6406, 1996.
5 J. Thoergersen, P. U. Jepsen, C. L. Thomsen, J. A. Poulsen, J. R. Buyberg, and S. R. Keiding, J. Phys. Chem.-A **101** , 3317, 1997.
6 M. J. Philpott, S. Charalambous, and P. J. Reid, Chem. Phys. Lett. **281,** 1, 1997.
7 M. J. Philpott, S. C. Hayes, and P. J. Reid, Chem. Phys. , in press.
8 S. C. Hayes, M. J. Philpott, and P. J. Reid, J. Chem. Phys., in press.
9 A. P. Esposito, C. E. Foster, R. Beckman, and P. J. Reid, J. Phys. Chem. A **101,** 5309, 1997. C. E. Foster and P. J. Reid, J. Phys. Chem.-A **102,** 3514, 1998.
10 M. J. Philpott, S. C. Hayes, and P. J. Reid, manuscript in preparation.

Instantaneous Product Anisotropies in the fs-Photolysis of Triiodide

Stephan Hess and Peter Vöhringer

Biomolecular and Chemical Dynamics, Max-Planck-Institute for Biophysical Chemistry, Am Fassberg 11, D-37077 Göttingen, Germany
E-mail: pvoehri@gwdg.de

Abstract. Multiple-pulse fs-polarization experiments are presented in order to explore exit channel dynamics in the femtosecond photodissociation of triiodide ions in ethanol solution. The recoil dynamics can be modeled quite accurately with a simple classical description of motion along the reaction coordinate. The corresponding instantaneous anisotropies induced in the system exhibit coherent contributions that arise from degeneracies in the electronic structure at very early stages of the bond breakage. The coherence decay reflects dephasing and lifting of the degeneracy as the fragments separate.

1. Introduction

The photodissociation of triiodide in liquid solution has been shown to occur on a variety of time scales reflecting different elementary dynamical processes that compose the chemical reaction [1]. The initial evolution of the system towards the transition state region of the corresponding I_2^-/I full collision can be observed in real time through wavelength selective probing of coherent wave packet motion along the nominal reaction coordinate [2]. Subsequent bond breakage with coherent product formation takes place within the first few hundred femtoseconds after impulsive photolysis [1,2]. As evidenced by the band shape stability of the diiodide photoproduct, excess vibrational energy is fully dissipated after 3 ps [3]. The reaction can be considered complete after approximately 20 ps. This is the time required for diffusive orientational randomization as monitored through time-resolved transient anisotropies with direct probing of the diatomic fragment [4]. In addition, the inertial component to the product anisotropy reveals a significant amount of excess rotational energy in the product [4]. However, the time scale for rotational relaxation and the time scales for recoil of the fragments and product solvation (i.e. motion towards the asymptotic limit of the reaction with negligible interactions between the fragments) are entirely unclear. We present new multiple-pulse experiments with polarization control in order to elucidate the time scale and the nature of these exit channel dynamics.

A 30-fs, 400 nm pulse photolyzes triiodide ions thereby generating I_2^- product ions. A linearly polarized 30-fs, 800-nm pump pulse excites the system at well defined delays T after photolysis. The anisotropic response of the product ions is detected for variable pump-probe delays τ with an 800-nm probe pulse polarized at 45° with respect to the pump. A Wollaston prism polarizer selects the parallel and perpendicular components of the pump-induced optical density from which the instantaneous anisotropy r(T,τ) can be determined.

2. Results and Discussion

Fig 1. shows the magic angle response of diiodide observed at 100 fs after photolysis of triiodide. At this photolysis-pump-delay T, the signal essentially displays a transient bleach. The impulsive stimulated Raman response of diiodide is clearly visible as periodic modulations [5]. For T < 1ps, the RISRS contribution contains basically three oscillatory components, two of which can be assigned as the fundamental and the second overtone of I_2^- around 112 and 220 cm^{-1}. The third component shows frequencies around 70 cm^{-1}. The T-dependence of the RISRS frequencies is displayed in Fig. 2. We model the recoil dynamics using an empirical LEPS potential described in Ref. [5]. The reaction coordinate for the I_3^- photodissociation is identified on this surface as the direction of steepest descent that leads to motion away from the Franck-Condon region toward the product channels. Motion along this 1-dimensional potential is modeled classically and damping is introduced phenomenologically using the zero frequency viscosity of the neat solvent. For each interfragment separation, the fundamental diiodide stretching frequency can be determined. The resulting T-dependence of the diiodide stretching frequency that results from motion along this model reaction coordinate is reproduced in Fig. 2. For comparison, the results of pure quantum dynamics calculations on the full LEPS potential are also shown. According to the classical model, the low-frequency component of the RISRS-response around 70 cm^{-1} arises from regions of the potential energy surface that are in close proximity to the transition state of the I_2^-/I full collision where the diiodide bond becomes unbound. The appearance of the fundamental frequency even at the earliest stages of the bond breakage may be explained by the enormous dispersion that the wavepacket experiences during the photolysis pulse. According to this

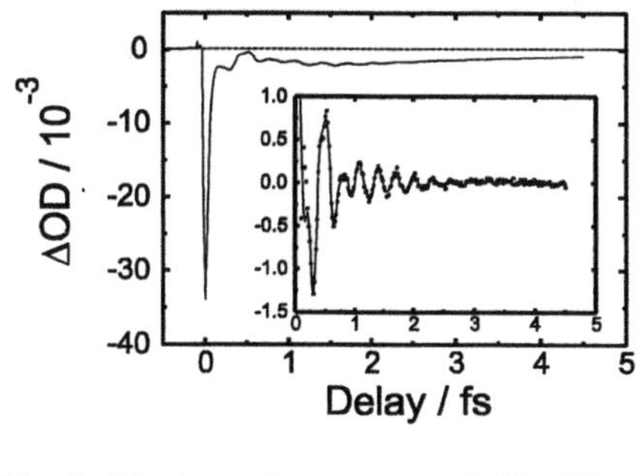

Fig. 1: Magic angle response of diiodide at a delay of 100 fs after impulsive photolysis of triiodide. Pump and probe wavelengths were 800 nm. Inset emphasizes the RISRS contribution (circles) together with a fit to a sum of exponentially damped cosines.

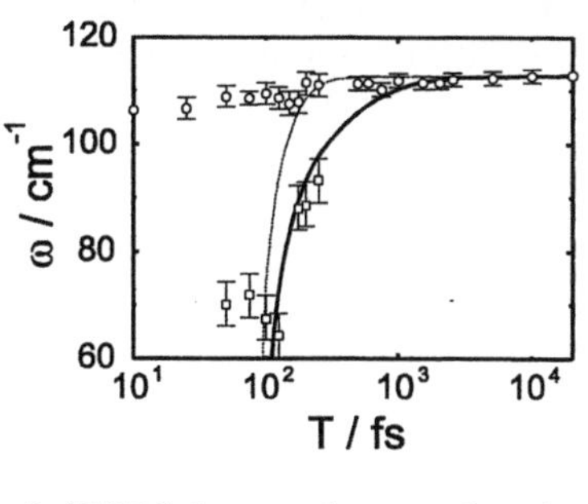

Fig. 2: RISRS frequencies as a function of photolysis-pump delay (circles). Solid curve: 1-dimensional classical model calculations. Dashed curve: pure quantum dynamics simulations for the isolated system

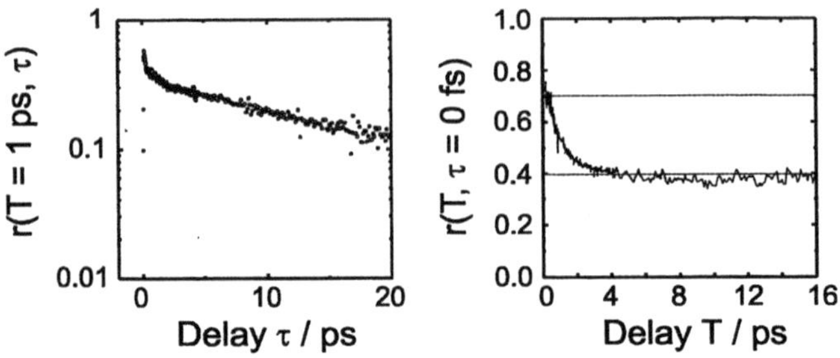

Fig. 3: Instantaneous anisotropy of triiodide/diiodide as a function of pump-probe time delay (left) and photolysis-pump time delay (right).

classical analysis, it takes about 1 ps until the fundamental frequency of diiodide of 112 cm^{-1} is recovered and residual interactions between the fragments vanish.

The instantaneous anisotropy decay corresponding to a photolysis pump delay of 1 ps is shown in Fig. 3. A triphasic behavior is clearly observed. An exponential component with a time constant of 11 ps reflects the rotational diffusion of the diiodide photofragment. An additional Gaussian component is reminiscent of the inertial decay of r(T=1ps,τ). The observed correlation time of 1 ps matches the free rotor rotational correlation time of diiodide at a temperature of 300 K. Apparently, at a delay of 1 ps after impulsive photolysis of triiodide, rotational relaxation of the diatomic fragment is already complete as expected from classical MD-simulations. Finally, r(T=1ps,τ=0) has a value of 0.6. The anisotropy decays in less than 100 fs to the value of 0.4 expected for parallel pump and probe transition dipoles. The T-dependence of the initial anisotropy is also diplayed in Fig. 3. It can be seen, that immediately after photolysis of the parent ion, the pump induced anisotropy has a maximum value of 0.7 which is indicative of higher lying electronic degeneracies in I_3^-. [6] The 800-nm pump creates a coherence in this degenerate electronic manifold which dephases along τ in less than 100 fs. As the fragments begin to seperate, this electronic degeneracy is lifted. The decay of r(T,τ=0) along the T-delay simply reflects an increasing level splitting with increasing interfragment seperation. The time constant for the decay of r(T,τ=0) is in qualitative agreement with the RISRS experiments which reveal a time scale of about 1 ps for recovery of the fundamental frequency of the equilibrated product ions. Hence, we conclude that it takes about 1 ps before the fragments of the I_3^--photodissociation appear in the asymptotic limit.

References

1 U. Banin and S. Ruhman, J. Chem. Phys. **98**, 4391, 1993.
2 T. Kühne, R. Küster, and P. Vöhringer, Chem. Phys. **233**, 161, 1998.
3 T. Kühne and P. Vöhringer, J. Chem. Phys. **105**, 10788, 1996.
4 T. Kühne and P. Vöhringer, J. Phys. Chem. A **102**, 4177, 1998..
5 U. Banin, R. Kosloff, and S. Ruhman, Chem. Phys. **183**, 289, 1994.
6 K. Wynne and R.M. Hochstrasser, Chem. Phys. **171**, 179, 1993.

Photoinduced Release of Nitric Oxide from S-nitrosoglutathione

Chengfei Wang, Brian K. Mohney, Rojana Leecharoen, and Gilbert C. Walker[1]

[1] Department of Chemistry, University of Pittsburgh, Pittsburgh, PA 15260, USA
E-mail: gilbertw@vm2.cis.pitt.edu

Abstract. A study has been undertaken to characterize the picosecond optical and infrared spectral kinetics associated with photoinduced bond cleavage of NO from S-nitrosoglutathione in water. Solvated NO formation and geminate recombination are observed.

1. Introduction

Nitric oxide (NO) has been found to perform many roles in intracellular signaling.[1] In the bloodstream, NO is found primarily bound to cysteine residues. A study of the photodynamics of S-N bond cleavage may reveal details of the mechanism by which the protein harbors the ligand. Glutathione (GSH) has been studied because it is both a naturally occurring NO-carrier and a model system for the principal bloodstream host, albumin.[2]

2. Experimental Methods

The time-resolved data were collected using a pump-probe technique, pumping with 400 nm and probing with either 400 nm or a tunable IR probe. The apparatus has been described in detail elsewhere.[3] The 400 nm pump-400 nm probe measurements were done in a 1mm pathlength flowing cell. The 400 nm pump-IR probe measurements were done at in a 100 μm pathlength CaF_2 flowing cell or in a 100 μm thick free jet. Flowing the sample prevented detrimental burning/bleaching of the sample and provided a fresh sample volume for each laser pulse. It also eliminated problems from NO and/or O_2 bubbles which formed in a closed, spinning sample cell.

3. Results and Discussion

Figure 1 shows the FTIR difference spectrum of GS-NO minus GSH in deuterated water solution. The spectrum shows a (cysteine)S-NO resonance at circa 1530 cm^{-1}, and perturbations in the Amide I' region around 1650 cm^{-1}. The NO resonance is quite wide (67 cm^{-1}) which may indicate a distribution of conformers. Photoexcitation of nitrosothiols and preparing either S_1 or S_2, followed by infrared probe provides the time-dependent difference spectrum which reveals population and relaxation dynamics for both the NO and the peptide or protein from which it was cleaved. Figure 1 also shows the infrared spectral changes resulting from 400 nm excitation, which prepares S_2. An

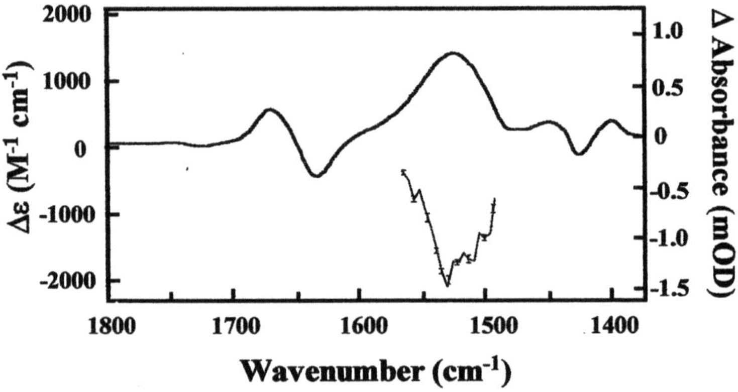

Fig. 1. IR difference spectrum (GS-NO minus GSH) (solid line) and 10 ps difference spectrum (crosses around 1530 cm^{-1}).

induced bleach is observed from ca 1490-1570 cm^{-1}, corresponding to loss of NO population from GS-NO. The spectral kinetics in this wavelength region show that the bleach recovers on two time-scales, 2.6 and 30 ps, respectively. The early time bleach amplitude is consistent with the predicted bleach signal, based on the excitation pulse energy. The fast part of the absorbance recovery is attributed to geminate recombination and the slow part to thermal relaxation. The bleach recovery does not show a pronounced frequency dependence.

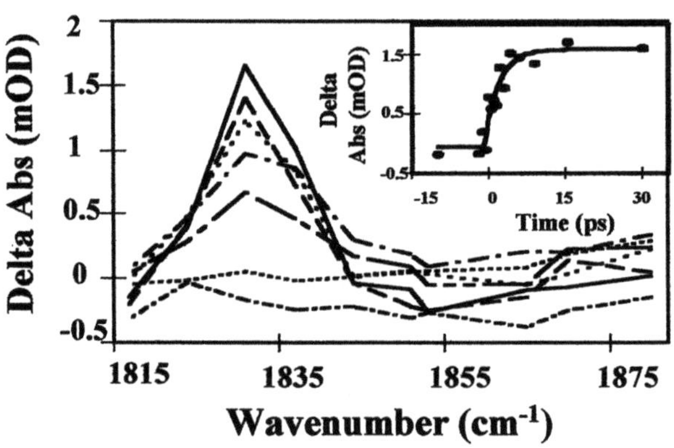

Fig. 2 Photoinduced formation of NO radical.

Figure 2 shows the time-dependence of the infrared absorbance in the region where the nitric oxide radical is expected to absorb. The gas phase NO absorbs around 1870 cm^{-1} while NO radical in water absorbs around 1835 cm^{-1}. In this study the formation time of the solvated NO species is circa 2.7 ps. There is no pronounced evidence of NO vibrational relaxation. The solvation process is not pronounced in the transient IR spectra, although there is a detectable transient absorbance between 1870 cm^{-1} and 1835 cm^{-1} during the first few picoseconds.

Rotational diffusion of photocleaved NO in the vicinity of the thiyl radical and the surrounding water cage provides a crude explanation for the similarity of the time-scales for rebinding and solvation. The rotational diffusion time of NO in water is estimated to be 1-2 ps. Future examination of the solvent and parent dependence of the rebinding rate, as well as the time-dependence of the anisotropy of the unsolvated radical will test this simple model, which neglects details of the photocleavage environment.

400nm pump/400nm probe experiments have been made, to study the kinetics of the optically prepared electronic states. These data are dominated by a signal from the geminately rebound species that is in a conformation that is not the equilibrium ground state conformation, and is the subject of ongoing study. Anisotropy data were obtained by measuring the polarized probe absorbances parallel ($||$) and perpendicular ($\perp$) to the polarized pump pulse and determining the mean angle between the pumped and probed transition dipole moments using the relation

$$\langle \cos^2[\Theta(t)] \rangle = (1+5r(t))/3, \text{ where } r(t) = (I_{||} - I_{\perp})/(I_{||} + 2I_{\perp}). \qquad (1)$$

The anisotropy for 400 nm pump/400 nm probe reveal a time-dependent anisotropy that is circa 0.2 +/- 0.04 between 0.4 and 3 ps. This signal is caused by both the anisotropy of the bleach and a single new absorbing species, and thus the new species has a transition moment that is at a 45° angle relative to the originally excited transition.

4. Conclusions

In summary, the first ultrafast studies of photoinduced NO deligation dynamics from S-nitrosothiol have been presented. These data may serve as a basis for modeling the structure and photodynamics in larger proteins that are S-nitrosylated.

Acknowledgments. We gratefully acknowledge the Petroleum Research Fund, the CRDF fund of the University of Pittsburgh and 3M Company.

References

1 J. S. Stamler *Cell,* **1994,** *78*, 931-936

2 B. K. Mohney and G. C. Walker *J. Am. Chem. Soc.*,1997, *119,* 9311-9312

3 B. Akhremitchev, C. Wang and G. C. Walker *Rev. Sci. Inst.*, **1996,** *67*, 3799.

Aromatic solute photoionization studied by femtosecond transient absorption spectroscopy

Jorge Peon, Tetsuro Yuzawa*, and Bern Kohler

Department of Chemistry, The Ohio State University, 100 W. 18th Ave, Columbus, OH 43210 USA
*Present address: Advanced Research Institute of Science and Engineering, Waseda University, Tokyo, Japan

E-mail: kohler@chemistry.ohio-state.edu

Abstract. Femtosecond pump-probe techniques are used to study the photoionization of aromatic molecules in polar solution.

1. Introduction

The photoionization of aromatic molecules by ultraviolet light is an important process in photochemistry and photobiology. Despite our detailed understanding of the photoionization of isolated molecules, there is still no comprehensive, microscopic theory to describe light-induced charge separation in the condensed phase. To construct such a theory, more information is required on (1) the nature of the ionizing electronic states of the solute, and (2) the elementary dynamical events that control charge separation and recombination. The minimum energy needed to ionize a molecule dissolved in a polar solvent can be reduced by more than 3 eV relative to the gas-phase, allowing experiments to be carried out with ultraviolet photons of modest energy. Such photons are not absorbed by the solvent at the one-photon level, effectively restricting the ionization process to the solute. In addition to changes in the energetics, mechanisms for charge separation exist in the condensed phase which have no counterpart in isolated molecules [1]. Dynamical studies of these mechanisms offer the possibility of new insights into the solute-solvent interactions that are central to much of solution-phase chemistry. Here we describe the use of femtosecond transient absorption spectroscopy to study N,N,N',N'-tetramethyl-p-phenylenediamine (TMPD), in several polar solvents.

2. Experimental Methods

Pump-probe transient absorption measurements were performed at room temperature on 2.0 millimolar solutions of TMPD in acetonitrile, ethanol, and methanol using a regeneratively amplified Ti:sapphire femtosecond laser system. By tripling the laser output in BBO, a 260 nm (4.77 eV) ~150-fs pump pulse was generated and used to photoexcite TMPD in its S_2 absorption band. Solutions were recirculated through a 0.5 mm pathlength flow cell. Probing was done with a white-light continuum generated in water. Transient absorbance signals were recorded either by using interference filters to spectrally select a portion of the continuum in conjunction with phase-sensitive detection, or by sending the entire continuum into an imaging spectrometer equipped with a CCD detector. The latter technique was used to record transient spectra at fixed pump-probe delays. All of the signals presented here are linear in the pump pulse energy, confirming single-photon excitation.

3. Results and Discussion

The spectrum of the radical cation of TMPD ("Wurster's Blue") is shown by the dashed line in Figure 1. Detecting photoionization by monitoring absorbance in

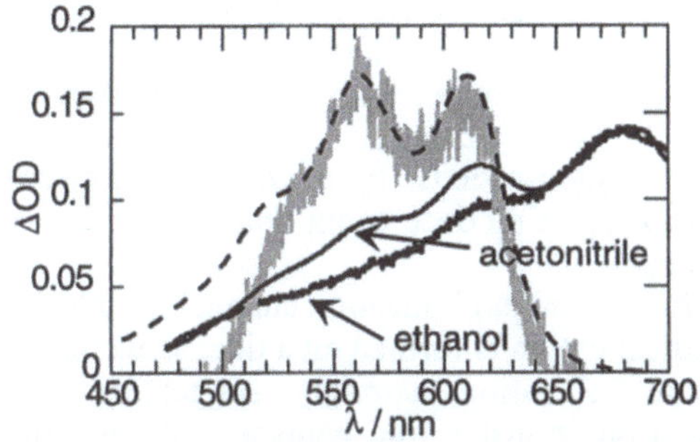

Fig 1. Transient absorption of TMPD at 30 ps delay in acetonitrile and ethanol (labelled curves), steady-state absorption spectrum of chemically-prepared Wurster's blue (dashed curve), and difference spectrum between acetonitrile and ethanol spectra (gray curve).

this band is difficult due to the presence of a very broad, overlapping band arising from excited state absorption of neutral TMPD [2,3]. The latter band accounts for the signal seen in ethanol at 30 ps delay (Fig. 1) since ionization is not expected in this solvent at our photon energy. In acetonitrile there may be some ionization by 30 ps, since the difference spectrum (gray curve in Fig. 1) between the data in acetonitrile and ethanol strongly resembles the radical cation band. Slow growth of the optical density at wavelengths corresponding to Wurster's blue in both acetonitrile and methanol (Fig. 2), provide evidence for a slow ionization process. We speculate that this slow process reflects ionization by a photoinduced electron transfer mechanism.

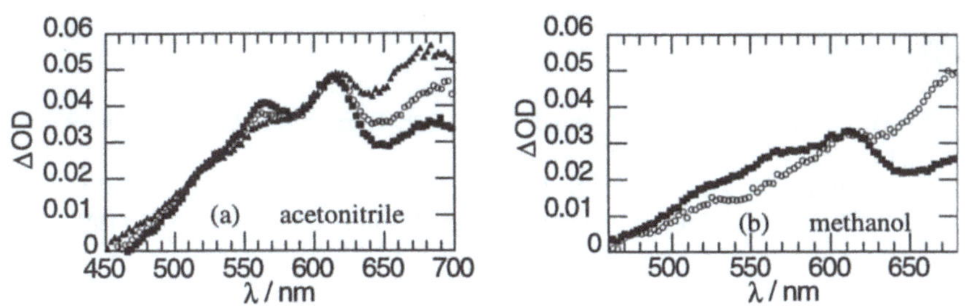

Fig. 2. (a) Transient absorption in acetonitrile at 20 ps (triangles), 300 ps (circles), and 600 ps (squares). (b) Transient absorption in methanol at 30 ps (circles) and 4.5 ns (squares).

Further evidence that ionization does not occur to a significant extent on an ultrafast timescale upon photoexciting TMPD at 260 nm comes from transient signals measured at discrete wavelengths. Data in Fig. 3 are for acetonitrile, but the results in methanol and ethanol are qualitatively similar. At long wavelengths (850 nm), the signal shows an initial decay followed by constant absorbance, while the signal at blue-edge wavelengths (570 nm) appears immediately within our time resolution (~150 fs) and shows no decay. At intermediate wavelengths (740 nm), the signal rises slightly in less than 1 picosecond. These data indicate ultrafast spectral reshaping of the $S_1 \rightarrow S_n$ band due to vibrational relaxation, solvation or to both processes. Following internal conversion, neutral TMPD molecules will have 9,000 cm^{-1} of excess vibrational energy. The dissipation of this energy requires several picoseconds. No evidence of electron-ion geminate recombination is observed, suggesting once again that single-photon photoionization of TMPD is a slow process.

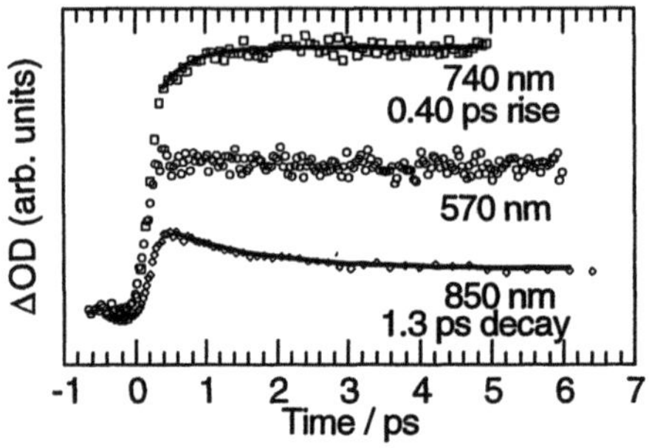

Fig. 3. Transients at three probe wavelengths for TMPD in acetonitrile

In conclusion, we have observed a slow (>100 ps) ionization channel for TMPD photoexcited at 260 nm. The TMPD radical cation is formed on a time scale that is much slower than that of vibrational relaxation, strongly suggesting that photoionization proceeds by an electron transfer mechanism without the intermediate formation of the delocalized electron. There is no evidence of ultrafast geminate recombination in the solvents examined. Further work is underway to better understand the dynamics of charge ejection from aromatic molecules in polar solvents.

Acknowledgments. This work was supported in part by the Ohio Board of Regents' Research Challenge Program.

References

1 M. U. Sander, K. Luther, and J. Troe, Ber. Bunsenges. Phys. Chem. **97**, 953 (1993).
2 Y. Hirata, Y. Tanaka, and N. Mataga, J. Phys. Chem. **89**, 4031 (1985).
3 Y. Hirata, M. Ichikawa, and N. Mataga, J. Phys. Chem. **94**, 3872 (1990).

Ultrafast Photomodulation of Conformation in Azobenzene Containing Cyclic Peptides

J. Wachtveitl[1], T. Nägele[1], A.J. Wurzer[1], M. Schenk[2] and L. Moroder[2]

[1] Institut für Medizinische Optik, Oettingenstr. 67, Ludwig-Maximilians-Universität München, 80538 München, Germany
[2] Max-Planck-Institut für Biochemie, am Klopferspitz 18a, 82152 Martinsried, Germany
E-mail: Josef.Wachtveitl@physik.uni-muenchen.de

Abstract. Real-time observation of conformational dynamics is described in a cyclic peptide. The reaction is triggered by subpicosecond photoisomerization of an integral azobenzene compound.

1. Introduction

The photoinitiation of conformational changes in peptides is a promising approach to study the mechanisms underlying the fast reactions in protein folding. The photoisomerization of azobenzene can be utilized to switch between structural motifs in cyclic peptides. The ultrafast photoreaction of azobenzene was recently subject to extensive investigations [1-3]. Here, we present a reaction model based on semiempirical calculations of the potential enegy surfaces (PES) and describe the main steps from a fast, reversible photoswitch to a photoresponsive peptide.

2. Methods

The transient absorption changes were recorded in a fs pump-probe setup including a multi-channel detection system as described in detail elsewhere [1,4]. For the PES calculation the MNDO-PM3-RHF method with CI and single excitation for the excited state was applied with HyperChem 5.0. The *trans*-ground state conformation was optimized with a conjugate gradient method. The torsion of the phenyl rings according to the C-N=N-C plane is fixed at an angle of 45°. Two degrees of freedom - rotation ψ (0°-180°) and inversion φ (120°-240°) - were considered in 10° steps (Figure 1a). The synthesis of the cyclic azo-bis-cysteinyl-peptides and of their linear precursors is described elsewhere [5].

3. Results and Discussion

The ability of azobenzene (Fig.1a) to act as fast photoswitch was tested by fs transient absorption spectroscopy. The dynamics of the photoisomerization of azobenzene excited at 435 nm in the long wavelength ($n\pi^*$)-band were analyzed over a wide spectral range for both isomers. Extremely fast photoproduct formation was

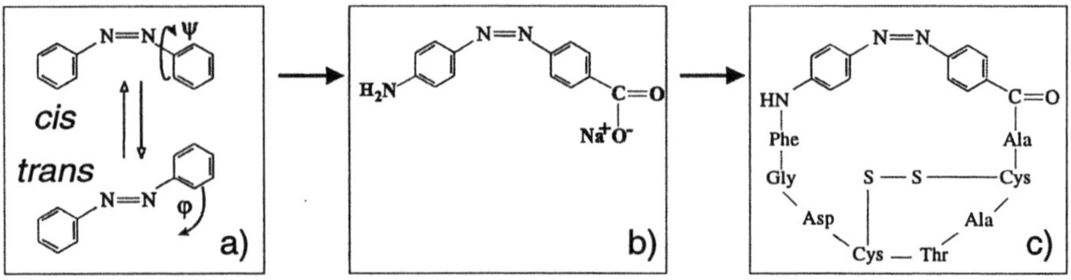

Fig. 1. Main steps in the synthesis of a photoresponsive cyclic peptide. Azobenzene is a fast, stable light switch and undergoes strong configurational changes upon isomerization (a). Addition of an amino- and a carboxyl-group leads to the pseudo-amino-acid (4-amino)phenylazobenzoic acid (APB), which can be inserted into peptides (b). Combination with an active-site fragment of thioredoxin reductase leads to a photoresponsive (bi-)cyclic peptide (c).

observed with dominating 170 fs (for *cis*-azobenzene) and 320 ps (for *trans*-azobenzene) components [1]. At early delay times the photoreaction clearly exhibits nonexponential and oscillatory features. The further temporal evolution is dominated by orientational and cooling effects: hot photoproducts cool via intra- and intermolecular energy redistribution processes on a timescale of 10 ps, the orientation depolarization time was found to be 14 ps (in ethanol). The experiments are consistent with the idea of a reaction pathway along a barrierless PES of the S_1 state and are well explained by the S_0 and S_1 PES obtained in semi-empirical calculations (Figure 2). Energy barriers separate *cis*-and *trans*- isomers in the lowest electronic state along the inversion and the rotation coordinate. The transition area between the two surfaces is clearly visible, however, due to the limits imposed by the few degrees of freedom considered and the accuracy of the semiempirical calculations, no conical intersection can be found. The photoinduced conformational changes can be described by a process termed "hula twist" mechanism [6]: after photoexcitation an initial movement of the N=N-C bond out of the molecular plane occurs and allows an inversional motion of both phenyl rings. The motion appears in our definition along the rotational coordinate (ψ). Subsequently the wavepacket reaches the transition area, i.e. the region with the smallest S_0/S_1 energy difference (ψ=85°, φ =140°). Here the radiationless electronic relaxation from the S_1 to the S_0 state is most likely. Finally the ground state relaxation to the *cis*-conformation (ψ=180°, φ=120°) follows.

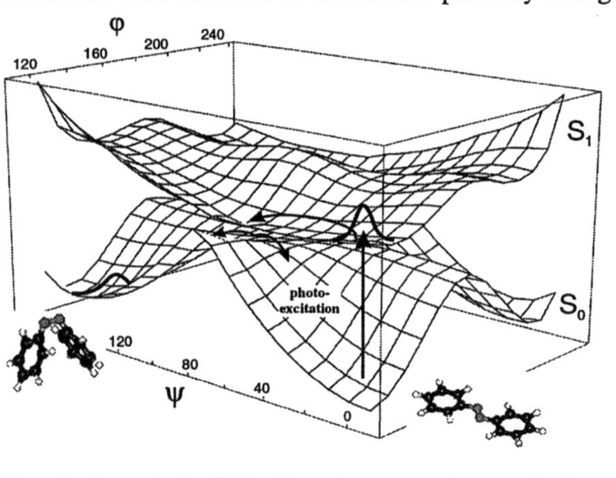

Fig. 2. S_0 and S_1 PES of azobenzene. Arrows indicate *trans*-photoexcitation and the most likely trajectory of a wavepacket along the steepest surface gradient.

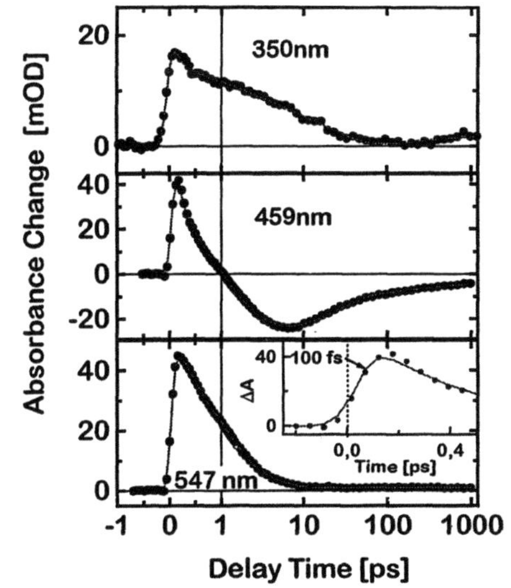

Fig. 3. Transient absorbance changes of *trans*-APB in ethanol after $\pi\pi^*$-excitation. The probing wavelengths shown represent spectral regions with dominating *cis*-APB (350 nm), *trans*-APB (459 nm) absorption or without significant ground state absorption (547 nm). The signal increase at λ_{pr}=547 nm exhibits a 100 fs delay. The expected time zero without delay is indicated by an arrow.

Substitution of the two phenyl rings of azobenzene with an amino and a carboxy group (Fig. 1b) leads to a red shift of the strong $\pi \rightarrow \pi^*$ transition of the *trans*-APB isomer, which allows its excitation at 435 nm [4]. Characteristic for the main reaction pathway of the APB photoreaction (Fig. 3) is a fast movement of the prepared wavepacket away from the initially populated FC regime on the S_2 potential energy surface within 100 fs, a fast, radiationless $S_2 \rightarrow S_1$ transition, a diffusion type wavepacket motion towards the S_1 minimum and a fast photoproduct formation within ≈ 1 ps. The orientational depolarization for *trans*-APB is 15 times slower than for azobenzene.

First FTIR and NMR experiments show that cyclic peptides linked to this photoswitch (Fig. 1c) can be reversibly modified in their conformation, dependent on the isomerisation state of the azo-compound. The *cis/trans* isomer equilibria indicate conformational preferences of the peptide chain, which are currently subject to fs experiments.

In conclusion, we observed excited state and ground state dynamics of azobenzene compounds and showed that isomerization occurs on a timescale faster than 250 fs. Azo-peptide compounds should be a highly suitable system for the study of fast conformational changes.

References

1 T. Nägele, R. Hoche, W. Zinth & J. Wachtveitl, *Chem. Phys. Lett.* **272,** 489, 1997.
2 I.K. Lednev, T.Q. Ye, R.E. Hester & J.N. Moore, *J. Phys. Chem.*, **100,** 13338, 1996.
3 I.K. Lednev, T.Q. Ye, M. Matousek, M. Towrie, P. Foggi, F.V.R. Neuwahl, S. Umapathy, R.E. Hester & J.N. Moore, *Chem. Phys. Lett.*, **290,** 68, 1998.
4 M. Seel, E. Wildermuth and W. Zinth, *Meas. Sci. Technol.* **8,** 449, 1997.
5 M. Schenk, S. Rudolph-Böhner, J. Wachtveitl, T. Nägele, D. Oesterhelt & L. Moroder, in *Peptides: Chemistry, Structure and Biology*, (in press), ESCOM, Leiden, 1998.
6 A.M. Müller, S. Lochbrunner, W.E. Schmid & W. Fuß, *Angew. Chem. Int. Ed.*, **37,** 505, 1998.
7 J. Wachtveitl, T. Nägele, B. Puell, W. Zinth, M. Krüger, S. Rudolph-Böhner, D. Oesterhelt & L. Moroder, *Photochem. Photobiol.*, **105,** 283, 1997.

Potential Energy Surfaces for Ultrafast Photochemistry: Short Chain Polyenes and Protonated Schiff Bases

Marco Garavelli[1], Fernando Bernardi[2], Massimo Olivucci[2], and Michael A. Robb[1]

[1] Department of Chemistry, King's College London Strand, London WC2R 2LS, United Kingdom
E-mail: marco@CHEM.ch.kcl.ac.uk
[2] Dipartimento di Chimica "G. Ciamician", Universitá di Bologna, via Selmi n. 2, Bologna I-40126, Italy

Abstract. The minimum energy paths (MEP) for the photochemical processes involving short chain linear protonated Schiff bases (PSB) and short chain polyenes have been computed. The results show that the photochemistry of the studied PSB and polyenes is driven by different S_1 excited states (ionic $1B_u$-like and covalent $2A_g$ respectively). The structures of the corresponding energy surfaces rationalise the observed ultrafast photochemistry and dynamics.

1. Introduction

Ultrafast excited state dynamics has been reported for short chain polyenes (butadienes, hexatrienes) [1]. These small molecules undergo a sub-picosecond decay to the ground state (S_0) which has been assigned to the opening of a $S_1 \rightarrow S_0$ radiationless decay channel leading to single/double-bond isomerization, reactant back-formation, and ring-closure. Fast, picosecond, excited state dynamics has been also reported for the photoisomerization of larger systems such as the 11-*cis* retinal protonated Schiff base (PSB11), the chromophore of rhodopsin (the human retina visual pigment), and its isomers [2]. These observations suggest the involvement of *barrierless* reactive processes where the molecules are not thermally equilibrated at the outset of the reaction. Thus, *the chemical change occurs on the same time-scale as intramolecular vibrational relaxation (IVR).*

A *qualitative* insight into conjugated systems photochemistry can be obtained using *structural* (i.e. *non dynamic*) information of the involved excited states. The shape of the excited surfaces is expected to play a dominant role in determining the observed short-lived dynamics which will be determined by: (i) the direction of the *initial* acceleration (i.e. the direction of the gradient) on the Frank-Condon (FC) region, which results in the population of only one or few selected vibrational modes of the system, and (ii) the structural details (i.e. the shape) of the excited state surface which control the redistribution of the initially localised vibrational energy among other modes and thus to the reactive mode leading to $S_1 \rightarrow S_0$ decay.

This analysis has been accomplished by using a new developed gradient-driven tool, which facilitates the computation of the down-hill minimum energy paths (MEP) starting from non-stationary points such as FC points and crossing points.

We have applied this technique to investigate the photochemical processes involving PSB of different length (2-*cis* pentadieniminium, *all-trans* heptatrieniminium, and 4-*cis* nonatetraeniminium cations) [2,3] and short chain polyenes (*s-cis* butadienes, and *all-trans* hexatriene) [1,4], using MC-SCF energy and gradient, and MP2 computations to improve the energetics.

2. Results and Discussion

The results show that the excited state evolution of the studied PSB and polyenes is driven by electronically different S_1 excited states: the spectroscopic $1B_u$-like ionic state and the covalent $2A_g$ excited state respectively. The different nature of the electronic states involved in the photochemistry of this class of compounds induces different electronic and geometric features in the structure of the $S_1 \rightarrow S_0$ decay channel. S_1/S_0 Conical Intersections (CI) in polyenes have a covalent tetraradicaloid character and involve a triangular arrangement of three carbon centres corresponding to a -$(CH)_3$- kink of the carbon skeleton [1,4], while S_1/S_0 CI in PSB have a ~90° twisted C-C "central" double-bond and correspond (at least partially) to a type of twisted intramolecular charge transfer (TICT) state [2,3] (see figure 1).

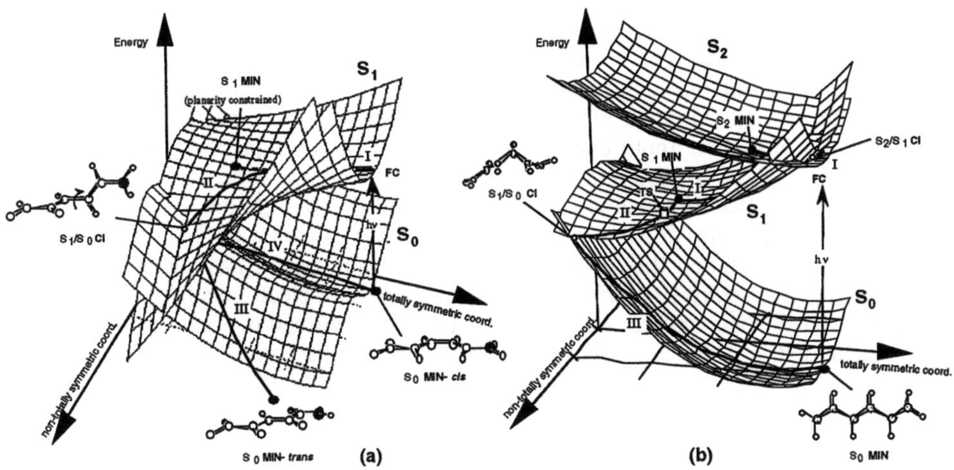

Fig. 1. Three-dimensional representation of model S_0, S_1 (and S_2 for polyene) potential energy surfaces of two isoelectronic systems: (a) 2-*cis* pentadieniminium cation and (b) *all-trans* hexatriene. The bold lines correspond to the computed MEP (I, II, III, and IV). Note that the S_1 symmetric critical point is located on the MEP only in polyenes, while in PSB the asymmetric path II starts before such a point is reached).

Also different relaxation paths occur in S_0: in polyenes we have found competitive relaxations corresponding to different recoupling schemes of the four unpaired electrons [1,4], while in the studied PSB only very efficient *cis→trans* isomerization channels for the "central" C-C double bonds were located [2,3].

Nevertheless, some common features may be detected in polyenes and PSB photochemistry. *(a)* The computed MEP demonstrates that the relaxation from the FC structure occurs along a barrierless path which ends at a twisted S_1/S_0 CI where an efficient and fast $S_1 \rightarrow S_0$ radiationless decay occurs. *(b)* The MEP co-ordinate suggests that the initial acceleration (point i above) occurs along a stretching mode (path I in figure 1) which *has no component* along the torsional deformation required to access the CI (point ii above). These results suggest that *IVR from the initially populated (stretching) mode to the reactive (torsional) mode* (path II in figure 1) *must take place in order for the reaction to occur.* In other words, the computation suggests that IVR is, at least in part, controlled by the structure of two different domains of the S_1 surface. *(c)* Geometries and energies of crossing points and ground state relaxation pathways (paths III-IV in figure 1) account (qualitatively) for the observed quantum yields and stereochemistry both in polyenes and PSB.

3. Conclusions

Above we have reported the structure of the barrierless S_1 potential energy surfaces which control ultrafast photochemical processes in the studied short chain PSB and polyenes. Such structure is characterised by the shape of two domains: the first (which develops along a *totally symmetric* co-ordinate corresponding to a *stretching* deformation) controls the initial excited state relaxation of the system, while the second (along a *non totally symmetric co-ordinate* dominated by torsional deformation about one or more bonds of the original reactant) is a low lying region of the S_1 potential energy surface and controls the evolution of the excited state molecule towards fully efficient decay to the ground state. For PSB this second domain defines a very efficient *cis→trans* C-C double bond isomerization channel: this may be one of the reason nature has chosen PSB as very efficient *cis-trans* isomerization switches. Moreover electronically different S_1 excited states (ionic and covalent respectively) are involved in PSB and polyenes photochemistry. *Thus we conclude that polyenes are not suitable models for the investigation of the retinal chromophore or other PSB photochemistry.*

References

1 F. Bernardi, M. Olivucci, and M. A. Robb, *Chem. Soc. Rev.* **25**, 321, 1996 and references cited therein.

2 M. Garavelli, P. Celani, F. Bernardi, M. A. Robb and M. Olivucci, *J. Am. Chem. Soc.* **119**, 6891, 1997 and references cited therein.

3 M. Garavelli, T. Vreven, P. Celani, F. Bernardi, M. A. Robb and M. Olivucci, *J. Am. Chem. Soc.* **120**, 1285, 1998.

4 M. Garavelli, P. Celani, F. Bernardi, M. A. Robb and M. Olivucci, *J. Am. Chem. Soc.* **119**, 11487, 1997.

Ultrafast photoisomerizations and S_1/S_0 internal conversions in organic and inorganic molecules: evidence for easily accessible conical intersections

W. Fuß, K. L. Kompa, T. Schikarski, W. E. Schmid, and S. A. Trushin

Max-Planck-Institut für Quantenoptik, D-85740 Garching, Germany,
E-mail: w.fuss@mpq.mpg.de

Abstract. Probing by intense-laser field dissociative ionization provides an unprecedented quantity of information about the motion of molecules on potential energy surfaces. Even the direction of motion and location of conical intersections can sometimes be deduced. In the molecules investigated, all processes are ultrafast, comparable or faster than the period of appropriate vibrations. Hence there must be a continuous pathway from the excited state down to the ground state of the product through easily accessible conical intersections.

1. Introduction

Probing ultrafast photochemical reactions by ionization by the electric field of intense laser radiation has the advantage that any electronically excited state is expected to give an enhanced ion signal, so that there is no dark state [1]. Fragment ions usually show a time behavior different from that of the parent ion and are thus an additional source of information. The probability to produce each ion depends on the position on the potential energy surface, since the ionization is more or less vertical. Thus monitoring the parent and various fragment ions as functions of the pump-probe delay in some *pericyclic reactions* as well as in the *decomposition of metal hexacarbonyls*, we could follow the molecules on the potential energy surface all along their path to the product ground state, obtaining a most detailed picture about the dynamics.

Photochemical *pericyclic reactions* have potential surfaces very similar to those of cis-trans isomerizations: A two-electron excited state (2A) has a minimum halfway between educt and product and collects the population of all the other excited states. But the distance of this minimum to the ground state 1A (which has a maximum near this geometry) is much too large (> 1 eV) to allow an ultrafast reaction. Recent ab-initio calculations (e.g. [2]) have shown that with an additional deformation the two surfaces intersect in the form of a conical intersection (CI), which is not only the leak of the „collection well“ (2A) or exit of the „funnel“, but also acts as a point of branching between reaction and internal conversion.

2. Results and Discussion

We found time constants $t_{2A} < 80$ fs for the ring opening of *cyclohexadiene* to hexatriene and for the H migration in *cycloheptatriene* (Fig. 1, 2). Both reactions are pericyclic. Such short times prove the existence and easy accessibility of CIs. Furthermore, since these times are shorter than the periods of relevant vibrations, they prove that the CIs are situated very near the lines of steepest descent and that almost no redistribution of vibrational energy is needed. The first of the time constants (5-10 fs) can be assigned to the time for departure from the Franck-Condon (FC) region. The deuterium effect in the cycloheptatriene derivative shows that the hydrogen begins to move immediately after this region, but still in the 1A'' state. So the 1A''/2A' intersection is displaced from the FC region in direction of FC active coordinates and a (a'') H shift. The second CI (2A'/1A') involves the same coordinates, but probably larger displacements.

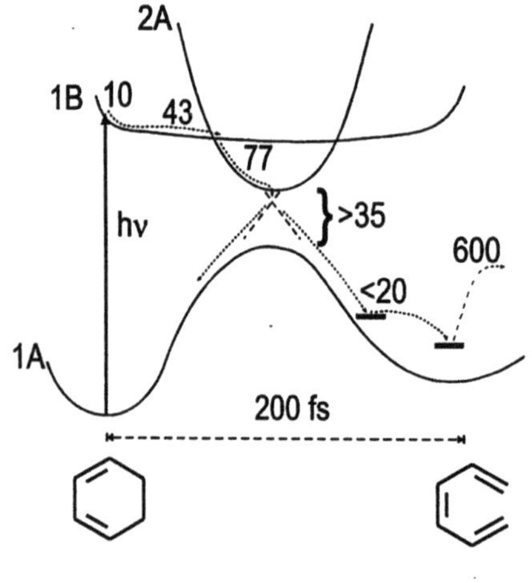

Fig. 1. Schematic potential energy surfaces for the electrocyclic photochemical ring opening of 1,3-cyclohexadiene → cZc-1,3,5-hexatriene and lifetimes (fs) for different intermediate levels. CI: conical intersection, situated outside the drawing plane.
Features general for photochemical pericyclic reactions are: a spectroscopic state („1B"), a collecting well (pericyclic minimum, „2A"), a high maximum in the ground state and a conical intersection (CI) connecting minimum and maximum.

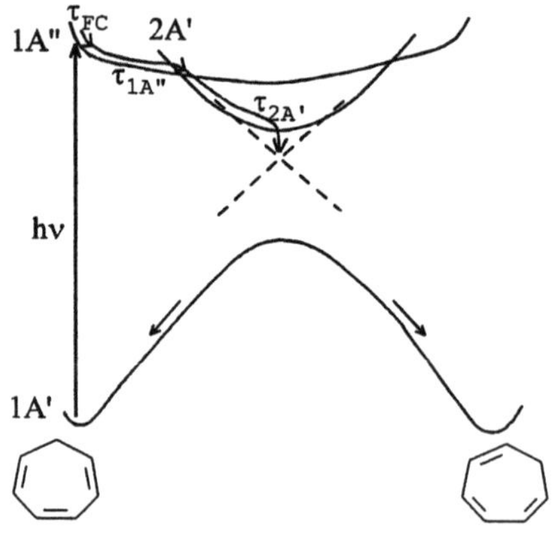

Fig. 2. Schematic potential energy surfaces for the [1,7]-sigmatropic hydrogen migration in cycloheptatriene, its 7-methyl derivative and the perdeuterated form of the latter.

	τ_{FC}/fs	$\tau_{1A''}$/fs	$\tau_{2A'}$/fs
C_7H_7-H	10	60	70
C_7H_7-CH_3	5	66	95
C_7D_7-CD_3	5	95	150

UV photodissociation of *metal hexacarbonyls* $M(CO)_6$, M = Cr, Mo, W proceeds within 33-69 fs, depending on M, via the electronically excited $M(CO)_5$ in the C_{4v} 1E state. Fig. 3 shows the case of M = Cr [3], and Table 1 gives the analogous numbers for the other carbonyls. The product $M(CO)_5$ then exhibits two consecutive ultrafast internal conversions via the D_{3h} 1E' state to produce finally the ground state C_{4v} $1A_1$ within additional 60-100 fs. On the ground state surface the molecules show coherent oscillations in a C_{ax}-M-C_{eq} vibration (Berry pseudorotation). The latter observation is strong support for the $C_{4v} \rightarrow D_{3h} \rightarrow C_{4v}$ pathway of the pentacarbonyl, and since in the intermediate the Jahn-Teller effect will induce a conical intersection (the branching space involves the coordinate mentioned), we again have a continuous pathway through a funnel just as in the organic reactions studied.

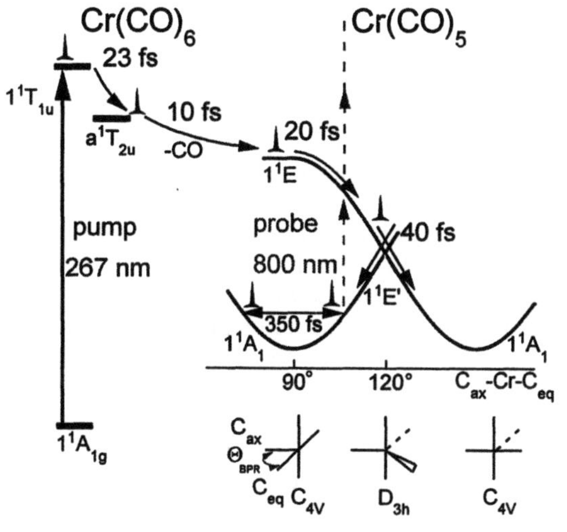

Fig. 3. Schematic representation of the early-time dynamics of single-photon UV decomposition of $Cr(CO)_6$. The Jahn-Teller effect splits all the degenerate states (T, E), giving rise to conical intersections. Such a splitting tunes the absorption through resonance with a probe at 800 nm, but not with one at 400 nm, so that the coherent oscillation (350 fs) can be observed in the former case, but not in the latter.

Table 1. Time constants (in fs) for $M(CO)_6$ analogous to those given in Fig. 3 for $Cr(CO)_6$.

M	τ_1	τ_2	τ_3	τ_4	τ_{BPR}
Cr	23	10	20	40	350
Mo	30	20	30	46	396
W	46	23	30	70	418

References

1 S.A. Trushin, W. Fuß, T. Schikarski, W.E. Schmid, K.L. Kompa, J. Chem. Phys. **106,** 9386, 1997; SPIE Proc. **3271**, 114, 1998.

2 F. Bernardi, M. Olivucci, Chem. Soc. Rev. **1996**, 321.

3 S.A. Trushin, W. Fuß, W.E. Schmid, K.L. Kompa, J. Phys. Chem. **102**, 4129, 1998.

Femtosecond Dynamics of Quantum Dissipative Systems: Localization of Electrons at Interfaces

C. B. Harris, N.-H. Ge, C. M. Wong, J. D. McNeill, and K. J. Gaffney

Department of Chemistry, University of California, Berkeley, CA 94720 and Chemical Sciences Division, Lawrence Berkeley National Laboratory, Berkeley, CA 94720, USA

Abstract. Using a recently developed electron transfer theory based on a path integral approach, we can model the femtosecond dynamics of electron self-trapping processes at interfaces.

Many physical and chemical processes can be described by the crossing of two potential energy levels spanned by a generalized nuclear coordinate. Familiar examples are Franck-Condon effects in electronic transition, spin-forbidden processes in unimolecular reactions, and electron transfer processes [1]. Traditional theoretical treatments have focused on transitions between weakly coupled initial and final states. In this non-adiabatic limit, lowest order perturbation theory in the electronic coupling yields the well-known golden rule. In strongly interacting systems, one often resorts to semiclassical treatments, such as the Landau-Zener formula for the transition probability in the adiabatic limit. A recent paper on electron transfer theory by Stuchebrukhov and Song [2] provides a unified theory which includes the two extreme cases (Fig. 1) on equal footing and is able to treat the bath modes quantum-mechanically. In their approach, the path integral technique is utilized to sum over all perturbation orders in electronic coupling, H_{fs}. Therefore, this formalism can work continuously from the non-adiabatic to the adiabatic regimes. Here we present the application of this theory to model the dynamics of electron self-trapping in 2-D.

Femtosecond time- and angle-resolved two-photon photoemission (TPPE) studies of excited electrons at bilayer n-heptane/Ag(111) interfaces revealed the time scale for electron localization [3]. The initially delocalized image electrons become localized as small polarons due to the strong electron-lattice interaction with the alkane layer. The localization time is found to be a strong function of the parallel momentum ($\hbar k_{\|}$) of the initially delocalized electrons with higher $k_{\|}$ electrons self-trapping faster. This behavior can be understood by making an analogy between the electron self-trapping processes and electron transfer reactions, as shown in Figure 2. Curves in the V_f manifold represent the different initial delocalized electronic states with translational symmetry labelled by $k_{\|}$, V_s represents the final self-trapped state, and the lattice distortion Q is the generalized nuclear coordinate for self-trapping. A self-trapping process starting from an initial state of particular $k_{\|}$ corresponds to an electron transfer reaction with exothermicity of $-\Delta\mathcal{E}$. As $k_{\|}$ and $-\Delta\mathcal{E}$ increases, the self-trapping rate will increase and go through an inverted region [1].

For small polaron formation in a molecular lattice, the electron can interact with both intra- and intermolecular vibrations. At our experimental temperature, 120 K, intermolecular vibrations are thermally excited but high-frequency intramolecular modes are frozen. We make the assumption that self-trapping involves a single quantum-mechanical intramolecular mode with a frequency ω_q and a reorganization energy E_{rq}, and all other intermolecular modes are classical with a reorganization energy E_{rc}. Adopting Stuchebrukhov and Song's formula [2] under this assumption, the rate expression reads

$$k_{st} = \frac{1}{\hbar\beta}\sum_{j=1}^{\infty}\frac{\pi^{1/2}}{2^{2j-2}j!(j-1)!}(\beta H_{fs})^{2j}\sum_{m=-\infty}^{\infty}\left[e^{-jS\coth\frac{\beta\hbar\omega_q}{2j}}\mathrm{I}_{|m|}(\alpha(j))e^{\frac{m\beta\hbar\omega_q}{2j}}\right]$$
$$\times\left\{e^{-\beta(\Delta\mathcal{E}+E_{rc}+\frac{m\hbar\omega_q}{j})^2/4E_{rc}}\sum_{l=0}^{j-1}\frac{\mathrm{C}_l^{j-1}(-1)^l}{(\beta E_{rc})^{\frac{2l+1}{2}}}\mathrm{H}_{2l}(\gamma(j,m))\right\}. \quad (1)$$

The quantity $S = E_{rq}/\hbar\omega_q$ is a measure of the electron-vibration coupling strength. $\mathrm{I}_{|m|}(x)$ is the modified Bessel function, $\alpha(j) = jS/\sinh(\beta\hbar\omega_q/2j)$, C_l^{j-1} is the binomial coefficient, $\gamma(j,m) = [\beta(\Delta\mathcal{E} + m\hbar\omega_q/j)^2/4E_{rc}]^{\frac{1}{2}}$, and $\mathrm{H}_{2l}(x)$ is the Hermite polynomial. The sum over j is a sum over all perturbation orders in the electronic coupling, H_{fs}, of the delocalized and self-trapped states. When only the $j = 1$ term is taken into account, the exact quantum mechanical expression for the non-adiabatic limit [1] is recovered. With this formalism, the rate can be calculated without presuming the coupling regime, weak or strong, the system is. One can determine the degree of a process being non-adiabatic or adiabatic by examining the relative contribution from each j term to the total rate.

Figure 3 shows the analysis of the self-trapping rate as a function of $-\Delta\mathcal{E}$. The parameters E_{rc} and E_{rq} extracted from the fit are consistent with the interaction energies of lattice and molecular polarons in organic molecular solids [4]. It is important to note that the $j = 1$ term accounts for nearly 96% of the fit in Fig. 3. The self-trapping process at 120 K is well described by the non-adiabatic quantum mechanical limit. The non-Arrhenius temperature dependence of the self-trapping

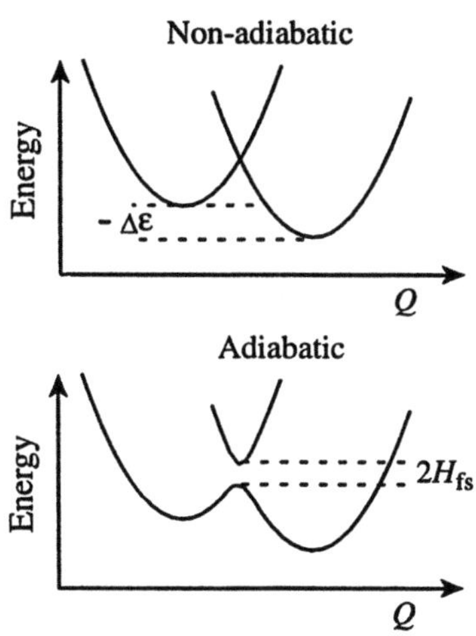

Fig. 1: Energy surfaces for the non-adiabatic and adiabatic limits.

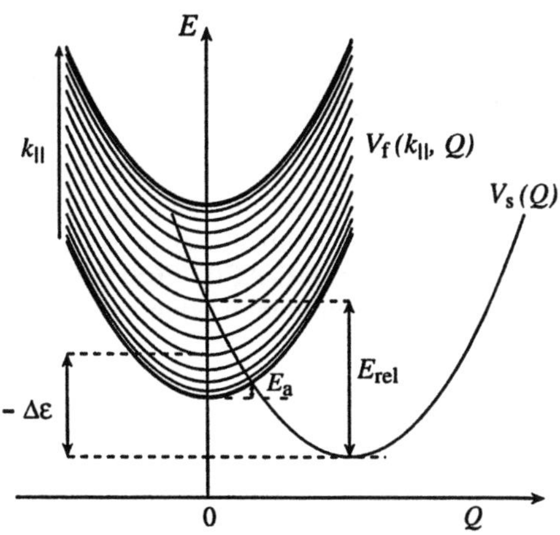

Fig. 2: Configuration coordinate diagram for electron self-trapping. E_{rel} is the lattice relaxation energy. E_a is the activation energy for $k_{||} = 0$.

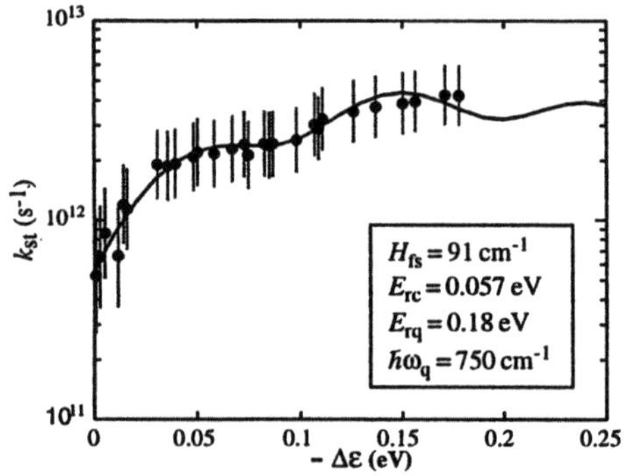

Fig. 3: Self-trapping rate versus exothermicity at 120 K.

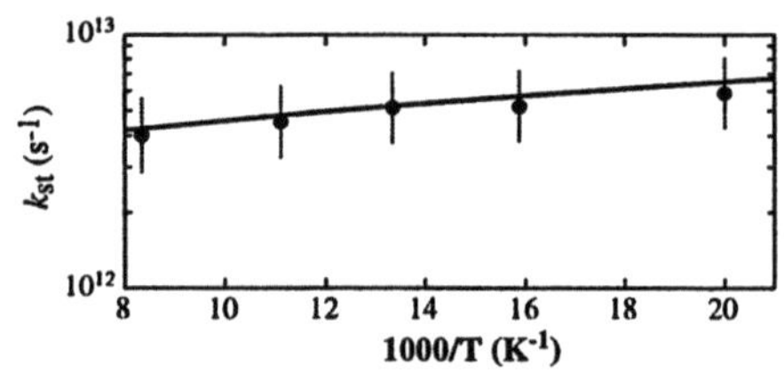

Fig. 4: Temperature dependence of self-trapping rate at $-\Delta\mathcal{E} = 0.14$ eV.

rate at $-\Delta\mathcal{E} = 0.14$ eV is reproduced with no additional adjustment of parameters (Fig. 4). The $j = 1$ term accounts for only 83% of the rate for the data point at 50 K. This shows that the higher order terms become more important at lower temperatures.

In conclusion, the dynamics of 2-D electron self-trapping processes as measured by femtosecond TPPE spectroscopy can be understood by analogy to electron transfer reactions. The theory we apply here is capable of accounting for all orders of electronic coupling and allows a determination of the adiabaticity of the self-trapping process.

Acknowledgements. We thank Dr. X. Song for very helpful discussions. This work was supported by the Director, Office of Energy Research, Office of Basic Energy Sciences, Chemical Sciences Division of the U. S. Department of Energy, under Contract No. DE-AC03-76SF00098. The authors acknowledge NSF support for specialized equipment used in some experiments described herein.

References

1 R. A. Marcus and N. Sutin, Biochim. Biophys. Acta **811**, 265, 1985.
2 A. A. Stuchebrukhov and X. Song, J. Chem. Phys. **101**, 9354, 1994.
3 N.-H. Ge, C. M. Wong, R. L. Lingle, Jr., J. D. McNeill, K. J. Gaffney and C. B. Harris, Science **279**, 202, 1998.
4 E. A. Silinsh and V. Capek, *Organic Molecular Crystals. Interaction, Localization, and Transport Phenomena*, AIP Press, New York, 1994.

Ultrafast reaction in solvated metal atom clusters: A dynamic study in the visible and near IR spectral range

Claus Peter Schulz[1], Axel Scholz[1], and Ingolf V. Hertel[1,2]

[1] Max-Born-Institute, Rudower Chaussee 6, D-12489 Berlin, Germany
E-mail: cps@mbi-berlin.de
[2] also: Fachbereich Physik, Freie Universität Berlin, Arnimallee 14, D-14195 Berlin, Germany

Abstract. The lifetime of the first electronically excited state of sodium-ammonia ($Na(NH_3)_n$) clusters is investigated with femtosecond laser pulses in the visible and near IR spectral range. The decay time of the pump-probe ion signal decreases drastically with increasing cluster size from 1000 ps (n = 1) to 0.1 ps for n = 15. Different decay mechanism are discussed. A purely statistical process can be excluded to be responsible for the observed decay times.

1. Introduction

Spectroscopic and dynamical properties of solvated electrons in polar liquids have been studied for many years. The development of ultrashort laser pulses has given a new impetus to this research field. Our approach to understand the formation and dynamics of the solvated electron differs from the typical experiments with liquid samples: Free clusters containing one sodium metal atom and an arbitrary number of ammonia molecules as solvent are studied in a molecular beam.

In our previous spectroscopic investigations it was found, that the *3s* → *3p* transition of the sodium atom is shifted from the visible to near IR spectral range with the number of solvent molecules bound to the metal atom [1,2] and that the excited state of sodium-ammonia complexes is short lived. For a detailed study of the excitation and decay dynamics we have set up a pump-probe experiment with femtosecond laser pulses tunable from the visible to the near IR spectral range.

2. Experimental Methods

Free sodium ammonia clusters $Na(NH_3)_n$ are prepared in a pickup source, where sodium atoms from an effusive beam are picked up in the expansion region of a neat ammonia jet [3]. Downstream the clusters are excited, ionized and mass separated in a time-of-flight mass spectrometer.

The femtosecond laser pulses are generated by a commercial, regenerative amplified Ti:Sapphire laser system. The output pulses with a duration of 200 fs and 440 μJ pulse energy, are tunable in a wavelength range from 770 nm to 840 nm. To extend the wavelength to the near IR spectral region an optical parametric generator/amplifier (OPG/OPA) was build based on the design of Petrov et al. [4]. The setup provides tunable laser pulses between 1200-1500 nm (signal) and 1700-2000 nm (idler) with a pulse width of 140 fs and pulse energies of 6 μJ at the cluster beam. For the probe pulse the second harmonic of the Ti:Sapphire fundamental was used for small $Na(NH_3)_n$ clusters up to n = 3. Due to the low ionization potential of the larger clusters we had to use probe pulses (500-590 nm), generated by sum frequency mixing of the fundamental wavelength and either the signal or the idler output of the OPG/OPA. Pulses of 220 fs duration and 5 μJ pulse energy were obtained.

3. Results and Discussion

With this experimental setup we were able to determine to the lifetime of the first electronically excited state of $Na(NH_3)_n$ clusters up to n = 15. The transition corresponds asymptotically to the $3s \rightarrow 3p$ in the sodium atom, which is shifted from the yellow atomic line to the near IR spectral region when ammonia molecules are bound to the sodium atom [1,2]. To measure the lifetimes, the light pulses have to be tuned to the absorption of the corresponding clusters. For $NaNH_3$ and $Na(NH_3)_2$ we used 820 nm to excite and 410 nm to ionize. For $Na(NH_3)_3$ 1300nm and 410 nm were pump and probe wave length. For all $Na(NH_3)_n$ with $n \geq 4$ the pump wavelength was 1690 nm and probe was 507 nm.

Figure 1 shows the measured lifetimes for $Na(NH_3)_n$ clusters (full squares) up n = 15 as a function of the inverse cluster radius. The lifetime decreases drastically with increasing cluster size. While the first excited state of $NaNH_3$ still has a lifetime of 1 ns, the lifetime drops to 1 ps with only two additionally bound ammonia molecules ($Na(NH_3)_3$). The first solvation shell is closed when four ammonia molecules are bound to the sodium atom forming a nearly tetrahedral complex. The higher symmetry leads to slightly longer lifetime compared to n = 3 and 5. Towards larger clusters the lifetime decreases to less than 0.2 ps.

In order to understand the observed lifetimes different decay processes have to be considered. The simplest process would be a direct dissociation, where the cluster finds itself on a repulsive potential surface after photo excitation. This process can be excluded to play a role here since the direct dissociation leads to very short decay times with only a small dependence on the cluster size. A second mechanism one can think of is a statistical redistribution of the excitation energy with subsequent evaporation of one or more ammonia molecules. In this case the decay time can be calculated by the RRK formula [5]. In Fig. 1 the calculated lifetimes are shown (open circles) with the parameters chosen a way, that the lifetime for n = 5 is reproduced. It is clear from Fig. 1 that a statistical redistribution cannot describe the decay process after photo excitation of the $Na(NH_3)_n$ clusters. At present, we assume that the energy of the electronically excited state is converted into vibrations of the surrounding ammonia molecules through an internal conversion process. This vibrationally excited cluster cannot be ionized with the wavelength used in the experiment. This interpretation is also consistent with the observed lifetime for deuterated clusters. For further discussion we refer to a future publication [6].

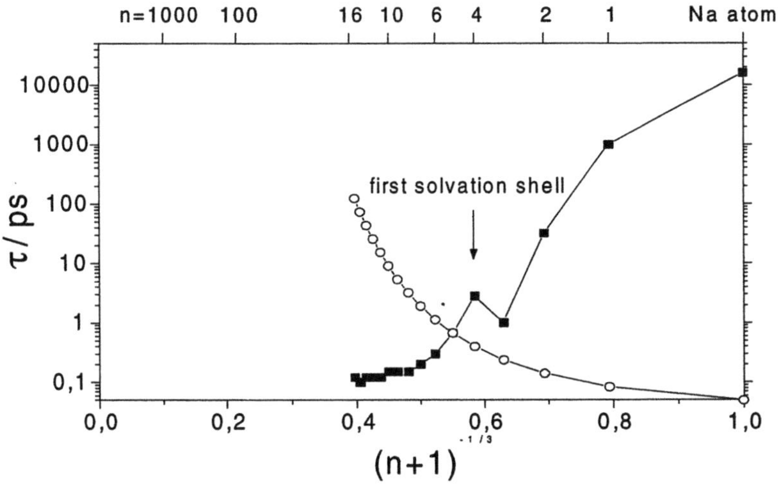

Fig. 1. Measured lifetimes of the first electronically excited state of $Na(NH_3)_n$ cluster (full squares). The observed decay cannot be described by a statistical RRK process (open circles).

4. Conclusions

The observed lifetimes for the first electronically excited state of $Na(NH_3)_n$ clusters change over 5 orders of magnitude from 1 ns for n = 1 to less than 0.2 ps for n = 15. A pure statistical process can be excluded to be responsible for the observed decay times. It is suggested that an internal conversion process can describe the observed decay times. Quantum chemical and dynamical calculations are needed for a rigorous understanding of the complex interplay between metal atoms and the polar environment.

Acknowledgments. Financial support by the Deutsche Forschungsgemeinschaft through SFB 337, Project A11, is gratefully acknowledge.

References

1 C.P. Schulz, and C. Nitsch, J. Chem. Phys. **107**, 9784, 1997.
2 P. Brockhaus, I.V. Hertel, and C.P. Schulz, J. Chem. Phys. submitted.
3 C.P. Schulz, R. Haugstätter, H.-U. Tittes, and I.V. Hertel, Z. Phys. D. **10**, 279, 1988.
4 V. Petrov, F. Seifert, and F. Noack, Appl. Phys. Lett. **65**, 268, 1994.
5 L.S. Kassel, J. Chem. Phys. **32**, 225, 1928.
6 A. Scholz, I.V. Hertel, and C.P. Schulz, to be published.

Photodissociation of CH_2I_2 and the Subsequent Electron Transfer in Its Cluster Formed in Solution

Yukito Naitoh[1], Ken-ichi Saitow[2], and Keitaro Yoshihara[3]

[1]Inoue Photochirogenesis Project, JST, Kamishinden, Toyonaka 565-0085, Japan
E-mail: naito@inoue.jst.go.jp
[2]Department of Diversity Science, Graduate School of Science and Technology, Chiba University, Chiba 263-8522, Japan
E-mail: saitoh@scichem.c.chiba-u.ac.jp
[3]School of Materials Science, Japan Advanced Institute of Science and Technology, Tatsunokuchi 923-1292, Japan
E-mail: yosihara@jaist.ac.jp

Abstract. We have studied photodissociation and the subsequent reactions of methylene iodide (CH_2I_2) upon excitation at 268 nm in acetonitrile and hexane solutions by the femtosecond transient absorption spectroscopy. In addition to dissociation and geminate recombination in the femtosecond region, formation of $CH_2I_2^+$ is observed in the early picosecond region. The reaction is proposed to occur by electron transfer between photofragment I and CH_2I_2. This indicates that the solute molecules aggregate even in a dilute solution and they are responsible for the cation formation.

Photodissociation in a solution phase shows different features from an isolated system through various interactions with solvent molecules. The solvent cage effect, which causes geminate recombination between photofragments, is one example. However, there are only few studies on how a solute-solute interaction influences a post-photodissociation process and such a secondary reaction mechanism remains unclear.

The photoinduced reaction of methylene iodide (CH_2I_2) upon excitation at 268 nm in acetonitrile and hexane solutions has been investigated by the transient absorption spectroscopy based on a regeneratively amplified femtosecond Ti:Sapphire laser. The transients probed over a wide range from 340 to 740 nm show a common profile having a fast rise, a fast decay, and a slow rise components. The transient absorption spectra in picosecond to nanosecond region show two peaks at around 400 nm and 540 nm. The intensity of blue band is greater than the red one by ca. four times. We have measured the concentration dependence of solute at 400 nm as shown in Figure 1. This shows that the two fast time constants, a 200 fs rise and a 500 fs decay, are independent of concentration. On the other hand, the slow rise becomes faster, 13 to 7 ps in acetonitrile and 50 to 13 ps in hexane, by increasing concentration. The intensity of transient absorption probed at 400 nm also has shown a linear pump-power dependence at the delay times of 500 fs and 20 ps.

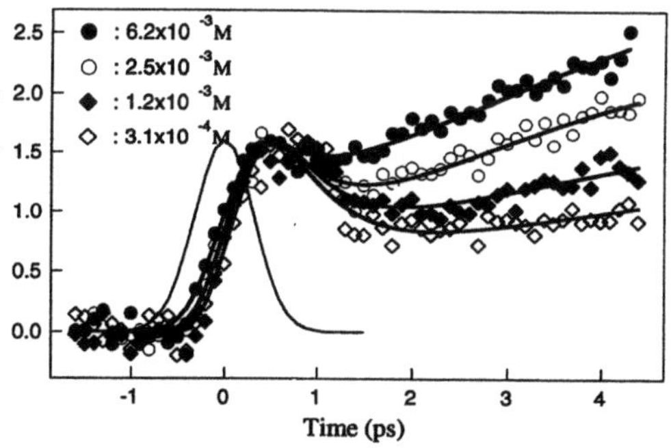

Figure1. Transient absorptions probed at 400 nm in CH_3CN. The absorbance is normalized at the peak of the first rise.

We assign the fast two components to generation of CH_2I radical by direct dissociation of the photoexcited CH_2I_2 and its disappearance by the subsequent geminate recombination. This supports the results observed upon excitation of CH_2I_2 at 310 nm [1]. The third slow component is not due to the vibrational relaxation of CH_2I radical as proposed in ref. [1], but to a transient induced by solute-solute interaction, because of widely observed rise behavior and the concentration dependence.

Figure 2 shows time-resolved absorption spectra of the blue band in picosecond time region. This absorption band consists of two bands; the band peaked at 430 nm initially appears and the band at 385 nm follows to grow up. The band at 385 nm is assigned to be a charge transfer (CT) complex, $CH_2I_2^{\delta+}\cdots I^{\delta-}$, formed between a photofragment I atom and another parent molecule CH_2I_2. This wavelength is in accordance with the estimation of the absorption peak based on the linear relationship between the CT-band and ionization potential of the donor molecule. The CT-

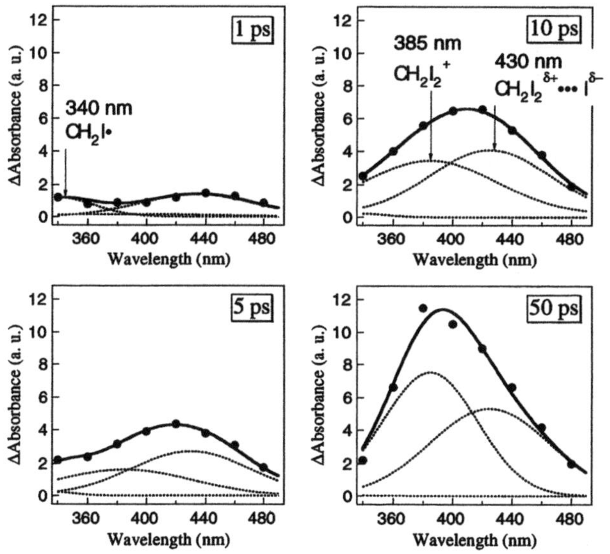

Figure2. Time-resolved absorption spectra in picosecond region, which are reconstructed from the single wavelength measurements.

complex is followed by a full electron transfer from CH_2I_2 to I to produce $CH_2I_2^+$, of which assignment is supported by the γ-radiolysis in a low temperature matrix [2], the photoelectron spectroscopy [3], and nanosecond flash photolysis [4].

Accordingly, we propose the following reaction mechanism upon photoexcitation of CH_2I_2;

$$CH_2I_2 \rightarrow CH_2I + I \quad (1)$$
$$CH_2I + I \rightarrow CH_2I_2 \quad (2)$$
$$CH_2I_2 + I \rightarrow CH_2I_2^{\delta+} \bullet\bullet\bullet I^{\delta-} \quad (3)$$
$$CH_2I_2^{\delta+} \bullet\bullet\bullet I^{\delta-} \rightarrow CH_2I_2^+ + I^- \quad (4)$$

After direct dissociation (1), photofragments suffer the geminate recombination in the solvent cage, (2). A part of I fragment escapes the geminate recombination and forms the CT-complex with a solute CH_2I_2 as shown in (3). The full electron transfer (4) occurs. This process is possible because the free energy gap (ΔG) is estimated to be almost 0 eV.

The observed CT-complex formation rate is much greater than that assuming a translational diffusion in a homogeneous solution at the experimental concentration range (10^{-4}~10^{-2}M). This means the formation occurs in the pre-existing solute clusters. Since the formation rate is in close connection with the degree of solute aggregation, we have estimated a relative local concentration, which reflects the number of CH_2I_2 around the fragment I in the cluster, using reaction rate equations. While the local concentration in hexane monotonically increases relative to the bulk concentration, it shows a gradual saturation in acetonitrile (Figure 3). The number of CH_2I_2 which can participate in CT-complex formation has an upper-limit due to the size limitation of the cluster and/or the restriction in the neighboring region around initially photoexcited CH_2I_2. Such a condition is achieved at a lower concentration in acetonitrile than in hexane.

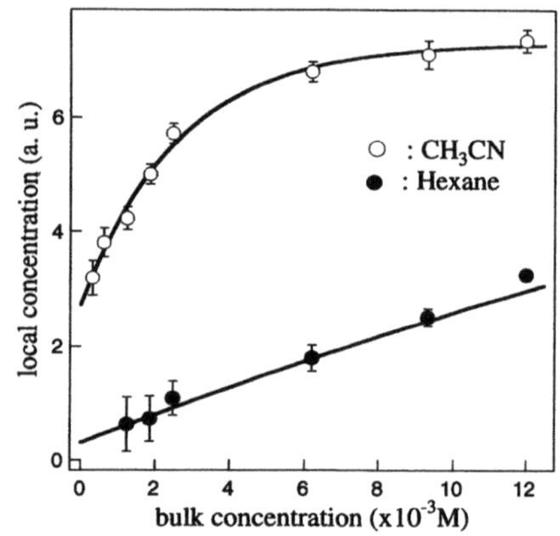

Figure3. Relationship between the local and bulk concentrations.

References

[1] B.J. Schwartz, J.C. King, J.Z. Zhang, and C.B. Harris, Chem. Phys. Lett., **203**, 503 (1993).

[2] H. Mohan and R.M. Iyer, Radiat. Effects, **39**, 97 (1978).

[3] L. Andrews, F.T. Prochaska, and B.S. Ault. J. Am. Chem. Soc., **101**, 9 (1978).

[4] K. Saitow, Y. Naitoh, K. Tominaga, and K. Yoshihara, Chem. Phys. Lett., **262**, 621 (1996).

Dynamics of Excited-State Evolution in Transition Metal Complexes: Time Resolving Electron Delocalization

Niels H. Damrauer[1], Giulio Cerullo[2], Alvin Yeh[1], and James K. McCusker[1]

[1] Department of Chemistry, University of California at Berkeley, Berkeley, CA 94720-1460, USA
E-mail: mccusker@socrates.berkeley.edu
[2] Permanent Address: Dipartimento di Fisica del Politecnico, P.za L. Da Vinci 32, 20133, Milano, Italia

Abstract. The excited-state dynamics of a series of aryl-substituted polypyridyl complexes of Ru^{II} have been examined. A combination of experimental work and theoretical calculations provide a cohesive physical model for the chromophore dynamics of these compounds in which changes in intraligand electron delocalization is coupled to motion of the peripheral aryl rings during excited-state thermalization. Preliminary femtosecond time-resolved absorption measurements in room-temperature CH_3CN solution indicate a time scale of ca. 200 fs for evolution of this delocalized state, and variable-temperature static emission spectra provide additional insights into how solvation dynamics may be coupled to this process.

1. Introduction

The excited-state dynamics of transition metal complexes represents an area that has received little attention from the ultrafast community as compared to small molecule or biological systems. However, the rich photophysics and photochemistry characteristic of metal-containing complexes [1] suggests that such studies may provide fertile ground for exploring new paradigms of excited-state dynamics relevant to a wide range of both fundamental and applied chemical processes. Our recent report [2] of the femtosecond dynamics of $[Ru(bpy)_3]^{2+}$,

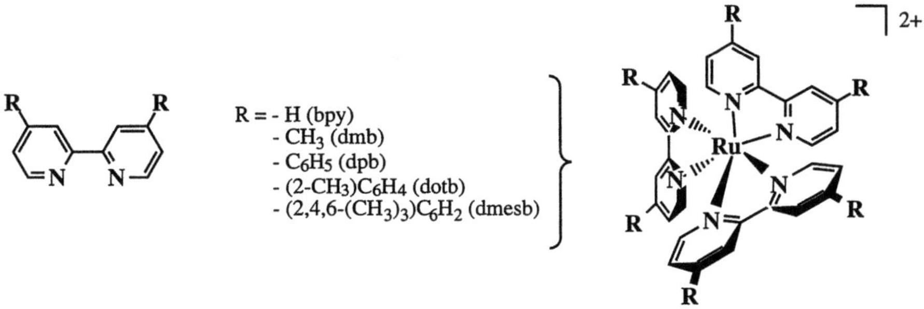

Fig. 1. Structures of 4,4'-substituted bipyridyl complexes of Ru^{II}

where bpy is 2,2'-bipyridine (Figure 1), represented the first step in characterizing the initial excited-state dynamics of this important class of molecules.

The aryl-substituted compounds indicated in Figure 1 comprise the focus of the work reported herein. Of particular interest is the change in the extent of delocalization of the photoexcited electron that occurs upon excited-state thermalization. A study of the dynamics of this process allows us in principal to probe electron delocalization in these compounds as it evolves in time.

2. Structural Model for Intraligand Delocalization

The basic model for the excited-state dynamics in this system has been described [3]. A large disparity in the non-radiative rate for excited-state decay (k_{nr}) between $[Ru(dmb)_3]^{2+}$ and $[Ru(dpb)_3]^{2+}$ was proposed to be due to differences in the extent to which the photoexcited electron could delocalize within the π system of the bipyridyl ligand. It was suggested that the peripheral phenyl rings of dpb rotated from a canted ground/Franck-Condon-state geometry to one in which they became coplanar with the bipyridyl fragment in the thermalized excited state. Photophysical properties observed for related complexes designed to increase steric hindrance for coplanarity were consistent with expectations of this model.

Theoretical studies on models for the ground-, Franck-Condon-, and thermalized excited-states [4] indicate canted geometries for the ground states of all of the aryl-substituted ligands, but a strong preference for planarity upon one-electron reduction. A comparison of the electron densities between the Franck-Condon state and the thermalized monoanions reveal the difference in intraligand electron delocalization that accompanies excited-state relaxation, raising the possibility of time-resolving this change in electron delocalization using ultrafast spectroscopy.

3. Ultrafast Spectroscopic Studies

Preliminary femtosecond time-resolved absorption measurements carried out on $[Ru(dpb)_3]^{2+}$ in CH_3CN reveal the presence of a transient bleach in the region of the $^1A_1 \rightarrow {}^1MLCT$ absorption band, consistent with depopulation of the ground state. A partial recovery of this initial signal to a steady-state bleach that persists well past Δt = 100 ps occurs with a time constant on the order of 200 fs. The long-lived signal is clearly that of the ^3MLCT excited state, known from nanosecond time-resolved measurements [3] to have a lifetime of 1.60(5) μs at room temperature in CH_3CN. The rapidly decaying initial transient is tentatively assigned as the formation of the ^3MLCT state. Based on the model described above, this signal defines the time scale for the change in intraligand electron delocalization concomitant with rotation of the peripheral phenyl ring and thermalization of the long-lived excited state.

4. Solvation Dynamics

We have examined the temperature dependence of emission through solvent glass-to-fluid transitions to gain insight into solvation processes that accompany excited-state relaxation. Spectra collected on all compounds in the "dpb series" reveal substantial differences in the nature of the solvent response of 4:1 EtOH/MeOH versus 2-MeTHF (Figure 2). The details of the solvent response

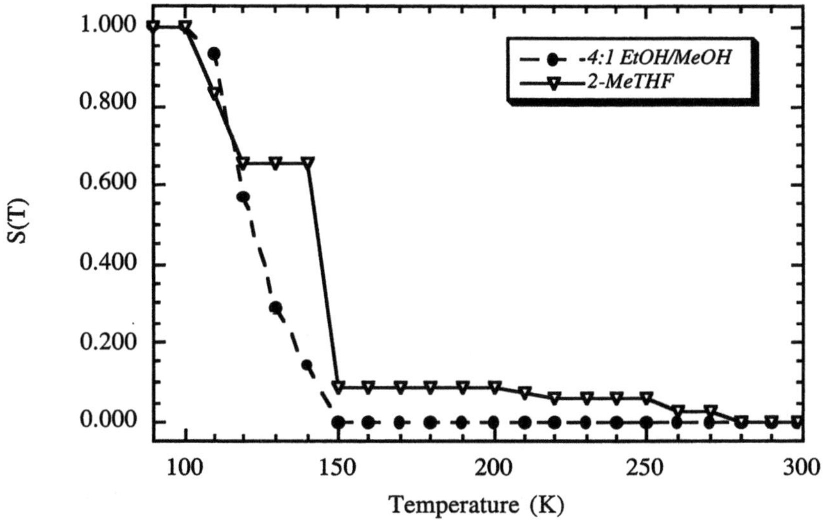

Fig. 2. Normalized temperature-dependent Stokes shift (S(T)) for $[Ru(dmesb)_3]^{2+}$ in 4:1 EtOH/MeOH and 2-MeTHF.

indicates the potential for strong coupling between intramolecular and solvation dynamics in these complexes. Further studies of this system are underway.

Acknowledgments. The authors thank Professor C. V. Shank for assistance.

References

1 D. M. Roundhill, *Photochemistry and Photophysics of Metal Complexes*, Plenum Press, New York, 1994.

2 N. H. Damrauer, G. Cerullo, A. Yeh, T. R. Boussie, C. V. Shank, and J. K. McCusker, Science **275**, 54, 1997.

3 N. H. Damrauer, T. R. Boussie, M. Devenney, and J. K. McCusker, J. Am. Chem. Soc. **119**, 8253, 1997.

4 N. H. Damrauer, B. T. Weldon, and J. K. McCusker, J. Phys. Chem. A **102**, 3382, 1998.

Modulation of Ultrafast Electron Transfer Dynamics by Wavepacket Motion in Oxazine 1

S. Engleitner[1], M. Seel[1], B. Wolfseder[2], G. Stock[3], W. Domcke[2] and W. Zinth[1]

[1] Institut für Medizinische Optik, Universität München
Oettingenstraße 67, D-80538 München, Germany
[2] Institut für Theoretische Chemie, Heinrich-Heine Universität
D-40225 Düsseldorf, Germany
[3] Theoretische Quantendynamik, Fakultät für Physik, Universität Freiburg
D-79104 Freiburg, Germany

Abstract. Experiments with highest temporal resolution on the electron transfer in oxazine 1 show pronounced nonexponential reaction dynamics and wavepacket motion in the reaction product.

Ultrafast photochemical reactions attract much attention since they seem to violate the restrictions imposed by conventional theory. As an example, the fastest electron transfer (ET) reactions observed so far are much faster than vibrational relaxation and well beyond the limits of non-adiabatic ET-theory. Theoretical considerations suggest that situations may be found where intramolecular vibrational modes become very important for the ET, thus resulting in non-exponential electronic relaxation dynamics [1]. Consequently in experiments non-exponential kinetics are expected: a fast component for the initial ET prior to the onset of dissipation which could reflect the vibrational motions. A slower component would account for ET after the onset of dissipation [1].

The system oxazine 1 in N,N-dimethylaniline is well suited to study the connection of electron transfer and vibrational motion since it shows a very fast electron transfer [2, 3] and pronounced wavepacket motion in both ground and excited state. While the initial ET occurs faster than 100 fs, the period of the dominating vibrational motion - observed as a wavepacket - is approximately 50 fs. Here we discuss a more detailed analysis of the absorption region of oxazine 1 in the red (600 nm to 700 nm) in addition to new pump-probe-experiments covering the absorption bands of the electron transfer products in the blue part of the spectrum (440 nm to 500 nm) with a time-resolution of better than 18 fs. The basic features of the experimental system used to study the fast reaction dynamics has been described in [3]. In short: we use amplified pulses from a CPM laser system after fiber compression, frequency shifting and recompression.

Analysis of the absorption kinetics at longer delay times between 600 nm and 700 nm laeds to the separation of ground and excited-state contributions. Extrapolation of this long-time dependence to shorter delay times allows to discriminate between the initial fast processes and the slower absorption transients.

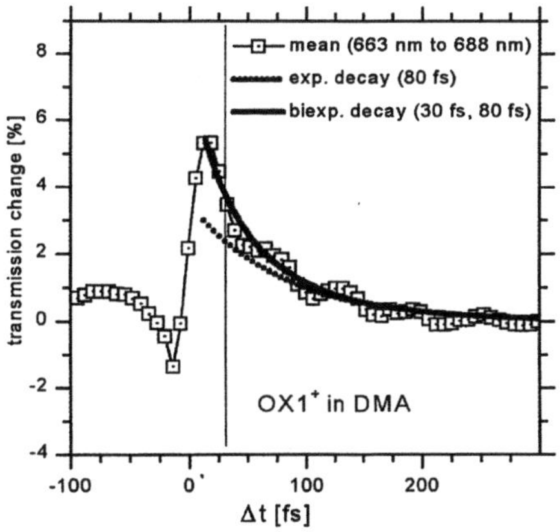

Fig. 1: Excited state contributions to the transient transmission change: Comparison of the averaged signal (663 nm to 688 nm) with a mono-exponential decay (80 fs) and a bi-exponential decay (30 fs and 80 fs).

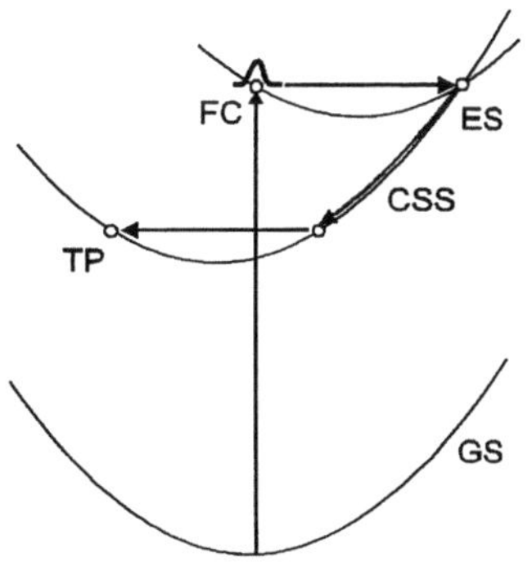

Fig. 2: Motion of the excited state vibrational wavepacket. The relative position of the potential curves is reconstructed from the phase of the oscillations showing up in the transient transmission at different wavelengths.

Fig. 1 shows the excited state contribution to the signal extracted in this way: The curve exhibits a very fast initial decay of the transient transmission on a time scale of about 30 fs. This process is followed by a slower decay with a time constant of about 80 fs. The existence of various features due to wavepacket motion causes absorption changes in this time domain and does not allow to exclude the existence of a weak 200 fs component as found in time resolved emission experiments [4]. Superimposed to the multiexponential decay is an oscillatory feature, which can be assigned to the spectral modulation of the stimulated emission by wavepacket motion in the excited state. This modulation is also well seen in the upper trace of Fig. 3, recorded close to the peak of the oxazine absorption at 657 nm. These findings are supplemented by the results of the experiments in the product absorption region (see Fig.3, lower trace). Pronounced oscillations are found in the spectral range of the product absorption. These oscillations can be observed at delay times between 0.2.and 1ps, in a period after the end of ET. Therefore these oscillations originate from wavepacket motion in the electron transfer products. Detailed analysis of the spectral behavior shows that this wavepacket is localized at oxazine 1. Experimental data of spectral ranges for ET-reactive and non-reactive systems (reference system: oxazine 1 in 1-chloronaphthaline) can be combined in a simple, phenomenological reaction model (Fig. 2). The model considers the structure of the observed time-dependent spectra (including the amplitudes and phases of the oscillations) and a stepwise decay of the excited state. After excitation to the Franck-Condon state FC a first nuclear motion leads the system to the intersection of the excited state potential surface ES with the product potential surface CSS.

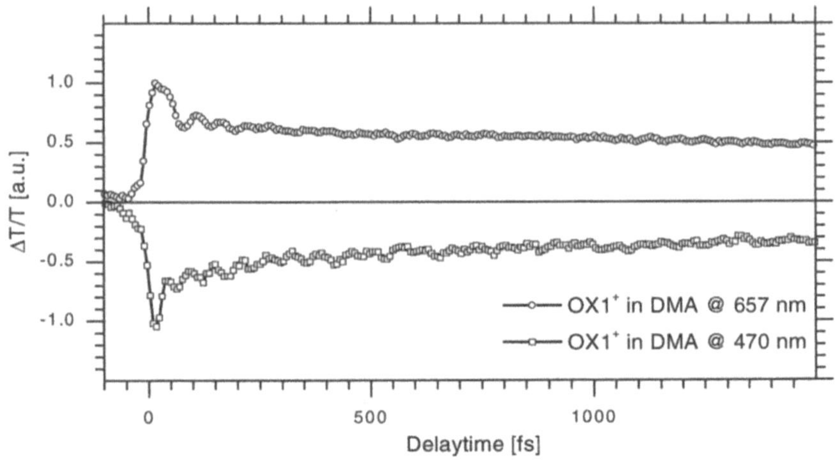

Fig. 3: Transient transmission dynamics for the reactive system $OX1^+$ in DMA at 657 nm (open circles) and 470 nm (open squares).

During this initial reaction part a large fraction of the molecules undergo ET. Subsequently the wavepacket remains oscillating on the potential surface of the product (charge separated) state. The 80fs component of the ET reaction is caused by multidimensional vibronic relaxation dynamics [5].

Quantum mechanical calculations based on a simplified model of the reactive system [5] qualitatively reproduce the experimental results and lead to the same relative positions of the potential curves for the ground state, the excited state and the charge-separated state.

References

1 W. Domcke and G. Stock, *Adv. Chem. Phys.* Vol. 100, 1, 1997.
2 K. Yoshihara, K. Tominaga and Y. Nagasawa, *Bull. Chem. Soc. Jpn.* Vol. 68, 696, 1995.
3 M. Seel, S. Engleitner and W. Zinth, *Chem. Phys. Lett.* Vol. 275, 363, 1997.
4 I. V. Rubtsov, H. Shirota and K. Yoshihara, *XI International Conference on Ultrafast Phenomena, Technical Digest*, ThP19, 1998.
5 B. Wolfseder, L. Seidner, W. Domcke, G. Stock, M. Seel, S. Engleitner and W. Zinth, *Chem. Phys.* 1998, in press.

Site Dependent Electron Transfer Reaction in Solution

Igor V. Rubtsov[1,2], Hideaki Shirota[3] and Keitaro Yoshihara[1]

[1] Japan Advanced Institute for Science and Technology, Hokuriku, Tatsunokuchi 923-1292, Japan, tel.: +81-761-51-1523
E-mail: ivr@jaist.ac.jp, yosihara@jaist.ac.jp
[2] Institute for Chemical Physics Research, Chernogolovka, Moscow Region, Russia
[3] Department of Chemistry and Biotechnology, Graduate School of Engineering, University of Tokyo, Tokyo 113-8656, Japan

Abstract. Highly nonexponential fluorescence decay due to electron transfer (ET) between photoexcited oxazine-1 (OX1) dye and different anilines was observed in a broad spectral range. The fastest component of ET observed in N,N-dimethylaniline (DMA) is about 60 fs. A strong dependence of fluorescence signal on the wavelength of observation was observed for the first time. The effect is very large in DMA and smaller in aniline (AN). The wavelength dependence is assigned to fluorescence from different configurations of the reactants surrounded by solvent molecules.

The mechanism of the electron transfer (ET) in condensed phase attracts a great theoretical and experimental interest. Recently, the system with intermolecular ET as fast as ~100 fs were found [1,2]. In this work we show that electron transfer rate in different configurations of reactants and solvent molecules could be measured separately.

Experimental details. The electron transfer from solvent molecules (different anilines) to excited oxazine-1 dye (OX1) was studied by fluorescence up-conversion technique with S/N ratio more than 10^3. The fluorescence up-conversion setup is based on femtosecond chromium-forsterite laser (1260 nm, 55 fs, 25 nJ/pulse). The second harmonic (630 nm, 40 fs, 10 nJ) was used for excitation of dye. The fluorescence was gated by a fundamental pulse. The cross-correlation function reaches 70 fs (FWHM). To control accurately the instrument response function on a particular wavelength of observation the Raman scattering signal of pure solvent was measured, which appears to be 85-90 fs depending on the wavelength.

Results

The ET process quenches the fluorescence of excited OX1 in electron donating solvents like anilines and determines a fluorescence decay rate. The fluorescence decay of OX1 in AN and DMA was measured in a wide range of emission wavelengths from 680 to 986 nm. As a control sample, the inert solvent 1-chloronaphthalene (ClNP) with a similar dielectric constant as anilines was used. The ET rate is very fast in DMA and more slow in AN (Fig. 1). The fastest

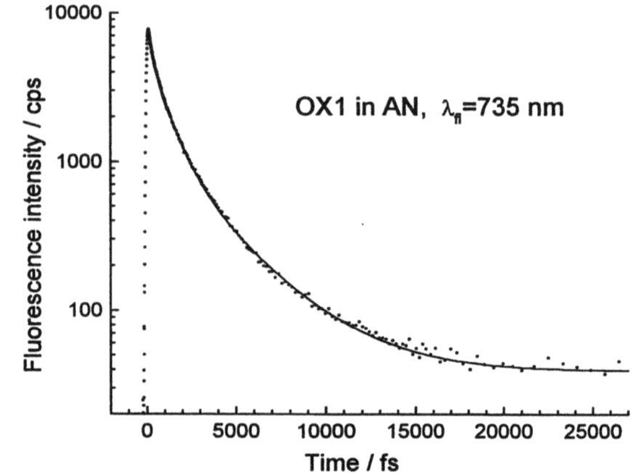

Fig. 1. Fluorescence decay of OX1 in AN at 735 nm in semilogarithmic scale

component observed at blue side of the measured range is about 60 fs in DMA (Fig. 2), and more than 3 times slower in AN. The fluorescence decay in both solvents is strongly nonexponential at all wavelengths observed (Fig. 1-3). The strong dependence of fluorescence decay on the wavelength of observation was detected in DMA (Fig. 2), the initial slope change is more than one order, and moderate effect was observed in AN (not shown). Almost no dependence on

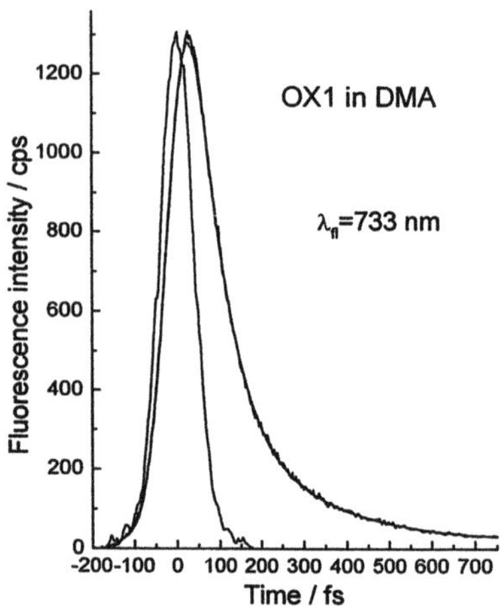

Fig. 2. Fluorescence decay of OX1 in DMA measured at 733 nm. The Raman scattering from pure DMA at the same wavelength is also shown. A fit with 3 exponentials (59 fs (80%), 200 fs (19%), 3 ps (1%)), convoluted with instrument response is given

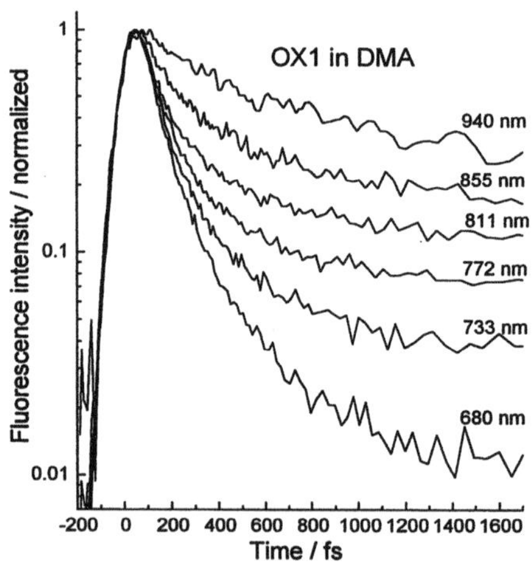

Fig. 3. Normalized fluorescence decay signals of OX1 in DMA at different wavelength of observation, 680, 733, 772, 811, 855, and 940 nm

wavelength (only very small difference due to solvation) was observed in ClNP where no ET takes place.

Discussion

The fluorescence at all wavelengths studied comes from the same excited electronic state, as no fast components in anisotropy decay was observed for different wavelengths in all solvents studied; and no change in time course at well separated wavelengths was observed for ClNP.

The nonexponential fluorescence signal without any definite predominant components suggests that ET occurs from different mutual configurations of reactants with a wide distribution of reaction rates. The total ET rate is composed of the terms with different decay rates and gives nonexponential decay. The configurations depends on the distance and angle between reagents as well as on solvent configuration. Initially there is some distribution of reactants configurations (consider only configurations where OX1 was excited). The fastest configurations ($OX1^*$-aniline) react quickly and slower configurations remains. Thus, the ET process changes distribution of configurations. Contrary, the motions of the reagents and solvent randomize the distribution of configurations. In the case of very fast ET (like in DMA), the distribution changes considerably by ET. Different configurations have different ΔG^0 of the reaction and different barriers. In the configurations corresponding to a higher energy of 0-0 transition (the absorption spectrum of OX1 is broad in DMA) the ΔG^0 of the reaction is larger and barrier is smaller. So, in that configurations the reaction is fast and the fluorescence of these configurations is blue shifted. Configurations whose absorption spectrum is red shifted reacts slower. As a result we observe the different fluorescence decays at different wavelengths, as they corresponds to the ET rate from different configurations. In AN case, where ET rate is intrinsically much slower than in DMA, the ET rate depends more on the mixing rate of different configurations. The initial distribution of configurations changes by ET not so strongly as in DMA. As a result the wavelength dependence is much smaller in AN.

Conclusions. Highly nonexponential fluorescence decay of dye in electron donating solvent with the fastest component of about 60 fs was observed. The ultrafast quenching is due to intermolecular electron transfer from solvent (various anilines) to the excited dye OX1. The calculations of reorganization energy showed that ET in DMA is almost barrierless. A strong wavelength dependence of fluorescence decay of OX1 in DMA was assigned to fluorescence highlighted by ultrafast ET reaction. The ultrafast ET allow us to measure at room temperature the ET rates in different configurations separately.

References

1 Y. Nagasawa, A. Yartsev, K. Tominaga, A. E. Johnson, K. Yoshihara, *J. Chem. Phys.*, **101** (1994) 5717

2 M. Seel, S. Engleitner, W. Zinth, *Chem. Phys. Lett.*, 275 (1997) 363

Measurements and Theoretical Modelling of Ultrafast Heterogeneous Spacer-Controlled Electron Transfer

B. Burfeindt[2], S. Ramakrishna[2], C. Zimmermann[1], B. Meißner[1], Th. Hannappel[1], W. Storck[2], F. Willig[1]

[1]Hahn-Meitner-Institut, Dept. CD, Glienickerstr. 100, D-14109 Berlin, Germany, Email: willig@hmi.de
[2]Fritz-Haber-Institut der MPG, Dept. PC, Faradayweg 4-6, D-14195 Berlin, Germany

Abstract. Femtosecond interfacial electron transfer from a chemically modified perylene molecule into empty electronic states of a sponge-like TiO_2 semiconductor was measured as transient absorption signal. A CH_2 -spacer group between anchor group and chromophore ensured weak electronic coupling with the electronic and optical properties of the chromophore retained.

1. Introduction

Electron transfer from the excited electronic singlet state of an anchored dye molecule into empty electronic states of the semiconductor TiO_2 was reported to occur on the femtosecond time scale [1-2]. For perylene as electron injecting molecule the decay of the reactant and rise of the product state was measured with identical time constants. Perylene is a suitable model molecule for such studies since the transient spectra of the excited and oxidized states are spectrally well separated [1], intersystem crossing is negligible, and the vibrational structure of the spectra is pronounced and well established. The perylene chromophore was modified by covalently attaching three molecular groups (Figure 1). The -CH_2-PO_3-group provides a spacer and anchors the molecule to the surface, and the tertiary CH_3-groups prevent dimerisation. The electronic interaction between the excited state of the chromophore and the conduction band levels is of the order meV. In ultra-high-vacuum the electron transfer process is similar to a 'solvent-free' redox-electrode reaction at extremly high forward bias.

2. Experimental Methods

A sponge-like anatase TiO_2 film (thickness about 1μm) was employed as the electron accepting solid. Strong signals were observed due to the dye adsorption at the huge internal surface. The sample was mounted in ultra-high-vacuum after the perylene dye was adsorbed from solution. The Ti:Sapphire laser system provided pulses of 6μJ energy, 100kHz repetition rate, and 60fs pulse width (fwhm). Ten percent of the light was split off and utilized for white light generation to provide the probe beam. An OPA was pumped with the remaining light intensity. The pump pulse at 450nm was obtained from the frequency doubled idler wave of the OPA. The typical pump pulse energy was about 20nJ. Further experimental details are described in Ref [1].

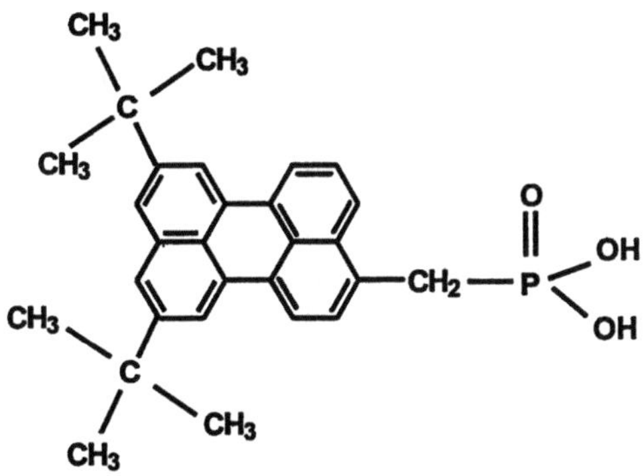

Fig. 1. Modified perylene molecule with two tert.-methyl-groups to prevent dimerisation. The CH_2-group acts as a spacer and the phosphonate-group anchors the dye molecule on the TiO_2 surface.

3. Results and Discussion

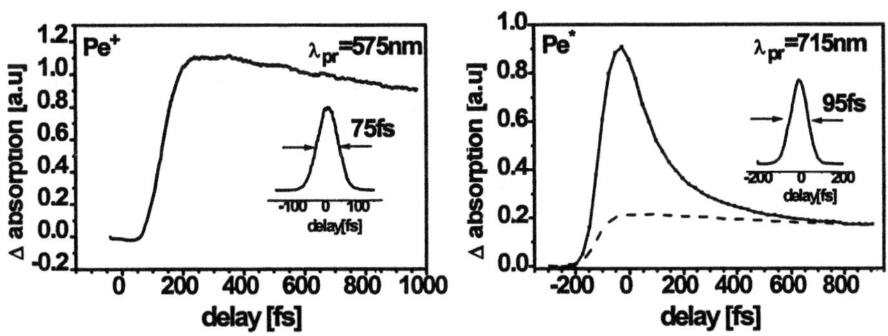

Fig. 2. Transient absorption signals of TiO_2 electrode covered with modified perylene detected at 575nm and 715nm indicating the cation absorption (left) and excited state absorption (right), respectively. The dashed line is a rescaled signal detected at 575nm. It indicates the contribution of cation absorption at 715nm. The instrumental functions are shown in the insets.

An electron injection time of 40fs was obtained from two independent signals (Figure 2). The rise of the signal at 575nm corresponds to the cation absorption (product state). The fast component of the decay at 715nm arises from the excited state absorption (reactant state). Cation absorption due to the spectral overlap leads to the slower decay component. It was observed that variations in the preparation of the TiO_2-perylene interface led to different electron transfer times in the fs-range. This can be attributed to variations in the electronic overlap between the excited dye state and the conduction band levels. This aspect is currently being investigated in some detail.

A theoretical model has been employed to obtain a formula for the lifetime of the excited state of the perylene chromophore in the above system. A modified Fano-Anderson model Hamiltonian is used to describe the excited state and its interactions with the continuum of the conduction band levels and with a dominant molecular vibrational mode. This minimal model explains the temperature independence and the ultrafast electron transfer time reported [1,2]. In the wide band limit, relevant for the above experimental system, the expression for the lifetime of the excited molecular state reduces to a simple exponential decay in time. For typical parameters valid for the above perylene-TiO_2 system, the lifetime of the excited state $M^*(t) \sim \exp(-\Delta t)$, wherein Δ is proportional to the square of the modulus of the electronic coupling between the excited state and the conduction band levels. Figure 3 indicates electron transfer occurs in fs time scales for relatively weak electronic interaction energies in the range of meV. Small variations in Δ below 10meV can lead to large changes in the lifetime. The observed sensitivity of electron transfer time to different surface preparations can be attributed to small variations in the electronic coupling.

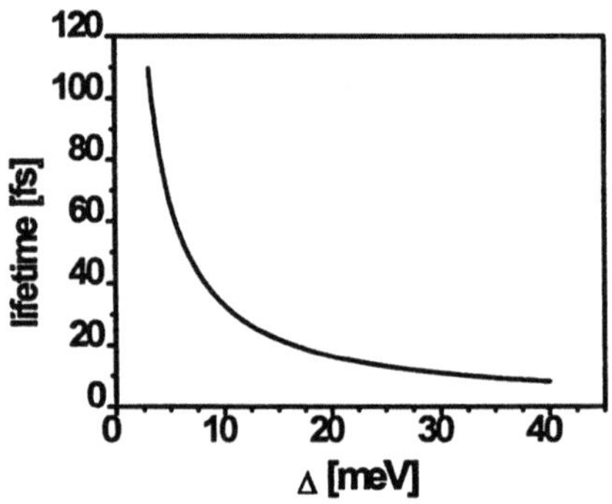

Fig. 3. Calculated lifetime of the excited state of the dye versus the electronic interaction energy.

Acknowledgements. The authors thank the Joule III program for financial support

References

1 B. Burfeindt, Th. Hannappel, W. Storck, F. Willig, in *Journal of Physical Chemistry* Vol.100, 16463, 1996.

2 Th. Hannappel, B. Burfeindt, W. Storck, F. Willig, in *Journal of Physical Chemistry B*, Vol. 101, 6799, 1997.

Femtosecond IR Study of $Ru(II)(SCN)_2(dcbpy)_2$ Sensitized Nanocrystalline TiO_2 Thin Films: Ultrafast Electron Injection and Relaxation Dynamics

John B. Asbury[a], Hirendra N. Ghosh[a], Randy J. Ellingson[b], Suzanne Ferrere[b], Arthur J. Nozik[b], and Tianquan Lian[a]

a: Department of Chemistry, Emory University, Atlanta, GA 30322, USA
E-mail: tlian@emory.edu
b: National Renewable Energy Laboratory, 1617 Cole Boulevard, Golden, CO 80401, USA

Abstract: Ultrafast electron injection(ca. 50 fs) and subsequent relaxation of electrons in $Ru(II)(SCN)_2L_2$[L=2,2'bupyridyl-4,4'-dicarboxylate] (also called N3) dye sensitized nanocrystalline TiO_2 thin films were studied by directly probing electrons using femtosecond IR spectroscopy.

1. Introduction

Photoelectrochemical solar cells based on dye sensitized TiO_2 films have received much attention in recent years because of their potential applications as a cost effective alternative to silicon based cells. Using ruthenium dye sensitizers, a solar to electricity conversion efficiency of about 10% has been reported[1]. This high efficiency is attributed to the fast electron injection and very slow back electron transfer(ET) to the sensitizer. The exact nature of the electron injection step has been a subject of intense recent debates[2-4]. For the Ru N3 dye sensitized nanocrystalline TiO_2 thin films, a biphasic electron injection process with time constant of <150 fs and 1.2 ps was first reported in a transient absorption study in the visible region[2]. A more recent study[4] reported a different assignment of the transient spectra and a much faster electron injection time of about 25 fs. In the latter study, the ET process was assigned based on an absorption signal at 1100 nm, which was attributed to hot electrons in the TiO_2 conduction band. Unfortunately, in the near IR region, the dye itself also has non-negligible absorption after visible excitation[6]. It is therefore more desirable to probe the injected electrons in the mid to far IR region, as demonstrated in recent mid IR spectroscopic studies of interfacial ET between semiconductor nanoparticles and various molecular adsorbates, such as coumarin 343[5a], $Fe(II)(CN)_6^{4-}$[5b], and SCN^-.

Recently, we carried out a femtosecond IR study of electron transfer and relaxation dynamics in N3 dye sensitized nanocrystalline TiO_2 thin films[6]. We have characterized the absorption of electrons in the mid-IR region and studied the forward ET and subsequent electron relaxation dynamics in the sensitized thin films. These spectra and dynamics were also compared to those of unsensitized thin films and bulk crystals, as well as colloidal nanoparticles. The details of the femtosecond IR spectrometer[5] and sample preparation[6] have been published elsewhere.

2. Results and Discussions

As shown in Fig. 1, after 400 nm excitation, a transient absorption at 2040 cm^{-1} was observed in N3 sensitized TiO_2 films. No noticeable signal was present in N3 sensitized Al_2O_3 films, and a 10 times smaller signal was observed in the naked TiO_2 film. The observed mid IR signal in the sensitized film is assigned to injected electrons in TiO_2 thin film. This assignment is consistent with the observation that in Al_2O_3 and ZrO_2 films, in which the conduction band energy is higher than the excited state of the dye, no injected electrons were detected. The observed signal in the naked film is attributed to electrons generated as a result of direct band gap transition. The spectral dependence of the observed signal in the N3 sensitized film and a bulk crystal are shown in Fig. 2. The qualitative similarity between these two spectra further supports the assignment of the IR absorption to electrons in TiO_2. At present, the detailed shape of the spectrum can not be fully analyzed until a spectrum at a wider wavelength region is obtained. Furthermore, the spectrum for the sensitized films is also complicated by sample inhomogeneity and spatial chirp of the probe pulse.

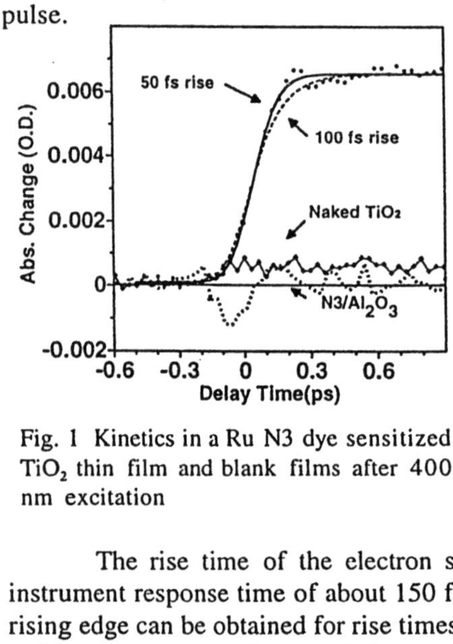

Fig. 1 Kinetics in a Ru N3 dye sensitized TiO_2 thin film and blank films after 400 nm excitation

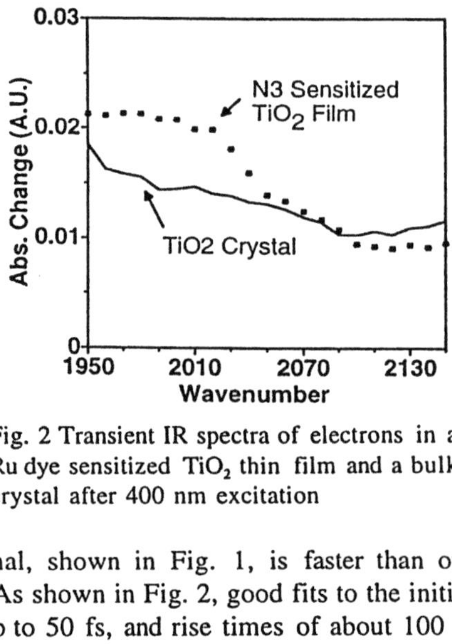

Fig. 2 Transient IR spectra of electrons in a Ru dye sensitized TiO_2 thin film and a bulk crystal after 400 nm excitation

The rise time of the electron signal, shown in Fig. 1, is faster than our instrument response time of about 150 fs. As shown in Fig. 2, good fits to the initial rising edge can be obtained for rise times up to 50 fs, and rise times of about 100 fs and longer can not give a satisfactory fit to the data. Based on this analysis, the electron injection time from the sensitizer to TiO_2 is estimated to be about 50 fs, consistent with the recent report of 25 fs ET time[4]. The same injection time has been obtained with kinetics in the 1900 to 2200 cm^{-1} region. The total amplitude of the transient absorption signal in sensitized films is similar to that of the bulk crystal under same excitation power. Since both the sensitized film and crystal have OD>2 at

400 nm, the quantum yield for electron injection can be estimated to be close to unity, assuming that the absorption strength of electrons in crystals and in films are similar in the mid IR region. This high injection yield is consistent with the observed fast electron injection time and estimated yields[2,7].

The subsequent dynamics of the injected electrons are very different from those of the naked film and nanoparticles. Shown in Fig. 3 is a comparison of long time decay kinetics of electrons in a N3 sensitized film, naked film, bulk crystal, and coumarin-343(C343) sensitized nano-particles in D_2O, each scaled to have similar peak amplitudes for comparison. The best fit to the signal in the sensitized film yields an instantaneous rise followed by a 6 ps rise which decays exponentially with a time constant of >> 1 ns. Because back ET has not been observed for the sensitized film on this time scale[3], the decay was assigned to electron trapping. The faster decay in the naked film indicated electron hole recombination in addition to trapping. It is interesting to note that these trapping dynamics appear to be much slower than those of the colloidal nanoparticles and slightly faster than those in bulk single crystals. This trend agrees with the trend in the density of surface defect states in these different TiO_2 materials.

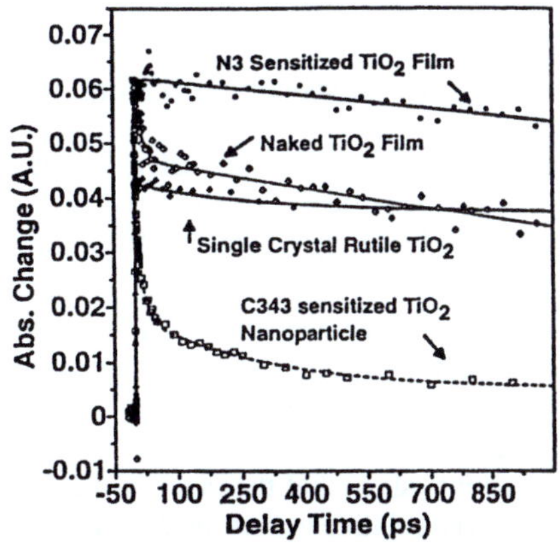

Fig. 3 Comparison of electron decay kinetics in a bulk crystal, thin films, and colloidal nanoparticles

In summary, the mid IR absorption of injected electrons in N3 dye sensitized nanocrystalline TiO_2 thin films have been characterized. The injection time from N3 sensitizer to TiO_2 is measured to be about 50 fs. The comparison of the decay kinetics in thin films, bulk crystals as well as colloidal nanoparticles indicates that the electron trapping dynamics in thin films is faster than those in bulk crystals and slower than those in colloidal nanoparticles, agreeing with the trend of defect densities.

References

1. B. Oregan, M. Gratzel, *Nature* **353**, 737-740 (1991).
2. Y. Tachibana, J. E. Moser, M. Gratzel, D. R. Klug, J. R. Durrant, *J. Phys. Chem.* **100**, 20056-20062 (1996).
3. J. E. Moser, D. Noukakis, U. Bach, Y. Tachibana, D. Klug, J. R. Durrant, R. Humphry-Baker, M. Gratzel, *J. Phys. Chem. B* **102**, 3649 (1998).
4. T. Hannappel, B. Burfeindt, W. Storck, F. Willig, *J. Phys. Chem. B* **101**, 6799 (1997).
5. a) H. N. Ghosh, J. B. Asbury, T. Lian, *J. Phys. Chem. B* , **in press** (1998); b) H. N. Ghosh, J. B. Asbury, T. Lian, submitted to *J. Phys. Chem. B*.
6. R. J. Ellingson, J. B. Asbury, S. Ferrere, H. N. Ghosh, T. Lian, A. J. Nozik, *J. Phys. Chem. B* . in press.
7. T. Heimer, E. J. Heilweil, *J. Phys. Chem. B* **101**, 10990-10993 (1997).

Femtosecond Photoselectivity in Intramolecular Double Proton Transfer in [2,2'-bipyridyl]-3,3'-diol

D. Marks, P. Prosposito, H. Zhang, and M. Glasbeek

Laboratory for Physical Chemistry, University of Amsterdam, Nieuwe Achtergracht 129, 1018 WS Amsterdam, The Netherlands
E-mail: glasbeek@fys.chem.uva.nl

Abstract. For photoexcited [2,2'-bipyridyl]-3,3'-diol in liquid solution two alternate double proton transfer processes were found to occur simultaneously: a concerted double proton transfer (one-step process), in less than 100 fs, and the sequential double proton transfer (two-step process), with time constants of less than 100 fs and 10 ps. The experiments show that the yield of the diketo tautomer (formed in the concerted process) is enhanced relative to that for the monoketo intermediate product, when laser pulses with longer wavelengths are used for photoexcitation. From the results, an energy barrier in the dienol-to-monoketo tautomerization reaction of about 600 cm^{-1} is inferred.

1. Introduction

Recently, the study of ultrafast enol-keto tautomerization reactions has attracted great interest [1-6]. For the single proton transfer in photoexcited 2-(2'-hydroxy-5'-ethylphenyl)benzotriazole in liquid solution a barrierless proton transfer, along a low-frequency coordinate, in about 700 fs was reported [1]. Intermolecular double proton transfer was studied for the model base pair, 7-azaindole dimer, in the supersonic beam [2] and in dendrites [3]. We have recently reported on our femtosecond fluorescence upconversion studies of the intramolecular double proton transfer dynamics in photoexcited [2,2'-bipyridyl]-3,3'-diol (hereafter referred to as $BP(OH)_2$, for structure see scheme) in liquid solution [4-6]. A detailed analysis of the spectra reconstructed from the fluorescence transients showed that the intramolecular excited-state double proton transfer in $BP(OH)_2$ develops along either of two alternate pathways: a concerted (one-step) process or a sequential (two-step) process [4]. The concerted reaction and the first step in the sequential process were faster than 100 fs, whereas the mono-to-diketo reaction in the two-step mechanism occurred with a typical time of 10 ps (see scheme). Stretching and bending vibrational modes within the $BP(OH)_2$ molecule were considered to promote the one-step and two-step proton transfer reactions, respectively [5, 6]. In this paper, we demonstrate that the yield of the monoketo tautomeric intermediate, in the two-step

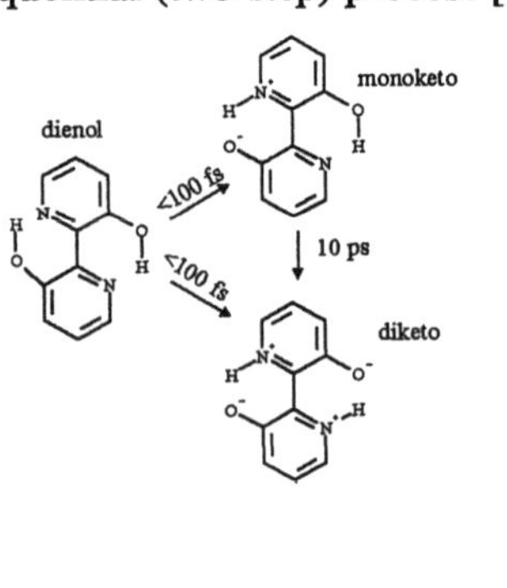

reaction dienol → monoketo → diketo, is significantly reduced when photoexcitation is to the lower vibrational levels of the dienol photoexcited S_1 state. It is noted that the excitation wavelength dependence of the two-step reaction yield manifests itself only because of the femtosecond probing of the proton transfer.

2. Experimental

[2,2'-bipyridyl]-3,3'-diol was purchased from Aldrich and dissolved without further purification in cyclohexane (Fluka, Microselect) in a concentration of about 10^{-3} M. Femtosecond fluorescence upconversion experiments and picosecond fluorescence decay measurements were performed as described previously [4-6]. The laser pulses delivered by a Quantronix regenerative amplifier laser system were split into two beams, one of which was used to pump an OPA system. The fourth harmonic of the signal pulses of the OPA, selected by a 2 mm thick optical filter DUG 11 (from 300 nm to 400 nm), were used to photoexcite the $BP(OH)_2$ solution in a flow cell at ambient temperature. Further details regarding excitation and upconversion signal detection are given elsewhere [6]. From the cross-correlation function of the excitation and gating pulses at 400 nm and 800 nm, the instrumental time response was estimated to be approximately 300 fs (FWHM).

3. Results and Discussion

Figure 1 shows a few characteristic fluorescence upconversion transients observed for $BP(OH)_2$, dissolved in cyclohexane, when the excitation wavelength is varied from 300 to 380 nm, while the detection wavelength is maintained at 480 nm. At this wavelength, the emission is almost completely due to the diketo tautomeric form [4]. The instantaneous rise component (for t ≤ 100 fs) in all transients corresponds to the ultrafast formation, directly from the dienol form, of the diketo tautomer [4]. The additional rise component (with a time constant of

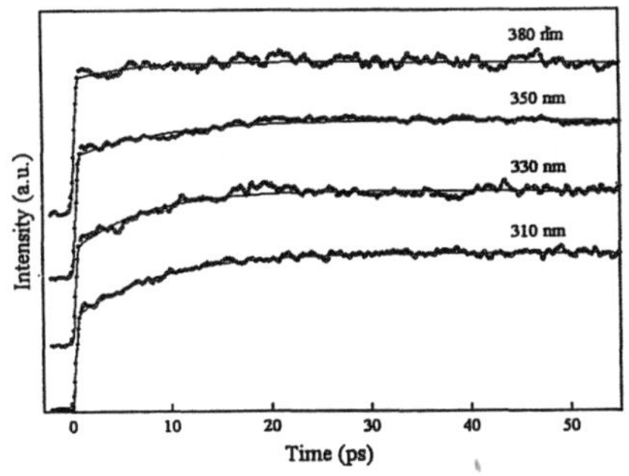

Fig. 1. Fluorescence upconversion transients of $BP(OH)_2$ in cyclohexane

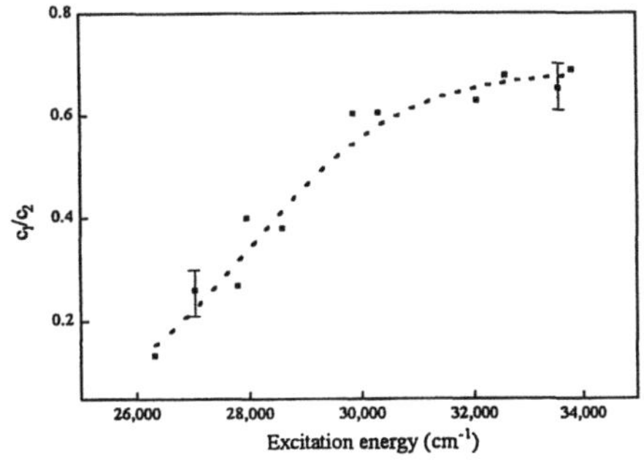

Fig. 2. Concentration ratio of the instantaneously formed *m*- and *d*- tautomers

about 10 ps) is representative of the diketo tautomer formed from the short-living (~ 10 ps) monoketo precursor [4] (see also scheme). As is seen from Figure 1, the ratio of the amplitudes of the instantaneous and the rising components is found to vary with the excitation wavelength: the relative yield of the diketo component formed from the monoketo intermediate becomes less for lower excitation energies. From this finding we conclude the presence of an energy barrier in the reaction pathway for the dienol-to-monoketo conversion. Best fittings to a biexponential function convoluted with the system response function are included in Figure 1. The ratio of the weights of the two exponential contributions, c_1/c_2, is representative of the ratio of the initially formed monoketo and diketo components. This ratio is plotted for a series of photon excitation energies in Figure 2. In a simplified approach, we now assume that the dienol-to monoketo proton transfer involves tunneling of the proton through a parabolically shaped barrier, of the form, $V= V_0 - \frac{1}{2}Ax^2$, with $A>0$. Then the permeability, G, becomes, $G= \{1+\exp[2\pi m^{\frac{1}{2}}(V_0 - W)/\hbar A^{\frac{1}{2}}]\}^{-1}$, where W is the excitation energy [6]. The plot of the best-fit function, in accordance with the expression for G, is included as the dotted curve in Figure 2. The best-fit values of the parameters are: V_0=28000 cm^{-1} and A/m =2.6x10^{30} s^{-2}. Since the 0-0 electronic transition has been estimated to be at 27400 cm^{-1} [6], it follows that in this approach the activation barrier height is about 600 cm^{-1}. Note that the curve in Figure 2 levels off for $c_1/c_2 \approx 0.7$, implying that the statistical limit value of 2 for the formation ratio of the monoketo and diketo tautomers at the higher excitation energies is not reached. The reduction in the quantum yield for the double proton transfer process is indicative for additional relaxation processes at the higher excitation energies. Candidates for such relaxation channels are IVR and energy relaxation that are very fast (less than 100 fs) in order to be competitive to the initial steps in the proton transfer processes.

Acknowledgements. This work was supported in part by The Netherlands Foundation for Chemical Research (SON) with the financial aid from The Netherlands Organization for Scientific Research (NWO). One of us (PP) gratefully acknowledges a Marie Curie Research Training Grant of the EC Training and Mobility of Researchers (TMR) Programme.

References

1 C. Chudoba, E. Riedle, M. Pfeiffer, and T. Elsaesser, *Chem. Phys. Letters*, **263**, 622, 1996.
2 A. Douhal, S. K. Kim, and A. H. Zewail, *Nature*, **378**, 260, 1995.
3 A. Douhal, T. Fiebig, M. Chachisvilis, and A. H. Zewail, *J. Phys. Chem. A*, **102**, 1657, 1998.
4 H. Zhang, P. v. d. Meulen, and M. Glasbeek, *Chem. Phys. Letters*, **253**, 97, 1996.
5 D. Marks, H. Zhang, M. Glasbeek, P. Borowicz, and A. Grabowska, *Chem. Phys. Letters*, **275**, 370, 1997.
6 D. Marks, P. Prosposito, H. Zhang, and M. Glasbeek, *Chem. Phys. Letters*, **289**, 535, 1998.

Feedback Quantum Control of Population Transfer Using Shaped Femtosecond Pulses

Christopher Bardeen[1], Vladislav V. Yakovlev[1], Kent R. Wilson[1], Scott D. Carpenter[2], Peter M. Weber[2], and Warren S. Warren[3]

[1] Department of Chemistry and Biochemistry, University of California, San Diego, La Jolla, CA 92093-0339, USA
[2] Box H, Department of Chemistry, Brown University, Providence, RI 02912, USA
[3] Department of Chemistry and Princeton Center for Photonics and Optoelectronic Materials, Princeton University, Princeton, NJ 08544-1009, USA

Abstract. Quantum control is extended to complex molecular systems by using experimental feedback to control the acousto-optic tailoring of ultrashort pulses.

1. Introduction

The extension of quantum control to large molecules in liquid solution, the domain of interest for most chemists, presents a challenge. The impossibility, given our lack of knowledge about molecular potential energy surfaces, of exactly calculating the optimal laser pulse to control such complicated systems has led to the proposal that one use the sample itself to derive the optimal field [1,2]. In other words, a computer algorithm searches for the optimal field using feedback from the experiment to vary and improve the laser pulse. We present here an experimental implementation of this concept, applied to the maximization of electronic population transfer for the laser dye IR125 in methanol solution [3]. This is, to our knowledge, the first experimental application of feedback control to a chemical system.

2. Experimental Methods

The experiment is outlined in Fig. 1. 40 fs, 800 nm pulses from a kHz amplified Ti:Sapphire laser system are tailored using a pulse shaper based on the design of Warren et al. [4] in which a RF pulse modulates both the amplitude and phase of the laser pulse spectrum through the acousto-optic effect. The shaped pulse excites fluorescence from a cell containing the dye solution, and the pulse energy and the fluorescence are detected simultaneously. The pulse shaper is controlled by a computer program which also reads the experimental measurements and uses a genetic algorithm to optimize the pulse shape. We use this algorithm to optimize two different quantities: the fluorescence *efficiency* (the ratio of fluorescence power to laser power) and the fluorescence *effectiveness* (the fluorescence power only), which is related to the total population inversion. By changing the

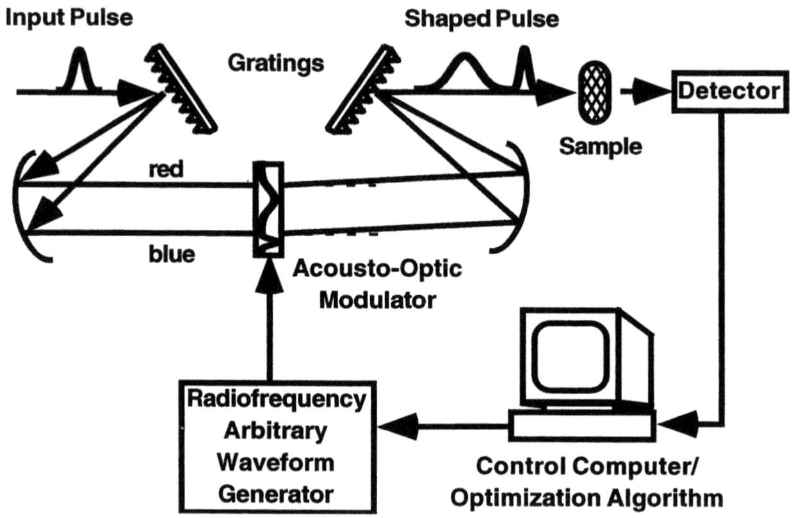

Fig. 1. Schematic of feedback experiment.

parameters of a Gaussian RF pulse, we vary five laser pulse parameters: the amplitude, central frequency, spectral width, and linear and quadratic chirp.

3. Results and Discussion

Using the experimental feedback, the algorithm converges to the optimum pulse shape within roughly 30 minutes in both cases. Fig. 2 shows the convergence of the genetic algorithm populations in the three dimensional space defined by the frequency, width, and linear chirp. The most *efficient* pulse is converged in both central frequency and spectral width, but the large spread in chirp implies that chirp is not an important parameter in this optimization. The optimal solution, shown in Fig. 3a, is simply a narrow bandwidth pulse tuned toward the peak of the absorption spectrum. The most *effective* pulse is converged in all three parameters and corresponds to a full power, full spectrum pulse with a strong positive chirp, which is shown in Fig. 3b. This has been observed before and also analyzed theoretically [5,6], with the conclusion that such chirped, strong field control is an effective and robust way to achieve nearly 100% transfer of the electronic population.

The fact that the optimization can find very different pulses depending on the quantity to be optimized shows that it is robust. We also find that the genetic algorithm is quite stable with respect to experimental non-idealities such as laser power noise and drift. The rapid convergence observed in the present experiments suggests that it should be possible to extend this method to more complex optimizations with a larger number of variable parameters .

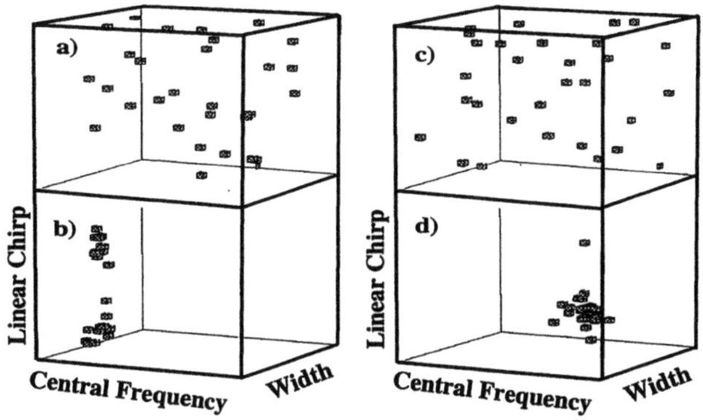

Figure 2. Evolution of genetic algorithm populations in the space defined by three pulse shaping parameters: the center frequency, the spectral width, and the linear chirp. Each square corresponds to an individual pulse shape generated by the algorithm. **a)** the first generation for fluorescence *efficiency*, **b)** the 17th generation, after convergence. **c)** the first generation for fluorescence *effectiveness*, **d)** the 17th generation, after convergence.

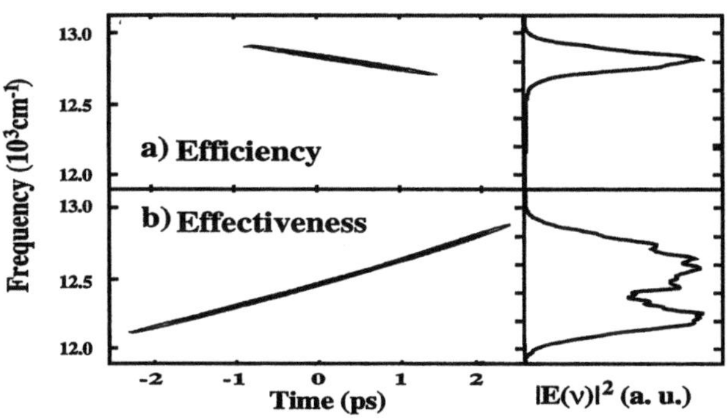

Figure 3. **a)** Optimal pulse for fluorescence efficiency determined by the feedback program. The left side is the experimentally measured Wigner transform, while the right side shows the intensity projection on the frequency axis. **b)** Optimal pulse for fluorescence effectiveness.

References

1. R.S. Judson and H. Rabitz, *Phys. Rev. Lett.* **68**, 1500 (1992).
2. Y.J. Yan, B.E. Kohler, R.E. Gillilan, R.M. Whitnell, K.R. Wilson, and S. Mukamel, *Ultrafast Phenomena VIII* (Springer-Verlag, Berlin, 1993) p.8.
3. C.J. Bardeen, V.V. Yakovlev, K.R. Wilson, S.D. Carpenter, P.M. Weber, and W.S. Warren, *Chem. Phys. Lett.* **280**, 151 (1997).
4. C.W. Hillegas, J.X. Tull, D. Goswami, D. Strickland, and W.S. Warren, *Opt. Lett.* **19**, 737 (1994).
5. G. Cerullo, C.J. Bardeen, Q. Wang, and C.V. Shank, *Chem. Phys. Lett.* **262**, 362 (1996).
6. J. Cao, C.J. Bardeen, and K.R. Wilson, *Phys. Rev. Lett.* **80**, 1406 (1998).

Control of Condensed Phase Molecular Excitation with Intense Infrared and Visible Chirped Pulses

V. D. Kleiman[1,2], E. J. Heilweil[1], D. McMorrow[2], and J. S. Melinger[2]

[1]National Institute of Standards and Technology, Gaithersburg, MD 20899, USA
[2]Naval Research Laboratory, Washington, DC 20375, USA
Tel: 202-767-5461 E-mail: valeriak@ccsalpha.nrl.navy.mil

Abstract. Chirped ultrashort laser pulses in the mid-infrared and visible are used to experimentally control population transfer in condensed phase molecules.

Recently, there has been much interest in using specially tailored ultrafast laser fields to produce a greater degree of control in molecular systems excitation. Both theoretical and experimental work has shown how appropriate variations in the phase and amplitude structure of a laser field can produce both more highly excited and selectively excited atomic and molecular systems. [1-3] In this work control over the phase structure of the laser field is used in the mid-infrared (mid-IR) to gain control of vibrational up-pumping in $W(CO)_6$, and in the visible to control population transfer between electronic manifolds in dye molecules. In each case, pulses with a (near) quadratic phase sweep (both positive and negative in sign) are produced with an appropriate time duration. For the visible excitation case we examine the role of laser pulse intensity on the excitation.

In the first example, the T_{1u} CO-stretching manifold of $W(CO)_6$ in n-hexane (298 K) is excited with chirped IR pulses near 1980 cm^{-1}. The experimental configuration is for IR pump-probe spectroscopy [4]. The IR pulse is produced by difference frequency mixing two visible laser pulses generated from independent dye lasers. The chirp is generated at the visible level - one of the dye laser pulses is chirped using an optical fiber/grating apparatus - and then transferred to the IR pulse in the down-conversion process. The chirp in the pump pulse was measured using the frequency resolved optical gating technique for ultrashort IR pulses [5]. IR pulse-widths were 1-2 ps with FWHM bandwidths of about 15 cm^{-1}. The pulses exhibited near-quadratic phase sweeps extending over the main part of the pulse. Figure 1 shows the transient spectra obtained with a weak broadband probe pulse ($\approx$ 120 cm^{-1} FWHM) after excitation by an intense pump pulse. Assuming a transition dipole of 1 Debye, the Rabi frequency is estimated to be about 5 cm^{-1}, and represents a significant fraction of both the 15 cm^{-1} anharmonicity of the $W(CO)_6$ potential, and the spectral bandwidth of the laser pulse. The middle panel corresponds to excitation with a near transform limited (TL) pulse and shows removal of population from the v = 0 ground state (at 1982 cm^{-1}). Two new features at 1967 cm^{-1} and 1952 cm^{-1} are generated with IR excitation and assigned to the v = 1 $\rightarrow$ 2 and v = 2 $\rightarrow$ 3 overtone absorptions, respectively [6]. The top and lower panel depicts transient difference spectra using positively and negatively chirped pump pulse excitation, respectively. The striking difference between the transient IR spectra for the three types of excitation pulses is the amount of v = 2 vs. v = 1 population transferred. Positive chirp excitation yields an almost completely diminished absorption at the v = 2 $\rightarrow$ 3 transition. The suppression of overtone transitions by the positively chirped pulse was checked at several choices of pump center frequency in the 1980 cm^{-1} region. For each choice of pump frequency a strong suppression of overtone excitation was observed. In contrast to the positively chirped excitation, negatively chirped excitation clearly enhances the up-pumping process by transferring significantly more of the population to the v = 2 vibrational state and, accordingly, maximizing the change in absorption for the v = 2 $\rightarrow$ 3 transition. We have modeled the ladder climbing process using a density matrix treatment of a simple one dimensional anharmonic oscillator without relaxation effects. The calculations show reasonable qualitative agreement between experiment and theory [4].

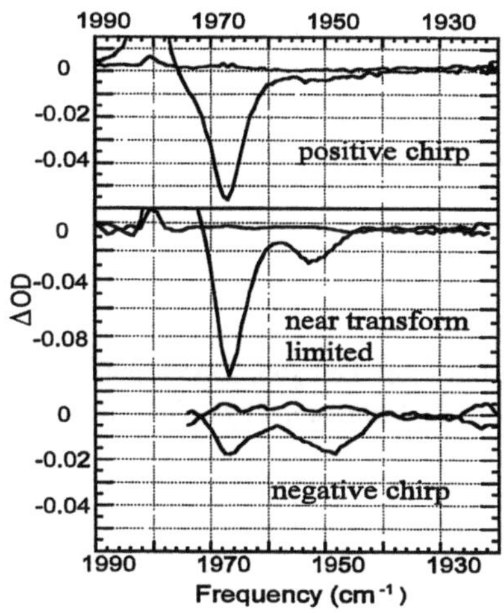

Fig. 1. Infrared transient absorption spectra of $W(CO)_6$ for different laser pulse chirps.

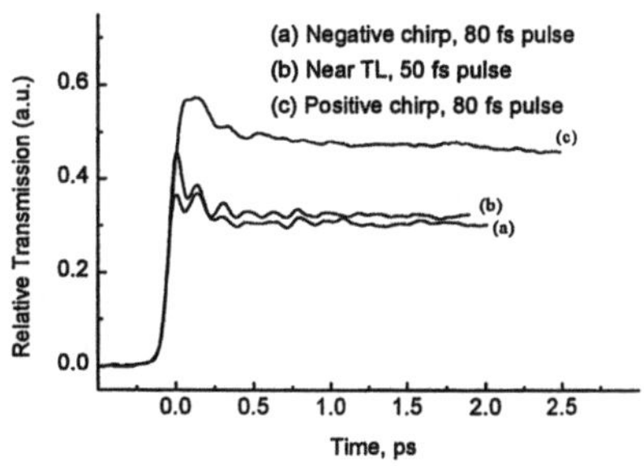

Fig. 2. DODCI transmission for different laser pulse chirps

In the second case, intense sub-100 fs laser pulses near 605 nm were used to excite and probe the dye molecule DODCI in ethylene glycol . The experiment here is similar to previous work that investigated the role of phase modulation in condensed phase excitation of a dye [2], with the addition that we examine the effect of a strong field. Figure 2 shows the transmission of a weak probe pulse (50 fs) following excitation by either a (near) TL pump pulse, or a bandwidth-equivalent chirped pump pulse made by passage through a prism pair. The pump fluence in each case was about 200 uJ/cm^2. The middle trace shows the probe transmission following excitation from a near-TL pulse of about 50 fs. Analysis of the oscillations reveals a pair of mode frequencies at 92 cm^{-1} and 215 cm^{-1}, which is consistent with previous ultrafast studies of DODCI [7]. The oscillations are indicative of coherent vibrational motion originating in either the S_0 or S_1 manifold (or both). The bottom and top traces show the effect of a negative and positive chirp. The negatively chirped pump pulse ($\approx$80 fs) gives rise to a probe transmission somewhat weaker than the near-TL case, while the positively chirped pump pulse ($\approx$80 fs) yields a significantly higher (by approximately a factor of 1.4) probe transmission. For each chirped pulse case the oscillations in the trace are attenuated. The same experiment performed at lower pump fluence ($\approx$20 uJ/cm^2) yields probe transmissions for each choice of chirp that are more nearly similar in magnitude. The results suggest that the intense negatively chirped pump pulse enhances a Raman-type process, which cycles population back down to the S_0 manifold, while the intense positively chirped pump is more effective in quantitatively exciting population to the S_1 manifold.

In conclusion, we have shown experimentally that chirped ultrashort laser pulses can provide enhanced control of population transfer in complex condensed phase systems.

References

1 S. Chelkowski, A. D. Bandrauk, P. B. Corkum, Phys Rev Lett., **65**, 2355 (1990).

2 B. Broers, H.B. van Linden van den Heuvell, L.D. Noordam, Phys. Rev. Lett., **69**, 2062 (1992).

3 (a) C.J. Bardeen, Q. Wang, C. V. Shank, Phys. Rev. Lett., **75**, 3410 (1995), (b) C.J. Bardeen, Q. Wang, C. V. Shank, J. Chem. Phys., 102, 2759 (1998).

4 V. D. Kleiman, S. M. Arrivo, J. S. Melinger, E. J. Heilweil, Chem. Phys., in press.

5 D. J. Kane, R. Trebino, IEEE J. Quant. Elect. **29**, 571 (1993).

6 S. M. Arrivo, T. P. Dougherty, W. T. Grubbs, E. J. Heilweil, Chem. Phys. Lett., **235**, 247 (1995).

7 F.W. Wise, M.J. Rosker, and C.L. Tang, J. Chem. Phys., **86**, 2827 (1986).

Part XIV

Biology and Medicine

Femtosecond Dynamics, Two Dimensional Infrared Spectroscopy and Echoes of Protein Vibrations

Robin M. Hochstrasser, Peter Hamm, and Manho Lim

Department of Chemistry, University of Pennsylvania, Philadelphia PA 19104-6323, USA
E-mail: hochstra@sas.upenn.edu

Abstract. Two approaches that use femtosecond infrared pulses to examine protein dynamics by means of nonlinear IR spectroscopy are described. In the first we dissect the featureless peptide amide-I band and introduce two-dimensional IR spectra [1]. In the second, the correlation function of the frequency fluctuations of a vibrational mode in the active site of carbonic anhydrase and heme pocket of hemoglobin is determined by three-pulse IR photon echoes [2]. These experiments permit the study of ultrafast dynamics without the requirement of chromophores.

1. Infrared Laser System

The IR-setup [3] is a white light seeded, two stage BBO-OPA followed by difference frequency mixing in $AgGaS_2$. The pulses near 1650 or 2000 cm^{-1} (bandwidth 130 cm^{-1}, pulse duration 180 fs, energy ca. 1 μJ) were essentially chirp free. The spectral range for the peptides (1700 cm^{-1} - 1580 cm^{-1}) was covered by a 32 channel MCT detector-array mounted on the focal plane of a 270 mm spectrograph. The overall sensitivity was 0.01 mOD (rms.) and the cross correlation width was typically 200-250 fs. In the peptide experiments either the whole band was excited with a short pulse (300 nJ) or spectrally narrow holes were burned into the amide I band by tunable pump pulses (25 nJ) from an IR Fabry-Perot filter. The echo experiments involved three equal intensity beams (1, 2 and 3) arranged in a box configuration to allow measurement of the $-k_1+k_2+k_3$ and $k_1 -k_2 +k_3$ signals as a function of times τ and T between the centers of pulses 2 and 1, and 3 and 2, respectively [4].

2. Ultrafast IR spectra of the amide-I band [1]

Femtosecond infrared (IR) pump probe and dynamic hole burning experiments were used to examine the ultrafast response of the modes in the 1600 cm^{-1} to 1700 cm^{-1} region (the so-called amide I modes) of N-methylacetamide (NMA), a model pentapeptide [5], and three small globular peptides, apamin, scyllatoxin and bovine pancreatic trypsin inhibitor (BPTI). These signals occur because there is anharmonicity in the amide I vibrational motion, measured in this work to be 16

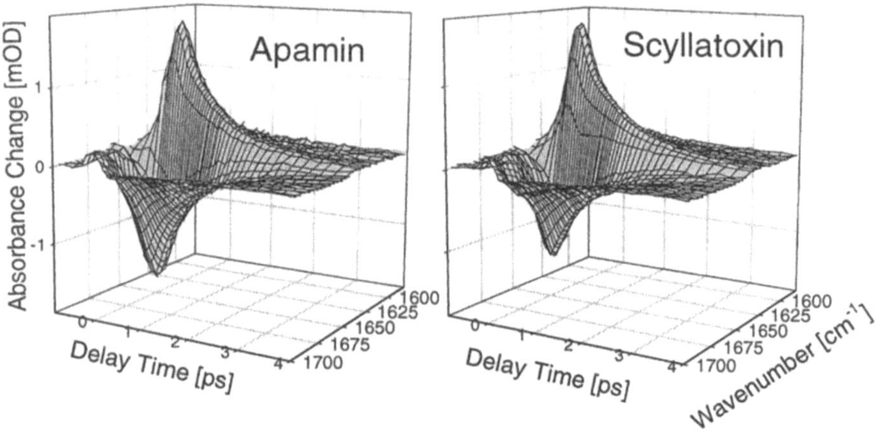

Fig. 1. 3D plot of the frequency dependent magic angle absorption change of apamin and scyllatoxin after exciting the sample with a broad band IR pump pulse

cm^{-1}, even when the dipole operator is nonlinear in displacement. Vibrational relaxation of the amide I modes of all investigated peptides occurs in ca. 1.2 ps (Figure 1). An even faster value of 450 fs is obtained for NMA, a model for the peptide unit. The vibrational relaxation is assumed to be dominated by Fermi resonances (IVR) and is an intrinsic property of the peptide group in any environment. Dynamic hole burning experiments (Figure 2) with a narrow band pump pulse which selectively excites only a subset of the amide I eigenstates, reveal

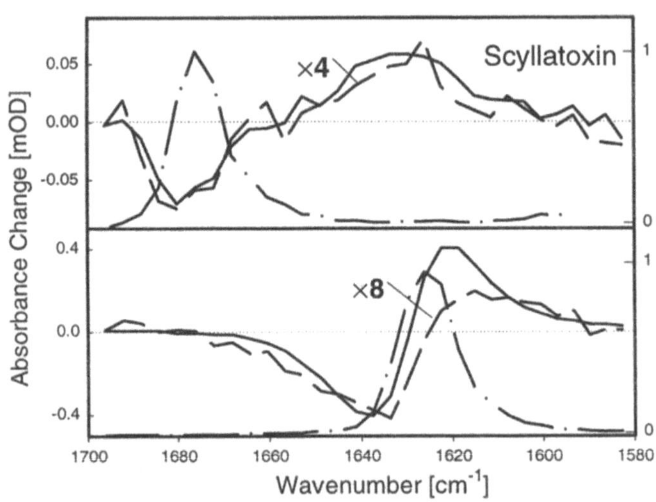

Fig. 2. Representative dynamic hole burning experiments where the samples were pumped with a narrow band pump pulse (dashed-dotted line). The scaled responses at "delay zero" (solid line) and 2.4 ps (dashed lines) are shown.

that energy migration between different amide I states is slow compared with vibrational relaxation. Two-dimensional pump-probe (2D-IR) spectra that display the spectral response of the amide I band as a function of the frequency of the narrow band pump pulse show that the amide I states are nevertheless delocalized along the peptide back bone (Figure 2). A model in which the protein structures define extended amide transition dipole arrays can be treated by exciton theory which adequately describes the nonlinear pump-probe spectrum, and after averaging over multiple structures it reproduces the experimental 2D-IR spectra. The third order nonlinear spectroscopies can provide enough information from the one and two exciton transitions to determine the ca. $n^2/2$ elements of this model although this has not yet been accomplished uniquely. It is estimated that the coherent spatial extent of the vibrational states in the peptide is ca. 8 Å. The differences between the experimental 2D-IR spectrum of apamin and scyllatoxin

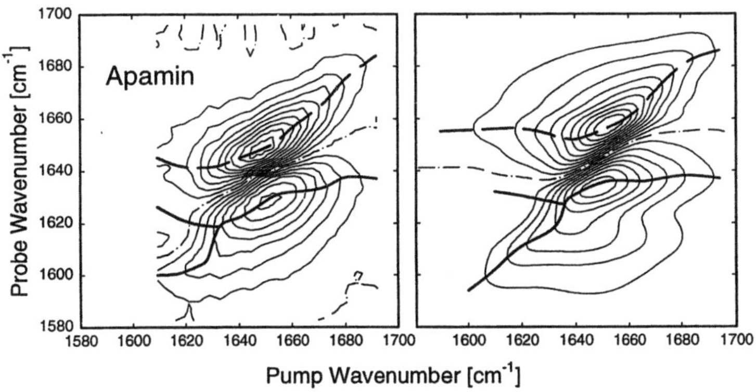

Fig. 3. Calculated (right) and observed (left) 2D IR spectra of apamin

are small, and the spectra do not resolve into well separated bands. However, we were able to distinguish between the localized versus delocalized character of the states and observe dephasing and equilibration via anisotropy measurements. A hydrogen bond dependent frequency shift was needed to fit the calculated and observed 2D-IR spectra of the apamin and the scyllatoxin, indicating that this technique is sensitive to the structure of the coupling Hamiltonian and that the 2D-IR spectra yield more information on the structure of a peptide than can a linear spectrum.

The 2D IR data for the model pentapeptide, whose linear spectrum is also shown in Figure 4, clearly manifests strong correlations between a number of the transitions. In this case a calculation based on the known geometric structure [5] of the peptide is employed to find the diagonal energies needed to reproduce the linear, pump/probe and three pulse echo decays. This procedure, which has worked reasonably well, yields the relationship between the structure and the spectra but its uniqueness requires to be tested on other model systems. In the amide-I region the first moment of the echo signal versus τ for various values of T

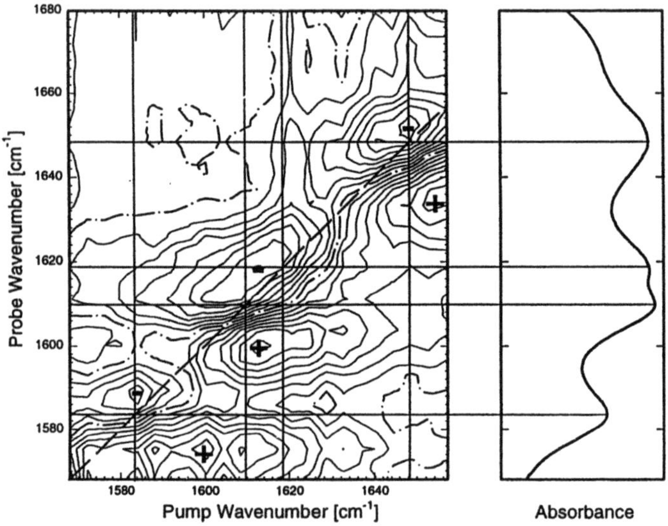

Fig. 4. Linear (right) and 2D IR spectrum (left) with perpendicular polarization of the model pentapeptide, cyclo (D-Abu-N-MeArg-Gly-Asp-Mamb) (see Ref. 5)

(Figure 5) shows slight decay, indicating that the inhomogeneous distribution is only slowly changing over the time scale of the two pulse echo (T=0) which yields an average dephasing time of ca. 1 ps for the various amide group transitions.

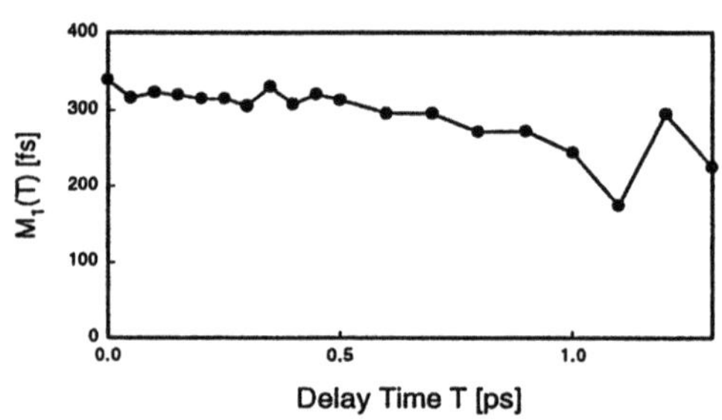

Fig. 5. First moment of the three pulse photon echo of the pentapeptide [5]

3. Fluctuations in Carbonic Anhydrase (CA) and Hemoglobin (Hb)

CA, the enzyme that interconverts CO_2 with bicarbonate, has a known structure which is uncompromised when the inhibitor azide (N_3^-) is bound [6]. The azide vibration at ca. 2050 cm^{-1}, which has known vibrational dynamics for the unbound ion [4,7], was studied in the proteins by three pulse photon echoes. The first

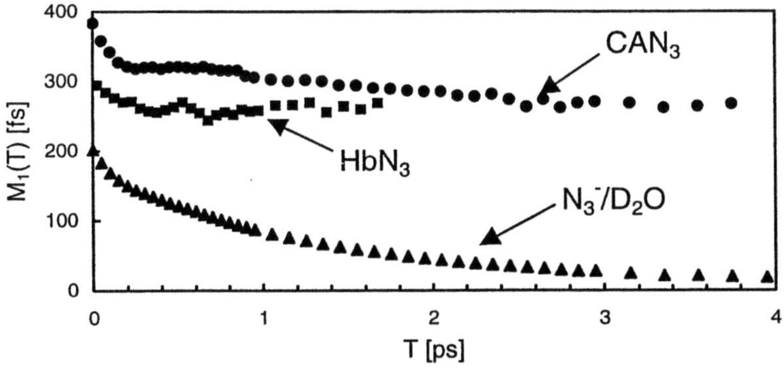

Fig 6. The three pulse photon echo of CA-N_3^-, HbN_3 and N_3^-/H_2O

moments of these signals shown in Figure 6 provide an approximate representation of the correlation function of the fluctuations in frequency of the vibrational mode. The dynamics are distinctly non-Markovian and the protein inhomogeneous distributions decay nonexponentially (see Figure 6) over time periods that exceed the lifetime of the vibrational state and the aqueous ion spectral diffusion [4]. The CA fluctuations are suggested to be from motions of the zinc cation to which the anion is bound and two nearby threonines that can influence the ground state potential surface of N_3^- [8]. Clearly the Hb pocket is the most rigid structure of the three on this time scale.

Acknowledgements. This research was supported by the NIH and NSF and used instrumentation developed under NIHRR 01348.

References

1 P. Hamm, M. Lim, and R. M. Hochstrasser, J. Phys. Chem. **102**, 28, 1998.
2 M. Lim, P. Hamm, and R. M. Hochstrasser, in preparation.
3 P. Hamm, M. Lim, and R. M. Hochstrasser, J. Chem. Phys. **107**, 10523, 1997.
4 P. Hamm, M. Lim and R. M. Hochstrasser, Phys. Rev. Lett., Submitted 1998.
5 A. C. Bach, C. J. Eyerman, J. D. Gross, M. J. Bower, R. L. Harlow, P. C. Weber, and W. F. DeGrado, J. Amer. Chem. Soc. **116** 3207, 1993.
6 S. K. Nair and D. W. Christenson, Eur. J. Biochem. **213**, 507, 1993.
7 M. Li, J. Owrutsky, M. Sarisky, J. P. Culver, A. Yodh, and R. M. Hochstrasser, J. Chem. Phys. **98**, 5499, 1993.
8 K. M. Merz and L. Banci, J. Phys. Chem. **100**, 17414, 1996.

Three Pulse Echo Peak Shift Measurements and *ab initio* Excited State Calculations of Photosynthetic Light Harvesting Function

G. D. Scholes,[1] J.-Y. Yu,[1] Y. Nagasawa,[1] I. R. Gould,[2] and G. R. Fleming[1]

[1] Department of Chemistry, University of California and Physical Biosciences Division, Lawrence Berkeley National Laboratory, Berkeley CA 94720-1460, U.S.A.
E-mail: fleming@cchem.berkeley.edu
[2] Department of Chemistry, Imperial College of Science, Technology and Medicine, London SW7 2AY, U.K.

Abstract. The analysis of three pulse echo peak shift (3PEPS) data for LH1 of *Rhodobacter sphaeroides* is discussed in comparison with that of the B820 subunit of *Rhodospirillum rubrum*. We report also *ab initio* molecular orbital (CIS/6-31G*) calculations of electronic couplings for supermolecule subunits of the light harvesting 2 complex (LH2) of *Rhodopseudomonas acidophila*.

1. Introduction

The key to understanding the primary processes of photosynthesis – light harvesting and charge separation – is to unravel the subtle interplay between the electronic states of the chromophores, the way these states are modified by the protein environment, the electronic couplings between the chromophores, and the nuclear motions of the protein. In natural systems the net result is almost perfectly efficient energy and electron transfer. These processes generally have strong vectorial character – presumably as a result of chromophore–protein interactions (as, for example, in the unidirectional primary charge separation step in purple bacterial reaction centers and in the directed energy transfer from antennas to reaction centers). An elucidation of the design principles at work in these supermolecular systems requires a combination of experiment and theory. In particular, we are using a combination of femtosecond nonlinear spectroscopy, theory, and *ab initio* electronic structure calculations to investigate the fundamental dynamics of purple bacterial light harvesting complexes. *Ab initio* electronic structure calculations are used here to define electronic couplings and the role of individual amino acids in modifying the electronic states.

The separation between photosynthetic cofactors is often less than 15Å; indeed individual chromophores may be in van der Waals contact. Thus simple models for electronic coupling (which gives rise to energy transfer) are unlikely to be valid. With the extraordinary increases in computing power available in the past few years it has now become possible to carry out *ab initio* quantum chemical calculations of clusters of chromophores and even include selected amino acids from the protein. We report here an analysis of the three pulse echo peak shift (3PEPS) data for LH1 of *Rhodobacter sphaeroides* in comparison with the B820 subunit of *Rhodospirillum rubrum*. We report also *ab initio* molecular orbital calculations of electronic couplings for the B850 ring of the light harvesting 2 complex (LH2) of *Rhodopseudomonas acidophila*.

2. Experimental Methods

3PEPS measurements were carried out as described elsewhere [1]. 37 fs FWHM laser pulses (cavity-dumped mode-locked Ti:sapphire) at 837 nm (250 kHz) were used at a maximum energy of 570 pJ/pulse.

In the absence of direct interchromophore orbital overlap, two chromophores (one in an excited electronic state) interact via the long-range *Coulombic* mechanism [2,3]. This interaction is mediated by an interaction between the transition moments of the two molecules, and dominates the coupling in many instances of singlet-singlet energy transfer. The Coulombic potential may be represented as a multipolar expansion of the interaction potential and it is usual to retain only the dipole-dipole term (hence the signature R^{-3} distance-dependence). Determination of the Coulombic contribution to the couplings was undertaken here using the recently-developed transition density cube (TDC) method [4].

When the electron density distributions (i.e. orbitals) on the two chromophores overlap, additional effects arise (as a consequence of the Pauli principle) which can dramatically increase the electronic coupling. Such a mechanism for energy transfer between forbidden transitions was first described by Dexter [5]. Broadly speaking, it is correct to attribute the primary overlap-dependent interactions to exchange "corrections", however, it has been shown [6] that by defining appropriate reactant and product states the coupling at close separations (or for forbidden transitions) is promoted primarily by terms other than the Dexter two-electron, four-center exchange integral. However, the exponential distance-dependence for the interaction is analogous to that introduced by Dexter (because the orbital-overlap-dependence is the same). The detailed elucidation of the mechanism of this coupling is not merely a matter of semantics; it serves to establish a connection with charge transfer configurations, and hence implications for spectral shifts, oscillator strength changes, superradiance, and polarisabilities of molecular aggregates.

The coupling is estimated from CI-singles (CIS) "supermolecule" calculations of the excited states. Thus we base our analysis on the premise that the CIS coupling is reliably-represented in the calculated splitting of the dimer excitation energy. At the CIS level of theory this is a good approximation, but the influence of higher-order CI has not been clearly established. We have found the calculated couplings to be reasonably precise, although they bear a poor relationship to experiment owing mainly to the overestimated Coulombic coupling (*ca.* a factor of four). To remedy this, we have attempted to dissect the coupling into contributions from the Coulombic coupling (which we then rescale) and the short-range (i.e. orbital overlap-dependent) coupling (which we take as a reasonable zero-order estimate) [7].

3. Results and Discussion

A vanishing value for the long-time peak shift is observed in liquids, which indicates that the solute can sample the full range of environments (i.e. loses the memory of its initial transition frequency completely) on the timescale that the peak shift approaches zero. With this in mind, the result of 3PEPS measurements on the light harvesting 1 (LH1) complex of *Rb. sphaeroides* shown in Fig 1 (open circles) is surprising. The peak shift on the timescale shown decays to a small value and on a 10 ps timescale decays to almost zero [1]. The reason for the extra decay component in LH1 is that energy transfer between the 32 bacteriochlorophyll (BChl) molecules comprising LH1 allows the excitation to "average" over the site energy distribution (sufficiently that the energy gap correlation function goes to zero [8]).

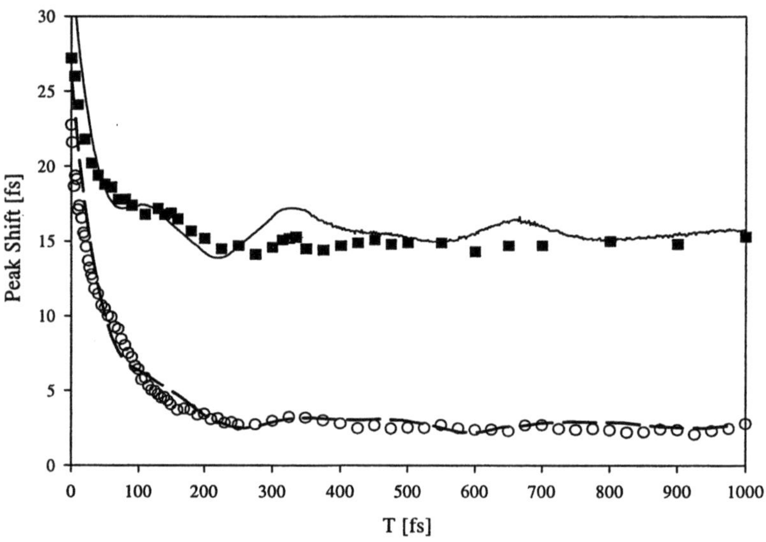

Fig. 1. 3PEPS data for LH1 of *Rb. sphaeroides* (open circles) and the B820 dimeric subunit of *Rs. rubrum* (solid squares). The dashed line is a simulation of the LH1 data, whilst the solid line is a simulation of the B820 data using identical parameters, but without the energy transfer term.

Striking confirmation of this picture appears in the upper curve of Fig. 1 (solid squares). Here 3PEPS data for the dimeric subunit of LH2 (B820) are presented [9]. Energy transfer outside of the dimer of BChl molecules cannot occur in B820, and thus the averaging effect of energy migration is absent. In other words, the lineshape function describing the LH1 data, in addition to containing the usual intramolecular and solvent (protein) spectral densities, also contains a term describing the energy transfer dynamics whose amplitude is related to the inhomogeneous broadening [10]. The full lineshape function was determined by fitting, and contains an energy transfer term with a 90 fs time constant and a static disorder of 160 cm^{-1}. This function gives the dashed curve in Fig. 1. Following the same logic, if no other properties of LH1 are changed in making B820, (other than prohibiting energy migration), then it should be possible to predict the B820 3PEPS result from the LH1 fit. The solid line in Fig. 1 is the prediction obtained by simply setting the energy transfer rate to zero. The degree of agreement between the actual and predicted curves is remarkable. The finite long-time value gives the static inhomogeneous width; now no longer averaged over by the energy migration process. Similarly, the increased initial value of the peak shift observed (and predicted) arises from the change in the ratio of static to dynamic line broadening mechanisms.

Table 1. Calculated couplings (CIS/6-31G*) for the intrapolypeptide and interpolypeptide dimers of the B850 ring of LH2 (*Rps. acidophila*): V(total), dissected into the Coulombic and short-range contributions. The dipole-dipole coupling, V(d-d), is given for comparison. Couplimgs are quoted in cm^{-1}.

Dimer	V(Coul)	V(short)	V(total)	V(d-d)
Intra	265	55	320	415
Inter	195	60	255	330

The analyses of the couplings between Bchl *a* chromophores which comprise the B850 ring of LH2 (*Rps. acidophila*) are summarised in Table 1. The Coulombic coupling, V^{Coul}, provides the largest contribution to the total couplings. It is found to be fairly insensitive to basis set. As anticipated, the dipole approximation does not reproduce the V^{Coul} for the B850 couplings. It is evident from Table 1 that the contribution to the couplings from orbital overlap-dependent interactions (short range coupling, V^{short}) is small compared to V^{Coul}, in accord with the conclusions of Alden, et al [11] (who did not, however, quantify V^{short}). Similar short range couplings contribute to both the interpolypeptide coupling and to the intrapolypeptide coupling. We conclude that the orbital overlap-dependent contribution to the coupling for both the intra- and interpolypeptide dimers is at least 60 cm^{-1}. This represents approximately 20% of the total coupling. To put this in perspective, it is at least twice the magnitude of the B800-B800 coupling.

It is known for B850 (and LH1) that the couplings are of the same magnitude as the disorder (inhomogenous broadening) and the homogenous linewidth. Hence a simple Förster model is inapplicable. In previous work it was found that the fluorescence depolarisation kinetics could be modelled by assuming an intrapolypeptide coupling of 230 cm^{-1}, an interpolypeptide coupling of 110 cm^{-1}, and a site disorder of 200 cm^{-1}. It was suggested that the resultant dynamics could be described as incoherent hopping between dimers (prompted by the strong interactions evident in the B820 dimeric subunit of LH1), however the authors emphasised that electronic structure calculations were necessary to confirm some of the speculation. Pullerits, et al [12] have simulated fs transient absorption experiments of the B850 region after excitation of the B800 band and subsequent energy transfer to B850. Their results suggest a delocalisation length of 4 ± 2 molecules. Studies and simulations of superradiance have led Monshouwer, et al [13] to conclude that the delocalisation length is approximately 2 to 3 molecules.

A key parameter required for the detailed analysis of spectroscopic data of molecular aggregates such as LH1 and LH2 is the electronic coupling. Mukamel and coworkers have recently described a method of analysis of 3PE data based on the density matrix, which requires input of disorder, site spectral densities, geometry of the aggregate and electronic couplings [14]. Consideration of the LH1/B820 results and analysis suggests that a number of open questions regarding the elucidation of energy transfer dynamics via 3PEPS data remain. With the availability of the crystal structure [15] and good estimates of the electronic couplings, more rigorous modelling of the data is possible, enabling us to address these issues in detail.

Acknowledgements. Professor Richard Cogdell is thanked for providing the crystal structure coordinates of LH2.

References

1 R. Jimenez, F. van Mourik, J.-Y. Yu, G. R. Fleming, J. Phys. Chem. B **101**, 7350, 1997.
2 Th. Förster, Ann. Phys. **2**, 55, 1947.
3 G. D. Scholes, D. L. Andrews, J. Chem. Phys. **107**, 5374, 1997.
4 B. P. Krueger, G. D. Scholes, G. R. Fleming, J. Phys. Chem. B, **102**, 5378, 1998.
5 D. L. Dexter, J. Chem. Phys. **21**, 836, 1953.
6 R. D. Harcourt, G. D. Scholes, K. P. Ghiggino, J. Chem. Phys. **101**, 10521, 1994.
7 G. D. Scholes, I. R. Gould, R. J. Cogdell, G. R. Fleming, in preparation.
8 T. Joo, Y. Jia, J.-Y. Yu, M. J. Lang, G. R. Fleming, J. Chem. Phys. **104**, 6089, 1996.
9 J.-Y. Yu, Y. Nagasawa, R. van Grondelle, G. R. Fleming, Chem. Phys. Lett. **280**, 404, 1997.
10 Y. Nagasawa, J.-Y. Yu, M. Cho, G. R. Fleming, Faraday Discuss. **108**, 23, 1997.
11 R. G. Alden, E. Johnson, V. Nagarajan, W. W. Parson, C. J. Law, R. G. Cogdell, J. Phys. Chem. B. **101**, 4667, 1997.
12 T. Pullerits, M. Chachisvilis, V. Sundström, J. Phys. Chem. **100**, 10787, 1996.
13 R. Monshouwer, M. Abrahamsson, F. van Mourik, R. van Grondelle, J. Phys. Chem. B **101**, 7241, 1997..
14 W. M. Zhang, T. Meier, V. Chernyak, S. Mukamel, J. Chem. Phys. **108**, 7763, 1998.
15 G. McDermott, S. M. Prince, A. A. Freer, A. M. Hawthorne-Lawless, M. Z. Papiz, R. J. Cogdell, N. W. Isaacs, Nature **374**, 517 (1995).

Two-Dimensional Femtosecond Spectroscopies of Coupled Chromophores

Wei Min Zhang, Vladimir Chernyak, and Shaul Mukamel

Department of Chemistry and Rochester Theory Center for Optical Science and Engineering, University of Rochester, Rochester, New York 14627, USA

Abstract. The coupling pattern of vibrational or electronic excitons as probed by multidimensional techniques provides a unique view into structure and dynamics of proteins and molecular aggregates.

1. Introduction

The elementary excitations of interacting localized chromophores may be described by the Frenkel exciton model. This applies to infrared vibrational transitions (e.g. the amide band of polypeptides [1,2]) as well as to optical electronic transitions (e.g. photosynthetic antenna complexes [3]). Two-dimensional four wave mixing spectroscopies of these systems, based on the application of sequences of carefully shaped and timed femtosecond pulses provide a novel multidimensional view of molecular structure as well as vibrational motions, interactions, and relaxation processes. Spreading the spectroscopic information in more than one dimension makes it possible to resolve congested spectra, selectively eliminates certain static broadening mechanisms, and provide structural and dynamical information unavailable from one-dimensional measurements. The basic ideas are analogous to multiple-pulse NMR techniques [4]. However, the resulting information is very different and complimentary [5,6]. The possible 2D resonant four wave mixing pulse sequences are studied.

2. Exciton Model of Localized Vibrations or Molecular Aggregates

The third order optical susceptibility of a system of coupled chromophores may be calculated by solving the nonlinear exciton equations (NEE) [6-8]

$$\chi^{(3)}(-\omega_s;\omega_a,\omega_b,\omega_c)=\frac{1}{3!}\sum_{p}\sum_{nm_1m_2m_3}\sum_{n'n''}\mu_n\mu_{m_1}\mu_{m_2}\mu_{m_3}$$

$$\times G_{n'm_1}(\omega_a)G_{n'm_2}(\omega_b)G^{+}_{m_3n''}(\omega_c)G_{nn''}(\omega_a+\omega_b-\omega_c)\overline{\Gamma}_{n''n'}(\omega_a+\omega_b)+c.c.$$

where G is the one exciton Green function and $\overline{\Gamma}$ is the exciton-exciton scattering matrix. The p summation represents all possible frequency permutations.

In a two-dimensional four wave mixing experiment, two of the three pulses are time-coincident and only differ by their wavevectors. Figure 1 shows the double-sided Feynman diagrams for all four possible techniques which survive the rotating wave approximation. Each technique is represented by two diagrams, with and without the participation of the two-exciton states.

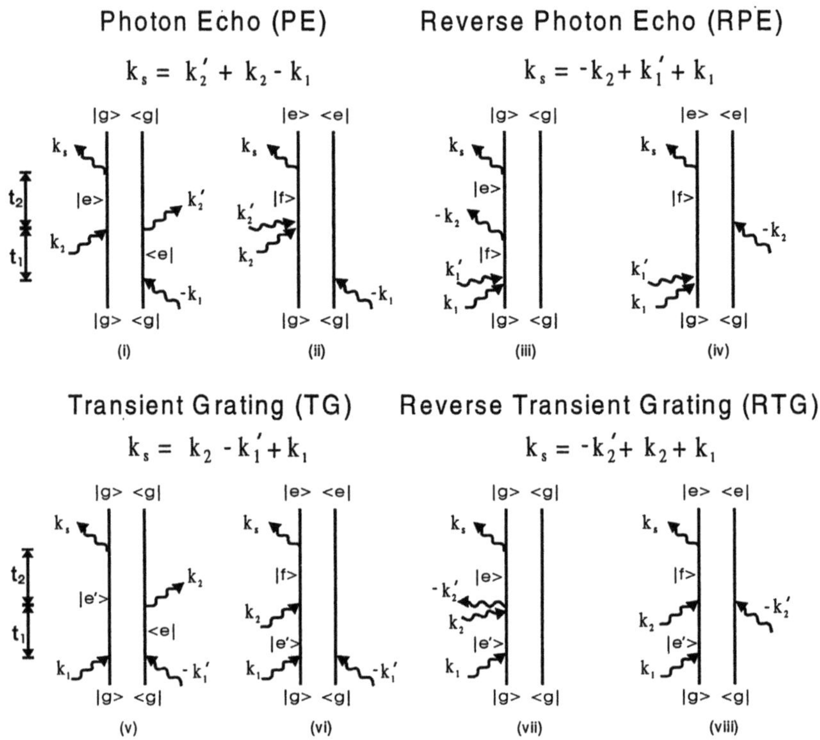

Fig. 1. Feynman diagrams of the four possible resonant 2D techniques. |g>, |e>, and |f> represent the ground, the one-exciton, and the two-exciton states, respectively.

Figure 2 shows model calculations for a system made out of 8 interacting vibrations [6]. Frequencies are given in units of the lowest transition frequency. Fig. 2(a) shows the linear absorption. Fig. 2(b), (c), (d) give the corresponding two-dimensional PE, TG and RTG infrared spectroscopies, respectively. Shown is the double Fourier transform of the heterodyne signal (Ω_1 and Ω_2 are the frequency variables conjugate to t_1 and t_2 respectively). Cross off-diagonal peaks reflect intermolecular couplings. The reverse photon echo (not shown) vanishes for non-interacting two level systems. This calculation demonstrates how two-dimensional spectroscopy can be used to directly probe the coupling pattern of interacting chromophores through the two-exciton manifold. It should be possible to extract useful structural and dynamical information by inverting such 2D signals.

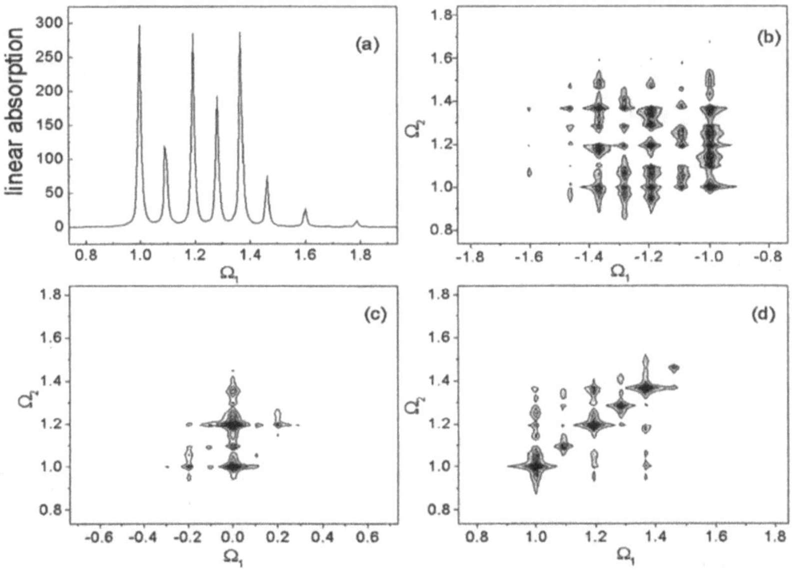

Fig. 2. (a) linear absorption of a coupled 8 chromophore aggregate. Panel (b) (c) (d) display the 2D spectroscopies of this aggregate. (b) PE, (c) TG, (d) RTG.

Acknowledgements. We wish to thank the support of the National Science Funcdation and the United States Air Force Office of Scientific Research.

References

1 H. Torii and M. Tasumi, *J. Chem. Phys.,* **96**, 3379 (1992).

2 P. Hamm, M. Lim, and R. M. Hochstrasser, "The Structure of the Amid I Band of Peptides Measured by Femtosecond Nonlinear IR Spectroscopy", *JCP*, preprint.

3 R. van Grondelle, J. P. Dekker, T. Gillbro, and V. Sundström, *Biochim. Biophys. Acta.* **1187**, 1 (1994).

4 R. R. Ernst and G. Bodenhausen and A. Wokaun, *Principles of Nuclear Magnetic Resonance in One and Two Dimensions*, Clarendon Press, Oxford, 1987.

5 Y. Tanimura and S. Mukamel, *J. Chem. Phys.* **99**, 9496 (1993)

6 W. M. Zhang, V. Chernyak, and S. Mukamel, "Multidimensional Femtosecond Correlation Spectroscopies of Electronic and Vibrational Excitons", *J. Chem. Phys.*, submitted.

7 S. Mukamel, *Principles of Nonlinear Optical Spectroscopy,* Oxford, New York, 1995.

8 V. Chernyak, W. M. Zhang, and S. Mukamel, "Multidimensional Femtosecond Spectroscopies of Molecular Aggregates and Semiconductor Nanostructures; the Nonlinear Exciton Equations", *J. Chem. Phys.*, submitted.

Fluorescence Upconversion and *ab initio* Studies of the Light-Harvesting Function of Carotenoids in Bacterial Light-Harvesting Antenna

Brent P. Krueger[1], Gregory D. Scholes[1], Ian R. Gould[2] and Graham R. Fleming[1]

[1] Department of Chemistry, University of California, Berkeley and Physical Biosciences Division, Lawrence Berkeley National Laboratory, Berkeley, California 94720-1460
[2] Department of Chemistry, Imperial College of Science, Technology and Medicine, London SW7 2AY, United Kingdom

Abstract. Fluorescence upconversion experiments and *ab initio* calculations show that S_2-Q_x electronic excitation transfer (to several Bchl Q_x transitions) is the primary pathway for carotenoid to Bchl energy transfer. Calculations on a carotenoid-B850-B850 aggregate suggest that the carotenoid may mediate B800-B850 EET by modifying the B850 transition density.

1. Introduction

Carotenoids have been the subject of increasing study as their photophysical properties and vital importance to photosynthetic organisms have become better understood [1]. One focus of recent work has been the function of carotenoids in the peripheral light-harvesting complex (LH2) of purple bacteria in which carotenoids harvest light and efficiently transfer excitation energy to nearby bacteriochlorophyll (Bchl). Recent high resolution crystal structures of LH2 from several species have spurred detailed analysis of structure-function relationships in the complexes [2]. In particular, we will consider the crystal structure of *Rhodopseudomonas (Rps.) acidophila* [3] which reveals a symmetric ring structure of nine repeating protomer units, each consisting of two closely spaced B850 molecules, one B800 molecule and one rhodopin-glucoside (RG) carotenoid which closely approaches both B850 and B800 pigments. We have performed fluorescence upconversion experiments [4] and novel *ab initio* calculations of coupling strengths [5] and orbital overlap effects [6] between RG and nearby Bchl pigments in the LH2 of *Rps. acidophila* to determine the dominant pathways of electronic excitation transfer (EET).

2. Results and Discussion

The upconversion data (Fig. 1) yield an *in situ* lifetime of 54±8fs for the carotenoid S_2 state and a corresponding wavelength dependent rise in the Bchl Q_y state that is 110±25fs at the peak of the steady-state emission (840nm). Modeling

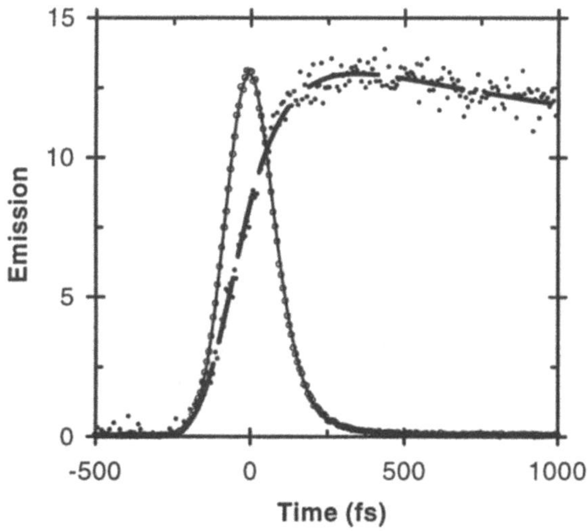

Figure 1. Fluorescence upconversion data and fits. Solid line and open circles are RG S_2 emission. Dashed line and solid circles are Bchl Q_y emission.

the data using estimates of the internal conversion timescales [4] and the S_1 to Q_y transfer rate [7] suggests a RG S_2 to Bchl Q_x transfer time of 90fs. This 90fs estimate represents an average of transfer from RG to any of several nearby Bchl pigments. Furthermore, any involvement of B800 Bchl or direct S_2-Q_y transfer was ignored in the model. Knowledge of the coupling strengths between the various chromophores is needed to extract more detailed information from the experimental data.

Accurate estimation of coupling strengths is difficult in LH2 because the molecules are close together (<15Å) and irregularly shaped (e.g. RG is twisted) so use of the ideal dipole approximation (IDA) is questionable. We have calculated Coulombic couplings between RG and all nearby pigments through the use of a new *ab initio* transition density cube (TDC) method which avoids the IDA and gives accurate Coulombic coupling strengths even for distorted molecules at very small separations [5]. Combining these coupling strengths with spectral overlap values allows us to estimate rates of energy transfer (in the weak coupling limit [8]) which are summarized in Fig. 2. These estimates suggest that significant amounts of energy flow from RG into several neighboring Bchl pigments and that interaction between pigments of different protomer units is important to light harvesting function. Along with estimates of the S_1-Q_y coupling [7], these calculations

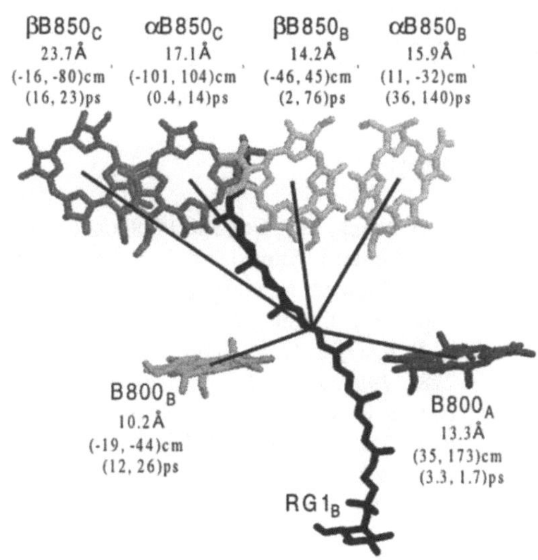

Figure 2. Center to center separations, coupling strengths and energy transfer times between RG S_2 and neighboring Bchl (Q_x, Q_y) transitions. A, B and C labels denote different protomers.

show that S_2-Q_x EET (to Q_x transitions on several Bchl pigments) is, by far, the primary pathway for carotenoid-Bchl transfer.

The TDC method allows accurate estimation of the Coulombic coupling between pigments which is the primary component of the electronic coupling at donor-acceptor separations beyond their van der Waals radii. However for pigments which approach very closely such as RG and B850, short range interactions (orbital overlap dependent) may become important [9]. For example, large *ab initio* MO calculations of an aggregate of two B850 Bchls, one B800 Bchl and one RG show that substantial electron density is shared between B850 and RG [6]. Thus the transition density of each is likely perturbed by the presence of the other. To examine this effect further, we have generated TDCs for an inter-protomer dimer ($\beta B850_B$ and $\alpha B850_C$) of Bchl and for a multimer of this dimer with $RG1_B$. Examination of these TDCs shows that the carotenoid does influence the transition density of the B850. Interacting these TDCs with B800 TDCs to determine the B800-'B850' coupling both with and without the carotenoid will quantify this effect [6].

3. Conclusions

Fluorescence upconversion experiments suggest <100fs EET times for the S_2-Q_x energy transfer, making it the primary pathway for carotenoid-Bchl EET. This is corroborated by *ab initio* TDC calculations that find strong coupling between the S_2 and Q_x transitions. TDC calculations also suggest that there are many minor pathways, acting both within and between protomer units, which each channel a few percent of the carotenoid excitation to Bchl. In addition, calculations carried out on LH2 pigment aggregates indicate that the carotenoid may mediate B800-B850 EET by modifying the transition density of the nearby B850.

References

1 H. A. Frank, R. J. Cogdell, Photochem. Photobio. **63**, 257, 1996.
2 T. Pullerits, V. Sundström, Acc. Chem. Res. **29**, 381, 1996.
3 G. McDermott, S. M. Prince, A. A. Freer, A. M. Hawthornthwaite-Lawless, M. Z. Papiz, R. J. Cogdell, N. W. Isaacs, Nature **374**, 517, 1995.
4 B. P. Krueger, G. D. Scholes, R. Jimenez, G. R. Fleming, J. Phys. Chem. B. **102**, 2284, 1998.
5 B. P. Krueger, G. D. Scholes, G. R. Fleming, J. Phys. Chem. B. **102**, 5378, 1998.
6 G. D. Scholes, I. R. Gould, B. P. Krueger, G. R. Fleming, in preparation.
7 G. D. Scholes, R. D. Harcourt, G. R. Fleming, J. Phys. Chem. B. **101**, 7302, 1997.
8 T. Förster, in *Modern Quantum Chemistry*, Edited by O. Sinanoglu, Vol. III, 93, 1965.
9 R. D. Harcourt, G. D. Scholes, K. P. Ghiggino, J. Chem. Phys. **101**, 10521, 1994.

Dynamics of Energy Transfer in the LH2 Antenna Complex of the Purple Bacterium *Rhodobacter sphaeroides*

Tomáš Polívka[1], Jennifer L. Herek[1], Tõnu Pullerits[1], Mats Dahlbom[1], C. Neil Hunter[2] and Villy Sundström[1]

[1]Department of Chemical Physics, Lund University, P.O. Box 124, S-221 00 Lund, Sweden;
[2]Department of Molecular Biology and Biotechnology, University of Sheffield, Western Bank, Sheffield S102UH, UK.

Abstract. Transient absorption spectra and kinetics are used for detailed studies of energy transfer dynamics within the B850 ring of the LH2 antenna of the purple bacterium *Rb. sphaeroides*. The initial exciton relaxation was found to occur within the first 100 fs and no temperature dependence of this process was observed. Transient absorption spectra of the LH2 antenna complex at 6 K revealed a new spectral band around 880 nm as a result of energy distribution of the lowest exciton states of B850 rings at cryogenic temperatures.

1. Introduction

Since the establishment of the three-dimensional structure of the B800-850 (LH2) antenna complex of purple bacteria [1,2], these systems play an increasingly important role in understanding the mechanism of light harvesting processes in photosynthesis. The LH2 antenna consists of two separate rings of pigments absorbing around 800 nm (B800) and 840-860 nm (B850). In *Rb. sphaeroides*, the B800 ring is formed by 9 BChl-a molecules, while B850 is comprised of a circular aggregate of 18 closely spaced BChl-a molecules. This packing leads to short distances between BChl-a molecules in B850 and suggests a strong excitonic interaction, which significantly influences the energy transfer characteristics within the ring.

2. Experimental Methods

Membranes of the LH2-only mutant strain DD13 [3] were dissolved in buffer solution and contained in a 1 mm optical path cuvette. For low temperature measurements, glycerol was added to achieve a glass-like sample of good optical quality. The femtosecond spectrometer consists of an amplified Ti:Sapphire laser system (Spectra Physics) that provides pulses with ~100 fs duration at a frequency of 5 kHz. The ~300 μJ pulses were divided into a pump pulse, which seeded an optical parametric generator (TOPAS, Light Conversion) that enabled tuning over a wide spectral range, and a probe pulse, which was focused into a 1 cm sapphire plate to generate a white-light continuum.

3. Results and Discussion

Both isotropic and anisotropic decays were calculated from traces measured for parallel and perpendicular polarizations. The isotropic decays consist of an instantaneous rise component and a very fast decay (~100 fs) that reflects ultrafast exciton dynamics within the B850 ring. After this ultrafast decay, the signal stays at a constant level as a result of trapping of energy within the B850 ring. The anisotropy decays from an initial value around 0.4 to a final value between 0.05 and 0.15 (depending on pump and probe wavelengths) within 80-90 fs at both room temperature and 6 K (Fig. 1), suggesting that the initial ultrafast decay of anisotropy does not depend significantly on temperature. Similar dynamics of excitonic states of B850 were observed recently at room temperature [4].

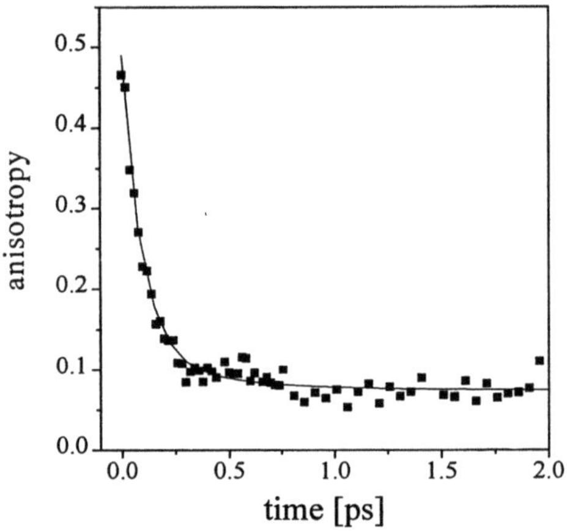

Fig. 1. Transient anisotropy of B850 at 6 K. Probed at 860 nm after 848 nm excitation.

Transient absorption spectra were measured at both temperatures for three different excitation wavelengths and were detected at different delay times extending to 30 ps. The spectra corresponding to 850 nm excitation are presented in Fig. 2. The spectra for both temperatures have a similar characteristic shape: a positive peak centered at 830-840 nm due to excited state absorption (ESA), and a negative peak centered at ~855 nm due to ground state bleaching (BL) and stimulated emission (SE). The transient absorption spectra reveal a very fast decay process within the first 150 fs. This decay can be ascribed to an excitonic relaxation to an equilibrium state of the system. A very distinct and unusual behavior of the transient absorption spectrum occurs at low temperature. While at room temperature the equilibration is finished within ~2 ps, at 6 K a new spectral band around 880 nm appears and the entire spectrum shifts to the red with time. There is no equilibration, even after 30 ps, and a significant shift of the zero-crossing (isosbestic) point is observed. Similar observations have been reported recently [5,6], and here we explain it as follows. At room temperature, after the initial exciton relaxation, the system reaches an equilibrium state which due to thermal excitations, is approximately the same for all B850 rings in the sample. At very low temperature, however, after the initial fast relaxation almost all excitations are in the lowest exciton state (LES), since thermal excitations are negligible. Due to the energy distribution of LESs among B850 rings in the sample, there is a probability for a given excitation to find a B850 ring with a lower-energy LES. This process results in hopping of excitations among different B850 rings toward the rings with the lowest-energy LESs. Such hopping is reflected by an increase of stimulated emission from the red-most B850 rings around 880 nm and by the shift of the

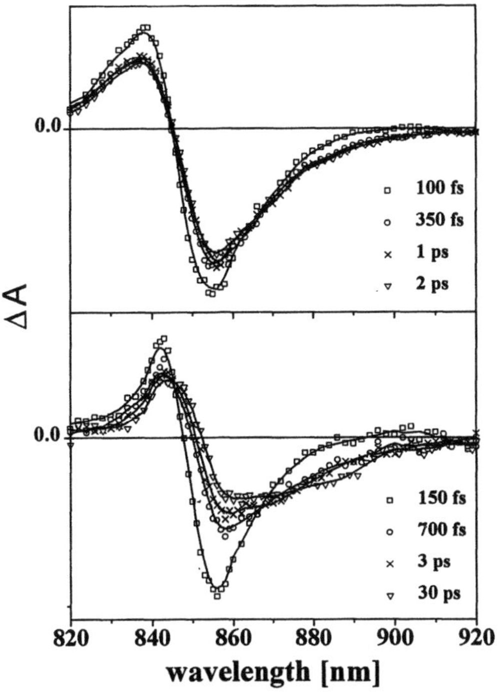

Fig. 2. Transient absorption spectra of B850 at four different time delays after excitation at 850 nm. Room temperature (top) and 6 K (bottom).

whole spectrum to the red as well. Kinetics probed at 880-890 nm reveal a complicated rise time characterized by at least three components with time constants of ~0.7 ps, 3-8 ps and more than 20 ps (the latter two depend on probing wavelength). The 3-8 ps component coincides well with previous measurements of inter-ring energy transfer in LH2-LH1 antenna complexes [7] and can therefore be ascribed to energy transfer between neighboring LH2 complexes, while the slowest component most likely reflects long-distance hopping between different LH2 rings.

References:

1 G. McDermott, S. M. Prince, A. A. Freer, A. M. Hawthornwaite-Lawless, M. Z. Papiz, R. J. Cogdell and N. W. Isaacs, Nature **374**, 517, 1995.

2 J. Koepke, X. Hu, C. Muenke, K. Schulten and H. Michel, Structure **4**, 21, 1996.

3 M. R. Jones, G. J. S. Fowler, L. C. D. Gibson, G. G. Grief, J. D. Olsen, W. Crielaard and C. N. Hunter, Molecular Biology **6**, 1173, 1992.

4 V. N. Nagarajan, R. G. Alden, J. C. Wiliams and W. W. Parson, Proc. Natl. Acad. Sci. USA **93**, 13774, 1996.

5 M. Chachisvilis, O. Kühn. T. Pullerits and V. Sundström, J. Phys. Chem. B **101**, 7275, 1997.

6 Freiberg, J. A. Jackson, S. Lin and N. W. Woodbury, J. Phys. Chem. A **102**, 4372, 1998.

7 S. Hess, M. Chachisvilis, K. Timpmann, M. R. Jones, G. J. S. Fowler, C. N. Hunter and V. Sundström, Proc. Natl. Acad. Sci USA **92**, 12333, 1995.

Ultrafast Singlet Energy Transfer in a Caroteno pyropheophorbide Dyad: A Biomimetic Model for Photosynthetic Light Harvesting?

Alisdair N. Macpherson and Tomas Gillbro

Department of Physical Chemistry, University of Umeå, S-901 87 Umeå, Sweden,
E-mail: alisdair@chem.umu.se

Abstract. Ultrafast singlet-singlet energy transfer with a ~60% yield has been directly measured from the S_2 state of a carotenoid to a covalently-linked pyropheophorbide by fluorescence upconversion. Transient absorption measurements indicate there is no energy transfer from the S_1 state of the carotenoid to the pyropheophorbide.

1. Introduction

Carotenoids are important light-harvesting pigments in photosynthesis because they absorb light in the blue-green spectral region and efficiently transfer energy to neighbouring (bacterio)chlorophylls [1,2]. At least two different pathways of singlet-singlet energy transfer are possible: one from the forbidden low-lying S_1 ($2A_g^-$) state of the carotenoid to the Q_y (S_1) level of the (bacterio)chlorophyll and another from the strongly-allowed S_2 (B_u^+) state of the carotenoid to the Q_x (S_2) level of the (bacterio)chlorophyll. Despite the short lifetime of the S_2 state of carotenoids, recent measurements suggest that energy transfer does indeed occur from the S_2 state [3,4] and furthermore, this pathway is, by far, the most important in the LH2 light-harvesting antenna of purple bacteria [4].

Efficient singlet energy transfer has also been reported in a caroteno-pyropheophorbide dyad model system designed to mimic the antenna functions of carotenoids [5]. The dynamics of this system were probed by picosecond transient absorption techniques [6]. In view of the development of laser flash techniques with much higher time resolution and the resulting discovery of the role of the carotenoid S_2 state in the natural system, we have reexamined the kinetics of the dyad excited singlet states using both femtosecond transient absorption and fluorescence techniques.

2. Experimental Methods

To accurately determine both the decay kinetics of the S_2 state of the carotenoids and the rise kinetics of the pyropheophorbide emission, a fluorescence upconversion spectrometer with a *ca.* 120 fs instrument response function was employed. Excitation close to 480 nm from a frequency doubled Ti:sapphire laser operated at 82 MHz was used to ensure selective excitation of the carotenoids.

The lifetime of the S_1 state of the carotenoids was determined near the excited state absorption maximum with white light probing and excitation at 485 nm, provided by the fourth harmonic of an OPA pumped at 5 kHz by a regeneratively amplified Ti:sapphire laser system, with a ≥ 150 fs instrumental response. The caroteno-pyropheophorbide and appropriate reference carotenoid were a generous gift from Arizona State University [5].

3. Results and Discussion

The fluorescence decay kinetics of the caroteno-pyropheophorbide (inset Fig. 1B) and an appropriate reference carotenoid in toluene are shown in Fig. 1A. There is a dramatic decrease in the lifetime of the S_2 state of the carotenoid from 150 fs (98.6%) to 54 fs (98.6%) upon linking to the pyropheophorbide. Concomitant with the decay of the carotenoid excited state, an ultrafast multiexponential rise ~60 fs (80%) in the excited state (S_1) emission of the pyropheophorbide moiety (Fig. 1B) is detected at 682 nm, the steady-state emission maximum.

These results suggest that the quenching of the S_2 state of the carotenoid can be attributed to singlet-singlet energy transfer ($k_{ET} \sim 1.2 \pm 0.3 \times 10^{13}$ s^{-1}). The efficiency of the energy transfer via the S_2 pathway is calculated to be ca. 60%, a little higher than the 50% quantum yield previously reported by steady-state fluorescence excitation measurements [5]. The pyropheophorbide rise kinetics (Fig. 1B) are complicated, however, by some overlapping emission from the S_2 state of the carotenoid. An ultrafast decay in the anisotropy from >0.3 to ~0.1 was found in this region, because of the presence of emission from the carotenoid.

The observation of a slower rise component (~2 ps) at 682 nm (Fig. 1B) suggests that there could be some additional energy transfer from the S_1 state. However, it should be noted that the amplitude of this rise component is emission wavelength-dependent, and it was absent at shorter wavelengths.

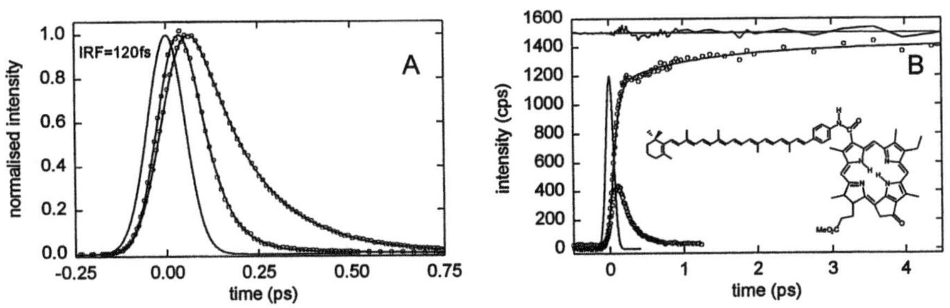

Fig. 1. (A) Decay kinetics of the reference carotenoid and dyad emission at 610 nm in toluene. (B) Isotropic rise kinetics of the pyropheophorbide dyad emission and decay kinetics of the reference carotenoid at 682 nm

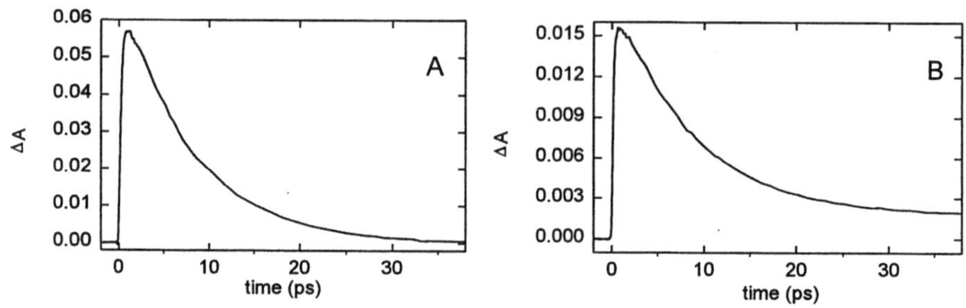

Fig. 2. Decay kinetics of the carotenoid S_1 state excited state absorption at 598 nm in toluene for (A) reference carotenoid, 7.8 ps and (B) dyad, 8.3 ps (89%), >3 ns (11%).

To determine if there is any energy transfer from the S_1 state, transient absorption measurements were made. Fig. 2 indicates that the S_1 lifetime of the carotenoid (~ 8 ps) is not quenched upon linking to the pyropheophorbide and therefore the slower rise component probably arises from an intermolecular relaxation process.

4. Conclusions

Efficient and ultrafast singlet-singlet energy transfer from the S_2 state of a carotenoid covalently-linked to a pyropheophorbide acceptor has been observed. This dyad successfully mimics the behaviour observed in some natural light-harvesting complexes, without the complication of the protein environment.

Acknowledgements. The dyad and carotenoid model were a generous gift from Arizona State University. We thank the Swedish Natural Science Research Council for financial support

References

1 Y. Koyama, M. Kuki, P. O. Andersson and T. Gillbro, Photochem. Photobiol., **63**, 243, 1996.
2 H. A. Frank and R. J. Cogdell, Photochem. Photobiol., **63**, 257, 1996.
3 M. Ricci, S. E. Bradforth, R. Jimenez and G. R. Fleming, Chem. Phys. Lett., **259**, 381, 1996.
4 A. N. Macpherson, J. B. Arellano, N. Fraser, R. J. Cogdell and T. Gillbro, to be published.
5 P. A. Liddell, D. Barrett, L. R. Makings, P. J. Pessiki, D. Gust and T. A. Moore, J. Am. Chem. Soc. **108**, 5350, 1986.
6 M. R. Wasielewski, P. A. Liddell, D. Barrett, T. A. Moore and D. Gust, Nature., **322**, 570, 1986.

Adiabatic Behaviour in the Primary Photosynthetic Electron Transfer

P. Huppmann, S. Spörlein, J. Wachtveitl and W. Zinth

Institut für Medizinische Optik, Ludwig-Maximilians-Universität München,
Oettingenstrasse 67, D-80538 München, Germany
E-mail: Petra.Huppmann@physik.uni-muenchen.de

Abstract. Pronounced modulations, extremely fast initial decay kinetics and unexpected temperature dependencies are observed for the initial electron transfer processes in mutated reaction centres (RC) of *Rhodopseudomonas viridis*. The observations indicate that at low temperatures adiabatic behaviour has to be considered for an adequate description of the electron transfer process.

1. Introduction

The application of the Marcus picture of electron transfer (ET) to the primary reaction in bacterial photosynthesis simplifies the description of the transfer by applying one effective nuclear coordinate coupling the donor (the excited special pair P^*) and the charge separated acceptor state P^+B^-, where the electron is transferred to the accessory bacteriochlorophyll (see Figure 1a, insert). In the non-adiabatic regime, the coupling is weak and the system passes the interaction region many times before ET takes place. From the temperature dependence of the reaction, one could deduce that the nuclear motion has a frequency in the 100 cm^{-1} range [1,2]. Surprisingly, similar oscillation frequencies were observed in absorption transients [3,4]. These features can either be assigned to a wavepacket motion in the excited electronic S_1 state modulating absorption and stimulated emission [5] or one could speculate from the coincidence of the frequencies, that the S_1 wavepacket motion is directly related to the ET-process. In that case the observed oscillation would reflect the modulated decay of the initial state P^*. At present a direct experimental proof for this interpretation is still missing.

2. Experimental Results

In *Rps. viridis* RCs, the resonant excitation of the special pair generates an occupation of P^*decaying biexponentially with a dominant fast component (Figure 1b). The oscillations superimposed to the biexponential decay could either be due to modulated electron transfer or due to pure wavepacket motion. The experimental results found so far do not allow to distinguish between these two possibilities. The situation for a modulated ET reaction is illustrated in Figure 1a: the ET leads to a stepwise decay of the donor state P^* and a stepwise formation of the acceptor state P^+B^-. In this case the relative height of a step is determined by the Landau-Zener factor or the ratio of the vibrational period τ_{vib} to the ET time τ_{ET}. In wild type reaction centres with typical values of $\tau_{ET}\approx 3$ ps and $\tau_{vib}\approx 250$ fs the step height is only $\approx 10\%$ of the total signal and a unique assignment of such a modulation to one specific process becomes difficult.

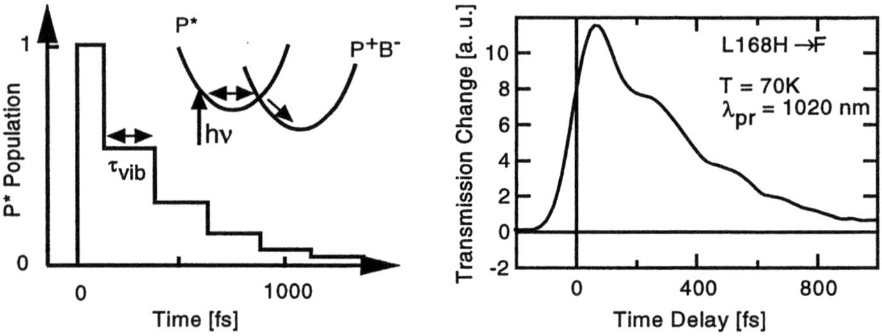

Fig. 1. a) left: potential energy curves (insert) and the stepwise decay of P^* expected for an electron transfer via the interaction region;
b) right: measured decay kinetics for RC of the mutant L168H→F at λ_{pr} =1020 nm.

An additional important observation: For wild type RCs at room temperature the vibrational relaxation time is faster than the ET time. So ET takes place from a thermally equilibrated P^*-state. For the L168H→F mutant at lower temperatures we have a completely different situation: the ET time ($\tau_{ET}\approx 250$ fs) is in the order of the vibrational period ($\tau_{vib}\approx 250$ fs). Besides, the relaxation time of this vibration is longer than the ET time indicating that the ET reaction in this mutant takes place from an excited vibrational state, which is a first indication for adiabatic ET.

Room temperature measurements of wild type and several mutant reaction centres can be explained in the frame of non-adiabatic ET theory. Extending the room temperature theory to lower temperatures and calculating the value of the Landau-Zener factor for the L168H→F mutant at lower temperatures one obtains a value in the order of unity. This gives a second clear indication that adiabatic ET behaviour is realized in these reaction centres.

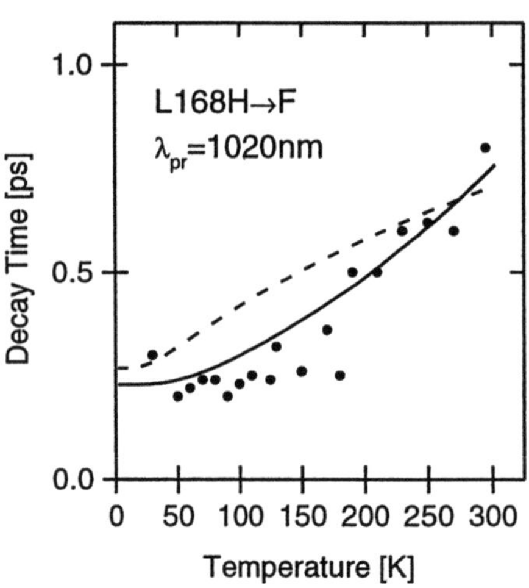

Fig. 2. Temperature dependence of the decay time of L168H→F; dotted line: semiclassical fit without heat expansion; solid line: with heat expansion

Figure 2 shows the observed temperature dependence of the primary ET time of the mutant L168H→F fitted according to non-adiabatic semiclassical theory. Heat expansion of the reaction centre according to the model given in [7] had to be considered to fit the data by the semiclassical formula of Hopfield [6]. Nevertheless the extremely fast increase of the ET rate with decreasing temperature from 300 K to 150 K and the constant reaction rate at low temperatures can not be fully explained. An attempt to extend the non-adiabatic Hopfield formula to consider adiabatic ET displaced the fit in Figure 2 by half a vibrational period to larger decay times and did not improve the coincidence with the experimental data. This observation agrees with a theory of Bixon and Jortner [8] who conclude that the influence of vibrational motion is reduced when the initially populated vibrational states of the donor are coupled to a Franck-Condon quasicontinuum of acceptor states.

3. Conclusions

Electron transfer reactions in the photosynthetic reaction centre show a wide range of reaction times where different molecular theories have to be applied: At room temperature, when the electron transfer times are larger or equal to 1 ps, the Landau-Zener factor is smaller than unity which indicates non-adiabatic electron transfer. Extremely fast processes exist at low temperatures for modified RCs in the mutant L168H→F. Here, the electron transfer starts from vibrationally excited states and the Landau-Zener factor is close to unity indicating that adiabatic electron transfer has to be applied.

References

1 G. R. Fleming, J. L. Martin, and J. Breton, *Nature*, 333, 190, 1988.
2 C. Lauterwasser, U. Finkele, H. Scheer, and W. Zinth, *Chem. Phys. Lett.*, 183, 471, 1991.
3 M. H. Vos, F. Rappaport, J.-C. Lambry, J. Breton, and J.L. Martin, *Nature,* 363, 320, 1993.
4 M. H. Vos, J.-C. Lambry, S. J. Robles, D. C. Youvan, J. Breton, and J. L. Martin, *Proc. Natl. Acad. Sci. USA*, 88, 8885, 1991.
5 S. Spörlein, W. Zinth, and J. Wachtveitl, *J. Phys. Chem.*, 102, (in press), 1998.
6 C. C. Moser, J. M. Keske, K. Warncke, R. S. Farid, and P. L. Dutton, *Nature*, 355, 796, 1992.
7 H. Frauenfelder, H. Hartmann, M. Karplus, I. D. Kuntz, Jr., J. Kuriyan, F. Parak, G. A. Petsko, D. Ringe, R. F. Tilton, Jr., M. L. Conolly, and N. Max, *Biochemistry*, 26, 254, 1987.
8 M. Bixon and J. Jortner, to be published.

Characterization of the Ultrafast Processes in the Trimeric Photosystem I Complex by Femtosecond Transient Absorption

Alfred R. Holzwarth[1], Marc G. Müller[1], Iris Martin[1], Navassard V. Karapetyan[2], and Gediminas Trinkunas[3]

[1] Max-Planck-Institut für Strahlenchemie, Stiftstr. 34-36, D-45470 Mülheim, Germany
[2] A.N. Bakh Institute of Biochemistry, Russian Academy of Sciences, Leninsky pr. 33, 117071 Moscow, Russia
[3] Institute of Physics, A. Gostauto 12, Vilnius 2600, Lithuania

Abstract. Energy transfer and charge separation kinetics in the photosystem I complex was investigated under low excitation intensities to characterize the functional organization of the complex.

1. Introduction

Photosystem I (PS I) is the antenna/reaction center complex which takes part in the electron transfer chain from PS II to NADP in oxygenic photosynthesis. The trimeric PS I core complex from the cyanobacterium *Spirulina platensis* contains approx. 90 antenna Chls/monomer [1]. The X-ray structure of a similar complex has been resolved recently [2]. Femtosecond transient absorption spectroscopy using 60-fs pulses from a Ti:Sa regenerative amplifier/OPA system and a high-speed camera detection [3] over a broad range in the visible and near-infrared region was performed at various excitation intensities from $1 \cdot 10^{13}$ to $2 \cdot 10^{14}$ ph/cm^2 to estimate possible annihilation effects (Fig. 1).

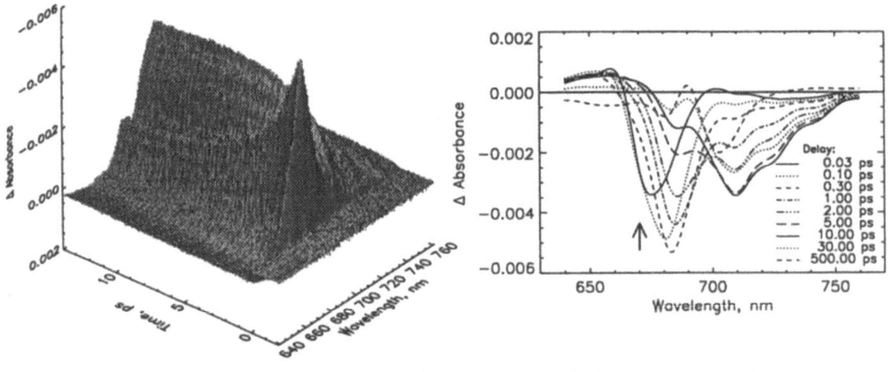

Fig. 1. Transient absorption data hypersurface (left) and difference spectra at selected delay times (right) of the trimeric complex of PS I at ≈15°C measured at low excitation intensity of $1.3 \cdot 10^{13}$ ph/cm^2 in a rotating cuvette. λ_{exc} = 670 nm

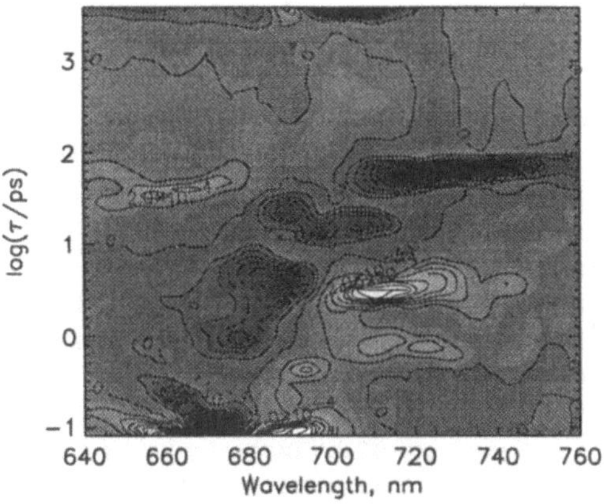

Fig. 2. Lifetime density map for the kinetics of the trimeric complex of PS I calculated from the low intensity data with 20-ps time range (shown in Fig. 1) and 500-ps range (not shown). The dark and white intense densities indicate individual lifetime components. Dark: negative, light: positive amplitude of the corresponding lifetime component

2. Lifetime Components

The intensity dependence of the kinetics reveals that annihilation effects predominantly result in an enhancement of amplitudes and shortening of lifetimes of some components from about 3-4 and 23 ps to about 2 and 10-15 ps, respectively. The lifetime shortening is observed both in the visible as well as the NIR range at ≈820 nm. The low-intensity data is consistent with the time-resolved fluorescence data which are essentially free of any non-linear effects [1].

The low intensity kinetics of the PS I trimers involves mostly distinct energy transfer and trapping components rather than broad lifetime distributions (Fig. 2). The lifetime component which appears to be most broadened is in the range 2-8 ps around 685 nm. However, the distinct rise term in the same lifetime range on the red wavelength side from 705 to 730 nm indicates one or two energy transfer components to red chlorophylls. Thus it is more likely that the 2-8 ps node is comprised of several unresolved single components. Further decay components in the picosecond range are about 13-23 ps at 690 to 715 nm and 50-80 ps from 705 to 755 nm. In addition several subpicosecond equilibration components were resolved. The most prominent are in the lifetime range of 0.7-1.0 ps seen as a decay of the bleaching at about 675 nm and a rise in the range from 705 to 730 nm. Furthermore energy equilibration in the bulk core proceeds on the very fast time scale with time constants of ≈100 fs and 400 fs.

3. Kinetic Modeling

In order to reveal the energy transfer pathways in PS I and the functional organization of the special long-wavelength absorbing chromophores (red pigments) with respect to the primary electron donor P700, as well as the intrinsic charge separation time of the system we used two different approaches of kinetic modeling. **I)** A compartment analysis limited to 6 pools (Chl a_{675}, Chl a_{680}, Chl a_{685}, Chl a_{720}, RP1 and RP2) recovered the most characteristic species-associated spectra and an estimate of a minimal kinetic model (Fig. 3). **II)** We employed genetic algorithms [4] for a detailed modeling taking into account the structure, the stationary absorption and the CD spectra. The genetic algorithm search (GA) is a combinatorial search for an optimal solution of different kind of parameters for 1 Chl site. These are i) the assignment of approx. 90 spectral forms (assuming only 10 are different) to the Chl sites in the structure, ii) 2 transition dipole orientations (1 bit/Chl site) and iii) 3 additional parameters for the intrinsic charge separation and recombination rate constant (6 bit/parameter). The energy transfer rate constants are calculated by the Förster formula. Since string order permutations are constrained to the spectral content conservation GA provide an efficient search procedure for an optimal (high fitness) solution in a multitude of possible bit combinations. An example for the convergence of the assignment of the red spectral form to one Chl site is outlined in Fig. 4.

Both kinetic models have in common that the *intrinsic* charge separation time is in the subpicosecond range ($\tau_{CS} \leq 0.5$ ps). All simulations indicate that the red pigment is *not* located in the direct vicinity of the primary electron donor.

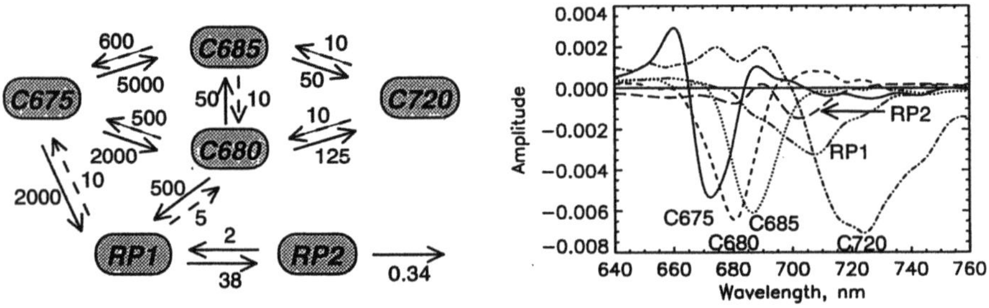

Fig. 3. Minimal kinetic scheme (left) and species-associated absorption difference spectra (right). Rate constants in units ns^{-1}. Calculated lifetimes: 106 fs, 890 fs, 3.7 ps, 23 ps, 69 ps, 3 ns. Assumed exc. probabilities at 670 nm: Chl a_{675}:Chl a_{680} = 77%:23%

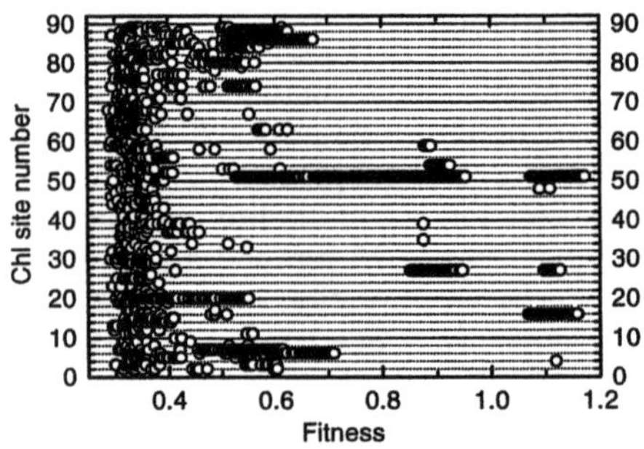

Fig. 4. Convergence of the genetic algorithm search for the assignment of the red spectral form to a Chl site. The Chl sites are renumbered continuously according to the sequence of Chl sites in the structure [2]. With elapsed compute time the fitness increases and only 2-3 sites are preserved

References

1 N. V. Karapetyan, D. Dorra, G. Schweitzer, I. N. Bezsmertnaya, and A. R. Holzwarth, Biochemistry **36**, 13830, 1997.

2 W.-D. Schubert, O. Klukas, N. Krauß, W. Saenger, P. Fromme, and H. T. Witt, J. Mol. Biol. **273**, 741, 1997.

3 A. R. Holzwarth, and M. G. Müller, Biochemistry **35**, 11820, 1996.

4 G. Trinkunas, and A. R. Holzwarth, Biophys. J. **71**, 351, 1996.

Single and Double Mutants of Bacteriorhodopsin and their Impact on Photoisomerization

Karsten Heyne, Reiner Dziewior, and Rolf Diller
Ulrike Alexiev

Institut f. Experimentalphysik, Freie Universität Berlin, Arnimallee 14, 14195 Berlin, Germany. E-mail: diller@physik.fu-berlin.de

Abstract. Optical transient absorption experiments on the photoisomerization in bacteriorhodopsin wild type and several mutants were performed in order to study the impact of the charge distribution in the active site on the isomerization dynamics.

1. Introduction

The photoinduced all-trans to 13-cis isomerization of the retinal chromophore in bacteriorhodopsin (bR) initiates an efficient proton transfer across the cell membrane. From comparison of retinal photoisomerization in retinal proteins it must be concluded that the specificity of the isomerization (100 % all-trans to 13-cis with a quantum yield of ≈ 64 %) and the high reaction rate is fine tuned by the constituents forming the chromophore binding pocket. We have performed a series of optical transient absorption experiments on the retinal photoisomerization in bR wild type (wt), the single mutants R82A, G231C and the double mutant R82A/G231C. In R82A, the positively charged arginine in the vicinity of the chromophore is replaced by the neutral alanine. In G231C the neutral glycine, located on the cytoplasmatic surface, is replaced by a cysteine. The results shed new light on the mechanism that might control the isomerization dynamics.

2. Experimental Methods

The transient absorption experiments were performed by means of a pump/probe spectrometer, based on a Titanium-Sapphire CPA-laser system (Clark). Excitation at 570 nm and probe light were extracted from a white light continuum at a repetition rate of 1 kHz, providing an instrumental response function of typically 450 fs. The following samples and conditions were used: wt, R82A, G231C and R82A/G231C in an aqueous solution of 15 mM KCl, 1 mM Tris pH 7.0 at room temperature with chromophore absorption maxima (λ_{max}) of about 568, 600, 568 and 567 nm, respectively.

3. Results and Discussion

Fig.1 shows absorption difference spectra at various delay times of R82A and G231C. The difference spectra of wt and R82A/G231C are indistinguishable from the spectrum shown for G231C within the experimental error. The spectral features are dominated by the bleach of the electronic ground state absorption band (520-600nm), the absorption band of the excited state (420 - 520 nm) and the photoproduct absorption band (600 - 700 nm). In Fig. 2, transient absorption kinetics of the four samples are shown at 480, 640 and 900 nm, respectively. Within the experimental error, no differences were found for the isomerization kinetics of wt, G231C and R82A/G231C. The excited state dynamics are characterized by a life time of 0.5 ps as expected for wt from the literature (1,2). However, the mutant R82A shows two components with substantially prolonged excited state life times (1.7 and 7 ps) as compared to wt. Similar results were described in (3) for R82Q.

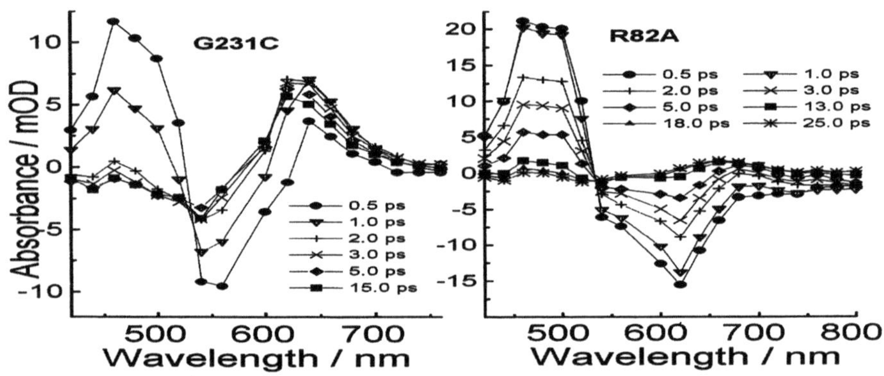

Fig. 1. Difference spectra of R82A and G231C at various delay times.

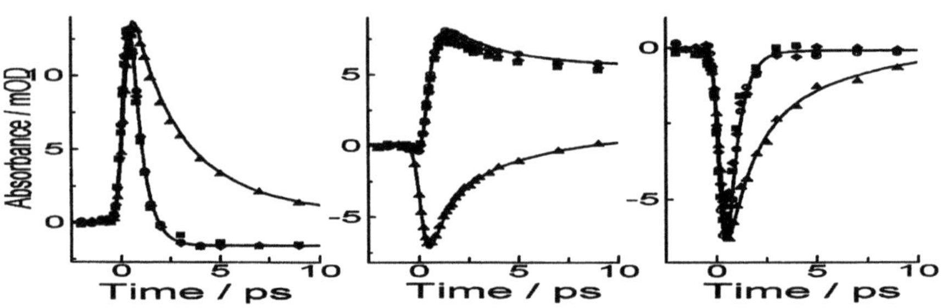

Fig. 2. Optical transients of bacteriorhodopsin wt (circle), R82A (triangle), G231C (diamond), R82A/G231C (square). From left to right: probe at 480, 640, 900 nm.

El-Sayed and coworkers (3) suggested a valence bond model in order to describe the impact of the charge distribution in the retinal binding pocket on the isomerization kinetics. According to this model the charge distribution stabilizes a specific isomer in the electronic ground state and increases the rate for photoisomerization around a specific C-C double bond. In R82A the charge distribution is altered as indicated by the 40 nm redshift of the λ_{max}. Also, higher dark isomerization rates (4) and lower photoisomerization rates (Fig. 2) occur. The single mutant G231C shows wild type behaviour ((5), Fig.2). This seems understandable considering the large distance between the chromophore and position 231. Surprisingly, the wild type photoisomerization kinetics are re-established when adding the mutation G231C to R82A. The rescue of the wild type electronic ground state properties (ground state absorption spectrum) and the nanosecond to second time scale kinetics of the photocycle were already reported (6). Following the valence bond model the results imply that in R82A/G231C the net charge of the diffuse counterion of the protonated Schiff base (SBH) is fully restored as compared to R82A. This could be facilitated by an altered involvement of the negatively charged D212. D212 compensates for the positive R82 in wt but remains as a negative net charge in R82A. The mechanism for the charge restauration in the double mutant could be a large scale conformational change induced by G231C in R82A/G231C that changes apparent pK_a-values near the chromophore or shields the negative charge of D212 by sterical means (7).

4. Conclusions

Our results suggest that the critical parameter in the valence bond model is the net charge of the diffuse counter ion of the SBH. The net charge can be realized by different secondary structures of the retinal binding pocket, involving charged residues (wt and R82A/G231C). From that follows that the oppositely charged residues R82 and D212 in the active site might not be important for the isomerization as long as they neutralize each other.

Acknowledgements. We thank D. Stehlik and M. P. Heyn for support and encouragement. This work was partly supported by DFG (Di 405/-1,2 and Sfb 312/B1).

References

1 J. Dobler, W. Zinth, W. Kaiser, and D. Oesterhelt, Chem. Phys. Lett. **144**, 215, 1988.
2 R. A. Mathies, C. H. Brito Cruz, W. T. Pollard, and C. V. Shank, Science **240**, 777, 1988.
3 L. Song, M. A. El-Sayed, and J. K. Lanyi, Science **261**, 891, 1993.
4 S. P. Balashov, R. Govindjee, M. Kono, E. Imasheva, E. Lukashev. T. G. Ebrey, R. K. Crouch, D. R. Menick, and Y. Feng, Biochem. **32**, 10331, 1993.
5 U. Alexiev, R. Mollaaghababa, P. Scherrer, H. G. Khorana, and M. P. Heyn, Proc. Natl. Acad. Sci. USA **92**, 372, 1995.
6 U. Alexiev, M. P. Krebs, R. Mollaaghababa, H. G. Khorana, and M. P. Heyn, Biophys. J. **70**, (2) A107, 1996.
7 U. Alexiev, R. Mollaaghababa, H. G. Khorana, and M. P. Heyn, in preparation.

Femtosecond - Picosecond Fluorescence Studies on Excited State Dynamics of Photoactive Yellow Protein (PYP) and It's Site - Directed Mutants

Haik Chosrowjan[1], Yutaka Shibata[1], Noboru Mataga[1], Yasushi Imamoto[2], and Fumio Tokunaga[2]

[1] Institute for Laser Technology, 3 - 11 - 20, Nakoji, Amagasaki, Hyogo 6610974, JAPAN
E-mail: haik@ile.osaka-u.ac.jp
[2] Department of Earth and Space Science, Osaka University, Toyonaka, Osaka 5600043, JAPAN
E-mail: imamoto@ess.sci.osaka-u.ac.jp

Abstract. Femtosecond fluorescence studies on PYP, it's chromophore and four site-directed mutants are performed. Effects of the protein environment on the photoisomerization of the chromophore are discussed.

1. Introduction

In photosensory or photoactive proteins with chromophores absorbing in the visible region such as rhodopsin for vision and photoactive proteins of various photoresponsive bacteria, light absorption usually leads to ultrafast and highly efficient reactions. PYP is one of such proteins. It is a small soluble protein found in a halophobic photosynthetic purple sulfur bacterium, *Ectothiorhodospira halophila*. Since the action spectrum of the photophobic response agrees with the absorption spectrum of PYP (446 nm), PYP is considered to function as a photoreceptor for the negative phototaxis of this organism. PYP is one of the simplest photoreceptor proteins discovered so far and as such holds great promise for the detailed investigation of its photophysical properties and mechanisms of reaction dynamics. Our first investigations on fluorescence dynamics of wild type PYP with 100 fs time resolution have shown that the fluorescence decay curves were nonexponential and initial main components have time constants of several hundreds fs to a few ps [1]. The decay process were interpreted as due to the barrierless twisting around vinyl bond of p-coumaric acid chromophore leading to trans/cis isomerization. However, whether this twisting of the chromophore starts by large scale rotational motion of phenyl ring around vinyl bond or by flipping thioester linkage bound to the protein at Cys69 is yet to be established. Moreover, effects of the modifications of amino acid residues surrounding the chromophore (protein nanospace) on the photoinduced ultrafast twisting of the chromophore in the PYP are still unchallenged. We have applied fluorescence up-conversion technique and carried out detailed fs time resolved fluorescence decay measurements on wild type PYP, it's chromophore (thiol-ester linked *p* -coumaric acid) and four site-directed mutants in a broad spectral region between 460 - 650 nm.

2. Experimental Methods

The fluorescence dynamics measurements were carried out by fluorescence up-conversion apparatus based on Ar ion laser pumped Ti-Sapphire laser (800 mW, 76 MHz, 100 fs) described in [1]. The FWHM of the instrumental response was 210 fs. The sample solutions were made flow through a 1 mm light path optical cell. The preparations of wild type PYP and its site-directed mutants were described elsewhere [2]. We have examined R52Q, T50V, E46Q, E46Q/T50V mutants, where the amino acid residues Arg52, Thr50, and Glu46 contiguously placed to the chromophore and rather strongly interacting with it are replaced with Gln, Val and Gln, respectively. For the measurements of the fluorescence dynamics of the chromophore in solution, PYP was denatured by addition of guanidinium hydrochloride at pH 9.5, where the protein was unfolded and the deprotonated chromophore was exposed to water environment. All measurements were made at room temperature.

3. Results and Discussion

The fluorescence dynamics of the free chromophore as shown in Fig.1 A, is dominated by well-known *dynamic solvation effect* (faster decay in shorter - and rise in longer wavelength regions) of an excited molecule with a non-zero dipole moment in a polar solvent, water in this case. The fluorescence decay of the free chromophore at 498 nm was monoexponential with 11 ps time constant. Contrary to these observations, with our current time resolution no detectable rise was observed for wild type PYP at any wavelength measured (Fig.1, B), indicating strong chromophore-protein interaction. Fluorescence decay curves were nonexponential and can be well fitted as a sum of three exponential decay components with time constants in the range of 130 fs - 350 fs (50-70%), 1.2 ps - 2 ps (25-40%), and 10 ps - 50 ps (5-10 %).

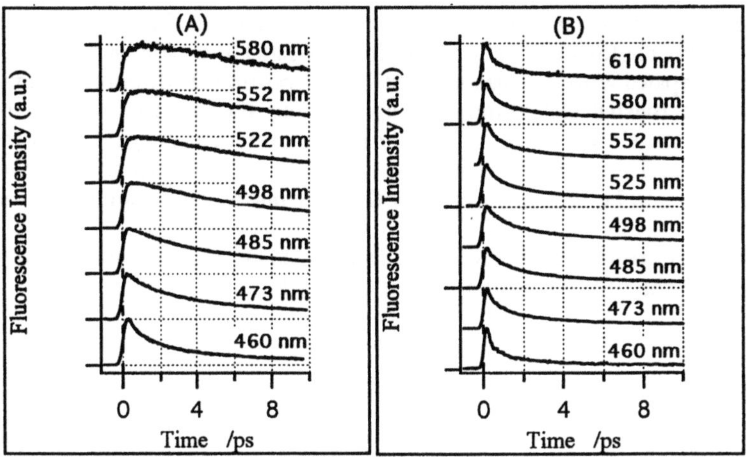

Fig. 1. (A) - Fluorescence decay kinetics of p-coumaric acid thioester chromophore (B) - Kinetics of wild type PYP. In both measurements the excitation was at 410 nm.

Hence, the rate of twisted state formation by photo-excitation is much enhanced in protein nanospace compared with the aqueous solution of the chromophore. Possible mechanisms for it might be: (a) In both cases, the twisting takes place by flipping thioester linkage and this process is accelerated in protein nanospace. (b) In aqueous solution, where the chromophore can move more freely, a large scale rotational motion around vinyl bond of the chromophore takes place, while the more rapid twisting by flipping thioester linkage occurs in the protein nanospace for the PYP. In this respect, effects of the modification of the protein nanospace by site-directed mutagenesis on the fluorescence spectra and dynamics of PYP are very interesting and important. In the wild type PYP, the $-O^-$ substituent of the chromophore is surrounded by proton donating amino acid residues Glu46, Tyr42 and Thr50 which make H-bonding network and undergo short range H-bonding interaction with $-O^-$ group. In the mutants E46Q, T50V and E46Q/T50V, H-bonding interactions with $-O^-$ group are weakened and the H-bonding network may be partly destroyed because Glu46 and Thr50 are replaced respectively by Gln and Val with much weaker or no function as proton donor in H-bonding. On the other hand, Arg52 is replaced by Gln leading to the loss of positive charge in R52Q where, however, the H-bonding network seems to remain unchanged. These changes affect not only the short range H-bonding interactions but also the structures and functions of protein nanospace that contains the chromophore, which in turn, affect seriously the ultrafast relaxation processes from the excited Franck-Condon to fluorescence and twisted state. In the case of the mutants, the short range H-bonding interactions and the H-bonding network are more or less weakened leading to the smaller reorganization energies associated with the transitions and slowing down the twisting process. The largest effect in slowing down the fluorescence decay dynamics due to the retardation of the twisting reaction among the mutants examined here was observed in E46Q/T50V.

4. Conclusions

The rate of the twisted state formation of the chromophore of PYP (the primary step of the trans→cis photoisomerization) was demonstrated to be considerably enhanced in protein compared with the aqueous solution supporting the flipping mechanism. Moreover, the twisted state formation by flipping the thioester linkage of the chromophore was slowed down by modifying the protein nanospace structure to a little looser one. The more restricted structure of the protein nanospace in the wild type PYP seems to be best engineered for the twisting by the flipping mechanism.

Acknowledgements. We thank Dr. F. Tanaka and Dr. H. Furukawa for providing and supporting the fitting program used in our work.

References

1 H. Chosrowjan, N. Mataga, N. Nakashima, Y. Imamoto and F. Tokunaga, *Chem. Phys. Lett.*, Vol. 270, 267, 1997.
2 K. Mihara, O. Hisatomi, Y. Imamoto, M. Kataoka and F. Tokunaga, *J. Biochemistry*, Vol. 121, 876, 1997.

Femtosecond Lasers for Ultra-Accurate Surgery: Application to Corneal Surgery

T. Juhasz[1,2], F.H. Loesel[1,3], C. Horvath[1,3], R.M. Kurtz[1,2], and G. Mourou[1]

1 Center for Ultrafast Optical Science, University of Michigan, 1006 IST Building, 2200 Bonisteel Blvd., Ann Arbor, MI 48109-2099, USA
2 Kellogg Eye Center, Department of Ophthalmology, University of Michigan, 1000 Wall Street, Ann Arbor, MI 48105, USA
3 Institute of Applied Physics, University of Heidelberg, Albert-Ueberle-Str. 3-5, D-69120 Heidelberg, Germany

Abstract. We investigated refractive corneal surgical procedures utilizing amplified femtosecond laser pulses from a reliable and compact femtosecond laser system suited for medical applications. Procedures such as corneal flap cutting and laser keratomileusis have been demonstrated in vivo and in vitro. Cutting was found to be well confined and highly precise.

1. Introduction

In recent years, ultrashort pulsed lasers have received increased attention for various biomedical applications, including refractive corneal surgery. In this work, we demonstrate that the femtosecond laser has the potential of being the corneal scalpel of the twenty-first century and may enable entirely new refractive approaches.

Different techniques can be used to surgically correct for instance for myopia (short-sightedness). Most of them rely on reducing the curvature of the cornea by removing corneal tissue. In contrast to direct photoablation of tissue with ultraviolet light, such as in excimer laser photorefractive keratectomy (PRK), the laser-tissue interaction with ultrashort pulses at transparent wavelengths is based on laser-induced optical breakdown (LIOB), resulting in the generation of a microplasma [1]. Absorption of laser pulse energy by the plasma leads to rapid temperature and pressure increases in the focal spot. The expansion of the hot plasma generates a shock wave and a cavitation bubble. As a result of this process, termed photodisruption, tissue in the focal volume is destroyed. Historically, ophthalmic surgical procedures which utilized photodisruption have relied on the Q-switched Nd:YAG laser, which produces pulses of approximately 10 nanosecond width at 1.06 μm wavelength. At this pulse duration, the energy required for initiating LIOB is relatively high and shows stochastic behavior [2]. This energy is then transmitted to the surrounding tissue in the form of shock and cavitation effects. The relatively large secondary shock waves associated with these relatively long-pulsed Nd:YAG lasers limit their ophthalmic applications. In contrast, ultrashort-pulsed lasers operating in the femtosecond time regime are associated with significantly smaller [3] and deterministic [2] threshold energies, as well as reduced shock waves and smaller cavitation bubbles [4] than the nanosecond or picosecond lasers. Femtosecond lasers may therefore be capable of minimally invasive and well localized corneal surgical procedures.

2. Experimental Methods

We designed a compact and reliable femtosecond laser system suited to perform specific real world medical applications. This all-solid-state Nd:glass laser system generates amplified pulses as short as ≈ 500 fs at 1.06 μm wavelength. Figure 1. shows a schematic overview of the laser system.

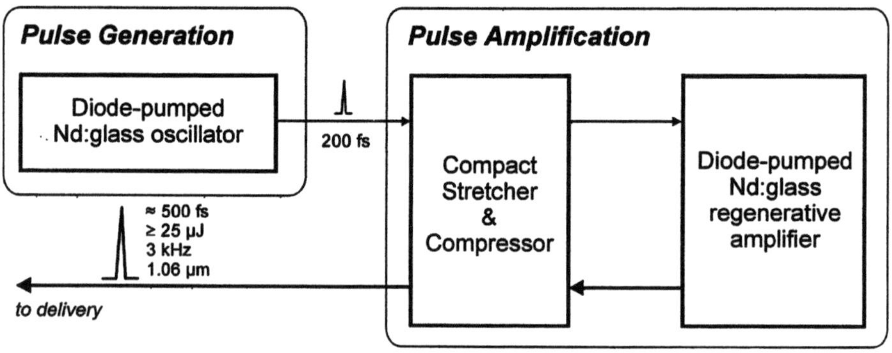

Fig. 1. Schematic overview of the all-solid-state femtosecond medical laser system

The system consists of a laser oscillator, a pulse stretcher, a regenerative amplifier, and a pulse compressor. The Nd:glass oscillator is pumped by only a single laser diode and mode-locked by a solid-state saturable absorber. Intracavity dispersion compensation is provided by Gires-Tournois mirrors and the laser produces a train of 200-fs pulses at a repetition rate of 80 MHz [5]. For chirped pulse amplification, pulses from the output of the oscillator are stretched to ≈ 60 ps pulse width in a very compact stretcher unit, based on a single holographic transmission grating [6]. Selected pulses are then seeded at a few kHz repetition rate into the regenerative amplifier using a Pockel's cell. Two 2W-diodes are used as pump sources for the Nd:glass gain medium, in order to achieve the required energy range. After maximum amplification, the pulses are dumped out of the amplifier laser at the polarizer by switching the voltage applied to the electrooptical cell. Figure 2. shows the measured pulse energy after the regenerative amplifier in dependence of the repetition rate of the seeding/dumping process. We have achieved pulse energies as high as 70 μJ at a pulse repetition rate of 1 kHz. The amplified pulses are then re-compressed in a compressor unit which uses the holographic transmission grating of the stretcher [6]. After compression, the laser pulse energy is measured to be as high as 25 μJ. Losses are primarily due to the non-optimized optics in the compressor. According to our calculations, pulse energies of 50 μJ could be obtained by using higher quality optics in the compressor unit. The spatial intensity profile of the output beam from the Nd:glass laser system was measured to be close to the ideal TEM_{00} mode. Our femtosecond laser system is very compact. The amplifier unit measures only 9 inch by 12 inch, and the oscillator and the combined stretcher/compressor unit have footprints of similar size.

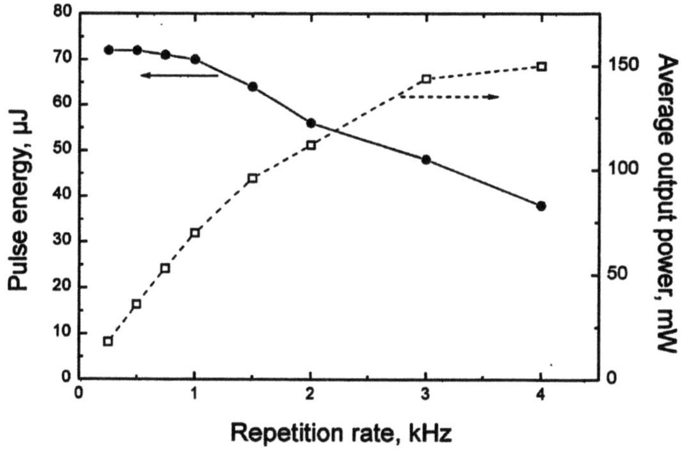

Fig. 2. Measured dumped pulse energies and average output powers from the regenerative amplifier in dependence of the repetition rate of the Pockel's cell

The amplified femtosecond laser pulses are delivered to the tissue via a computer-controlled optical system. A high numerical aperture lens unit can focus the beam down to a spot diameter of 5 µm within the cornea (corneal stroma). We use a flat contact glass as an interface between eye and laser to establish a fixed reference plane for reproducible intrastromal photodisruption. The current optical system allows scanning of the beam within a disc of 9 mm in diameter and within a depth of approximately 1 mm.

Enucleated porcine and primate eyes, as well as live rabbits and live primates were used in our study. In vivo experiments were followed up to 6 weeks. After placing the eyes at the contact glass, their position was fixed by generating a weak vacuum with a suction ring. We then performed various intrastromal scanning patterns with different refractive approaches. After irradiation, we evaluated the achieved ablation, using SEM, histopathological analysis, slit-lamp photography, corneal topography, and pachymetry for measuring the corneal thickness.

3. Results and Discussion

By scanning the focus spot of the laser below the corneal surface, the overlapping small ablation volumes of single pulses resulted in contiguous tissue cutting and vaporization. Only a very thin layer of tissue was removed and we found no sign of collateral damage to adjacent tissue. The ablating beam was scanned in various patterns to create corneal flaps, lenticules, and layered intrastromal disruption. As an example, Figure 3. demonstrates the principle of the femtosecond laser assisted keratomileusis (fs-LASIK) procedure. In a first step, the laser focus is scanned along a spiral pattern inside the corneal stroma, which results in a disc-type dissection. This cut forms the lower side of a lens-like volume (lenticule). After moving the focus closer to the cornea surface, a curved disc pattern completes the lenticule cut. Then, several ring-type cuts at decreasing depth in the cornea connect the lenticule cut to the cornea surface. In a second step, the eye is removed from the contact glass and the tissue above the lenticule can be lifted like a flap (a) and the lenticule can be extracted with a forceps (b). Finally, the flap can be easily repositioned on the cornea. Due to the missing volume of the lenticule, the surface of the cornea flattens and a change in refractive power is achieved.

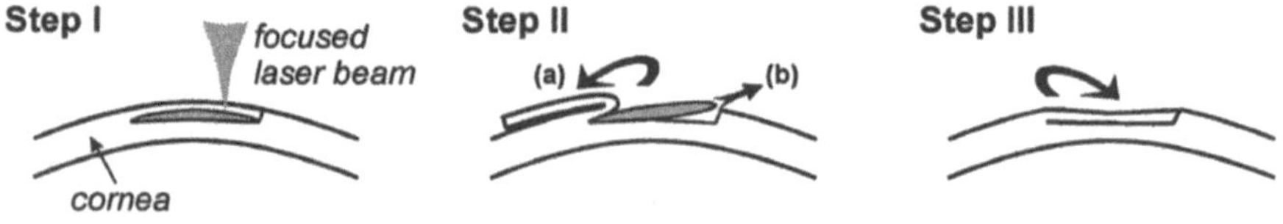

Fig. 3. Schematic principle of femtosecond laser assisted keratomileusis (fs-LASIK)

In our experiments, performing the fs-LASIK procedure with a pulse energy of 3 μJ and 10 μm spot separation resulted in excellent dissection of the flap and the lenticule. The lenticule could be removed with gentle traction on the anterior surface. We observed superior surface quality of the femtosecond laser cuts, when compared to microkeratome cuts or to results obtained with longer laser pulse widths. The SEM pictures in Figure 4. show a comparison of the LASIK procedure performed with femtosecond and picosecond laser pulses, respectively.

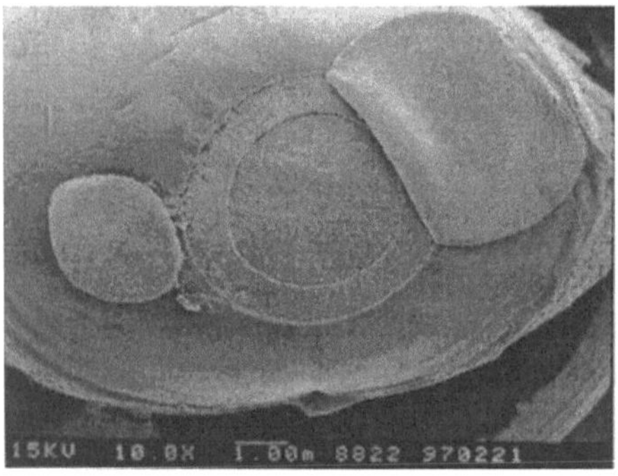

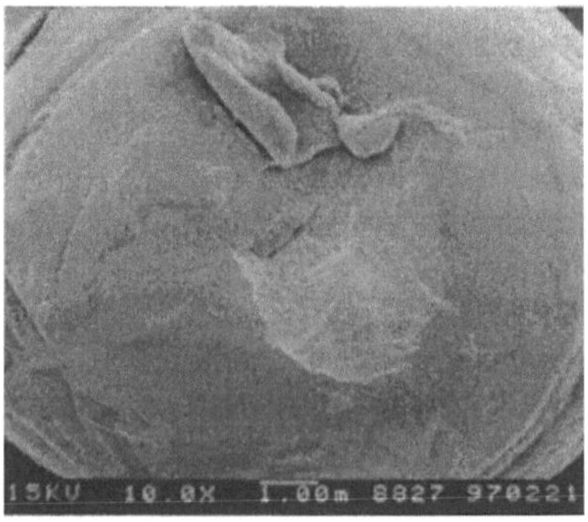

Fig. 4. SEMs of in vitro short-pulse LASIK in the primate eye. *Left:* Lifted corneal flap and extracted lenticule, generated with the femtosecond laser. *Right:* Unsatisfactory dissection when using longer 60 ps pulses of a Nd:YLF laser

The ablation parameters were comparable for both pulse widths and the pulse energies were adjusted to the same level above the respective threshold of ablation.

In live animal studies, corneal clarity returned within 3 days of the procedure, without significant corneal haze (Figure 5.). Preliminary data for refractive effects showed consistent changes after performing fs-LASIK and layered intrastromal disruption.

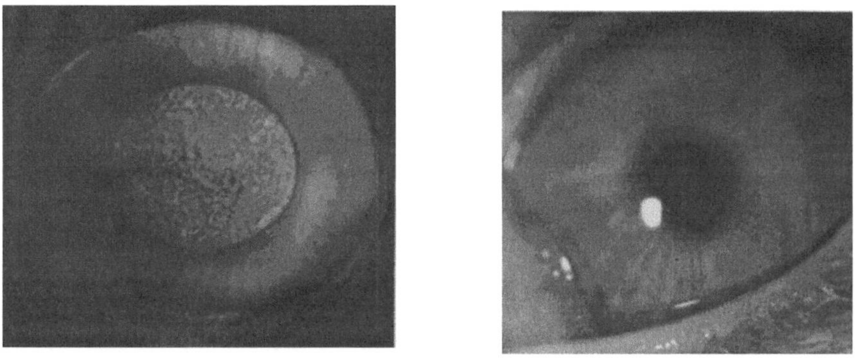

Fig. 5. Slit lamp photographs of rabbit eye. Intraoperative (*left*) and 1-week post-operative (*right*) view of rabbit eye treated with femtosecond laser pulses in a spiral pattern. Strong scattering at the cavitytion bubbles is observed (*left*). Transparency returns to normal (*right*) after approximately 3 days

4. Conclusions

We conclude that femtosecond laser pulses can perform a variety of high precision corneal refractive procedures, offering significant advantages over current mechanical and laser devices and possessing potential for entirely new refractive approaches.

References

1 N. Bloembergen, IEEE JQE 10 375-386 (1974).

2 D. Du, X. Liu, G. Korn, J. Squier, and G. Mourou, Appl. Phys. Lett. 64 3071 (1994).

3 F.H. Loesel, M.H. Niemz, J.F. Bille, and T. Juhasz, IEEE JQE 32 1717-1722 (1996).

4 T. Juhasz, G.A. Kastis, C. Suarez, Z. Bor, and W.E. Bron, Lasers Surg. Med. 10 23-29 (1996).

5 C. Horvath, F.H. Loesel, L. Turi, and T. Juhasz, in *CLEO Technical Digest*, 1998.

6 C. Horvath, A. Braun, H. Liu, T. Juhasz, and G. Mourou, Opt. Lett. 22 1790-1792 (1997).

Index of Contributors

Springer Series in Chemical Physics

Editors: Vitalii I. Goldanskii Fritz P. Schäfer J. Peter Toennies

1 **Atomic Spectra and Radiative Transitions** By I. I. Sobelman
2 **Surface Crystallography by LEED** Theory, Computation and Structural Results. By M. A. Van Hove, S. Y. Tong
3 **Advances in Laser Chemistry** Editor: A. H. Zewail
4 **Picosecond Phenomena** Editors: C. V. Shank, E. P. Ippen, S. L. Shapiro
5 **Laser Spectroscopy** Basic Concepts and Instrumentation By W. Demtröder 3rd Printing
6 **Laser-Induced Processes in Molecules** Physics and Chemistry Editors: K. L. Kompa, S. D. Smith
7 **Excitation of Atoms and Broadening of Spectral Lines** By I. I. Sobelman, L. A. Vainshtein, E. A. Yukov
8 **Spin Exchange** Principles and Applications in Chemistry and Biology By Yu. N. Molin, K. M. Salikhov, K. I. Zamaraev
9 **Secondary Ion Mass Spectrometry SIMS II** Editors: A. Benninghoven, C. A. Evans, Jr., R. A. Powell, R. Shimizu, H. A. Storms
10 **Lasers and Chemical Change** By A. Ben-Shaul, Y. Haas, K. L. Kompa, R. D. Levine
11 **Liquid Crystals of One- and Two-Dimensional Order** Editors: W. Helfrich, G. Heppke
12 **Gasdynamic Laser** By S. A. Losev
13 **Atomic Many-Body Theory** By I. Lindgren, J. Morrison
14 **Picosecond Phenomena II** Editors: R. M. Hochstrasser, W. Kaiser, C. V. Shank
15 **Vibrational Spectroscopy of Adsorbates** Editor: R. F. Willis
16 **Spectroscopy of Molecular Excitions** By V. L. Broude, E. I. Rashba, E. F. Sheka
17 **Inelastic Particle-Surface Collisions** Editors: E. Taglauer, W. Heiland
18 **Modelling of Chemical Reaction Systems** Editors: K. H. Ebert, P. Deuflhard, W. Jäger
19 **Secondary Ion Mass Spectrometry SIMS III** Editors: A. Benninghoven, J. Giber, J. Lászl6, M. Riedel, H. W. Werner
20 **Chemistry and Physics of Solid Surfaces IV** Editors: R. Vanselow, R. Howe
21 **Dynamics of Gas-Surface Interaction** Editors: G. Benedek, U. Valbusa
22 **Nonlinear Laser Chemistry** Multiple-Photon Excitation By V. S. Letokhov
23 **Picosecond Phenomena III** Editors: K. B. Eisenthal, R. M. Hochstrasser, W. Kaiser, A. Laubereau
24 **Desorption Induced by Electronic Transitions DIET I** Editors: N. H. Tolk, M. M. Traum, J. C. Tully, T. E. Madey
25 **Ion Formation from Organic Solids** Editor: A. Benninghoven
26 **Semiclassical Theories of Molecular Scattering** By B. C. Eu
27 **EXAFS and Near Edge Structures** Editors: A. Bianconi, L. Incoccia, S. Stipcich
28 **Atoms in Strong Light Fields** By N. B. Delone, V. P. Krainov
29 **Gas Flow in Nozzles** By U. G. Pirumov, G. S. Roslyakov
30 **Theory of Slow Atomic Collisions** By E. E. Nikitin, S. Ya. Umanskii
31 **Reference Data on Atoms, Molecules, and Ions** By A. A. Radzig, B. M. Smirnov
32 **Adsorption Processes on Semiconductor and Dielectric Surfaces I** By V. F. Kiselev, O. V. Krylov
33 **Surface Studies with Lasers** Editors: F.R. Aussenegg, A. Leitner, M.E. Lippitsch
34 **Inert Gases** Potentials, Dynamics, and Energy Transfer in Doped Crystals. Editor: M. L. Klein
35 **Chemistry and Physics of Solid Surfaces V** Editors: R. Vanselow, R. Howe
36 **Secondary Ion Mass Spectrometry, SIMS IV** Editors: A. Benninghoven, J. Okano, R. Shimizu, H. W. Werner
37 **X-Ray Spectra and Chemical Binding** By A. Meisel, G. Leonhardt, R. Szargan
38 **Ultrafast Phenomena IV** By D. H. Auston, K. B. Eisenthal
39 **Laser Processing and Diagnostics** Editor: D. Bäuerle

Springer Series in Chemical Physics

Editors: Vitalii I. Goldanskii Fritz P. Schäfer J. Peter Toennies

Managing Editor: H. K.V. Lotsch

40 **High-Resolution Spectroscopy of Transient Molecules**
By E. Hirota

41 **High Resolution Spectral Atlas of Nitrogen Dioxide 559–597 nm**
By K. Uehara and H. Sasada

42 **Antennas and Reaction Centers of Photosynthetic Bacteria**
Structure, Interactions, and Dynamics
Editor: M. E. Michel-Beyerle

43 **The Atom-Atom Potential Method**
Applications to Organic Molecular Solids
By A. J. Pertsin and A. I. Kitaigorodsky

44 **Secondary Ion Mass Spectrometry SIMS V**
Editors: A. Benninghoven, R.J. Colton, D.S. Simons, and H.W. Werner

45 **Thermotropic Liquid Crystals, Fundamentals**
By G. Vertogen and W. H. de Jeu

46 **Ultrafast Phenomena V**
Editors: G.R. Fleming and A.E. Siegman

47 **Complex Chemical Reaction Systems**
Mathematical Modelling and Simulation
Editors: J. Warnatz and W. Jäger

48 **Ultrafast Phenomena VI**
Editors: T. Yajima, K. Yoshihara, C. B. Harris, and S. Shionoya

49 **Vibronic Interactions in Molecules and Crystals**
By I. B. Bersuker and V.Z. Polinger

50 **Molecular and Laser Spectroscopy**
By Zu-Geng Wang and Hui-Rong Xia

51 **Space-Time Organization in Macro molecular Fluids**
Editors: F. Tanaka, M. Doi, and T. Ohta

52 **Multiple-Photon Laser Chemistry**
By. R.V. Ambartzumian, C.D. Cantrell, and A. Puretzky

53 **Ultrafast Phenomena VII**
Editors: C. B. Harris, E. P. Ippen, G.A. Mourou, and A.H. Zewail

54 **Physics of Ion Impact Phenomena**
Editor: D. Mathur

55 **Ultrafast Phenomena VIII**
Editors: J.-L. Martin, A. Migus, G.A. Mourou, and A.H. Zewail

56 **Clusters of Atoms and Molecules**
Solvation and Chemistry of Free Clusters, and Embedded, Supported and Compressed Clusters
Editor: H. Haberland

57 **Radiationless Transitions in Polyatomic Molecules**
By E. S. Medvedev and V. I. Osherov

58 **Positron Annihilation in Chemistry**
By O. E. Mogensen

59 **Soot Formation in Combustion**
Mechanisms and Models
Editor: H. Bockhorn

60 **Ultrafast Phenomena IX**
Editors: P. Barbara, W. H. Knox, G. A. Mourou, and A. H. Zewail

61 **Gas Phase Chemical Reaction Systems**
Experiments and Models 100 Years After Max Bodenstein
Editors: J. Wolfrum, H.-R. Volpp, R. Rannacher, and J. Warnatz

62 **Ultrafast Phenomena X**
Editors: P. F. Barbara, J. G. Fujimoto, W. H. Knox, and W. Zinth

Springer and the environment

At Springer we firmly believe that an international science publisher has a special obligation to the environment, and our corporate policies consistently reflect this conviction.

We also expect our business partners – paper mills, printers, packaging manufacturers, etc. – to commit themselves to using materials and production processes that do not harm the environment. The paper in this book is made from low- or no-chlorine pulp and is acid free, in conformance with international standards for paper permanency.

Springer

MIX
Papier aus verantwortungsvollen Quellen
Paper from responsible sources
FSC® C105338

If you have any concerns about our products,
you can contact us on
ProductSafety@springernature.com

In case Publisher is established outside the EU,
the EU authorized representative is:
Springer Nature Customer Service Center GmbH
Europaplatz 3, 69115 Heidelberg, Germany

Printed by Libri Plureos GmbH
in Hamburg, Germany